Evolution in the Genus *Bufo*

Evolution in the Genus *Bufo*

Edited by W. FRANK BLAIR

UNIVERSITY OF TEXAS PRESS • AUSTIN AND LONDON

International Standard Book Number 0-292-72001-7
Library of Congress Catalog Card Number 75-185242

Type set by Southwestern Typographics, Inc., Dallas, Texas
Printed and Bound by TJM Corporation, Baton Rouge, Louisiana
Color plates produced by Gulf Printing Company, Houston, Texas

CONTENTS

COLOR PLATES *Facing page*

PREFACE

Recent advances in scientific techniques enable the student of animal phylogeny to solve problems of evolutionary relationship that have been essentially unsolvable solely by use of morphological comparisons. The diversity of methods that one can now utilize makes it almost a necessity that a team approach be utilized if all pertinent evidence is to be gathered.

This book is a report and synthesis of the results of team research and of international, cooperative research. It reports much of my own research over the past fifteen years. Seven of the chapters are contributed by my present or former students, most of whose work was done on the same individuals and species as my own. One chapter is coauthored by a former postdoctoral research associate, Murray J. Littlejohn, whose work is also based on the same material. In addition, close liaison, including the sharing and exchange of material, has been maintained over several years between my group and the team of Jose M. Cei, at Mendoza, Argentina, and V. Erspamer and M. Roseghini, at Rome, Italy.

The fate of a particular specimen of toad illustrates how many of the data were gathered. After being received in my laboratory, a male toad would have its release call recorded by Littlejohn or Brown, and in some instances it might have its mating call recorded as well. Low would squeeze out the parotoid-gland secretions for her biochemical analyses. After this male was sacrificed for use in one of my *in vitro* crosses, Guttman removed the blood for his work, and Bogart removed an eye for his karyotype study. W. F. Martin made vocal measurements on the recently dead animal. The skeleton might be preserved for R. F. Martin, and in some instances the animal would be skinned and the dried skin sent to the team of Cei, Erspamer, and Roseghini for their biochemical work. The success of this cooperative approach must be judged from the contents of the book.

The additional chapters to round out the book have been contributed by R. F. Inger, Osvaldo Reig, Joe Tihen, and Henryk Szarski, each working in a field of his own high competence.

The kind and magnitude of research that forms the basis for much of this book would not have been possible without the support for basic research that is provided by the National Science Foundation. Through the years, my own research contributory to this volume has had support from NSF Grants G–328, G–2215, G–4956, G–12335, GB–790, GB–3483, and GB–5406X. Most of the authors who did their research as graduate students in my laboratory have had assistance from NSF Grants GB–4128 or GB–6914.

The volunteer efforts of scores of people have made it possible to assemble live toads from all the major continental areas, and it is not feasible to list all who have sent material for our work. We are indeed indebted to every person whose sending of animals made this work possible. A few may be mentioned to illustrate the cooperation we have had. Dr. Avelino Barrio of Buenos Aires has responded to every request and has sent such hard-to-get toads as *B. variegatus* of the southern Andes. Dr. A. K. Mondal, who has the intriguing title of "Frog Development Officer," has supplied Indian species on many occasions. Dr. Eviatar Nevo has sent *B. virdis* from Israel whenever asked. Mr. R. Guibert of Deux-Sevres, France, has helped us and many other workers obtain exotic species for our research. Mrs. Kay Mehren, Peace Corp volunteer in Kenya, sent us a large stock of *B. kerinyagae* (Keith 1968), so that we knew a great deal about the biochemistry, karyotype, and genetic compatibility of this toad even before it had a scientific name. Mills Tandy's contribution of materials during and after his tour as a Peace Corp volunteer in Nigeria was largely responsible for the present thorough treatment of African toads.

Sing Sioe Liem made valuable suggestions concerning limb musculature and permitted use of his data on nonbufonids. Preparation of chapter 8 benefitted from the technical assistance of Michelle D. Grieser, Thomas Olechowski, and Peter Solt.

For permission to study specimens in their care in connection with preparation of chapter 8, we are grateful to Charles M. Bogert, American Museum of Natural History (AMNH); Alice G. C. Grandison, British Museum (Natural History) (BM); Konrad Klemmer, Natur-Museum und Forschungs-Institut "Senkenberg" (SNF); Alan Leviton, California Academy of Sciences (CAS); Günther Peters, Institut für Spezielle Zoologie and Zoologisches Museum, Berlin (ZMB); James A. Peters, United States National Museum (USNM); and Heinz Wermuth, Staatliches Museum für Naturkunde in Stuttgart (MNS).

Jill Brown generously provided help during many phases of study on which chapter 16 is based. Acoustical analysis was carried out at the University of Melbourne. Data were processed at Illinois State University, utilizing the computer services made available by R. Reiter. G. C. Ramseyer and K. Tcheng gave valuable advice concerning statistical analysis. Partial support of the work of Brown and Littlejohn was provided by research grants from Illinois State University and from the University of Melbourne. During the initial stages of this study, M. J. Littlejohn received a travel grant from the Australian-American Education Foundation and was granted sabbatical leave from the University of Melbourne. We thank A. A. Martin for critically reading the manuscript of chapter 16.

W. Frank Blair
Austin, Texas

Evolution in the Genus *Bufo*

1. Introduction

W. FRANK BLAIR

The University of Texas at Austin
Austin, Texas

The genus *Bufo* is an ideal group of vertebrates for a multidisciplinary effort to reconstruct evolutionary history and phylogenetic relations. Its distribution is nearly worldwide, involving the Americas, Eurasia, and Africa. Furthermore, there are enough but not too many species for reasonable analysis. Also, the members of this genus have probably more systems (genetical, behavioral, biochemical, bioacoustical) that are amenable to analysis than almost any other group of animals.

There are about two hundred species of *Bufo*. It will not be possible to state the exact number in the foreseeable future. There are two reasons for this. First, there is at present no possibility of estimating the number of cryptic species that have not been revealed by previous, purely morphological studies. A case in point is the recent (1960) discovery by Bogert that what had been considered a single species, *B. compactilis*, was really two quite distinct species. Great differences in the mating call provided the primary evidence. A second example is presented in this book. In the *regularis* group of Africa, there is differentiation of mating call and there is genetic incompatibility between populations that, on strictly morphological grounds, had been previously referred to as a single species, *B. regularis*. One can only guess that there will be other examples when the entire genus is subjected to the kind of analysis that has been applied heretofore to only a modest part of it.

Second, it is highly probable that additional species remain to be discovered in addition to the cryptic species mentioned above. Only North America and Europe have been adequately explored for *Bufo*. The largest *Bufo*, *B. blombergi*, from Colombia, was described as recently as 1951 by Myers and Funkhouser. The spectacular endemic *B. periglenes*, from Costa Rica, was described only in 1966 by Savage. In South America, the complex of species that has been treated as a single polytypic species, *B. granulosus*, by Gallardo (1965) is actually comprised of several species (Chap. 6). The complex of variously disjunct populations that has been treated previously as a polytypic species, *B. spinulosus* (see Cei 1962), is clearly comprised of several quite differentiated species (Chaps. 6, 11), but additional work remains to be done to clarify the degree of differentiation of the various populations. In Africa, the *regularis* group needs much more study before the number of species is satisfactorily determined.

Our state of knowledge of the genus *Bufo* prior to the studies on which the following chapters are based, like that of any other widely distributed taxon, has resulted from many disconnected and individual efforts varying from local to provincial in scope. Also, the knowledge about the genus varies greatly from country to country and from continent to continent. The three species of the depauperate European fauna have been thoroughly investigated, but the relations of these to closely related forms that range across temperate Asia remain uncertain. The Asiatic toads, especially those of tropical Asia, remain the poorest known of any. The works of Pope and Boring (1940–1941) on Chinese amphibia, of Liu (1950) on the amphibians of western China, and of Taylor (1962) on the amphibians of Thailand are very useful, but all are geographically limited in scope and all reflect the bare bones of alpha taxonomy. A much more modern treatment, although for a limited area, is Inger's (1966) work on Bornean Amphibia.

The evolutionary relations, distribution, and ecological life history of the United States *Bufo* are better known than those of the toads of any other country. Check lists, such as that of Schmidt (1953) for North American amphibians and reptiles north of Mexico, manuals, such as Wright's (1949) "Handbook of Frogs and Toads" or Stebbin's (1951) "Amphibians of Western North America," or field guides, such as Conant's (1958) "Field Guide to Reptiles and Amphibians of Eastern North America" provide a wealth of general information about the various species. For Mexico and Central America, the situation is very different. For Mexico, there is the bare check list and key of Smith and Taylor (1948), now badly outdated. For Central American countries we have few comprehensive works that include *Bufo*. Most important are Stuart (1948 and 1963) on Guatemala and Taylor (1952) on Costa Rica. Some groups of Mexican and Central American *Bufo* have received study by modern methods of analysis, particularly studies by Porter (1964*a*, 1964*b*, 1965), Porter and Porter (1967), and W. F. Blair (1966).

For South America, there is no continent-wide account of *Bufo* aside from the valuable summary by Cei (1968). The most valuable regional papers are those of Cei (1960, 1962), "Batracios de Chile"; Rivero (1961), "Salientia of Venezuela"; Vellard (1959) on Andean *Bufo;* Cochran (1955), "Frogs of Southeastern Brazil," and Cochran and Goin (1970), "Frogs of Colombia."

Knowledge of African *Bufo* is comparable to that of the South American ones. There is no continent-wide treatment of the genus, and there has been undue confusion of the taxonomy because most studies have been restricted by political boundaries. Important recent works are those of Poynton (1964) for South Africa, of Schiøtz (1964*a*, 1964*b*, 1964*c*, 1966) for West Africa, of Schmidt and Inger (1959) for the Congo, and of Stewart (1967) for Malawi.

The *Bufo* Pattern

The fossil record attests that the genus *Bufo* has had a long existence. Tihen (1962) interpreted a fossil from the lower Oligocene of Argentina as being the oldest known *Bufo* as well as a member of an extant group. This implies an evolutionary conservatism that is also supported by other aspects of the fossil record and by evidence from extant forms. This suggests that *Bufo* evolved a very successful mode of existence that has survived through millions of years. This mode of existence may be best characterized as one of generalism, and it may be that this generalism accounts for the long persistence of the genus through time and for the apparent conservatism of species and species groups.

The toad body form departs little from one of short, stout body and relatively short legs. Locomotion is by short hops, or in some species by walking, which means that *Bufo* essentially lacks the ability to escape by leaping, as do hylids, many leptodactylids, and other anurans. This lack of leaping ability is probably compensated for by the presence of skin venoms. Protective skin secretions are concentrated in the usually prominent parotoid glands, and in some large species there are accessory glands of this type on the legs as well.

The terrestrial habit of *Bufo* is one of the important features of the generalism of the group. Some are remarkable burrowers, using their sharp, keratinized metatarsal "spade." Others do not burrow but use holes made by other animals. The life history is a generalized anuran one, with many eggs laid in strands in water, with a normal larval stage in water, followed by metamorphosis. There is inconsequential variation on this theme, for instance the single, rather than stranded, eggs of such species as *B. punctatus* and *B. marmoreus* and the specialized larvae of *B. perreti* mentioned by Schiøtz (1963). This adherence to a generalized life history contrasts strongly with the specializations, such as abbreviated life history, terrestrial deposit of eggs, foam nests, or specialized modes of transport of larvae by one or the other parent in various genera of the related leptodactylids, hylids, and atelopodids.

Another feature of *Bufo* life is the general occurrence of opportunistic breeding, reproductive activities being generally coordinated with patterns of rainfall. Exceptions to this breeding pattern do occur in very arid regions, where some species breed in permanent water and independent of the scarce rainfall.

The male mating call provides an attractant for the females, and, hence, its species specificity is an isolating mechanism of paramount importance (Blair 1956, 1958, 1962, 1963, 1964; Bogert 1960; Mecham 1961, and others). The male release call provides a sex identificatory signal when one male clasps another.

Although body form and mode of locomotion remain relatively constant in the genus, body size shows great variation. The evolutionary pressures that might account for the range of body sizes remain to be elucidated, although one might guess a variety of reasons having to do with temperature, water balance, or even interspecies relations. Great variation in body size is found among the *Bufo* species of all continents. In South America the range is from the tiny *B. pygmaeus*, in which females measure from 31 to 42 mm. in snout-vent length (data from Myers and Carvalho 1952), to *B. paracnemis*, in which the huge females are the size of a large dinner plate. In North America, the size range encompasses the tiny *B. quercicus*, in which females reach slightly more than 30 mm. in snout-vent length, and *B. alvarius*, in which females reach more than 150 mm. In Africa, the range encompasses *B. rosei*, roughly the size of *B. quercicus*, and *B. superciliaris*, approximately the size of *B. alvarius*. The females are almost always larger than the males in *Bufo*.

Among sensory modalities, vocalization seems to provide the main attractant for a mate, but various species among the narrow-skulled toads have lost the mating call and have substituted other, presently unknown, mechanisms. Vision functions importantly in food-getting (Brower, Brower, and Westcott 1960; Brower and Brower 1965). Vision and olfaction are important in orientation to the environment and in homing (Grubb 1968). Solar, lunar, and celestial orientation have been demonstrated in *B. woodhousei* (Ferguson and Landreth 1966; Landreth and Ferguson 1968).

Toads are generally arthropod feeders, but the larger ones, such as *B. marinus*, have become generalized in that they will feed on a variety of invertebrates and vertebrates of appropriate size.

Toads have been able to adapt to a wide range of environmental conditions. The fossil record suggests that there was a very early dichotomy into warmth-adapted and cold-adapted toads (W. F. Blair 1970). The warmth-adapted forms are limited today to the tropics and their borders. The cold-adapted toads have distributions that follow the cordilleras of the western parts of the Americas and the temperate latitudes across Eurasia to North Africa. On the whole, the genus represents adaptation to semixeric conditions, but the group is excluded from the very dry deserts of the world. In the broad-skulled toads, a few, like members of the *guttatus* group of South America and *B. superciliaris* of Africa, are restricted to the rain forest. Others, being highly eurytopic, for instance, *B. marinus*, occur commonly in the rain forest but also occupy other tropical biomes.

Objectives

This book attempts to bring to bear many lines of evidence on several questions about the evolution within a single genus. The main questions to which this book is addressed are:

1. Where did the genus *Bufo* originate?
2. What are the evolutionary history and phylogenetic relations of the approximately two hundred species of the nearly cosmopolitan genus *Bufo*? How did the genus spread to its almost world-wide distribution? How many continental exchanges have occurred? By what routes have they been accomplished?
3. What is the age of the genus, and when in its history have major dichotomies occurred?
4. What are the boundaries of the various species groups?
5. What are the relations of the species groups to one another?
6. What are the relations of the various single species that seem to have no close relatives to other species of *Bufo*?

REFERENCES

Blair, W. F. 1956*a*. Call difference as an isolation mechanism in southwestern toads (genus *Bufo*). Texas J. Sci. 8:87–106.

———. 1956*b*. The mating call of hybrid toads. Texas J. Sci. 8:350–355.

———. 1956*c*. Comparative survival of hybrid toads (*B. woodhousei* x *B. valliceps*) in nature. Copeia 1956: 259–260.

———. 1958. Mating call in the speciation of anuran amphibians. Amer. Nat. 92:27–51.

———. 1962. Non-morphological data in anuran classification. Syst. Zool. 11:72–84.

———. 1963*a*. Evolutionary relationships in North American toads of the genus *Bufo*: A progress report. Evolution 17:1–16.

———. 1963*b*. Intragroup genetic compatibility in the *Bufo americanus* species group of toads. Texas J. Sci. 15:15–34.

———. 1964*a*. Isolating mechanisms and interspecies interactions in anuran amphibians. Quart. Rev. Biol. 39:334–344.

———. 1964*b*. Evidence bearing on relationships of the *Bufo boreas* group of toads. Texas J. Sci. 16:181–192.

———. 1964*c*. Evolution at populational and interpopulational levels, isolating mechanisms and interspecies interactions in anuran amphibians. Quart. Rev. Biol. 39:333–344.

———. 1966. Genetic compatibility in the *Bufo valliceps* and closely related groups of toads. Texas J. Sci. 18:333–351.

Bogert, C. M. 1960. The influence of sound on the behavior of amphibians and reptiles, p. 137–320. *In* W. W. Lanyon and W. N. Tavolga (eds.), Animal sounds and communication. A.I.B.S. Publ. 7.

Brower, Jane Van Zandt, and L. P. Brower. 1965. Experimental studies of mimicry. 8. Further investigations of honeybees (*Apix mellifera*) and their dronefly mimics (*Eristalis* spp.). Amer. Nat. 99:173–188.

Brower, Lincoln P., and J. van Z. Brower. 1960. Experimental studies of mimicry: Reactions of toads to bumble bees and their asilid-fly mimics. Proc. 11th Int. Congr. Ent. 3(4):258.

———, ———, and Peter W. Westcott. 1960. Experimental studies of mimicry. 5. The reactions of toads (*Bufo terrestris*) to bumblebees (*Bombus americanorum*) and their robberfly mimics (*Mallophora bomboides*), with a discussion of aggressive mimicry. Amer. Nat. 94:343–355.

Cei, Jose M. 1960. Geographic variation of *Bufo spinulosus* in Chile. Herpetologica 16:243–250.

———. 1962. Batracios de Chile. F. Bruckmann KG, Graph. Kunstanst, Munich.

———. 1968. Remarks on the geographical distribution and phyletic trends of South American toads. Texas Mem. Mus. Publ. Pearce-Sellards Series, 13:1–20.

Cochran, Doris M. 1955. Frogs of southeastern Brazil. U.S. Nat. Mus. Bull. 206:423.

———, and C. J. Goin. 1970. Frogs of Colombia. U.S. Nat. Mus. Bull. 228.

Conant, Roger. 1958. A field guide to reptiles and amphibians of the United States and Canada east of the 100th meridian. Houghton Mifflin, Boston.

Ferguson, Denzel E., and Hobart F. Landreth. 1966. Celestial orientation of Fowler's toad *Bufo fowleri*. Behaviour 26:105–123.

Gallardo, J. M. 1965. The species *Bufo granulosus* Spix (Salientia: Bufonidae) and its geographic variation. Bull. Mus. Zool. (Harvard Univ.) 134(4):107–138.

Grubb, Jerry Carl. 1970. Homing behavior in post-reproductive *Bufo valliceps*: The importance of certain sensory mechanisms. M.A. Thesis, Univ. Texas, Austin.

Inger, R. F. 1966. The systematics and zoogeography of the Amphibia of Borneo. Fieldiana Zool. 52:1–402.

Landreth, Hobart F., and Denzel E. Ferguson. 1968. The sun compass of Fowler's toad *Bufo woodhousei fowleri*. Behavior 30:27–43.

Liu, Cheng-chao. 1950. Amphibians of western China. Fieldiana Zool. (Mem.) 2:1–400.

Mecham, J. S. 1961. Isolating mechanisms in anuran amphibians, p. 24–61. *In* W. F. Blair (ed.), Vertebrate speciation. Univ. Texas Press, Austin.

Myers, G. S., and J. W. Funkhouser. 1951. A new giant toad from southwestern Colombia. Zoologica 36:279–282.

———, and A. Leitao de Carvalho. 1952. A new dwarf toad from southeastern Brasil. Zoologica 37(1)1–3.

Pope, C. H., and A. M. Boring. 1940–1941. A survey of Chinese amphibia. Peking Bull. Nat. Hist. 15:13–86.

Porter, Kenneth R. 1964*a*. Distribution and taxonomic status of seven species of Mexican *Bufo*. Herpetologica 19:229–247.

———. 1964*b*. Morphological and mating call comparisons in the *Bufo valliceps* complex. Amer. Midl. Nat. 71:232–245.

———. 1965. Intraspecific variation in mating call of *Bufo coccifer* Cope. Amer. Midl. Nat. 74:350–356.

———, and Wendy F. Porter. 1967. Venom comparisons and relationships of twenty species of New World toads (genus *Bufo*). Copeia 1967:298–307.

Poynton, J. C. 1964. The Amphibia of southern Africa. Ann. Natal Mus. 17:1–334.

Rivero, J. A. 1961. Salientia of Venezuela. Bull. Mus. Comp. Zool. 126(1):5–207.

Savage, Jay M. 1966. An extraordinary new toad (*Bufo*) from Costa Rica. Rev. Biol. Trop. 14:153–167.

Schiøtz, A. 1963. The amphibians of Nigeria. Vidensk. Medd. dansk naturh. Foren. (Copenhagen) 126:1–92.

———. 1964*a*. A preliminary list of amphibians collected in Ghana. Vidensk. Medd. dansk naturh. Foren. (Copenhagen) 127:1–17.

———. 1964*b*. A preliminary list of amphibians collected in Sierra Leone. Vidensk. Medd. dansk naturh. Foren. (Copenhagen) 127:19–34.

———. 1964*c*. The voices of some West African amphibians. Vidensk. Medd. dansk naturh. Foren (Copenhagen) 127:35–83.

———. 1966. On a collection of Amphibia from Nigeria. Vidensk. Medd. dansk naturh. Foren. (Copenhagen) 129:43–48.

Schmidt, Karl P. 1953. A check list of North American amphibians and reptiles. Sixth Edition. Univ. Chicago Press, Chicago.

———, and R. F. Inger. 1959. Amphibians. Explor. Parc Nat. Upemba. 56:1–264.

Smith, Hobart M., and Edward H. Taylor. 1948. An annotated checklist and key to the Amphibia of Mexico. U.S. Nat. Mus. Bull. 194:1–118.

Stebbins, R. C. 1951. Amphibians of western North America. Univ. California Press, Berkeley.

Stewart, Margaret. 1967. Amphibians of Malawi. State Univ. New York Press, New York.

Stuart, L. C. 1948. The amphibians and reptiles of Alta Verapaz Guatemala. Univ. Mich. Mus. Zool. Bull. 69: 1–109.

———. 1963. A checklist of the herpetofauna of Guatemala. Misc. Publ. Mus. Zool. (Univ. Mich.) 122: 1–150.

Taylor, Edward H. 1952. The frogs and toads of Costa Rica. Univ. Kansas Sci. Bull. 35:577–942.

———. 1962. The amphibian fauna of Thailand. Univ. Kansas Sci. Bull. 43:265–599.

Tihen, J. A. 1962. A review of new world fossil bufonids. Amer. Midl. Nat. 68:1–50.

Vellard, J. 1959. Estudios sobre batracios andinos. 5. El genero *Bufo*. Mem. Mus. Hist. Nat. Xavier Prado 8:1–48.

Wright, Albert Hazen, and Anna Allen Wright. 1949. Handbooks of frogs and toads of the United States and Canada. Comstock Publishing Co., Ithaca, N.Y.

2. The Fossil Record

JOSEPH A. TIHEN

University of Notre Dame
Notre Dame, Indiana

Introduction

When a group of vertebrates has had a long geological history, has been widely distributed geographically for a long period of time, and also exhibits extensive intragroup ecological diversity, one might reasonably anticipate that an extensive and informative fossil record exists. These conditions obtain for *Bufo* and the bufonids, but the anticipation is unfortunately not realized. Fossils are not uncommon in the later Tertiary, but these tell us nothing of the origin of, or the major phylogenetic relationships within, the genus. Attempts to reconstruct phylogenies must still rely almost entirely on comparative studies of living forms. The known fossil record does serve, however, to provide some sort of a framework in geologic time into which any hypothesized phylogenies must be fitted.

The present discussion will concern itself solely with the genus *Bufo,* not only because of the scope of the present compilation, but also because all fossils that can be unequivocally recognized as bufonids can also be assigned with confidence to this genus. In a few cases that will be discussed, the generic reference is questionable, as well as the family reference. A number of actual or purported fossil genera have from time to time, by various workers, been assigned to the Bufonidae. In some instances (e.g., *Indobatrachus* Noble 1930) this resulted only from accepting a broad concept of the family that included most or all of the forms now considered Leptodactylidae (*sensu lato*). In many other cases, however, the assignment was simply an error, due to misinterpreting of the fossil specimens. Kuhn (1962) included several toothed genera in the Bufonidae; all these can be referred with reasonable certainty to some other extant family.

One questionable group merits more extensive comment. Fejérváry (1917) proposed recognition of a subfamily Platosphinae within the Bufonidae. He referred the type genus *Platosphus* de L'Isle (1877) and a new genus, *Pliobatrachus,* to this subfamily. He also placed in it, with reservations, the genera *Diplopelturus* Depéret (1897) and *Bufavus* Portis (1885). The major distinguishing characteristics of the subfamily are purportedly the incorporation of more than one vertebra into the sacrum and the consistent presence of horizontal laminae on the urostyle, which is in some forms fused with the "synsacrum."

All these forms are from France, *Platosphus* and *Bufavus* from the Miocene, the other two from the Pliocene. Młynarski (1960, 1961) records *Pliobat-*

rachus from the Pliocene of Poland and continues to refer the genus to the Bufonidae; the 1961 paper explicitly recognizes the Platosphinae, but makes no mention of *Bufavus* and suggests that *Diplopelturus* (there written *Diploplecturus*) is a synonym of *Bufo*. Kuhn (1962) refers all these genera, plus several others, to the Bufonidae. Despite these recent treatments, all the genera (excepting, of course, *Bufo* itself, and the inadequately known *Diplopelturus*) included in the Bufonidae by Kuhn and these previous authors can be referred with reasonable confidence to other known families; at the very least, all can be removed from the Bufonidae, even if the proper familial assignment is not entirely clear.

Hecht and Hoffstetter (1962) report recovery of bufonid material, to which they do not assign any name, from deposits representing approximately the Paleocene-Eocene transition in Belgium. Vertebral fusions are common in this material, including a usual incorporation of three vertebrae into a "synsacrum." Association of other skeletal elements with these vertebrae was difficult, however; they point out that no edentulous maxillae were recovered from these deposits, either because of accident of collection or because this was a toothed form. In any event, the numbers of vertebrae collected demonstrate conclusively that fusions were the normal situation in the population represented. The specimens can scarcely represent the genus *Bufo*, but it is not conclusively demonstrated that they are actually bufonid. In any event, it is unlikely that these are identical with the later "Platosphinae." Hecht and Hoffstetter point out the probable inadequacy of the bases for recognizing a platosphine subfamily, but by inference allow the supposed platosphines to remain within the bufonids. If the form represented by their specimens is indeed related to, even though not identical with, later "platosphines," then it is not a bufonid; in any event, regardless of relationships, assignment of these specimens to the Bufonidae is highly speculative.

Oligocene Fossils

The genus *Bufo* has been reported to occur in the Oligocene. These reports are based on only two specimens, one from Europe and the other from South America; the accuracy of each identification is open to serious question.

Filhol (1877) proposed the name *Bufo servatus* for a specimen from the Quercy phosphorites in France. These deposits range in age from late Eocene to early Oligocene, and dating of particular specimens within this range is uncertain, but, if this specimen is indeed a *Bufo*, it would represent the earliest known record of the genus. This specimen, like others from the Quercy phosphorites, is not a skeleton, but simply a partial reproduction of the external body form in phosphate. Filhol based his generic assignment on the purported presence of parotoid glands, but the accuracy of this interpretation is open to question (see Piveteau 1927, and Tihen 1962 for comment). Subsequent workers have continued to employ this name, largely because there is no adequate basis for allocating it properly. Thus, there is no real evidence that it is not a *Bufo*; the specimen simply appears unidentifiable on the basis of present knowledge and cannot be used as direct evidence for the existence of *Bufo* in the early Oligocene.

A second Oligocene fossil that has been referred to *Bufo*, but again questionably, is from the lower Oligocene (Deseadan) of Patagonia, in Argentina. This specimen, a fairly complete but fragmented and somewhat distorted skeleton, was described by Schaeffer (1949) under the name *Neoprocoela edentata*. He considered it an edentulous leptodactylid, but noted a number of *Bufo*-like features. Because of these numerous features, Tihen (1962) tentatively referred this species to the genus *Bufo*. However, Mr. John D. Lynch (personal communication), in a comparative study of leptodactylid osteology, concluded that this is indeed a leptodactylid, probably related to *Batrachophrynus*.

Therefore, as with *Bufo servatus*, this Patagonian fossil offers no real evidence that *Bufo* existed in the Oligocene. If it is truly a bufonid, its morphological affinities seem to be with the Eurasian *B. calamita–B. raddei* complex, as evidenced by the presence of a frontoparietal fontanelle. If, as is equally likely, it is a leptodactylid, no close relationship to the bufonids can be inferred. It does, however, demonstrate once again the already well recognized fact that bufonid-like osteological characteristics are readily derivable from a leptodactylid type.

Miocene Fossils

The earliest unquestioned fossil occurrence of *Bufo* is *B. praevius* Tihen (1951) from the Thomas Farm deposits of Gilchrist County, Florida. The Thomas Farm deposits are generally considered to be of late Arikareean age, that is, late Lower Miocene (Bader 1956; Romer 1966). Although the specimens recovered consist almost entirely of dissociated, often fragmentary, individual bones, the number and variety of the specimens are adequate to provide a firm basis for identification, at least to the generic level. The closest affinities of this species within the genus are somewhat less clear — partly because we are not yet certain of the systematic significance of various osteological features even among recent forms. Auffenberg (1956), primarily because the fossil form possesses nearly parallel supraorbital crests that form a sharp angle with the postorbital crests, suggested a relationship to "western" members of the *americanus group*. But Tihen (1962) felt that a number of features, notably the fusion of the frontoparietal with the proötic or otoccipital, were not consistent with this interpretation. He believed that it belongs to the same general group as *B. punctatus, B. debilis* and relatives, and *B. quercicus*. These species share certain osteological features suggesting relationship, as do certain other characteristics, although some other lines of evidence do not indicate that this "group" constitutes a natural unit. Osteologically, *B. praevius* does correspond in all respects except the disposition of the cranial crests (in which *B. quercicus* is also divergent) with this putative group, which probably also includes all or most Antillean species and perhaps some South American ones. There is thus a possibility that this species or its precursor arrived in Florida via an Antillean island-hopping route, rather than by way of the mainland.

Whatever the actual status of the *punctatus-quercicus* complex as a natural group, *Bufo praevius* seems certain to be allied with the general section of the genus in which one or more of these species belongs; it is therefore already well advanced along a line of evolution that has taken place entirely within the New World. No particular resemblances to any Old World toads can be discerned, nor to any hypothetical ancestral New World type. In fact, some of the living New World species are more similar to certain Old World forms than is *B. praevius*. The genus must have been established in the New World (whether orginating there or immigrating) considerably prior to the time of the formation of the Thomas Farm deposits and must have already undergone great diversification in this hemisphere.

The earliest reasonably well confirmed appearance of the genus in Eurasia is in the Middle Miocene deposits of Oeningen, Germany, whence Tschudi (1838) described *Paleophrynus gessneri*. European workers concur that this is a *Bufo* (see, e.g., Kuhn 1962). Published descriptions and figures are inadequate to demonstrate unequivocally the correctness of this allocation, but they are at least consistent with such an interpretation, and there is no apparent reason to question it seriously. Available information is also consistent with the interpretation that this fossil is similar to the present *B. bufo* in many respects, but postulation of real affinity with that species would be highly speculative in our present state of knowledge. At least, *Bufo gessneri* is a narrow-skulled form, without obvious crests or prominent dermal ornamentation.

In the Upper Miocene the fossil record is extended and includes the other two major land masses now inhabited by *Bufo* — South America and Africa. The South American form is represented by a single specimen, comprising the posterior portion of the skull, eight articulated vertebrae, and several girdle and appendicular elements, and constitutes a part of the La Venta Fauna (Savage 1951) from the Upper Miocene of Colombia. Estes and Wassersug (1963) described this specimen in detail and provided illustrations, referring it to the extant species *B. marinus*. As they point out, the extent of osteological variation within the modern *B. marinus*, and the possible osteological differences between that species and other closely related ones, are not known. The fossil seems to differ from recent specimens of *B. marinus* only in minor details involving the extent of ossification of a few elements; the taxonomic significance, if any, of these minor differences is not clear. Certainly this Miocene form is extremely similar to, if not completely identical with, the present-day *B. marinus*. However, the modern *B. arenarum* is also very similar osteologically to *B. marinus*, yet hybridization experiments (Chap. 10) suggest that the relationship between those two species may not be as close as the osteological similarity would imply. In any event, the fossil very clearly exhibits certain diagnostic features (notably the broad overlap of the pterygoid onto the ventral surface of the parasphenoid) unknown in the genus except in the *marinus-arenarum* complex.

Hecht, Hoffstetter, and Vergnaud (1961) reported briefly that bones referable to *Bufo* had been recovered from the Upper Miocene of Beni-Mellal, Morocco. Vergnaud-Grazzini subsequently (1966) described and illustrated these specimens in greater detail. They consist almost entirely of postcranial elements, and exact specific identification is perhaps not feasible at present. Nor did Vergnaud-Grazzini attempt specific identification of the bones, referring them only to "*Bufo* sp.". She pointed out that there are apparently two species present and that one of them (the one represented by the greatest number of individual bones) closely resembles the modern *B. regularis* in all discernible aspects. These characteristics suggest, but by no means prove, that the 20-chromosome African group of toads was already in existence by the Upper Miocene. The second species from Beni-Mellal shows some resemblances to *B. bufo*, particularly the ilium, but the question of its relationships must remain completely open for the present.

No other *Bufo* have been reported in the fossil record before the end of the Miocene. However, a few limb and girdle elements and vertebrae referable to the genus, but not identifiable to species or species group, have been recovered from scattered Middle to Late Miocene deposits of western Nebraska and South Dakota. Size, proportions, and other observable characteristics of these specimens are not inconsistent with their tentative reference to *Bufo valentinensis* (see below), but neither are they inconsistent with a number of other possible interpretations.

Estes and Tihen (1964) reported two species of *Bufo* from the lower Valentine Formation of north central Nebraska. These date from either the end of the Miocene or the beginning of the Pliocene; they are essentially transitional between the two epochs. Both species appear related to either the *americanus* or the *cognatus* group. These two groups are not separable from each other osteologically, although individual species within each group possess individually diagnostic features. Both are characterized by a narrow skull, frontoparietal independent of the proötic and usually prominent crests or ornamentation, and, in most species, an exceptionally high ilial prominence. Although the available fossil specimens consist entirely of dissociated individual bones, they demonstrate conclusively that the Valentine species possessed the cranial features mentioned. One of the two species has also a relatively high ilial prominence. The other does not, but this is not a universal characteristic of modern members of the group; *B. terrestris*, for example, is fully comparable to the fossil form.

The smaller of the two species, *B. valentinensis*, is not definitely known other than from the Valentine Formation, but, as mentioned above, a few other elements from various Middle and Upper Miocene deposits of the same general area may be referable to this or to a similar species. The other species, *B. hibbardi*, was first described from the Edson Beds, Middle Pliocene, of Kansas (Taylor 1936).

The fossil record of pre-Pliocene toads, although revealing nothing about the ancestry or orgin of the genus, does make it clear that it had attained essentially world-wide distribution before the end of the Miocene. The relationships of the Old World fossil forms are less clear, but there is some evidence suggesting that at least one of the extant groups in each major landmass was present before the end of the Miocene and in the same major area in which it now occurs. The wide distribution and considerable diversification that had been attained by the Upper Miocene indicate a long prior history for the genus. Even though the purported pre-Miocene occurrences of the genus reported to date are unconvincing, there is strong reason to believe that *Bufo* existed in the Oligocene and possibly even earlier.

Pliocene Fossils

Surprisingly perhaps, no *Bufo* have been reported from the Pliocene of Africa, South America, or Asia, and few from Europe, which surely reflects the lack of extensive use of appropriate collecting techniques in these areas, rather than any absence or scarcity of representatives of the genus during that epoch.

In Europe, both *B. bufo* and *B. viridis* have been reported from the Pliocene of Hungary (Bolkay 1913). Młynarski (1960, 1961) has reported a number of bufonid remains from the Pliocene of Poland; some of these were not specifically identified; others were referred to *Pliobatrachus langhae*, which, as previously noted, is probably not a bufonid. Still others were referred to a new species, to which he gave the name of *Bufo tarloi*. He pointed out a resemblance between some of the elements referred to *B. tarloi* and the corresponding elements in the poorly known *Diplopelturus ruscinensis* from the Lower Pliocene of France, suggesting that the latter is a *Bufo* and should be known as *Bufo ruscinensis*. If his suggestion is correct, four species of the genus are known from the European Pliocene. In my opinion, however, the available specimens do not yet demonstrate that four different species actually existed; "*B.*" *ruscinensis* is so inadequately represented that even its family affinities are not unequivocally demonstrable. What does seem clear is that the genus was present and was represented by one or more species that at least did not differ drastically from the living species, *B. bufo* and *B. viridis*.

The Pliocene record in North America is more

extensive. The existence of *B. hibbardi* and *B. valentinensis* in the Miocene-Pliocene transitional period has already been mentioned. One of these, *B. hibbardi,* and three other species appear in the Middle Pliocene (some perhaps in the Lower Pliocene) of Kansas. The other species include the extant *B. cognatus* and the exclusively fossil forms *B. spongifrons* and *B. alienus* (Tihen 1962). *B. spongifrons,* like *B. hibbardi* and (obviously) *B. cognatus,* belongs to the *americanus* group–*cognatus* group complex. *B. alienus* is known only from a single ilium, markedly different from the ilia of any other known North American toad. Morphologically, it is more similar to the ilia of some of the European species, notably *B. calamita,* but no taxonomic conclusions can be drawn from such meager information.

Two fossil species are known from the Middle Pliocene of areas other than the Great Plains. *Bufo campi* Brattstrom (1955), from the Yepomara Formation of Chihuahua, was described on the basis of a single tibiofibula. This bone bears a unique ridge along its postaxial border, but neither this characteristic nor any other provides clues about the affinities of the species. The Middle Pliocene Floridian *B. tiheni* Auffenberg (1957) is scarcely better represented: two sacra and five ilia have been recovered. Because the ilial prominence is relatively high, Tihen (1962) referred the species to the *americanus* group. However, the sacral vertebrae are markedly depressed, and this seems to have been a small species. Both of these characteristics suggest the possibility of relationship to *B. quercicus,* a species now inhabiting the area in which the fossils were found.

From the Upper Pliocene (Blancan), Tihen (1962) reported three species, all occurring in the Rexroad Formation of southwestern Kansas. These included the extant "*B. compactilis*" (probably better referred to *B. speciosus*) and the fossil forms, *B. rexroadensis* and *B. suspectus.* The last species is based on a single ilium; morphological features suggest affinity with *B. punctatus,* but the suggestion remains highly speculative. *B. rexroadensis* is appreciably similar to the modern *B. woodhousei;* the two species probably represent temporal stages in a single population line, and it is highly possible that *B. hibbardi* represents an earlier stage in this same line.

Quaternary Fossils

The Pleistocene-Holocene record of the genus scarcely merits extensive discussion. Although in a few instances specific names have been provided for Pleistocene specimens, such action has probably been ill advised; all known Pleistocene specimens are probably referable to living species. As is true for the Pliocene, Africa remains *terra incognita* in the fossil record of toads. Only *B. arenarum* has been reported from the South American Pleistocene (Tihen 1962). Asia is scarcely better represented, with *B. raddei* and *B. bufo* from Choukoutien (Pei 1940) and the latter species also from Japan (Shikama and Okafuji 1958 and others). Both *B. bufo* and *B. viridis* have been reported by several authors from a number of localities in the Pleistocene of Europe. It is interesting, but probably of no special significance, that the third common European species, *B. calamita,* has not been recorded as a fossil, insofar as I am aware (although no extensive literature search has been made to verify this lack). At least nine extant species have been reported from the Pleistocene of North America; Gehlbach (1965) provides a recent summary. In all known cases, from all continents concerned, Pleistocene specimens appear to be properly referable to extant species that now live either in the same locale as the fossil occurrence, or at least sufficiently close so that their presence at the fossil site is readily explained by the shifts in the borders of ranges, which must have occurred during the strong climatic fluctuations of the Pleistocene.

Summary

The fossil record as currently known seems to represent only the later stages in the phylogenetic history of the genus. No clear record exists before the Middle Miocene. From that time to the present, all fossils about which we have adequate information seem to have affinities with species groups now inhabiting the same general area. Some Pliocene and at least one Upper Miocene form are indistinguishable from living species. In other words, by no later than Middle or Upper Miocene, the genus had already become widely distributed and undergone diversification in each of the major geographic areas where it occurred, so that many species groups, or at least major complexes deriving from several closely related species groups, existed. Despite provocative possibilities suggested by a few inadequately known forms, there is no concrete evidence of any fossil representative being geographically "out of place" when compared with modern distribution of the groups and species. If the known fossil record does primarily represent evolution only at the level of species within a species group, one assumes that the species groups themselves, and more inclusive complexes, as well as the genus *Bufo* itself, developed at appreciably earlier dates. It is to be hoped that these earlier evolutionary developments have also left a fossil record, which needs only to be found.

REFERENCES

Auffenberg, W. 1956. Remarks on some Miocene anurans from Florida with a description of a new species of *Hyla*. Breviora, no. 52,.pp. 1–11.

———. 1957. A new species of *Bufo* from Pliocene of Florida. Quart. J. Florida Acad. Sci. 20:14–20.

Bader, R. S. 1956. A quantitative study of the Equidae of the Thomas Farm Miocene. Bull. Mus. Comp. Zool. 115 (2):1–78.

Bolkay, S. J. 1913. Additions to the fossil herpetology of Hungary from the Pannonian and Praeglacial Periods. Mitteil. Jahrb. kön ungar. geol. Reichsanst. 21(7):217–230, pls. 11, 12.

Brattstrom, B. H. 1955. Records of some Pliocene and Pleistocene reptiles and amphibians from Mexico. Bull. So. Calif. Acad. Sci. 54:1–4.

De L'Isle, A. 1877. Note sur un genre nouveau de batraciens bufoniformes du terrain á *Elephas meridionalis* de Durfort (Gard) (*Platosphus gervaisii*). J. Zool. (Paris) 6:472–478.

Depéret, Ch. 1897. Les animaux pliocènes du Roussillon. Mém. Soc. Géol. France (Paléontol.) 7(3):1–180.

Estes, R., and J. A. Tihen. 1964. Lower vertebrates from the Valentine Formation of Nebraska. Amer. Mid. Nat. 72:453–472.

———, and R. Wassersug, 1963. A Miocene toad from Colombia, South America. Breviora, no. 193, pp. 1–13.

Fejérváry, G. J. 1917. Anoures fossiles des couches préglaciaires de Püspökfürdö en Hongrie. Földtani Közlöny (Budapest) 47:141–172.

Filhol, H. 1877. Recherches sur les phosphorites du Quercy. Ann. Sci. Géol. (Paris) 8:1–340.

Gehlbach, F. R. 1965. Amphibians and reptiles from the Pliocene and Pleistocene of North America: A chronological summary and selected bibliography. Texas J. Sci. 17:56–70.

Hecht, M., and R. Hoffstetter. 1962. Note preliminaire sur les amphibiens et les squamates du Landenien Supérieur et du Tongrien de Belgique. Inst. R. Sci. Nat. Belg. 38:1–30.

———, ———, and C. Vergnaud. 1961. Amphibiens, p. 103. *In* G. Choubert, et al. (eds.), Le gisement de vertébrés miocènes de Beni-Mellal (Morocco). Notes & Mém. Serv. Géol. Maroc. (Rabat), no. 155.

Kuhn, O. 1962. Die vorzeitlichen Frösche und Salamander: Ihre Gattungen und Familien. Jahrb. Ver. vaterl. Naturk. (Württemberg) 117:327–372.

Młynarski, M. 1960. Pliocene amphibians and reptiles from Rebielice Królewskie (Poland). Acta Zool. Cracoviensis (Cracow) 5(4):131–153.

———. 1961. Amphibians from the Pliocene of Poland [in Polish, English and Russian summaries]. Acta Palaeontol. Polonica 6:261–282.

Noble, G. K. 1930. The fossil frogs of the Intertrappean beds of Bombay, India. Amer. Mus. Novitates, no. 401, pp. 1–13.

Pei, Wen-Chung, 1940. The upper cave fauna of Choukoutien. Paleontol. Sinica. Ser. C, 10:1–84.

Piveteau, J. 1927. Études sur quelques amphibiens et reptiles fossiles. Ann. Paléontol. 16:59–99.

Portis, A. 1885. Appunti paleontologici. 2. Resti di batraci fossili Italiani. Atti. R. Accad. Torino (Turin) 20:1178–1201, pl. 13.

Romer, A. S. 1966. Vertebrate Paleontology. 3rd ed. Univ. of Chicago Press, Chicago and London.

Savage, D. E. 1951. Report on fossil vertebrates from the upper Magdalena Valley, Colombia. Science 114: 186-187.

Schaeffer, B. 1949. Anurans from the early Tertiary of Patagonia. Bull. Amer. Mus. Nat. Hist. 93:47–68.

Shikama, T., and G. Okafuji. 1958. Quaternary cave and fissure deposits and their fossils in Akiyosi district Yamagati prefecture. Sci. Rep. Yokohama Nat. Univ. 2(7):43–104.

Taylor, E. H. 1936. Una nueva fauna de batracios anuros del Plioceno Medio de Kansas. Ann. Inst. Biol. México 7:513–529.

Tihen, J. A. 1951. Anuran remains from the Miocene of Florida, with the description of a new species of *Bufo*. Copeia 1951:230–235.

———. 1962. A review of New World fossil bufonids. Amer. Midl. Nat. 68:1–50.

Tschudi, J. J. von. 1838. Classification der Batrachier, mit Berücksichtigung der fossilen Thiere dieser Abteilung der Reptilien. [A preprint, later appearing in] Mém. Soc. Sci. Nat. Neuchâtel 1839:1–99.

Vergnaud-Grazzini, C. 1966. Les amphibiens du Miocène de Beni-Mellal. Notes du Serv. Géol. Maroc (Rabat) 27:43–74.

3. *Macrogenioglottus* and the South American Bufonid Toads

OSVALDO A. REIG[*]

Universidad Central de Venezuela
Caracas, Venezuela

Introduction

South America has a remarkably rich anuran fauna showing its own pattern of diversity and including many endemic groups. The evolutionary relationships among the taxa of this fauna are, however, rather poorly known. Most of the available information about the frogs and toads of this continent comes from the descriptive approach of classical taxonomy, which is hardly fruitful for the purposes of modern evolutionary systematics.

This limitation is being overcome by the results of more modern research, as exemplified by the various contributions to this volume on experimental hybridization, comparative biochemistry, bioacoustics, and comparative karyology. The more traditional approach of comparative anatomy, however, has only been preliminarily used to infer evolutionary relationships among South American anurans, even though it has been repeatedly proven that it provides fruitful information for modern taxonomy and evolution.

Among the South American anurans, there are several monotypic genera of rather obscure relationships, whose internal anatomy is almost completely unknown. This is the case of *Telmatobufo* from Chile, *Strobomantis* from Venezuela, and the Colombian *Amblyphrynus, Niceforonia*, and *Tachyphrynus.* Some Brazilian genera, such as *Basanitia, Notaden,* and *Proceratophrys,* are in the same situation, and, among these, *Macrogenioglottus* is exceptional since it has been the object of some anatomical research.

When improvements are made in the knowledge of the anatomy of these poorly known genera of dubious relationships, a better picture of their relationships and classification will arise. New information can also produce radical changes in the currently held conclusions about classification and evolutionary relationships of the pertinent genera and even of the major taxa. We shall see in this chapter that this is what has happened in the case of *Macrogenioglottus,* the investigation of whose anatomy not only clarified its family allocation, but also cast new light on the problem of bufonid history and bufonid diversity in South America.

[*]Present address: Department of Zoology, University College London, London, England.

The Problem

Macrogenioglottus alipioi was described by Carvalho (1946) as the only known species of a genus of Brazilian toads related to *Odontophrynus* and *Bufo*. The animal is extremely rare, only three specimens being known: the two original individuals described by Carvalho that were found in 1945 at Ilheus, state of Bahia, and a third one that was caught at Itapeví, near the city of São Paulo, by Dr. R. Hoge. The three specimens belong to the collection of the Museum Nacional of Rio de Janeiro. Information from Dr. Bertha Lutz on the discovery of some additional specimens in the state of São Paulo has not been confirmed, but another individual from that provenance has been available for chromosome work by Beçak *et al.* (1970).

Carvalho described *Macrogenioglottus* as a member of the family "Ceratophrydidae." This family name was first used by Miranda Ribeiro (1920, 1926), but wrongly formed, the correct spelling being Ceratophrynidae, as emendated by Reig and Limeses (1963), Limeses (1963), and Barrio (1963). However, the informal taxon-name "Ceratophrydes" was first used by Tschudi in 1838 (see Kuhn 1965) and later (1886) by Cope as a division of the "Cystignathidae" (= Leptodactylidae), and the name Ceratophrynidae as a subfamily of Leptodactylidae was erected in 1935 by Parker. Laurent (1942) also uses the misspelled family name "Ceratophryidae" as a mere equivalent of the whole Leptodactylidae. This and other misspellings have been confusedly used by different authors, as summarized by Limeses (1963). The taxonomic concept was not adequately defined; neither was its extension. The corresponding name was used by various authors to include a differing array of broad-headed, short-legged, and clumsy toads and frogs carelessly supposed to be related. Miranda Ribeiro (1926) included in this family the genera *Ceratophrys, Odontophrynus, Stombus, Zachaenus,* and *Proceratophrys.* The first three have been generally recognized as making up the core of the Ceratophrynidae, but even Miranda Ribeiro (1923) cast some doubt that natural relationships existed among them. As already mentioned, *Macrogenioglottus* was added to the family later, as were *Lepidobatrachus* and *Chacophrys.* These latter were previously confused under *Ceratophrys.* Moreover, various authors suggested also that *Cyclorhamphus, Amblyphrynus,* and even *Craspedoglossa* might belong to the same taxon.

Studies undertaken by myself and associates on the skeleton (Reig 1960*a*; Reig and Cei 1963; Reig and Limeses 1963), the musculature (Reig and Limeses 1963; Limeses 1963, 1964, 1965*a*, 1965*b*), and the behavior and voice (Barrio 1963) gradually bolstered the conclusion that the name Ceratophrynidae should be applied to a natural taxon of family rank clearly distinguished from the Leptodactylidae and comprising only the genera *Ceratophrys, Lepidobatrachus,* and *Chacophrys* (Reig and Limeses 1963: fn. p. 125; Barrio 1963: p. 147). This conclusion has been more recently supported by biochemical studies (Cei, Erspamer, and Roseghini 1967) and karyological data (Bogart 1967). The allocation of the remaining broad-headed genera usually referred to the Ceratophrynidae remains vague after this restatement of the extension of the family, although it is generally agreed that they can be included in the Leptodactylidae. Most probably *Odontophrynus* and *Stombus,* which are closely related in skeletal and muscle characters, are to be placed in the family Leptodactylidae, making a subfamily of their own, which might also include *Proceratophrys. Amblyphrynus* and *Zachaenus* have been reported (Cochran and Goin 1961) as sharing many character states, but their suprageneric allocation within the Leptodactylidae would require a better knowledge of the characteristics of that taxon's many designated subfamilies. *Strobomantis cornuta,* usually referred to as a broad-headed species of *Eleutherodactylus,* is a taxon to take into account when considering the relationships of this array of *Ceratophrys*-like genera. We shall discuss the situation of *Macrogenioglottus* later.

One interesting conclusion of the myological work of Limeses (see especially Limeses 1964 and 1965*b*) was that *Ceratophrys, Lepidobatrachus,* and *Chacophrys* seemed more closely related to the Bufonidae than to the Leptodactylidae. This conclusion is supported by the detailed studies of the thigh and jaw muscles, especially by the jaw muscle studies. As in *Bufo,* in the above-mentioned genera the mandibular branch of the trigeminus nerve is laterally covered by the muscle levator mandibulae externus (= mandibularis externus), whereas in all the examined genera of the Leptodactylidae and also in the ranids and hylids (Stadie 1960) the main stem of the nerve passes over the outer surface of that muscle. Moreover, the depressor mandibulae, which has two well-developed parts of the leptodactylids, namely a pars scapularis and a pars tympanica, has only rudiments of the pars scapularis (*Ceratophrys*, *Chacophrys*) or completely lacks that portion (*Lepidobatrachus*) in the ceratophrynids. This agrees with the situation in *Bufo*, which has only the pars tympanica. Other characteristics, for instance the extension of muscle levator mandibulae subexternus and the lack of "temporalis" position of m. levator mandibulae posterior, are character states shared in common by ceratophrynids and bufonids. Furthermore, members of these two taxa show an open squamosal angle (Griffiths 1954, 1963) measuring more than 55°; they also have an extensive dermal ossification in the skull roof but they lack the omosternum, or have it in a very reduced condition.

It could be alleged that similar states in the above characters might have arisen by convergence and that they do not afford conclusive proof that bufonid and ceratophrynid toads are closely related. In this sense, it can be advocated that *Odontophrynus* and *Stombus* also agree with *Bufo* in several characters, especially the thigh musculature, despite their general agreement with the leptodactylids in other character states. It is precisely because some genera of broad-headed and clumsy leptodactylids match the bufonids in a few, but not in all, character states shared in common by ceratophrynids and bufonids which reinforce the conclusion that they are leptodactylids that converged in some respects with the bufonid-ceratophrynid complex, whereas the overall similarity of members of this latter complex is better thought of as an indication of true evolutionary relationships.

If this hypothesis is true, we have to admit that a bufonid stock existed in South America from a very early date and that it underwent a process of diversification in this continent that led to the separation of at least three different families, namely Bufonidae, Atelopodidae, and Ceratophrynidae, the latter two being endemic to South America. Needless to say, the hypothesis would receive additional support if the above prediction were confirmed by the fossil record or by the discovery in the living South American fauna of another group of toads of bufonid relationships. It is interesting to investigate, thus, what the relationships are of *Macrogenioglottus*, a genus originally described as a ceratophrynid, which shows many similarities with *Bufo*.

After the original paper of Carvalho, *Macrogenioglottus* was studied by Limeses (1964, 1965*b*) in her surveys of thigh and jaw musculature. Her observations stimulated further examination of the anatomical characteristics of this rare toad. Thanks to the assistance of Dr. Antenor Leitao de Carvalho and Dr. Bertha Lutz, I had the chance to examine part of the musculature and the skeleton of the female allotype of *M. alipioi*, although a thorough preparation of the skeleton was avoided in order to keep the anatomical parts of this specimen as complete as possible. We are therefore able to offer a survey of several relevant morphological characteristics of this genus.

External Characters

Macrogenioglottus alipioi is a large toad, comparable in size to *Bufo arenarum*. Its general aspect is clumsy and very *Bufo*-like, differing especially from the species of that genus by its larger head, which is, however, not so large and broad as in the ceratophrynids. As in *Bufo* and the ceratophrynids, the posterior legs are short and the anterior ones long, and the skin is thick and glandular.

The head is very deep and wider than it is long, its length amounting to a little less than one-third of the snout-vent length, whereas its breadth covers almost one-half of that length. The snout is very *Bufo*-like: short, truncate in profile, with a well-defined canthus rostralis. The nostrils are a little nearer to the eyes than to the tip of the snout and are placed below the canthus; the openings are directed upward and forward and are separated from each other by an interval shorter than their distance from the eyes. The loreal region is convex in profile but markedly concave in transverse section because of the two conspicuous ridges at the canthus that diverge backward up to the front of the eyes and continue above the eyes to the rear as two subparallel and sharp supraorbital ridges. Neither postorbital nor orbitotympanic ridges are evident, but a narrow descending preorbital ridge goes downward from the angle of the canthal and supraorbital ridges.

The eyes are large, moderately projecting, and placed forward. The eyelids are conspicuous and have a thickened and granular border, but no hornlike projections. A prominent tubercle is present in the skin between the eye and the border of the mouth. The skin in the postocular region is thickened and elevated, resembling a parotoid gland. If there is a true gland in this region, it is not, however, as large and well defined as in *Bufo*. The tympanum is fairly large and distinct but it is covered by the skin. The mouth is large and broad, the maxillary border not projecting beyond the lower jaw. Small teeth are present on the maxillary, premaxillary, and vomer (see below).

A peculiar characteristic of this genus is the morphology of its tongue, previously described by Carvalho, which provided the origin of the generic name. It is much more reduced than in the species of *Bufo* and the ceratophrynids, being subcircular in outline and nearly one-third the width of the mouth opening. The lateral and posterior borders are free, and the anterior border is far from the front border of the mouth. Thus, the genioglossus muscle is visible through the skin of the floor of the mouth, forming a broad ridge from the mandibular symphysis to the anterior border of the tongue.

The hands are large and have distinct palmar and subarticular tubercles. The fingers are free and have simple terminal phalanges; their lateral borders are distinctly ridged. The third finger is longer than the first, and the latter is longer than the fourth, the second being the shortest. The toes are fringed and almost completely free, a vestigial web being present between the second and the third and between the third and the fourth. The underside of the feet is not tuberculated, but an elongated inner and a round lateral tubercle are evident, the tarsal ridge not being prominent. Between the phalanges of the toes, single and smooth subarticular tubercles are present. The fourth toe is the longest, the third being longer than the fifth and reaching the base of the antepenultimate phalanx of the fourth.

The skin of the dorsum is rather smooth, thick, and glandular, without warts. There is no bony shield under the skin, as is present in most ceratophrynids. On the flanks, three ridges of large, glandular, and oblong tubercles extend on each side from above the tympanum to the groin. The skin of the underside has no tubercles, but its entire surface is covered by small granules. The skin of the anterior and posterior legs is covered with elongated tubercles on the dorsal side, but the underside is smooth. There is no distinct tibial gland on the posterior leg.

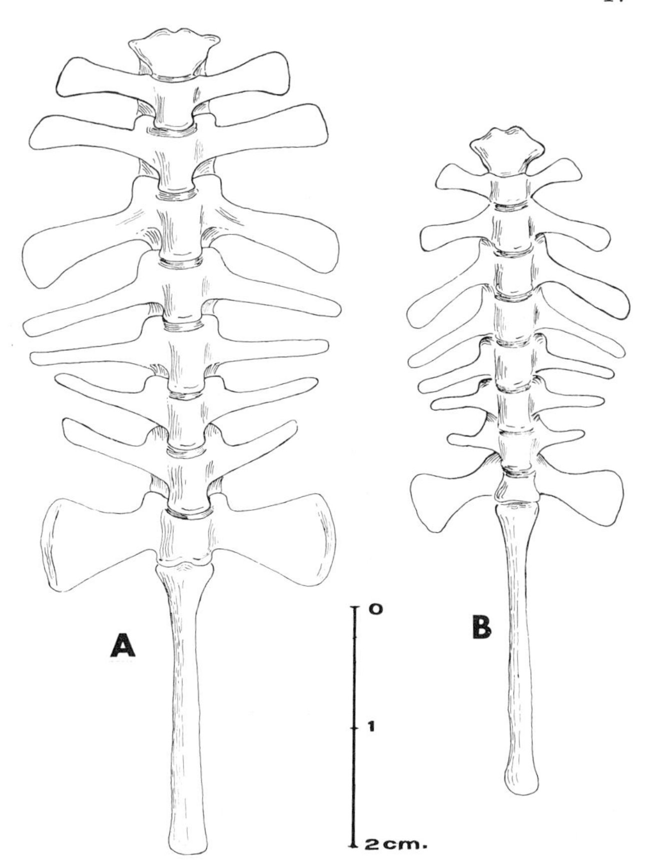

Fig. 3–1. Ventral view of the vertebral column of *Macrogenioglottus alipioi (A)* and *Bufo ictericus (B).*

Skeleton

Vertebral Column

The vertebral column of *Macrogenioglottus* (Fig. 3-1) shows the widespread combination of eight presacral independent procoelous vertebrae and one sacral vertebra with expanded diapophyses, united to the urostyle by two condyles. The vertebral centra are elongated as in *Bufo*, being twice as long as wide. The atlantal cotyles are closely approximated ventrally and are separated by a narrow notch, as in most species of *Bufo*, in *Ansonia*, and in most of the ceratophrynids. The form and proportion of the transverse processes of vertebrae two to eight are much more bufonid than ceratophrynid or leptodactylid. As in bufonids, the transverse processes of the fourth vertebra are longer than those of the third and are oriented posteriorly, whereas those of vertebrae two and three are oriented anteriorly. Moreover, there is no sharp difference in development between those of the vertebrae two to five and those of the last four vertebrae, the latter becoming progressively slightly shorter. In the ceratophrynids, the transverse processes of the third vertebra are much longer than those of the fourth, and there is a sharp reduction in size in those of the four last vertebrae.

The sacral diapophyses are moderately expanded, their anterior corners being distinctly anterior to the prezygapophyses. The expansion of the borders of the diapophyses is as in *Bufo mazatlanensis* and is less developed than in *Bufo spinulosus.* The length of the diapophyses is equal to that of the transverse processes of the third vertebra. They are clearly much more expanded than in *Ceratophrys, Lepidobatrachus,* and the broad-headed leptodactylids *Odontophrynus* and *Stombus.*

The urostyle is short. It is not, however, as short as in ceratophrynids but is markedly shorter than in *Odontophrynus.* Its length is a little less than the width of the diapophyses of the sacral vertebra and equals the combined length of the six first vertebral centra. The spine of the ossis coccygis reaches a little beyond the middle of the bone and it gradually descends backward; no lateral spines are indicated. The posterior portion of the bone is broad and slightly flattened. Generally speaking, the urostyle is similar to that of the bufonids.

Briefly, the vertebral column of *Macrogenioglottus* is similar to that of *Bufo* in all the observed characters, differing strikingly from that of the ceratophrynids and of *Odontophrynus* and *Stombus.*

Shoulder Girdle

Like most bufonids and ceratophrynids, *Macrogenioglottus* (Fig. 3-2) lacks an ossified omosternum; this element is also not developed as cartilage. *Odontophrynus* and other broad-headed leptodactylids, on the other hand, typically have a well-developed omosternum. The xiphisternum, on the contrary, is extensive in *Macrogenioglottus.* It is longer than it is wide and has a notched posterior edge. Its anterior portion is more or less calcified, in the form of two contiguous plates to which the pectoralis sternalis musculature is attached. The epicoracoids overlap in a typical arciferous manner, but a precoracoid bridge, representing around 10 percent of the total epicoracoid length, is noticeable.

The clavicles are strongly forward oriented and have an extensive union with the scapulae. They abut each other with their inner extremities at the precoracoid bridge, and they have a concave anterior border. The coracoids are short and, as in *Bufo* and *Ceratophrys,* have a moderately expanded medial border.

The scapula is a very large bone, similar in relative length to that of ceratophrynids, *Odontophrynus,* and *Stombus,* and longer than in bufonids and most leptodactylids. Its length is twice that of the coracoid. It has a clefted and broadened medial portion and a curved shaft ending at a flat and expanded distal border. Its anterior edge is evenly curved to form a concavity, but the posterior edge shows this to a lesser degree. The major axis of the bone is oriented at approximately right angles with reference to the medial line, so that its distal border opposes both the pars acromialis and the pars glenoidalis. In *Bufo* I have observed a backward orientation of the main axis of the scapula. This results in a more posterior position of its distal border, which faces only the glenoid capsule, while the condition in the ceratophrynids is like that in *Macrogenioglottus.* As in *Bufo,* the ventral side of pars acromialis projects forward to form with the proximal portion of the clavicle a moderately developed preglenoidal projecting surface for the attachment of the scapulohumeralis and part of the deltoid musculature. The arrangement of this region is similar in the ceratophrynids, but in the leptodactylids so far examined, including *Odontophrynus* and *Stombus,* there is a more extensive development of this projection, which is formed by the single or prevailing participation of the scapula.

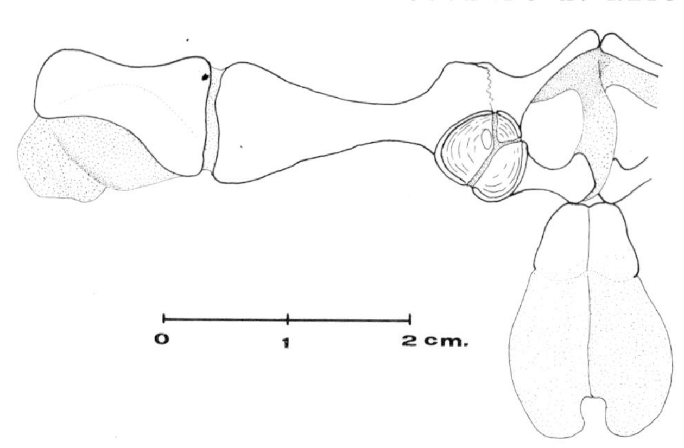

Fig. 3–2. Shoulder girdle of *Macrogenioglottus alipioi*, in ventral view. The scapula and suprascapula are artificially straightened to show their outer and upper faces.

The suprascapula shows a well-ossified anteroproximal portion and a posterior cartilaginous part, which is only poorly calcified. The bony portion is very similar to that of *Bufo* and has no special characteristics.

In brief, the shoulder girdle of *Macrogenioglottus* is bufonid-ceratophrynid in structure, more closely resembling the ceratophrynids than the bufonids in the elongation and orientation of the scapula. Since this resemblance also extends to some deep-bodied leptodactylids, it is safe to conclude that it is due to convergence.

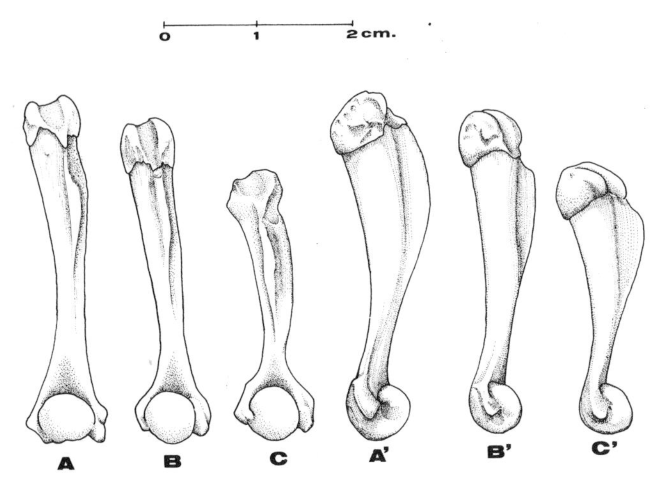

Fig. 3–3. Ventral (*A,B,C*) and lateral (*A′,B′,C′*) views of the humerus of *Macrogenioglottus alipioi* (*A*), *Bufo marinus* (*B*), and *Ceratophrys dorsata* (*C*).

Humerus

The humerus of the female *Macrogenioglottus* is quite similar to that of female specimens of *Bufo marinus* (Fig. 3-3), in that it is almost impossible to describe the differences other than those which can be found within a sample of various humeri of the latter. Both in *Bufo* and in *Macrogenioglottus* the bone is more elongated than in the ceratophrynids and shows a less prominent crista ventralis (= deltoid crest), but it is stouter and less slender than in *Odontophrynus*, *Stombus*, and most of the leptodactylids.

Other bones of the anterior leg, as well as those of the posterior leg and pelvic girdle, have not been studied, in order to conserve complete anatomical pieces for the observation of musculature.

Bony skull

Macrogenioglottus has a well-ossified skull, much like that of some species of *Bufo* in general appearance and even in structural details. The osteocranium of *Bufo* is, however, highly variable in the different species and species groups, and it is unfortunately little known in bufonids other than *Bufo* (see Tihen 1960). In the following description, we have followed both Tihen's (1962*a*) and Sanders' (1953) accounts, and our own observations on South American bufonids, ceratophrynids, and leptodactylids.

The osteocranium is distinctly elevated (Fig. 3-4C). Its height at the posterior end (from the ventral borders of quadratojugals to the dorsal surface above the foramen magnum, in the midline) is 76 percent of the length of the skull (from occipital condyles to the edge of the snout, in the midline). Because of this feature, it resembles the toads of the *Bufo americanus* group. Neither in any of the ceratophrynids, nor in the broad-headed leptodactylids, *Odontophrynus* and *Stombus*, does the skull reach such a remarkable height, although in them it is distinctly deeper than in the flat-headed *Leptodactylus*, *Caudiverbera* (= *Calyptocephalella*), *Telmatobius*, and most of the remaining leptodactylids. Skull depth is an adaptation in amphibians to a dominantly

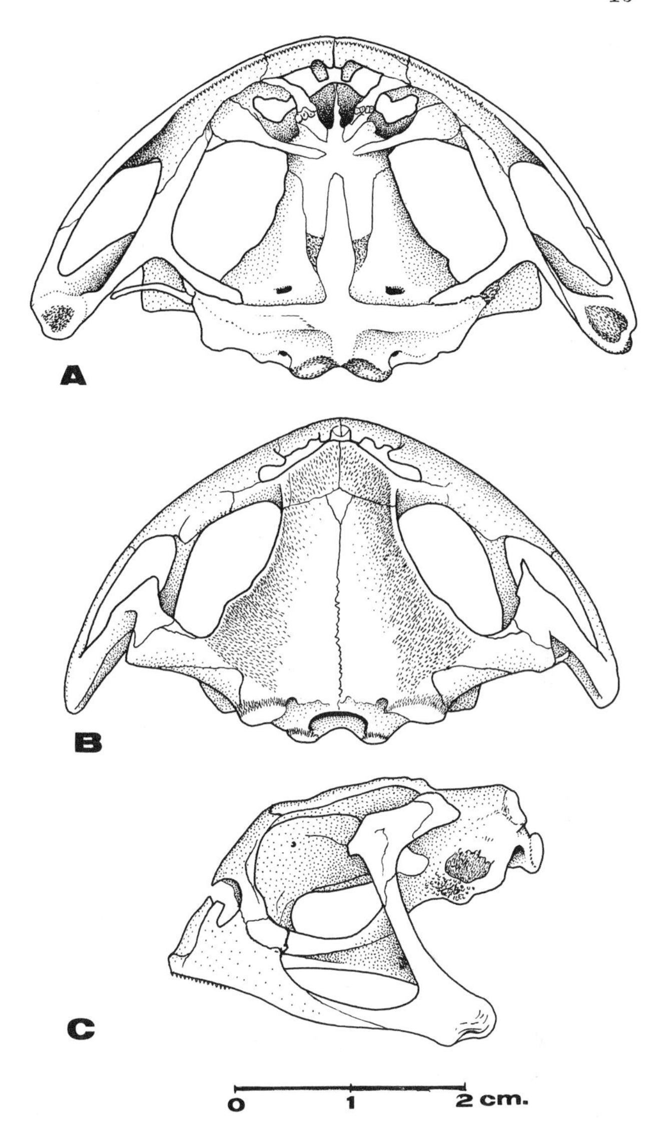

Fig. 3–4. Ventral (*A*), dorsal (*B*), and lateral view (*C*) of the skull of *Macrogenioglottus alipioi*.

terrestrial life and improves the mechanism of lung-breathing (Schmalhausen 1957).

The outer border of the osteocranium describes a nearly perfect semicircle (Fig. 3-4 A, B). The width of the skull at the mandibular articulation is only a little shorter than twice the length of the skull, and the quadrates are in a line with the occipital condyles, as in *Odontophrynus*, *Chacophrys*, and in the North American *Bufo cognatus*.

The dermal covering of the roof bones is well developed, but not as extensive as in *Bufo marinus, B. valliceps*, or *B. guentheri*, or in the ceratophrynids *Lepidobatrachus* and *Ceratophrys*. There is no true dermal ornamentation, but the surfaces of the frontoparietals are distinctly roughened.

The frontoparietals are very broad when compared with those of *Odontophrynus*, *Stombus*, and *Chacophrys* but are narrow by the standards of some species of *Bufo*. The development of the frontoparietals is comparable to the development in *Bufo mauritanicus* and *Bufo terrestris*, being definitely wider than in *Bufo calamita* or *Bufo spinulosus*. Their lateral borders strongly diverge posteriorly and are produced into well-developed and upturned supraorbital crests. As in most species of *Bufo*, except the *calamita* and *spinulosus* groups, the posterior part of the frontoparietals extends as a shelf over the posteromedian part of the orbit. In the underside of this shelf, the frontoparietals show a small aperture for the occipital artery, which passes through a canal that opens to the occipital surface at the medial corner of the epiotic eminences. Therefore, there is an enclosed occipital canal for the artery, as is true in many species of *Bufo* and the ceratophrynids instead of the open groove shown in the *calamita, spinulosus, americanus*, and other groups of the *Bufo* species and the leptodactylids *Odontophrynus, Stombus, Leptodactylus*, and others. The frontoparietals are in contact along most of their medial border. However, a very small portion of these medial borders diverges anteriorly, and, since the posterior border of the nasals diverges posteriorly, the sphenethmoid is exposed in the dorsal view in a rhomboidal area between nasals and frontoparietals. A similar exposure of the sphenethmoid is seen in many species of *Bufo*, in the ceratophrynid *Chacophrys*, and in *Odontophrynus* and many other Leptodactylids.

The nasals are very *Bufo*-like. They are somewhat swollen and they slant downward sharply, being concave in transverse section. Their posterior borders form an angle with the apex directed anteriorly; they are slightly emarginated in the middle in connection with the dorsal exposure of the sphenethmoid.

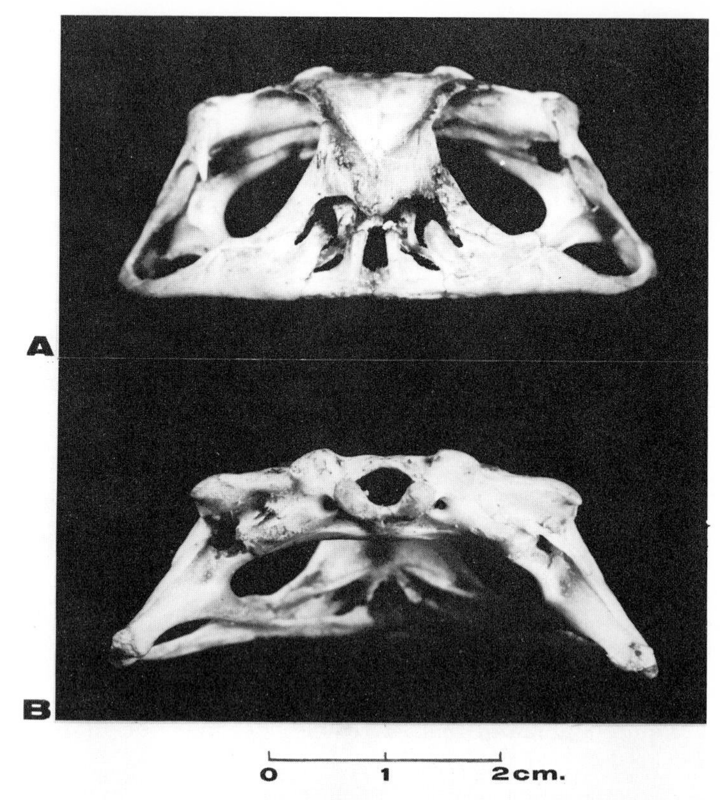

Fig. 3–5. Frontal (*A*) and occipital (*B*) view of the skull of *Macrogenioglottus alipioi*.

A significant feature of the skull of *Macrogenioglottus* is the development of an otoparietal plate (Sanders 1953) covering most of the dorsal surface of the proötic and having an extensive union with the squamosal. This is characteristic of the bufonid skull and also occurs in the ceratophrynids, but is absent in *Odontophrynus* and *Stombus*. *Macrogenioglottus* differs, however, from *Bufo* in that the dermal covering of the proötic is almost exclusively contributed by this otoparietal extension of the frontoparietals whose transverse portion encircles the whole posterior margin of the orbit and overlaps the whole anterior and lateral portion of the dorsal surface of the proötics. The development of a temporal plate of the squamosal is not great and it does not overlap the proötic as in most species of *Bufo* and in the ceratophrynids. This plate is merely an expanded part of the posterior head of the squamosal, which has a broad dorsal contact with the transverse portion of the otoparietal plate. The latter also has a longitudinal portion, as in *Bufo,* which covers dorsally a part of the epiotic eminences, the proötic being free from dermal covering at a small area outward from each eminence. There is no postorbital ridge, the anterior edge of the otoparietal plate being very thin and not projecting, which results in an emargination that interrupts the line of the diverging lateral edges of the frontoparietals.

The posterior border of the proötics forms a sharp ridge proceeding laterally from the epiotic eminences. The undersurface of the proötics anterior to this ridge forms the roof of the otic capsules. The fenestra ovalis is oblong in outline and large, having a well-defined, projecting posterior border. It is covered by a cartilaginous operculum to which the long, rodlike, and curved columella attaches by an enlarged footplate. The epiotic eminences are well developed in the form of a swollen plug projecting backward and upward. They have a blunt posterior surface, which does not show the sharp posterior border seen in *Ceratophrys* and *Lepidobatrachus.* The foramen magnum is ovoid, broader than high.

The stem of the squamosal is elongated and oriented backward and downward, forming an angle of about 55° with the maxilloquadratojugal border. Its internal border is intimately united with the posterior branch of the pterygoid. This branch of the pterygoid is nearly as long as its anterior branch, which is thin and rather wide and reaches forward to the planum antorbitale, contacting the expanded lateral portion of the palatines. The same contact of pterygoids and palatines occurs in many species of *Bufo* and in *Ceratophrys,* as well as in *Stombus* and *Odontophrynus,* but it does not occur in *Chacophrys* and *Lepidobatrachus.* The zygomatic process (anterior branch) of the squamosal is short and acute and slants sharply downward. As in *Odontophrynus* and in most species of *Bufo,* it does not reach the maxillary to form a posterolateral fenestra as it does in *Stombus* and the ceratophrynids.

The quadratojugals are well developed and have a broad contact with the maxillaries. Both bones overlap along the posterior two-thirds of the pterygoid fossa, the quadratojugals being extensively exposed dorsally to the maxillaries and having a ventral exposure much shorter than the dorsal one. The pterygoid fossa is extensive, as in *Lepidobatrachus* and *Odontophrynus,* extending farther forward than in *Bufo* to reach the level of the anterior end of the parasphenoid.

The maxillaries have a rodlike zygomatic posterior portion in oblique contact with the quadratojugals, and there is a somewhat expanded anterior portion that lacks roughenings or ornamentation. This latter portion contributes to the lower margin of the orbit and has a broad contact with the descending process of the nasal, which forms the anterior border of the orbit. Anterior to the descending process of the nasal, each maxillary develops an ascending process, which contributes to the definition of the anterior border of the external nares. The maxillary border bears minute teeth (see below) on the anterior one-third of its length. The premaxillaries are moderately developed and they also bear small teeth along their entire border. Their ascending processes do not reach the nasals.

The vomers are long and narrow bones that diverge anteriorly and do not contact each other. They have a dentigerous posterior part and an anterior prechoanal process that extends forward to contact the maxillaries and premaxillaries at their junction. No lateral process or postchoanal element is developed. There are four vomerine teeth on one side and three on the other. They are larger than the maxillary and premaxillary teeth and are implanted in two small and distinct, widely separated patches on the inner border or the choanae. This dentigerous posterior portion of the vomers does not come in contact with the palatines. The latter are well developed and are composed of an inner slender part and an expanded lateral portion contacting with the maxillary and with the anterior end of the parasphenoid, without contacting one another. *Macrogenioglottus* differs from all known species of *Bufo* in possessing vomerine teeth and lacking lateral processes in the vomer. Vomerine teeth are also reduced or absent in the ceratophrynids, but they are well developed in *Odontophrynus* and *Chacophrys.* Palatines are well developed in *Bufo,* as well as in the above-mentioned broad-headed leptodactylids and in the ceratophrynids, with the exception of *Lepidobatrachus,* which lacks any trace of these bones, as is the case in some genera of bufonids other than *Bufo* (see Tihen 1960).

The sphenethmoid is well ossified and has an extensive exposure anterior to the orbit and, as already indicated, a small exposure on the dorsal side of the skull. The stem of the parasphenoid is acuminated and weakly keeled and it is much shorter than the combined length of the alate portions of the bones. These portions are long and broad and are obliquely truncated at their lateral end underlying the fenestra ovalis. The medial arm of the pterygoids overlaps with the anterior border of the alate processes at their distal part. This region of the skull is not particularly characteristic but represents character-states that are wide-spread among various groups of anurans.

In overall structure, the skull of *Macrogenioglottus* is very close to that of *Bufo* and is less similar to that of the ceratophrynids. It resembles only very generally the skulls of the leptodactylids, including *Odontophrynus* and *Stombus*. The only important differences from *Bufo* are in the presence of maxillary, premaxillary, and vomerine teeth and in the exclusive participation of the otoparietal plate in the dermal covering of the proötic. The first can be ascribed to the retention of an ancestral character state, and the second might be regarded as a peculiar development from a common ancestral state lacking exostosic disposition on the proötic.

Musculature

Pectoral

The pectoral musculature of *Macrogenioglottus* is much like that of *Bufo*, which is known best from the contributions of Bigalke (1927) and Bhati (1955). Jones (1933) is the author of a comparative survey of the pectoral musculature, in which he claims that *Bufo* is peculiar in the absence of "episternohumeralis" muscle (=m. deltoideus, pars epicoracoidea, in this paper). This absence could provide an interesting clue for comparing other genera of supposed bufonid relationships, but, unfortunately, Jones's observations are rather superficial and disagree with the excellent studies of Bigalke and Bhati, who described a well-developed deltoideus epicoracoideus in *Bufo vulgaris* (= *B. bufo*) and *Bufo andersoni*. Nevertheless, the accounts of the above authors and our own observations in *Bufo marinus* and *Bufo arenarum* show that the pectoral musculature of *Bufo* is fairly characteristic in several other peculiarities, thus affording a solid ground for inferring taxonomic relationships. The lack of accurate and extensive studies on the musculature of leptodactylids and ceratophrynids is, however, a challenge to the validity of this inference. We tried to overcome this challenge through comparisons with *Ceratophrys* and *Leptodactylus* from our own studies and with *Lepidobatrachus* (Limeses 1968), but failed to include in this report comparative observations of the broad-headed leptodactylids. In the following description the nomenclature of Bhati (1955) has been used with only minor changes. (See Fig. 3-6.)

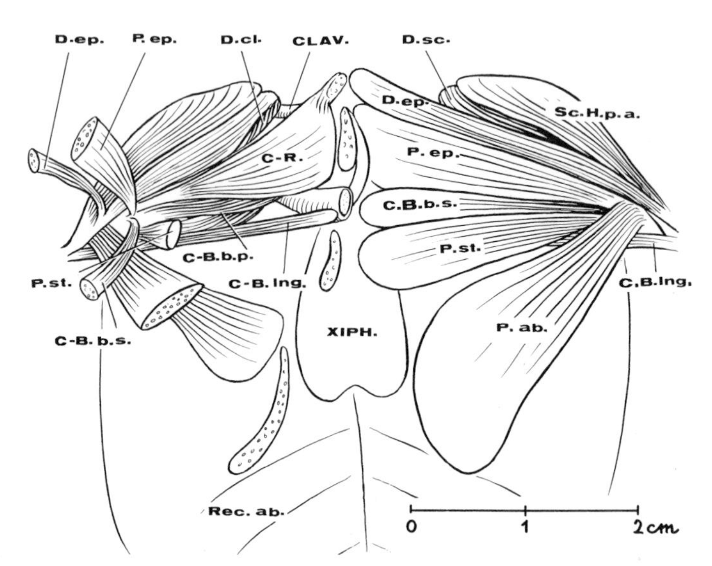

Fig. 3–6. The pectoral muscles of *Macrogenioglottus alipioi*, seen from the ventral side. *C-B,b,p*, coracobrachialis brevis profundus; *C-B,b,s*, coracobrachialis brevis superficialis; *C-B, lng.*, coracobrachialis longus; *CLAV*, clavicle; *C-R*, coracoradialis; *D. cl.*, deltoideus clavicularis; *D.ep.*, deltoideus epicoracoideus; *D.sc.*, deltoideus scapularis; *P.ab.*, pectoralis abdominalis; *P.ep.*, pectoralis epicoracoideus; *P.st.*, pectoralis sternalis; *Rec. ab.*, rectus abdominalis; *Sc.H.p.a.*, scapulohumeralis profundus anterior; *XIPH.*: xiphisternum.

The deltoideus complex is complete in *Macrogenioglottus*, and its three portions resemble those of *Bufo*. In ventral view, the deltoideus scapularis is visible between the scapulohumeralis profundus anterior and the deltoideus episternalis, being partially covered by the latter. In *Leptodactylus*, however, this muscle is broadly exposed on the ventral surface, and in *Ceratophrys* and *Lepidobatrachus*, only the deltoideus episternalis is partially visible in ventral view. In *Macrogenioglottus*, the deltoideus scapularis has its origin on the dorsal surface of the acromial portion of the clavicle and extends outward and backward to insert in the crista ventralis humeri, as it does in *Bufo* and *Ceratophrys*, while in *Leptodactylus* it goes farther, to insert in the crista medialis humeri.

The deltoideus episternalis (= deltoideus cleidohumeralis longus, episternohumeralis) is a long, narrow muscle lying in front of and in close contact with the pectoralis epicoracoideus. It arises at the medial extremity of the clavicle and extends laterally and slightly backward to insert, as in *Bufo*, on the crista ventralis humeri, near to the attachment of the scapulohumeralis profundus. In *Ceratophrys* it has a similar insertion, but it is only partially exposed in the ventral surface. In *Lepidobatrachus*, inferring from the illustration given by Limeses (1968), who confused this muscle with the anterior portion of the coracoradialis, it is like *Ceratophrys* and *Macrogenioglottus* in insertion, but it is more exposed ventrally than in the former. In *Leptodactylus* it takes its proximal insertion on the lateral surface of the omosternum and has extensive exposure on the ventral surface, running more obliquely backward to its insertion on the crista ventralis humeri.

Anterior to the deltoideus episternalis, a portion of the deltoideus scapularis is visible in ventral view, but the main muscle in this region is the scapulohumeralis profundus, which has its origin on the anterior border of the dorsal surface of the pars acromialis scapulae and runs obliquely outward and backward to insert, after passing over the shoulder joint, on the crista ventralis humeri. It is not visibly divided into an anterior and a posterior portion as described in *Bufo* and *Rana* (Bhati *op.* 1955). It is a strong muscle in *Leptodactylus*, in close contact with the deltoideus scapularis. These two muscles, together with the deltoideus clavicularis, insert in *Leptodactylus* on the crista medialis humeri. In *Ceratophrys*, neither the scapulohumeralis profundus, nor the deltoideus scapularis is visible in the ventral view, and the two attach to the humerus as in *Bufo*, namely, on the crista ventralis.

The three pectoralis muscles are well developed and broad, forming most of the pectoral musculature exposed on the ventral side. The pectoralis epicoracoideus is a broad triangular muscle situated between the deltoideus episternalis and the coracobrachialis brevis superficialis, as in *Bufo*, *Ceratophrys*, and *Lepidobatrachus*, while in *Leptodactylus* it only contacts the pectoralis sternalis posteriorly. It has its proximal insertion in the epicoracoidal cartilage and runs outward over the coracoradialis to insert on the proximal part of the crista ventralis humeri, before the insertion of the deltoideus episternalis and beside the attachment of the coracobrachialis brevis superficialis.

Noble (1926) pointed out that possession of a "supracoracoideus profundus" is a characteristic feature of his bufonids. Jones (1933) referred to the same muscle as lying at the posterior border of the "supracoracoideus superficialis" (= pectoralis epicoracoideus) and between it and the pectoralis sternalis, being "sometimes overlapped by one or the other of these muscles" and arising from "the medial end of the coracoid and the adjoining epicoracoid." This is obviously the pars superficialis of the muscle coracobrachialis brevis, as described by Bhati and previous authors. This muscle is visible on the ventral surface in *Bufo*, having the characteristic feature of intervening between the pars epicoracoidea and the pars sternalis of the pectoralis, leaving a portion of the coracoradialis visible in front of its proximal half. Exactly the same disposition is found in *Ceratophrys* and *Lepidobatrachus* (though in the latter genus the coracoradialis is completely covered by coracobrachialis brevis superficialis, referred to by Limeses, 1968, as the supracoracoideus profundus), while in *Leptodactylus ocellatus* this muscle is thin and is completely covered by the pectoralis epicoracoideus, the two pectoralis muscles being in broad contact and not leaving room for a ventral exposure of the coracoradialis. In *Macrogenioglottus*, the coracobrachialis superficialis brevis is a narrow muscle situated between the pectoralis epicoracoideus and the pectoralis sternalis. As in *Lepidobatrachus*, it completely covers the posterior portion of the coracoradialis and leaves no portion of the latter exposed ventrally. It arises, as in *Bufo*, from the posterior part of the medial expansion of the ventral surface of the coracoid and runs outward to attach on the crista ventralis humeri, beside and anterior to the insertion of the pectoralis epicoracoideus.

The pectoralis sternalis is a broad, triangular, and flat muscle adjacent to and posterior to the coracobrachialis superficialis brevis. It covers completely the coracobrachialis longus and arises from the anterior calcified portion of the xiphisternum to extend outward to its insertion on the shaft of the humerus. Near its distal end, it is overlapped by the tendon of the pectoralis abdominalis, as is true in *Bufo*, *Lepidobatrachus*, and *Ceratophrys*. In the latter and in *Leptodactylus* the muscle is broader and fan-shaped, with a longer attachment on the medial line than in *Bufo*.

The pectoralis abdominalis of *Macrogenioglottus* is the broadest and the strongest of the muscles of the pectorohumeral set. It is fan-shaped and has a medial broad origin in the sheath of the rectus abdominis, lateral and posterior to the cartilaginous xiphisternum, and running forward obliquely and laterally. The fibers converge distally into a broad tendon that overlaps the coracobrachialis longus, the pectoralis sternalis and epicoracoideus, and the coracobrachialis brevis superficialis to insert on the crista ventralis humeri. This tendon is partially overlapped laterally by the deltoideus episternalis. Most of the fibers of the pectoralis abdominalis cover the oblique externus scapularis. The described morphology of the pectoralis abdominalis agrees with that of *Bufo* in detail. In *Ceratophrys* this muscle is generally similar, but it is stronger, and its distal insertion does not reach beyond the xiphisternum as it does in *Bufo*, *Lepidobatrachus*, and *Macrogenioglottus*. In *Leptodactylus* it is much more slender and extends strongly backward to reach a very posterior insertion on the sheath of the rectus abdominis.

Dissecting the pectoralis epicoracoideus, the coracoradialis becomes visible. This is a fairly broad, triangular, and fan-shaped muscle covered by the pectoralis epicoracoideus and partially by the coracobrachialis brevis superficialis. As in *Bufo*, it arises from the medial line of the epicoracoid and the medial portion of the clavicle, being divided into one anterior and one posterior portion that pass on distally into a single narrow and strong tendon. This tendon runs through a canal adjacent to the crista ventralis humeri, passes into a groove between the two distal condyles of the humerus, and inserts on the olecranon process of the radioulna. After dissecting the coracobrachialis brevis superficialis, the inner elements of the coracobrachialis complex are visible. The coracobrachialis brevis profundus forms a bulky mass of muscle partially covered by the posterior portion of the coracoradialis and the coracobrachialis longus. As in *Bufo*, it is divided into anterior, medialis, and posterior portions. The former two arise from the distal part of the coracoid, close to the glenoid cavity. The coracobrachialis brevis profundus anterior is inserted on the ventral surface of the coracoid, whereas the medialis portion arises from its dorsal surface. The two portions have distinct attachments on the lateral side of the crista ventralis humeri. The posterior portion is covered by the coracobrachialis longus, which is a narrow and long muscle running obliquely outward and a little backward underneath the coracobrachialis brevis superficialis, the pectoralis sternalis, and the distal portion of the pectoralis abdominalis. It originates in the posterior part of the medial border of the coracoid, just behind the attachment for the coracobrachialis brevis superficialis, and inserts on the ventral surface of the humerus at a fairly distal point. The same morphology has been observed in *Bufo* and *Ceratophrys*.

After the pectoralis abdominalis is dissected, the obliquus externus scapularis is visible. This is a flat fan-shaped muscle originating, as in *Bufo*, on the sheath of the rectus abdominis in front of and outside the origin of the pectoralis abdominalis, which completely covers its ventral portion. It runs forward and outward and narrows into a feeble tendon abutting at the middle of the dorsal surface of the scapula. The morphology and disposition of this muscle is quite similar in *Bufo*. In *Ceratophrys* and *Lepidobatrachus* it is much stronger and broader, while in *Leptodactylus* it is a very reduced strip, scarcely, if at all, covered by the pectoralis abdominalis. Other muscles of the pectoral region are not described, because of the difficulty of observing the single available specimen, which is badly preserved.

The above comparative description demonstrates that *Macrogenioglottus* greatly resembles *Bufo* in pectoral musculature. From the fourteen studied muscles, thirteen are similar in both genera in insertions, relations with other muscles, and relative development. The only two appreciable differences are the apparent lack of division in the scapulohumeralis profundus and in the position of the coracobrachialis brevis superficialis, which intervenes between the pectoralis epicoracoideus and the pectoralis sternalis, thus leaving no space for the ventral exposure of the coracoradialis. These are clearly minor differences of the type found in *Bufo* at the intrageneric level. *Macrogenioglottus* and *Bufo* are also closely related to *Ceratophrys* and *Lepidobatrachus* in pectoral musculature, but the relationships with the latter are not so close as between the former. Finally, *Leptodactylus* must be placed in a separate group.

Thigh

The thigh musculature of *Macrogenioglottus* has been studied by Limeses (1964) using the same specimen we redescribe here. (See Fig. 3-7.)

Because of the weak development of the posterior leg, the thigh of *Macrogenioglottus* is not very bulkily built, but somewhat slender, with narrow, long muscles. The cruralis is a little more developed than the gluteus and makes up most of the anterior surface of the thigh. The tensor fasciae latae is, as in *Bufo* and *Ceratophrys*, a short muscle originating from the ilium at a posterior position, behind the insertion to the iliacus externus and slightly in front of the iliacus internus. Distally, the tensor is widely broadened, originating from the cruralis and the gluteus. In *Odontophrynus americanus* this muscle

I. Broad-skulled toads: (1) *B. poeppigi*, (2) *B. marinus*, (3) *B. paracnemis*, (4) *B. ictericus*, (5) *B. arenarum*, and (6) *B. crucifer* of South America; (7) *B. humboldti* of the intermediate-skulled South American *granulosus* group; the broad-skulled (8) *B. typhonius* and (9) *B. haematiticus* of South America; (10) *B. superciliaris* of Africa; (11) *B. blombergi* of South America; and (12) *B. melanostictus* of Asia.

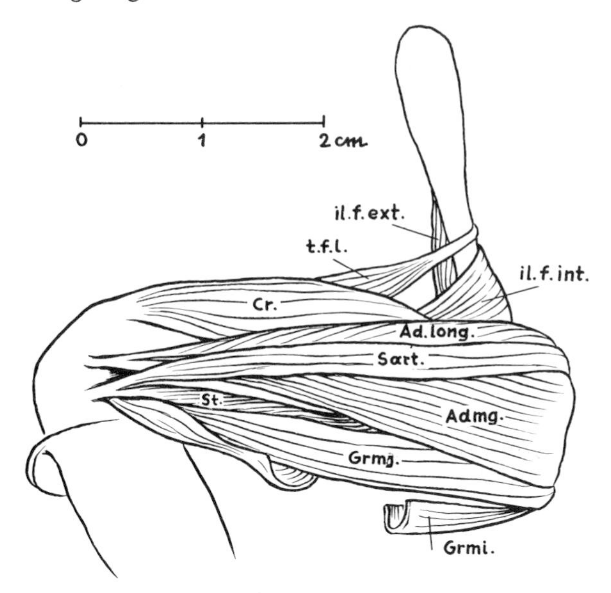

Fig. 3–7. *Macrogenioglottus alipioi*, thigh muscles in ventral view. *Ad. long.*, adductor longus; *Admg.*, adductor magnus; *Cr.*, cruralis; *Grma.*, gracilis major; *Grmi.*, gracilis minor; *il.f.ext.*, iliofemoralis externus; *il.f.int.*, iliofemoralis internus; *sart.*, sartorius; *st.*, semitendinosus; *t.f.l.*, tensor fasciae latae.

is completely absent, while in two other species of the same genus, *O. occidentalis* and *O. cultripes*, it is present but much reduced.

The adductor longus is a distinct, flat, and long muscle. It is half the width of the sartorius and runs parallel to it through most of its length. It is not overlapped by the sartorius and, laterally, it covers a part of the posterior portion of the cruralis. This situation is similar to *Bufo*, *Telmatobius*, *Leptodactylus*, and *Craspedoglossa*. In *Odontophrynus* the muscle is slender and is overlapped extensively by the sartorius. The distal tendon of the adductor longus unites with the distal tendon of the ventral head of the adductor magnus.

The sartorius is narrow, as in the species of *Bufo* that have been studied and in the ceratophrynids, while in *Leptodactylus* it is distinctly broad. *Odontophrynus* shows an intermediate development. The sartorius overlaps the distal portion of the adductor magnus and inserts on the knee by a tendon united with the tendon of the semitendinosus and passes over the tendons of the graciles in the typical "bufonid" way.

The semitendinosus has two portions — dorsal and ventral. The ventral portion has a medial elongated belly and two tendinous ends, while the dorsal portion has a proximal belly and a long distal tendon that unites with the tendon of the ventral portion. This single distal tendon lies ventral to the tendons of the gracilis and joins the tendon of the sartorius before reaching the knee. Before joining, the tendon of the sartorius and that of the semitendinosus run side by side in the same plane, as is the case in *Bufo*, *Stombus*, *Lepidobatrachus*, and *Odontophrynus*.

The adductor magnus caput ventralis is the main muscle of the ventral surface of the thigh. It has a broad origin on the ventral edge of the puboischiadic plate and runs distally, narrowing gradually to end in a strong tendon before reaching the knee, to which the tendon of the adductor longus unites. It is partially overlapped by the sartorius in its proximal half and distally it is obliquely crossed by that muscle, so that the tendon of the sartorius and semitendinosus insert posterior to the tendon of adductor longus and adductor magnus. A well-developed accessory head of the adductor magnus is distinct on its distal portion. This accessory head is composed of several fibers originating both from the tendon and the belly of the ventral portion of the semitendinosus, as it also does in *Bufo* and *Leptodactylus*, whereas in *Ceratophrys* and *Odontophrynus* their fibers only originate from the tendon of the semitendinosus.

The gracilis major and gracilis minor are both well developed. The latter almost completely overlaps the former. The two gracilis muscles end distally as tendons passing dorsal to the tendons of the sartorius and semitendinosus. The iliacus internus is a broad, well-developed muscle with a short attachment to the ilium. The iliacus externus has a bulky belly and attaches to the dorsal surface of the ilium at a level anterior to the insertion of the tensor fasciae latae.

On the dorsal side of the thigh, the distal half of the iliofemoralis is visible between the gluteus and the semimembranosus. The latter is a fairly broad muscle, with an extensive insertion on the dorsal and posterior edge of the puboischiadic plate, contacting a strong and narrow tendon inserted on the condyles of the femur, and it has a long contact behind with the gracilis minor.

As stated by Limeses, *Macrogenioglottus* matches *Bufo* in thigh musculature in many relevant character states, namely (1) the complex of distal tendons, with the tendon of semitendinosus lying ventral to the tendons of the graciles and joining the sartorius before reaching the knee; (2) tendons of sartorius and semitendinosus running side by side in the same plane to their insertion in the knee; (3) iliacus externus is short with a bulky belly; (4) gracilis minor is well developed; (5) semitendinosus has two bellies of different size; (6) accessory head of adductor magnus is well developed, rising from both the tendon and the belly of the ventral semitendinosus; (7) sartorius is long and narrow; (8) adductor longus is distinct, flat, and long; (9) tensor fasciae latae is well developed, but short. The ceratophrynids *sensu strictu* agree with most of the above character states but show for a few of them, for instance, (1), (2), (3), and (6), different states in different genera so that it seems safe to conclude that *Macrogenioglottus* is more *Bufo*-like than ceratophrynidlike in thigh musculature. *Odontophrynus* and *Stombus*, to judge from Limeses's paper, also agree considerably in thigh musculature with *Bufo*, *Macrogenioglottus*, and the ceratophrynids, but, taking other features into consideration, it is wise to ascribe those similarities to convergence from a primitive leptodactylid condition.

It is interesting to point out, moreover, that *Macrogenioglottus* is even more *Bufo*-like in thigh musculature than other bufonid genera, such as *Pseudobufo*, *Ansonia*, *Wolterstorffina*, *Pelophryne*, *Mertensophryne*, *Laurentophryne*, and *Nectophryne*, which lack the adductor longus and have an elongate tensor fasciae latae (Tihen 1960).

Jaw

The muscles involved in the jaw movements have also been reported for *Macrogenioglottus* by Limeses (1965). We have notes on this part of the musculature that confirm that description. The following account is based mostly on the paper of Limeses. The nomenclature of Edgeworth (1931) is used with the minor changes proposed by Limeses (1965).

The depressor mandibulae is composed of two parts (Fig. 3–8). The anterior part (pars tympanica) is the stronger and is attached to the ridge of the posterior border of the proötics, the posterior head of the squamosal, and the posterior and lower border of the tympanum. The posterior part (pars scapularis) is proximally well differentiated from the belly of the pars tympanica and arises from the dorsal fascia above the suprascapula. The morphology of this muscle resembles that of *Ceratophrys* and *Stombus*, which also have a reduced pars scapularis and a strong pars tympanica, whereas in *Bufo* and *Lepidobatrachus* the pars scapularis is completely absent. Unfortunately, other genera of the bufonids have not

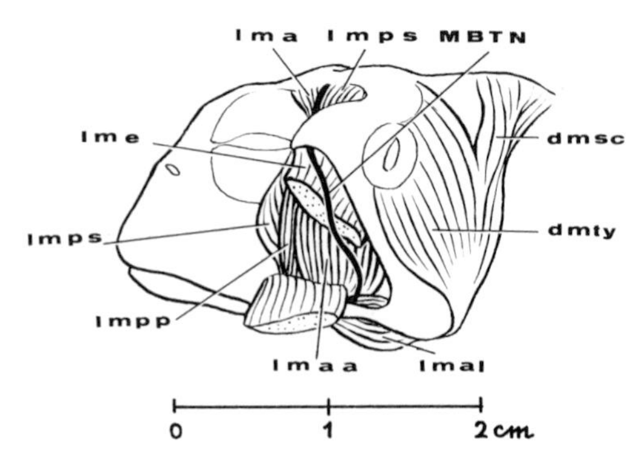

Fig. 3–8. Jaw muscles of *Macrogenioglottus alipioi*. *dmsc*., depressor mandibulae, pars scapularis; *dmty*, depressor mandibulae, pars tympanica; *lma*, levator mandibulae anterior; *lmaa*, levator mandibulae anterior articularis; *lmal*, levator mandibulae anterior lateralis; *lme*, levator mandibulae externus; *lmpp*, levator mandibulae posterior profundis; *lmps*, levator mandibulae posterior superficialis; *MBTN*, ramus mandibularis of trigeminus nerve.

been accounted for, but it is known that the condition in *Bufo* is repeated in the atelopodids (Griffiths 1959, 1963).

The levator mandibulae anterior forms the posterointernal border of the eye cavity; it arises from the ventral surface of the postorbital shelf and is not visible in the dorsal view of the skull. The levator mandibulae posterior superficialis is a bulky and broad muscle that constitutes most of the posterior wall of the eye cavity, bulging into a belly down to the eye. It arises from the ventral surface of the frontoparietal and, contrary to *Ceratophrys*, but agreeing with *Lepidobatrachus*, *Chacophrys*, and *Bufo*, it is not exposed above the proötic.

The levator mandibulae externus is a broad and flat muscle arising from the zygomatic process and the anterior border of the stem of the squamosal. It covers most of the lateral exposure of the levator mandibulae anterior articularis and levator mandibulae posterior profundus. The latter is situated between the levator mandibulae posterior superficialis and the levator mandibulae anterior articularis. It is a flat muscle, as in *Bufo*, with its major breadth in the transverse plane. The levator mandibulae anterior articularis is a voluminous, broad, and rather flat muscle, much more developed than the levator mandibulae posterior superficialis. It is more similar to that of *Ceratophrys* than to that of *Bufo*. The levator mandibulae subexternus, well developed in *Bufo, Rana,* and *Leptodactylus,* is absent in *Macrogenioglottus,* as it is in *Chacophrys, Stombus, Odontophrynus,* and in two of three studied species of *Ceratophrys.* The levator mandibulae anterior lateralis attaches to the inner side of the quadratojugal and extends forward, being visible in lateral view.

In the passage of the main stem of the mandibular branch of the trigeminus, *Macrogenioglottus* belongs to the leptodactylid type, as defined by Limeses (1965*b*; p. 55). As in *Leptodactylus, Stombus, Odontophrynus, Cyclorhamphus, Craspedoglossa, Zachaenus* (Limeses 1965*b*), *Telmatobius* (Stadie 1960), and *Hyla* (Lubosch 1915), the main stem of the nerve, after going over the levator mandibulae posterior superficialis in its proximal section, passes backward between the levator mandibulae externus and the integument. Alternatively, in *Bufo* and *Rana* (Lubosch 1915; Stadie 1960) and in the typical ceratophrynids *Ceratophrys, Chacophrys,* and *Lepidobatrachus* (Reig and Limeses 1963; Limeses 1965*b*), it passes between the masses of the levator mandibular externus and the more internal adductors, that is, the levator mandibulae posterior profundus and levator mandibulae anterior articularis. It is covered laterally by the fibers of the levator mandibulae externus. In its proximal course, however, the nerve has a path similar to that of *Bufo.* It goes from the foramen proöticum forward, passes between the levator mandibulae anterior and the levator mandibulae posterior superficialis to the lateral surface of the latter, and turns downward to follow its course downward and backward over the levator mandibulae externus.

Herre (1960) and Stadie (1960) stressed the probable systematic value of the differences in the passage of the main stem of the mandibular branch of the trigeminus, and Limeses (1965) inferred that *Chacophrys, Lepidobatrachus,* and *Ceratophrys* were more closely related among themselves and with *Bufo* than with *Odontophrynus, Stombus,* and the leptodactylids because of the characteristics of the mandibularis nerve relations. By the same reasoning, *Macrogenioglottus* is more leptodactylidlike than bufonid-ceratophrynidlike in the characteristics of the nerve. This opposes the pattern of relationships inferred from all the other analyzed characters, even those of the jaw musculature, by which the advanced reduction of the pars scapularis of depressor mandibulae and the lack of "temporalis" exposure of the adductors are typical bufonid-ceratophrynid character states. Unfortunately, we have no information about the characteristics of the nerve in the other genera of bufonids. This information could be relevant to the problem of what the primitive situation of the nerve is in the bufonids as a whole. *Macrogenioglottus* could be viewed as having retained a primitive bufonid-leptodactyloid condition, either from an already modified common ancestor, or independently acquired from one unmodified common ancestor, which was eventually modified in the same way in *Bufo* and the ceratophrynids.

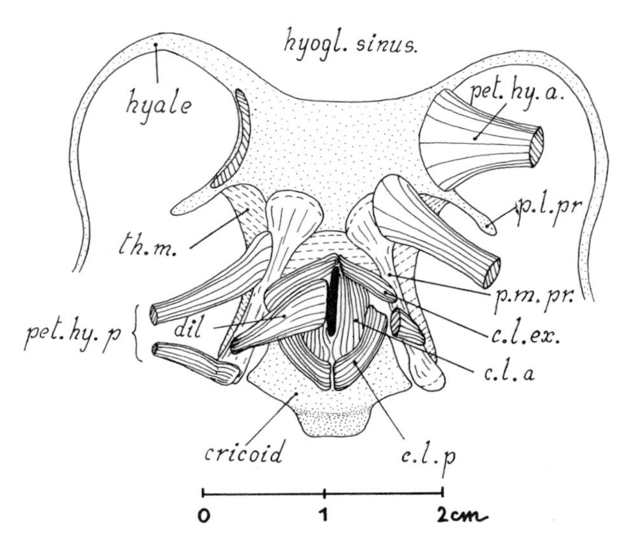

Fig. 3–9. Ventral view of hyolaryngeal apparatus of *Macrogenioglottus alipioi. c.l.a.,* constrictor laryngis anterior; *c.l.ex.,* constrictor laryngis externus; *c.l.p.,* constrictor laryngis posterior; *dil.,* dilator laryngis; *pet.hy.a.,* petrohyoideus anterior; *pet. hy.p.,* petrohyoidei posteriores; *p.l.pr.,* posterolateral process; *p.m. pr.,* posteromedial process; *th. m.,* thyroid membrane.

Hyolaryngeal Apparatus

The hyolaryngeal apparatus of *Macrogenioglottus* is significant in showing a mixture of bufonid-atelopodid and ceratophrynid features. For comparisons with *Bufo* and the Atelopodidae we have followed the papers of Trewavas (1933) and Griffiths (1963). The basis of the comparative data on the ceratophrynids comes from our own observations of *Ceratophrys ornata* and *Lepidobatrachus asper* and the information on *Ceratophrys cornuta* in the paper of Blume (1930). Unfortunately, the single available specimen of *Macrogenioglottus* was poorly preserved and somewhat macerated in the delicate hyolaryngeal muscles and cartilages, so that some anatomical features could not be adequately studied, and others, such as the presence of pulvinaria vocalia, could not be checked at all. (See Fig. 3–9.)

The hyoid plate is very short and wide, its width being about one and one-half the median length, and it is remarkable in its total absence of alary processes. In these two characters it agrees with *Ceratophrys* and *Lepidobatrachus*. In *Leptodactylus* (Trewavas 1963) the hyoid plate is also short and wide, but there are well-developed alary processes. In *Bufo* it is longer than wide and also has conspicuous alary processes. The hyoglossal sinus is very shallow, and the hyale are slender and long. The posterolateral processes are also slender and acuminated, as in *Ceratophrys* and *Leptodactylus*. The bony posteromedial processes are well developed. They have an expanded proximal end not as wide as in *Ceratophrys* and *Lepidobatrachus* (in which they are exceedingly expanded), and a posterior rodlike shaft that scarcely expands at its distal end. A narrow thyroid membrane extends from the posterior border of the posterolateral processes to the lateral border of the posteromedial processes.

The petroomohyoid musculature is weak, as it is in *Ceratophrys* and especially in the *Bufo*-atelopodid complex, contrary to the condition in the leptodactylids. The petrohyoideus anterior inserts on the dorsal surface of the lateral edge of the hyoid plate. The omohyoideus is slender and attaches on the dorsal surface of the lateral part of the proximal expansion of the posterolateral process. Only two petrohyoidei posterioris are present; the more anterior one is inserted on the thyroid membrane and on the lateral edge of the shaft of the posteromedial process, and the more posterior one is attached to the edge of the distal end of the same process. It is obvious, therefore, that they correspond to the first and third of the petrohyoidei posterioris muscles of *Leptodactylus*, *Ceratophrys*, and most anurans. Consequently, *Macrogenioglottus* agrees with *Bufo* and the atelopodids in the absence of m. petrohyoideus posterior medius.

The set of laryngeal muscles is complete, and the individual muscles are similar to those of *Ceratophrys* and *Leptodactylus*. The dilator laryngis is more slender than in those genera and in *Bufo* and is attached to the posteromedial process anterior to the insertion of the third petrohyoideus posterior. The constrictor posterior is present, contrary to the condition in *Bufo* and the atelopodids, but it is weak and comparatively slender. The relations of the distal halves of the hyoglossus with the posteromedial process, which also characterize the *Bufo*-atelopodid complex, were impossible to determine because the region was badly preserved.

The observed characters show that *Macrogenioglottus* is characterized by a somewhat modified and ceratophrynidlike hyolaryngeal apparatus. In the reduction of the petroomohyoid musculature it resembles *Bufo* and atelopodids more closely than *Ceratophrys*.

Teeth

It has been already stated that *Macrogenioglottus* is a toothed toad, possessing teeth on the maxilla,

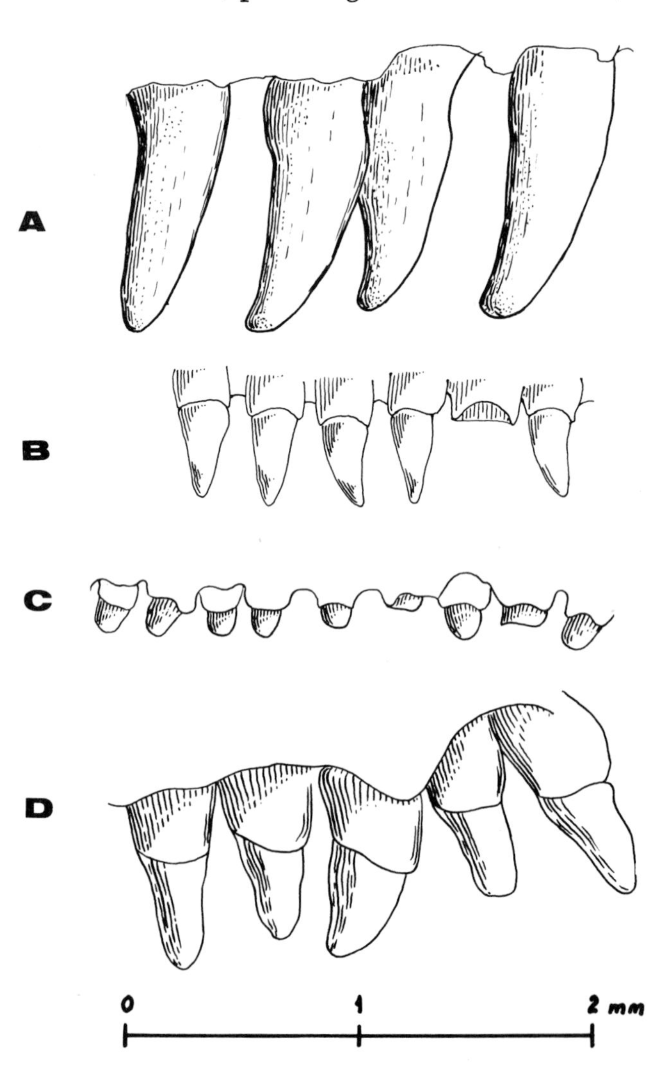

Fig. 3–10. Lateral view of the maxillary teeth of *Ceratophrys dorsata* (A), *Macrogenioglottus alipioi* (B), *Odontophrynus americanus* (C), and *Leptodactylus pentadactylus* (D).

premaxilla, and vomer. The vomerine teeth have already been described.

There are twenty five teeth on each maxilla and eighteen on each premaxilla (Fig. 3-5A). They are very tiny, sharp structures about 0.6 mm. in length and 0.2 mm. in diameter (Fig. 3–10), which are typically divided, as in most anurans (Parsons and Williams 1962) into a crown and a pedicel. They differ in this respect from the ceratophrynids, which do not show such a division, at least in microscopic observation (Reig and Limeses 1963, and the critical observations of Peters 1967). The pedicel is short, about 0.2 mm. in depth, while the crown is a long and acuminate cone, as in *Leptodactylus*, ending in a sharp point that is curved distinctly inward. The toothed portion of the maxilla is limited to the anterior third, and the teeth gradually decrease in size from the maxillary-premaxillary junction backward.

Compared with other anurans, the teeth of *Macrogenioglottus* are very small. They are about half the size of the teeth of *Leptodactylus pentadactylus* and about one-third the size of the teeth of *Lepidobatrachus llanensis*. These two species have a skull of a size similar to that of *Macrogenioglottus*. The teeth are about the same size as those of *Stombus* and *Odontophrynus*, genera that are distinctly smaller in skull and body size, but the number of teeth is fewer than in those genera. Moreover, the teeth of *Odontophrynus* and *Stombus* differ greatly from those of *Macrogenioglottus* in morphology, since they have a very short and blunt crown.

Both the minute size and the reduced number of teeth suggest that the dentition of *Macrogenioglottus* is in a regressive state. This is to be expected in a genus so closely related to the edentulous bufonids in other characters.

Classification of *Macrogenioglottus*

The above comparative described characters of the skeleton, the muscles, the hyolaryngeal apparatus, and other features of *Macrogenioglottus* indicate that this genus is phenetically more closely related to the bufonids (as represented by *Bufo*) than to the ceratophrynids. At the same time, cogent arguments have been mustered that conclude that *Macrogenioglottus*, the bufonid-atelopodid complex, and the ceratophrynids resemble each other more closely than any one of them resembles the leptodactylids, or such broad-headed genera as *Stombus* and *Odontophrynus*.

This last conclusion supports the idea of the Bufonoidea constituting a taxon concept[1] of superfamily rank, which would include the families Bufonidae, Atelopodidae, Ceratophrynidae, and *Macrogenioglottus* and would be distinct from a superfamily Leptodactyloidea, whose extension covers the families Leptodactylidae, Hylidae, Centrolenidae, Pseudidae (if these two latter are really worth family status), and probably the Dendrobatidae. What are the relevant characteristics that define each of these two taxon concepts of superfamily rank? In the process of improving the system of classifying a group of organisms, it is easier to assess the limits of the taxon concepts that emerge from the taxonomic elucidation than to assess what the new concepts mean. It is, thus, much easier to define extensively the Bufonidea and the Leptodactyloidea, as we have done, than to look for diagnostic character states that define intensively each of those taxon concepts. In order to reach a fully intelligible concept of the involved taxa, it is, of course, necessary to achieve a precise definition of them. In our case, however, this task is untimely, as we need to gather much more information about the anatomy and other relevant features of the whole array of the Neobatrachia (Reig 1958) in order to reliably define their subordinate taxa. In the meantime, definitions by statement of the limits are not to be avoided, since they have a high heuristic value.

What is the position of *Macrogenioglottus* within the Bufonoidea? I have already pointed out that this genus shows a higher degree of affinity with the Bufonidae and the related Atelopodidae than with the Ceratophrynidae and that in many relevant characteristics it is very closely related to *Bufo* itself. The inclusion of *Macrogenioglottus* in the Leptodactylidae is completely rejected on the basis of its overall morphological traits, despite some limited convergences with the broad-headed *Odontophrynus* and its ally *Stombus*.

The inclusion of *Macrogenioglottus* in the Ceratophrynidae, as originally proposed by Carvalho (1946), was only tenable according to a vague concept of that taxon and previous to the recent research redefining this taxon in extension and, by implication, in intension. *Macrogenioglottus* cannot be a ceratophrynid as presently understood, because of many relevant characters, for instance, the number, form,

[1] For a better understanding of this paragraph, see my treatment of the attributes of taxon concepts (Reig 1970).

and structure of teeth, the passage of the mandibular branch of the trigeminus nerve, the extent of the dermal ossification on the skull, the absence of a temporal plate of the squamosal covering the proötic, the deltoid and scapulocoracoid musculature, the absence of the medial portion of muscle petrohyoideus posterior, and the relative development of the transverse processes of the vertebrae, to name only a few.

In spite of the high degree of affinity shared by *Macrogenioglottus* and *Bufo* in pectoral musculature, structure of the skull, axial skeleton, humerus, thigh musculature, and some features of the hyolaryngeal apparatus, *Macrogenioglottus* can hardly be included in either the Bufonidae or the Atelopodidae (Brachycephalidae). The presence of teeth and the passage of the ramus mandibularis of the trigeminus nerve, not to mention some peculiar features of the shoulder girdle, the hyoid plate, the laryngeal musculature, and the tongue, preclude its inclusion in the Bufonidae. Most of these characteristics, together with the arciferal shoulder girdle, also prevent its allocation within the Atelopodidae. We are thus forced to conclude that *Macrogenioglottus* belongs to a family of its own and that it is necessary to establish the new family name Macrogenioglottidae to designate that family. The system of the Bufonoidea should therefore be as follows:

Class Amphibia
  Subclass Lissamphibia
    Superorder Salientia
      Order Anura
        Suborder Neobatrachia
          Superfamily Bufonoidea
            Family Macrogenioglottidae
            Family Bufonidae
            Family Atelopodidae
              (= Brachycephalidae)
            Family Ceratophrynidae

The Macrogenioglottidae is, to our present knowledge, a monotypic family, which could be defined by the following set of characters:

Bufonoid neobatrachians with a well-ossified, broad, deep, and moderately exostosic osteocranium, with minute teeth distinctly divided into crown and pedicel in the maxillae and premaxillae and a few larger teeth on the vomer. Broad frontoparietals with a well-developed otoparietal plate overlapping the proötics. Temporal plate of the squamosal poorly developed, with a broad contact with the otoparietal plate, but scarcely covering the proötic. Quadratojugals and palatines well developed. Operculum and columella regularly developed; tympanum distinct, covered by the skin. Vertebral centra elongated, with no fusion among them. Transverse processes of the fourth vertebra the longest; sacral diapophyses moderately expanded. Shoulder girdle arciferous, with a short precoracoid bridge. Omosternum completely absent; xiphisternum long and broad, cartilaginous, moderately calcified anteriorly. Scapula very long, twice the length of the coracoid. Humerus regular, not shortened. Deltoideus episternalis long and narrow, visible in ventral view. Coracobrachialis brevis superficialis placed between pars epicoracoidea and pars sternalis of the pectoralis, covering all the space between them and completely overlapping the coracoradialis. Tensor fasciae latae rather short. Adductor longus distinct, not overlapped by the sartorius. Pars scapularis of depressor mandibulae present, but much reduced; levator mandibulae subexternus absent. Ramus mandibularis of trigeminus passing over the levator mandibulae externus. Alary processes of hyoid plate absent; hyoid plate short and wide. M. petrohyoideus posterior medius absent; constrictor laryngis posterior present, not strong. Tongue subcircular, reduced; m. genioglossus visible on the floor of the mouth in front of the tongue. Skin moderately thick, glandular. A glandular thickening of the skin at the parotoid region.

Phylogenetic and Biogeographic Generalizations

The fossil record of the bufonoid toads hardly affords a sound basis for inferring phylogenetic relationships among their families. There are no known atelopodid and macrogenioglottid fossils. The paleontological information on the bufonids is analyzed by Tihen in chapter 2 of this book. Tihen is conclusive in demonstrating the incompleteness of bufonid fossil record. For the ceratophrynids, the only certain information comes from the Upper Pliocene of Argentina, where a form of *Ceratophrys*, hardly separable from the living *C. ornata*, has been found often (see Rovereto 1914). The older (Friasean, late Miocene) *Wawelia gerholdi* has recently been referred to this family by Casamiquela (1963). If this form is a ceratophrynid, the material is too incomplete to give any meaningful evidence for the history of that family. Attempts to infer phylogenetic relationships within the Bufonidae are therefore almost exclusively dependent upon the data afforded by the comparative anatomy of living forms.

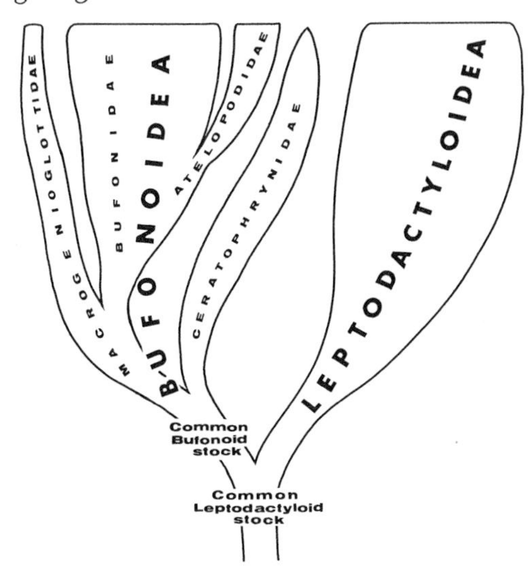

Figure 3–11. Phylogenetic diagram showing the probable relations of bufonid and leptodactyloid neobatrachian anurans, and the probable relations among the families of the Bufonoidea.

Our comparative studies provide a hypothetical reconstruction of the probable relationships among the different groups of bufonoids (Fig. 3-11). In accordance with the available evidence, the bufonoids are presumably an offshoot of a primitive leptodactyloid stock. This has been recently suggested by Griffiths (1959, 1963) for the bufonid-atelopodid complex and also by Reig (1960). The general morphological evidence suggests that the ancestral bufonoid stock might have derived from a primitive terrestrial group of leptodactylids, the further evolution of the group being characterized by an improvement in progressive terrestrial trends.

This primitive bufonoid ancestral stock might be thought of as a taxon with reduced omosternum, deep, well-ossified, and moderately exostosic skull, teeth of generalized type, unmodified hyolaryngeal apparatus, a distinct coracobrachialis brevis superficialis behind the pectoralis epicoracoideus, the bufonid pattern of thigh musculature, a trend toward the reduction of pars scapularis of depressor mandibulae, and the retention of the leptodactylid type of passage of ramus mandibularis. *Macrogenioglottus* matches this hypothetical taxon in most characters, and it could be considered a relict of such an original stock that shows some progressive features in the hyolaryngeal apparatus, the shoulder girdle, and the reduced dentition. It is also probable that it represents an advanced form of the bufonoid ancestral taxon on itself, and that the latter belongs to the family Macrogenioglottidae.

The ceratophrynids are reasonably thought of as an early offshoot of this bufonoid ancestral taxon. They retain some primitive features in the hyolaryngeal apparatus, for instance the full development of the constrictor muscles and of the three petrohyoidei posteriores, but they have developed several specializations of their own for a predatory way of life, including a very large, broad head with sharp teeth, an aggressive pattern of behavior, and specialized predaceous larvae. They are also very modified in the advanced exostosis and ornamentation of the roof bones of the skull, the elongation of the scapula, the powerful pectoral musculature, and the correlated strengthening of the humerus. In the passage of the mandibular branch of the trigeminus, they have reached a fully bufonid state. This may be interpreted in two ways: Either bufonids and ceratophrynids derived from a common ancestor with the bufonid pattern of nerve passage, or this latter character state arose independently in both groups from an ancestral stock retaining the leptodactylid pattern. It is well known that *Rana* has the same kind of nerve passage as *Bufo* (Lubosch 1915), which suggests that this state could have been attained independently in different groups of neobatrachians. It is also improbable that bufonids and ceratophrynids had a common ancestor at an advanced stage of bufonoid evolution, since they diverge completely in the evolution of the dentition. *Macrogenioglottus*, with its numerically reduced and minute teeth, represents perhaps the beginning of the trend toward the edentulous condition in bufonids and atelopodids, and it retains the leptodactylid pattern of nerve passage. All this seems to indicate, therefore, that bufonids and ceratophrynids reached independently the similar state of the nerve passage.

The bufonids might have been derived from the macrogenioglottids at an early stage in the evolution of the family, from a genus less specialized than *Macrogenioglottus* in the elongation of the scapula, the shortening of the hyoid plate, the absence of alary processes, and the reduction of the tongue. It is assumed that in the evolution from the macrogenioglottid ancestor to bufonids, the nerve passage changed from a leptodactylid to a bufonid condition. It is highly suggestive that, aside from the above features, *Macrogenioglottus* as a genus might be considered a good ancestor of *Bufo* in all the remaining studied characters, which emphasizes that it is a living derivative of the immediate ancestor of the bufonids.

The atelopodids have been shown to be bufonid derivatives with modifications in shoulder girdle morphogenesis resulting in a firmisternal condition. Griffiths pointed out (1959, 1963) that they might be a branch of the bufonids that accentuated some paedormorphic trends occurring in the latter. Though they seem to deserve classification as a distinct family (see also Laurent 1942), they are merely a well-differentiated branch in the internal process of bufonid diversification. This process is still poorly known, though a good point of departure for understanding it has been presented by Tihen (1960).

One important conclusion of Tihen's paper is that the radiation of the bufonids might have arisen from the genus *Bufo* itself, so that this is not only the most diversified and widespread bufonid, but also the ancestral genus from which several other genera evolved. It is suggestive, in this connection, that *Macrogenioglottus*, a genus that is probably very close to the ancestral stock of the bufonids, shows more affinity with *Bufo* than with any of the remaining bufonids. Another of Tihen's meaningful conclusions is that the early members of the genus *Bufo* are not to be sought among the thin-skulled forms with narrow frontoparietals, such as the members of the *calamita*, *viridis*, or *spinulosus* groups, but in forms having strongly ossified and moderately exostosic skulls with rather broad frontoparietals, as in *Bufo mauritanicus*. This is also in agreement with the assumed macrogenioglottid derivation of *Bufo*, since *Macrogenioglottus* shows the characteristics postulated for the early members of *Bufo*.

A significant consequence of the above facts and arguments is that the geographical origin of the bufonids (and then of *Bufo*, as the primitive genus of this family) was South America. The evidence shows that the major radiation of the bufonoid group is best represented by purely South American taxa, for instance, the ceratophrynids, the atelopodids, and the macrogenioglottids, and that the group most likely to be ancestor of the bufonids is to be sought in the South American macrogenioglottids. This conclusion is in close agreement with the generalization reached by Blair (1968 and Chaps. 11 and 18 of this volume) in his study of hybridization, which indicates a neotropical origin of the genus *Bufo*. The strongest evidence comes from the existence in South America and in southern North America of generalized immediate forms, in crossability and morphology, and the absence of such intermediate types in Eurasia and Africa. Blair's hypothesis is reinforced by the discovery in South America of a relict member of the taxon from which the bufonids probably evolved.

When did *Bufo* originate as an early genus of the bufonids? Although the question cannot be answered with certainty, it would be useful to analyze available evidence that bears on this problem.

As stated by Tihen (Chap. 2 of this volume), no clear record of *Bufo* exists before the Middle Miocene. Previous assignment of *Bufo* of the Deseadan form *Neoprocoela edentata* by Tihen (1962*b*) is placed in doubt on the basis of Lynch's unpublished studies, and the so-called *Bufo servatus* from the European Quercy phosphorites is, according to Tihen, unidentifiable. Tihen has pointed out that the available paleontological evidence of *Bufo* is likely to represent only the latter stages of the phylogenetic history of the genus and that *Bufo* as a genus and its included species group might have evolved at a time appreciably earlier than the Middle Miocene. The certainty of this conclusion is supported by the discovery of *Bufo marinus* from the Upper Miocene (Friasean) of Colombia (Estes and Wassersug 1963). This fossil is meaningful since it suggests that, by Friasean times, the closely related species of the *marinus* group (*B. marinus, B. ictericus, B. paracnemis*, etc.) were already differentiated (Estes, personal communication). *Bufo pisanoi* (Casamiquela 1967), a Pliocene form closely related (if not conspecific) to *B. paracnemis*, gives additional support to the same conclusion. Moreover, specimens now under study by Estes (personal communication) show that *Bufo* itself was present in the South American Paleocene, represented by a species hardly separable from the living *B. spinulosus*.

It was customary to speculate about the phylogeny of modern anuran families as if their histories occurred during the span of the Cenozoic. The recent discussions on this subject, however, accept an earlier history for the Neobatrachia (see especially Hecht 1963), which includes probably most of the Cretaceous. Relevant evidence exists, moreover, indicating that the group was well established in Morrison and the late Cretaceous (Estes 1964, Hecht and Estes 1960). Though the question is still controversial (see Tihen 1965 and Griffiths 1963 for a dissident view), there seem to be cogent reasons to maintain that the neobatrachians evolved by the early Cretaceous probably from a discoglossid-ascaphid basal group (Hecht 1963). The early presence in Patagonia of the first-recorded representatives of this group, namely *Vieraella* (Lower Jurassic) and *Notobatrachus* (Middle Jurassic), suggests that the neobatrachians had their origins in a southern landmass. It has been advocated elsewhere (see for instance Hecht 1963: p. 31) that the leptodactyloids can be considered the ancestral group from which the radiation of the Neobatrachia originated. We can assume that the first representatives of the neobatrachians were primitive leptodactylids similar to the living Australian Cycloraninae.

As suggested by Laurent (1942, 1951), this early leptodactylid stock could have been the origin of the ranoids, which apparently evolved in southern Asia, differentiating later into the Australian leptodactylid radiation and the array of South American leptodactyloids and bufonoids. The complete absence in Australia of ceratophrynids and bufonids, other than the introduced *Bufo marinus*, suggests that the differentiation of the early bufonoid ancestral stock occurred in South America after the disappearance of the Antarctic migrational route connecting Patagonia and Australia, thus placing the time at the end of the Cretaceous (Reig 1968). A somewhat earlier origin of the bufonoids cannot, however, be completely excluded, since the group could have originated by early or middle Cretaceous times and could have been impeded from entering Australia by ecological obstacles. In any case, since the separation of the bufonids from the bufonoid stock necessarily took place at a later time than the period of the first origin of the bufonoid ancestral stock, it is necessary to think of the origin of *Bufo* and the bufonoids at a time earlier than the Lower Paleocene or, less probably, the late Cretaceous.

This is perhaps all that can reasonably be said at present about the time of bufonid origins, and even this vague conclusion is highly speculative. South America has been rich in suggestive paleobatrachological discoveries in the last decade, and it is to be expected that new discoveries in the early Cenozoic and late Mesozoic will eventually clarify the question of the time of bufonid origins. The very recent finding of fossil anurans from the Upper Cretaceous of Peru (Sige 1968) will probably be relevant.

ADDENDUM

After this paper was submitted for publication, the chromosomes of *Macrogenioglottus alipioi* were described and chromosome information was published on several taxa of South American frogs and toads connected with my discussion. It is of interest to attempt a brief revision of this new evidence and to discuss its bearing on the hypotheses stated in this paper.

It has been confirmed that bufonids have a widespread karyotype of 2n=22 chromosomes, with the sole exception of the members of the African *regularis* group (Morescalchi and Garguilo 1969). Within the 2n=22 general karyotype, chromosome multiformity in size and centromeric index was found in several South American species (Brum-Zorrilla and Saez 1971).

The ceratophrynids as conceived in this paper have a basic chromosome complement of n=13 chromosomes, expressed by the regular diploid number in the species of *Lepidobatrachus,* in *Chacophrys,* in *Ceratophrys calcarata,* and in the Chacoan *Ceratophrys ornata* (Bogart 1966; Morescalchi 1967; Barrio and Rinaldi de Chieri 1970*b*). *C. ornata* from the pampean region, as well as *Ceratophrys varia,* have been found to be octaploid (Beçak *et al.*, 1967; Barrio and Rinaldi de Chieri, 1970*b*).

Among the Leptodactylidae, most of the studied species have either 2n=22 or 2n=26 chromosomes. Members of the subfamily Leptodactylinae (*Leptodactylus, Pleurodema, Physalaemus,* etc.), as well as *Stombus* and the diploid species of *Odontophyrynus*, are characterized by 2n=22 chromosomes. The latter genus is also known to comprise tetraploid species with 2n=44 chromosomes (Bogart 1967; Beçak *et al.*, 1966, 1970; Saez and Brum 1966). On the other hand, taxa grouped within the Telmatobiinae show 2n=26 chromosomes (Barrio and Rinaldi de Chieri 1971), and the same happens with species of *Oocormus* and *Cycloramphus* (Beçak *et al.*, 1970).

The karyotype of *Macrogenioglottus alipioi*, as described by Beçak *et al.* (1968; in Beçak *et al.*, 1970), is composed of 2n=22 chromosomes, quite similar to that of the diploid *Odontophrynus* and the species of *Stombus.* From this evidence, Barrio and Rinaldi de Chieri (1971:684) maintain that *Macrogenioglottus* and *Odontophrynus* must be classified within the Leptodactylinae. However, we see that the chromosomes of *Macrogenioglottus alipioi* are very similar, in number and structure, to many species of *Bufo.* In fact, the karyotype of *Macrogenioglottus* is the one to be expected in the ancestral stock of the Bufonidae, and it is therefore in full agreement with my own results, which support the view that that genus might be considered closely related to the ancestry of the Bufonidae. It is true that the chromosomes of *M. alipioi* agree in details with those of *Odontophrynus carvalhoi*, *O. occidentalis*, and *Stombus boiei*. But it is also true that they likewise agree in relative size and centromeric indices with those of other leptodactylids and species of *Bufo* (cf. Beçak *et al.*, 1970, Table II; Brum-Zorrilla and Saez 1971, Table II). Moreover, it is to be assumed that repatterning of the centromere position by pericentric inversions might have arisen repeatedly during chromosome evolution within the 2n=22 karyotype, leading to some convergent results. Changes in chromosome numbers seem to have been rarer events in anuran evolution, and they probably deserve more weight in phylogenetic inference.

The general picture of the karyotypes in the Bufonoid-Leptodactyloid anurans seems to indicate that the chromosome number in the common ancestor of the whole group was 26. This number was maintained in the ceratophrynids, in which further changes in the genome were obtained by polyploidy. It was also maintained in the ancestral stock of South American leptodactylids, as represented by the Telmatobiinae. Some offshoots of the primitive leptodactyloids experienced reduction in number to 24 (Cycloraniinae and Miobatrachinae; see Morescalchi *et al.*, 1968) and to 22 (most Leptodactylinae and the broad-headed *Stombus* and *Odontophrynus*), with further modifications by tetraploidy in *Odontophrynus* and in *Pleurodema* (for the latter, see Barrio and Rinaldi de Chieri 1970). Species of *Eupsophus s. s.* (excluding the forms now assigned to *Alsodes*), *Eleutherodactylus*, and *Syrrophus* (Bogart 1970) showing numbers higher and lower than 22 and 26 would be interpreted as examples of exceptional mechanisms of chromosome evolution within the family. Within the Bufonoids, *Macrogenioglottus* and bufonids share the synapomorphic occurrence of 2n=22 chromosomes, arisen independently from the origin of the same number in some Leptodactylids. Among the bufonids, modification of this number to 2n=20 occurred only in one group of species limited to Africa.

Previous suggestions of close relationships between *Macrogenioglottus* and *Odontophrynus* are not, therefore, necessarily corroborated by chromosome number. Nor is the assumption that *Odontophrynus*, because it possesses 22 chromosomes, be postulated as close to the ancestry of bufonids. The anatomical evidence against those views seems to be overwhelming, and both anatomy and chromosome evidence agree in supporting my view of the close affinity between *Macrogenioglottus* and the bufonids as well as the hypothesis holding that the former is a living representative of the ancestral stock of the latter.

REFERENCES

Barrio, A. 1963. Consideraciones sobre comportamiento y "grito agresivo" propio de algunas especies de Ceratophrynidae (Anura). Physis 24: 143–148.

———. 1968. Revisión del género *Lepidobatrachus* Budget (Anura, Ceratophrynidae). Physis 27:445–454; 28:95–106.

———, and P. Rinaldi de Chieri. 1970*a*. Estudio citogenético del género *Pleurodema* y sus consecuencias evolutivas (Amphibia, Anura, Leptodactylidae). Physis 30:309–319.

———. 1970*b*. Relaciones cariosistemáticas de los Ceratophryidae de la Argentina. Physis 30:321–329.

———. 1971. Contribución al esclarecimiento de la posición de algunos batracios patagónicos de la familia Leptodactylidae mediante el análisis cariotípico. Physis 30:673–685.

Beçak, M. L., W. Beçak, and M. N. Rabello. 1966. Cytological evidence of constant tetraploidy in the bisexual South American frog *Odontophrynus americanus*. Chromosoma 19:188–193.

———. 1967. Further studies in polyploid amphibians (Ceratophrydidae). I. Miotic and meiotic aspects. Chromosoma 22:192–201.

Beçak, M. L., L. Denaro and W. Beçak. 1970. Polyploidy and mechanisms of karyotypic diversification in Amphibia. Cytogenetics 9:225–238.

Bhati, D. P. S. 1955. The pectoral musclature of *Rana tigrina* Daud and *Bufo andersoni* Boulenger. Ann. Zool. 1(2):23–78.

Bigalke, R. 1927. Zur Myologie der Erdkröte (*Bufo vulgaris* Laurenti). Z. Anat. Entw.-Gesch. (Berlin) 6: 236–253.

Blair, W. F. 1968. Evolución del género Bufo en América del Sur. IV Congreso Latinoamericano de Zoología (Caracas, Nov. 10–16, 1968).

Blume, W. 1930. Studien am Anurenlarynx. Gegenbaurs. Morph. Jahrb. 65:307–464.

Bogart, J. P. 1967. Chromosomes of the South American amphibian family Ceratophrydae with a reconsideration of the taxonomic status of *Odontophrynus americanus*. Canad. J. Genet. Cytol. 9:531–542.

———. 1970. Systematic problems in the Amphibian family Leptodactylidae as indicated by karyotypic analysis. Cytogenetics 9:369–383.

Brum-Zorrilla, N., and F. A. Saez. 1971. Chromosomes of South American Bufonidae (Amphibia, Anura). Experientia 27:470–471.

Carvalho, A. L. de. 1946. Um novo gênero de Ceratofridídeo do Sudeste Balano. Bol. Museu Nac. Zool., no. 73, pp. 1–5.

Casamiquela, R. M. 1963. Sobre un par de anuros del Mioceno de Rio Negro (Pantagonia): *Wawelia gerholdi*, n. gen. et sp. (Ceratophrydidae) y *Gigantobatrachus parodii* (Leptodactylidae). Ameghiniana 3(5):141–160.

———. 1967. Sobre un nuevo *Bufo* fosil de la Provincia de Buenos Aires (Argentina). Ameghiniana 5(5): 161–168.

Cei, J. M., V. Erspamer, and M. Roseghini. 1967. Taxonomic and evolutionary significance of biogenic amines and polypeptides occurring in amphibian skin. I. Neotropical Leptodactylid frogs. Syst. Zool. 16: 328–342.

Cochran, D. M., and C. J. Goin. 1961. A new genus and species of frog (Leptodactylidae) from Colombia. Fieldiana Zool. 39:543–546.

Cope, E. D. 1889. The Batrachia of North America. U.S. Nat. Museum Bull., no. 34.

Edgeworth, F. H. 1935. The cranial muscles of vertebrates. Cambridge Univ. Press, London and New York.

Estes, R. 1964. Fossil vertebrates from the late Cretaceous Lance Formation eastern Wyoming. Univ. Calif. Publ. Geol. Sci. 49:1-180.

———, and R. Wassersug. 1963. A Miocene toad from Colombia, South America. Brevoria, no. 193, pp. 1–13.

Griffiths, I. 1954. On the otic element in Amphibia Salientia. Proc. Zool. Soc. (London) 123:781–792.

———. 1959. The phylogeny of *Sminthillus limbatus* and the status of the Brachycephalidae (Amphibia Salientia). Proc. Zool. Soc. (London) 132:457–487.

———. 1963.The phylogeny of the Salientia. Biol. Rev. 38:241–292.

Hecht, M. K. 1963. A reevaluation of the early history of the frogs. II. Syst. Zool. 12:20–35.

———, and R. Estes. 1960. Fossil amphibians from quarry nine. Postilla (Yale Peabody Mus.), no. 46, pp. 1–19.

Herre, W. 1960. Zur Problematik der Taxonomie der Anuren. Zool. Anz. 164:394–400.

Jones, E. J. 1933. Observations on the pectoral musculature of Amphibia Salientia. Ann. Mag. Nat. Hist. (London) 12:403–420.

Kuhn, O. 1965. Die Amphibien. System und Stammesgeschitchte. Verlag Oeben, Krailling (Munich).

Laurent, R. 1942. Note sur les procoellens firmisternes (Batrachia Anura). Bull. Mus. R. d'Hist. Nat. Belgique 18(43):1–20.

———. 1951. Sur la nécessité de supprimer la famille des Rhacophoridae mais de créer celle des Hyperoliidae. Rev. Zool. Bot. Afr. 45:116–122.

Limeses, C. E. 1963. La musculatura del muslo de las especies del género *Lepidobatrachus* (Anura-Ceratophrynidae). Physis 24:205–218.

———. 1964. La musculatura del muslo en los Ceratofrínidos y formas afines, con un análisis crítico sobre la significación de los caracteres miológicos en la sistemática de los Anuros superiores. Contr. Cient. Univ. Buenos Aires (Zool.) 1(4):193–245.

———. 1965*a*. Musculatura del muslo de los Ceratofrínidos. Anais do II Congreso Latinoamericano de Zoología (São Paulo, July 16–21, 1962). 2:249-260.

———. 1965*b*. La musculatura mandibular en los Ceratofrínidos y formas afines (Anura, Ceratophrynidae). Physis 25:41–58.

———. 1968. *Lepidobatrachus* Budget (Anura, Ceratophrynidae). Nota miológica complementaria. Physis 28:127–134.

Lubosch, W. 1915. Die Kaumuskeln der Amphibien. Jena Z. Naturwiss. 35:51–188.

Miranda Ribeiro, A. 1920. Algunas consideraçoes sobre o genero *Ceratophrys* e suas especies. Rev. Mus. Paul. 22:291–304.

———. 1926. Notas para servirem ao estudo dos Gymnobatrachios (Anura) Brasileiros. Arch. Mus. Nac. Rio de Janeiro, 27:1–227.

Morescalchi, A. 1967. The close karyological affinities between *Ceratophrys* and *Pelobates* (Amphibia, Salientia). Experientia 23:1–4.

———, and G. Garguilo. 1969. Cytotaxonomic remarks on the genus *Bufo* (Amphibia, Salientia). Archivos Soc. Biol. Montevideo 27:88–91.

———, and E. Olmo. 1968. Note citotassonomiche sui Leptodactylidae (Amphibia, Salientia). Boll. Zool. 35.

Noble, G. K. 1922. The phylogeny of the Salientia. Bull. Amer. Mus. Nat. Hist. 46:1-57.

———. 1926. The pectoral girdle of the Brachycephalid frogs. Amer. Mus. Novitates, no. 230, pp. 1–14.

Parker, H. W. 1935. The frogs, lizards and snakes of British Guiana. Proc. Zool. Soc. (London) 1935:505–530.

Parsons, T. S., and E. E. Williams. 1962. The teeth of Amphibia and their relation to amphibian phylogeny. J. Morph 110:375–390.

Peters, J. A. 1967. The generic allocation of the frog *Ceratophrys stolzmanni* Steindachner, with a description of a new subspecies from Ecuador. Proc. Biol. Soc. (Washington) 80:105–112.

Reig, O. A. 1958. Proposiciones para una nueva macrosistemática de los anuros. Physis 21:109–118.

———. 1960*a*. La anatomía esquelética del género *Lepidobatrachus* Budget (Anura-Leptodactylidae) comparada con la de otros Ceratofrininos. Actas y Trab. Primer Congr. Sudamer. Zool. (La Plata 1959) 4:133–147.

———. 1960*b*. Lineamientos generales da la historia biogeográfica de los Anuros. Actas 1° Congr. Sudamer. Zool. (La Plata 1959) 1:270–278.

———. 1968. Le peuplement en vertébrés Tétrapodes de l'Amérique du Sud, IV, 185–229. *In* C. Delamare Debouteville and E. Rapoport, (ed.), Biologie de L'Amérique Australe. C.N.R.S., Paris.

———. 1970. The Proterosuchia and the early evolution of the archosaurs. An essay on the origin of a major taxon. Bull. Mus. Comp. Zool. 139:229–292.

———, and J. M. Cei. 1963. Elucidación morfológico-esta-dística de las entidades del género *Lepidobatrachus* Budget (Anura-Ceratophrynidae) con consideraciones sobre la extensión del Distrito Chaqueño del Dominio Zoogeográfico Subtropical. Physis 24:181–204.

———, and C. E. Limeses. 1963. Un nuevo género de Anuros Ceratofrínidos del Distrito Chaqueño. Physis 24:113–128.

Rovereto, C. 1914. Los estratos araucanos y sus fósiles. Anal. Mus. Nac. Hist. (Buenos Aires) 25:1–247.

Sanders, O. 1953. A new species of toad, with a discussion of morphology of the bufonid skull. Herpetologica 9:25–47.

Saez, F. A. and N. Brum. 1966. Citogenética de anfibios anuros de América del Sur. Los chromosomas de *Odontophrynus americanus* y *Ceratophrys ornata*. Anal. Fac. Med. Montevideo 44:414–423.

———. 1966. Karyotype variation in some species of the genus *Odontophrynus* (Amphibia, Anura). Caryologia 19:55–63.

Schmalhausen, I. I. 1957. Biologicheskie osnovy vozniknovenila nazemnykh pozonochnykh. Izd. Akad. Nauk (Biol.) (Moscow) 1:3–30.

Sige, B. 1968. Dents de micromammifères et fragments de coquilles d'oeufs de dinosauriens dans la faune de vertébrés du Crétacé supérieur de Laguna Umayo (Andes péruviennes). C. R. Acad. Sc. 267:1495–1498.

Stadie, C. 1960. Einige vergleichend-anatomische Bemerkungen zur Kritik von Taxonomie und Systematik der Anuren. Zool. Anz. 166:245–257.

Tihen, J. A. 1960. Two new genera of African bufonids, with remarks on the phylogeny of related genera. Copeia 1960:225–233.

———. 1962*a*. Osteological observations on New World *Bufo*. Amer. Midl. Nat. 67:157–183.

———. 1962*b*. A review of New World fossil bufonids. Amer. Midl. Nat. 68:1–50.

———. 1965. Evolutionary trends in frogs. Amer. Zoologist 5:309–318.

Trewavas, E. 1933. The hyoid and larynx of the Anura. Phil. Trans. Royal Soc. (London) B, 222:410–527.

4. Evidence from Osteology

ROBERT F. MARTIN

The University of Texas at Austin
Austin, Texas

Introduction

In the past decade, research on evolution within the genus *Bufo* has proceeded with increasing energy. Baldauf (1955, 1958, 1959), Tihen (1960, 1962*a*, *b*), and Martin (1964) have investigated the osteology of New World species and have constructed various phylogenetic assemblages based primarily on cranial morphology. Porter (1964*a*, *b*, 1965, 1966) and Porter and Porter (1967) have studied primarily Mexican and Central American species and have applied parotoid secretion chromatography, mating call analysis, and morphology toward an understanding of their evolutionary relationships. Other researches, as reported in this book, have also contributed much toward a comprehensive understanding of the genus. Nevertheless, many problem areas remain. It is hoped that these poblems will be partially resolved by this study.

Materials and Methods

Sixty-one species of *Bufo* have been examined; minimal samples were available for some. Only the description of major osteological details is presented. The species examined, sample size, and locality data are presented in Appendix A.

Skeletal preparations were accomplished by a modification (Martin 1964) of the technique described by Sanders (1953) and also by use of dermestid beetles. During preparation, specimens were examined for species characters; many had been previously identified prior to hybridization experiments by W. F. Blair.

Figure 4-1 illustrates the osteological nomenclature used in the text and depicts major skull types, cristation patterns, and relationships between bones of the cranium.

The dorsal roof of the skull is composed primarily of three paired dermal bones: the nasals (anterior), the frontoparietals (medial and posterior), and the laterally placed squamosals. Any or all may bear crests. When crests are present, two major categories exist: the *valliceps* and *americanus*-type patterns (Fig. 4-1). Conditions in which the lateral edges of the frontoparietals are slanted upward but are not produced into elevated crests also exist. If no mention is made of a particular type of crest, it is not present.

The nasals may be produced anteriorly beyond the line of the jaw or may terminate posterior to or level with it. The latter condition is most common, and only deviations from it are discussed. The junction of the nasals with the frontoparietals is extremely variable intraspecifically, usually incompletely

NARROW FRONTOPARIETAL TYPE
AMERICANUS TYPE CRISTATION

BROAD FRONTOPARIETAL TYPE
VALLICEPS TYPE CRISTATION

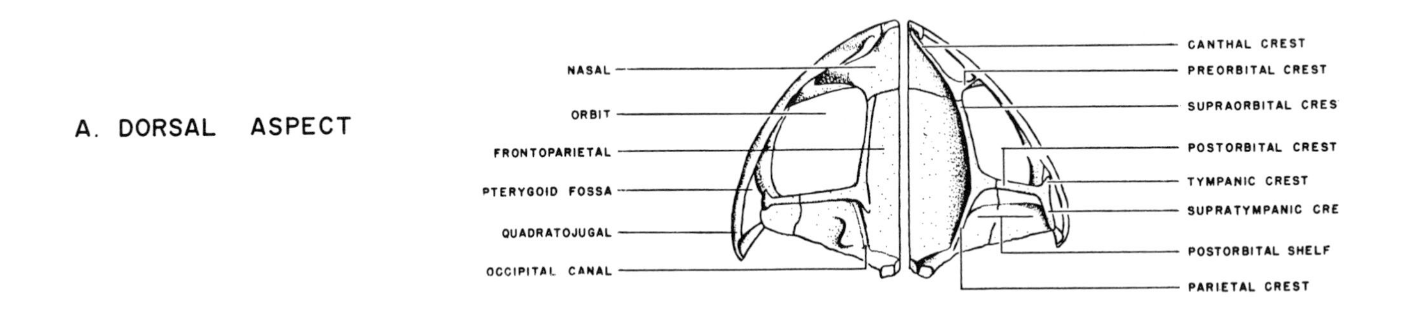

SUPRAPTERYGOID FENESTRA
NOT OCCLUDED

SUPRAPTERYGOID FENESTRA
MARKEDLY OCCLUDED

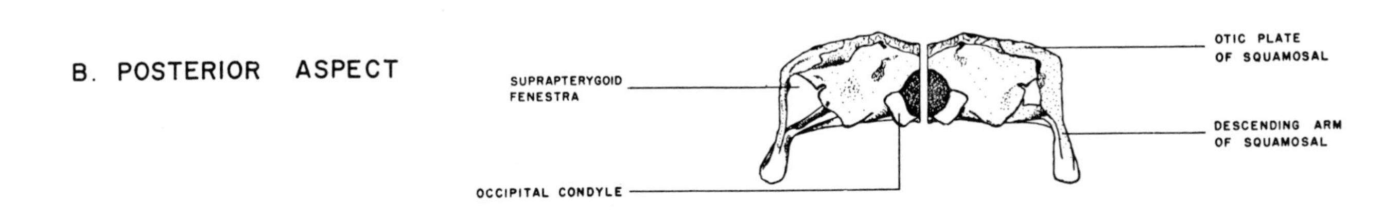

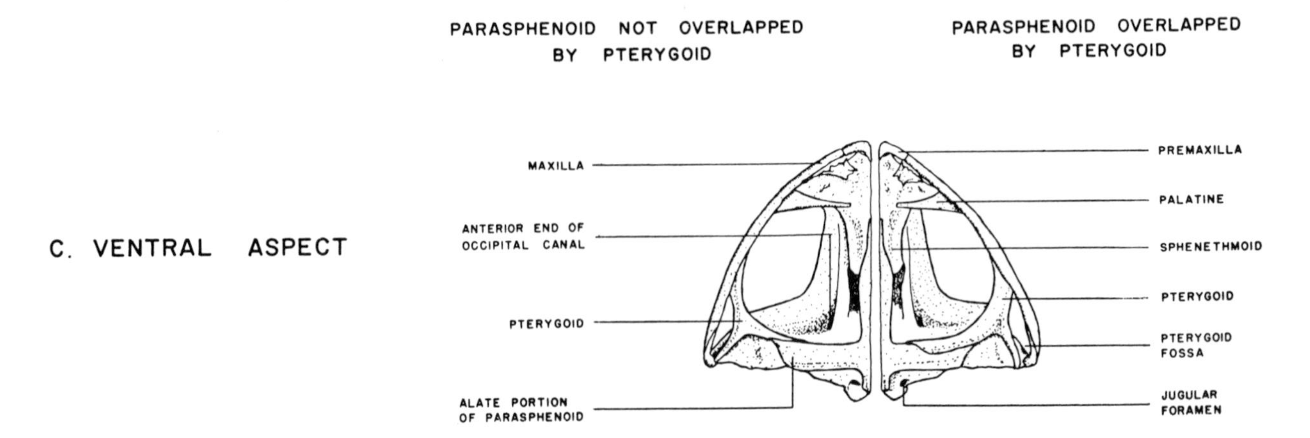

Fig. 4–1. Nomenclature of cranial characters.

exposing the underlying sphenethmoid bone. The extent to which the posterior portions of the horizontal laminae of the frontoparietals project above the posteromedial quadrant of the orbit (Fig. 4-1) has been treated as a character of considerable importance by Tihen (1962*a*, *b*,) and Blair (1963*a*, 1964). The tentative phylogeny of Blair (1963*a*) divided North American *Bufo* into two major evolutionary lines, one (southern line) generally possessing broad, the other (northern line), narrow frontoparietals. The occipital canals pass over the frontoparietals posterior to their entrance to the orbit, and the extent to which they are covered is discussed. Sutural contact of the frontoparietal with the squamosal may be present or absent, and bones may occur within the orbit fusion between proötic and frontoparietal.

The surfaces of the dermal cranial elements may be smooth or thickened by varying amounts of cancellous or spiculate dermal bone. At the rear of the skull, the suprapterygoid fenestrae may be open or occluded to varying degrees by flanges of the pterygoid or squamosal (Fig. 4-1). Ventrally, the medial arm of the pterygoid may not contact the lateral wings of the parasphenoid, may be tangent with them, or may overlap them to varying degrees (Fig. 4-1). The quadratojugal bone may be short and not in contact with the maxilla, or may extend more anteriorly from the angle of the jaw, overlapping the maxilla and forming a large part of the lateral wall of the pterygoid fossa. The anterior-posterior location of the angle of the jaw is usually discussed in relation to the transverse plane including the jugular foramen (Baldauf 1959; Tihen 1962*a*) and varies from far anterior to far posterior to this plane. This character is quite variable intraspecifically, however (Martin 1964), and only extreme deviations from this plane will be discussed.

Measurements were made with vernier calipers or with an ocular micrometer fitted to a dissecting microscope and were recorded to the nearest one-tenth millimeter. The following characters were measured:

1. *Skull length:* the distance from the posterior end of the occipital condyles to the tips of the nasal bones.
2. *Skull width:* the distance between the lateral edges of the maxillae (or quadratojugals) at their widest breadth.
3. *Skull height:* the perpendicular distance between a plane upon which the ventral surfaces of a skull are tangent and a parallel plane including the most dorsal midline point of the frontoparietal bones.
4. *Transorbital width:* the distance between the posteromedial corners of the orbit measured at the dorsal skull roof. If an arc of bone rather than a corner existed in this area, the measurement was taken at a point (subjectively) judged to be its midpoint.
5. *Occipital canal width:* the distance between the centers of the anterior openings of the occipital canals (or grooves).
6. *Neurocranial height:* the distance between the most dorsal median point of the frontoparietal bones and a line determined by the lowest anterior and posterior points of the parasphenoid bone, along which the skull was viewed in occipital aspect.
7. *Foramen magnum height:* the distance between midline roof and floor of the foramen magnum when viewed in occipital aspect.
8. *Foramen magnum width:* the greatest distance between lateral edges of the foramen magnum.
9. *Vertebral column length:* the distance from the most anterior edge of the articular facets of the first to the most posterior edge of the articular facets of the ninth vertebral centrum.
10. *Vertebral column width:* the greatest width of the anterior end of the centrum of the fourth vertebra.
11. *Humeral length:* the greatest length of the humerus when the humeral crest is perpendicular to the plane of the calipers.
12. *Humeral width:* the width of the humerus at the most elevated portion of and including the humeral crest.
13. *Femoral length:* the greatest length of the femur.
14. *Femoral width:* the width of the femur at the midpoint of and including the femoral crest.
15. *Tibiofibula length:* the greatest length of the tibiofibula when the median longitudinal depression of that bone is perpendicular to the plane of the caliper arm.
16. *Tibiofibula width:* the shortest distance across the neck of the tibiofibula taken parallel to the broadest plane of the bone.

The dextral bone of the specimen was used when possible for nonmedian measurements. The frequent occurrence of anomalous structure and breakage explains variation in sample size among the various measurements within species.

Since amphibians grow throughout life, recourse to ratios (indices) of continuous variates is generally made when sampling necessitates comparison between unbalanced size groups. Underhill (1961) has indicated an isometric relationship of characters in ratios constructed for adult *B. woodhousei.* This relationship is assumed to be true for the characters I have measured.

The previous measurements were used to construct the following indices:

1. Skull length/width × 1,000 (Fig. 4–2).
2. Occipital canal width/transorbital width × 1,000 (Fig. 4-3). This index is an attempt to quantify the terms "broad" and "narrow" when applied to the frontoparietal bones. In this research (and general usage) these terms apply not to the actual width of these bones at any point, but to the amount of horizontal projection of the frontoparietal bone above the dorsal posteromedial corner of the orbit (and projection beyond the dorsolateral portion of the braincase at the rear of the orbit). The most notable landmark in this corner of the orbit is the anterior end of the occipital canal.
3. Neurocranial height/occipital canal width × 1,000 (Fig. 4-4). This index describes the shape of the braincase: elevated, depressed or intermediate.
4. Neurocranial height/skull height × 1,000 (Fig. 4-5). This index indirectly describes the degree of elevation of the base of the braincase above the most ventral portion of the skull and is a function of the length of the descending arm of the squamosal and also the angle this arm forms with the vertical.
5. Vertebral column length/width × 1,000 (Fig. 4-6).
6. Humeral width/length × 1,000 (Fig. 4-7).
7. Femoral width/length × 1,000 (Fig. 4-8).
8. Tibiofibula width/length × 10,000 (Fig. 4-9).

The range, sample mean, standard deviation, standard error, and 95 percent confidence intervals were calculated for each ratio in the species examined. In the calculation of confidence intervals, the computer program utilized estimates of the value of *t*. When sample size is in the vicinity of 10, this estimator produces intervals that compare to six significant figures with those calculated by use of a *t* value extracted from a Table of *t*. Below this sample size, the accuracy decreases gradually with sample size until correspondence is only to three significant figures at N = 3. In figures 4-2 to 4-9 the horizontal line indicates the range; the long vertical line intersecting it, the mean. For samples larger than 5, the 95 percent confidence intervals are indicated by short vertical lines intersecting the range lateral to the mean. Sample size is indicated in parentheses.

Osteological Characters

North and Central American *Bufo*

The approaches of authors attacking evolutionary problems of New World *Bufo* have yielded various results. Although there are differences, general areas of agreement are present. Two major evolutionary lines appear to exist. Blair (1963*a*) previously recognized these as northern and southern lines, as did Low (1967); Tihen (1962*a, b*), as *americanus* and *valliceps* groups; Baldauf (1959), as groups I and II. The term "group" is applied differently by the authors. In this book it is used in the sense of a biological species group.

Alvarius Group. *B. alvarius.* The skull is quite wide and short (Figs. 4-2, 4-10). The frontoparietals are broad (Fig. 4-3), their lateral edges meeting in a smooth arc with the anterior edge of the postorbital shelf. Prominent supraorbital and postorbital crests are present. The otic plate of the squamosal is large, as is the entire postorbital shelf. The nasals terminate anteriorly in a transverse plane that is posterior to the anterior edges of the bases of the premaxillae. The dorsal surfaces of frontoparietals and squamosals are very heavily ornamented, the nasals much less so. The anterior edges of the frontoparietals overlap the nasals. The occipital canals are closed and the frontoparietals are fused to the proötics. The braincase is depressed (Fig. 4-4). The shafts of the squamosals are proportionately long (Fig. 4-5). The suprapterygoid fenestra is not occluded. The wings of the parasphenoid are not overlapped by the pterygoid in four of the specimens examined, but one specimen of average size shows a small amount of anteroventral overlap. The quadratojugals are thick and extend forward for approximately two-thirds the length of the pterygoid fossae. The vertebral column is of average dimensions (Fig. 4-6). The humerus and tibiofibula are of

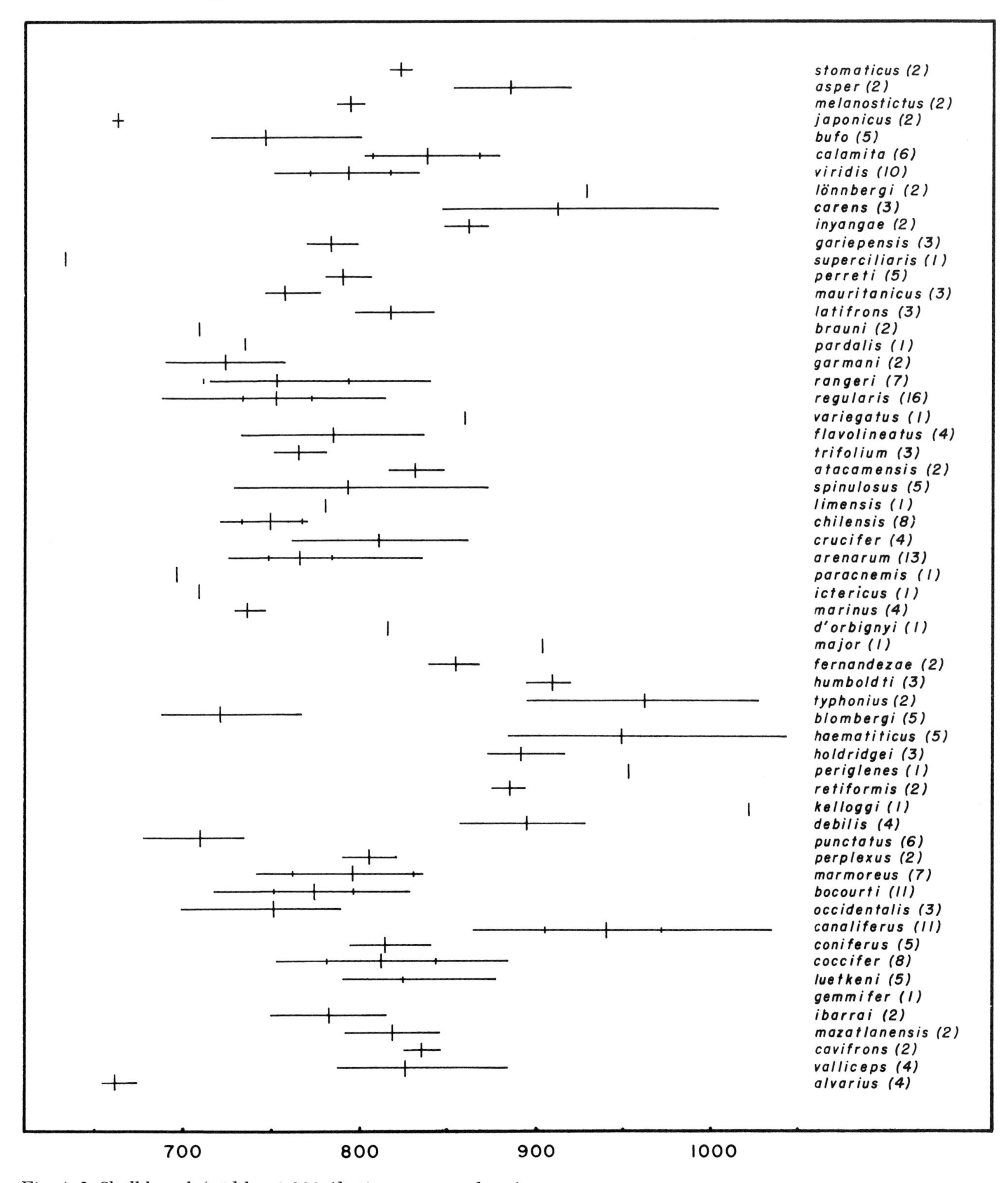

Fig. 4–2. Skull length/width x 1,000 (*latifrons* = *maculatus*).

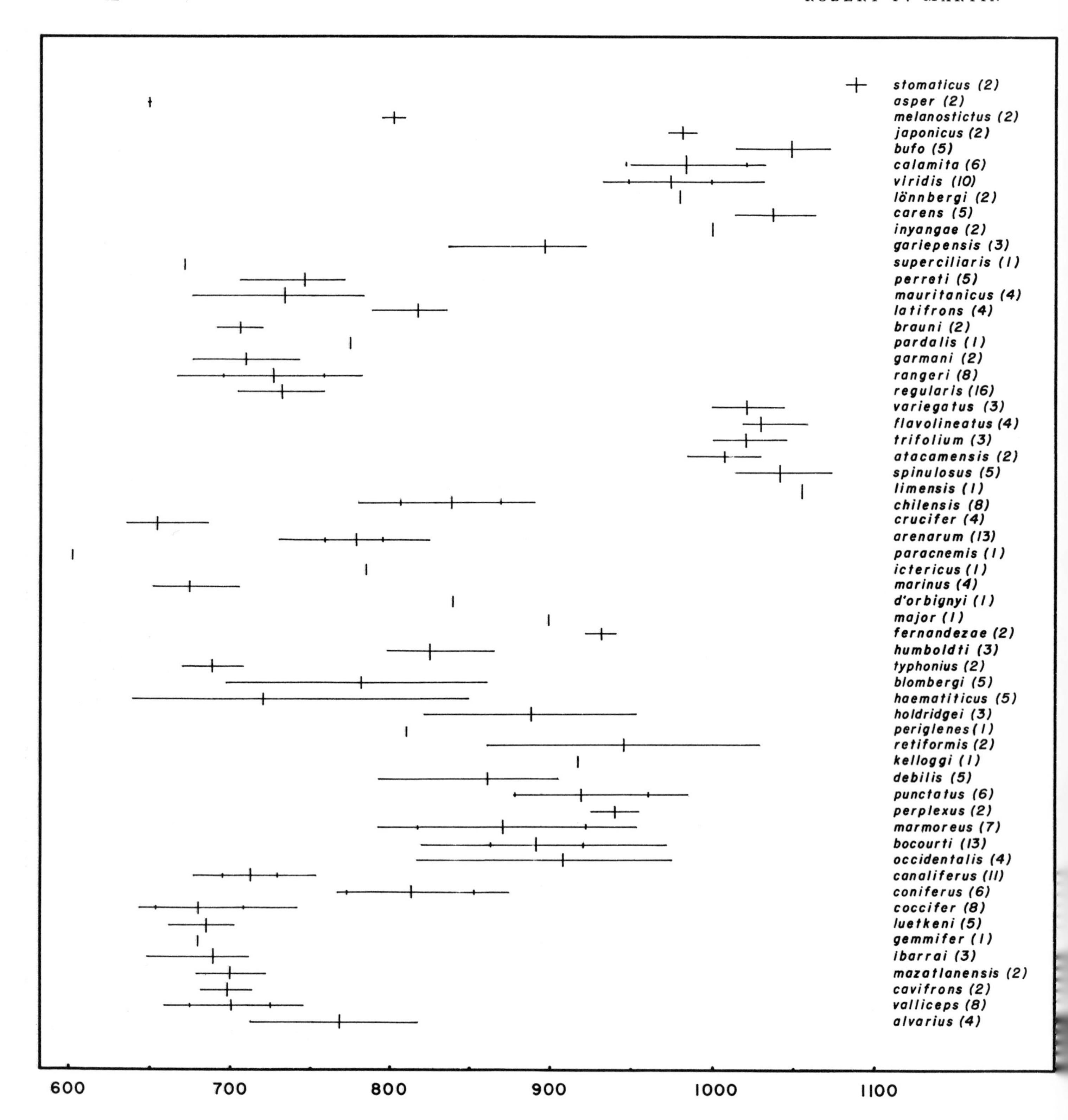

Fig. 4–3. Occipital canal width/transorbital width x 1,000 (*latifrons* = *maculatus*).

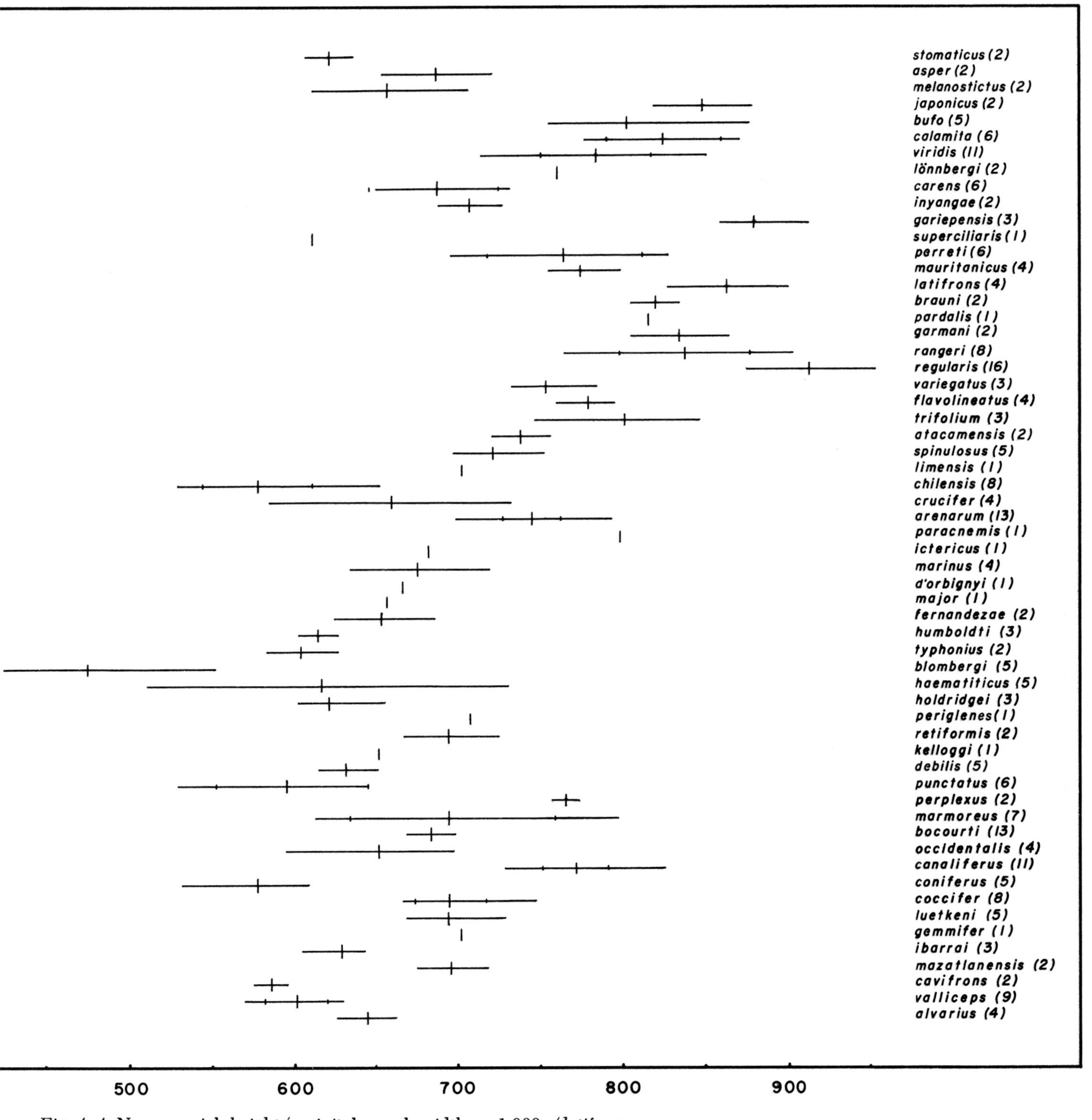

Fig. 4–4. Neurocranial height/occipital canal width x 1,000 (*latifrons* = *maculatus*).

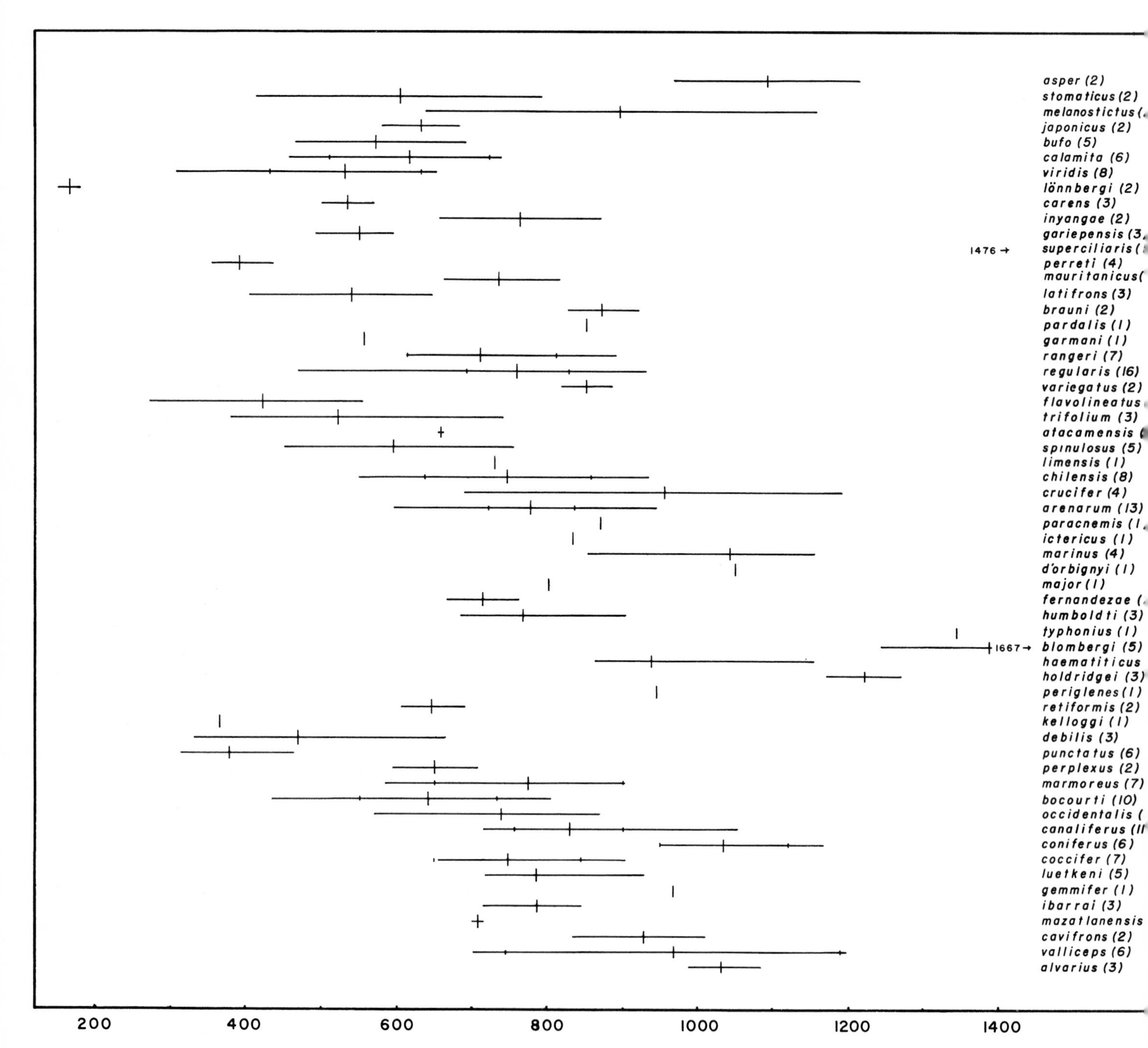

Fig. 4–5. Total height – neurocranial height/neurocranial height x 1,000 (*latifrons* = *maculatus*).

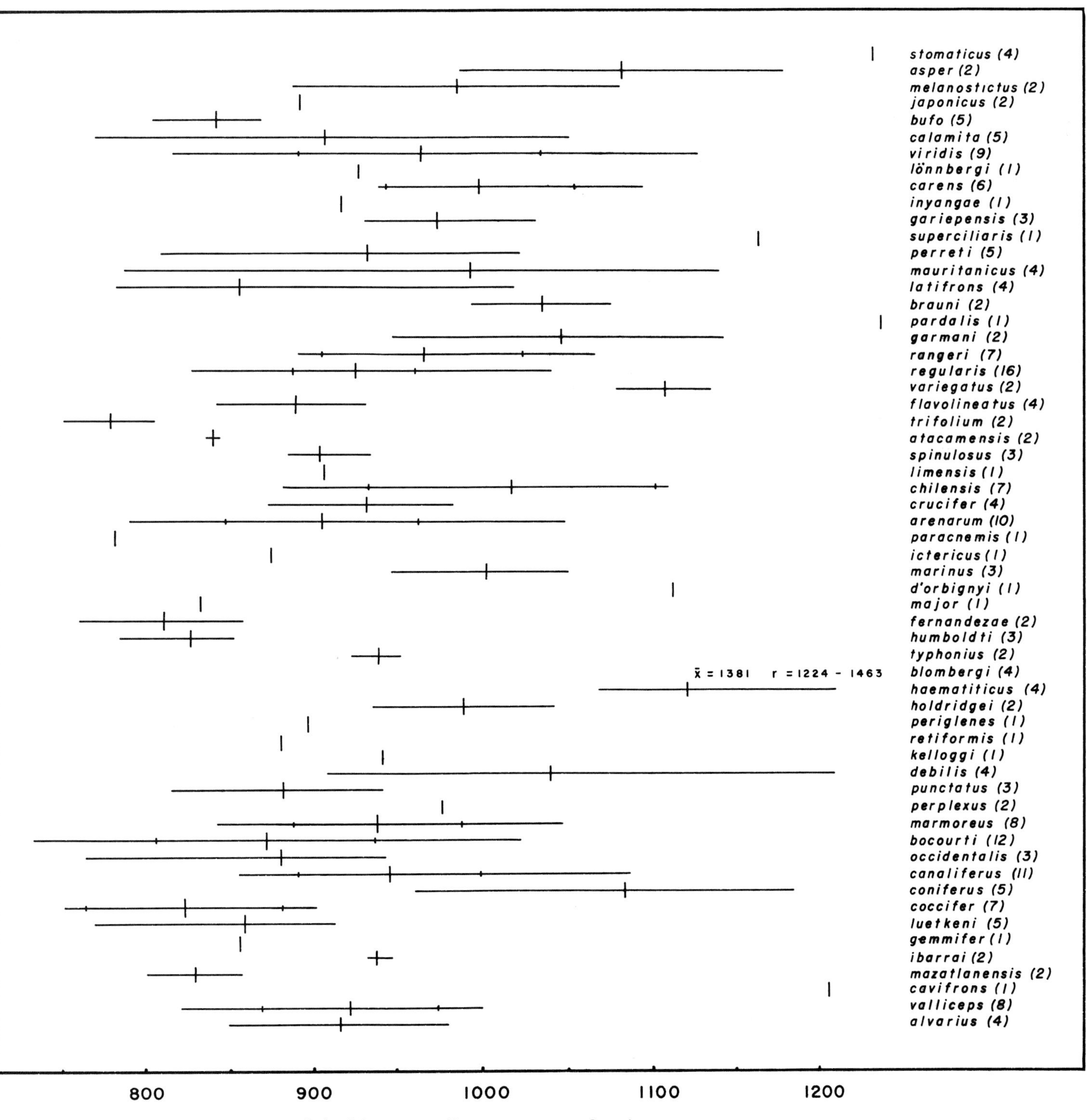

Fig. 4–6. Vertebral column length/width x 1,000 (*latifrons* = *maculatus*).

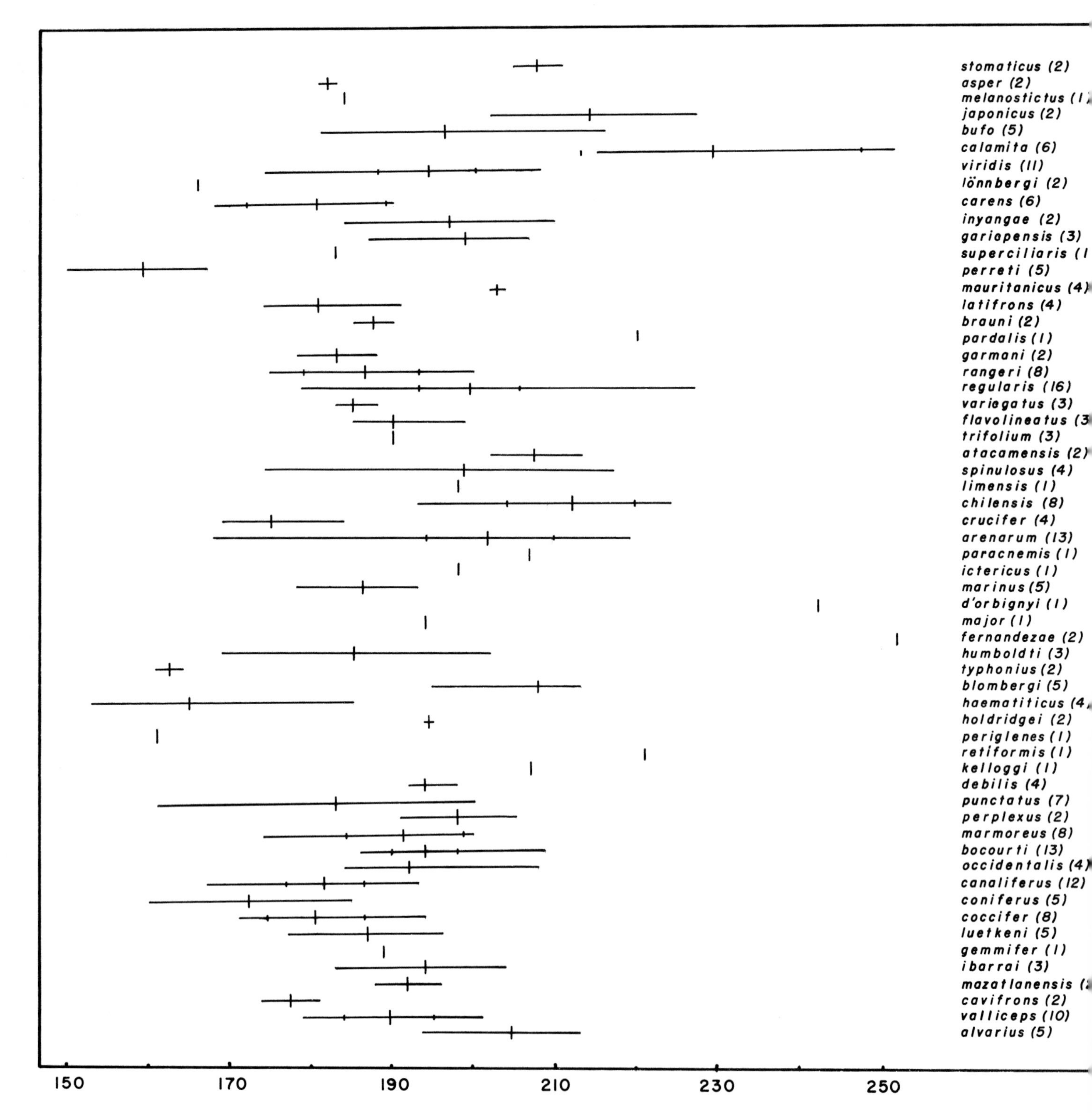

Fig. 4-7. Humeral width/length x 1,000 (*latifrons* = *maculatus*).

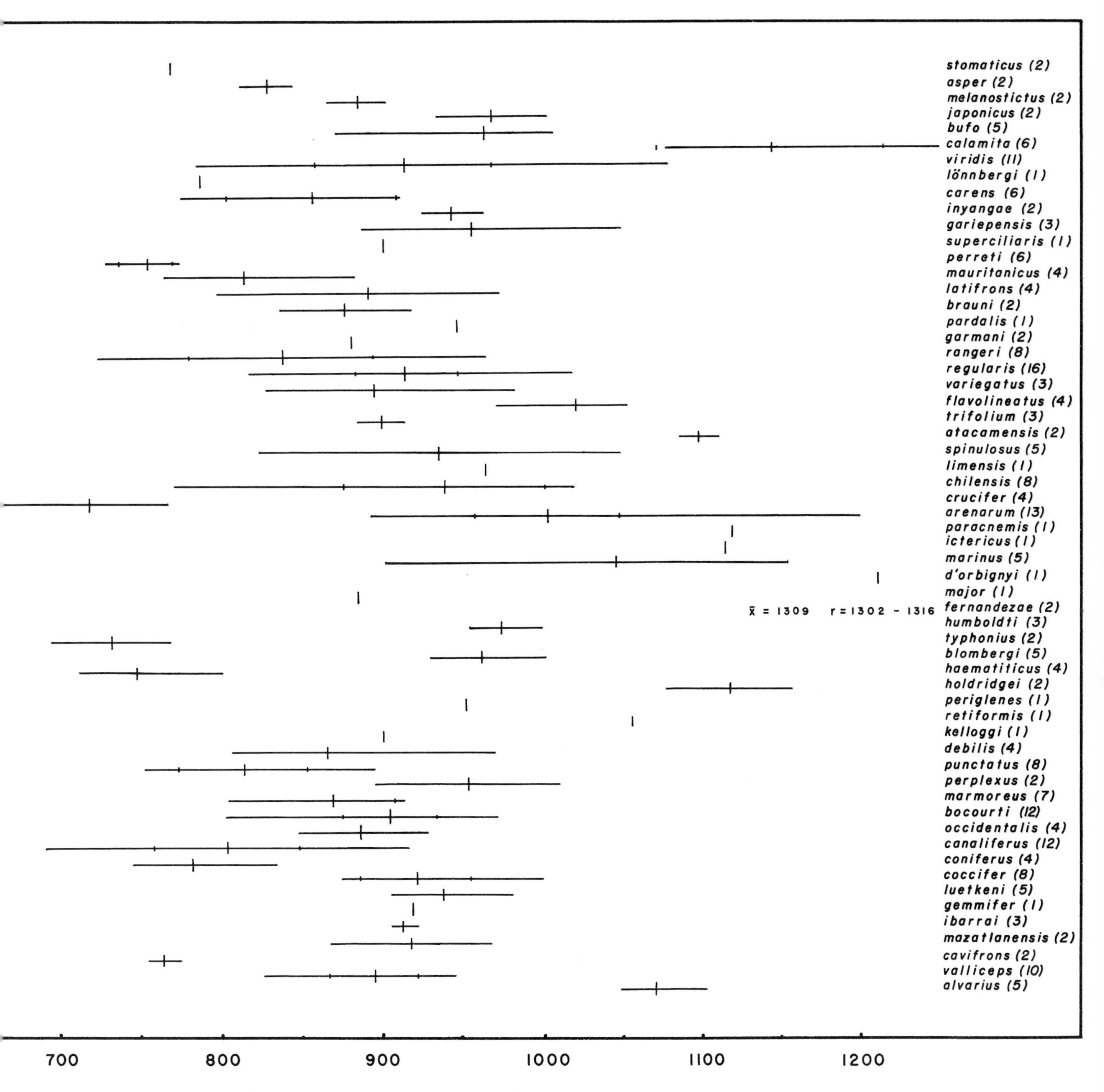

Fig. 4–8. Femoral width/length x 1,000 (*latifrons* = *maculatus*).

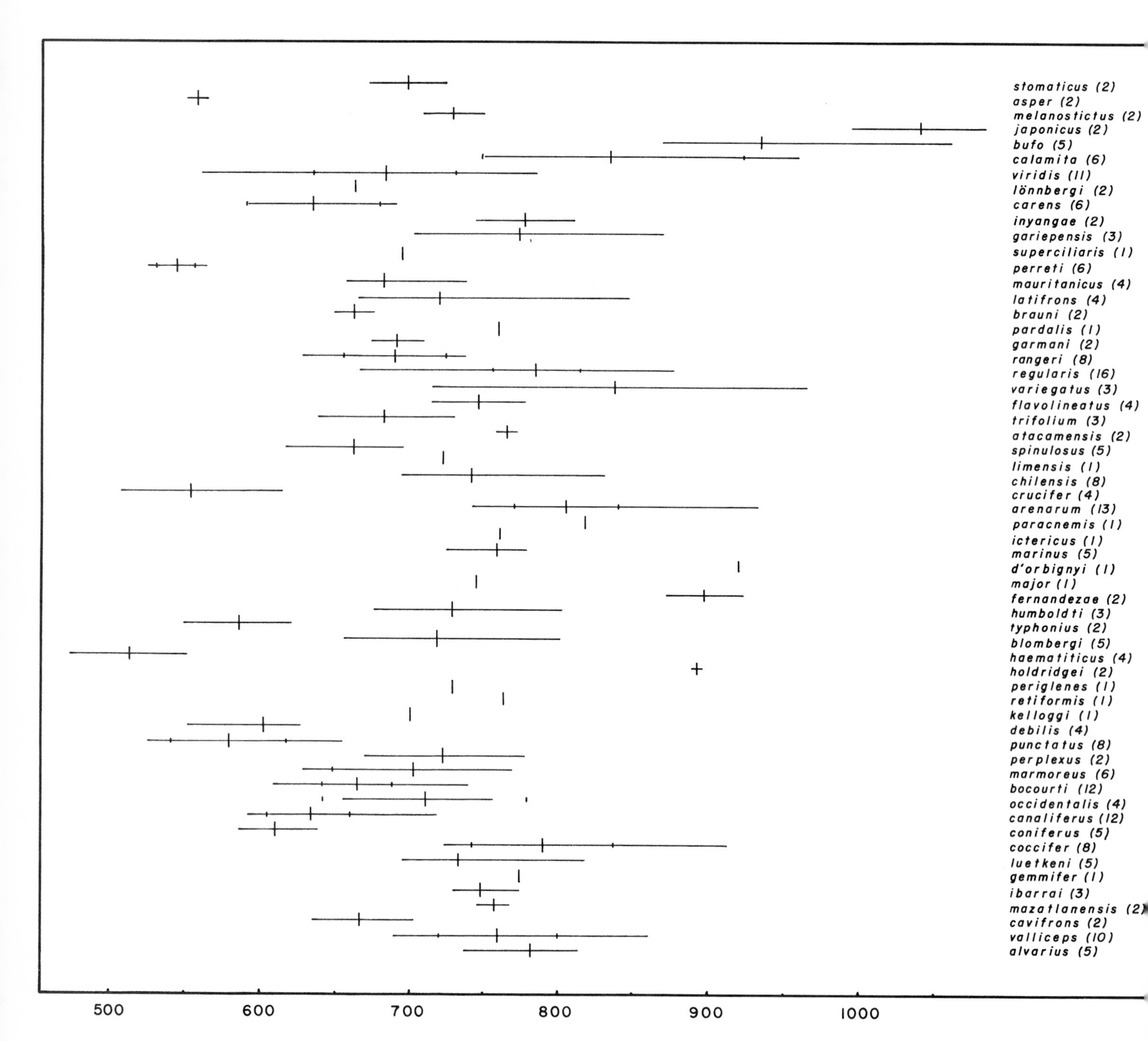

Fig. 4–9. Tibiofibula width/length x 10,000 (*latifrons* = *maculatus*).

average dimensions, the femur, quite thick (Figs. 4-7, 4-8, 4-9).

Osteologically, *B. alvarius* is quite unlike *boreas*-group toads (see Martin, 1964) but very similar to members of the *valliceps* and *marinus* groups.

Valliceps Group. *B. valliceps, B. cavifrons, B. luetkeni, B. ibarrai, B. mazatlanensis, B. gemmifer, B. coniferus.* In dorsal aspect the skull is generally of average proportions (Figs. 4-2, 4-10). The frontoparietals are broad (Fig. 4-3) and bear prominent dorsolaterally directed supraorbital crests that join posteriorly in an arc with the anteriorly projecting postorbital crests to roof the posteromedial portion of the orbit. The nasal bone contributes extensively to the supraorbital crest, which continues anteriorly upon this bone to form the canthal crest. At the anterior end of the orbit, a sharp ridge of the nasal projects downward to form the anterolateral border of the orbit. The dorsolateral border of the squamosal bone bears the supratympanic crest, which anteriorly joins the postorbital crest; the tympanic crest projects anteroventrally on the squamosal from this juncture. The parietal crest projects posteromedially across the dorsal surface of the skull from the juncture of the supra- and postorbital crests. The frontoparietals are fused with the proötic bones. The occipital canals are closed. The otic plate of the squamosal is extensive and forms a rectangular or trapezoidal-shaped area that comprises at least two-thirds of the dorsal surface of the postorbital shelf. The neurocranium is not elevated (Fig. 4-4), and the shafts of the squamosal are long or of average length (Fig. 4-5). The quadratojugal bones are extensive and project anteriorly for at least two-thirds the length of the pterygoid fossae. The medial arms of the pterygoid bones abut anteriorly the wings of the parasphenoid bone but do not overlap them to any marked extent. The suprapterygoid fenestra is not occluded.

The supraorbital, postorbital, and parietal crests are generally hypertrophied in *B. cavifrons* and are slightly reduced in *B. mazatlanensis*. The parietal crests are nearly absent in *B. gemmifer*. Sculpturing of the dorsomedial surface of the frontoparietals is nearly lacking in *B. luetkeni*, lacking completely in *B. coniferus*, and present but variable in the other species. Although the braincase is quite broad in *B. coniferus* (Figs. 4-4, 4-10), the juncture of supra- and postorbital crests is more angular than in other species of the group (Figs. 4-3, 4-10). The postorbital shelf of *B. luetkeni* is rectangular and more extended laterally than in the other species; that of *B. coniferus* is rounded in the posterolateral quadrant. Although this character is subject to a great deal of individual variation (see Martin 1964), it appears that the angle of the jaw is normally behind the jugular foramen only in *B. cavifrons*. Associated with this feature, the shaft of the squamosal in *B. cavifrons* is more horizontal. Although apparently not closely related (restricted sense) to each other, both *B. cavifrons* and *B. coniferus* are differentiated from the remaining species in possessing more attenuated vertebral centra and limb bones (Figs. 4-6, 4-7, 4-8, 4-9).

Coccifer Group. *B. coccifer.* The osteological description of the *valliceps* group applies also to this species (Figs. 4-2 to 4-10). Evidence from other disciplines, discussed elsewhere in this book, however, suggests erection of a monotypic group for it (Chap. 11). This evidence seems to suggest an earlier separation from the line that later gave rise to the *valliceps* group.

Canaliferus Group. *B. canaliferus.* The skull is long and narrow (Figs. 4-2, 4-10). The frontoparietals are broad (Fig. 4-3) and bear supraorbital crests that join the postorbital crests in a smooth curve above the posteromedial corner of the orbit. Parietal crests are lacking. The roofing bones of the skull are smooth. The roofing of dermal bone above the sphenethmoid bone at the juncture of nasals and frontoparietals is incomplete, lending a crescentric shape to the nasals. The braincase is moderately elevated (Fig. 4-4), the squamosal shafts moderately long (Fig. 4-5). The frontoparietals are fused to the proötics and the occipital canals are closed. The otic plate of the squamosal is reduced. The quadratojugal usually extends anteriorly for at least two-thirds the length of the pterygoid fossa. The medial arms of the pterygoid abut but do not overlap the lateral wings of the parasphenoid. The suprapterygoid fenestra is not occluded. The vertebral column is of average proportions (Fig. 4-6); the limb bones slender (Figs. 4-7, 4-8, 4-9). Osteological similarity to the *valliceps* group is not great.

Occidentalis Group. *B. occidentalis.* The skull is short and broad (Figs. 4-2, 4-11). The frontoparietals are of intermediate width (Fig. 4-3), their lateral edges turned upward to form a relatively upright supraorbital crest. The junction with

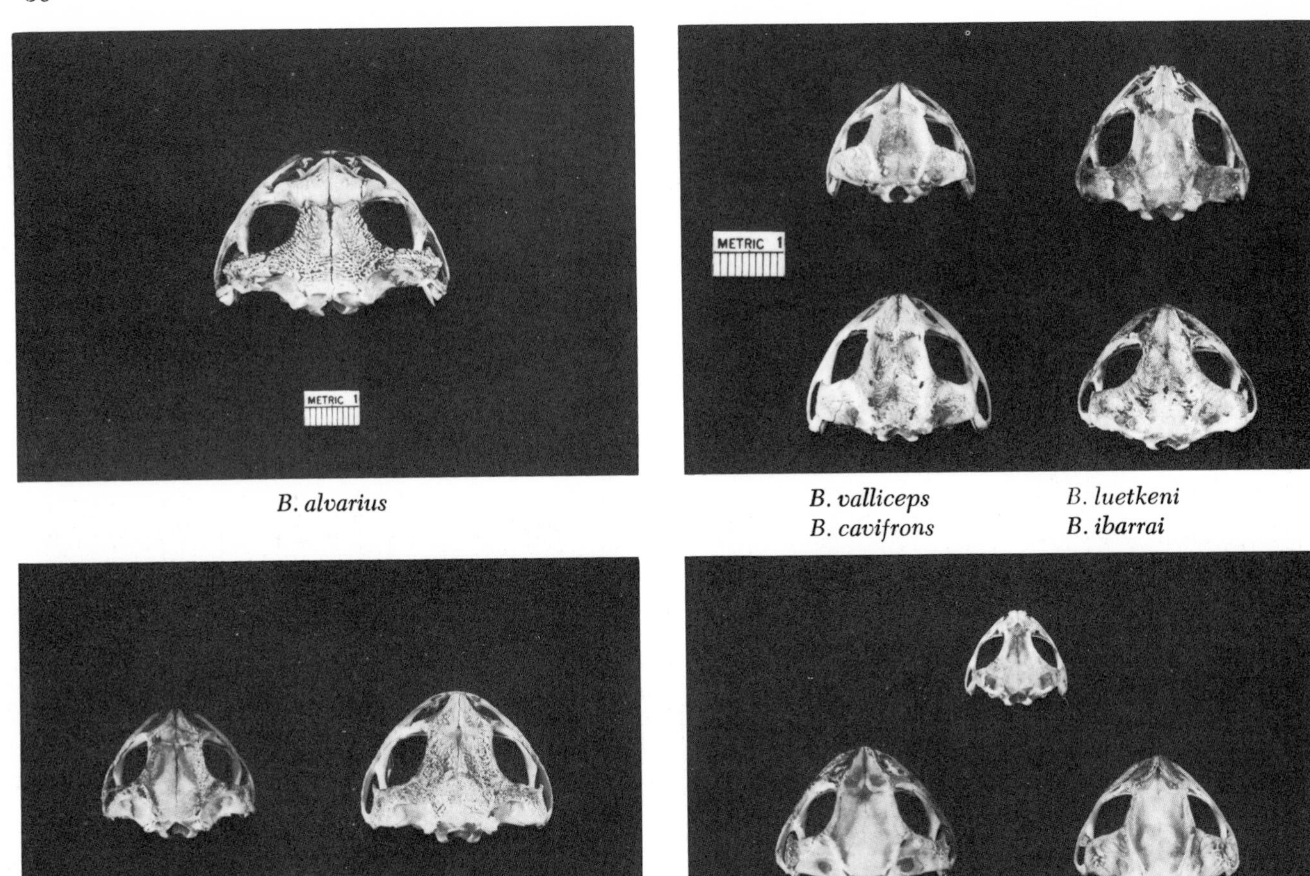

B. alvarius

B. valliceps *B. luetkeni*
B. cavifrons *B. ibarrai*

B. mazatlanensis *B. gemmifer*

B. canaliferus *B. coccifer* *B. coniferus*

Fig. 4–10. *Bufo* skulls

the postorbital ridge varies from arclike to nearly angular. A parietal spur projects posteromedially from this junction, only partially roofing the occipital canal. The postorbital shelf is subtriangular, and the otic plate of the squamosal is only moderately developed. Dermal bone sculpturing is very light, appearing only on the crests. The premaxillary bones extend anteriorly beyond the nasals. The frontoparietals are fused to the proötics. The braincase is moderately depressed (Fig. 4-4), and squamosal shafts are of average length (Fig. 4-5). The quadratojugal is thin but extends anteriorly at least one-half the length of the pterygoid fossa. The medial arms of the pterygoid do not occlude the suprapterygoid fenestra. The alate portions of the parasphenoid are only slightly (if at all) overlapped by the medial arms of the pterygoid. The vertebral column is moderately broad and limbs are of average proportions (Figs. 4-6, 4-7, 4-8, 4-9).

Osteologically, *B. occidentalis* shows several features characteristic of narrow frontoparietal types: the postorbital shelf is reduced, dermal ornamentation is reduced and the occipital canals are only partly roofed (narrow). The pattern of cristation is intermediate. The braincase is flattened, as in most broad frontoparietal species. At present, the above information argues against a definite allocation.

Bocourti Group. *B. bocourti*. The skull is moderately short and broad (Figs. 4-2, 4-11). The frontoparietals are of moderate width (Fig. 4-3). The

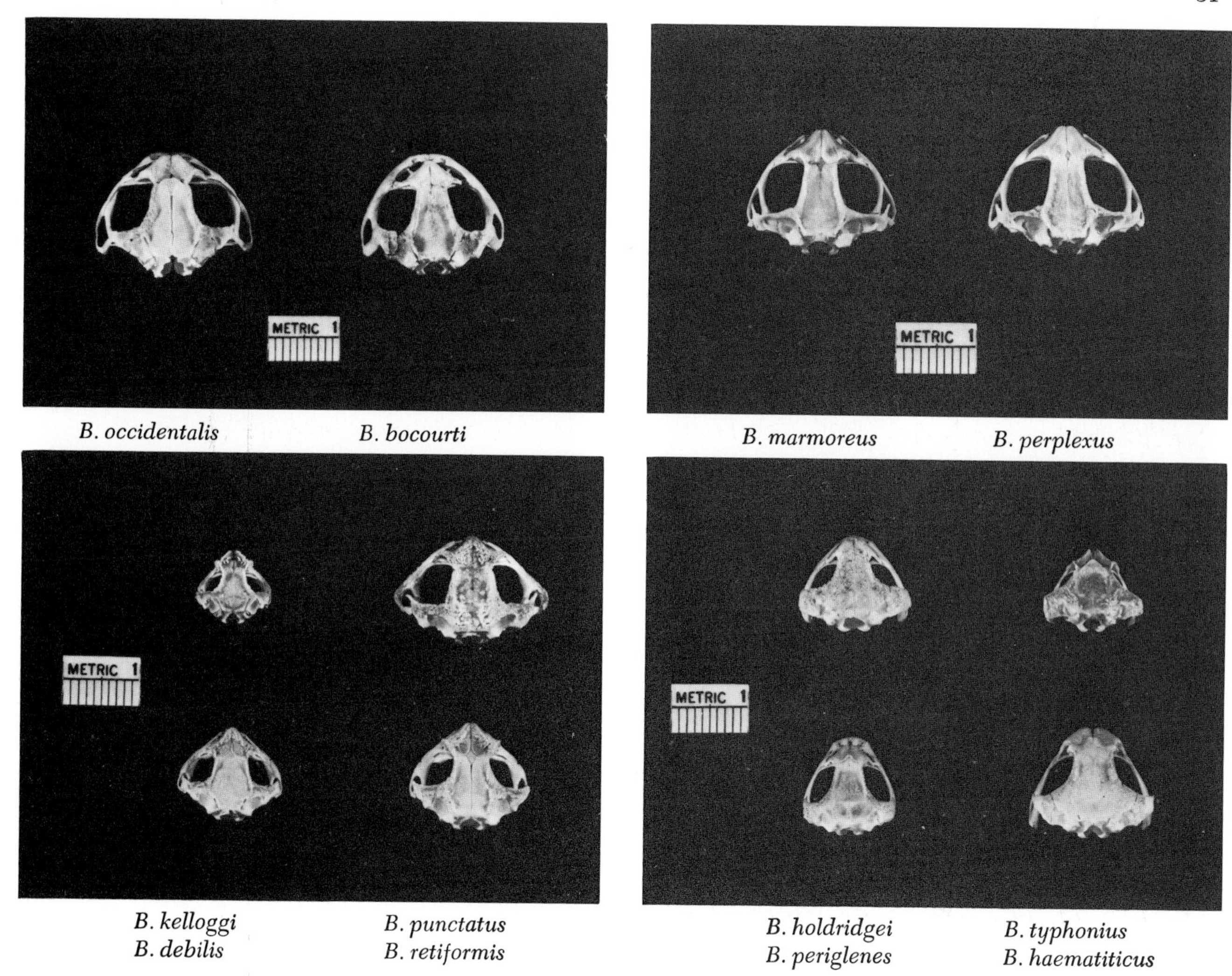

B. occidentalis *B. bocourti* *B. marmoreus* *B. perplexus*

B. kelloggi *B. debilis* *B. punctatus* *B. retiformis* *B. holdridgei* *B. periglenes* *B. typhonius* *B. haematiticus*

Fig. 4–11. *Bufo* skulls

supraorbital crests are pronounced but not greatly elevated and meet with the postorbital ridge in a somewhat smoothed obtuse angle. In some specimens the parietal ridge is a short spur, leaving the occipital canal open posteriorly; in others, it forms a longer parietal crest, roofing the canal completely. The frontoparietals are fused to the proötics. The otic plate of the squamosal is of moderate extent and bears a variable supratympanic crest. The postorbital shelf is subrectangular. Light ornamentation is present upon the crests. The braincase is moderately depressed (Fig. 4-4), and the squamosals are of average length (Fig. 4-5). The premaxillae extend anteriorly past the nasals; their ascending processes slant backward appreciably from their bases. The ventral surfaces of the maxillae are considerably curved. In most *Bufo*, these surfaces are largely tangent to the horizontal plane; in *B. bocourti* the midpoint of the ventral surface of the maxilla is appreciably dorsal to its ends. The medial arms of the pterygoid overlap the wings of the parasphenoid only slightly, if at all, and do not occlude the suprapterygoid fenestra. No columella auris is present. The vertebral column is short (Fig. 4-6). Humerus and femur are of average proportions; the tibiofibula is slender (Figs. 4-7, 4-8, 4-9). Although no definite allocation is strongly indicated, osteological evidence suggests derivation from the narrow frontoparietal line rather than from the broad line.

Marmoreus Group. *B. marmoreus, B. perplexus.* The skull is of average proportions (Figs. 4-2,

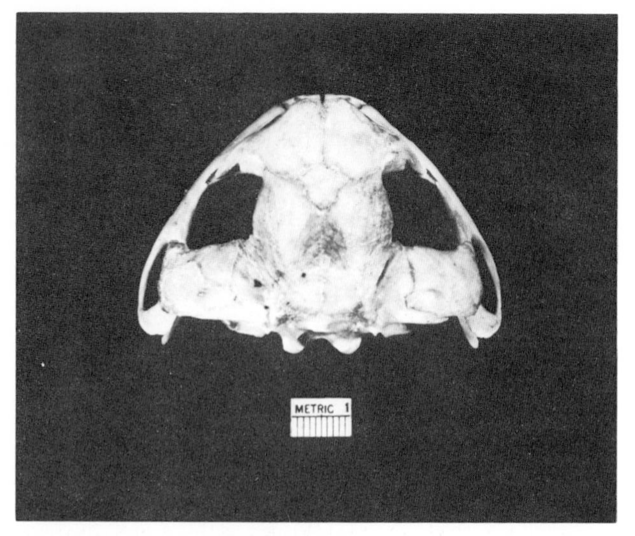

B. blombergi

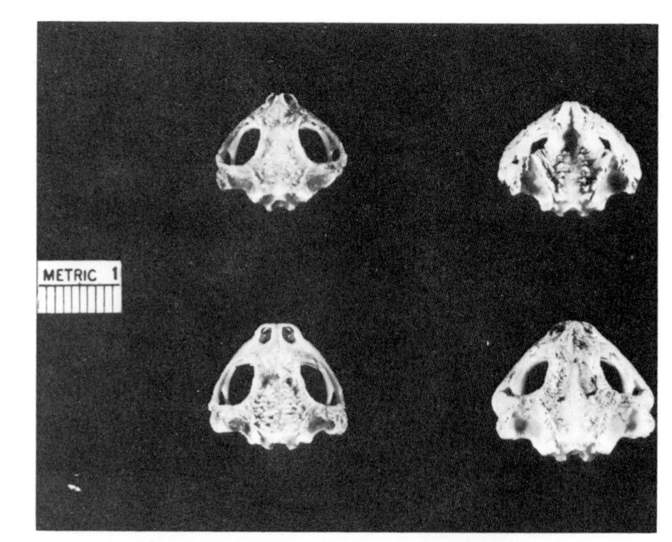

B. fernandezae *B. d'orbignyi*
B. humboldti *B. major*

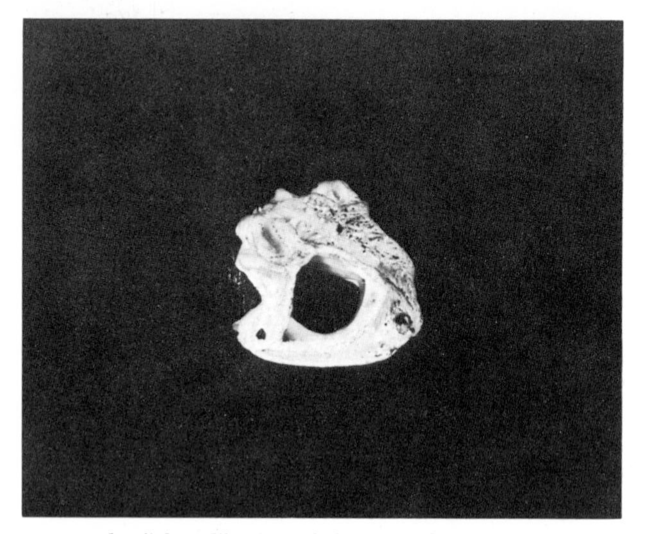

The "closed" orbit of the *granulosus* group

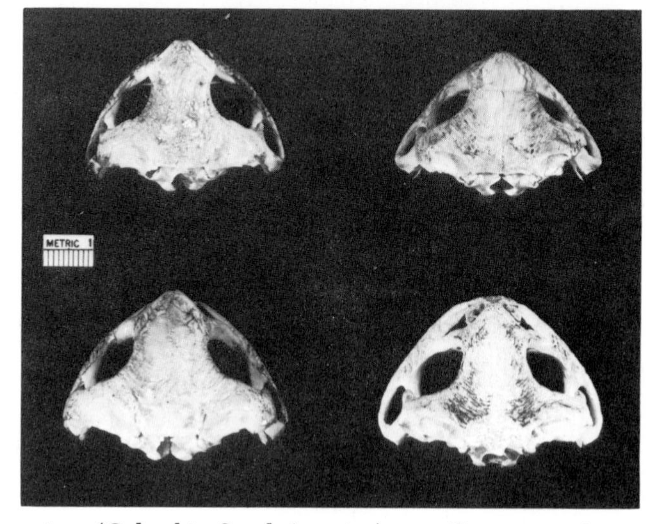

B. marinus (Colombia, South America) *B. paracnemis*
B. marinus (Mexico) *B. ictericus*

Fig. 4–12. *Bufo* skulls

4-11). The frontoparietals are of only moderate width (Fig. 4-3), their lateral edges directed dorsally and laterally to form low supraorbital crests. A low postorbital crest joins the supraorbital crest at the posteromedial corner of the orbit; in my sample this junction is slightly more angular in *B. perplexus*, more arclike in *B. marmoreus*. A low and short supratympanic crest is present in some specimens. The parietal crest is low and variable in length; accordingly, the occipital canals are either covered or partially exposed. (In these species increased development of the crest is correlated with increased size.) The anterior ends of nasals and premaxillae approximate the same vertical plane. The otic plate of the squamosal varies from moderate to fairly extensive; the shape of the postorbital shelf varies concurrently from subtriangular to nearly rectangular. Light ornamentation is present only upon the crests of *B. marmoreus*; in *B. perplexus*, the horizontal laminae of the frontoparietals are also lightly ornamented. The frontoparietals are fused with the proötics.

The braincase and squamosal arms are of average height (Figs. 4-4, 4-5). The alate portions of the parasphenoid are not overlapped by the medial arms of the pterygoid, and the suprapterygoid fenestrae are not occluded. The vertebral column and limbs are of average proportions (Figs. 4-6, 4-7, 4-8, 4-9).

The osteological description indicates intermediacy between major lines but stronger affinity with the narrow rather than the broad line.

Punctatus Group. *B. punctatus.* The skull is short and broad (Figs. 4-2, 4-11). The frontoparietals are of moderate width (Fig. 4-3). The lateral edges of the frontoparietals join the anterior edges of the postorbital shelves in an angle or a short arc. In most specimens, crests are nondescript, but some Mexican populations display low supra and postorbital crests. The nasals are acuminate and usually project slightly anterior to the premaxillae. The surfaces of nasals, frontoparietals, squamosals, and premaxillae are heavily ornamented. The occipital canal is closed. The frontoparietals are fused to the proötics. The postorbital shelf is reduced and subtriangular in shape. The lateral (orbital) arm of the squamosal is not broad but is considerably longer than the descending arm, in some specimens nearly reaching the maxilla. The descending arm is nearly absent and, in my specimens, contacts neither pterygoid nor quadratojugal. The braincase is quite depressed (Fig. 4-4). The wings of the parasphenoid are not overlapped by the medial arms of the pterygoid. The occlusion of the suprapterygoid fenestra varies from moderate to absent. The articulation of the jaw is considerably anterior to the jugular foramen. The vertebral centra are broad (Fig. 4-6), the limbs slender (Figs. 4-7, 4-8, 4-9).

The only strong osteological similarities are with *B. debilis*, *B. kelloggi*, *B. retiformis*, and to a lesser extent with some members of the *granulosus* group.

Debilis Group. *B. debilis*, *B. kelloggi*, *B. retiformis.* The skull is long (Figs. 4-2, 4-11). The frontoparietals are of moderate width (Fig. 4-3), and their lateral edges meet with the anterior edges of the postorbital shelf in a junction intermediate between an arc and an angle. In *B. debilis* and *B. retiformis*, the supraorbital and postorbital crests are low, in *B. kelloggi*, distinctly elevated. In *B. retiformis* and in some specimens of *B. debilis*, a low parietal crest is present; in *B. kelloggi*, it is absent. Ornamentation is conspicuous toward the edges of the frontoparietals, squamosals, and nasals. The nasals are acuminate and project anteriorly well beyond the line of the jaw. The occipital canals are closed or partially open in mid-length. The frontoparietals are fused with the proötics. The postorbital shelf is somewhat reduced and subtriangular. The lateral arm of the squamosal is extensive, nearly contacting the maxilla in some specimens of *B. debilis.* The descending arm (shaft) of the squamosal is reduced in *B. debilis* and *B. kelloggi* (Fig. 4-5), its length being equal to or shorter than the lateral arm. In *B. retiformis* the shaft is of average length (Fig. 4-5). Contact of this arm with the quadratojugal is variable, being present or absent in *B. debilis* and *B. retiformis* and absent in the one damaged specimen of *B. kelloggi* examined. The braincase is flattened (Fig. 4-4). The suprapterygoid fenestra is nearly occluded and the pterygoid fossa is very short. The medial arms of the pterygoid do not overlap the alae of the parasphenoid. The angle of the jaw is far anterior to the jugular foramen. The vertebral centra of *B. retiformis* are short and broad, those of *debilis* are moderately long, and those of *B. kelloggi* are of average proportions (Fig. 4-6). The limbs of *B. retiformis* are more robust than those of *B. kelloggi* and *B. debilis* (Figs. 4-7, 4-8, 4-9). The humerus and femur of *B. debilis* and *B. kelloggi* are of average proportions, the tibiofibulae, more slender (Figs. 4-7, 4-8, 4-9).

Strong osteological similarity with the *punctatus* group is supported by chromosomal evidence (Chap. 10).

Periglenes Group. *B. periglenes.* The skull is long (Figs. 4-2, 4-11). The frontoparietals are moderately broad (Fig. 4-3), and the posteromedial quadrant of the orbit is roofed by a smooth arc of bone. Dorsal surface ornamentation is reduced, and the postorbital shelf is moderately extensive. The nasals are short, terminating anteriorly, considerably posterior to the most anterior projections of the premaxillae. The occipital canals are closed and the frontoparietals are fused with the proötics. The lateral bar of the squamosal is moderately enlarged and trapezoidal in shape. The braincase is of average proportions (Fig. 4-4); the squamosal shafts are long (Fig. 4-5). The alae of the parasphenoid are only slightly overlapped by the medial arm of the pterygoid, but the suprapterygoid fenestra is considerably occluded by it. No columella auris is present. The vertebral column is of average length (Fig. 4-6). The humerus is thin, the femur and tibiofibula, of average proportions (Figs. 4-7, 4-8, 4-9).

Little is known of the biology of this species (R. Sage, personal communication). It inhabits the montane cloud forest of Costa Rica. The sexes are dimorphic; the male is bright orange, the female, bluish green. This plus the reduced auditory apparatus strongly suggest breeding on visual cues. There is general ecological and osteological similarity with *B. holdridgei* but little additional information of real systematic value.

Holdridgei Group. *B. holdridgei.* The skull is moderately long (Figs. 4-2, 4-11). The frontoparietals are of moderate width (Fig. 4-3), their lateral edges joining the anterior edge of the postorbital shelf in a smooth arc. The lateral edges of the frontoparietals are only slightly elevated. The postorbital shelf is extensive, its lateral edge sloping upward posteriorly to form a distinct eminence at the posterolateral corner of the shelf. No parietal crests are present, but the occipital canal is closed. The frontoparietals are fused to the proötics. The dorsal surface of the skull is conspicuously granular in texture. The lateral process of the squamosal is conspicuously enlarged and granular, terminating ventrally in a blunt square. The squamosal shafts are long (Fig. 4-5) and upright. The braincase is depressed (Fig. 4-4). The wings of the parasphenoid are not overlapped by the medial arms of the pterygoid, and the suprapterygoid fenestra is not occluded. The quadratojugal is moderately long, extending anteriorly from one-half to two-thirds the length of the pterygoid fossa. A columella auris is absent. The vertebral centra are of average proportions (Fig. 4-6), but the urostyle bears unique and conspicuous lateral flanges, extending in width beyond its sacral articular facets. The humerus is of average proportions (Fig. 4-7), but both femur and tibiofibula are extremely short and thick (Figs. 4-8, 4-9).

This species cannot be conveniently placed with any other grouping. It shares small size, granular skull roof, and extensive lateral plate of the squamosal with *B. punctatus*, the *debilis* group, and the *granulosus* group, but the squamosal shafts are long, and the suprapterygoid fenestra is not occluded. The peculiar urostylar flange and loss of columella are further specializations. Savage and Kluge (1961) have described a similar specialization of the urostyle in *Crepidius epioticus*, a bufonid from a similar habitat in Costa Rica. Morphological characters of *Crepidius* argue strongly for its retention in a separate taxon, but speculation may be entertained about the derivation of this genus from *B. holdridgei* or an ancester.

South American *Bufo*

Guttatus Group. *B. haematiticus.* The skull is long and narrow (Figs. 4-2, 4-11). The frontoparietals are broad (Fig. 4-3), their supraorbital and postorbital edges upturned and extending as a roof above the orbit. In larger specimens, these edges appear as low crests. The lateral edge of the squamosal bears a low supratympanic crest. The dorsal surfaces of the nasals bulge slightly above those of the frontoparietals in adult specimens, markedly so in smaller specimens, giving the bones an "inflated" appearance. The dorsal margins of these bones are rounded, rather than sharply edged as in the *valliceps* group. The dorsal surfaces of the frontoparietals and nasals are only lightly ornamented; the lateral edges of the squamosal are ornamented to a greater degree. No parietal crests are present. The occipital canals are enclosed. The frontoparietals are fused to the proötics. The otic plate of the squamosal is extensive and trapezoidal in shape. The neurocranium is depressed (Fig. 4-4), but the squamosal shafts are long (Fig. 4-5). The quadratojugals are thin but moderately extensive, projecting anteriorly from the angle of the jaw for approximately half the pterygoid fossa. The medial arms of the pterygoid overlap the wings of the parasphenoid only slightly. The medial arms of the pterygoid bear thin dorsal flanges that slightly occlude the suprapterygoid fenestra. The vertebral centra are elongate (Fig. 4-6). The transverse processes of vertebrae 5, 6, 7, and 8 are long and bear prominent horizontal flanges, and the neural spines of all vertebrae are flattened into horizontal plates. The limb bones are long and slender (Figs. 4-7, 4-8, 4-9). Tihen (1962*a*) has considered this species sufficiently unique to erect a monotypic group for it. Unquestionably, it possesses several peculiar specializations; however, some of the features described by Tihen are not characteristic of my specimens. Only one of my specimens displays the extremely narrow maxillary ellipse figured in Tihen's research. My specimen is small, poorly ossified, and it is possible that its maxillary configuration was damaged in preparation. The largest of my specimens does not

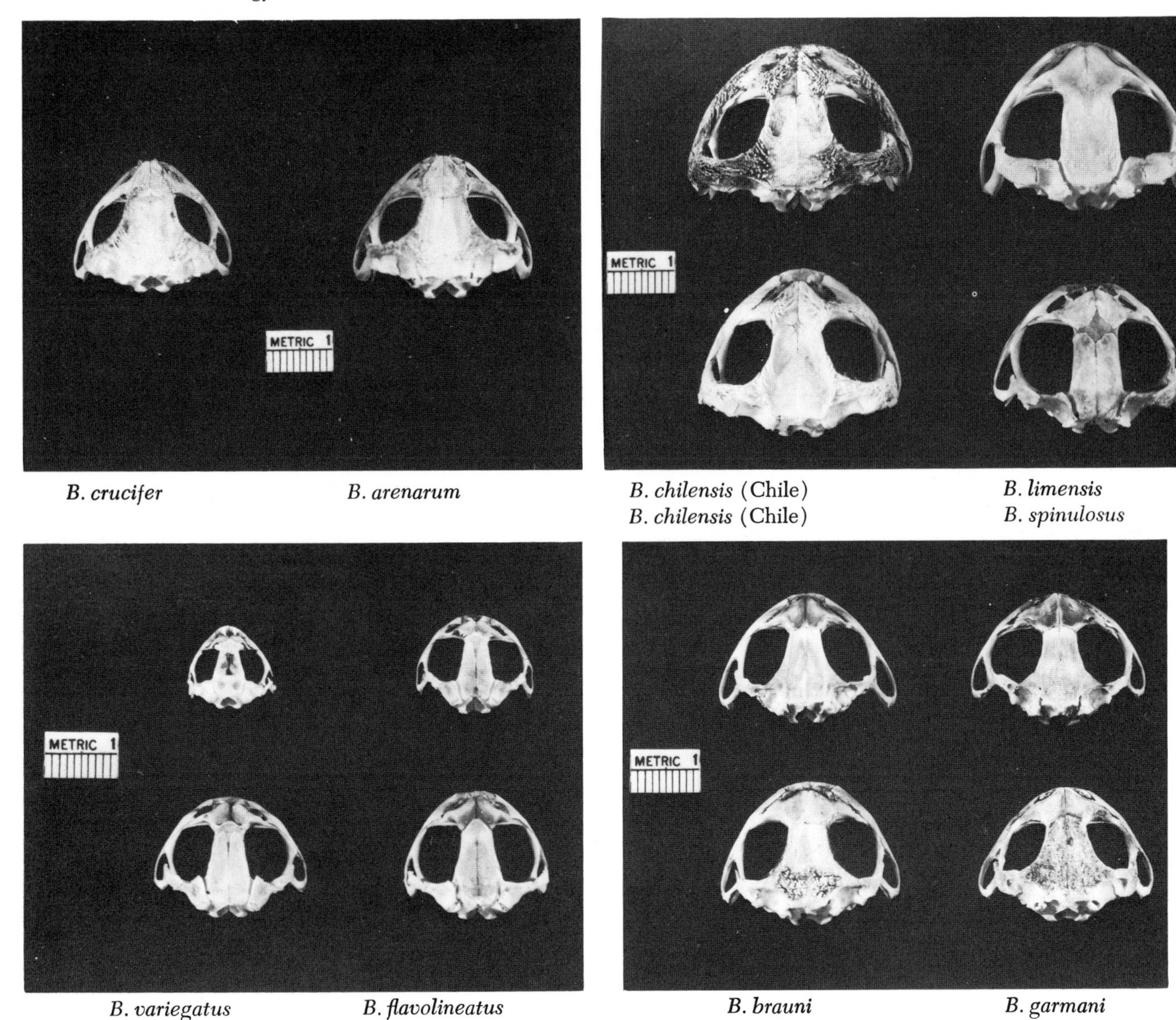

B. crucifer *B. arenarum*

B. chilensis (Chile) *B. limensis*
B. chilensis (Chile) *B. spinulosus*

B. variegatus *B. flavolineatus*
B. trifolium *B. atacamensis*

B. brauni *B. garmani*
B. rangeri *B. regularis*

Fig. 4–13. *Bufo* skulls

display the poorly ossified sphenethmoid characteristic of Tihen's description and of my smaller individuals. Tihen also described this species as possessing unfused frontoparietal-proötic contact (a character of considerable weight in his scheme of classification). In my specimens, this character varies with size, from absence of fusion in small specimens, to complete fusion in large individuals. Thus, the condition varies ontogenetically (as in other fused species), but differs in that the state of fusion appears to be reached later than usual in the life of the individual.

B. blombergi. The skull is broad and short (Figs. 4-2, 4-12). The frontoparietals are broad (Fig. 4-3); their lateral edges roof a considerable portion of the orbit, but are only slightly upturned. The postorbital shelves are nearly rectangular, their anterior edges forming an obtuse angle with the lateral edges of the frontoparietals. The dorsal surface of the skull is moderately sculptured in both spiculate and cancellous pattern. Parietal crests are prominent but low. The occipital canal is closed. All dorsal corners of the nasals are rounded. Although my specimens are adult, fusion of the frontoparietal and proötic is completely present in only one, incomplete in three, and absent in one. The size range of the species (Myers and Funkhouser 1951) extends beyond that of my sample, and, since Tihen (1962*a*) describes the species as fused, it appears that, as in *B. haematiticus*, fusion occurs late in ontogeny. The neurocranium is very depressed (Fig. 4-4), and the descending arms of the squamosals moderately long (Fig. 4-5). The quadratojugal is of moderate proportions, extending forward for slightly more than half the length of the pterygoid fossa. Although Tihen (1962*a*) has indicated only slight occlusion, in three of my five specimens prominent flanges borne upon the medial arms of the pterygoid markedly occlude the suprapterygoid fenestra. The wings of the parasphenoid are broadly overlapped by the medial arms of the pterygoid. The sphenethmoid bone is well ossified and extensive, projecting anteriorly beyond the medial arms of the palatines nearly as far as posteriorly. The foramen magnum is much broader than it is tall; its height divided by its width averages .56. The vertebral centra are quite long (Fig. 4-6), as are the transverse processes. The neural spines of the vertebrae are flattened and roughened, but not to the extent that occurs in *B. haematiticus*. The limbs are of average proportions (Figs. 4-7, 4-8, 4-9).

W. F. Blair (Chap. 11) has grouped *B. haematiticus* and *B. blombergi* in his *guttatus* group partly because of the peculiar testicular morphology that appears involved with an inhibitory secretion produced in the seminal fluid.

In the light of the osteological similarities mentioned earlier, as well as the gonadal similarity, it seem logical to group *B. blombergi* with *B. haematiticus*.

Typhonius Group. *B. typhonius*. The skull is narrow, (Figs. 4-2, 4-11). The frontoparietals are broad (Fig. 4-3) and their laterally directed supraorbital crests merge in a smooth arc with the postorbital crests. Canthal, antorbital, tympanic, and supratympanic crests are present. The postorbital shelf is extensive and trapezoidal in shape. Parietal crests are not evident, but the area usually occupied by them is raised by ridges and granulations that roof the occipital canal. The dorsal surface of the skull is lightly ornamented. The nasals are acuminate and their anterior edges project considerably anterior to the bases of the premaxillae. In conjunction with this condition, the ascending processes of the premaxillae project conspicuously forward as they rise toward the nasals. The frontoparietals are fused to the proötics. The braincase is low (Fig. 4-4), but the squamosals are quite long (Fig. 4-5) and parallel to each other. The wings of the parasphenoid are considerably overlapped by the medial arm of the pterygoid, and the suprapterygoid fenestra is moderately occluded by flanges of the pterygoid and squamosal. The quadratojugal is of moderate extent, extending anteriorly for approximately half the pterygoid fossa. The fossa itself is short and does not extend anteriorly beyond a transverse plane tangent to the anterolateral edge of the squamosal. The vertebral centra are of average proportions (Fig. 4-6). The limb bones are long and slender (Figs. 4-7, 4-8, 4-9).

The "dead-leaf" pattern is similar to that of *B. haematiticus* and *B. blombergi*, and general skull and limb proportions are similar to those of *B. haematiticus*. It does not share their peculiar testicular or vertebral morphology, however, and its relationships to them and other species must remain uncertain.

Granulosus Group. *B. humboldti, B. major, B. fernandezae, B. d'orbignyi*. The skull proportions of the various forms range from broad to average (Figs. 4-2, 4-12). All possess broad or moderately wide frontoparietal bones that bear conspicuous supraorbital crests. Less conspicuous canthal and preorbital crests are borne upon the nasals. The supraorbital crests join pronounced postorbital crests in a smooth arc. Supratympanic crests of varying height are present. Parietal crests are prominent in *B. major, B. fernandezae*, and *B. d'orbignyi*, but vary from present to nearly absent in *B. humboldti*. The occipital canal is completely roofed. The surfaces of frontoparietals, nasals, squamosals, maxillae, and premaxillae are heavily ornamented, and dermal ossification obscures many of the sutures. The nasals and ascending processes of the premaxillae are intimately united by dermal ossification and project forward beyond the line of the jaw in a very pronounced snout. The postorbital shelf is extensive and rectangular or trapezoidal in shape; the otic plate of the squamosal is large. The frontoparietals are fused to the proötics. The lateral and ascending arms of the squamosal are fused together into a single bar that joins the maxilla and closes the orbit (Fig. 4-12). The braincase is somewhat depressed (Fig. 4-4). The length of squamosal shaft is average in *B. fernandezae*, *B. major*, and *B. humboldti*, quite long in *B. d'orbignyi*. The suprapterygoid fenestra is occluded, and the wings of the parasphenoid are overlapped anteroventrally for at least half their length by the medial arms of the pterygoid. The quadratojugal-maxillary suture is obscured by ornamentation, and the articulation of the jaw is anterior to the jugular foramen. The vertebral centra of *B. major*, *B. humboldti*, and *B. fernandezae* are all short and broad; those of *B. d'orbignyi*, long (Fig. 4-6). The limbs of *B. humboldti* and *B. major* are of normal proportions; those of *B. fernandezae* and *B. d'orbignyi* are very short and broad (Figs. 4-7, 4-8, 4-9).

Gallardo (1965) has treated these forms as subspecies in his morphological study. Cei (Chap. 6) has indicated that *fernandezae* and *major* maintain their identity in sympatry and believes that several species are included under the name *B. granulosus*. Significant osteological differences are present between populations. I am using the subspecific names of Gallardo as if they applied to species. Panamanian specimens (*B. humboldti*) display very reduced parietal crests; in others the crests are conspicuous. Specimens of *B. major* and *B. fernandezae* collected sympatrically in Resistencia, Argentina, display great disparity in limb proportions (Figs. 4-7, 4-8, 4-9), and Cei and Erspamer (1966) demonstrated that these populations also show considerable differentiation in the concentration of 6-indolealkylamine in their integument. Although this differentiation is present, all forms share the unique "closed orbit" and heavy ornamentation. These facts lead to the conclusion that species rather than subspecies are represented here; the range of the group is, however, considerable, and the available material is in no way an adequate sampling of it.

Marinus Group. *B. marinus*, *B. ictericus*, *B. paracnemis*, *B. arenarum*. The skull is short and broad (Figs. 4-2, 4-12, 4-13). The frontoparietals are broad (Fig. 4-3), their lateral edges produced into prominent supraorbital crests that continue anteriorly onto the nasals and bifurcate to form prominent canthal and preorbital crests. Posteriorly, the supraorbital crests are joined in a smooth arc by the postorbital crests, which are in turn joined laterally by the tympanic and supratympanic crests. The frontoparietal-nasal suture is nearly transverse. In *B. ictericus* the anterior edges of the premaxillae extend anteriorly beyond the anterior ends of the nasals, but in other members of this group these bones terminate in approximately the same transverse plane. The dermal roofing bones are generally heavily striated and granulated. Distinct parietal crests are not present, but the area that they usually occupy is elevated by parallel ridges of bone. The postorbital shelf is extensive and trapezoidal in shape; the otic plate of the squamosal is large. The frontoparietals are fused to the proötics. The occipital canal is closed. The braincases range from moderately depressed to slightly elevated (Fig. 4-4). The shafts of the squamosal are of average or less than average length (Fig. 4-5). The medial arms of the pterygoid overlap the wings of the parasphenoid for over half their length. The suprapterygoid fenestra is nearly completely occluded by flanges borne on the shaft of the squamosal and the medial arm of the pterygoid in all group members. The vertebral columns of *B. paracnemis* and *B. ictericus* are short and broad, those of the remaining members of average proportions (Fig. 4-6). The femora of *B. ictericus*, *B. marinus*, *B. paracnemis*, and *B. arenarum* are short and thick, as are the tibiofibulae of *B. paracnemis* and *B. arenarum*. Other limb bones are of average proportions (Figs. 4-7, 4-8, 4-9).

Crucifer Group. *B. crucifer*. This species is very similar in osteology to members of the *marinus* group. The following differences are present: its skull is narrower and longer (Figs. 4-2, 4-12, 4-13), the suprapterygoid fenestra is only moderately occluded, and the limb bones are more slender (Figs. 4-7, 4-8, 4-9).

Spinulosus Group. *B. spinulosus*, *B. atacamensis*, *B. trifolium*, *B. flavolineatus*, *B. limensis*, *B. chilensis*.

B. chilensis. The skull is broad (Figs. 4-2, 4-13). The frontoparietals are relatively broad (Fig. 4-3) and their lateral edges are upturned to form supraorbital crests that meet with the anterior edges of the postorbital shelves in a smooth arc. Low parietal, supratympanic, and preorbital crests are present. The surfaces of premaxillae, maxillae, nasals, squamosals, and frontoparietals are heavily ornamented. The occipital canals are completely covered in some specimens, but the roofing is incomplete posteriorly in others. The nasals terminate anteriorly posterior to the anterior ends of the premaxillae. The otic plate of the squamosal is long, broad, and cancellous; the descending arm is of average length (Fig. 4-5). The braincase is quite depressed (Fig. 4-4). The suprapterygoid fenestra is not occluded. The medial arm of the pterygoid overlaps the anteroventral portions of the parasphenoid alae for one-third to one-half their lengths. The quadratojugal extends forward for at least three-fourths the length of the pterygoid fossa. Vertebral centra are of average proportions (Fig. 4-6). The humeri are short and broad; other limb proportions are average (Figs. 4-7, 4-8, 4-9).

The entire skull of *B. limensis* is moderately broad (Figs. 4-2, 4-13). The frontoparietals are narrow (Fig. 4-3), and their lateral edges meet with the anterior edges of the postorbital shelves in a sharp right angle. No crests are present. The anterior edges of the nasals terminate slightly posterior to the bases of the premaxillae. The otic plate of the squamosal is well developed, and the postorbital shelf is large. There is only very light sculpturing of the roofing bones. The occipital canals are open but are flanked laterally and medially by low ridges of bone. The frontoparietals are fused to the proötics. The lateral arm of the squamosal is of only moderate extent; the descending arm is of average length (Fig. 4-5). The braincase is of average proportions (Fig. 4-4). The suprapterygoid fenestra is only slightly occluded. The medial arms of the pterygoid overlap the anteroventral portions of the parasphenoid wings for approximately one-third their length. The quadratojugals are slender and extend anteriorly nearly the entire length of the pterygoid fossa. The vertebral centra are slender (Fig. 4-6), and the limbs are of average proportions (Figs. 4-7, 4-8, 4-9).

The skull of *B. spinulosus* is moderately broad (Figs. 4-2, 4-13). The frontoparietals are very narrow (Figs. 4-3, 4-13). The irregularity of the anterior edge of the postorbital shelf tends to obscure the angular nature of the inner corner of the orbit. This irregularity also imparts a somewhat crescentric appearance to the entire postorbital shelf. Anteriorly, the frontoparietals are in only partial contact with the nasals; in some specimens, contact is not present. The nasals terminate anteriorly posterior to the bases of the premaxillae. The otic plate and lateral arm of the squamosal are reduced. No crests are present. The occipital canals are open anteriorly and either partially or totally roofed for a short distance at their posterior ends. The frontoparietals are fused to the proötics. The braincase is of average proportions (Fig. 4-4), as are the descending arms of the squamosals (Fig. 4-5). The suprapterygoid fenestra is not occluded. The medial arms of the pterygoid overlap the wings of the parasphenoid anteroventrally for less than one-third of their length. The quadratojugal is moderately extensive, from one-half to two-thirds the length of the pterygoid fossa. The vertebral column is of average dimensions (Fig. 4-6). Humerus and femur are of average proportions, but the tibiofibula is relatively slender (Figs. 4-7, 4-8, 4-9).

Table 4–1. Condition of Frontoparietal-Proötic Fusion in *B. atacamensis*, *B. trifolium*, and *B. flavolineatus*

Species	Cranial Length in mm.	Condition of Fusion
B. atacamensis (sample below	17.0	fused
species mean in body size)	16.9	partially fused
B. trifolium (all in sample	17.0	fused
at least 10 mm. below	15.9	partially fused
species mean in body size)	14.1	partially fused
B. flavolineatus (all in	13.8	unilaterally fused
sample at least 6 mm.	14.0	partially fused
below species mean in body	12.9	partially fused
size)	13.0	unfused

B. atacamensis, *B. trifolium*, and *B. flavolineatus* are all quite similar to *B. spinulosus*. None of my specimens reach even the mean size reported for these species by Vellard (1959) and Cei (1962), which may introduce error into the description. The skulls are of average or broader than average proportions (Figs. 4-2, 4-13). The frontoparietals are narrow (Figs. 4-3, 4-13). No crests are present. The nasals are not in sutural contact with the frontoparietals. The mid-frontoparietal suture is incompletely closed for most of its length. No ornamentation of roofing bones is present. The postorbital shelf and otic plate of the squamosal are reduced. Although variable (probably ontogenetically), the terminal relationship of frontoparietal and proötic appears to be that of fusion (Table 4-1). The braincase is moderately elevated (Fig. 4-4). The descending arms of the squamosal

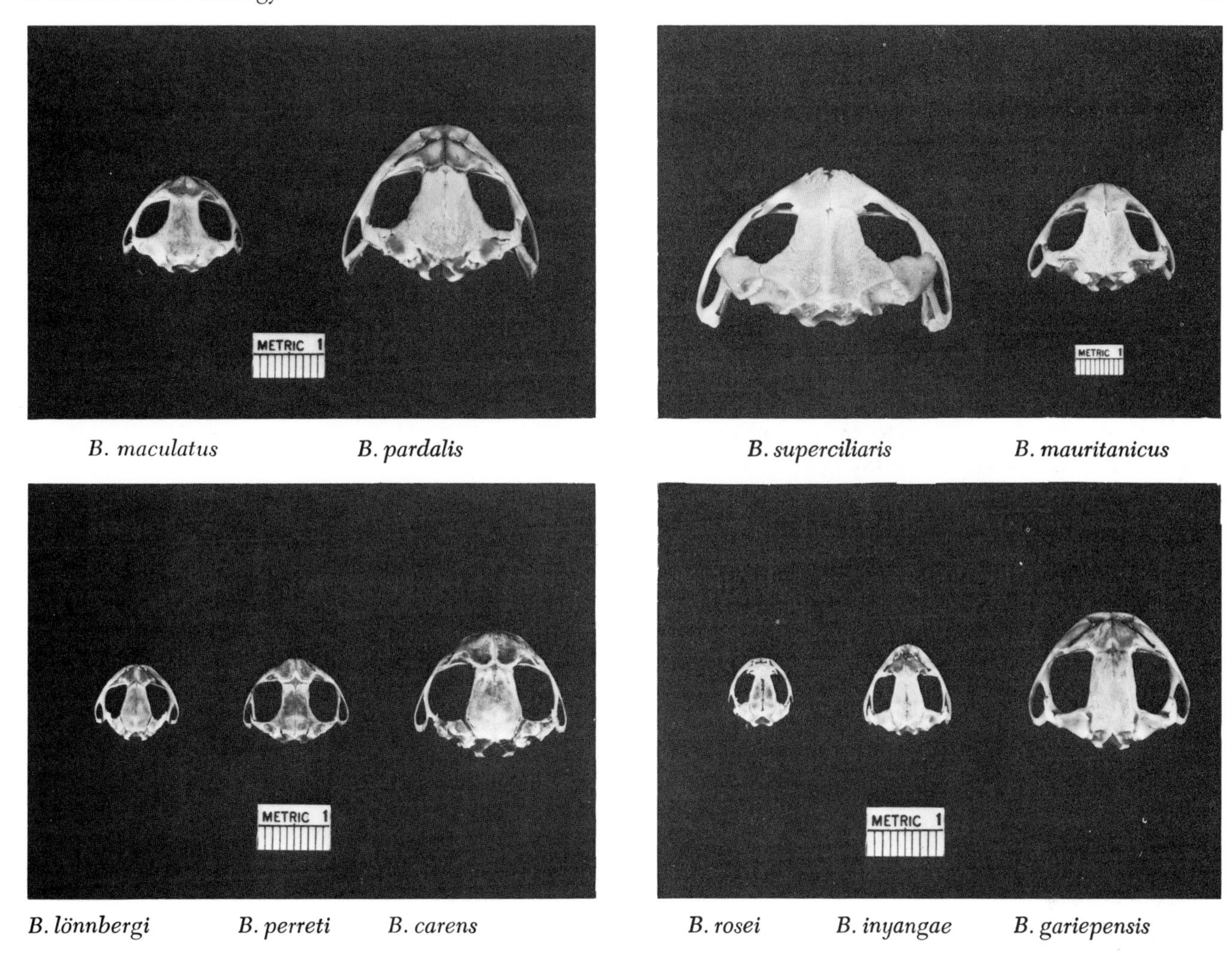

B. maculatus *B. pardalis* *B. superciliaris* *B. mauritanicus*

B. lönnbergi *B. perreti* *B. carens* *B. rosei* *B. inyangae* *B. gariepensis*

Fig. 4–14. *Bufo* skulls

are of average length in *B. atacamensis*, proportionately shorter in *B. trifolium* and *B. flavolineatus*. The suprapterygoid fenestra is slightly occluded in *B. flavolineatus*, but not occluded in *B. trifolium* and *B. atacamensis*. Overlap of the wings of the parasphenoid is present to only a limited extent or is entirely absent. The vertebral centra are short (Fig. 4-6). Humeri and tibiofibulae are of average proportions except in *B. trifolium*; its tibiofibula is moderately slender (Figs. 4-7, 4-9). The femur of *B. trifolium* is of average dimensions; those of *B. flavolineatus* and *B. atacamensis* are more short and robust (Figs. 4-8, 4-9).

The morphology and ecology of members of this group have been discussed in moderate detail by Vellard (1959) and Cei (1962). When *B. chilensis* is compared to the other Andean forms, it appears at first greatly dissimilar due to the increased amount of dermal bone. More intimate relationship with these, however, is suggested by the osteologically intermediate *B. limensis*.

Variegatus Group. *B. variegatus.* All my specimens are below the dimensions Cei (1962) reported as average body size for this species, and the reduced nature of the dermal bones may be partly ontogenetic. The skull is of average proportions (Figs. 4-2, 4-13). The frontoparietals are narrow (Fig. 4-3) and do not contact the nasals anteriorly. The nasals are reduced and terminate far posteriorly to the bases of the premaxillae. The frontoparietals are broadly separated anteriorly; posterior to the sphenethmoid, a fontanelle is present.

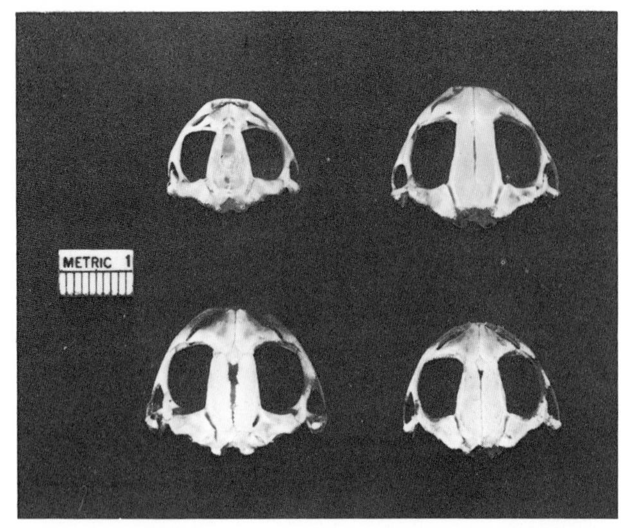

B. calamita *B. viridis* (Europe)
B. viridis (Israel) *B. viridis* (Israel)

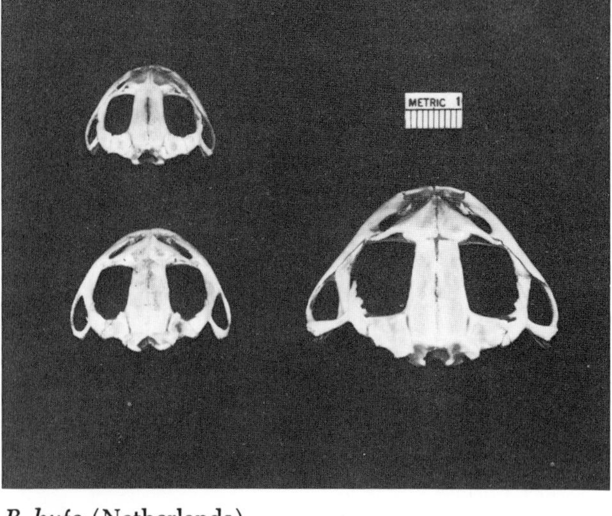

B. bufo (Netherlands)
B. bufo (Italy) *B. b. japonicus* (Japan)

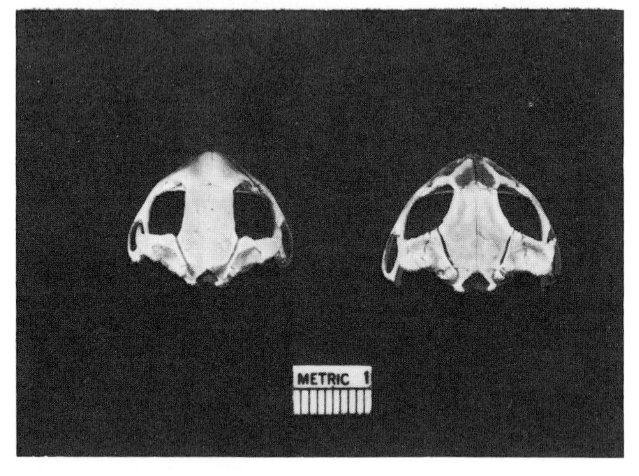

B. stomaticus *B. melanostictus*

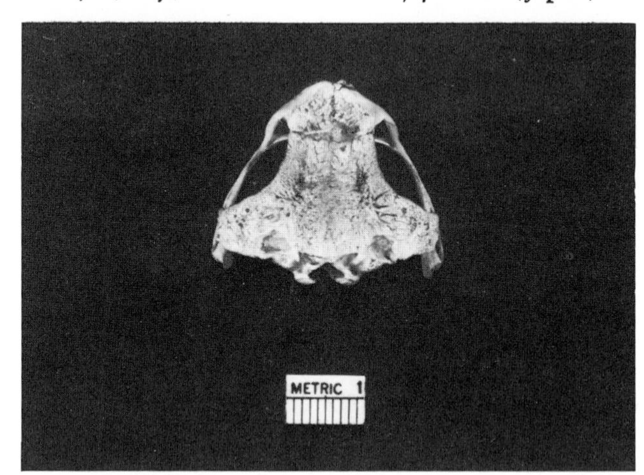

B. asper

Fig. 4–15. *Bufo* skulls

Posteriorly, midline contact of these bones is only approximated in my sample. The occipital canals are open. The otic plate of the squamosal is reduced to a narrow bar of bone applied only laterally to the proötic. The frontoparietals are not fused to the proötics in the specimens I have examined. The braincase is moderately elevated (Fig. 4-4), and the descending arms of the squamosal are long (Fig. 4-5). The medial arm of the pterygoid abuts the proötic but does not contact the parasphenoid. The suprapterygoid fenestra is not occluded. The quadratojugal is very short and does not reach the maxilla. No columella auris is present. The vertebral centra are narrow (Fig. 4-6). Humeri and femora are of average dimensions; the tibiofibulae are short (Figs. 4-7, 4-8, 4-9). Although placed separately, *B. variegatus* is similar to the members of the *spinulosus* group with narrow frontoparietals and is probably closely related to that group.

African *Bufo*

Regularis Group. *B. regularis, B. rangeri, B. garmani, B. pardalis, B. brauni.* The skull is broad in all species (Figs. 4-2, 4-13, 4-14). The frontoparietals are broad (Fig. 4-3), their lateral edges joining the anterior edge of the postorbital shelf in an arc that roofs a considerable portion of the orbit.

Cranial crests are absent. The dorsal surfaces of the frontoparietals vary from slightly concave to slightly convex. Dermal ornamentation is variable within the species but is particularly reduced in *B. brauni*. The occipital canals are closed. The postorbital shelf is of moderate extent and varies from subrectangular to merely triangular. The specimens of *B. garmani* and *B. brauni* that I have examined show postorbital shelf reduction in excess of that of the other species. In *B. pardalis* the anterior ends of the nasals terminate considerably posterior to the anterior edge of the bases of the premaxillae; in the other species these terminate in approximately the same transverse plane. The frontoparietals are not fused to the proötics. The braincase is elevated (Fig. 4-4). In *B. regularis*, *B. rangeri*. *B. brauni*, and *B. pardalis*, the shafts of the squamosal are moderately long, in *B. garmani*, shorter (Fig. 4-5). The medial arms of the pterygoid do not overlap the wings of the parasphenoid, nor do they occlude the suprapterygoid fenestrae. The quadratojugals are moderately long, extending anteriorly at least two-thirds the length of the pterygoid fossa. The vertebral columns of *B. pardalis* and *B. brauni* are long; those of the others are of average proportions (Fig. 4-6). Humeri range from somewhat slender in *B. garmani*, *B. rangeri*, and *B. brauni* to average in *B. regularis* and short in *B. pardalis* (Fig. 4-7). The femora of *B. garmani*, *B. brauni*, *B. pardalis*, and *B. regularis* are of average dimensions, those of *rangeri*, moderately slender (Fig. 4-8). Tibiofibulae of *B. regularis* and *B. pardalis* are of average proportions, while those of the other species are slender (Fig. 4-9).

Since my sample sizes of several of the species listed are quite small, and because the osteology of these species is quite similar, I do not feel that I can make any statements concerning intragroup relationships. I have examined only one specimen of each of the species that seem to be most divergent (*B. brauni* and *B. pardalis*) from each other and from other members of the group. Inger (1959) regarded *rangeri* and *pardalis* as subspecies of *B. regularis*, but this seems unrealistic in view of Poynton's (1964) description of their disparate premating behavior in the area of sympatry. Guttman (1967) noted that the single hemoglobin bands of *B. rangeri* and *B. garmani* were alike and identical with one of the bands of *B. brauni*. The data of Low (1967) offer no clear proximities but suggest that *B. garmani* is more biochemically distinct than are the other species. As mentioned elsewhere in this book, there is great geographic variation and possibly speciation among the toads to which the name *B. regularis* is currently applied. Surprisingly, although osteological variation is present, it is of no greater magnitude than that displayed by species of more narrow range.

Maculatus Group. *B. maculatus*. The osteology of this species is similar to that of the *regularis* group, with the following exceptions: its skull is of more average proportions (Figs. 4-2, 4-13, 4-14); the squamosal shafts are not as long (Fig. 4-5); the vertebral column is shorter and broader (Fig. 4-6); and the humeri and tibiofibulae are more slender (Figs. 4-7, 4-9).

Perreti Group. *B. perreti*. The skull is moderately broad (Figs. 4-2, 4-14). The frontoparietals are narrow anteriorly (Fig. 4-3) but broaden rapidly at the rear of the orbit to join the anterior edges of the postorbital shelves in a smooth arc. No crests are present. The anterior edges of the nasals approximate or extend slightly beyond the transverse plane including the anterior ends of the bases of the premaxillae. The otic plate of the squamosal and postorbital shelf are reduced in area. The dorsal surfaces of the roofing bones possess only very light sculpturing. The occipital canals are either completely closed or open for a short distance posteriorly. The frontoparietals are not fused to the proötics. The braincase is moderately elevated (Fig. 4-4), but the squamosal shafts are short (Fig. 4-5). The medial arm of the pterygoid does not occlude the suprapterygoid fenestra. The wings of the parasphenoid are not, or are only slightly, overlapped by the medial arm of the pterygoid. The quadratojugal is extensive and extends anteriorly at least three-fourths the length of the pterygoid fossa. The vertebral centra are of average proportions (Fig. 4-6). All limb bones are long and delicate (Figs. 4-7, 4-8, 4-9). *B. perreti* seems a very specialized derivative of the *regularis* group.

Mauritanicus Group. *B. mauritanicus*. The skull is broad (Figs. 4-2, 4-14). The frontoparietals are broad (Fig. 4-3), and their lateral edges are considerably upturned, imparting a concave appearance to the roof of the skull. The posteromedial quadrant of the orbit is roofed in a smooth arc. Ornamentation of the roofing bones varies from light to heavy. The anterior edges of the nasals lie in approximately the same transverse plane as the bases of the premaxillae. The postorbital shelf is moderately extensive and varies in shape from subtriangular to trapezoidal. The occipital canal is completely roofed and the frontoparietals are fused to the proötics. The braincase is elevated (Fig. 4-4), and the squamosal shafts are moderately long (Fig. 4-5). The medial arms of the pterygoids do not occlude the suprapterygoid fenestrae, nor do they overlap the wings of the parasphenoid. The quadratojugal is moderately extensive, extending anteriorly at least two-thirds the length of the pterygoid fossa. The vertebral column and humerus are of average proportions (Figs. 4-6, 4-7). The hind limb bones are slender (Figs. 4-8, 4-9).

Superciliaris Group. *B. superciliaris.* The skull is very broad (Figs. 4-2, 4-14). The frontoparietals are broad (Fig. 4-13) and concave, their lateral edges curving smoothly upward laterally and joining in an arc with the anterior edges of the postorbital shelf. The most anterior point of the nasals lies anterior to the transverse plane that includes the bases of the premaxillae. The postorbital shelf is fairly extensive, and the otic plate is large and trapezoidal in shape. The dorsal dermal bones are moderately ornamented. The occipital canals are completely roofed. A small area of fusion is present unilaterally between one frontoparietal and proötic. The specimen is adult and large; it is possible that fusion occurs late in the ontogeny of the species. The braincase is flattened (Fig. 4-4), and the descending arms of the squamosal are very long (Fig. 4-5). The medial arms of the pterygoid neither occlude the suprapterygoid fenestra nor overlap the wings of the parasphenoid. The quadratojugals extend anteriorly for approximately two-thirds the length of the pterygoid fossa. The angle of the jaw is far posterior, located considerably behind the occipital condyles. The vertebral centra are long (Fig. 4-6). The humerus and tibiofibula are elongate; the femur is of average proportions (Figs. 4-7, 4-8, 4-9).

Although distinct osteologically, *B. superciliaris* shows more similarity with *B. mauritanicus* and with the *regularis* group than with other African species.

Gariepensis Group. *B. gariepensis, B. inyangae,* and *B. rosei.* The skull of *B. gariepensis* is moderately broad (Figs. 4-2, 4-14). The frontoparietals are narrow (Fig. 4-2), their lateral edges approximating straight lines that diverge posteriorly and meet with the posterior wall of the orbit in slightly obtuse angles. No crests are present. The occipital canal is open and its anterior end is deflected medially, which causes the occipital artery to course inward over the dorsal surface of the frontoparietal bone rather than lateral to it. This modification results in a frontoparietal ratio (Fig. 4-3) that could be misleading. The otic plate of the squamosal is greatly reduced and covers only a small lateral portion of the proötic. Concurrently, the postorbital shelf is also reduced and the posterior wall of the orbit is broadly exposed in dorsal aspect. A lateral contribution of the frontoparietal projects dorsally as a spur at the posteromedial corner of the orbit. The posterolateral portions of the frontoparietals are lightly sculptured. The braincase is quite elevated (Fig. 4-4), but the squamosal shafts are moderately short (Fig. 4-5). The suprapterygoid fenestra is slightly occluded by the medial arm of the pterygoid. The arms of the parasphenoid are not overlapped. The quardratojugal is extensive and projects anteriorly for approximately three-fourths of the pterygoid fossa. The vertebral centra are of average porportions (Fig. 4-6), but the transverse processes are short. The limb bones are of average proportions.

In dorsal aspect, the small *B. inyangae* (max. size = 46.5 mm; Poynton, 1964) is only superficially similar to *B. gariepensis.* The skull is more narrow (Figs. 4-2, 4-14), as are the frontoparietals (Fig. 4-3). No ornamentation is present. The anterior termination of the nasals is far posterior to the bases of the premaxillae. The braincase is less elevated (Fig. 4-4), and the squamosal shafts are proportionately longer. Vertebral and limb proportions are similar (Figs. 4-6, 4-7, 4-8, 4-9).

B. rosei is a very small species; the maximum size given by Poynton (1964) is 38.5 mm. Most of my specimens were damaged or modified by preparation and were not measured. It is generally similar to *B. inyangae.* Differences are as follows: a mid-frontoparietal fontanelle is present in some of the specimens; its presence is not correlated with sex or size. The quadratojugals are shorter, not extending anteriorly for greater than one-half the length of the pterygoid fossa and frequently not reaching the maxilla. The medial arms of the pterygoid are short and do not contact the wings of the parasphenoid. No columella auris is present in the members of the Capetown population that I have examined.

The three forms included here are members of the *angusticeps* group of Poynton (1964), who described *inyangae* as a subspecies of *B. gariepensis.* Osteologically, both differences in size and in proportion suggest specific status; *B. rosei* appears most closely related to *B. inyangae.*

Lönnbergi Group. *B. lönnbergi.* The skull is long and narrow (Figs. 4-2, 4-14). The frontoparietals are narrow (Fig. 4-3) and meet with the postorbital shelf in an angle or shallow arc. The arclike juncture is not due to the width of the frontoparietal but to emargination of the anterior edge of the postorbital shelf. No crests are present. The postorbital shelf appears in dorsal view as an anteriorly directed crescent. The otic plate of the squamosal is reduced. No sculpturing of the cranial bones is evident. The frontoparietals are fused to the proötics. The occipital canals are open. The braincase is elevated (Fig. 4-4). The squamosal shafts are very short (Fig. 4-5) and do not contact the quadratojugal. The medial arms of the pterygoid moderately occlude the suprapterygoid fenestra but do not overlap the wings of the parasphenoid. The quadratojugals are very long, continuing anteriorly from the angle of the jaw for the entire length of the pterygoid fossa. The angle of the jaw is considerably anterior to the level of the jugular foramen. The vertebral centra are of average proportions (Fig. 4-6), the limbs long and delicate (Figs. 4-7, 4-8, 4-9).

Osteological similarity is greater with *B. rosei* and *B. inyangae* than with other African species examined.

Carens Group. *B. carens.* The skull is narrow (Figs. 4-2, 4-14). The lateral edges of the narrow (Fig. 4-3) frontoparietals transverse the orbit in straight lines and join the anterior edge of the postorbital shelf in an obtuse angle. No crests are present. The junction of the nasals with the frontoparietals is extensive, and in some specimens the frontoparietals overlap the nasals slightly. The nasals are produced anteriorly to or beyond the level of the bases of the premaxillae. Sculpturing of the roofing bones is lacking. The postorbital shelf is reduced and varies from subrectangular to subtriangular in shape. Its lateral portion is not bony but appears to be calcified cartilage. The very reduced (splintlike) otic portion of the squamosal is applied to this edge. The occipital canals are open, but in some specimens bony ridges flank and nearly cover them posteriorly. The frontoparietals are fused to the proötics. The braincase is of average proportions (Fig. 4-4), but the shafts of the squamosals are thin and short (Fig. 4-5). The wings of the parasphenoid are not overlapped by the medial arms of the pterygoid. The base of the parasphenoid is unique; upon reaching the sphenethmoid anteriorly, it terminates bluntly and abruptly, not tapering and continuing anteriorly as in all other *Bufo* examined. The quadratojugal extends anteriorly for less than two-thirds of the pterygoid fossa. The angle of the jaw is well anterior to the level of the jugular foramen. The vertebral column is of average proportions (Fig. 4-6); the limb bones are slender (Fig. 4-7, 4-8, 4-9).

B. carens is distinct osteologically, but affinity is suggested with the narrow frontoparietal line.

Eurasian *Bufo*

Viridis Group. *B. viridis.* The skull is of average proportions (Figs. 4-2, 4-15). The frontoparietals are narrow (Fig. 4-3); their straight lateral edges meet the postorbital shelf in a slightly obtuse angle. The frontoparietals diverge in midline anteriorly: the separation may be broad or narrow, for only a short distance, or for nearly the entire length of the frontoparietals. Its degree is not correlated with size. No ornamentation is present. The postorbital shelf is reduced and variable in shape. The otic plate of the squamosal is greatly reduced and splintlike. The occipital canals are usually open but are flanked laterally and medially by ridges of bone that have joined to roof the posterior portion of the canal in one of my specimens. The frontoparietals are fused to the proötics. The braincase is tropibasic (Fig. 4-4); the lateral arms of the squamosal are relatively short (Fig. 4-5). The medial arms of the pterygoid do not occlude the suprapterygoid fenestra, nor do they overlap the wings of the parasphenoid. Tihen (1962*b*) has described this species as possessing a parasphenoid that does not contact the pterygoid, but contact is present in several of my specimens from Israel, and it is felt that a larger sample of European specimens would also display instances of tangency. The quadratojugals are thin and extend anteriorly less than half the length of the pterygoid fossa. The vertebral centra are of average length (Fig. 4-6), and the transverse processes of vertebrae 5, 6, 7, and 8 are short. Humerus and femur are of average dimensions, but the tibiofibula is moderately slender (Figs. 4-7, 4-8, 4-9).

Calamita Group. *B. calamita.* This species is similar to *B. viridis* (Figs. 4-2, 4-3, 4-4, 4-5, 4-15). The major differences are that the skull is less broad, and the frontoparietals are broadly separated for more than half their length, forming a large frontoparietal fontanelle. Pterygoid-parasphenoid contact, described as absent by Tihen (1962*b*), is present in all the specimens I have examined. The quadratojugals are somewhat more reduced than in *viridis* and do not contact the maxilla in several of the specimens examined. The limb bones measured are extremely short and thick (Figs. 4-7, 4-8, 4-9).

Tihen (1962*b*) included *B. carens, B. raddei, B. calamita,* and *B. viridis,* as well as several fossil forms, in his *calamita* group. *B. raddei* has not been examined in this study; the specializations of *B. carens* are sufficient to exclude it from this group.

Bufo Group. *B. bufo.* The skull is broad (Figs. 4-2, 4-15). The frontoparietals are narrow (Fig. 4-3); their straight lateral edges join the anterior edges of the postorbital shelf in approximately a right angle. The transverse plane including the anterior edges of the nasals is behind the plane that includes the bases of the premaxillae. The postorbital shelf and the otic plate of the squamosal are moderately extensive. Ornamentation of the roofing bones is absent or reduced. The occipital canals are open. The frontoparietals are fused to the proötics. The braincase is tropibasic (Fig. 4-4); the squamosal shafts are proportionally short (Fig. 4-5). The medial arms of the pterygoids usually overlap slightly the anteroventral edges of the alae of the parasphenoid. The suprapterygoid fenestra is not occluded. The quadratojugals are long, extending anteriorly for at least three-fourths the length of the pterygoid fossa. The vertebral centra are short (Fig. 4-6). Humeri and femora are of average length (Figs. 4-7, 4-8); the tibiofibulae are short and thick (Fig. 4-9).

B. b. japonicus, a large form, departs from the preceding in several characters. The frontoparietals are fused to the proötics. The skull is somewhat broader (Fig. 4-2). The otic plate of the squamosal is more extensive, and the humerus is more robust (Fig. 4-7).

In the one disarticulate damaged specimen of *B. b. asiaticus* examined, the cranium is similar to that of *B. b. japonicus,* but the otic plate of the squamosal is more reduced.

Guttman (1967) found the hemoglobin and transferrin patterns of *japonicus* to be completely divergent from those of other specimens of *B. bufo* he had examined. Consequently, although these forms are usually recognized as subspecies, he suggested species recognition. Osteological evidence tentatively supports this conclusion and also suggests similar treatment for *asiaticus.*

Melanostictus Group. *B. melanostictus.* The skull is of average proportions (Figs. 4-2, 4-15). The frontoparietals are broad (Fig. 4-3), roofing a considerable portion of the orbit and joining in a broad arc with the anterior edge of the postorbital shelf. Conspicuous supraorbital, postorbital, tympanic, supratympanic, canthal, and preorbital crests are present. Inconspicuous bony striations of the supraorbital crests are present. The otic plate of the squamosal is large; the postorbital shelf is trapezoidal in shape. The occipital canals are roofed for a short distance anteriorly and posteriorly but are open for much of their length. It is possible that a larger sample would display both open and closed canals. The frontoparietals are fused to the proötics. The braincase is moderately depressed (Fig. 4-4); the squamosal shafts are proportionately long (Fig. 4-5). The suprapterygoid fenestrae are not occluded, nor are the alate portions of the parasphenoid overlapped by the medial arms of the pterygoid. The quadratojugals are moderately extensive, reaching anteriorly for at least two-thirds the length of the pterygoid fossa. The vertebral column and limbs are of average proportions (Figs. 4-6, 4-7, 4-8, 4-9).

This species shows no close osteological affinity with any other Eurasian species.

Stomaticus Group. *B. stomaticus.* The skull is of average proportions (Figs. 4-2, 4-15). The frontoparietals are narrow (Fig. 4-3), do not roof the posteromedial quadrant of the orbit, and meet the anterior edge of the postorbital shelf at an angle. No crests are present. The otic plate of the squamosal is moderately extensive. The posteromedial portions of the frontoparietals are lightly ornamented. The occipital canals are open, and the frontoparietals are fused to the proötics. The braincase is depressed (Fig. 4-4). The squamosal shafts are moderately short (Fig. 4-5). The medial arms of the pterygoid partially occlude the suprapterygoid fenestra but do not overlap the wings of the parasphenoid. The quadratojugals are long and extend the entire length of the pterygoid fos-

Table 4–2. Frontoparietal Type and Relationship with Proötic in *Bufo*

	Broad	Intermediate	Narrow
Unfused	*americanus* group (part) *regularis* group *maculatus* group (M) *mauritanicus* group (M) *superciliaris* group (M) *perreti* group (M) *asper* group (M)	*cognatus* group (part)	*americanus group* (part) *boreas* group *cognatus* group (part) *bufo* group (part)
Fused	*granulosus* group (part) *spinulosus* group (part) *guttatus* group *marinus* group *crucifer* group (M) *typhonius* group (M) *alvarius* group (M) *valliceps* group *coccifer* group (M) *melanostictus* group (M) *canaliferus* group (M)	*granulosus* group (part) *punctatus* group (M) *debilis* group *occidentalis* group (M) *bocourti* group (M) *marmoreus* group *holdridgei* group (M) *periglenes* group (M)	*bufo* group (part) *spinulosus* group (part) *calamita* group (M) *viridis* group (M) *gariepensis* group *stomaticus* group (M) *carens* group (M) *lönnbergi* group (M)

(M) = monotypic group

sa. The vertebral centra are long (Fig. 4-6). The humerus is moderately robust (Fig. 4-7); the femur and tibiofibula are slender (Figs. 4-8, 4-9).

Osteology indicates greater affinity with narrow rather than broad frontoparietal species.

Asper Group. *B. asper.* The skull is narrow (Figs. 4-2, 4-15). The frontoparietals are very broad (Fig. 4-3), roof a considerable portion of the orbit, and join in a broad arc with the anterior edges of the postorbital shelf. Low supraorbital, postorbital, and supratympanic crests are present. The anterior edges of the nasals project anteriorly well beyond the bases of the premaxillae. The postorbital shelf is extensive and trapezoidal in shape; concurrently, the otic plate of the squamosal is large. Granular ornamentation of the roofing bones is extensive. The occipital canals are completely roofed. The frontoparietals are not fused with the proötics. The braincase is neither elevated nor depressed (Fig. 4-4). The shafts of the squamosals are long (Fig. 4-5). The medial arms of the pterygoid overlap the wings of the parasphenoid but do not occlude the suprapterygoid fenestra. The quadratojugals are moderately long, reaching anteriorly approximately two-thirds the length of the pterygoid fossa. The vertebral centra are long, the limbs slender (Figs. 4-6, 4-7, 4-8, 4-9).

The osteological data do not indicate close relationship to any other of the Eurasian species examined.

Phylogenetic Considerations

In the preceding section, a brief description of osteological characters preceded a short discussion of possible immediate relationships of the species concerned. The following discussion develops a broader scheme of relationships that attempts to reconcile existing lines of evidence. Many species examined are absent from this tentative phylogeny; for these, systematic data are inadequate, inconclusive, or conflicting.

The most convenient osteological division, and possibly that which represents an early major evolutionary dichotomy in *Bufo*, is to separate species possessing broad frontoparietal bones from those whose frontoparietals are narrow. As might be expected, intermediate conditions exist, but these are not as common as either extreme. This character reflects the amount of dermal bone in the dorsal roof of the skull and correlates well with other characters (such as extensiveness of the postorbital shelf, and degree of closure of the occipital canals) that are also dependent upon extensiveness of dermal bone. All continents inhabited by *Bufo* have representatives of both frontoparietal types; the broad generally occur in lowland species; the narrow, in upland environ-

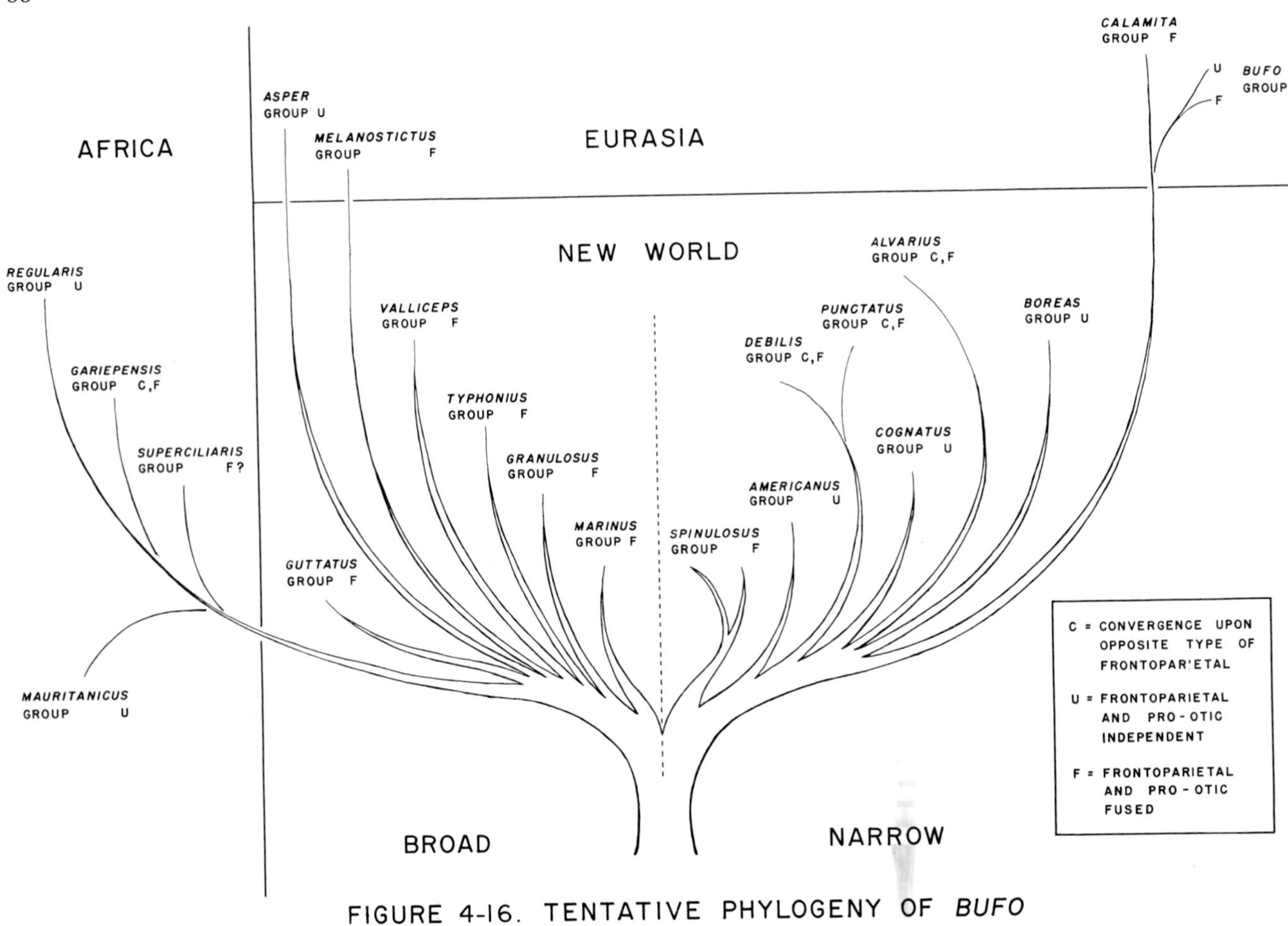

Fig. 4–16. Tentative phylogeny of *Bufo*.

ments. There has been a reduction of dermal bone in the elements of the cranium in the evolution of the Amphibia; protection lost has apparently been offset by the advantages of increased mobility. Tihen (1960, 1962*a*) has hypothesized that primitive members of the genus *Bufo* possessed well-ossified skulls. Although difficulties in interpretation exist, representatives of both narrow (advanced) and broad (primitive) frontoparietal types appear to be present in the Miocene (Tihen, Chap. 2; also 1962*a*, *b*). This fact, as well as the apparently broad distribution of the genus at that time, suggests a considerable history of the genus earlier in the Tertiary. Both major lines have undergone parallel reduction in other dermal cranial elements as well as in the frontoparietals, and it is likely that several instances of secondary addition of dermal bone have occurred in the narrow line.

Crania can be further divided into those in which fusion occurs between frontoparietal and proötic and those in which these bones remain independent, the latter condition being hypothetically more primitive (Tihen 1960, 1965). The species examined and their skull types are shown in table 4-2. Five of the thirteen nonmonotypic species groups examined fall into more than one contingency. Of these five, only three divide into extreme contingencies; this suggests that the characters listed are relatively conservative.

Figure 4-16 depicts a possible evolutionary scheme involving major species groups and transcontinental links within the genus. Low (1967) has thoroughly reviewed the problems of intercontinental transport as they relate to *Bufo* and concluded that the hypothesis of rafting from South America to Africa is quite acceptable. Blair (1963*a*) and Tihen (1962*a*) have considered the Bering land bridge an acceptable route for northern crossing. Bogart (1967, 1968), W. F. Martin (Chap. 15), Blair (Chap. 18), and Low (1967) have evaluated the merits of various continents as sites of origin, and, although the question is not completely settled, the greatest amount of evidence favors South America.

Broad-Skulled Toads

The frontoparietals of members of the South American *marinus* group are broad, crested, fused with the proötics, and generally heavily ornamented. The pterygoid and squamosal are well developed. A closely allied fossil form exists from the Miocene of Colombia (Estes and Wassersug 1963). Members of the Mexican and Central American *valliceps* group are similar in cristation pattern to the *marinus* group, but the thickness of dermal ornamentation is generally more reduced. The flanges of the pterygoid and squamosal are also less extensively developed; consequently, the wings of the parasphenoid are not overlapped nor is the suprapterygoid fenestra occluded as in the *marinus* group. Thus, the *valliceps* group appears more osteologically advanced. Crosses between this group and *B. arenarum* of the *marinus* group have yielded high reciprocal metamorphosis (Blair 1966; also Chap. 11). No positive pre-Pleistocene fossil record exists for the *valliceps* group.

The Asian *B. melanostictus* shows great similarity to the *valliceps* group in crest morphology and other cranial characters. Dermal ornamentation is more reduced than is generally the case among members of the *valliceps* group, and the occipital canal is open in mid-length. These osteological similarities strongly support a northern crossing to Asia by an ancestral member of the *valliceps* group.

The breadth of the frontoparietals varies from broad to medium in the South American *granulosus* group, and cristation patterns are similar to the above groups. The frontoparietals are fused to the proötics, and the flanges of the pterygoid are extensive. The extreme degree of dermal ossification, particularly at the snout and squamosal, is possibly not primitive but secondarily evolved.

Although not greatly similar in dorsal aspect, *B. blombergi* and *B. haematiticus* of the South American *guttatus* group share peculiar vertebral and testicular morphology. Both bear broad frontoparietals that fuse with the proötics relatively late in ontogeny.

Members of the African *regularis* group display broad acristate, independent frontoparietals and are unique in possessing only twenty chromosomes (Chap. 10). They show one-way metamorphosis with some members of the broad-skulled line of New World species, but little genetic compatibility or biochemical similarity with palearctic species. This evidence led to the search for an ancestral group in South America. Since members of the *regularis* group possess two primitive skull conditions, the *guttatus* group is a possible candidate. These toads do not possess definite crests, their frontoparietals are broad, and the frontoparietals fuse with the proötics, although apparently late in ontogeny. *Bufo mauritanicus* is very similar to the *regularis* group in osteology but shows the greater degree of dorsal concavity of the frontoparietals characteristic of South American forms and possesses twenty-two chromosomes (Chap. 10). The large African species *B. superciliaris* shows osteological and other similarities to the *regularis group*. In addition, it displays the cryptic leaf coloration of the *guttatus* group and has been reported by Noble (1923) to have large folded testes, a character shared by males of the *guttatus* group. With this information it is possible to postulate that either *B. mauritanicus* or *B. superciliaris* or both are remnants of a stock that crossed from the New World and later gave rise to the twenty-chromosome *regularis* group.

The Asian *B. asper* possesses broad unfused frontoparietals and appears quite similar to the Central American *B. haematiticus* in many osteological characters. Both have long narrow skulls, projecting nasals, closed occipital canals, long, vertical squamosal shafts, greatly flattened neural spines, and long transverse processes and vertebral centra. The frontoparietals are fused with the proötics in *B. haematiticus*, but this does not strongly argue against affinity.

Narrow-Skulled Toads

The frontoparietals in the *americanus* group are narrow (except in *B. terrestris*)[1] and independent from the proötics. Dermal bone is only moderately reduced, however, and well-developed cranial crests that join at an angle above the orbit (Fig. 4-1) are present. Ornamentation of the roofing bones is variable within the group, and the occipital canals are partially open posteriorly.

In the *boreas* group of western North America (see

[1] Osteological (Sanders 1961; Tihen 1962*b*; R. F. Martin 1964) and biochemical characters (Low 1967) of *B. terrestris* that display similarities to southern line (*valliceps* group) toads have been described. Alternative explanations of convergence or genetic exchange between the groups exist.

Karlstrom 1962; Tihen 1962*a*; Martin 1964), dermal bone is reduced and the frontoparietals are narrow and independent of the proötics. Ornamentation is absent and the occipital canals are open. Except for lack of frontoparietal-proötic fusion, members of this group are quite similar to the narrow frontoparietal members of the *spinulosus* group. European *B. bufo* of the Eurasian *bufo* group are extremely similar osteologically to members of the *boreas* group; in the Asian members of the *bufo* group, the frontoparietals are fused with the proötics. In the palearctic *calamita* group the dermal roofing bones are further reduced in size and proötic-frontoparietal fusion is present. The osteological data strongly suggests greatest proximity between *B. bufo* and *B. boreas.*

Intermediate or Problematic Groups

Both extremes of dermal ossification are represented in the Andean *spinulosus* group, and several interpretations may be considered. Osteology and other evidence suggest that several species that have been treated as subspecies are present in this group. Also, this evidence does not conclusively preclude the possibility that several species groups are actually present. Under these circumstances, branches from both hypothetical major lines may be represented. A more conservative treatment would locate the group as an offshoot of the broad line in which reduction parallel to the narrow line had occurred. Since most of the species possess narrow frontoparietals, the group may be treated as a branch of the narrow line, with secondary addition of dermal bone occurring in *B. chilensis.* The latter treatment is followed in figure 4-16.

The members of the North American and Mexican *cognatus* group (Blair 1963*a*; Bogert 1960; Martin 1964) are generally similar to the *americanus* group. In *B. compactilis*, the frontoparietals are narrow and the occipital canal is open, but in *B. cognatus* and *B. speciosus* the frontoparietals roof a portion of the orbit, and the occipital canals are closed anteriorly (Martin 1964).

The frontoparietals of the North American and Mexican *B. punctatus* are of intermediate width, but, although the posteromedial portion of the orbit is partially roofed, the juncture of crests at this point is more angular than arclike. The depressed braincase, a character common in broad-line species, is possibly a secondary adaptation to crevice dwelling. Dermal ornamentation is heavy, and the occipital canals are closed. Blair (Chap. 11), however, has found reciprocal metamorphosis only with the *boreas* group and with the Asian *B. stomaticus.* Thus, he believes its affinities lie with the narrow line. In addition, Bogart (Chap. 10) considers *punctatus* close to the *americanus* group on the basis of chromosomal evidence. To accommodate this, secondary addition of dermal bone must be postulated.

Members of the *debilis* group are very similar to *B. punctatus*, but the roofing of the orbit is slightly more arclike. Savage (1954) suggested derivation of this group from an ancestor similar to *B. valliceps.* The osteological similarity with the *punctatus* group suggests common ancestry. Since, in this situation, secondary addition of dermal bone must again be postulated, the very real possibility of broad-line ancestry cannot be avoided.

The Mexican *marmoreus* group is intermediate between narrow and broad frontoparietal types.

The frontoparietals of the Mexican *B. occidentalis* are of intermediate width and are fused with the proötics. Although most other osteological characters are typical of the narrow line, other evidence strongly indicates affinity with broad frontoparietal forms.

The Guatemalan *B. bocourti* displays intermediate frontoparietals that are fused with the proötics. Osteological characters of both lines are present but suggest greater affinity with the narrow.

The frontoparietals of *B. alvarius* are broad, highly ornamented, and fused with the proötics. Other osteological characters are also typical of the broad line. Other evidence implies derivation from the narrow line, and secondary addition of dermal bone accompanied by fusion between frontoparietal and proötic must be considered possible.

The African *gariepensis* group possesses narrow, fused frontoparietals and reduced dermal bone. Dermal ornamentation is light or lacking, and the occipital canals are open. Chromatographic data indicate strong affinity with the osteologically dissimilar *regularis* group and with broad frontoparietal South American species. If this group was derived from the stock that is hypothesized to have crossed from South America and to have given rise the the *regularis* group, subsequent loss of dermal bone must have been accompanied by frontoparietal-proötic fusion.

The frontoparietals of *B. stomaticus* of Asia are narrow and fused with the proötics. Dermal ornamentation is lacking and the occipital canal is open. Although somewhat intermediate in character osteologically, *stomaticus* appears to be a member of the narrow rather than the broad line.

Summary

The osteology of approximately sixty-five species of *Bufo* was examined in an attempt to elucidate evolutionary relationships in this genus by integrating this type of evidence with existing systematic data. Although small samples were used, considerable intraspecific variation was found.

Groupings suggested by a series of osteological characters usually correspond closely with assemblages based upon other disciplines; exceptions occur, however, and advise against a completely osteological approach. Several changes in taxonomic level suggested by other evidence are also indicated by osteology.

It is likely that osteological evolution usually proceeded with reduction in size of dermal cranial elements; however, certain instances of secondary addition of dermal bone must be postulated. Two major skull types of *Bufo* exist; these possibly represent an early evolutionary dichotomy in the genus. The *marinus, valliceps, canaliferus, melanostictus, granulosus, typhonius, guttatus, regularis,* and *asper* groups possess broad frontoparietals. Members of these groups generally possess well-developed dermal roofing bones with conspicuous ornamentation. Narrower frontoparietals and reduction in dermal bone are generally characteristic of the *boreas, americanus, spinulosus, bufo, calamita, stomaticus, gariepensis, carens,* and *lönnbergi* groups. Several groups show intermediacy in frontoparietal width.

A phylogeny is presented that attempts to reconcile major lines of evidence. Osteological evidence supports three previously proposed intercontinental links: the first, between the narrow frontoparietal North American *boreas* group and the Eurasian *bufo* and *calamita* groups; the second, between the Central American and Mexican *valliceps* group and the Asian *B. melanostictus*; the third, between the South American *guttatus* group and the African *mauritanicus, superciliaris,* and *regularis* groups. Considerable osteological similarity is present between the *guttatus group* and the Asian *B. asper*, which possibly indicates a fourth link.

REFERENCES

Baldauf, R. J. 1955. Contributions to the cranial morphology of *Bufo w. woodhousei* Girard. Texas J. Sci. 7:275–311.

———. 1958. Contributions to the cranial morphology of *Bufo valliceps* Wiegmann. Texas J. Sci. 10:172–186.

———. 1959. Morphological criteria and their use in showing bufonid phylogeny. J. Morph. 104:527–560.

Blair, W. F. 1963. Evolutionary relationships of North American toads of the genus *Bufo*: A progress report. Evolution 17:1–16.

———. 1964. Evidence bearing on relationships of the *Bufo boreas* group of toads. Texas J. Sci. 16:181–192.

———. 1966. Genetic compatibility in the *Bufo valliceps* and closely related groups of toads. Texas J. Sci. 18:333–351.

Bogart, J. P. 1967. Chromosomes of the South American amphibian family Ceratophridae with a reconsideration of the taxonomic status of *Odontophrynus americanus*. Canad. J. Genet. and Cytol. 9:531-542.

———. 1968. Chromosome number difference in the amphibian genus *Bufo*: The *Bufo regularis* species group. Evolution 22:42–45.

Bogert, C. 1960. The influence of sound on the behavior of amphibians and reptiles, p. 137–320. *In* W. E. Lanyon and W. N. Tavolga (eds.), Animal sounds and communication. A.I.B.S. Publ. 7.

Cei, J. M. 1962. Batracios de Chile. Universidad de Chile, Santiago.

——— and V. Erspamer. 1966. Biochemical taxonomy of South American amphibians by means of skin amines and polypeptides. Copeia 1966:74–78.

Estes, Richard, and R. Wassersug. 1963. A Miocene toad from Colombia, South America. Breviora 193:1–13.

Gallardo, J. M. 1965. The species *Bufo granulosus* Spix (Salientia: Bufonidae) and its geographic variation. Bull. Mus. Comp. Zool. 134 (4): 107–138.

Guttman, S. I. 1967. Evolution of blood proteins within the cosmopolitan toad genus *Bufo*. Ph. D. Thesis, Univ. Texas.

Inger, R. F. 1959. Amphibia, VI, 510–533. *In*, Hanstrom *et al.* (eds.), South African animal life: Results of the Lund University Expedition in 1950–1951. Almquist & Wiksell, Stockholm.

Karlstrom, E. L. 1962. The toad genus *Bufo* in the Sierra Nevada of California. Univ. Calif. Publ. Zool. 62:1–104.

Low, B. S. 1967. Evolution in the genus *Bufo*: Evidence from parotoid secretions. Ph.D. Thesis, Univ. Texas.

Martin, R. F. 1964. Osteological morphology and the

phylogeny of certain North American toads (genus *Bufo*). Unpublished M.A. Thesis, Univ. Texas.

Myers, G. S., and J. W. Funkhouser. 1951. A new giant toad from southwestern Colombia. Zoologica 36:279–283.

Noble, G. K. 1923. Amphibia, p. 147–347. *In*, Contributions to the herpetology of the Belgian Congo based on the collection of the American Museum Congo Expedition 1909–1915. Bull. Amer. Mus. Nat. Hist. 49.

Porter, K. R. 1964*a*. Distribution and taxonomic status of seven species of Mexican *Bufo*. Herpetologica 19: 229–247.

———. 1964*b*. Morphological and mating call comparisons in the *Bufo valliceps* complex. Amer. Midl. Nat. 71:232–245.

———. 1965. Intraspecific variations in mating call of *Bufo coccifer* Cope. Amer. Midl. Nat. 74:350–356.

———. 1966. Mating calls of six Mexican and Central American toads (genus *Bufo*). Herpetologica 22:60–67.

———, and W. F. Porter. 1967. Venom comparisons and relationships of twenty species of New World toads. (Genus *Bufo*). Copeia 1967:298–307.

Poynton, J. C. 1964. The Amphibia of southern Africa: A faunal study. Ann. Natal Mus. 17:1–334.

Sanders, O. 1953. A rapid method for preparing skeletons from preserved Salientia. Herpetologica 9:48.

———. 1961. Indications for the hybrid origin of *Bufo terrestris* Bonaterre. Herpetologica 17:145–156.

Savage, J. M. 1954. A revision of the toads of the *Bufo debilis* complex. Texas J. Sci. 6:83–112.

———, and A. G. Kluge. 1961. Rediscovery of the strange Costa Rican toad, *Crepidius epioticus* Cope. Rev. Biol. Trop. 9:39–51.

Tihen, J. A. 1960. Two new genera of African bufonids, with remarks on the phylogeny of related genera. Copeia 1960:225–233.

———. 1962*a*. Osteological observations on New World *Bufo*. Amer. Midl. Nat. 67:157–183.

———. 1962*b*. A review of New World fossil bufonids. Amer. Midl. Nat. 68:1–50.

———. 1965. Evolutionary trends in frogs. Amer. Zoologist 5:309–318.

Underhill, J. C. 1961. Variation in Woodhouse's toad, *Bufo woodhousei* Girard in South Dakota. Copeia 1961:333–336.

Vellard, J. 1959. Estudios sobre batracios andinos. V. El género *Bufo*. Mem. Mus. Hist. Nat. Xavier Prado 8:1–48.

5. Integument and Soft Parts

*HENRYK SZARSKI**

Copernicus University
Torum, Poland

Introduction

Various species of the genus *Bufo* have served as material for anatomical and physiological studies. The literature describing the organs of these animals is therefore very extensive. The majority of investigators were not concerned, however, with the implications that their observations could have for the systematics within the genus or for the problem of affinities of *Bufo* with other genera, so that in consequence it is very difficult to evaluate that information from the point of view of this book. It is impossible to review all such papers here, although the facts they contain may become important in the future when more comparable data have accumulated.

Inger (1958) postulated that the definitions of genera should be based on characters that reflect the principal adaptive trends of a group of species and he used as an example a description of the genus *Bufo*. It is, however, often difficult to demonstrate the adaptive value of a morphological trait. It is perhaps easier to fulfill this postulate for the soft-part characters, but it will be seen below that in some cases the explanatory hypotheses are constructed with great difficulty.

* Present address: Jagellonian University, Cracow, Poland

Integument

Every description of a toad underlines its warty skin. Elias and Shapiro (1957) introduced a detailed nomenclature of the different "tubercles, warts, ridges, papillae, cones and spines" in their description of the skin of sixteen species of Salientia. The skin of *Bufo* is by far the richest in these structures. The principal warts in *Bufo* are huge verrucae glandulares, containing big glandular masses. The large skin glands of *Bufo* have their own capillary nets (Fig. 5-1; J. Czopek 1955; G. Czopek and J. Czopek 1959; Bieniak and Watka 1962). This is probably a generic character, as in other Salientia so far examined, and even in those possessing very large glands, these glands are supplied only by capillaries of the subepidermal and dermal nets. Special nets around glands are present in some Caudata (e.g., in *Salamandra*, J. Czopek, personal communication). The presence of vascular nets around glands suggests a high level of metabolism and accordingly an active production of secretion.

The presence of skin glands influences the thickness of the skin, which reaches considerable dimensions in *Bufo*. Contrary to popular belief this does not mean that the body fluids are especially well protected from desiccation. The net of capillaries is situated directly under the epidermis, and the thick-

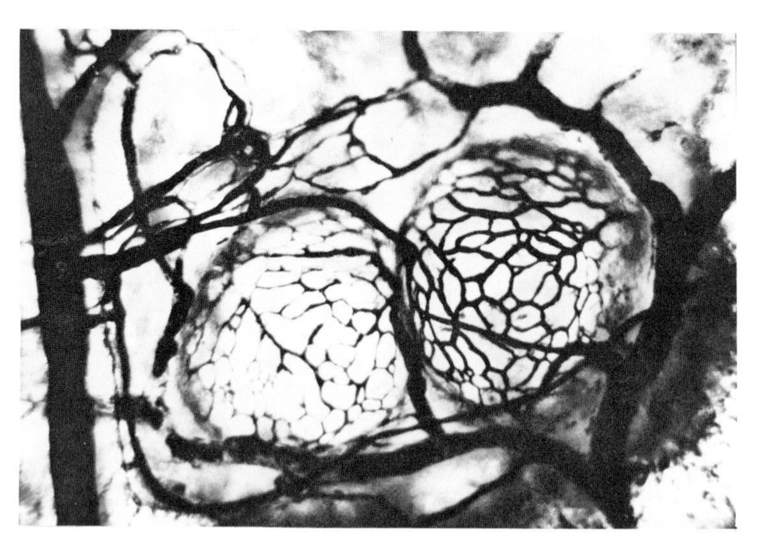

Fig. 5-1. Injected vascular nets on the granular glands of *Bufo bufo*. After Czopek (1955).

ness of this last layer does not differ in *Bufo* from that of other Salientia (Table 5-1).

Elias and Shapiro (1957) gave the thickness of epidermis in warts of three more species of *Bufo*, but they did not calculate the mean value. Their data generally agree, however, with Czopek's observations. It is apparent that some Salientian species that live near water (*B. variegata, R. esculenta*) have a far stronger epithelium than that of some terrestrial forms (e.g., *S. holbrooki*).

The thickness of the epidermis is not the only factor that can influence the permeability of the epidermis. Other elements are the degree of cornification and the nature and amount of mucus. There is some controversy about the protection offered by the skin of frogs and toads against evaporation. According to Adolph (1932), frogs without skin lose water with the same speed as intact animals. Thorson (1955) claimed that *Bufo boreas* loses water in experimental conditions with a rapidity similar to that of other

Table 5-1. Mean Thickness of the Epidermis*

Bufo		Other Anura	
Bufo bufo	51.9μ	*Bombina variegata*	65.2μ
B. calamita	56.0μ	*B. bombina*	22.8μ
B. cognatus	30.3μ	*Scaphiopus holbrooki*	19.2μ
B. speciosus	49.0μ	*Rana esculenta*	62.3μ
B. viridis	46.0μ	*R. pipiens*	37.0μ

*Czopek 1965.

Salientia, which are less resistant to aridity. Dinesman (1948) demonstrated that the loss of water of skinned salientians is greater than in intact ones and that the protection offered by the skin is especially apparent in *Bufo viridis*. Bentley (1966) has shown that *Bufo bufo* loses water about 25 percent less rapidly than *Rana esculenta*. According to Dinesman (1948) the mucus covering the epithelium has a great influence on the permeability of the skin to water; thefore, the conditions to which the animal was subjected immediately before the experiment influence the results. In Warburg's experiments (1965) *Bufo marinus* had a lower rate of water loss when compared with Australian native salientians. According to Cloudsley-Thompson (1967), the evaporative water loss has a vital importance in the African species *B. regularis*, since it decreases the body temperature below the critical level that is often met in its environment.

The subepidermal vascular net has been studied in five *Bufo* species. Results were summarized by Czopek (1965). The lowest number of meshes per one mm.^{2} is found in *B. speciosus* (154), the highest in *B. viridis* (181). In most salientians the subepidermal net is less developed (e.g., in *Bombina variegata*, 42/mm.^{2}, in *Leiopelma hochstetteri*, 48/mm.^{2}). Denser nets were found only in *Rana temporaria* and in *Rana esculenta* (up to 225/mm.^{2}). Czopek concluded that the toads have an ability to exchange gas through skin. This was confirmed by Johansen and Ditadi (1966), who found that the cutaneous effluent blood of *Bufo paracnemis* is saturated up to 80 percent with oxygen.

Hutchison, Whitford, and Kohl (1968) compared the cutaneous and lung exchange in several species of anurans under different temperatures and found that the four species of *Bufo* do not differ significantly from other anurans in this respect. Thus, the lowest ratio of pulmonary oxygen uptake, namely 32.8 percent, is found among *Bufo* in *B. americanus* at 5°C, among remaining genera, 30.9 percent in *Hyla gratiosa* at 5°C. The highest ratio of pulmonary to skin oxygen uptake is found among *Bufo* in *B. marinus* at 15°C, namely 72.3 percent, among remaining genera in *Eleutherodactylus portoricensis* at 15°C, 76.8 percent (Table 5-2).

Salée and Vidrequin-Deliège (1967) have demonstrated that nerve-skin preparation of *B. bufo* alters the skin potential after an electrical stimulation of the brachial plexus like a preparation of *Rana temporaria*. In contrast to this species, however,

Table 5–2. Oxygen Uptake in Various Anurans at 15° C*

Xenopus laevis	33.83
Bufo marinus	37.38
Rana pipiens	51.87
Ceratophrys calcrata	53.46
Bufo boreas	56.34
Rana catesbeiana	57.10
Microhyla carolinensis	66.13
Eleutherodactylus portoricensis	69.36
Bufo cognatus	76.20
Bufo alvarius	78.70
Bufo terrestris	80.48
Scaphiopus holbrooki	83.66
Scaphiopus couchi	84.06
Hyla gratiosa	89.03
Scaphiopus hammondi	93.91
Rana sylvatica	98.90
Hyla versicolor	104.15
Bufo americanus	110.18
Hyla cinerea	126.51
Scaphiopus bombifrons	139.17

*After Hutchison, Whitford, and Kohl (1968). Values in microliters per gram-hour.

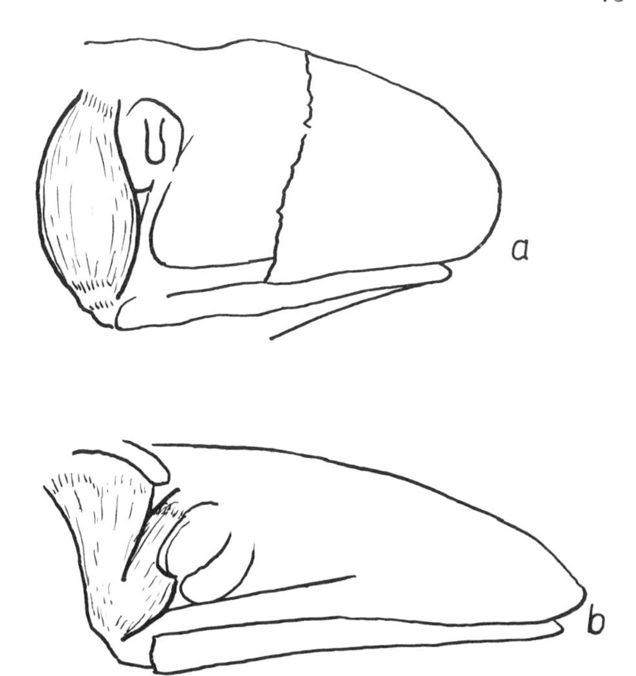

Fig. 5–2. Depressor mandibulae muscle in *Bufo arenarum (a)* and in *Rana pipiens (b)*. Simplified after Limeses (1965).

there is no relation between the level of skin polarization and the amplitude of the deviation of the potential in *Bufo*. Similar results were obtained by Gonzales, Sanchez, and Concha (1966) for *B. spinulosus*. The skin potential is related to the migration of ions across the skin, and it can be expected that it will behave differently in species adapted to different environments. The present data are, however, too scarce for further speculation.

Muscular System

The arrangement of muscles of the pelvic limb was one of the characters on which Noble (1922) based his systematics of Salientia. His observations were repeated and enlarged by Dunlap (1960). Both these authors have stated that the limb-muscle arrangement in *Bufo* is very similar to that in the family Leptodactylidae. According to Limeses (1964), who examined several neotropical forms, the thigh musculature of Bufonidae and Leptodactylidae demonstrates no constant differences from Ranidae and other families belonging to Noble's order Diplasiocoela. Limeses regarded, therefore, the last suborder and Procoela as one natural unit.

In the majority of Salientia, m. depressor mandibulae is composed of two parts. The anterior part originates from the posterior margin of the skull; the posterior part of the muscle arises from the dorsal fascia or from the suprascapula. In *Bufo* the posterior part is always absent (Fig. 5-2). Griffiths (1954, 1959, 1963) considers this feature to be of great significance. He has stressed that except for the Bufonidae the posterior part of m. depressor mandibulae is absent only in Atelopodidae, which are also similar to Bufonidae in other features. Recently, Limeses (1965) demonstrated the bufonid condition of the depressor mandibulae in the leptodactylid genus *Lepidobatrachus*. This and other evidence induced her to erect the family Ceratophrynidae, including *Lepidobatrachus* and some other neotropic genera, which, however, possess the posterior part of the depressor mandibulae. Griffiths (1963) considered the bufonid condition of the depressor mandibulae a paedomorphic feature, since it is repeated in the ontogeny of every salientian species.

The remaining somatic muscles have not been studied in a large number of species. In the literature there are only comparisons between *Bufo* and *Rana* (Bigalke 1927; Bhati 1961). If we agree with Tihen (1965) that a muscle once lost is never regenerated and that there is nevertheless a general trend toward increasing complexity of musculature, it can be concluded that *Bufo* is more advanced in muscle

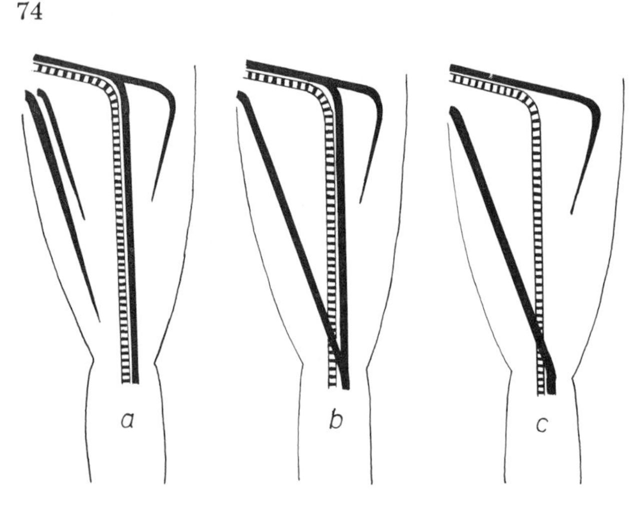

Fig. 5-3. The connections of the facial vein in *Rana esculenta* (*a*), *R. tigrina* (*b*), and *Bufo melanostictus* (*c*). After Bhaduri (1933).

anatomy than *Rana*. Several muscles that are uniform in *Rana* are composed of more or less separate parts in *Bufo*: m. temporalis externus, m. temporalis internus, m. rectus abdominis, m. rhomboideus posterior, m. serratus medius, m. ileolumbaris, m. dorsalis scapulae, m. flexor carpi radialis, m. palmaris longus, and m. extensor carpi ulnaris. An example of a lost muscle is the posterior part of m. depressor mandibulae.

G. Czopek (1963) studied the distribution of capillaries in the muscles of some amphibians, among them *Bufo calamita*. She found that, since the vascularization of muscles diminishes with the growth of the animal, it is advisable to compare the results obtained in specimens of similar weight. It can nevertheless be stated that the muscles of *Bufo calamita* are the most intensely vascularized among the muscles of all of the Amphibia examined. The mean value of capillaries per one mm.^{2} of cross surface area in *B. calamita* weighing 16 g. is 417, while in small specimen of *Rana esculenta* weighing only 2.65 g. this value is 362, and in a large specimen of *R. esculenta* weighing 250 g. it falls to 144. Other parameters have similarly high values in *Bufo*. Length of capillaries in a unit of muscle is 314 m/g. in *Bufo*, while in the smallest specimen of *Rana* it is 272 m/g, in the largest only 100 m/g. As in other species, the muscle best supplied with capillaries is m. submaxillaris, which has 768 capillaries per one mm.^{2} of cross surface area. G. Czopek stressed that the greatest difference in the vascularization of individual muscles is found in *Bufo*, while in other investigated species the muscles are vascularized more uniformly. These facts point to a high degree of specialization of the *Bufo* muscular system.

Vascular System

The problem of the distribution of blood in the amphibian heart has a voluminous literature. Some species of *Bufo* have served as experimental animals. Although Simons (1959) stated that the precise pattern of distribution varies between species, there is little knowledge of the differences in this matter between *Bufo* and the other Salientia. Buddenbrock (1967), however, reports the relative weight of the heart in *B. viridis* at 8.05, while in *Hyla arborea* it is 5.79, and in *Rana pipiens* 3.2 to 4.1. Simons has claimed (1957) that the blood pressure in the aortic arches is higher in *Bufo* than in *Rana*. Johansen and Ditadi (1966) demonstrated a high level of blood separation in a toad heart.

The arrangement of principal blood vessels in Salientia has been the subject of only a few studies (e.g., Millard 1941). The reported differences suggest that an examination of more genera and species may furnish interesting data. According to Szarski (1947–1948) the principal features of the vascular system of *Bufo bufo* are as follows. As in the majority of Salientia there are no posterior cardinal veins (present in Discoglossidae and Ascaphidae, Szarski 1951). Their presence has been noted in some aberrant specimens (Bhaduri 1944–1945). *Bufo* has a caudal vein, absent in *Rana*. In *Bufo* and *Rana* there is no ischiadic vein, and the femoral vein is the principal vessel of the thigh. In other Salientia so far studied the ischiadic vein is the main thigh vessel (Figs 5-3, 5-4). The facial vein in *Bufo* unites with

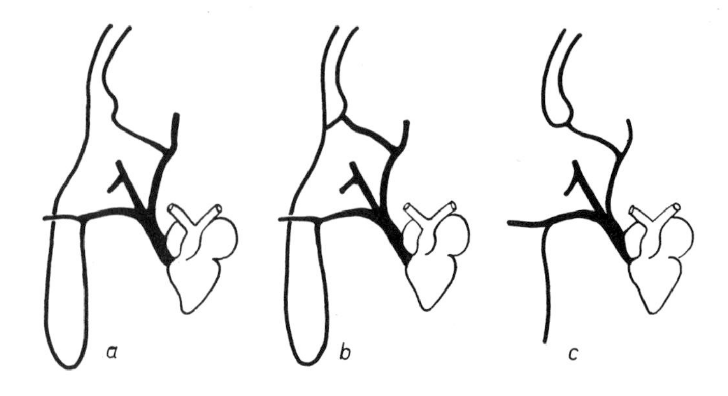

Fig. 5–4. The principal veins of the thigh in *Pelobates* (*a*), *Bombina* (*b*), and *Bufo* and *Rana* (*c*). After Szarski (1947-1948).

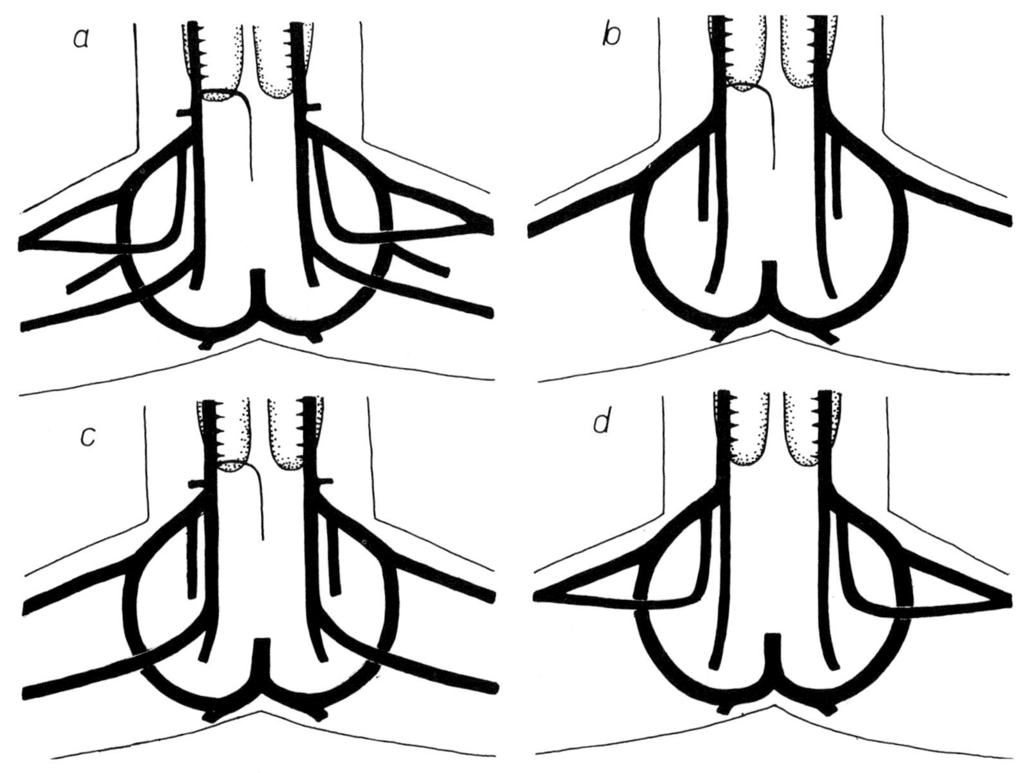

Fig. 5–5. The principal veins of the pelvic region in *Pelobates* (*a*), *Bufo* (*b*), *Bombina* (*c*), and *Rana* (*d*). After Szarski (1947-1948).

the internal mandibular vein to form the faciomandibular as in all other Salientia. Only in *Rana* does the facial vein continue as a large cutaneous vein – in European species without anastomosis with mandibular, or in American and Indian species, with anastomosis (Bhaduri 1933, 1938; Fig. 5-5). The venous anastomoses in the elbow region of Bufo are different from those of other Salientia. It is difficult to interpret the other peculiarities of the Bufo vascular system noted by Szarski (1947–1948). This will be possible when extensive comparative material can be studied.

The amphibian carotid labyrinth has been recently studied in several forms (e.g., Rogers 1966; Carman 1967*a, b*). These studies point up some pecularities of the labyrinth of *Bufo*. The small number of lymph hearts has been regarded as primitive (Tihen 1965). In all the Salientia there is only one anterior pair of these structures. According to Hoyer (1934), *Bufo* has only one posterior pair of lymph hearts, similar to *Bombina* and *Breviceps*. Several pairs are found in *Rana, Ceratophrys*, and *Pelobates*.

Recent interest in the evolution of immunological mechanisms has led several investigators to examine the structure of the thymus and of the lymph nodules in some species of Salientia. It has been found that the localization of lymphoid organs varies in the investigated forms. A comparative study of these structures may furnish future phylogenetic information. A review of present literature is given by Cooper (1967).

Respiratory Organs

The internal lung surface is very well developed in *Bufo* (Czopek 1965). The area of lung septa divided by the area of lung sacs without septa reaches 7.75 in *B. cognatus* (weight of the specimen 71.5 g.). In *B. bufo* it is 5.09 (weight 48 g.), in *B. speciosus* it is 4.80 (weight 48 g.), in *B. viridis* it is 4.30 (weight 33.5 g.), and in *B. calamita* it is 3.00 (weight 20.0 g.). A larger value than in *B. cognatus* was found among Amphibia only in a very large (250 g.) specimen of *Rana esculenta*, which reached 8.19. When comparing these values one must remember that the area of lung septa increases during growth (Andrzejewski *et al.* 1962). Thus, in a specimen of *Rana esculenta* weighing 53 g. the coefficient is 4.5. According to Marcus (1937) the best-developed lung surface among Salientia is found in the genus *Pipa*. Marcus presented, however, no quantitative data.

The well-developed lung surface in *Bufo* is moderately supplied with capillaries. The number of meshes on one mm.^{2} is also correlated with the size of the animal and with the number of lung septa. In *B. bufo* it is 520/mm.^{2}, in *B. calamita* 676/mm.^{2}, in *B. cognatus* 645/mm.^{2}, in. *B. speciosus* 655/mm.^{2}, and in *B. viridis* 729/mm.^{2} (weights of specimens as above). The densest net among Amphibia was found in a small specimen of *Rana pipiens* (weight 0.6 g.) whose lungs were devoid of septa. There were 1384 meshes per one mm.^{2} in this animal. Very low values characterize small and primitive species: *Leiopelma hochstetteri* (weight 5 g.), 396 and *Bombina bombina* (weight 6.6 g.), 280.

The total number of respiratory capillaries per gram of body weight is very high in all *Bufo* species. It is highest in *B. cognatus*: 25.6 m/g (weights of specimens as above), in *B. speciosus* it is 24.3 m/g, in *B. calamita* it is 23.9 m/g, in *B. viridis* it is 21.8 m/g, and in *B. bufo* 19.7 m/g. Higher values were found in small specimens of the genera *Hyla* and *Acris*, and in freshly metamorphosed specimens of *Rana*. Toads demonstrate the highest values of the amount of respiratory capillaries per unit of weight among Salientia weighing more than 20 g.

The regression equations for oxygen consumption at 15°C and body weight, calculated by Hutchison, Whitford, and Kohl (1968) are: Bufonidae: $M = 2.96W^{0.59}$; Hylidae: $M = 1.35W^{0.82}$; Ranidae: $M = 1.75W^{0.64}$; Pelobatidae: $M = 1.18W^{0.94}$. Although the exponent that determines the slope of the regression line is low in Bufonidae, the constant K, which reflects the general level of metabolism, is significantly higher in Bufonidae than that found in other families. It can be stated therefore that, other things being equal, *Bufo* has the highest oxygen consumption among Anura.

Alimentary Canal and Larynx

All species of *Bufo* possess a protrusible tongue, attached medioanteriorly. This is a very common character in Salientia. Lack of tongue is demonstrated by *Pipa* and *Xenopus*, and its different structure (*Werneria*, *Rhinophrynus*) is regarded as a secondary condition. The structure of the glands surrounding the nasal cavities was studied in detail in some Salientia. Müller (1932) divided the arrangement of glands and their ducts into five types. The ducts may open immediately to the mouth cavity or they may open into the prechoanal grooves or into prechoanal sacs. All species of *Bufo* studied by Müller belong to his Type V. Ramaswami (1936) demonstrated later that *B. melanostictus*, *B. parientalis*, *B. hololius*, and *B. beddomi* belong to Müller's Type II. According to Baldauf (1959), prechoanal sacs are present in *B. woodhousei*, *B. terrestris*, and *B. microscaphus*, are extremely short in *B. marinus* and *B. debilis*, while only prechoanal grooves are present in *B. punctatus* and *B. quercicus*. Both sacs and grooves are absent in *B. valliceps* and *B. angusticeps*. The gland ducts open into the sacs in *B. terrestris*, into the grooves in *B. microscaphus*, and both into the grooves and into the buccal cavity in *B. woodhousei*. The glands empty directly into the buccal cavity in *B. valliceps*, *B. marinus*, and *B. quercicus*. Michael (1961) studied the nasal region in the Egyptian species *B. regularis* and found that this species has reduced prechoanal sacs, while the glandular ducts open directly into the mouth cavity. After also considering the shape of olfactory cavities, Michael concluded that *B. regularis* was intermediate between different species groups. Proper evaluation of the systematic value of these structures will be possible only after their functional value is partly understood. Regal's (1966) considerations on the function of the tongue and palatal teeth in Caudata can furnish an example. Publications on the prechoanal sacs, ducts, and gland ducts of Salientia yield no hypotheses concerning function.

The structure of the larynx is considered one of the distinctive characters of Bufonidae (Trewavas 1933; Griffiths 1959 1963; Chacko 1965). In the Bufonidae, the apical and basal cartilages are absent; vocal cords are stretched dorsoventrally and are attached to arytenoids. Pulvinaria vocalia (masses of connective tissue with which the anterior and posterior ends of arytenoids are padded in other Salientia) are absent, as are the constrictor laryngis posterior and petrohyoideus posterior II muscles. According to Chacko (1965) all these peculiarities of the bufonid larynx are consequences of a dorsoventral flexure of laryngeal cartilages. A similar larynx is found in Atelopodidae, which is an argument in favor of deriving this family from Bufonidae.

Dornesco, Santa, and Zacharia (1965) attracted attention to the mode of entrance of the pancreatic duct to the duodenum in Salientia. According to their description, it is possible to discern two arrangements. In the first, the pancreatic duct opens to the bile duct in the wall of the intestine (*Bombina* and *Pelobates*) or in its immediate vicinity (*Bufo* and *Hyla*). In the second, found in *Rana,* the pancreatic duct joins the bile duct before this channel leaves the pancreas. Dornesco *et al.* examined only some European species; their results must be confirmed in a larger variety of forms to decide whether the rules formulated by them have a general application.

Van Dijk (1959) described in detail the cloacal region of *Ascaphus* and compared its structure to that of some of the other Salientia. He concluded that *Bufo* exhibits no marked features of the cloacal region that could distinguish it from the majority of genera.

Urogenital System

An extensive survey of the urogenital system in Salientia was conducted by Bhaduri (1953). He stressed the well-known fact that the genus *Bufo* is characterized by the presence of Bidder's organ, at least in young specimens. Its appearance is, however, irregular in older specimens. It is usually absent in adult females; it was not found by Bhaduri in the only male specimen of *B. w. fowleri* he examined, although its presence in this species was confirmed by Davis (1936). According to Griffiths (1963), the presence of Bidder's organ in a species was a proof of bufonid affinities. He has therefore strongly emphasized the presence of this organ in *Atelopus.*

Mullerian ducts are absent in male salientians in a majority of genera; they are, however, generally present in males of *Bufo.* According to Bhaduri, they are absent only in *B. quercicus.* The Wolffian ducts unite terminally in males of all species of *Bufo* except *B. boreas,* but the united segment of the ducts is short, less than one mm. long. The testes exhibit in *Bufo* a considerable variability between species in shape, size, color, and position (Chap. 17).

Development and Larval Characters

The majority of toads lay large quantities of small eggs. This is a primitive trait and an adaptation to egg laying in conditions where the mortality of tadpoles is high. The larvae development is generally short, but not as short as in *Scaphiopus* (Brown 1967), and the larvae do not attain large size. According to Moulton, Jurand, and Fox (1968), *Bufo* tadpoles have no Mauthner cells, which are present in tadpoles of remaining Anura. The presence and good development of Mauthner neurones is correlated with high swimming ability. Orton (1953) stressed the value of tadpole morphology for amphibian systematics and divided tadpoles into four types. *Bufo* tadpoles belong to Orton's Type four, that is they possess horny beaks, they have labial teeth in single rows, and their spiracle is single and asymmetrical. The anus is median. Such a tadpole structure speaks clearly against closer affinity of bufonids with the primitive salientians but gives little information on the position of the family within an "advanced" group. Griffiths (1961) has shown that bufonid tadpoles lack the "manicotto glandulare" present in the majority of tadpoles. He does not consider this character significant, because tadpoles devoid of manicotto are also found in a few species belonging to various salientian families.

A unique property of *Bufo* tadpoles is the ability to produce an alarm substance (Eibl-Eibesfeldt 1949; Hrbáček 1950; Pfeiffer 1966). The skin extracts of tadpoles of European species produce a flight reaction in tadpoles. Larvae of other genera do not react to this substance. These experiments are worth repeating on tadpoles of other continents. Voris and Bacon (1966) reported that fishes avoid *Bufo* tadpoles as food, presumably because of some skin secretions. Perhaps the same substance defends *Bufo* tadpoles against predation and serves as an alarm signal.

Physiological and Ecological Observations

There are many publications on the resistance of various species of *Bufo* to high temperature. The results of these studies can be summarized by stating that the resistance to heat, both of adult specimens and of tadpoles, is correlated with the climate of the area of distribution (e.g., Volpe 1957; Schmidt 1965; Hutchison, Whitford, and Kohl 1968). Different results were obtained by Kusakina (1965). She studied the heat stability of hemoglobin and of cholinesterase in three subspecies of *B. bufo* — one of which lives on the border of the Black Sea, the other in the vicinity of Leningrad, the third near Vladivostok — and found no differences. The heat stability of preparations obtained from *B. viridis,* a species reaching farther south, although sympatric in a large area, was somewhat higher.

Highest resistance to heat was described by Straw (1958) in a population of *B. exsul* from warm springs. These tadpoles tolerate temperatures up to 40°C; adult animals are less resistant to warm water but in dry air they can support even 44°C. The survey of literature gives the impression that numerous species of *Bufo* are resistant to high temperature (e.g., Inger 1959; Cloudsley-Thompson 1967).

Amphibian species highly resistant to heat are often also able to tolerate aridity. As mentioned earlier, the epidermis of *Bufo* is not very thick, and the protection offered by it to water loss is limited. Thorson and Svihla (1943) demonstrated that the resistance to aridity in Salientia is based on the ability to tolerate the water loss. According to Cloudsley-Thompson (1967), *B. regularis* survives a water loss that causes the body weight to drop to 50 percent of the original value. One of the mechanisms responsible for this is the ability to accumulate a large quantity of diluted urine in the urinary bladder, which acts as an organ of water storage (Ruibal 1962).

The highest resistance to salinity among amphibians was found in *Rana cancrivora* (Gordon, Schmidt-Nielsen, and Kelly 1961). Resistance to salinity has also been demonstrated among *Bufo* species, but it is never very high. The tadpoles of *B. melanostictus* can survive salinities reaching 0.75 percent, although not in the first days of life (Strahan 1957). *Bufo viridis* from Denmark is claimed to be most resistant, since it survives in sea water diluted to 50 percent, which has a salinity of 1.6 percent (Gordon 1965). The same species in Algeria tolerates 2.2 percent of salinity (Bons and Bons 1959). According to Schoffeniels and Tercafs (1965–1966) the mechanism of adaptation to high osmotic pressure in *B. viridis* is based on the ability to store urea, similar to that demonstrated in *R. cancrivora*. Gordon (1965) obtained, however, different results. According to him the osmotic concentration in *B. viridis* adapted to brackish water is raised almost entirely by changes in NaCl concentration.

Toads tolerate complete submersion less well than other amphibians except *Hyla* (Czopek 1962), although they possess the mechanism that induces bradycardia and a depression of metabolism in this circumstance (Leivestadt 1960). Hutchison and Dady (1964) immersed amphibians in a definite amount of water. They found that the level of oxygen at the time of death is higher in *B. terrestris* than in *Rana pipiens*.

Maher (1967) has reported that thyroxine raises the metabolism of *Rana pipiens* when the temperature of the animal is between 20 and 25°C. In *B. woodhousei* the effect of thyroxine was observed at temperatures ranging from 5 to 25°C. Maher concluded that this is correlated with the nocturnal life of the toad.

The numerous scattered observations on the food of toads are difficult to evaluate since they were collected differently. Inger and Marx (1961), during an extensive study of the food of amphibians in an African reserve, found that *Bufo* feeds exclusively on terrestrial organisms, that the toads show the greatest concentration to one, or to a few foods, and that they eat large quantities of small and very small prey, such as termites and ants. These observations prove that *Bufo* is comparatively highly specialized in collecting food. Regal (1966) gave a convincing picture of the evolution of tongue and palatal teeth of Caudata, which demonstrates that the successive improvements in catching and swallowing led to the ability to utilize very small food items.

Toads learn quickly to avoid stinging insects and their mimics (Brower 1960; Brower and Brower 1962). This ability is strongly contrasted by the behavior of *Xenopus*, which learns only after hundreds of experiments (Altner 1962). The considerable abilities of toads for homing and for orientation in space have often been noted (e.g., Goin and Goin 1960; Brattstrom 1962; Ferguson and Landreth 1966; Oldham 1966).

Discussion

Bufo is a very successful amphibian genus rich in species and nearly cosmopolitan in distribution. Some of its representatives are small, others reach considerable proportions. The different forms show various adaptations to many environments and modes of life, but it is tempting to speculate about the nature of the key adjustments that are responsible for its evolutionary radiation. Let us discuss some hypotheses.

Probably the first improvement in bufonid evolution was the extraordinary development of skin glands. This made the animal unpalatable to the majority of predators and permitted a decrease in jumping ability. Jumping and running are antagonistic; a jumper must possess long legs with very long thighs (Hinsche 1950), which makes running cumbersome. Many characters of *Bufo* are correlated with its ability for prolonged movement; the legs are comparatively short, muscles are diversified and very well supplied with blood. The lungs have a complicated structure; the amount of respiratory capillaries per unit of weight is very high, as is the blood pressure, and the heart is large.

The ability to cover long distances permitted active hunting. The active mode of life put a high premium on the perfection of learning ability and orientation in space. In consequence, the toad is an intelligent amphibian. *Bufo* does not passively wait for its food but it returns regularly to good hunting sites (Brattstrom 1962). The food selection probably results in part from returning regularly to successful feeding places, where a similar prey is collected night after night, but it may also be a consequence of discriminatory ability. Every observer of *Bufo* has noted that the animal bends the neck and fixes the prey before attacking it. Hinsche (1950) has demonstrated that the visual field of the toad comprises the ground in its' immediate vicinity but is limited from above, while a frog sees well the prey moving above it and is little attracted by animals moving on the ground.

Bending the head downward may explain the peculiar structure of m. depressor mandibulae. The portion of the muscle attached to the dorsal fascia would be inefficient when the neck region is flexed. The head mobility may be related to the fact that the toads are active chiefly during the night. Many animals hunted at night are wet and slimy, for instance slugs and earthworms; they will not stick to the tongue but must be caught with the jaws, which necessitates bending the neck.

The night wanderings may conduct animals far from humid hiding places, and the toads are consequently able to tolerate considerable heat and water loss. The ability to survive the hot and dry hours of the day is increased in some species by very big size, which changes surface-to-mass ratio (e.g., *B. paracnemis*, *B. marinus*, *B. blombergi*). Many smaller species have a considerable ability to dig.

Thus, many *Bufo* characters can be meaningfully correlated by assuming that the ancestor of the genus was specialized by being a highly venomous, terrestrial, active, and intelligent night hunter. There are, however, other important traits that are not explained by this supposition, namely, the presence of Bidder's organ and the peculiar structure of the larynx.

REFERENCES

Adolph, E. F. 1932. The vapour tension relation of frogs. Biol. Bull. 62:112–125.

Altner, H. 1962. Untersuchungen über Leistungen und Bau der Nase des südafrikanischen Krallenfrosches *Xenopus laevis* (Daudin). Z. vergl. Physiol. 45:272–306.

Andrzejewski, H., J. Czopek, L. Siankowa, and H. Szarski. 1962. The vascularization of respiratory surfaces in amphibians reaching large body size. Studia Soc. Sci. Torun. Sect. E (Zool.) 6:163–177.

Baldauf, R. J. 1959. Morphological criteria and their use in showing bufonid phylogeny. J. Morph.104:527–560.

Bentley, P. J. 1966. Adaptations of Amphibia to arid environments. Science 152:619–623.

Bhaduri, J. L. 1933. Observations on the course of the facial vein and the formation of the external jugular vein in some Indian frogs and toads. J. Proc. Asiat. Soc. (Bengal) 27:101–110.

———. 1938. Observations on the course of the facial vein and the formation of the external jugular vein in an American bullfrog *Rana catesbeiana*. Anat. Anz. 86:170–172.

———. 1944–1945. A further instance of the posterior cardinal vein in *Bufo melanostictus* Schneider. Sci. and Cult. 10:351–352.

———. 1953. A study of the urogenital system of Salientia. Proc. Zool. Soc. (Bengal) 6:1–111.

Bhati, D. P. S. 1961. Studies on the pectoral girdle and its musculature (with innervation) of *Rana tigrina* Daud. and *Bufo andersoni* Bouleng. Agra Univ. J. Res. Sci. 10:131–135.

Bieniak, A., and R. Watka. 1962. Vascularization of respiratory surfaces in *Bufo cognatus* Say and *Bufo compactilis* Wiegman. Bull. Acad. Polon. Sci. Ser. 2, 10:9–12.

Bigalke, R. 1927. Zur Myologie der Erdkröte (*Bufo vulgaris* Laurenti) Z. Anat. Entw.-Gesch. 82:236–253.

Bons, J., and N. Bons. 1959. Sur la faune herpétologique de Doukkala. Bul. Soc. Sci. Nat. Phys. (Morocco) 39:117 (quoted in Schoffeniels and Tercafs 1965–1966).

Brattstrom, B. H. 1962. Homing in the giant toad, *Bufo marinus*. Herpetologica 18:176–180.

Brower, J. V. Z. 1960. The reactions of southern toad (*Bufo terrestris*) to honey-bees (*Apis mellifica*) and their syrphid mimics (*Eristalis* sp.). Anat. Rec. 183: 338.

Brower, L. P., and J. V. Z. Brower. 1962. Investigations into the mimicry. Nat. Hist. 71:8–19.

Brown, H. A. 1967. Embryonic temperature adaptation and genetic compatibility in two allopatric populations of the spadefoot toad, *Scaphiopus hammondi*. Evolution 21:742–761.

Buddenbrock, W. V. 1967. Vergleichende Physiologie. 6. Blut und Herz. Birkhauser, Basel and Stuttgart.

Carman, J. B. 1967*a*. The carotid labyrinth in the anuran *Breviceps mossambicus*. Trans. R. Soc. New Zealand (Zool.) 10:1–15.

———. 1967*b*. The morphology of the carotid labyrinth in *Bufo bufo* and *Leiopelma hochstetteri*. Trans. R. Soc. New Zealand (Zool.) 10:71–76.

Chacko, T. 1965. The hyolaryngeal apparatus of two anurans. Acta Zool. (Stockholm) 46:83–108.

Cloudsley-Thompson, J. L. 1967. Diurnal rhythm, temperature and water relation of the African toad. *Bufo regularis*. J. Zool. (London) 152:43–54.

Cooper, E. L. 1967. Lympho-myeloid organs of Amphibia. I. Appearance during larval and adult stages of *Rana catesbeiana*. J. Morph. 122:381-398.

Czopek, G. 1963. The distribution of capillaries in muscles of some Amphibia. Studia Soc. Sci. Torun. Sect. E (Zool.) 7:61–98.

———, and J. Czopek. 1959. Vascularization of respiratory surfaces in *Bufo viridis* Laur. and *Bufo calamita* Laur. Bull. Acad. Polon. Sci. Ser. 2, 7:39–45.

Czopek, J. 1955. The vascularization of respiratory surfaces of some Salientia. Zool. Polon. 6:101–134.

———. 1962. Tolerance to submersion in water in amphibians. Acta Biol. Cracoviensia (Ser. Zool.) 5:241–251.

———. 1965. Quantitative studies on the morphology of respiratory surfaces in amphibians. Acta Anat. 62:296–323.

Davis, D. D. 1936. The distribution of Bidder's organ in the Bufonidae. Field Mus. Nat. Hist. (Zool.) 20:115–125.

Dinesman, L. G. 1948. Adaptatsia amfibii k razlichnim usloviam vlashnosti vozdukha. Zool. Sh. 27:231–239.

Dornesco, G. T., V. Santa, and F. Zacharia. 1965. Sur les canaux hépatiques, cholésdoques et pancréatiques des Anoures. Anat. Anz. 116:239–257.

Dunlap, D. G. 1960. The comparative myology of the pelvic appendage in the Salientia. J. Morph. 106:1–76.

Eibl-Eibesfeldt, I. 1949. Über Vorkommen von Schreckstoffen bei Erdkrötenquappen. Experientia 5:236.

Elias, H., and J. Shapiro. 1957. Histology of the skin of some toads and frogs. Amer. Mus. Novitates, no. 1819 pp. 1-27.

Ferguson, D. E., and H. F. Landbreth. 1966. Celestial orientation of Fowler's toad *Bufo fowleri*. Behaviour 26:105–123.

Goin, O. B., and C. J. Goin. 1960. Return of the toad *Bufo terrestris* to the breeding site. Herpetologica 16:276.

Gonzales, S. J., O. Sanchez, and B. Concha. 1966. Changes in potential difference and short circuit current produced by electrical stimulation in a nerve-skin preparation of toad. Biochim. Biophys. Acta 120:186–188.

Gordon, M. S. 1965. Intracellular osmoregulation in skeletal muscle during salinity adaptation in two species of toads. Biol. Bull. 128:218–229.

———, K. Schmidt-Nielsen, and H. M. Kelly. 1961. Osmotic regulation in the crab eating frog (*Rana cancrivora*). J. Exp. Biol. 39:659–678.

Griffiths, I. 1954. On the "otic element" in Amphibia Salientia. Proc. Zool. Soc. (London) 124:35–50.

———. 1961. The form and function of the fore gut in anuran larvae (Amphibia Salientia) with particular reference to the manicotto glandulare. Proc. Zool. Soc. (London) 137:249–283.

———. 1963. The phylogeny of the Salientia. Biol. Rev. 38:241–292.

Hinsche, G. 1950. Funktionskritische Zonen am Wirbeltierkörper. Zool. Anz. 145:170–180.

Hoyer, H. 1934. Das Lymphgefäßsystem der Wirbeltiere vom Standpunkte der vergleichenden Anatomie. Mém. Acad. Polon. Sci. Cl. Méd. 1934:1–205.

Hrbáček, J. 1950. On the flight reaction of tadpoles of the common toad caused by chemical substances. Experientia 6:100–102.

Hutchison, V. H., and J. Dady. 1964. The viability of *Rana pipiens* and *Bufo terrestris* submerged at different temperatures. Herpetologica 20:149–162.

———, W. C. Whitford, and M. Kohl. 1968. Relation of body size and surface area to gas exchange in anurans. Physiol. Zoöl. 41:65–85.

Inger, R. F. 1958. Comments on the definition of genera. Evolution 12:370–384.

———. 1959. Amphibia. South African Animal Life. Lund Univ. Exp. 6:510–553.

———, and H. Marx. 1961. The food of amphibians. Explor. Parc Nat. Upemba. 64:1–86.

Johansen, K., and A. S. F. Ditadi. 1966. Double circulation in the giant toad, *Bufo paracnemis*. Physiol. Zoöl. 39:140–150.

Kusakina, A. A. 1965. Teplostoichivost gemoglobina i kholinesterazy myshts i pechenii u predstavitelei trekh podvidov seroi zhaby (*Bufo bufo* L.). Sb. rab. Inst. Citol. (Moscow) 8:209–211.

Leivestadt, H. 1960. The effect of prolonged submersion on the metabolism and the heart rate in the toad (*Bufo bufo*). Arb. Univ. Bergen Mat. (Naturv. Ser.) 5:1–15.

Limeses, C. E. 1964. La musculatura del muslo en los Ceratofrinidos y formas afines. Con un análisis crítico sobre la significación de los caracters miológicos en la sistemática de los Anuros superiores. Contr. Cient. Univ. Buenos Aires (Zool.) 1(4):193–245.

———. 1965. La musculatura mandibular en los Ceratofrinidos y formas afines (Anura, Ceratophrynidae). Physis 25:41–58.

Maher, M. J. 1967. Response to thyroxine as a function of environmental temperature in the toad. *Bufo woodhousei* and the frog, *Rana pipiens*. Copeia 1967:361–365.

Marcus, H. 1937. Lungen, III, 909–988. *In* Bolk-Göppert-Kallius-Lubosch's Handbuch der vergleichenden Anatomie der Wirbeltiere, edited by Louis Bolk et al. Berlin and Vienna, Urban and Schwarzenberg.

Michael, M. J. 1961. The adult morphology of the olfactory organs of the Egyptian toad, *Bufo regularis* Reuss. J. Morph. 109:1–18.

Millard, M. 1941. The vascular anatomy of *Xenopus laevis*. Trans. R. Soc. S. Africa. 28:387–439.

Moulton, J. M., A. Jurand, and H. Fox. 1968. A cytological study of Mauthner cells in *Xenopus laevis* and *Rana temporaria* during metamorphosis. J. Embryol. Exp. Morphol. 19:415–431.

Müller, E. 1932 Untersuchungen über die Mundhöhlendrüsen der anuren Amphibien. Morphol. Jahrb. 70: 131–172.

Noble, G. K. 1922. The phylogeny of the Salientia. I. The osteology and thigh musculature; their bearing on classification and phylogeny. Bull. Amer. Mus. Nat. Hist. 46:1–87.

Oldham, R. S. 1966. Spring movements in the American toad, *Bufo americanus*. Canad. J. Zool. 44:63–100.

Orton, G. L. 1953. The systematics of vertebrate larvae. Syst. Zool. 2:63–75.

Pfeiffer, W. 1966. Die Verbreitung der Schreckreaktion bei Kaulquappen und die Herkunft des Schreckstoffes. Z. vergl. Physiol. 52:79–98.

Ramaswami, L. S. 1936. The morphology of the bufonid head. Proc. Zool. Soc. (London) 4:1157–1169.

Regal, P. J. 1966. Feeding specializations and the classification of terrestrial salamanders. Evolution 20:392–407.

Rogers, D. C. 1966. A histological and histochemical study of the carotid labyrinth in the anuran amphibians, *Bufo marinus*, *Hyla aurea* and *Neobatrachus pictus*. Acta Anat. 63:249–280.

Ruibal, R. 1962. The adaptive value of bladder water in the toad *Bufo cognatus*. Physiol. Zoöl. 35:218–223.

Salée, M.-L., and M. Vidrequin-Deliège. 1967. Nervous control of the permeability characteristics of the isolated skin of the toad *Bufo bufo* L. Comp. Biochem. Physiol. 23:583–597.

Schoffeniels, E., and R. R. Tercafs. 1965–1966. L'osmorégulation chez les Batraciens. Ann. Soc. R. Zool. (Belgium) 96:23–39.

Schmidt, W. D. 1965. High temperature tolerances of *Bufo hemiophrys and Bufo cognatus*. Ecology 46: 559–560.

Simons, J. R. 1957. The blood pressure and the pressure pulses in the arterial arches of the frog (*Rana temporaria*) and the toad (*Bufo bufo*). J. Physiol. 137:12–21.

———. 1959. The distribution of the blood from the heart in some Amphibia. Proc. Zool. Soc. (London) 132:51–64.

Strahan, R. 1957. The effects of salinity on the survival of larvae of *Bufo melanostictus* Schneider. Copeia 1957:146–147.

Straw, R. M. 1958. Experimental notes on deep springs toad *Bufo exsul*. Ecology 39:552–553.

Szarski, H. 1947–1948. On the blood-vascular system of the Salientia. Bull. Acad. Polon. Sci. Ser. B. 1947: 146–211.

———. 1951. Remarks on the blood vascular system of the frog *Leiopelma hochstetteri* Fitzinger. Trans. R. Soc. New Zealand 79:140–147.

Thorson, T. B. 1955. The relationship of water economy to terrestrialism in amphibians. Ecology 36:100–116.

———, and A. Svihla. 1943. Correlations of the habitat of amphibians with their ability to survive the loss of body water. Ecology 24:374–381.

Tihen, J. A. 1965. Evolutionary trends in frogs. Amer. Zoologist 5:309–318.

Trewavas, E. 1933. The hyoid and larynx of the Anura. Philos. Trans. Royal Soc. (London) B, 222:410–527.

Van Dijk, D. E. 1959. On the cloacal region of Anura in particular of larval *Ascaphus*. Ann. Univ. Stellenbosch. A, 35:169–249.

Volpe, E. P. 1957. Embryonic temperature tolerance and rate of development in *Bufo valliceps*. Physiol. Zoöl. 30:164–176.

Voris, H. K., and J. P. Bacon, Jr. 1966. Differential predation on tadpoles. Copeia 1966:594–598.

Warburg, M. R. 1965. Studies on the water economy of some Australian frogs. Austr. J. Zool. 13:317–330.

6. *Bufo* of South America

JOSE M. CEI

Universidad Nacional de Cuyo
Mendoza, Argentina

Introduction

The geographical distribution and ecological features of the South American toads of the genus *Bufo*, besides agreeing generally with the main phyletic lines and the ancient relationships of its major species groups, correspond also to the geomorphological and paleoclimatical trends of the ancient and present structural units of the continent. Various studies have successfully demonstrated some phenetic and biological affinities between supraspecific groups of neotropical and nearctic toads, for instance the claimed relationships between *B. marinus* and *B. valliceps*, or between the *boreas* and *spinulosus* complexes. A synthetic discussion on intrageneric groupings may be found in Tihen's osteological considerations (1962). Moreover, recent reports by Blair (1963, 1964, 1966), dealing with genetic compatibility and interspecific hybridization, provide an objective, experimental basis for a tentative arrangement and a suggestive evolutionary history of the most precocious stock of these anurans. On the other hand, the fundamental paleogeographic development of South America and the major tectonic events that have taken place since the close of the Cretaceous, such as the Tertiary transgressions and volcanic activity and later the uplift of the Andean cordilleras, have played an unquestionable role in determining the present dispersal of the supraspecific groups. The importance of these paleogeographic events as an important factor in speciation of these toads must be stressed.

Five main lines or sections of South American toads can easily be identified by many comparative morphophysiological characters. It is interesting to analyze how much these species-complexes agree with both the present topographic and climatic features of soil and habitat and with the alternate raisings and lowerings of the ancient emerging geotectonic units that led to the actual embossments of the landscape and to the continental hydrographic systems.

At first glance, the dominant structural element of South America is the Andean cordilleras, which spread longitudinally from the equatorial zone to the cold subantarctic Magellanian latitudes. A tentative reconstruction of the evolution of this geosynclinal area has been synthetized by Nygren (1950) and especially by Harrington (1962). The latter author stressed the differences between the westward "migrating" southern part of the geosynclinal axis and the so-called Bolivar geosyncline or Ecuadorian sea portal of northwestern South America where the presence of a postulated western land mass on the site of the present Galapagos-Malpelo submarine platform acted as a true "foreland" to the Mediterranean Colombian-Ecuadorian geosynclines. An almost independent evolution of the southern part of the continent can well be assumed up until the final foundering of the Malpelo landmass below the waters of the Pacific. This relatively recent event took place in late Miocene to early Pliocene time, contemporaneous with the greatest uplift of the cordilleras of Colombia, which followed the Upper Oligocene full subsidence of the Bolivar geosyncline and the definitive close of the Ecuadorian sea portal.

The *Spinulosus* Group

One supraspecific group of toads, the *spinulosus* group, is virtually restricted to the southern part of the cordilleran belt (Vellard 1959). These bufonids, very resistent to dryness and altitude, breeding at permanent waters, live in the Andean mountains from Loja province (quoted by Parker) to the south of Chile and Argentina, along the Pacific coast, splitting into a number of geographical units, which are sometimes separated by strong topographical or ecological barriers (Cei 1960). *B. spinulosus* Wiegmann (1835) has been considered a very polymorphic, polytypic species; but a careful, experimental and serological screening of its so-called complex of subspecies may suggest a high degree of physiological isolation among many of these allopatric populations, which have probably reached various levels of speciation. *B. spinulosus spinulosus* Wiegmann is found from the high mountains of Peru (Cuzco and Ayacucho provinces) to Lake Titicaca. It then spreads on the Bolivian uplands in the Tarapaca and Antofagasta mountains of Chile, southward along the cordilleran summits of Chile and Argentina, from 1,000–1,200 meters to 5,000 meters. It reaches at least 43° south latitude, where in the wooded districts of the cold austral forest it is known as *B. spinulosus papillosus* (Philippi). A subspecies, *B. spinulosus altiperuvianus* Gallardo (1961), is reported from the Oruro mountains (Bolivia: 3,700 meters). Another form, *B. spinulosus arequipensis* Vellard is known from the volcanic Arequipa mountains and their southwestern valleys (Moquegua, Tarata, Tacna), but its real taxonomic position in Wiegmann's type is still obscure. Perhaps additional subspecific divisions of *B. spinulosus* could be recognized along the widespread Andean range.

Specific status is apparent for *B. chilensis* (Tschudi) (= *arunco* [Molina] in Garnot and Lesson), which is distributed in the central Chilean arid steppe (*Acacia caven* association), from Coquimbo to Concepción, facing eastward at some 1,000–1,200 meters the neighboring *B. spinulosus* from the cordilleran slopes. The great ecological barrier of the stony desert of Atacama isolates another form, likewise a good species, *B. atacamensis* (Cei 1961), which is striking for its remarkable sexual dichromatism and which lives in the coastal rivers and streams of this very dry, almost azoic environment. On the other hand the cold Valdivian rain forest (up to 4,000 mm. of annual rainfall) contains a specialized and uncommon species, *B. rubropunctatus* Guichenot, which is characterized by its ventral color pattern (white spots on a velvet black belly) (Gallardo 1962), strongly reminiscent of the rare Californian *B. exsul*. Moreover, *B. variegatus* (Gunther), which lives there, is a sympatric bufonid form of problematic affinities (Gallardo 1962), erroneously placed in the *spinulosus* complex, as discussed by Gallardo (1965). This small toad, similar by its yellow dorsal stripes to the juvenile specimens of *B. flavolineatus* Vellard from Peru, is a very localized delicate inhabitant of the *Nothofagus* forest, where it may represent some relict of an ancient Tertiary Patagonian fauna, together with *Calyptocephalella* or *Eupsophus* frogs from the same biota. At any rate the present sympatry in the Valdivian forest near Llanquihue Lake of both *B. spinulosus* and *B. rubropunctatus*, as well as *B. variegatus*, must be pointed out (Silva *et al.* 1968).

North to 13° south latitude, from the sandy shores of Pisco, along the arid Peruvian coast to the Sechura Desert, a big form, *B. limensis* Werner, lives in the narrow valleys and creeks descending to the Pacific from the longitudinal steep Andean chains. It is replaced in the high Marañón valley by a closely related toad, *B. orientalis* Vellard, probably a subspecies of *B. limensis*, to which some still unknown populations from Cajamarca must also be referred. The exceptionally broken landscape of the Peruvian uplands also contributes to geographical isolating mechanisms of speciation, with its sharp parallel mountains surrounding, in many cases, enlarged, flat, wet "racers," with scattered, isolated, shallow lakes. The peculiar form *B. flavolineatus* Vellard is a typical example, inhabiting the high central Andean plateau, between Junín and Callejon de Huaylas, up to an altitude of 4,600 meters. The relations of *B. flavolineatus* with the adjacent *B. trifolium* Tschudi are still unclear where the mountain ranges of both approach one another. *B. trifolium* is related to *B. spinulosus* Wiegmann due to its general morphology and

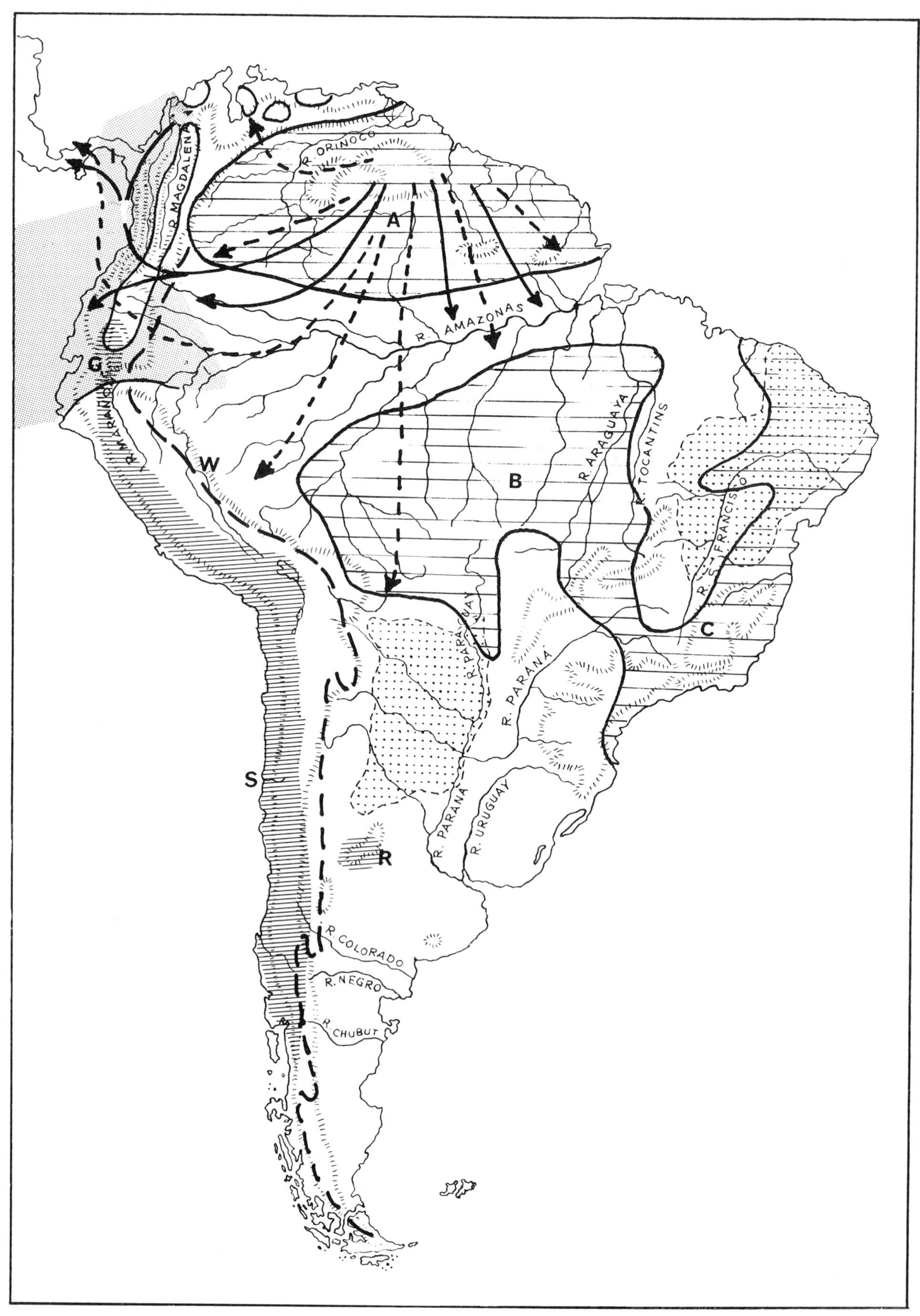

Fig. 6–1. Guiana (*A*), central Brazilian (*B*), and coastal Brazilian (*C*) shields (ancient cratonic or geotectonic units; see Harrington 1962). *I*, Cenozoic interruption by the Isthmian or Panamian portal; *G*, Bolivar geosyncline and Ecuadorian sea portal; *W*, western migrating Pacific geosynclines (broken line); *dashed arrows*, radiation of the *typhonius* group from an early Tertiary Guiana center; *solid arrows*, radiation of the *guttatus-haematiticus* group from an early Tertiary Guiana center; *S* (horizontally thin hatched area), present distribution of the Andean specific-subspecific *spinulosus* complex; *R*, extra-Andean relicts of *Bufo spinulosus* (Sierras of Cordoba).

biology, but its yellow median dorsal stripe is unmistakable. It extends from the Amazonian head of the Huallaga-Marañón basin, in the subtropical range of Huánuco (2,800 meters) to the Andean tops of Junín, Tarma, Ayacucho, Andahuaylas, and the eastern borders of Cajamarca. It is obvious that such a widespread and uneven range with its attendant isolation could have led to genetic incompatibility among many of these clearly disjunct populations. Thus, the real status of the not yet explained populational complex of *B. trifolium* needs further study. A final species, *B. cophotis* Boulenger, belongs to the northern toads of the *spinulosus* species group, because rare specimens have been found in Cajamarca, Libertad, and Ancash, between 2,200 and 3,500 meters. Although its distribution lies between the ranges of *B. limensis* and *B. trifolium*, sympatry with these toads is still unreported.

Some evident conclusions may be derived from our brief examination of the biogeographical patterns of the Andean toads. No species of *Bufo*, other than those of the *spinulosus* complex, occur in the Tertiary-uplifted cordilleran landmass, southward of 7° south latitude. Peripheral populations of *B. poeppigi*, entering the narrow coastal range up to 10° south, must be considered merely recent tropical immigrants. The clear-cut interruption of the geographical distribution of the *spinulosus* group near the Sechura Desert coincides with the area of the Bolivar geosyncline rising above sea level during the Pliocene. On the other hand the scattered populations of the same stock in the southern Ecuadorian Andean valleys of Loja could have filtered through to the north during the Pliocene or Pleistocene. A significant relationship seems evident between the very prolonged evolution of the *spinulosus* group and the geomorphological events of the Andean belt following the emersion of the southern part of the Pacific geosyncline. Some primitive line of *spinulosus*-like ancestors must be postulated at the beginning of the slowly rising Oligocene cordilleran fronts. Its supposed early affinities with the northern lines having *calamita*-like or *boreas*-like features is discussed elsewhere (Chap. 18). A widespread Tertiary range in the south and Patagonian relationships of such a primitive line are suggested by the limits of the group and by its unique presence (*B. rubropunctatus*, southernmost populations of *spinulosus*), together with the aberrant *B. variegatus*, in such a relict biocenotic community as the Chilote or Valdivian rainforest.

The present patterns of speciation at such different latitudinal levels could be attributed to further interventions of topographic and climatic factors, covering probably late Tertiary and Pleistocene time. That may be the case in the ecological isolation of the small-sized *B. flavolineatus*, which lives in the cold, green *páramos* of Junín, where the annual rainfall is 1,000 mm. This is very different from the coastal subtropical desert, home of the giant, stout *B. limensis*, where the annual rainfall is 10–24 mm. On the other hand, a reflection of the late Pleistocene climatic changes and their severe effects on the xeric north Chilean environments may be the present genetic isolation (genetic drift?) shown by *B. atacamensis* from the Atacama rivers, likewise a derivative species of *B. chilensis* from the central steppe zone. A very high incompatibility between these two similar forms is also indicated by serological distance, which is almost the same as the serological distances between *B. chilensis* and *B. spinulosus* from Mendoza, Argentina, or Cuzco, Peru (Cei 1968, 1969). A noticeable aspect of the present distribution of *B. spinulosus* is the marginal and very limited populations that occur in some extra-cordilleran biotypes, as in the Tunuyan Hills facing Mendoza, and the Pampa de Achala Sierras (2,200 meters) neighboring Córdoba, Argentina, in which some scanty floristic elements from the antarctic biocenosis are found (*Maitenus boaria, Pernettia.*) The problem is whether such extra-cordilleran relicts represent remnants of a much earlier stage of radiation, or whether the range of the species extended eastward during a more humid and cooler climatic condition, probably in the Pleistocene. On the other hand, recent morpho-physiological findings support a true specific status, at least for the isolated toads from the granite ravines of the Pampa de Achala summits (*Bufo achalensis*; Cei 1969).

Broad-Skulled Toads

The early and late history of basic South American geological units, such as the Guiana, central and coastal Brazilian shields, is the natural background for the slow evolutionary and adaptive history of the other four main lines of neotropical toads: the *marinus, granulosus, typhonius,* and *guttatus* groups. As indicated earlier, a major evolutionary line, the "thin-skulled line," has recently been postulated by Blair to explain remote affinities between the southern Andean toads and the northwestern American representatives of this group, in spite of a very huge gap in the present distribution from Ecuador to northern Baja California. Similarly, ancestral relationships

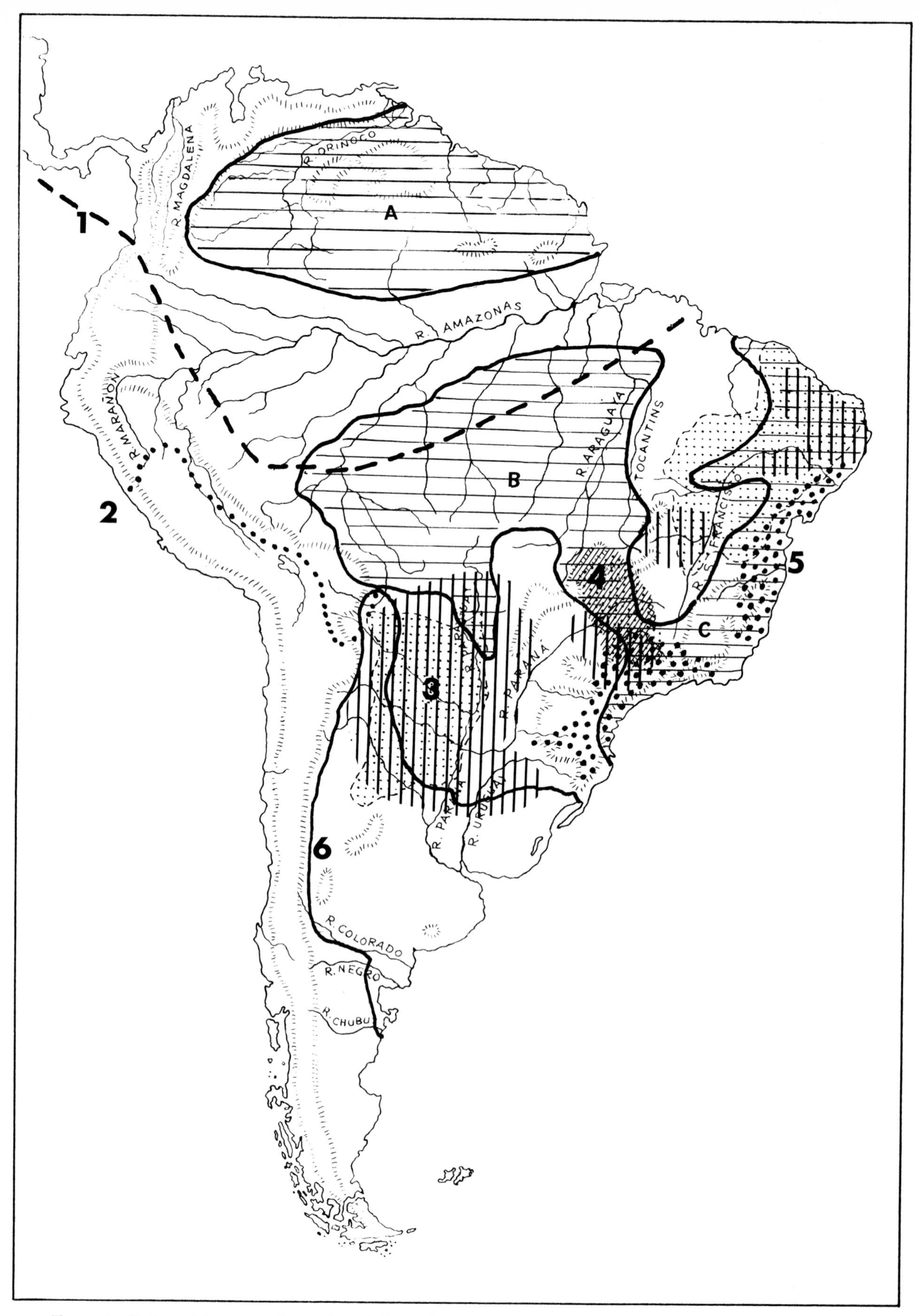

Fig. 6–2. Guiana (*A*), central Brazilian (*B*), and coastal Brazilian (*C*) shields. *1* (dashed line), tentative southern limits of *Bufo marinus marinus* (Guiana-Amazonian range); *2* (dotted line), southern limits of *Bufo (marinus) poeppigi* on the eastern forest-covered Andean slopes; *3* (vertically hatched zones), area of *Bufo paracnemis* covering the central xeric zones of the Chacoan belt and caatingas (faint stippled areas in both maps); *4* (obliquely faint hatched-stippled zone), area of *Bufo rufus* in the Brazilian *planaltos*; *5* (strongly stippled area), coastal forest range of *Bufo ictericus*; *6* (solid line), tentative limits *of Bufo arenarum* complex.

between the Central American toads of the *valliceps* group and the *marinus* group have been pointed out. Without any doubt, the occurrence of *B. marinus* in Central America, together with members of the *granulosus*, *typhonius*, and *guttatus* stocks, is explainable by a secondary Pliocene Panamanian connection. But the prolonged extensive marine flooding in the northwestern corner of the continent since Neocomian time (Lower Cretaceous) may represent the most impressive, or decisive, geotectonic event for any aspect of the phyletic radiation of any primeval neotropical bufonid line.

Parallel trends of speciation and continental dispersal are evident for the *marinus* and *granulosus* stocks, which extend widely, with their peripheral species, 42° southward, and 29° and 15° northward respectively, from the equatorial line. An older Guiana center could be indicated. *B. marinus* (Linne), reaching to the Rio Grande of Texas, is the most common toad in Colombian, Venezuelan, Guianan, Ecuadorian, and in general Amazonian mesic environments, but a smaller derivative, *B. poeppigi* Tschudi, morphologically related to *B. arenarum*, replaces it on the eastern Andean slopes, up to Bolivian foggy "Yungas," crossing the cordilleras between 0° and 10° south latitude. Its southwestern marginal distribution seems quite similar to that of another peripheral member of the group, *B. rufus* Garman, named for its reddish, warty, dorsal skin and known from the Brazilian *planaltos* (Goiás, Minas Gerais). *B. rufus* also approaches the southernmost species, *B. arenarum* Hensel, a very adaptive toad, which ranges with many regional morphs (see Gallardo 1965; Laurent, unpublished paper) from the warm subtropical Chacoan belt to the Pampean grasslands of Uruguay, Rio Grande do Sul, and Argentina, to the "creosote bush shrubs" of the desert area of Cuyo near the eastern Andean slopes. There, as *B. arenarum mendocinus* (Philippi), it is sympatric in some valleys with *B. spinulosus*. Populations of *B. arenarum* reach the ecotonal Patagonian associations (Río Negro, Chubut) and overlap the distributional range of *B. poeppigi* on the southern slopes of the Bolivian uplands (Cochabamba). The physiological polymorphism of these toads is remarkable as well as their great ecological adaptiveness (Cei 1956, 1959; Rodriguez, unpublished data). In the mountains of La Rioja, central Argentina, they are found up to an altitude of 2600 meters, which reflects the successful character of these late invasive elements.

A primitive, *marinus*-like central Guiana form can be assumed. It is probably related to the Upper Miocene fossil remains of La Venta, Colombia (Estes and Wassersug 1963), having split later by peripheral speciation into a number of more or less specialized elements, some of which, like the *arenarum* complex, may be still considered a dynamic step of incipient speciation. It is suggestive to observe that most of our so-called peripheral forms occur on the pericratonic and intercratonic areas of the continent, subjected by their structural characters to the more intensive geological crises, when periods of subsidence and deposition alternated with uplifts and erosion or sudden rampant volcanic activities. The two remaining species of the *marinus* complex reflect this same general process of geographic differentiation. *B. ictericus* Spix is a mesic forest toad living in the eastern coastal range, southward to the Misiones territory near the Paraná River. It is presently separated from the Amazonian form by the broad xeric zone of the Chacoan *monte* and the central caatingas, extending into the southern Chacoan *Schinopsis* woods and the northern dry provinces of the San Francisco River and Ceará, Brazil. The latter is the homeland of the giant toad, *B. paracnemis* Lutz, which shows the most striking characters of the group, such as very strong cephalic crests, bulky parotoid glands, and tibial glands. The isolation of this probably terminal evolutive species, derived from some ancient *marinus*-like undifferentiated ancestor (cf. *B. pisanoi* from Argentine Pliocene: Casamiquela, in press), may have taken place during the changing of the climate, flora, and ecological environments following the Pleistocene glacial and interglacial periods. However, the conservative physiognomy of *B. ictericus*, as a relict of a prolonged cratonic connection prior to the late Pliocene geological events, is suggested. According to Cochran (1955), *B. paracnemis* and *B. ictericus* are found in São Paulo and Rio de Janeiro without intergrading, but they occur in different ecological niches, such as in the streams (*B. ictericus*) and the drier mountain summits (*B. paracnemis*). Sympatry is known also in Misiones, Argentina, where *B. paracnemis* is frequent on the Paraná borders and *B. ictericus* in the mesic central forest.

The *Granulosus* Group

The *granulosus* complex shows the same general trends of distribution as the *marinus* group, displaying extreme geographic fragmentation from some ancient stock, closely tied to the now highly specialized species of the Caribbean *peltacephalus* group. In spite of its morphological affinities with the "southern line," hinted by the strong cranial crests or the high glandular activity, the still poorly known forms that radiated from the so-called central species of Spix (1825?) are not yet well delimited by their intra-group genetic compatibility (Chap. 11). Their present arrangement is unsatisfactory, having been regarded by Gallardo (1965) as a widespread network of subspecific geographic units distributed according to the drainage systems of the continent. A number of subspecies have been reported from the Guiana and Orinoco regions: *B. granulosus merianae* Gallardo (1965), *B. granulosus beebei* Gallardo (1965) (also in Trinidad), and *B. granulosus nattereri* Bokermann (1966), which is the latest from the Roraima Mountains. *B. granulosus humboldti* Gallardo (1965) lives in the Magdalena Valley, entering into the isthmian lands; *B. granulosus barbouri* Gallardo is an insular endemic from Margarita Island, Venezuela. A broad tropical, perhaps primitive, area of sympatric speciation may then be assumed for the Guiana shield and the neighboring countries. Gallardo's map (1965) reveals the overlapping of *humboldti* and *beebei* in the Colombian Santa Marta district, and in Maracay (El Paito), Falcón State, and Atabapo (Orinoco River) in Venezuela. Also *nattereri*, *humboldti*, and *merianae* may overlap near the Guiana-Brazil frontier. Only an adequate assessment of the genetic compatibility and ecophysiological relationships between such typologically named subspecies can reveal their true evolutionary status.

Similarly, with the distribution of a more eastern, mesic, equatorial form, *B. granulosus mirandariberoi* Gallardo (1965) from the Aguaraya-Tocantins basin), *B. granulosus goeldi* Gallardo (1965) spreads from Santarém into the Amazon basin to the Marañón River, reaching the upper limits of *B. granulosus minor* Gallardo (1965) in the Guaporé territory. This later form, from the Beni basin, is also touched northwards by the Chacoan toad, *major* Muller and Hellmich, an opportunistic breeder that is very adaptive to the dryness and seasonal differentiation of the central xeric environments, as just indicated for the specialized giant toad *B. paracnemis*. The form to which *major* belongs is probably a natural eastern group centering on the eastern coastal Brazilian shield and including *granulosus* Spix, *lutzi* Gallardo (1965), and perhaps *azarai* Gallardo (1965) from the Paraguay basin. Like that of the giant toad of the caatingas, the differentiation of *major*, indicated by biochemical and ethological evidence (for instance, seroprotein patterns, warning and mating calls), could have occurred in the above-mentioned late Tertiary and Pleistocene climatic and ecological crisis.

It is presently hard to evaluate relationships among the puzzling tropical members of the *granulosus* group. A classification of their status is needed for the southeastern populational groups: *fernandezae* Gallardo (1957) and *pygmaeus* Myers and Carvalho (1952) to which *d'orbignyi* Dumeril and Bibron approaches. Perhaps *d'orbignyi* is a hypertelic subspecies of *fernandezae* with prominent orbital crests, in spite of their cited sympatry in Montevideo or near the Chascomús lagoons (Buenos Aires). The Paraná basin is the home of *fernandezae*, which is sympatric in many localities with *pygmaeus* and *major*, as in the central Chacoan borders (Resistencia) or in the province of Santa Fe, Argentina. The dwarf toad *B. pygmaeus*, somewhat long nosed, like the Chacoan species *major*, was originally discovered near Rio de Janeiro, Brazil, but its scattered burrowing populations reach southward beyond 32° south latitude, along the Paraná River. In addition to other physiological differences, many kinds of isolating mechanisms separate the *granulosus* forms, for instance, mating calls or warning vibrations (Cei 1964). This suggests a high level of speciation in these southernmost forms, whose radiation from some unspecialized *granulosus*-like stock could have extended over a considerable period of time.

The conservative physiography of the Archeozoic Orinoco and Guiana landmass suggests its apparent role as a lasting genocenter (Rivero 1961, 1963). Repeated processes of speciation and peripheral radiation seem evident for the *marinus* and *granulosus* lines, as well as for the *typhonius* and *guttatus* groups. The closed tropical rain forest and the ignorance about much of northern South America prevent a satisfactory delimitation of the geographic boundaries of many of the rare or only systematically recognized species or races. Moreover, in accordance with the apparently low physical selective pressure and the extremely keen competition among the inhabitants of the quite stabilized mild tropical belt, the specific interrelationships are very intricate, in the same measure as the ecological "responses" to the complex environmental "challenges" (Dobzhansky 1950). This is undoubtedly a very effective and differential factor leading to a continuous creative process. Adaptive versatility, polymorphism, and multiplication of the species, but in a sharply reduced

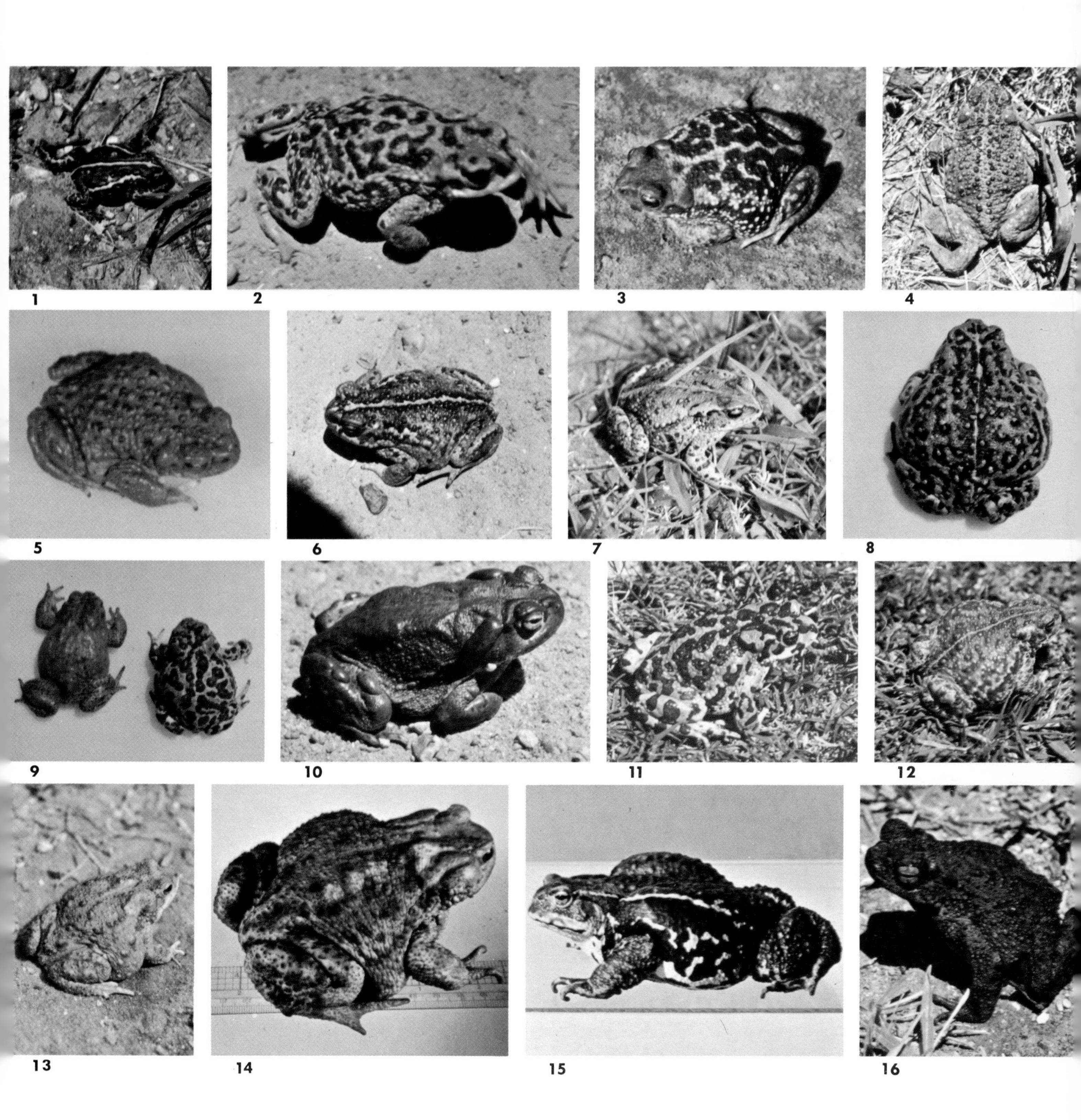

II. Narrow-skulled toads: (1) *B. variegatus*, (2) *B. chilensis*, (3) *B. atacamensis*, (4) *B. trifolium*, (5) *B. spinulosus*, and (6) *B. flavolineatus* of the South American *spinulosus* group; (7) *B. bocourti* of Guatemala; (8) *B. boreas* and (9) *B. canorus* (♂ left, ♀ right) of the *boreas* group of North America; (10) *B. alvarius*; (11) *B. viridis* of Eurasia; (12) *B. calamita*; (13) *B. stomaticus*; (14) ♀ *B. bufo* and (15) ♂ *B. bufo* of Japan; and (16) *B. asper*.

number of populations, are some of the most impressive features of tropical organisms, mediating largely in their dynamics and distribution.

The *Typhonius* Group and *B. Crucifer*

Among the *typhonius* group, a widespread, probably central species, *B. typhonius typhonius* (Linne), is known. Related forms border or overlap its Amazonian range: the hypertelic *B. typhonius alatus* Thominot from Venezuela (west of the delta of the Orinoco River), which exhibits enormous cephalic crests; *B. ockendeni* Boulenger, *B. inca* Steineger, *B. leptoscelis* Boulenger, and *B. fissipes* Boulenger, from the wild slopes of the Peruvian and Bolivian mountains (1,400–2,000 meters), between 2° and 17° south latitude. *B. quechua* Gallardo is reported from the Chapare, Yungas of Bolivia (Cochabamba: 2,600 meters); *B. coeruleostictus* Gunther is cited from Ecuador; *B. ceratophrys* Boulenger from Guiana and the Orinoco basin. *B. gnustae* Gallardo (1967) from northern Argentina cannot be seriously considered, because it has been described from a doubtful, poorly preserved specimen. The most remarkable, extreme representative of these smooth-skinned, "dead-leaf"-like, delicate toads, the long-snouted B. *dapsilis* Myers and Carvalho (1952), occurs in the Amazon basin. A number of other uncertain, presently museological species is concealed in the rain forest, such as the Venezuelan B. *sternosignatus* Gunther of unclear relationships (Rivero 1964), B. *intermedius* Gunther, and B. *manicorensis* Gallardo (1961) from Madeira River (Amazonia), or B. *ocellatus* Gunther from Minas Gerais. In the east Brazilian selva southward from the gallery forest of Misiones, along the Paraná River, the *typhonius* complex is ecologically superseded by the elegant, less specialized, *B. crucifer* Wied, whose subspecific or racial subdivision is scantily revealed from Lutz's previous works (1934).

The *Guttatus* Group

An important character of the big soft-skinned toads of the *guttatus* group is the absence of cephalic crests. Their phyletic traits differ decidedly from those of the other neotropical lines, and an independent origin from an unknown early Tertiary ancestor might be tentatively suggested. *B. guttatus* Schneider occurs in a broad area of the Guiana shield to the south of the Amazon River (Rivero 1964). *B. glaberrimus* Gunther, also indicated as a subspecies of *B. guttatus*, extends westward along the upper shores of the Amazon River up to the Ecuadorian selvas, in the Bobonaza and Pastaza basins. These large bufonids cross the Ecuadorian-Colombian uplands, spreading into the forest-covered Pacific coastal strands. *B. blombergi* Myers and Funkhauser (1951) is the largest form, living in the western rainforest from Esmeralda (Ecuador) to Nariño and Cauca (Colombia; Heredia 1968), but its relationship with other giant *glaberrimus*-like populations from the only partially explored Napo-Pastaza region remains a yet unexplained problem. In the tropical province of Chocó, northward to Panama, Costa Rica, and Cartagena on the Caribbean, *B. haematiticus* Cope occurs, morphologically resembling *B. guttatus* because of its "dead-leaf" color pattern. It is physiologically related to *B. blombergi* by the peculiar chemical factor of the testis, which prevents fertilization or normal development of the eggs in heterologous crosses (Chap. 11). Let us assume that a biologically well defined strain of remote evolutionary history, such as "*guttatus*," must have been especially moulded, in the ancient and present routes of its dispersal, by the sequence of the geological events that took place in the Tertiary, emphasized in this case by the consecutive orogenic uplifts of the longitudinal cordilleras of Colombia.

Discussion

The paleogeographic trends of the northeastern lands, west of the undeformable "positive" Guiana shield, indicate a long series of alternate periods of subsidence, continental deposits, and slow orogeny following the Upper Cretaceous ingressions. The beginning of the uplift of the eastern cordillera of Colombia is placed in the Eocene, contemporaneous with the opening of the "Bolivar geosyncline," one of the great structural barriers to the past dispersals of the main early evolutionary lines of the neotropical batrachians. If we consider that during the Middle and Upper Oligocene the renewed orogenic movements enlarged the Caribbean-Venezuelan mountains, the early Miocene witnessed the first important rising of the eastern cordillera of Colombia, in spite of the essentially unaltered sedimentary conditions of the Venezuelan troughs, the Bolivar channel, and the Colombian-Peruvian pericratonic foredeep. The Middle and Upper Miocene saw the definitive uplift of the eastern cordillera of Colombia and the orographic separation of the Llanos basin from the peripheral troughs. That is the age of the Huila formations (Upper Magdalena Valley) to which the *marinus*-like fossil studied by Estes and Wassersug is referred. An expanded period of land connections, westward, through the Middle Tertiary floodplains can be assumed, prior to the major tectonic events at the close of the Pliocene, which witnessed the last general uplift of the whole Andean range, the rampant volcanic activity, and the upthrow of the Panamian isthmian link. Such late Cenozoic and Pleistocene events can, therefore, explain the present disjunctive radiation of the *guttatus* group on the western and eastern side of the Colombian cordilleras. The isthmian bridge enabled the reciprocal further migration to and from adjacent Central America, such as that of *B. haematiticus* and *B. coniferus* Cope, a Costa Rican form belonging to the *valliceps* group that enters eastern Colombia. Also, the almost recent marine regression between the lower Atrato or lower Magdalena rivers, has probably been a biogeographic factor favoring or mediating the late dispersal of *B. haematiticus* along the shores of the Gulf of Darien, or the infiltration of *B. granulosus humboldti*, *B. typhonius*, and *B. marinus* toward the isthmian lands.

Summary

There are apparently two major centers of genetic differentiation and geographic radiation in the South American continent. A southern, probably early Patagonian, stock corresponds to the Andean orogenesis. Speciation of the *spinulosus* group, following different chronological steps, corresponds closely to the Tertiary and Pleistocene evolution of the cordilleran geological units, the result of which has been the past and present geographic and ecological barriers isolating the various populations. The adaptive trends and the evolutionary polymorphism of these toads are remarkable; the group has occupied a great number of ecological niches, from the cold relict *Nothofagus* forest in the south to the subtropical and tropical deserts, from the high Andean summits to the steppes of central Chile. A northern limiting barrier seems to have been established by the Oligocene-Miocene Bolivar geosyncline. No species of other neotropical lines appear to have entered the emerging cordilleras along the "western migrating" southern Pacific geosyncline. The marine ingressions and the climatic effects of the late Pleistocene have also been a predominantly negative factor for the present pattern of distribution in the austral biocenotic communities, which are now in equilibrium since their postglacial recuperation. Opposing the Patagonian-Andean stock, an early Tertiary tropical northwestern genocenter may be identified, radiating from the Guiana shield and connecting eastward by prolonged intercratonic coupling with the central and coastal Brazilian massifs. Such a broad emerging area has been the ancestral homeland of many relatively independent phyletic lines, whose interrelations remain in controversy. The *marinus, granulosus, typhonius*, and *guttatus* groups belong to this area, and they seem to have undergone some common general features of sympatric speciation and parallel geographic radiation during their phyletic history. A great number of species and a network of reduced populations are the rule under the push of the keen competition typical of the tropical environments. Consequently, taxonomic or specific-subspecific appraisal of most of the identified forms is very difficult in spite of the biological reality of the above-mentioned natural groups. Contrasting with the remarkable tectonic stability of the cratonic units, the peripheral basins and the Mediterranean geosynclines facing the postulated "Malpelo foreland" have been submitted during the late Mesozoic and Tertiary to prolonged geological crisis. The final uplift of the Colombian cordilleras represents the major regional event that took place in northwestern tropical America. The actual pattern of distribution of the bufonids, both in the Pacific and Amazonian slopes of the Ecuadorian-Colombian uplands, may be easily understood considering the late Miocene and Pliocene orogenesis and the recent emergence of the isthmian Panamanian bridge, a two-way migration route that has been continuously broken since early Paleocene time. Very suggestive examples are given here by the present trends of speciation of the *guttatus* group and by its northern invasion of Central America, together with some peripheral elements of the *marinus, granulosus*, and *typhonius* groups.

REFERENCES

Blair, W. F. 1963. Evolutionary relationships in North American toads of the genus *Bufo*: A progress report. Evolution 17:1–16.

———. 1964. Evidence bearing on relationships of the *Bufo boreas* group of toads. Texas J. Sci. 16:181–192.

———. 1966. Genetic compatibility in the *Bufo valliceps* and closely related groups of toads. Texas J. Sci. 18:333–351.

Bokermann, W. C. A. 1966. Notas sobre a distribuição de *Bufo granulosus* Spix na Amazônia e descrição de uma subespecie nova. Symposium on the Biota of the Amazon Basin, Belem, June 5, 1966 (Atas do Simposio sôbre a Biota Amazônica, Zoologia, Vol. 5, 1967).

Cei, J. M. 1956. Occurrence of the dwarf toad in Argentina. Herpetologica 12:324.

———. 1959. Ecological and physiological observations on polymorphic populations of the toad *Bufo arenarum* Hensel from Argentina. Evolution 13:532–536.

———. 1960. Geographic variation of *Bufo spinulosus* in Chile. Herpetologica 16:243–250.

———. 1961. *Bufo arunco* (Molina) y las formas chilenas de *Bufo spinulosus* Wiegmann. Inv. Zool. Chil. 7:59–81.

———. 1964. La vibración preventiva en poblaciones simpatridas chaquenas de *Bufo granulosus major* y *Bufo granulosus fernandezae*. Notas Biol. Fac. Cienc. Exac. Fis. Nat. Corrientes (Zool.) 4:15–21.

———. 1968. Remarks on the geographical distribution and phyletic trends of South American toads. Texas Mem. Mus. Publ. Pearce-Sellards Series 13:1–20.

———. Analisis sero-immunologico de diferentes niveles de especiación en *Bufo* del grupo *spinulosus*. (Acta II Jorn. Arg. Zool. S. Fé, 1969: Acta Zool. Lilloana, 28, 1971.)

———. Segregación corológïca y procesos de especiación por aislamiento en anfibios de la Pampa de Achala, Córdoba. (Acta II Jorn. Arg. Zool. S. Fé, 1969: Acta Zool. Lilloana, 28, 1971.)

Cochran, D. M. 1955. Frogs of southeastern Brasil. U.S. Nat. Mus. Bull. 206:1–422.

Dobzhansky, T. H. 1950. Evolution in the tropics. Amer. Scientist 38:209–221.

Estes, R., and R. Wassersug. 1963. A Miocene toad from Colombia, South America. Breviora 193:1–13.

Gallardo, J. M. 1957. Las subespecies Argentinas de *Bufo granulosus* Spix. Rev. Mus. Arg. Ciencias Nat. B. (Rivadavia) 6:337–374.

———. 1961. Three new toads from South America: *Bufo manicorensis*, *Bufo spinulosus altiperuvianus*, and *Bufo quechua*. Breviora 141:1–8.

———. 1962. Caracterización de *Bufo rubropunctatus* Guichenot y su presencia en la Argentina. Neotropica 8(25):28–30.

———. 1962. A proposito de *Bufo variegatus* (Gunther) sapo del bosque humedo Antartandico, y las otras especies de *Bufo* neotropicales. Physis 23:93–102.

———. 1965. The species *Bufo granulosus* Spix (Salientia: Bufonidae) and its geographic variation. Bull. Mus. Comp. Zool. 134(4):107–138.

———. 1965. Especiación en trés *Bufo* neotropicales (Amphibia, Anura). Papeis Avulsos. Dep. Zool. (São Paulo) 17:57–75.

———. 1967. *Bufo gnustae* Sp. n. del grupo de *Bufo ockendeni* Boulenger, hallado en la Prov. de Jujuy, Argentina. Neotropica 13(41):54–56.

———. 1967. Un nuevo nombre para *Bufo granulosus minor* Gallardo. Neotropica 13(41):56.

Harrington, H. J. 1962. Paleogeographic development of South America. Bull. Amer. Assn. Petrol. Geol. 46: 1773–1814.

Heredia, F. 1968. Estudios preliminares sobre el sapo gigante *Bufo blombergi*. Bol. Dep. Biol. Univ. del Valle (Colombia) 1(1):31–36.

Lutz, A. 1934. Notas sobre espécies brasileiras do genero *Bufo*. Mem. Inst. Oswaldo Cruz 28(1):111–133.

Myers, G. S., and J. W. Funkhouser. 1951 A new giant toad from southwestern Colombia. Zoologica 36:279–282.

———, and A. Leitao de Carvalho. 1952. A new dwarf toad from southeastern Brasil. Zoologica 37:1–3.

Nygren, W. E. 1950. Bolivar geosyncline of northwestern South America. Bull. Amer. Assn. Petrol. Geol. 34:1998–2006.

Rivero, J. A. 1961. Salientia of Venezuela. Bull. Mus. Comp. Zool. 126(1):5–207.

———. 1963. The distribution of Venezuelan frogs. 3. The Sierra de Perija and the Falcón region. Carib. J. Sci. 3:197–199.

———. 1964. The distribution of Venezuelan frogs. 4. The Coastal Range. Carib. J. Sci. 4:307–319.

———. 1964. The distribution of Venezuelan frogs. 5. The Venezuelan Guayana. Carib. J. Sci. 4:411–420.

———. 1964. The distribution of Venezuelan frogs. 6. The Llanos and the Delta Region. Carib. J. Sci. 4: 491–495.

Silva, F., A. Veloso, J. Solervicens, and J. C. Ortiz. 1968. Investigaciones Zoologicas en el Parque Nacional Vicente Perez Rosales y Zona de Pargua. Mus. Nac. Hist. Nat. (Santiago) 148:1–12.

Spix, J. B. von. 1825. Animalia nova, sive species novae Testudinum et Ranarum, quas in itinere Brasiliam, annis 1817–1820, collegit et descripsit. Munich.

Tihen, J. A. 1962. Osteological observations on New World *Bufo*. Amer. Midl. Nat. 67:157–183.

———. 1962. A review of New World fossil bufonids. Amer. Midl. Nat. 68:1–50.

Vellard, J. 1955. Répartition des Batraciens dans les Andes au Sud de l'Équateur. Trav. de l'Inst. Français d'Études Andines 5:141–161.

———. 1959. Estudios sobre Batracios Andinos. 5. El genero *Bufo*. Mem. Mus. Hist. Nat. Xavier Prado 8:1–48.

Wiegmann, A. F. A. 1835. Beiträge zur Zoologie, gesammelt auf einer Reise um die Erde, von Dr. F. J. F. Meyen. 7. Abhandl. Amphibien. Nova Acta Acad. Leop. Carol. 17:183–268.

7. *Bufo* of North and Central America

W. FRANK BLAIR

The University of Texas at Austin
Austin, Texas

Introduction

The greatest diversity of the genus *Bufo* is found in the Middle American tropics. The small country of Costa Rica, for example, has seven species: *B. coniferus*, *B. melanochloris*, *B. marinus*, *B. haematiticus*, *B. holdridgei*, *B. periglenes*, and *B. coccifer*. An eighth species, *epioticus* is regarded by Savage and Kluge (1961) as belonging to Cope's (1875) genus *Crepidius*, and three other named forms that are poorly known or of dubious validity also occur there. Mexico is rich in species, but the number decreases rapidly with increasing latitude, and only one species, *B. boreas*, reaches as far north as coastal Alaska. The number of species also decreases from west to east in North America, and only one or two species can be found at any particular locality in the eastern half of the United States and only one in eastern Canada.

Among the Central and North American *Bufo* there are seven species groups, each with two to eight species that presumably have resulted from late Tertiary or Pleistocene speciation. In addition, there are three species that apparently have no close relatives and that consequently seem to represent the results of early evolutionary dichotomies. There are seven species that represent the penetration into Central America or southern North America by groups that are distributed primarily in South America. The movement of these latter species into Central and North America presumably occurred in late Pleistocene or even post-Pleistocene.

The Thin-Skulled Line

In western North America there are toads that form a segment of a major evolutionary line that can be traced from the southern Andes through the western cordillera of North America to coastal Alaska and across Eurasia to the Middle East. These toads tend toward diurnalism, probably because they inhabit high and consequently cold montane habitats. There is also a tendency toward sexual dimorphism in color pattern, which may be related, in turn, to the diurnal habits. The voice is usually lost in the New World species. *B. bocourti* Brocchi (1881–1883) of the Guatemalan highlands is separated by a wide gap from the related *spinulosus* group of South America. This species not only has lost the mating call, but, in addition, has lost the columella from the middle ear (Chap. 15), hence is probably incapable of hearing airborne sounds. There is conspicuous sexual dimorphism in size and color pattern, with the males being smaller and more uniformly colored than the females. Another large gap exists between the northern limits of this species in Chiapas, Mexico, and the southern limits of the *boreas* group in the mountains of Baja California. *B. tacanensis* Smith (1952), a poorly known species from Volcan Tacana, Chiapas, is possibly related to *B. bocourti*.

B. granulosus Spix (1824) is actually a complex of several South American species that remains to be clarified. Gallardo (1965) has treated this as a single species, using only morphological evidence and a typological concept. The northern limit of these small toads is Panama.

The *boreas* group is distributed along the parallel coastal mountains of the western United States and Canada and in the Rocky Mountains. The species *B. boreas* Baird and Girard (1852*b*) ranges northward to coastal Alaska. The high altitude relict, *B. canorus* Camp (1916), is largely restricted to elevations between 8500 and 10,000 feet (Karlstrom 1962) in the central Sierra Nevada of California. This toad retains a strong mating call that is very similar to that of the related Eurasion *B. viridis*. The color pattern is highly dimorphic, with the small males showing a plain yellowish dorsum and the larger females showing a reticulated dorsum. *B. boreas*, which ranges very widely in western North America, has lost the mating call but it shows only slight sexual dimorphism in color pattern. The distribution of *B. boreas* and derivative species reflects the effect of Pleistocene climatic change. A population of small, dark-colored toads that is restricted to the vicinity of permanent water in Deep Springs Valley, Inyo County, California, has been given the name *B. exsul* Myers (1942). This population presumably became isolated from *B. boreas* with the late and post-Wisconsin drying of the Southwest. Another isolate, which has been given the name *B. nelsoni* Stejneger (1893), is found in Oasis Valley, Nye County, Nevada.

Four species or species groups of close affinity with the *boreas* group occur in western North America. *B. punctatus* Baird and Girard (1852*a*) is a desert-adapted species of moderately small size that ranges over the Mexican Plateau northward to southern Utah and Colorado and from the deserts of California eastward across the Edwards Plateau of Texas. The color pattern varies geographically, with the red-spotted pattern that gives the species its name becoming plain and pale gray on the limestones of the Edwards Plateau. This species is an opportunistic breeder, with a strong mating call, that breeds after rainstorms except in the Great Basin, where, like other anurans that are able to exist there, it breeds at permanent water irrespective of rainfall. This species lays single, adhesive eggs as a probable adaptation for stream breeding, for in many instances the eggs are laid in creeks (Tevis 1966; author's data). The only other toads that we know lay single eggs are the closely related *B. marmoreus* and the toads of the *B. debilis* group.

B. alvarius Girard (1859) is a large, aquatic species that is restricted to the Sonoran Desert. The green skin is smooth, but there are large skin glands on the hind legs comparable to the parotoid glands in their secretions. The mating call has been lost or remains vestigially in some individuals (Blair and Pettus 1954; Inger 1958). A loud male release call is retained. This species may breed independently of rainfall, with the males actively chasing females or any other moving object in the permanent pools in which they breed. This species occurs commonly in fields that are irrigated for agriculture.

The *debilis* group is a complex of three good biological species (Bogert 1962), *B. debilis* Girard (1854), *B. kelloggi* Taylor (1938), and *B. retiformis* Sanders and Smith (1951), which mainly occupy the Sonoran and Chihuahuan deserts, but one, *B. debilis*, extends eastward to central Texas and central Oklahoma. These are small toads, with *B. retiformis* being the largest. *B. kelloggi* is brown, the other two are green. Breeding is opportunistic with rainfall. The eggs of *B. debilis* and *B. kelloggi* are laid in very short strings or singly; those of *B. retiformis* are unknown to me. The embryonic development is peculiar but has not been described in detail.

The *cognatus* group consists of three species that have desert or grassland adaptations. *B. compactilis* Wiegmann (1833) occurs on the lower Mexican Plateau. Until the mating calls were recorded and recognized as taxonomic characters of great significance (Bogert 1960), it was confused with the morphologically similar *B. speciosus* Girard (1854) of the northern Mexican Plateau and southwestern United States. The calls of the two are enormously different in duration. The latter species ranges from northwestern Oklahoma and southern New Mexico down across the eastern part of the Mexican Plateau and the Gulf coastal plain of Mexico. It shows preference for sandy soils. *B. cognatus* Say (1823), with a mating call very similar to that of *B. compactilis*, is markedly set off by its green reticulated color pattern. *B. cognatus* ranges over the Mexican Plateau into the deserts of Arizona and New Mexico and northward through the central grasslands to western Minnesota and southern Canada. All three members of this group have highly developed mating calls and all are opportunistic breeders with rainfall. There is moderate overlap in the ranges of *B. cognatus* and *B. speciosus*, and natural hybrids are known. Natural hybrids between *B. woodhousei* and *B. speciosus* are also known (Ballinger 1966).

The *marmoreus* group is composed of a pair of species with fairly close relationship to the thin-skulled line of toads. *B. marmoreus* Wiegmann (1833) ranges from Sinaloa to Chiapas on the Pacific slope and from Veracruz to the Isthmus of Tehuantepec on the Atlantic slope of Mexico. *B. perplexus* Taylor (1943) has a restricted range on the Pacific slope of Mexico from Guerrero to Chiapas (Smith and Taylor 1948). Porter and Porter (1967) have suggested that the two are so similar biochemically in their skin secretions that they may be conspecific. However, this suggestion ignores that there are important biological differences setting these populations apart. The mating call is different. *B. marmoreus* shows striking sexual dimorphism in color pattern, with the males being plain yellowish, while the females are dorsally patterned. *B. perplexus* shows essentially no dimorphism. Furthermore, *B. marmoreus* lays single adhesive eggs, while *B. perplexus* lays strands of eggs, as in most species of *Bufo*. From this evidence it seems unquestionable that these are two species; the similarity in one kind of character, namely skin biochemistry, is easily attributable to parallel evolution of the skin compounds.

The *coccifer* group includes geographically variable toads that range through tropical scrub of the Pacific drainages of Central America from Costa Rica to southern Mexico. The size is fairly small, and the thick skin is presumably adaptive to the xeric environment. These toads do not seem closely related to any other living species. A disjunct population in southern Mexico has been named *B. cycladen* by Lynch and Smith (1966), with the name *B. coccifer* Cope (1866) retained by all the morphologically variable Central American populations. Porter (1967*a*), who (1965) pointed out the difference in mating call, which provides the only real justification for considering the Mexican populations a separate species, regards this naming as premature. He has listed locality records for the Mexican form (1964).

The Broad-Skulled Line

The *valliceps* group is the largest species group of *Bufo* in Central and North America with eight nominal allopatric species, most of which are well differentiated, but some taxonomic problems remain unsolved. The most widely distributed species is *B. valliceps* Wiegman (1833), which occupies the Gulf of Mexico lowlands from Louisiana to Yucatán and Honduras and crosses the Isthmus of Tehuantepec to the Pacific coast (Porter 1964). Although this ecologically tolerant species ranges from the coastal deciduous forest in Louisiana to the Central American tropics, much of its distribution coincides with that of lowland, tropical scrub. There is minor morphological variation from end to end of the range. The throat color of males changes from lemon yellow in the north to orange in the south. Baylor and Stuart (1961) have described a slightly differentiated population from the Grijalva Valley of Guatemala as *B. v. wilsoni.* Other populations could also be described if there were merit in giving subspecies names to them. One interesting, but confusing, phenomenon of southern populations is the tendency for hypertrophy of the parietal crests (Porter 1964; Blair 1966) at higher elevations. Firschein and Smith (1957) have described a supposedly high-crested form from Oaxaca as *B. v. macrocristatus*, but Duellman (1960) was unable to recognize it and Porter (1964) regards this form as unrecognizable. Natural hybrids between *B. valliceps* and *B. woodhousei* have been reported from various places in the overlap zone in the United States (Thornton 1955; Volpe 1956).

Two montane populations in which the parietal crests are hypertrophied have been named. One of these is *B. cristatus* Wiegmann (1833) from Jalapa, Veracruz, Mexico, which has rarely been collected since its description. The other is *B. cavifrons* Firschein (1950) from Volcan San Martin, Veracruz. Porter (1964) has discussed the relations of these high-crested species and has referred specimens from Chiapas and Oaxaca to *B. cavifrons* and specimens from Tezuitlan, Puebla, to *B. cristatus.* A live immature toad collected by Ron Altig near Tezuitlan showed similarities to the *guttatus* group but became more *valliceps*-group-like as an adult and is probably the toad to which the name *cristatus* applies. *B. cavifrons,* as known from Volcan San Martin, seems to be a good species. In addition to the hypertrophied crests, the color is more yellowish than in *B. valliceps*, the body and legs are more elongated, and the behavior is more active and froglike. Porter (1967*b*) has reported *B. cavifrons* from dense rainforest in Nicaragua.

B. ibarrai Stuart (1954) has a restricted distribution at intermediate elevations in Guatemala. Morphological characters have been interpreted as indicating relationship of this species to *B. coccifer* (Stuart 1954) or to *B. v. wilsoni* (Baylor and Stuart 1961). Porter and Porter (1967) have claimed that mating call, morphology, and skin venoms show *B. ibarri* to be more closely related to *B. coccifer* than to *B. valliceps*, but their claims are unconvincing because of the positive genetic compatibility tests with *B. valliceps*, and in my own opinion *B. ibarrai* shows more general resemblance to *B. valliceps* than to *B. coccifer*. Furthermore, several of their groupings of species from venom data alone are ones that are strongly inconsistent with other kinds of data.

B. mazatlanensis Taylor (1939) has an elongate distribution in the Pacific lowlands of Mexico from upper Colima to the middle Sonoran coast. It occupies xeric environments and, as with other members of the group, it is an opportunistic breeder with rainfall. The color is generally grayer than in *B. valliceps*; it is quite comparable in size to that species. The Mexican Plateau and the cordilleras separate this species from *B. valliceps*, and it is likely that this separation into eastern and western populations was the earliest speciational event in the *valliceps* group (Blair 1966).

B. gemmifer Taylor (1939) is a disjunct population of limited distribution in xeric environments near Acapulco, Mexico. It is apparently separated from the similar *B. mazatlanensis* by the Balsas Depression. The dorsal color is fairly uniform; the parietal crests are reduced by comparison with those of *B. mazatlanensis*. It is presently uncertain whether this is a species distinct from *B. mazatlanensis*.

B. luetkeni Boulenger (1891) is a Central American species that occurs at low and moderate elevations on the Pacific slope from Costa Rica to southeastern Guatemala (Stuart 1963). There is conspicuous sexual dichromatism that is much greater than in any other member of the *valliceps* group except *B. coniferus*. The males are usually yellow green and unpatterned, while the females are dorsally patterned. The body size is larger than in southern *B. valliceps* but it is equalled in northern ones.

B. coniferus Cope (1861) ranges from Costa Rica southward into Colombia and Ecuador (Taylor 1952). There is more polymorphism than in any other member of the *valliceps* group, with males ranging from yellow to green or gray. There is striking sexual dichromatism, with the males tending to be plain in color, the females patterned, as in *B. luetkeni*.

B. melanochloris Cope (1877) is found on the Pacific slope of Costa Rica. This is a medium-sized, poorly known species. The dorsal color pattern and general appearance are much as in *B. valliceps*, but the throat is gray as in *B. haematiticus* rather than yellow orange as in *B. valliceps*.

One Mexican and one Central American species are quite close to the *valliceps* group but show more divergence from it than do any of its members from one another. *B. canaliferus* Cope (1877) is a small, highly polychromatic species in which the dorsal colors range from lemon yellow to grays and browns and from plain to highly patterned. Its range extends along the Pacific lowlands from west of the Isthmus of Tehuantepec southeast to El Salvador. This is a lowland species but it ranges up to 400 meters (Porter 1964). *B. occidentalis* Camerano (1878) has departed ecologically from the basic *valliceps*-group pattern, for it is an upland species that ranges over the Mexican Plateau, northward to Chihuahua. The color pattern shows little variation except for the red-versus-gray dichromatism, which is also found in some other North American species (e.g., *B. americanus*).

The *marinus* group of large tropic-adapted toads is represented in the United States only by *B. marinus* (Linnaeus) (1758), which ranges northward through eastern Mexico to the lower Rio Grande Valley of Texas. This is one of the largest species of *Bufo* in the world in body size, as represented by populations in the Amazon basin. The parotoid glands are very large, and their secretions are capable of killing a predator the size of a dog. This is a highly euryceous species that ranges through tropical environments from those of the selva to those of tropical and subtropical scrub. *B. marinus* has been introduced into southern Florida, where it is now firmly established.

B. typhonius (Linnaeus) (1758) is a widely distributed species of small toads of the South American rain forest that range northward into Panama. The color pattern is the "dead-leaf" pattern that characterizes various species living in the selva. Some South American populations show greatly hypertrophied supraorbital crests, but the Panamanian representatives are low crested.

Central American Forms

Three Costa Rican species are possibly members of the *guttatus* group of northern South America. *B. haematiticus* Cope (1861) is a small member of this group. The "dead-leaf" pattern is comparable to that of the Colombian *B. blombergi* and the African *B. superciliaris.* As in *B. blombergi,* the hypertrophied testes contain some compound, as yet unidentified, that prevents fertilization or normal development of the eggs in attempted crosses with other species of *Bufo. B. holdridgei* Taylor (1952) is an even smaller, and in this instance sexually dimorphic, species known from Volcan Barba in Costa Rica. The hypertrophied testes have an effect similar to that of the testes of *B. haematiticus* in attempted crosses with other species. *B. periglenes* Savage (1966) is another small, montane species and is known only from Monte Verde, Costa Rica. Except for the color pattern, there is superficial resemblance to *B. holdridgei,* but the testes are not hypertrophied and do not inhibit fertilization. This species shows great sexual dichromatism, which is possibly associated with their diurnal habits and sexual selection in the high, cold cloud forest. The males are bright orange in color; the females are considerably larger than the males and are blue with yellow spots. The relations of this species are obscure, but the resemblance to members of the *guttatus* group is greater than to any other group, although the chemical specialization in the testes, characteristic of that group, is lacking here.

Except for those mentioned above, small montane isolates in Central America have not been investigated adequately to provide any useful estimate of their relations to the major radiation of the genus or to one another. These include such forms as *Crepidius epioticus* Cope of Savage and Klug (1961), which may be a *Bufo.* They also include such species listed by Taylor (1952) as *B. coerulescens, B. gabbi,* and *B. fastidiosus.* Taylor (1952) considered the last species to be close to *B. bocourti* in morphology.

The Americanus Group

The *americanus* group, with six species, is the most widely distributed group of *Bufo* in North America. Members of this group range from northern Chihuahua northward to James Bay in Canada, and eastward from the Imperial Valley of California and the Columbia River valley in Washington and Oregon to the Atlantic coast from Florida northward to southern Quebec. This group does not show very close affinity with any other group, but it is probably closer to the *boreas* group than to any other North American group. Tihen (1962) has pointed out the osteological similarity to *B. bufo* of Eurasia.

Relations within the *americanus* group have received the attention of many workers (A. P. Blair 1941, 1942, 1955; Stebbins 1951; W. F. Blair 1957*a*, 1957*b*, 1962, 1963 in particular). The most widely ranging species, *B. woodhousei* Girard (1854), has a range from the Imperial Valley of California and the Columbia River valley of Washington and Oregon eastward to the Atlantic coast and northward to southern New Hampshire. This species occurs sympatrically, in part at least, with four of the five other species of the *americanus* group. Only the northern *B. hemiophrys* fails to contact this species. Natural hybrids between this species and all four of the sympatric species of the *americanus* group have been reported (A. P. Blair 1941, 1942, 1955; W. F. Blair 1957*a*; Brown 1967; Cory and Manion 1955; Volpe 1952, 1959). However, there is evidence (W. F. Blair 1962) that this hybridization results in reinforcing the isolating mechanism of call difference, at least in *B. americanus* and *B. terrestris.*

B. woodhousei is a complex species that has a zone of secondary interbreeding – along the eastern forest-grassland ecotone – between a small eastern form and a large prairie form (A. P. Blair 1941, 1942; Meacham 1962). Body size in the prairie populations is the largest reached by any member of the *americanus* group; in the Great Basin the body size is small, and there is a distinctive boss formed by filling in the area between the supraorbital crests. Breeding is opportunistic with rainfall through much of the range of the species, but in the Great Basin there is independence of rainfall, with the breeding season presumably being timed by photoperiod or temperature. Both in its zone of sympatry with *B. microscaphus* in the west and with *B. americanus* in the east, *B. woodhousei* tends to breed in the larger rivers, whereas the sympatric member of the group usually breeds in the smaller tributary streams. In the east it also breeds in temporary rainpools. These partial ecological isolating mechanisms of difference in breeding site contribute to the isolating mechanism complex that is inadequate to prevent any sympatric hybridization.

The range of *B. americanus* Holbrook (1836) is exceeded only by *B. woodhousei* and extends from the Piedmont in Georgia and Mississippi northward to James Bay in Canada. It extends westward into the ecotone between the eastern deciduous forest and the central grasslands. The northernmost populations show bright coloration; populations at the southwestern limits of the range in Oklahoma show plain, dull coloration. In addition to breeding along small, headwater streams rather than along large rivers, *B. americanus* is partially isolated where sympatric with *B. woodhousei* by its habit of breeding at an earlier season and at colder temperatures.

A group of disjunct populations on the Gulf of Mexico coastal plain in Texas, from Bastrop to Houston, has been given the name *B. houstonensis* Sanders (1953). This is clearly a slightly differentiated group of relictual populations that reflect a former more southern extension of *B. americanus*. It is likely that these populations reached the Texas coastal plain in the Wisconsin stage of the Pleistocene and that their isolation from the main body of *B. americanus* occurred no more than ten thousand years ago. The populations are small, and hybridization occurs with both *B. woodhousei* and *B. valliceps* (Brown 1967); hybridization with the former could result in swamping of these small populations, since the hybrids are fertile, and hybridization with the latter could lead to extinction, since hybrids of one reciprocal are sterile and of the other, inviable. Destruction of the habitat by man is also contributing to the probable forthcoming extinction of these populations.

B. terrestris (Bonnaterre) (1789) replaces *B. americanus* on the coastal plain of the southeastern United States from the Pearl River in Louisiana to Florida and northward to southeastern Virginia. The most distinctive feature is the enlarged and somewhat swollen pair of parietal crests, especially in large females. The relations with *B. americanus*, where the ranges of the two approach one another, are unclear. There seems to be a temporal and altitudinal separation of the two breeding populations, and in some places there appears to be a gap in which neither occurs. Netting and Goin (1946) proposed, solely on morphological evidence, that the two are conspecific, but this seems doubtful on present evidence (W. F. Blair 1963).

B. hemiophrys Cope (1886) is a fairly small-sized member of the *americanus* group that inhabits the prairies of the north central United States and Canada. The range extends from western Minnesota and eastern North Dakota northward to northern Alberta. There is an essentially undifferentiated population on the Laramie Plateau of Wyoming, which, like *B. houstonensis*, is presumably a post-Pleistocene isolate. However, the absence of noticeable differentiation and the far northern locality suggest a shorter period of isolation than for that species. The most conspicuous morphological character of *B. hemiophrys* is the "boss" of keratin that fills the space between the supraorbital crests and in some specimens arches up beyond their plane. The range of *B. hemiophrys* in western Minnesota seems to be largely separated from that of *B. americanus* by an ecotonal strip in which neither occurs. However, Henrich (1968) has reported presumed natural hybrids between the two species in eastern North Dakota.

B. microscaphus Cope (1867) is the westernmost species of the *americanus* group with the exception of the wide-ranging *B. woodhousei*. Isolated populations of *B. microscaphus* are found from southwestern Utah and southeastern Nevada southward into western Chihuahua and eastward into western New Mexico. There are also isolated populations in coastal southern California and in northern Baja California, Mexico. The disjunct nature of the populations of *B. microscaphus* provides a clear example of the effects of environmental drying following the last Pleistocene "pluvial" in the Southwest. Only these isolated populations remain from what must have been a continuously distributed population. Distribution is limited to the places where there is permanent water, either in streams or in irrigated lands. Some of the latter have been under irrigation since prehistoric times, which may have been a factor in the survival of the disjunct populations (A. P. Blair 1955). Breeding occurs in permanent water. It is independent of rainfall in this xeric region and is presumably stimulated by photoperiod or temperature, or both. The general pattern of calling by the males is comparable to that in *B. americanus*, with the individuals scattered along the small streams rather than in large choruses. Larger groupings do occur in the artificial impoundments that exist on many of these streams. The harsh desert in which these toads live is severely limiting on adults, and they simply do not move away from the permanently watered areas.

Bufo quercicus

B. quercicus Holbrook (1840) is a tiny species and is the smallest of all known species of *Bufo*. The adult length rarely exceeds 30 mm. The distribution is limited to the Gulf of Mexico and Atlantic coastal plains from the Pearl River in Louisiana to North

Carolina. This miniature species is externally similar to toads of the *americanus* group and seems more closely related to it than to any other group. The reticulate patterning on the testes is duplicated in *Melanophryniscus* of South America, which is a *Bufo* or very near to the genus. The throat is black, as in *B. woodhousei*, and the color pattern is reminiscent of the pattern in that species. The eggs are laid in short bars of two to six, rather than in long strings, as is typical of most *Bufo*. *B. quercicus* is an opportunistic breeder that typically breeds after heavy rains.

Biogeographical Relations

The pattern of distribution of the species of *Bufo* in Central and North America is fairly simple. The great diversity of species in Central America is explainable on two grounds, one of which is the close proximity to South America, and thus the land and ecological continuity with northern South America, which permits the northward extension of primarily South American groups. The other reason for this diversity is the existence of great ecological diversity provided by mountains, lowland tropical forest, and xeric and subxeric tropical scrub. Among the montane forms, *B. bocourti* seems close to the thin-skulled line, which can be traced from the southern Andes through North America and Eurasia and which, with the exception of this species, has a huge gap in its distribution for the whole distance from Ecuador to northern Baja California in Mexico. *B. holdridgei* and possibly *B. periglenes* seem close to the *guttatus* group of northern South America. *B. tacanensis*, a volcanic-mountain form of the Mexican-Guatemalan border is possibly a relative of *B. bocourti*; another volcanic inhabitant, *B. cavifrons*, is a *valliceps*-group toad. The rain forest toads of Central America, with the exception of members of the *valliceps* group, belong to groups that have their main distributions in South America. These include *B. haematiticus* of the *guttatus group*, *B. typhonius*, and *B. marinus*. In the tropical scrub environments of Central America and southern Mexico, the species are mostly of the *valliceps* group, or a derivative, *B. canaliferus*, or apparently old branches from the thin-skulled line, *B. marmoreus* and *B. perplexus*. *B. coccifer* appears to have no close relative. *B. granulosus*, which reaches only to Panama, is representative of another South Amercan group.

The main feature affecting *Bufo* distribution in Mexico is the very extensive Mexican Plateau, which separates the coastal lowlands on each side of the country. Members of the *valliceps* group and *B. marinus* extend northward through these lowlands. The plateau itself is faunistically continuous with the deserts of the southwestern United States. The characteristic *Bufo* of this large, xeric area are derivatives from the thin-skulled line, which have probably been separate for a long time. These include the *cognatus* group, the *debilis* group, *B. punctatus*, and *B. alvarius*. *B. woodhousei* of the *americanus* group is the only other *Bufo* adapted to this xeric environment. Disjunct populations of the *boreas* group and of *B. microscaphus* are relicts and exist in the desert region only where local conditions permit.

The mountains of the West belong to the *boreas* group, which ranges from northern Baja California to coastal Alaska. The one species of *Bufo* that is endemic to the central grasslands is the far northern *B. hemiophrys* of the *americanus* group, but the most widely distributed species in the grasslands is the wide-ranging *B. woodhousei* of the same group. A third species, *B. cognatus*, ranges far northward through the grasslands from the deserts of the Southwest. The eastern deciduous forest, which occupies more than one-third the total area of the United States, has only members of the *americanus* group and, on the coastal plain, *B. quercicus*.

REFERENCES

Baird, S. F., and C. Girard. 1852*a*. Characteristics of some new reptiles in the museum of the Smithsonian Institution. Proc. Acad. Nat. Sci. (Philadelphia) 6: 173.

———. 1852*b*. Description of new species of reptiles, collected by the U. S. Exploring Expedition under the command of Capt. Charles Wilkes, U.S.N. Proc. Acad. Nat. Sci. (Philadelphia) 1852:174–177.

Ballinger, R. E. 1966. Natural hybridization of the toads *Bufo woodhousei* and *Bufo speciosus*. Copeia 1966: 366–368.

Baylor, E. R., and L. C. Stuart. 1961. A new race of *Bufo valliceps* from Guatemala. Proc. Biol. Soc. (Washington) 74:195–202.

Blair, A. P. 1941. Variation, isolation mechanisms, and hybridization in certain toads. Genetics 26:398–417.

———. 1942. Isolating mechanisms in a complex of four species of toads. Biol. Symposia 6:235–249.

———. 1955. Distribution, variation, and hybridization in a relict toad (*Bufo microscaphus*) in southwestern Utah. Amer. Mus. Novitates no. 1722, pp. 1–38.

Blair, W. F. 1957*a*. Mating call and relationships of *Bufo hemiophrys* Cope. Texas J. Sci. 9:99–108.

———. 1957*b*. Structure of the call and relationships of *Bufo microscaphus* Cope. Copeia 1957:208–212.

———. 1962. Non-morphological data in anuran classification. Syst. Zool. 11:72–84.

———. 1963. Intragroup genetic compatibility in the *Bufo americanus* species group of toads. Texas J. Sci. 15:15–34.

———. 1966. Genetic compatibility in the *Bufo valliceps* and closely related groups of toads. Texas J. Sci. 18: 333–351.

———, and D. Pettus. 1954. The mating call and its significance in the Colorado River toad (*Bufo alvarius* Girard). Texas J. Sci. 6:72–77.

Bogert, C. M. 1960. The influence of sound on the behavior of amphibians and reptiles, p. 137–320. *In* W. W. Lanyon and W. N. Tavolga (eds.), Animal sounds and communication. A.I.B.S. Publ. 7.

———. 1962. Isolation mechanisms in toads of the *Bufo debilis* group in Arizona and western Mexico. Amer. Mus. Novitates, no. 2100, pp. 1–37.

Bonnaterre. 1789. *B. terrestris.* Tabl. Encycl. Meth. Erp.

Boulenger, G. A. 1891. Notes on American batrachians. Ann. Mag. Brocchi Nat. Hist. Ser. 6, 8:453–458.

Brocchi, P. 1881–1883. Etudes des batraciens de L'Amérique Centrale. *In* Mission Scientifique au Méxique et dans l'Amérique Centrale. Pt. 3. 2:1–222.

Brown, L. C. 1967. The significance of natural hybridization in certain aspects of the speciation of some North American toads (genus *Bufo*). Unpubl. Ph.D. Thesis, Univ. Texas.

Camerano, L. 1878. Di alcune specie dí anfibii anuri eistenti nelle collezioni del R. Museo Zoologico di Torino. Attri. R. Accad. Torino (Turin) 14:866–897.

Camp, C. L. 1916. Description of *Bufo canorus,* a new toad from the Yosemite National Park. Univ. Calif. Publ. Zool. 17(3):11–14.

Cope, E. D. 1861. On some new and little known American Anura. Proc. Acad. Nat. Sci. (Philadelphia) 1861:151–159.

———. 1866. Fourth contribution to the herpetology of tropical America. Proc. Acad. Nat. Sci. (Philadelphia) 1866:123–232.

———. 1866. On the reptilia and batrachia of the Sonoran Province of the Nearctic Region. Proc. Acad. Nat. Sci. (Philadelphia) 1866:300–314.

———. 1875. On the batrachia and reptilia of Costa Rica. J. Acad. Nat. Sci. Philadelphia. Ser. 2, 8(4): 93–154.

———. 1877. Tenth contribution to the herpetology of tropical America. Proc. Amer. Philos. Soc. 17:85–98.

———. 1886. Synonymic list of North American species of *Bufo* and *Rana,* with descriptions of some new species of Batrachia, from specimens in the National Museum. Proc. Amer. Philos. Soc. 23:514–526.

Cory, L., and J. J. Manion. 1955. Ecology and hybridization in the genus *Bufo* in the Michigan-Indiana region. Evolution 9:42–51.

Duellman, W. E. 1960. A distributional study of the amphibians of the Isthmus of Tehuantepec, Mexico. Univ. Kan. Publ. Zool. (Mus. Nat. Hist.) 13(2):19–72.

Firschein, I. L. 1950. A new toad from Mexico with a redefinition of the *cristatus* group. Copeia 1950:81–87.

———, and H. M. Smith. 1957. A high-crested race of toad (*Bufo valliceps*) and other noteworthy reptiles and amphibians from southern Mexico. Herpetologica 13:219–222.

Gallardo, J. M. 1965. The species *Bufo granulosus* Spix (Salientia: Bufonidae) and its geographic variation. Bull. Mus. Comp. Zool. 134(4):107–138.

Girard, C. 1854. A list of North American bufonids, with diagnoses of new species. Proc. Acad. Nat. Sci. (Philadelphia) 7:86–88.

———. 1859. *In* Reptiles of the Boundary. S. F. Baird (ed.) United States and Mexico Boundary Survey. 2(2):1–35.

Henrich, T. W. 1968. Morphological evidence of secondary intergradation between *Bufo hemiophrys* Cope and *Bufo americanus* Holbrook in eastern South Dakota. Herpetologica 24:1–13.

Holbrook, J. E. 1836. North American Herpetology. Philadelphia.

———. 1840. North American Herpetology. 4. Philadelphia.

Inger, R. F. 1958. The vocal sac of the Colorado River toad (*Bufo alvarius* Girard). Texas J. Sci. 10:319–324.

Karlstrom, E. L. 1962. The toad genus *Bufo* in the Sierra Nevada of California. Ecological and systematic relationships. Univ. Calif. Publ. Zool. 62:1–104.

Linnaeus, C. 1758. *Systema Naturae.* 10th ed. Stockholm.

Lynch, J. D., and H. M. Smith. 1966. A new toad from western Mexico. Southwestern Naturalist 11:19–23.

Meacham, W. R. 1962. Factors affecting secondary intergradation between two allopatric populations in the *Bufo woodhousei* complex. Amer. Midl. Nat. 67:282–304.

Myers, G. S. 1942. The black toad of Deep Springs Valley, Inyo County, California. Occas. Papers Mus. Zool. (Univ. Mich.) 460:1–19.

Netting, M. G., and C. J. Goin. 1946. The correct names

of some toads from eastern United States. Copeia 1946:107.

Porter, K. R. 1964. Distribution and taxonomic status of seven species of Mexican Bufo. Herpetologica 19: 229–247.

———. 1965. Intraspecific variation in mating call of *Bufo coccifer* Cope. Amer. Midl. Nat. 74:350–356.

———. 1967*a*. *Bufo cycladen* (Bufonidae): A case of *nomen dubium*. Southwestern Naturalist 12:200–201.

———. 1967*b*. *Bufo cavifrons* Firschein collected in Nicaragua. Herpetologica 23:66.

———, and W. F. Porter. 1967. Venom comparisons and relationships of twenty species of New World toads (genus *Bufo*). Copeia 1967:298–307.

Sanders, O. 1953. A new species of toad with a discussion of morphology of the bufonid skull. Herpetologica 9:25–47.

———, and H. M. Smith. 1951. Geographic variation in toads of the *debilis* group of *Bufo*. Field and Lab. 19:141–160.

Savage, J. M. 1966. An extraordinary new toad (*Bufo*) from Costa Rica. Rev. Biol. Trop. 14:153–167.

———, and A. G. Kluge. 1961. Rediscovery of the strange Costa Rican toad, *Crepidius epioticus* Cope. Rev. Biol. Trop. 9:39–51.

Say, T. 1823. *In* Stephen F. Long, Account of an expedition from Pittsburgh to the Rocky Mountains performed in the years 1819 and 1820 by order of the Hon. J. C. Calhoun, Sec'y. of War; under the command of Major Stephen H. Long from the notes of Major Long, Mj. T. Say, and other gentlemen of the exploring party. Compiled by Edwin James, botanist and geologist for the expedition. Philadelphia, 2 vols. (+ atlas).

Smith, H. M., and E. H. Taylor. 1948. An annotated checklist and key to the Amphibia of Mexico. U.S. Nat. Mus. Bull. 192:1–118.

Smith, P. W. 1952. A new toad from the highlands of Guatemala and Chiapas. Copeia 1953:175–177.

Spix, J. B. 1824. Animalia nova sive species novae testudinum et ranarum, quas in itinere per Brasiliam, annis 1817–1820 jussu et auspiciis Maximiliani Josephi I. Bavarie Regis, suspecto collegit et discripsit J. D. Spix. Munich.

Stebbins, R. C. 1951. Amphibians of western North America. Univ. Calif. Press, Berkeley.

Stejneger, L. 1893. Annotated list of the reptiles and batrachians collected by the Death Valley expedition in 1891, with descriptions of new species. North Amer. Fauna 7:159–228.

Stuart, L. C.1954. Descriptions of some new amphibians and reptiles from Guatemala. Proc. Biol. Soc. (Washington) 67:159–178.

———. 1963. A checklist of the herpetofauna of Guatemala. Misc. Publ. Mus. Zool. (Univ. Mich.) 122:1–150.

Taylor, E. H. 1936 [1938]. Notes on the herpetological fauna of the Mexican state of Sinaloa. Univ. Kansas Sci. Bull. 24:505–530.

———. 1939. Herpetological miscellany No. 1. Univ. Kansas Sci. Bull. 26:489–571.

———. 1943. Herpetological novelties from Mexico. Univ. Kansas Sci. Bull. 29:343–360.

———. 1952. The frogs and toads of Costa Rica. Univ. Kansas Sci. Bull. 35:577–942.

Tevis, L., Jr. 1966. Unsuccessful breeding by desert toads (*Bufo punctatus*) at the limit of their ecological tolerance. Ecology 47:766–775.

Thornton, W. A. 1955. Interspecific hybridization in *Bufo woodhousei* and *Bufo valliceps*. Evolution 9: 455–468.

Tihen, J. A. 1962. Osteological observations of New World *Bufo*. Amer. Midl. Nat. 67:157–183.

Volpe, E. P. 1952. Physiological evidence for natural hybridization of *Bufo americanus* and *Bufo fowleri*. Evolution 6:393–406.

———. 1956. Experimental F_1 hybrids between *Bufo valliceps* and *Bufo fowleri*. Tulane Studies Zool. 4(2): 61–75.

———. 1959. Experimental and natural hybridization between *Bufo terrestris* and *Bufo fowleri*. Amer. Midl. Nat. 61:295–312.

Wiegmann, A. F. 1833. Herpetologische Beyträge. I. Ueber die Mexicanischen Kröten nebst Bemerkungen über ihnen vermandte Arten anderer Weltgegenden. Isis von Oken 7:651–662.

8. *Bufo* of Eurasia

ROBERT F. INGER

Field Museum of Natural History
Chicago, Illinois

Introduction

The area comprising Europe, Asia, the East Indies, and North Africa — the palearctic and Oriental zoogeographic regions — has at least forty-four species of *Bufo*. The relationships of some populations to one another make the exact number of species uncertain. Among such problematical groups are the West China populations related to *Bufo bufo*: *gargarizans, wrighti, tibetanus*, and *minshanicus*. The questions surrounding these populations in that topographically and climatically complex region are illustrated by comparing the differing treatments given them by Pope and Boring (1940) and Liu (1950). Similar difficulties are associated with *Bufo andersoni* Boulenger and *B. stomaticus* Lütken and one or two other groups of populations.

Some of these problems will not be resolved by examining a few preserved specimens in the comfort of a museum thousands of miles from the habitats of the animals. I have chosen, therefore, to ignore the problems of the species and subspecies levels. Instead, I will focus on the species group level, which can be studied profitably with preserved material. Subsequent field studies on the Chinese populations related to *Bufo bufo* will probably not affect the decisions at a higher categorical level. At most such studies would expand or contract the number of forms in a species group.

The species delimitations of the most recent regional works have been accepted. The results of overconfidence in establishing any particular set of species boundaries will be too many or too few species in a group. But, again, the pattern of relationships will not be affected.

In general, only the forms actually examined will be considered. In most cases the literature does not provide information on enough characters to permit reliable placement of a form in a species group.

The procedures followed were simple and conventional. The external morphology of sexually mature toads was examined with particular attention to those characters that have been cited frequently in the literature. Skeletons and limb musculature were also examined insofar as sample size permitted. The characters examined and their states are described in Appendix B. The species and number of specimens examined are listed in Appendix C.

Character Evaluation

The characters were weighted after the observations were made on all species. The weighting system is coarse and is in essence traditional. Characters showing excessive amounts of intraspecific variation are of little use in detecting relationships above the species level and were discarded. The effect is to give such characters no weight. Characters that vary only slightly or not at all at the interspecific level provide little information on relationships between species groups and consequently have little weight. Finally, there are characters that may show some intraspecific variation but that show moderate interspecific variation and correlation with variation in other characters. Characters of this sort are most likely to reveal the structure of interspecific relationships and are the kind taxonomists have traditionally weighted heavily.

Two weaknesses of this approach are obvious. First, the difference between the second and third types of characters is not sharp. There is usually a gradation of values within this range. Second, one cannot know in advance which characters will have great value in a new group.

To decide on the basis of these arguments that character weighting is unjustifiable and that all characters should be equated is overlooking three important criteria. First, objective, quantitative methods of evaluating characters in terms of their usefulness in taxonomy are being developed. Throckmorton (1968) has developed a measure of a character's capacity to "predict" other characters. Even without such methods, however, a taxonomist with much experience in working on a group learns which characters have a high probability of providing the kind of information required in phylogenetic taxonomy. Second, phenetic methods that utilize many unweighted characters produce dendrograms that can be duplicated by smaller subsets of characters; the remainder of the characters contribute noise rather than information (Voris 1969). Third, the analysis of character variation and correlation has important by-products, such as information on adaptation.

In the final analysis certain characters that had been recorded were deleted and, hence, were given no weight. These characters and the reasons for rejecting them follow. The numbering system is that used in Appendix B.

4. Shape of palatal fenestra. In a number of instances intraspecific variation exceeded interspecific variation.
6. Width/length of parasphenoid process. Rejected for the same reason as the preceding character.
14. Hyomandibular foramen. Almost all species examined have a notch in the lateral edge of the proötic. Intraspecific variation again exceeded interspecific.
15. Suprapterygoid fenestra. Rejected because it appears in all Eurasian species.
26. Extensor digitorum communis longus. Rejected because of excessive intraspecific variability. In terms of the toes on which distal slips insert, *B. viridis*, for example, has the following states: 2–3 (1 individual), 2–4 (3), 2–5 (4).
28. Tympanum. Only one species, *B. surdus*, lacks a visible tympanum. The character varies too little to reveal interspecific relationships.

This initial evaluation of characters was done before the formal, final analysis, since the weakness of the eliminated characters was obvious. The remaining characters were included in the analysis but they do not contribute equal amounts of information on phylogenetic relations, as will be shown below.

For analyzing character correlation and intragroup variation, character states were coded as follows:

2. Skull width/length

State:	1.05–1.20	1.21–1.35	1.36–1.50	1.51–1.65	1.66–1.80
Code:	0	1	2	3	4

3. Skull height/length

State:	0.351–0.450	0.451–0.550	0.551–0.650	0.651–0.750	0.751–0.850
Code:	0	1	2	3	4

5. Braincase depth/width

State:	0.451–0.550	0.551–0.600	0.551–0.650	0.601–0.700	0.651–0.750	0.701–0.800	0.751–0.850
Code:	0	1	2	3	4	5	6

7. Dorsal surface of skull

State:	smooth	smooth to pitted	rugose or rugose and pitted
Code:	0	1	2

8. Frontoparietals

State:	separated	not separated
Code:	0	1

9. Dorsal exposure of sphenethmoid

Descriptors: not exposed = a
exposed only in midline = b
exposed half width of snout = c
exposed full width of snout = d

State:	a	b	c	d
Code:	0	1	2	3

10. Occipital canal

State:	exposed	half-covered	completely covered
Code:	*0*	1	2

11. Dorsal otic plate of squamosal

State:	narrow	broad
Code:	0	*1*

12. Proötic-frontoparietal relationship

State:	separated	fused
Code:	0	1

13. Quadratojugal overlap of maxilla
Descriptors: not overlapping = a
overlapping in posterior half of pterygoid fenestra = b
overlapping in anterior half of fenestra = c
quadratojugal meeting pterygoid = d

State:	a	b	c	d
Code:	0	1	2	3

16. Pterygoid angulation
Descriptors: intrapterygoid angle > pterygoid-parasphenoid angle = a
intrapterygoid angle = pterygoid-parasphenoid angle = b
intrapterygoid angle < pterygoid-parasphenoid angle = c

State:	a	b	c
Code:	0	1	2

17. Transverse parasphenoid ridge
Descriptors: absent = a
shallow, obtuse = b
deep, obtuse = c
deep, sharp = d

State:	a or b	c	d
Code:	*0*	1	2

18. Palatine

State:	smooth	denticulate
Code:	*0*	1

19. Third transverse process/skull length

State:	< 0.900	0.901–1.060	1.061–1.220
Code:	0	1	2

20. Seventh transverse process/third

State:	0.576–0.725	0.726–0.875	0.876–1.025
Code:	0	*1*	2

21. Diapophysis width/skull length

State:	< 0.250	0.251–0.350	0.301–0.400	0.351–0.450
Code:	0	1	2	3

22. Third spinal nerve foramen

State:	< 11	11–15	11–20	16–25
Code:	0	1	2	3

23. Vertebral crests

State:	no crest	median crest	double crest
Code:	0	*1*	2

24. Humerodorsalis muscle, main slip to fourth finger

State:	absent	present
Code:	0	*1*

25. Supinator manus humeralis

State:	absent	present
Code:	0	*1*

27. Adductor longus

State:	absent	present
Code:	0	*1*

29. Cranial crests
Descriptors: none = a
supratympanic = b
orbitals = c
parietal = d

State:	a	b	c	bc	cd	bcd
Code:	0	1	2	2	4	5

30. Tibia gland

State:	absent	present
Code:	*0*	1

31. Tarsal ridge

State:	absent	present
Code:	0	1

32. Subarticular tubercles doubled

State:	no digits	3rd finger	4th toe	both
Code:	0	1	2	3

33–34. Vocal apparatus
Descriptors: no vocal sac = a
vocal sac present = b
gular pigment present = c

State:	a	b	bc
Code:	0	1	2

The distribution of these character states is given in table 8-1.

Attempts to assess cladistic relationships must be based on derived character states (Hennig 1966; Throckmorton 1968), which imply knowledge of direction of change. In the preceding tabulation of

Table 8–1. Character States of Mainly Eurasian Taxa

Taxa	Characters														
	2	3	5	7	8	9	10	11	12	13	16	17	18	19	20
biporcatus	1	3	0	2	1	1	2	1	1	2	2	2	0	2	1
divergens	1	2	2	2	1	1	2	1	1	1/2	0/2	1	0	1	2
philippinensis	1	2	1	2	1	1/2/3	2	1	1	2	0	2	1	1	1
parvus	0	2	2	2	1	2/3	2	1	1	1/2	2	2	0	1	1
quadriporcatus	1	2	5	2	1	2	2	1	1	1	0	2	0	2	2
claviger				2			2								
asper	1	2	5	2	1	0	2	1	0	1/2	0	0	0	0	2
juxtasper	1	2	5	2	1	2	2	1	0	1	0	0	0	1	2
celebensis							2								
macrotis							0							1	1
galeatus							2								
sumatranus															1
melanostictus	1	2	1	1	1	1/2	1	1	1	1/2	0/2	0	0/1	1	0
parietalis							2			1					
stuarti							0							0	1
himalayanus	2						1							0	1
sulphureus							2	1		2					
beddomi					1		0			1				0	0
latastei					0	3		0		0				0	0
abatus															
stomaticus	1	1	0	2	1	2	0	1	1	3	2	0	0	1	0
fergusoni	0	1	0	2	1	0/1	0	1	1	3	0	0	0	0	0
olivaceus							0			3					
dhufarensis							0							2	0
pentoni				2			2	1							
dodsoni	1	0		0	1	2	0	0	1	1	0	0	0	1	0
surdus	1				0		0	0		0				0	0
orientalis	1	1	5	0	1	1/3	1	0	0	1	0	0	0	0	1
viridis	2	1	3	1	0	3	0	0	1	0	2	0	0	0	0
calamita	3	3	4	0	0	3	0	0	1	0	2	0	1	1	0
raddei	1	2	4	0	0	3	0	0	1	0	2	0	0	0	0
mauritanicus	4	2	1	2	1	0	2	1	0	2	1	0	0	2	0
bufo	3	3	4	0	0/1	0/3	0	1	0	2	0	0	0	1	0
gargarizans	3	3	3	1	1	0/1/2	0	1	1	2	0	0	0	1	0
bankorensis	2	2	4	0	1	0	0	1	1	1/2	0	1	0	1	0
minshanicus	2	2		0	1		0	1			0	0		0	0
tibetanus	2	3	3	0	0/1	2/3	0	1	1	2		0	1	0	0
stejnegeri				0	1										
											Extraterritorial forms				
regularis	2	2	6	2	1	0	2	1	0	2	0	2	0	1	1
boreas	3	3	4	0	0	3	0	0	0	2	2	0	0	1	0
americanus	3	4	5	2	0	1/2	1	1	0	1	0/2	0	1	1	0
marinus	3	2	3	2	1	0	2	1	1	2	0	0	1	2	0
spinulosus	2	2	2	0	1	3	0	1	1	1	0	0	0	2	1

character states, the primitive state is italicized in the characters that later will be especially useful. For two of these characters (29, 31) direction of change cannot be determined on the basis of present information. The direction of change in the remainder of the significant characters is predicated on the following set of assumptions.

1. Primitive bufonids lacked teeth and had smooth palatines. These are almost universal characteristics within the family.

2. Primitive bufonids had moderately heavy skulls. This assumption avoids choosing an extreme, specialized condition as the primitive state. It follows that frontal fontanelles were lacking, the frontoparietals met the nasals, the quadratojugal overlapped the maxilla for moderate distances, the occipital canal was exposed, and the frontoparietal and proötic were not fused.

3. Primitive bufonids had moderate skeletal proportions. This assumption is also designed to avoid selecting a specialized ancestral form.

4. Primitive bufonids had a single median dorsal crest on the vertebral column formed by the neural spines. This is the common state in frogs of all families.

5. Primitive bufonids had vocal sacs and melanophores in the surrounding muscles and connective tissues. The first part of this assumption is reasonably well founded considering the distribution of vocal sacs in other families of anurans (see Liu 1935). The second part is less certain and depends on the occurrence of this gular pigment in species of *Bufo* from all parts of the generic range.

6. Primitive bufonids had two slips of the humerodorsalis muscle inserting on the fourth finger and a supinator manus humeralis. Both character states are common in the Ranidae, Leptodactylidae, and Pelobatidae (data from S. S. Liem).

7. Primitive bufonids had a distinguishable adductor longus. All the so-called higher families of frogs share this state.

In a subsequent section of this chapter species groups are designated and justified. One of the tests of species groups discussed later is within-group similarity in a number of characters. Consequently, one of the qualities necessary for a character to have weight in matters of interspecific relations is that it shows little within-group variation. Using the data of table 8-4, which lists the characters exhibiting little or no variation within groups, we can determine the number of species groups within which a character is relatively invariable (Table 8-2). Since there are only six species groups, the maximum frequency of within-group constancy is six. Three characters reach this level and eleven others are relatively constant within each of five groups. At the opposite extreme, four characters are invariable within fewer than three groups. Since the criterion for species groups is internal phenotypic homogeneity, the fourteen characters that are constant within five or six groups are the ones that contribute most to the process of recognizing species groups.

Table 8–2. Constancy of Characters of Eurasian *Bufo**

Character Number**	No. of Groups***	Character Number	No. of Groups
2	2	19	2
3	3	20	5
5	2	21	4
7	5	22	3
8	4	23	5
9	1	24	6
10	5	25	6
11	5	27	5
12	4	29	5
13	3	30	5
16	3	31	6
17	5	32	3
18	5	33–34	5

*Measured by lack of variation within species groups.

**Character identification numbers correspond to numbers in Appendix B.

***Number of species groups within which character is invariable. Maximum number is 6. See Table 8–4.

These fourteen characters span diverse portions of the morphology, as the following tabulation shows.

No.	*Character Name*
7	Rugosity of skull surface
10	Occipital canal
11	Dorsal otic plate of squamosal
17	Transverse parasphenoid ridge
18	Palatine denticulation
20	Ratio of seventh to third transverse process
23	Vertebral crests
24	Humerodorsalis
25	Supinator manus
27	Adductor longus
29	Cranial crests
30	Tibia gland
31	Tarsal ridge
33–34	Vocal apparatus

Table 8–3. Association between Pairs of Characters in Eurasian Species of *Bufo**

Char-acters	Characters													
	7	10	11	17	18	20	23	24	25	27	29	30	31	33–34
7		15.6	6.3	6.1	1.3	9.0	15.4	6.5	10.0	3.9	15.3	3.3	7.2	10.8
10	0.01		5.3	8.9	1.3	17.2	20.4	7.6	15.0	6.1	22.5	4.0	8.4	12.9
11	0.04	0.08		2.6	0	2.5	5.7	4.3	4.2	0.8	7.1	3.4	5.2	6.9
17	0.18	0.09	0.25		0.7	12.8	10.9	3.0	1.6	6.3	18.5	1.8	8.7	10.2
18	0.5	0.5	1.0	0.7		1.2	0.5	0.1	0.1	0.1	0.9	0.5	0.1	0.4
20	0.07	0.001	0.3	0.015	0.5		17.7	10.2	16.3	5.6	23.0	4.0	6.1	13.0
23	0.007	0.001	0.05	0.03	0.75	0.003		9.9	28.0	6.1	41.6	3.2	11.1	22.9
24	0.15	0.03	0.05	0.25	0.8	0.008	0.008		7.0	3.1	10.7	2.9	6.7	7.7
25	0.008	0.001	0.05	0.4	0.8	0.001	0.001	0.009		6.1	19.6	1.7	4.5	23.9
27	0.16	0.05	0.4	0.04	0.8	0.07	0.012	0.08	0.015		11.6	0.7	4.9	5.6
29	0.02	0.001	0.07	0.007	0.8	0.001	0.001	0.015	0.001	0.01		4.4	22.0	19.6
30	0.2	0.14	0.06	0.4	0.4	0.15	0.2	0.12	0.2	0.4	0.2		5.2	2.2
31	0.03	0.02	0.03	0.016	0.7	0.05	0.007	0.05	0.04	0.03	0.001	0.03		3.5
33–34	0.04	0.015	0.04	0.04	0.8	0.012	0.001	0.03	0.001	0.07	0.005	0.3	0.18	

*Chi-square values above the diagonal, probability values below.

We have, thus, five skull characters (out of 13 used), two vertebral characters (of 5), three muscle characters (of 3), three external characters (of 4), and one vocal apparatus character (the only one used).

Correlations among these fourteen characters were analyzed by means of chi-square tests (Table 8-3). The proportion of the character-by-character tests showing high degrees of association ($p \leqq 0.05$) is large (51/91). Only two characters, 18 and 30, have virtually no correlations with any other. The remaining twelve are correlated with from six to eleven of the others.

Associations of characters can be explained by pleiotropic genes, functional relations, or historical (phylogenetic) relations. Of the associations in the table, only the three involving characters 7, 10, and 29 seem to be attributable to either of the first two causes. Cranial crests (29) are usually accompanied by build-up of bones on the roof of the skull leading to rugosity of the dorsal surface (7) and to the covering of the occipital canal (10). The relationship among these characters appears to be a necessary correlation (Cain and Harrison 1960), although it is not perfect. *Bufo mauritanicus, pentoni*, and *regularis* lack cranial crests but have covered occipital canals and rugose skulls. *Bufo stomaticus* has a rugose skull but an exposed occipital canal and no cranial crests. Despite the apparent developmental and mechanical relationships among these three characters, they cannot be treated as a unit, because that would make some pairs of taxa seem more similar and others less similar than they are.

None of the other character associations appears to have a pleiotropic or functional foundation. They are best explained by a historical or phylogenetic relationship.

In retrospect since the fourteen stable characters were good indicators of species groups, they should also be useful for assigning incompletely known species to groups. If data are available for many of the fourteen characters for any of the unassigned species it should be possible to place it in the correct species group. It is assumed there is a high probability that the pattern of variation of these characters will not change significantly in other species of *Bufo* from Eurasia. This is in effect a formal rationale for using weighted characters in unstudied or incompletely known species within this segment of the genus.

The relationships of these incompletely known species, demonstrated by the fourteen characters, are what one would have predicted solely on the basis of geography. Since the characters were not selected with geographic criteria in mind, the relationships shown in table 8-6 strengthen confidence in them.

The coarse weights assigned to characters in this paper do not apply to all conceivable taxonomic studies involving the same set of species. The tympanum has been given no weight in the study of species groups. But if one were concerned with recognizing species boundaries, the absence of the tympanum in *B. surdus* would be very helpful in distinguishing it from related forms. Similarly, the tarsal spur that is peculiar to *B. pentoni* has not been used in forming species groups, but would be valuable if the problem were delimitation of species. At a higher level, if one were concerned with the relations among anuran families, the absence of maxillary teeth in these species of *Bufo* would be useful, though this character has been given no weight in this study of species groups within the genus *Bufo.*

Interspecific Relationships

As specimens were examined, an impression of similarity among species gradually formed in my mind. This process was not free of bias. The names used in the literature for some taxa indicate relationships, for example, the trinomials for *B. bufo bufo*, *B. bufo gargarizans*, and *B. bufo bankorensis*. In other cases my own previous experience had led me to associate certain taxa, for example, *B. asper* with *B. juxtasper* and the group of forms associated with *B. biporcatus* (Inger 1966). Nevertheless, this bias did not dominate my initial assessment, since it did not apply to most of the taxa examined and, when it did have influence, it was countered by examining characters, mainly muscular and skeletal, that had not been surveyed previously.

This subjective analysis formed the basis of the following preliminary species groups:

biporcatus group

biporcatus — *quadriporcatus*
divergens — *parvus*
philippinicus — *claviger*

asper group

asper — *juxtasper*

melanostictus group

melanostictus — *stuarti*
parietalis — *himalayanus*

stomaticus group

fergusoni — *dhufarensis*
olivaceus — *stomaticus*

viridis group

viridis — *orientalis*
calamita — *surdus*
raddei — *latastei*

bufo group

bufo — *bankorensis*
gargarizans — *minshanicus*
wrighti — *tibetanus*

A number of the taxa examined are not included in this listing and will be discussed later.

The immediate purpose of forming species groups is to associate taxa that are closely related genetically. Since the genotype was effectively inaccessible during this study, I had to make the assumption, which taxonomists regularly do, that a good survey of the phenotype will give us a reliable gauge of the genotype.

Thus, an objective test of these subjectively designated species groups is available. If the tentative groups are internally homogeneous in phenotype, then they do consist of genetically related taxa. We can examine each species group and ask how much of the phenotype is relatively invariable. In table 8-4 are the results of this examination. Three criteria had to be satisfied before a character was listed in table 8-4: (1) the character appears in no more than two states within a species group; (2) only one species within a group differs from the others; and (3) if there are only two species within a group, they are identical in the character under consideration.

The *melanostictus* group has the smallest number of invariable characters (10). Since data were available, however, for at least three of the four species in only thirteen characters, the degree of homogeneity is high. In the other species groups, eighteen to twenty-two of the twenty-six characters are invariable. Since these characters cover a wide spectrum of the phenotype, all the species groups satisfy the criterion of internal homogeneity of phenotype.

But we might ask whether any arrangement of these species would yield groups that had as much internal similarity. To answer this question the same twenty-six taxa were formed into groups of the same size using a table of random numbers to assign species to groups. Analysis of two such synthetic sets of species groups was carried out, applying the three criteria mentioned above. The species groups generated in this way have 1–13 and 5–10 characters invariable within groups in the first and second sets, respectively (Table 8-5). The average numbers of constant characters are 8 and 9 in these synthetic sets as compared to 18 in what we may call the real set. The "real" species groups are much more homogeneous internally, that is, they consist of more close-

Table 8–4. Phenotypic Homogeneity within Species Groups of Eurasian *Bufo*

Groups					
biporcatus	*asper*	*melanostictus*	*stomaticus*	*viridis*	*bufo*
Invariable Characters*					
(2)	2	(19)	3	(7)	(7)
(3)	3	(20)	5	(8)	10
7	5	(21)	7	9	11
8	7	23	8	(10)	(13)
10	8	24	10	11	16
11	10	25	11	(12)	(17)
12	11	27	12	(13)	(18)
(17)	12	(29)	13	(16)	20
(18)	16	30	17	17	(21)
(21)	17	31	18	(18)	(22)
(22)	18		20	(19)	23
23	20		22	(20)	24
24	21		24	23	25
25	23		25	(24)	27
29	24		27	25	29
30	25		(30)	27	(30)
31	27		(31)	29	(31)
(32)	29		33–34	31	32
33–34	30			33–34	33–34
	31				
	32				
	33–34				

*Character identification numbers correspond to those in Appendix B.

Characters in parentheses indicate that one species differs from the others within a species group. Characters without parentheses indicate that all the species within a group have the same state.

Table 8–5. Phenotypic Homogeneity within Synthetic Species Groups of Eurasian *Bufo**

Synthetic Groups					
I	II	III	IV	V	VI
Invariable Characters** First Set					
	3	2	(10)	12	7
(2)	5	3	12		16
7	10	8	18		(17)
(8)	11	11	(29)		(23)
11	12	17	30		(25)
(12)	13	24	31		(29)
(17)	17	(25)	(32)		(30)
18	20	27			(31)
24	23	30			
	29	32			
	30				
	32				
	33–34				
Second Set					
3	11	(2)	10	17	(7)
8	12	3	12	18	(8)
11	17	7	17	25	(9)
12	18	8	18	27	11
(19)	19	11	20	30	12
20	21	13	23		(16)
23	22	16	29		(18)
25	27	24	31		(23)
27	30	32			(24)
(30)	31				30

*Groups formed by random assortment of species used for Table 8–1.
**Conventions as in Table 8–4.

Table 8–6. Similarity of Certain Species of *Bufo* to Groups of Eurasian Species

	Number of Significant Characters*	Percentage of Potential Similarity*: Species Groups						Percentage of Potential Similarity*: Unassigned Species								Number of Shared, Derived States: Species Groups						Number of Shared, Derived States: Unassigned Species							
		biporcatus	*asper*	*melanostictus*	*stomaticus*	*viridis*	*bufo*	*celebensis*	*macrotis*	*galeatus*	*sulphureus*	*beddomi*	*pentoni*	*dodsoni*	*mauretanicus*	*biporcatus*	*asper*	*melanostictus*	*stomaticus*	*viridis*	*bufo*	*celebensis*	*macrotis*	*galeatus*	*sulphureus*	*beddomi*	*pentoni*	*dodsoni*	*mauretanicus*
Species with Incomplete Data																													
claviger	9	100	86	40	50	0	0	83	57	83	60	40	25	14	28	8	6	2	2	0	0	5	4	5	3	2	2	1	2
celebensis	7	83	83	50	33	0	0		83	83	40	40	33	16	16	5	5	3	1	0	0		5	5	2	2	2	1	1
macrotis	10	55	71	40	25	0	0			86	33	50	14	28	0	5	5	2	1	0	0			6	2	2	1	2	0
galeatus	8	75	75	60	66	0	0				50	40	14	28	14	6	6	3	2	0	0				3	2	1	2	1
sulphureus	8	50	33	80	33	0	33					50	50	16	25	3	2	4	1	0	1					2	2	1	1
beddomi	7	33	16	50	75	67	67						50	100	20	2	1	2	3	2	2						2	3	1
pentoni	11	33	25	40	50	25	25							14	75	3	2	2	2	1	1							1	3
Species of Uncertain Relations																													
dodsoni	13	20	33	16	80	100	80								37	2	3	1	4	5	4								3
mauretanicus	13	20	33	16	80	50	50									2	3	1	4	3	3								
Extraterritorial Species																													
regularis	13	40	37	40	50	20	25									4	3	2	2	1	1								
boreas	12	11	11	0	50	83	100									1	1	0	2	5	4								
americanus	12	33	37	60	75	20	25									3	3	3	3	1	1								
marinus	13	50	50	80	75	25	25									5	4	4	3	1	1								
spinulosus	12	22	25	20	50	40	50									2	2	1	2	2	2								
No. of characters in derived states																11	9	6	5	6	5								

*See explanation in text.

ly related species than do the synthetic species groups.

The species groups also satisfy another test of internal consistency — the geographic test. The rationale behind this test of natural groups is given elsewhere (Inger and Marx 1965; pp. 244–245). Each of the species groups has a cohesive geographic range. The *biporcatus* and *asper* groups are confined to Southeast Asia and the East Indies (Fig. 8-10, 8-11); the *bufo* and *viridis* groups to the Palearctic (Figs. 8-5, 8-6); the *stomaticus* group largely to southwest Asia and the northern part of the Indian subcontinent (Fig. 8-8), thus straddling the Palearctic and Oriental regions; and the *melanostictus* group to the Oriental region (Fig. 8-9) with most of the species in the Indian subregion.

Twelve of the species studied were not initially assigned to any of these groups. Data on the fourteen significant characters are reasonably complete for two of the twelve uncertain species. Data on the remaining ten are incomplete. Comparison of these twelve unassigned species to the six species groups and to each other should help in determining their relationship. Table 8-6 presents the comparisons for those species for which I have data on at least seven of the fourteen characters. Since only similarity in shared derived states provides information on cladistic relations, similarities are shown in those terms in table 8-6.

For the purpose of these comparisons, characters fall in one of the following three categories:

(1) No data for uncertain species
(2) Data for uncertain species, primitive state in compared species group
(3) Data for uncertain species, derived state in compared species group

Only the third category can contribute the kind of information we are seeking. The number of characters in this category gives the maximum potential similarity between taxa for table 8-6. The ratio of actual to potential similarity in characters of this category gives the percentage of potential similarity in table 8-6.

Similarity percentages over 70 percent are set off by a heavy line in table 8-6. Although this is an arbitrary standard, it is a high one. The unassigned taxa show such high similarity values only with species groups from their respective geographic regions. For example, the East Indian forms, *B. claviger* and *B. celebensis*, have strongest similarity to the two species groups, *biporcatus* and *asper*, centered in the East Indies and Southeast Asia. So do the taxa *B. macrotis* and *B. galeatus*, from Southeast Asia. *Bufo dodsoni* and *B. mauritanicus*, from northeast and North Africa respectively, show greatest similarity to a group or groups that occur in the adjacent Arabian area.

The relationships among these unassigned taxa follow the pattern set by their similarities to the six species groups, as would be expected. The cluster of four taxa from Southeast Asia and the East Indies shows higher similarities to one another than to any other species. A conspicuous exception is the low similarity between *B. claviger* and *B. macrotis*. But *B. macrotis* is also not very similar to the *biporcatus* group to which *B. claviger* is identical on the basis of this test.

Bufo sulphureus and *B. beddomi* are only moderately similar to each other and they show great similarities to different species groups. *Bufo beddomi* has a high similarity to *B. dodsoni* and both are similar to the *stomaticus* group.

The two surprises in this part of table 8-6 are the low similarity between *B. dodsoni* and *B. mauritanicus* and the high similarity between *B. pentoni* and *B. mauritanicus*. I am not sure what significance these two observations have.

On the basis of the numbers in table 8-6, *B. claviger* should be assigned to the *biporcatus* group. Doing so would not reduce the phenotypic homogeneity of this group shown in table 8-4. Although *B. celebensis* and *B. galeatus* are clearly related to the *biporcatus* and *asper* groups, the values in table 8-6 do not make a choice between these alternatives easy. *Bufo macrotis* and *B. galeatus* may be the nucleus of an additional species group allied to *B. biporcatus* and *B. asper*.

Bufo dodsoni probably should be placed in the *viridis* group, judging from the results in table 8-6. However, the *viridis* group would then be much less homogeneous phenotypically. The list of constant characters (Table 8-4) would be reduced to thirteen. Perhaps the best solution here is to remove *B. orientalis* from the *viridis* group, since it shows a number of differences from the other members (See Table 8-1), and place it with *B. dodsoni*. This would produce two closely allied species groups, the *orientalis* group having fifteen invariable characters, using the criteria given previously, and a new, restricted *viridis* group having twenty.

After the preceding analyses had been carried out, relationships within the entire set of Eurasian taxa listed in table 8-1 were investigated by means of the combinatorial method of Felsenstein and Sharrock (in manuscript). This computer method yields a list of nonredundant, monothetic combination[1] of taxa.

The fourteen significant characters were used in this analysis. Since phyletic relationships were sought, the presumed primitive states were deleted, leaving a total of twenty-four derived states that were coded as below.

Character Number	New State Code	State Description
7	1	skull smooth
	2	skull rugose and pitted
10	3	occipital canal partly covered
	4	occipital canal completely covered
11	5	otic plate narrow
17	6	parasphenoid ridge obtuse
	7	parasphenoid ridge sharp
18	8	palatine denticulate
20	9	seventh transverse process short
	10	seventh transverse process long
23	11	no vertebral crests
	12	double vertebral crests
24	13	humerodorsalis without fourth main slip
25	14	supinator manus humeralis absent
27	15	adductor longus absent
29	16	no cranial crest
	17	supratympanic crests only
	18	orbital crests only
	19	supratympanic, orbital, and parietal crests
30	20	tibia gland present
31	21	tarsal ridge absent
	22	tarsal ridge present
33–34	23	no vocal sac
	24	vocal sac present, no gular pigment

Taxa variable in one or more characters were represented in the data deck by two or more cards so that all their derived states could be considered. Taxa for which I lacked data on any characters were treated as though they lacked the corresponding states. This treatment has certainly made the dendrograms less efficient.

The combinatorial method assumes that a taxon has not only its derived states but also the states ancestral to them. That is, a phyletic relationship between taxa means not merely that the taxa agree in their present character states but also that they share evolutionary steps leading to those states (Felsenstein and Sharrock, in manuscript). Consequently, in a few instances in the dendrograms to be discussed below, some taxa are shown as having two derived states instead of just the one credited them in table 8-1.

A number of dendrograms were constructed from the monothetic combinations generated by this method. Four of the best, complete trees are shown in figures 8-1 to 8-4. The code numbers of the taxa used in the dendrograms are listed below.

Code No.	Taxon	Code No.	Taxon
1	*philippinensis*	21	*tibetanus*
2	*divergens*	22	*tibetanus*
3	*beddomi*	23	*mauritanicus*
4	*fergusoni*	24	*parvus*
5	*dhufarensis*	25	*claviger*
6	*stomaticus*	26	*biporcatus*
7	*stomaticus*	27	*quadriporcatus*
8	*olivaceus*	28	*calamita*
9	*viridis*	29	*celebensis*
10	*surdus*	30	*galeatus*
11	*raddei*	31	*macrotis*
12	*latastei*	32	*juxtasper*
13	*orientalis*	33	*asper*
14	*orientalis*	34	*stuarti*
15	*dodsoni*	35	*himalayanus*
16	*bufo bankorensis*	36	*melanostictus*
17	*bufo gargarizans*	37	*melanostictus*
18	*bufo gargarizans*	38	*sulphureus*
19	*bufo bufo*	39	*pentoni*
20	*minshanicus*		

Objective criteria for evaluating alternative dendrograms are discussed in detail elsewhere (Inger, manuscript). Briefly, dendrograms can be compared in terms of (1) the relative probability of the particular convergences implicit in each; (2) the total number of convergences; (3) the ratio of convergences between major lineages to total convergences within major lineages; (4) the number of states unique to single major lineages; and (5) the number of major lineages distinguished by unique states.

Only one of the character states involved in this study has attributes that suggest it has a lower probability for convergence than other states — the double vertebral crest (state 12). This state is rare in bufonids and apparently confined to species in Southeast Asia and the East Indies. In combination, these attributes indicate that the state arose just once. If this hypothesis is correct, dendrograms I and III, which require convergence in this state, are unacceptable.

Since we have little reason to differentiate among the other derived states in terms of probability of

[1] A monothetic combination is one whose member taxa share at least one character state. A redundant combination is one completely contained within a larger combination sharing the same character states.

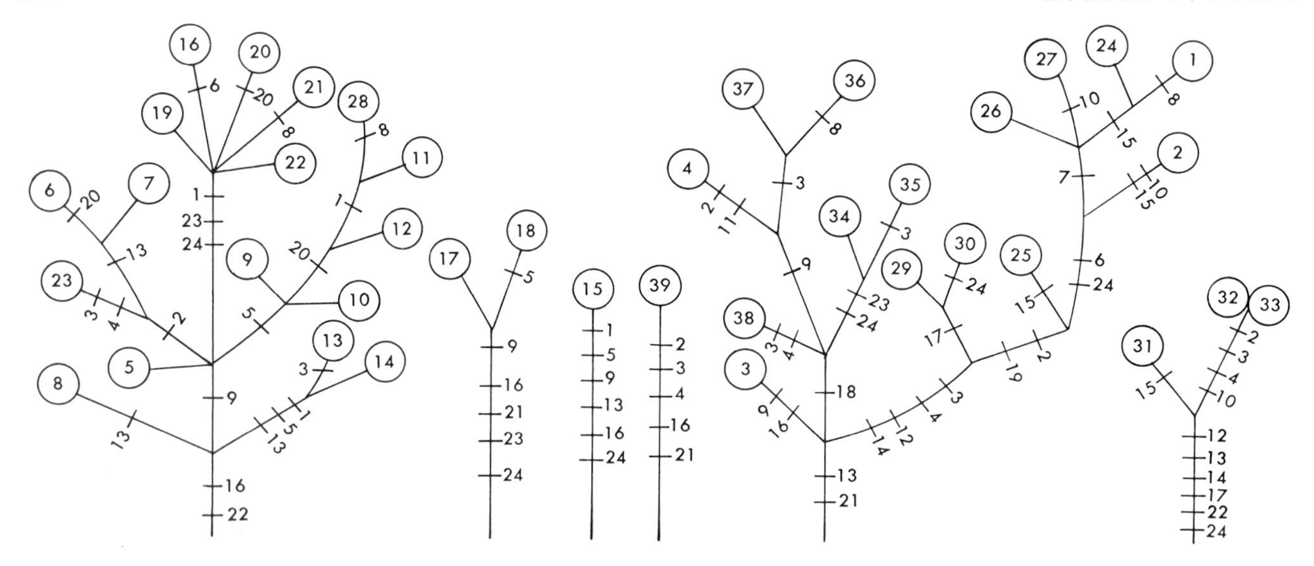

Fig. 8–1. Relationships among Eurasian forms of *Bufo*, alternate dendrogram I. Numbers in circles are code numbers of forms listed in text. Numbers next to small crossbars are code numbers of states listed in text.

convergence, they offer no help with the first criterion suggested for choosing among the alternative dendrograms.

In table 8-7 the dendrograms are compared on the basis of the other criteria mentioned above. In this case, because there is so little difference between states, the assumption that the probability of convergence is the same for all of them (except state 12) is reasonable. Therefore, we can use the argument that the fewer the total number of convergences, the more suitable the dendrogram as a probability estimate of the true phylogeny. On this basis dendrogram IV (Fig. 8-4) is preferable to the others.

The third criterion is based on the argument that the closer the phylogenetic relationship of two taxa, that is, the more similar the genomes, the more likely the convergence. One therefore expects a good dendrogram to have fewer convergences between major lineages than within major lineages. Comparison of the fourth and fifth columns of table 8-7 shows that dendrogram IV is the best in this criterion.

States confined to a single lineage are a measure of the distinctiveness of the genome of that lineage. Dendrogram IV has the most states confined to single major lineages (column 6 of Table 8-7), suggesting that its lineages are the most distinct. Since this is one of the features that an ideal dendrogram should have, IV is again the best of the series.

However, more than total number of unique states is involved in distinctiveness of lineages. Each major lineage should have at least one state unique to it. Only dendrogram IV has each of its major lineages distinguished by at least one unique state (column 7 versus column 2).

Table 8–7. Comparison of Four Dendrograms Relating 34 Eurasian Forms of *Bufo*

Dendrogram	Number of Major Lineages	Total Number of Convergences	Total Convergences within Major Lineages	Convergences between Major Lineages	States Unique to Single Major Lineages	Major Lineages with Unique States
I	6	59	21	38	6	3
II	6	56	28	28	10	3
III	3	55	33	22	8	2
IV	3	52	36	16	13	3

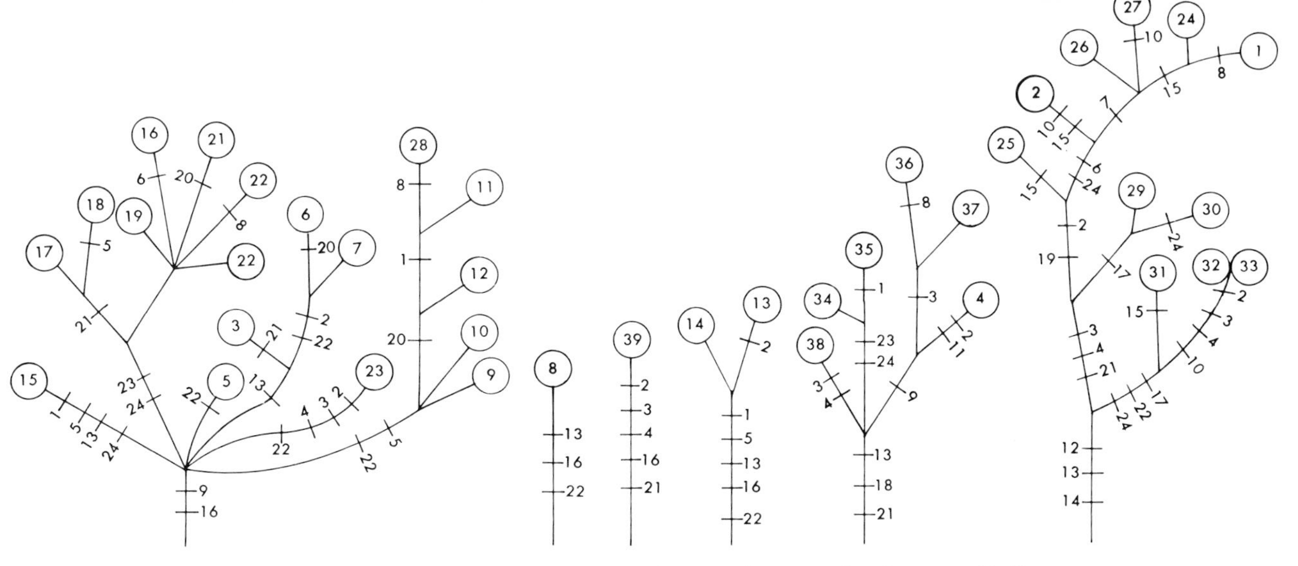

Fig. 8–2. Relationships among Eurasian forms of *Bufo*, alternate dendrogram II. Conventions as in Figure 8–1.

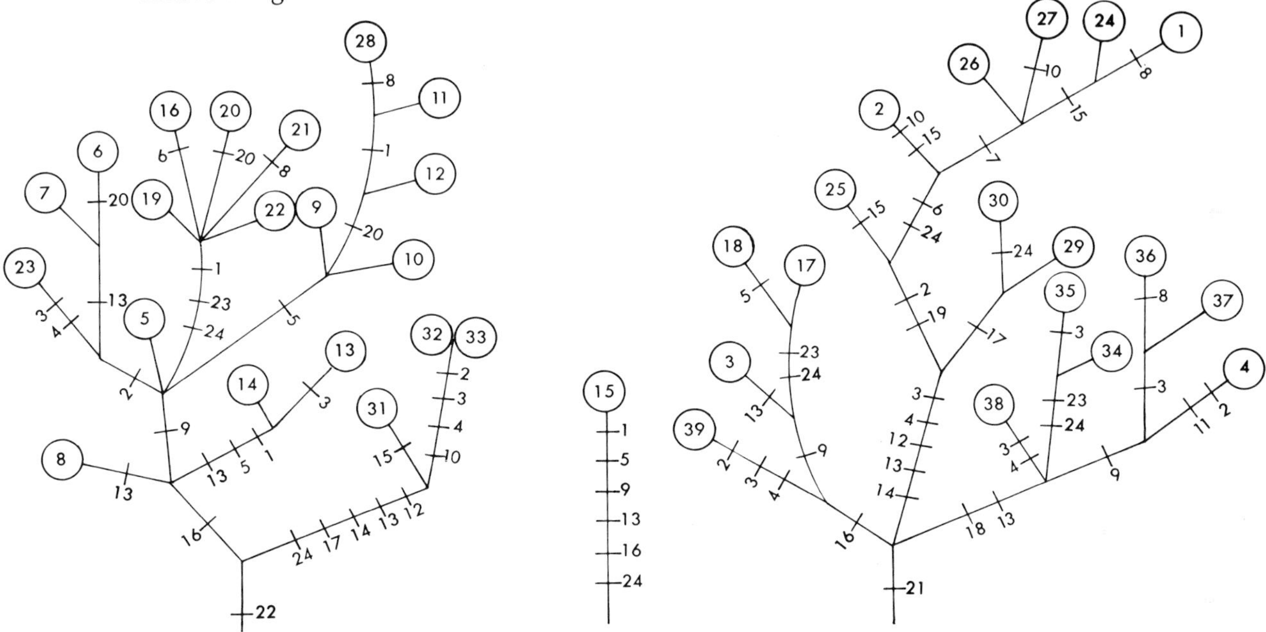

Fig. 8–3. Relationships among Eurasian forms of *Bufo*, alternate dendrogram III. Conventions as in Figure 8–1.

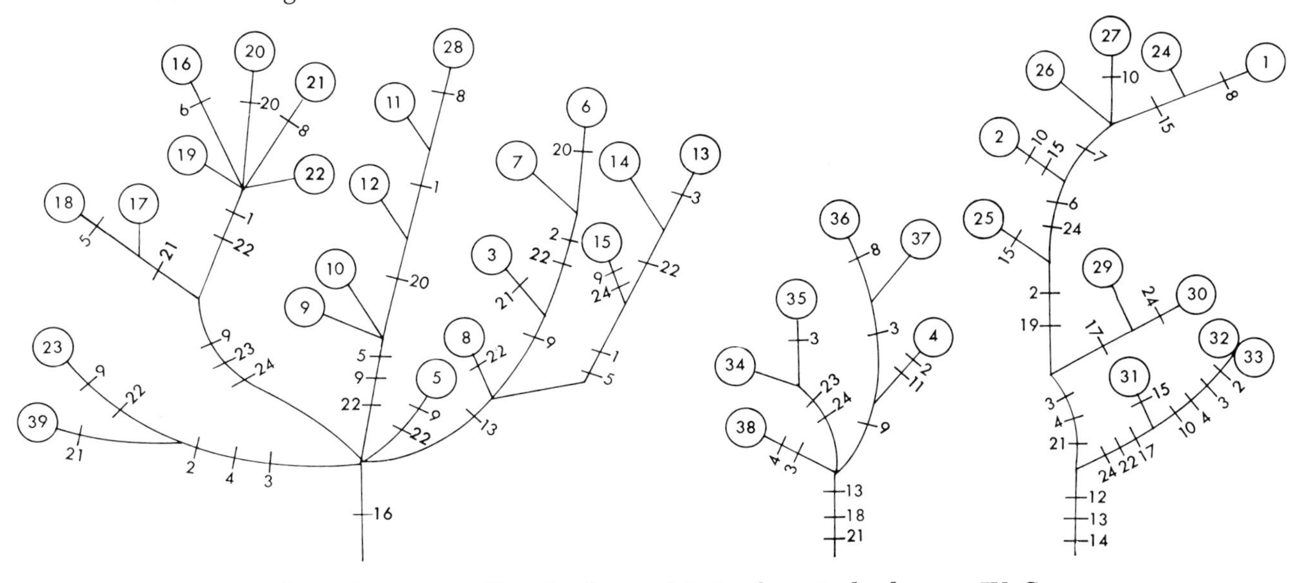

Fig. 8–4. Relationships among Eurasian forms of *Bufo*, alternate dendrogram IV. Conventions as in Figure 8–1.

Two other tests, which are not always appropriate, can be applied to these dendrograms. In this example, we have several taxa generally regarded as subspecies of *Bufo bufo*. A biologically sound dendrogram should place all these forms (code numbers 16–19) in the same major lineage. Only dendrograms II and IV satisfy this requirement.

The second of these supplementary tests is geographic consistency of lineages. Dendrograms I-III place the Palearctic forms in several major lineages and consequently are not as satisfactory as IV on these grounds.

Thus, on the basis of all these tests, dendrogram IV is the best estimate of the phylogenetic relationships of these toads. Now, how do the relationships in dendrogram IV match previous conclusions in this paper?

The preliminary series groups outlined above are supported with these exceptions: (1) *Bufo orientalis* is pulled out of the *viridis* group and joined by *B. dodsoni* in one line. This change was also suggested earlier for different reasons. (2) *Bufo fergusoni* is moved from the *stomaticus* group to the *melanostictus* group.

If we now make a table similar to table 8-4 for these rearranged groups, the number of invariable characters per species groups ranges from 11 to 22 with an average 17.4, or almost exactly the values in table 8-4.

This dendrogram (Fig. 8-4) also maintains the relationships brought out in table 8-6.

The Eurasian species groups have varying similarities to the extraterritorial species (Table 8-6), which were selected for these comparisons because they have been regarded in the past as representing species groups. The Southeast Asian and East Indian groups *biporcatus* and *asper* show the least similarity to the extraterritorial species. The *bufo* and *viridis* groups, especially the former, are very similar to *boreas*. In fact, including *boreas* in the *bufo* group leaves the list of invariable characters (Table 8-4) of that group virtually unchanged; only two of the nineteen characters are eliminated. The *stomaticus* group shows strong relationship with *marinus* and *americanus*, as does the *melanostictus* group with *marinus*. The similarity of any of these Eurasian groups, including the *orientalis* group, which is not shown separately in table 8-6, with *regularis* and *spinulosus* is low.

There is a coarse pattern in these relationships. Transhemispheric relationships are closest between taxa occupying the northernmost ranges. From this point, similarity decreases roughly as the ranges of the compared similar taxa are progressively more southern. The correlation is far from complete but it is at least suggestive of a time and distance relationship.

Artificial hybridization represents an independent test of the interspecific relationships suggested here. Unfortunately, not enough crosses have been made using the Eurasian species (Chap. 11) to provide a satisfactory test of the species groups. Blair obtained either very few or no hybrid larvae that completed metamorphosis, using *asper*, *melanostictus*, *stomaticus*, *bufo*, *viridis*, and *calamita*. Blair concludes these species are "quite divergent from one another." Since

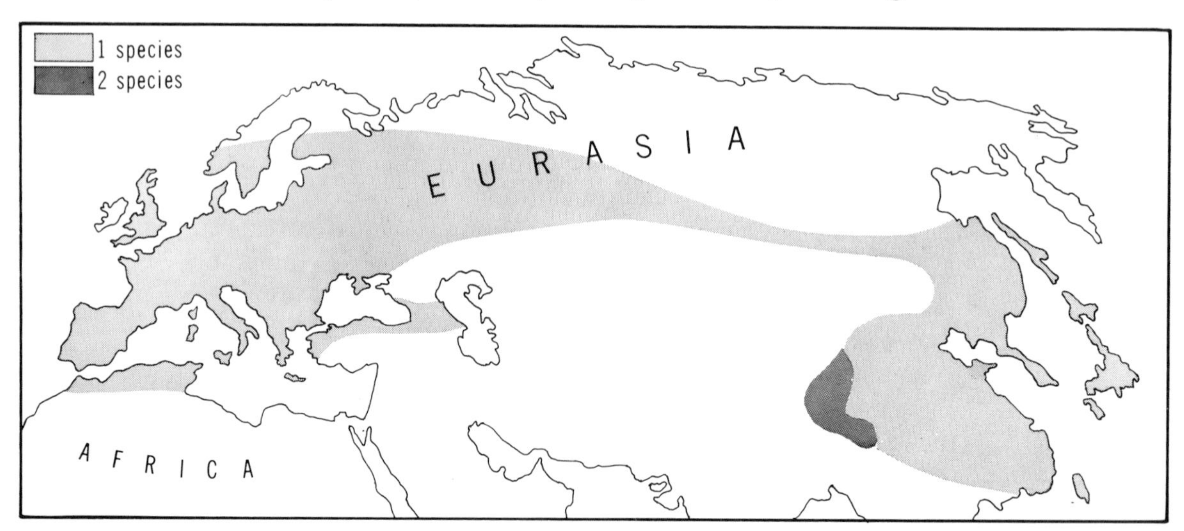

Fig. 8–5. Distribution of the *Bufo bufo* species group.

viridis and *calamita* have been placed in one group, Blair's experiments to date do not support the grouping.

Blood proteins (see Chap. 14) provide some evidence that *bufo*, *viridis*, and *calamita* are related. But the evidence is limited.

Clearly, the biochemical and breeding tests are important in evaluating species groups arrived at on the basis of gross morphology. In this case, however, we have to wait for more complete testing.

Diagnosis and Definition of Species Groups

Bufo Group

No cranial crests; tarsal ridge present, rarely absent; no vocal sac; tibia gland present or absent; supinator manus humeralis present; humerodorsalis with main slips to third and fourth fingers and a short accessory slip to fourth metacarpal; adductor longus present; vertebral column with a single median crest formed by neural spines; seventh transverse process 0.576–0.725 of third; occipital canal exposed; dorsal surface of skull smooth, pitted in a small percentage of individuals; squamosal with a moderately broad dorsal otic plate; transverse parasphenoid ridge absent or shallow and obtuse; palatine smooth.

Included forms: *B. bufo bufo, B. bufo bankorensis, B. bufo gargarizans, B. bufo wrighti, B. bufo japonicus, B. minshanicus, B. tibetanus.* The exact status of *B. minshanicus* is uncertain. The entire complex requires field study.

Range: Europe, western North Africa, northern Asia Minor, central Asia, northeastern Asia, Japan, Taiwan (Fig. 8-5).

Viridis Group

No cranial crests; tarsal ridge present; vocal sac present, surrounding muscle and connective tissue with melanophores; tibia gland present or absent; supinator manus humeralis present; humerodorsalis with main slips to third and fourth fingers and an accessory slip to fourth metacarpal; adductor longus present; vetrebral column with a single, median crest; seventh transverse process 0.576–0.725 of third; occipital canal exposed; dorsal surface of skull smooth or weakly pitted; squamosal without a dorsal otic plate; transverse parasphenoid ridge absent; palatine usually smooth.

Included forms: *B. viridis, B. calamita, B. raddei, B. surdus, B. latastei. Bufo luristanica* Schmidt is probably a member of this group.

Range: Europe, North Africa, southwest Asia except for the Arabian Peninsula, central Asia, Korea, and North China (Fig. 8-6).

Orientalis Group

No cranial crests; tarsal ridge present; vocal sac present, surrounding muscle with or without melanophores; tibia gland absent; supinator manus humeralis present; humerodorsalis with a long slip to the third finger only, a short, accessory slip to the fourth metacarpal; vertebral column with a single, median crest; seventh transverse process 0.575–0.875 of third; occipital canal exposed or half covered by bone; dorsal surface of skull smooth; squamosal without a dorsal otic plate; transverse parasphenoid ridge absent; palatine smooth.

Included species: *B. orientalis, B. dodsoni.*

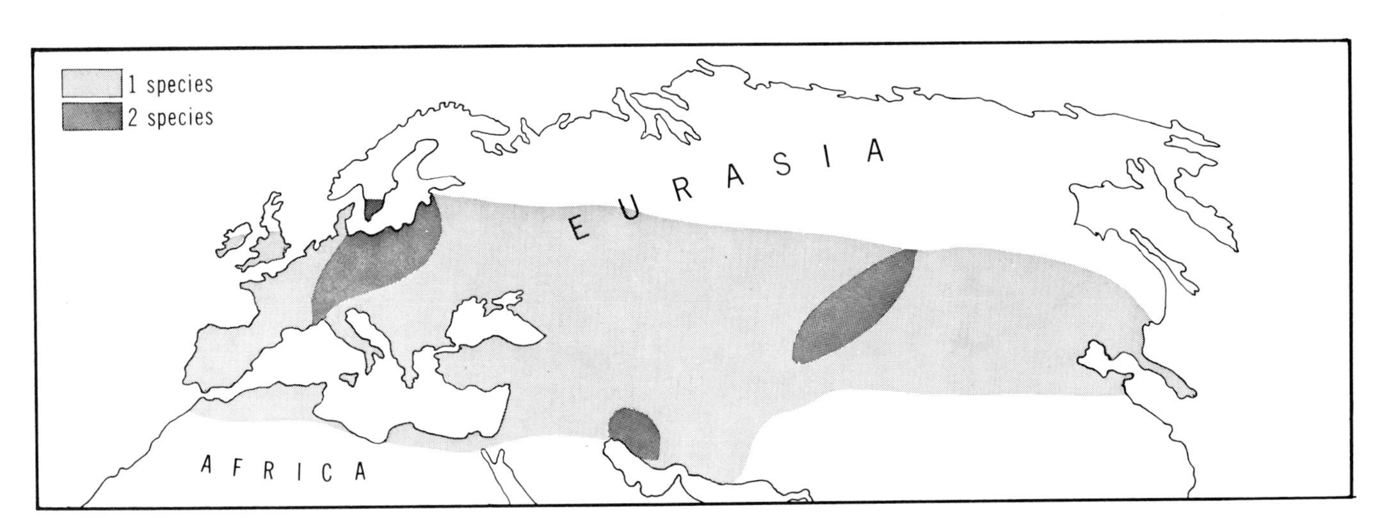

Fig. 8–6. Distribution of the *Bufo viridis* species group.

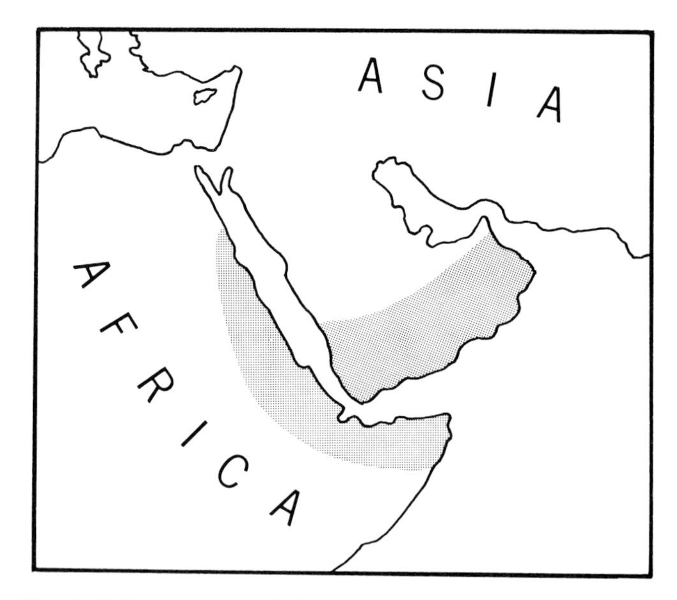

Fig. 8–7. Distribution of the *Bufo orientalis* species group.

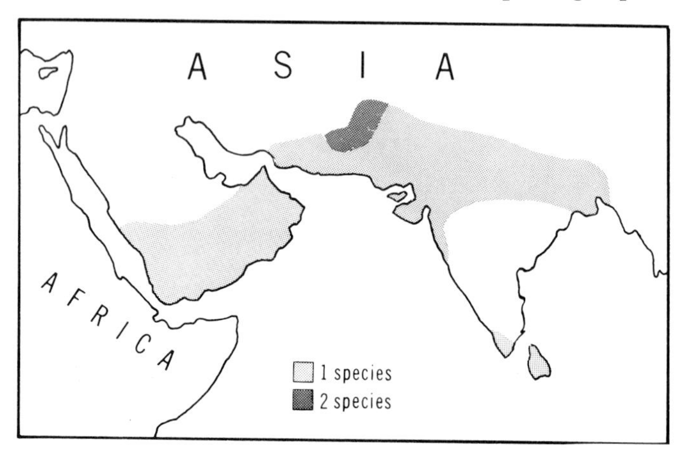

Fig. 8–8. Distribution of the *Bufo stomaticus* species group.

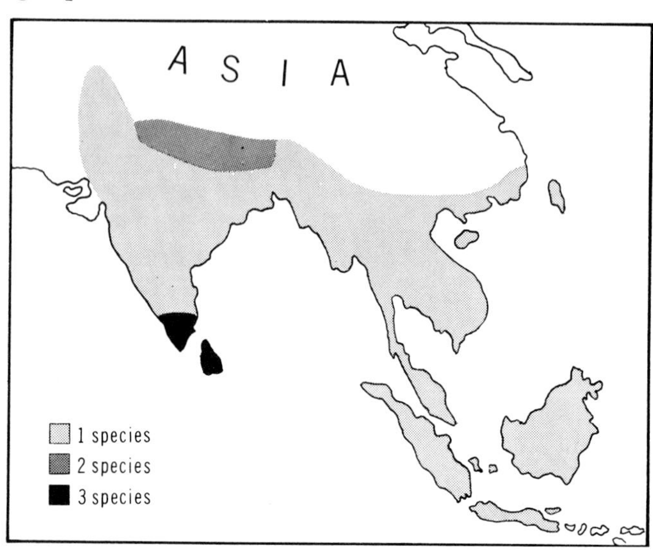

Fig. 8–9. Distribution of the *Bufo melanostictus* species group.

Range: Northeastern Africa and southern third of Arabian Peninsula (Fig. 8-7).

Stomaticus Group

Cranial crests absent; tarsal ridge usually present; tibia gland usually absent; vocal sac present, surrounding muscle with melanophores; supinator manus humeralis present; humerodorsalis with a long slip only to third finger, a short slip to fourth metacarpal; adductor longus present; vertebral column with a single median crest; seventh transverse process 0.575–0.725 of third; occipital canal exposed; dorsal surface of skull smooth or rugose; squamosal with a dorsal otic plate; transverse parasphenoid ridge absent; palatine smooth.

Included species: *B. stomaticus, B. olivaceus, B. dhufarensis, B. beddomi.* The last species is tentatively assigned to this group. On the basis of material from Nepal, India, Pakistan, and Afghanistan, I am unable to distinguish between *B. stomaticus* Lütken and *B. andersoni* Boulenger. If these are indeed separate species, both belong to this group.

Range: Southern half of Arabian Peninsula, southwestern Iran, Afghanistan, West Pakistan, northern India, Nepal, extreme southern India, and Ceylon (Fig. 8-8).

Melanostictus Group

Orbital cranial crests present, parietal crests present in one species; no tarsal ridge; tibia gland absent; vocal sac present or absent, surrounding muscle with melanophores when present; supinator manus humeralis present; humerodorsalis with a long slip only to third finger, a short accessory slip to fourth metacarpal; adductor longus present; vertebral column with a single median crest; seventh transverse process 0.575–0.875 of third; occipital canal exposed, partly covered, or completely covered by bone; dorsal surface of skull smooth or pitted; squamosal with a dorsal otic plate; transverse parasphenoid ridge absent; palatine smooth or denticulate.

Included species: *B. melanostictus, B. fergusoni, B. parietalis, B. stuarti, B. himalayanus.*

Range: All of Indian region except West Pakistan, one species (*B. melanostictus*) extending to Bali, Borneo, Taiwan, and all of Southeast Asia (Fig. 8-9).

Asper Group

Supratympanic crests present; tarsal ridge present; tibia gland absent; vocal sac present, surrounding muscle and connective tissue without melanophores;

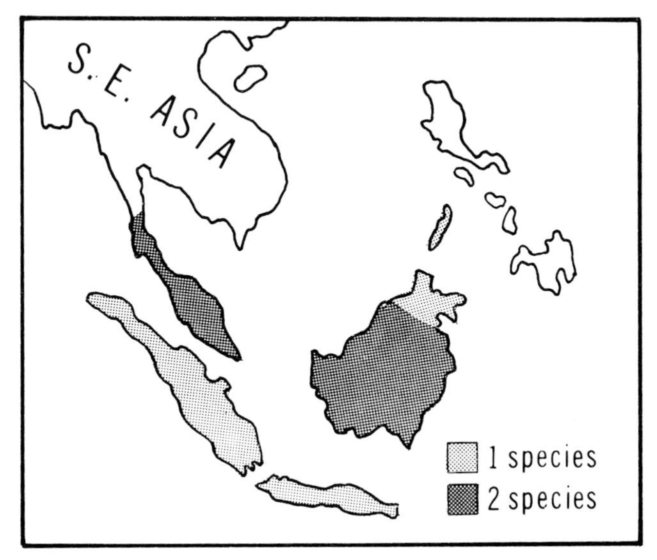

Fig. 8–10. Distribution of the *Bufo asper* species group.

Fig. 8–11. Distribution of the *Bufo biporcatus* species group.

supinator manus humeralis absent; humerodorsalis with long slip only to third finger, a short accessory slip to fourth metacarpal; adductor longus present; vertebral column with a pair of dorsal ridges or crests; seventh transverse process 0.876–1.025 of third; occipital canal completely covered by bone; dorsal surface of skull rugose; squamosal with a broad dorsal otic plate; transverse parasphenoid ridge absent; palatine smooth.

Included species: *B. asper, B. juxtasper.*

Range: Malay Peninsula, Borneo, Sumatra, Java (Fig. 8-10).

Biporcatus Group

Supraorbital, parietal, and supratympanic crests present; tarsal ridge absent; vocal sac present, but no melanophores in surrounding muscle and connective tissue; tibia gland absent; supinator manus humeralis absent; humerodorsalis with a long slip to third finger only, a short accessory slip to fourth metacarpal; adductor longus absent in most populations; vertebral column with a pair of dorsal ridges or crests; seventh transverse process 0.726–1.025 of third; occipital canal completely covered by bone; dorsal surface of skull rugose; squamosal with a broad dorsal otic plate; transverse parasphenoid ridge deep and sharp; palatine smooth in most populations.

Included forms: *B. biporcatus biporcatus, B. biporcatus divergens, B. biporcatus philippinicus, B. parvus, B. quadriporcatus, B. claviger.*

Range: Malay Peninsula, Sumatra and adjacent islets, Java, Bali, Borneo, and southwestern island chain of Philippine Islands (Fig. 8-11).

REFERENCES

Cain, A. J., and G. A. Harrison. 1960. Phyletic weighting. Proc. Zool. Soc. (London) 135:1–31.

Dunlap, D. G. 1960. The comparative myology of the pelvic appendage in the Salientia. J. Morph. 106: 1–76.

———. 1966. The development of the musculature of the hindlimb in the frog, *Rana pipiens.* J. Morph. 119:241–258.

Gaupp, Ernst. 1896. Anatomie des Frosches. Braunschweig, Friederick Vieweg u. Sohn.

Haines, R. W. 1939. A revision of the extensor muscles of the forearm in tetrapods. J. Anat. 73:211–233.

Hennig, Willi. 1966. Phylogenetic systematics. Univ. Illinois Press, Urbana.

Inger, R. F. 1966. The systematics and zoogeography of the Amphibia of Borneo. Fieldiana Zool. 52:1–402.

———, and Hymen Marx. 1965. The systematics and evolution of the Oriental colubrid snakes of the genus *Calamaria.* Fieldiana Zool. 49:1–304.

Kändler, Rudolf. 1924. Die sexuelle Ausgestaltung der

Vorderextremität der anuren Amphibien. Jena. Z. Naturwiss. 60:176–240.

Liu, C. C. 1935. Types of vocal sac in the Salientia. Proc. Boston Soc. Nat. Hist. 41:19–40.

———. 1950. Amphibians of western China. Fieldiana Zool. (Mem.) 2:1–400.

Pope, C. H., and A. M. Boring. 1940–1941. A survey of Chinese Amphibia. Peking Bull. Nat. Hist. 15:13–86.

Tihen, J. A. 1962. Osteological observations on New World *Bufo*. Amer. Midl. Nat. 67:157–183.

Throckmorton, L. H. 1968. Concordance and discordance of taxonomic characters in *Drosophila* classification. Syst. Zool. 17:355–387.

Voris, H. K. 1969. Evaluation of the Hydrophiidae with a critique on method. Ph.D. Thesis. Univ. Chicago.

9. *Bufo* of Africa

MILLS TANDY

The University of Texas at Austin
Austin, Texas,

and

RONALDA KEITH

The American Museum of Natural History
New York, New York

Introduction

The interest of men in African toads probably dates from the first time both organisms coexisted. African literature has many references to toads and usually attributes to them evil spectres — an attitude common throughout mankind. The Ibo story below illustrates this attitude and also reveals the manner in which mythology is used to explain a biological phenomenon — the pattern of inheritance.

The Toad

"When Death first entered the world, men sent a messenger to Chuku, asking him whether the dead could not be restored to life and sent back to their old homes. They chose the dog as their messenger.

The dog, however, did not go straight to Chuku, and dallied on the way. The toad had overheard the message, and, as he wished to punish mankind, he overtook the dog and reached Chuku first. He said he had been sent by men to say that after death they had no desire at all to return to the world. Chuku declared that he would respect their wishes, and when the dog arrived later with the true message he refused to alter his decision.

Thus, although a human being may be born again, he cannot return with the same body and the same personality."

An Ibo story (Nigeria)
[From Beier 1966]

Most Africans are aware of the toxic properties of toad venom, particularly its effects on the eyes. For this reason, and because of the often unsanitary microhabitat preferences of *Bufo*, as well as for mythological reasons, toads are avoided. Most African languages have a special name for *Bufo* as distinguished from *Rana* and other genera — *ọpolọ́* (Yoruba), *awọ̀* (Ibo), *ikwọt* (Efik), *fụ̀njọ́* (Bafut). Some languages have names for different species, e.g., *asọ́ngam mó-kwi ñgàm* for *Bufo superciliaris* in Bafut, but this is unusual. Stewart (1967) has discouraged the use of African names for amphibians because of the restricted geographical usage of many of the names. However, such names are useful for field workers, and, if their linguistic and geographic usages are specified, there should be no confusion. Also, experience in African educational institutions has shown that use of African languages in biology texts — even when many different languages are spoken within a country — is a great aid in giving students a feeling of identification with their study material and an incentive to relate this material to their own experiences (Savory 1963). Loveridge realized the utility of such names in fieldwork, and their documentation is to be encouraged.

The interest of European scientists in African toads probably dates from Aristotle's transliterations of the calls of anurans. *Bufo bufo* and *B. viridis*, primarily Eurasian forms, were described in the first works of Linné (1735) and Laurenti (1768). The first published reference to a member of the *regularis* group dates from the plate of *Grenouille ponctuée* from Egypt by Geoffroy St. Hilaire in 1827. Until the late nineteenth-century, descriptions of other species were based on studies in Europe and America of collections brought from Africa by wealthy adventurer-travelers. The end of the nineteenth century and the early part of the twentieth saw a turn in this European interest as museum curators competed feverishly to describe the world fauna and flora. Several distinctive African species of *Bufo* were described during the period 1880–1910, chiefly by Bocage, Boulenger, Nieden, Peters, and Werner. This competitive work by scientists from different European countries on different parts of the African continent resulted in much duplication and taxonomic confusion, which has persisted to the present. The genus *Bufo* suffered particularly because of lack of serious interest in it by many workers and because of the difficulty in distinguishing many species by the morphological criteria used by these workers. Much of the twentieth-century literature on African *Bufo* consists of reports of museum-sponsored collecting expeditions, notably those of the Museum of Comparative Zoology (Loveridge's work), the American Museum of Natural History (Noble 1924), the Senckenberg (Mertens), the Stockholm (Andersson), the Transvaal (FitzSimons), and more recently the Copenhagen (Schiøtz) museums. The British Museum has accumulated large quantities of herpetological material, mostly collected by governmental colonial officers. Workers in South Africa associated with local museums have contributed much information on the southern bufonid fauna. The works of Hewitt and of Power are among the very few biological studies of African toads. Another exceptional work is Sanderson's (1936) detailed study of the ecology of species of the Mamfe area of Cameroun. This work is also the only known study on the effects of vegetational succession on the anuran fauna of a tropical area. Hewitt and Power were apparently the first to note the importance of the mating call as a reproductive isolating mechanism among African *Bufo*, although they described several sympatric species, which they knew to be acoustically distinct, as subspecies! Rose also wrote much about the South African forms, but, as Schiøtz (1969) has noted, these writings tend to tell as much about Rose as about the anurans. Wager (1965) has collected much life history information on the South African species. Poynton published his revision of the South African forms in 1964. Perret similarly treated the Camerounian forms in 1966. A. A. Martin made the first serious acoustical investigation of African *Bufo* in 1961 in South Africa, although his work was not published at that time. Schiøtz published the first sound spectrograms of three species of African toads in 1964. Keith published spectrograms of calls for five additional species in 1968. Schneider's oscillograms, spectrograms, and descriptions of the calls of *B. bufo* and *B. viridis* appeared in 1966.

During 1961–1965 Ronalda Keith recorded extensively in eastern and southern Africa under the sponsorship of the American Museum of Natural History. Mills Tandy lived in Nigeria from 1963 to 1965 and made recordings and field observations during this time. He also traveled and collected extensively in southern and eastern Africa in the latter part of 1965, partially under the sponsorship of Dr. W. F. Blair. The present work concerns mainly Keith's and Tandy's investigations of the acoustics and ecology of African toads, but it is also a summary of everything that is known to date about the evolution of these species.

Breeding Behavior Patterns

Introduction

The details of the breeding behavior of African *Bufo* are largely unknown. The literature has various accounts of field observations on seven to twelve of the suspected fifty to sixty African species. Our work adds an additional eighteen species. Only two detailed field studies have been made — those by Chapman and Chapman (1958) and by Menzies (1963) on populations of *Bufo "regularis"* in Tanzania and Sierra Leone respectively. Complete developmental information has been compiled by Power (1925–

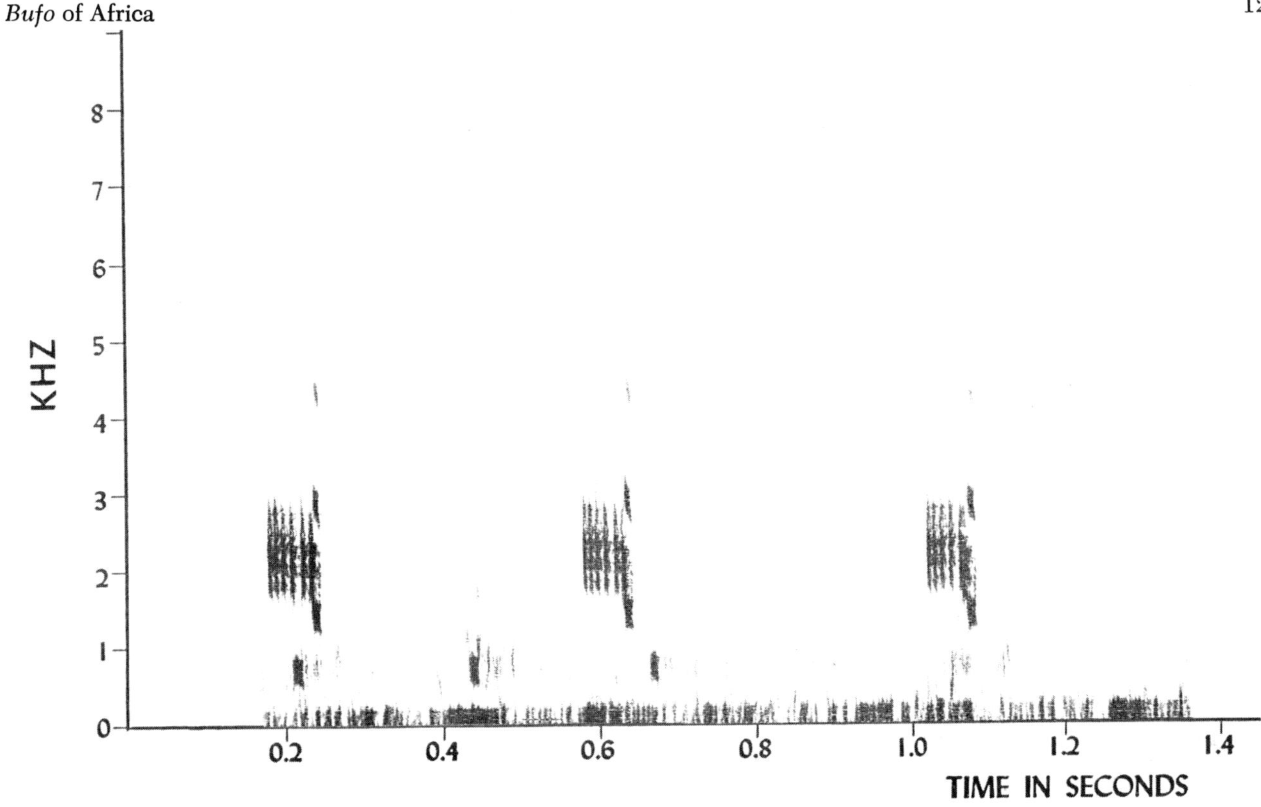

Figure 9–1. A sound spectrogram of what may be a release call of *B. lönnbergi,* Nanyuki, Kenya, by R. Keith. Filter band width 50Hz.

1932), Power and Rose (1929), de Villiers (1929*a*, 1929*b*), and Wager (1965) for several South African species. Prior to our work, the mating calls of only eight African species had been recorded on magnetic tape. We now have recordings of twenty-one or twenty-two species. Our acoustical data include analysis by sound spectrograph of approximately one thousand mating calls of three hundred individual animals from eighty-four populations among twenty-one to twenty-two species. The literature includes verbal descriptions of the calls of three to five additional species.

The species that are least known behaviorally are forest and montane forms. There are only fragmentary literary accounts of the behavior of three to four of the nine to sixteen species of earless toads with the exception of *B. rosei* (Power and Rose 1929). We have a recording of what may be a mating call or a release call for one of these earless species – *B. lönnbergi* (Fig. 9-1). One forest species that is reproductively unknown is *B. superciliaris,* although this species is fairly common in museum collections from its area of distribution.

Physical Structure of the Mating Calls

The mating calls of different species of African *Bufo* have temporal structures containing physical elements of four different grades of complexity (in terms of Broughton 1963). These are pulses (p), pulse trains (pt), complex pulse trains (cpt), and first order sequences of complex pulse trains (fcpt). This terminology is based on the definition of a pulse as "a train of sine waves." It is equivalent to "figure" of Davis (1964) and Schiøtz (1967). Martin (1967) has demonstrated that among different *Bufo* species there are two different types of pulses based on different physiological mechanisms. One is produced passively by the arytenoid valves of the larynx acting as relaxation oscillators during the flow of air from the lungs to the vocal sac. He terms the mechanism passive amplitude modulation (pam). Such pulses are the commonest type in calls of African *Bufo.* A second type of amplitude modulation is produced actively by the thoracic musculature, permitting reversal of the direction of air flow between the lungs and the vocal sac. This mechanism is termed

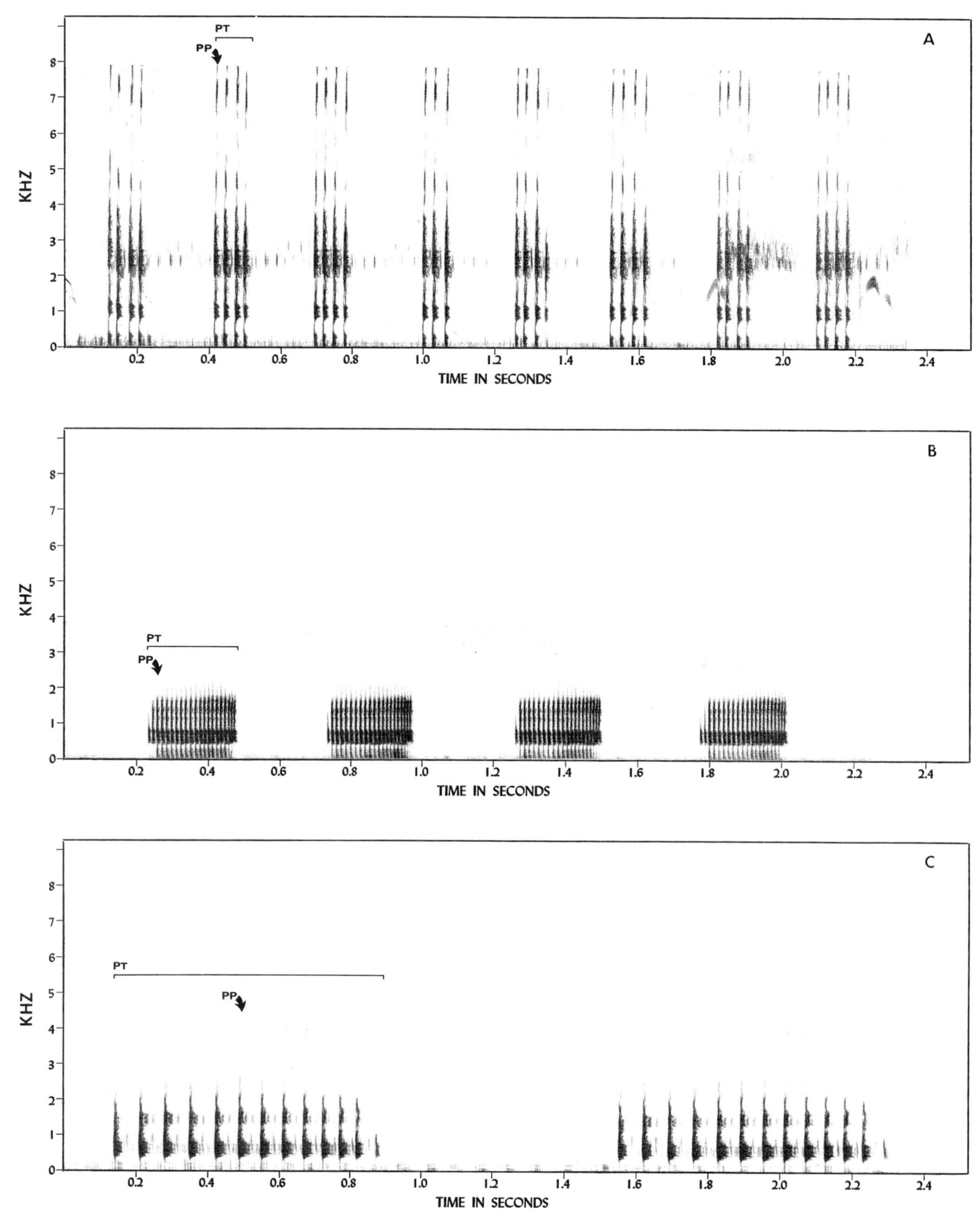

Fig. 9–2. Sound spectrograms of parts of the mating calls of three species of African *Bufo* consisting of complex pulse trains containing active and passive amplitude modulation. *(A) B. steindachneri*, near Butiaba, Lake Alberta, Uganda, by R. Keith, August 3, 1963. (*B*) *B. garmani*, Letaba, Kruger National Park, South Africa, by M. Tandy, October 8, 1965, 9:00 P.M. Air temperature, 25.5° C. (*C*) *B. regularis* (E), Ol Tukai, Masai, Amboseli Game Reserve, Kenya, by R. Keith, March 14, 1961. Effective filter band width 50 Hz.

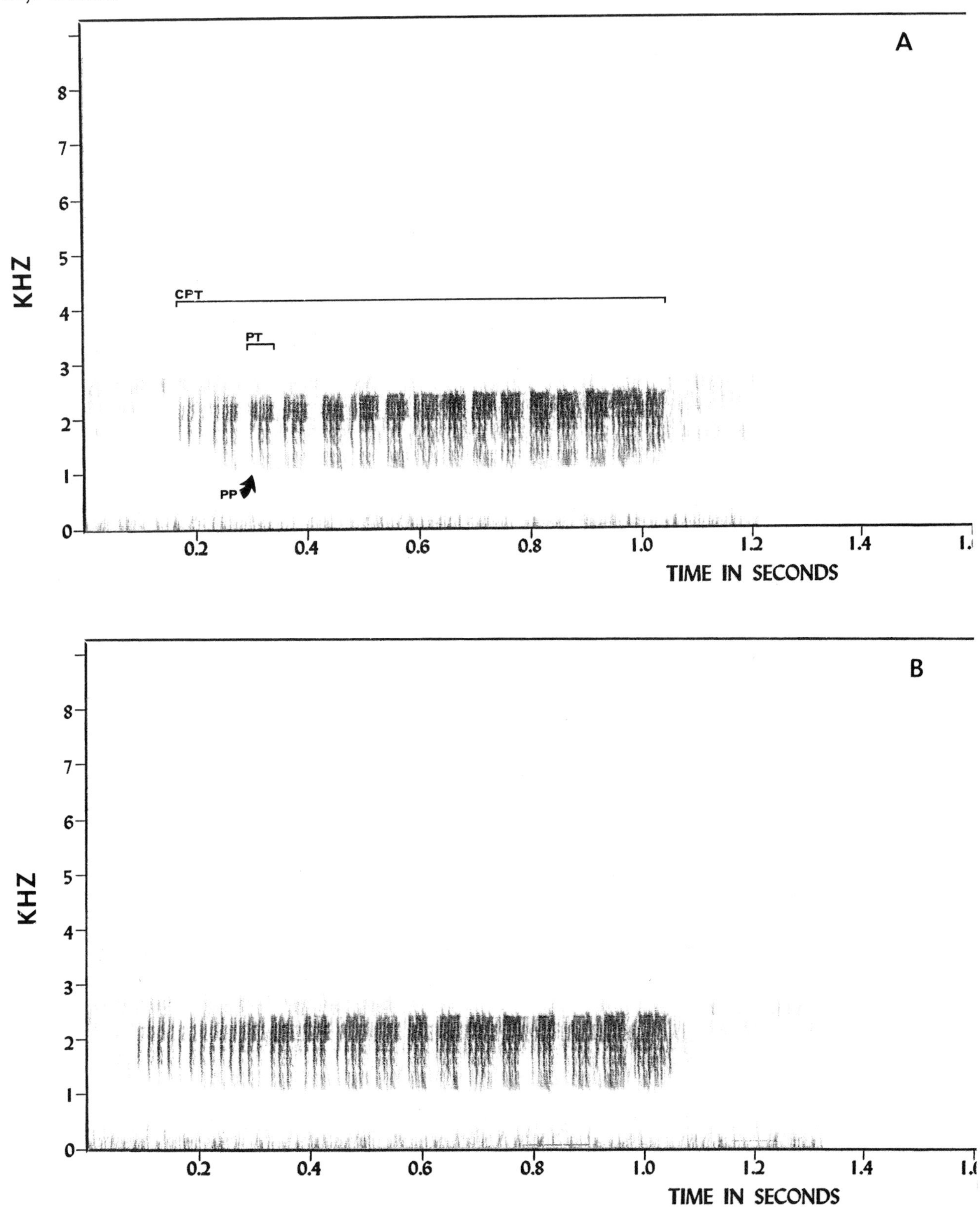

Fig. 9–3. Sound spectrograms of parts of the mating call of *B. perreti*, which consists of a first-order sequence of complex pulse trains containing active and passive amplitude modulation. Note that in (*A*) the entire complex pulse train is broken up into brief simple pulse trains, but in (*B*) the first part of the complex pulse train is not so divided. Also note that this first part of the complex pulse train in (*B*) very much resembles the simple pulse trains of *B. garmani* shown in Fig. 9–2(B). Idanre, in western Nigeria, by M. Tandy, June 26, 1965. Air temperature, 23.3° C. Effective filter band width 50 Hz.

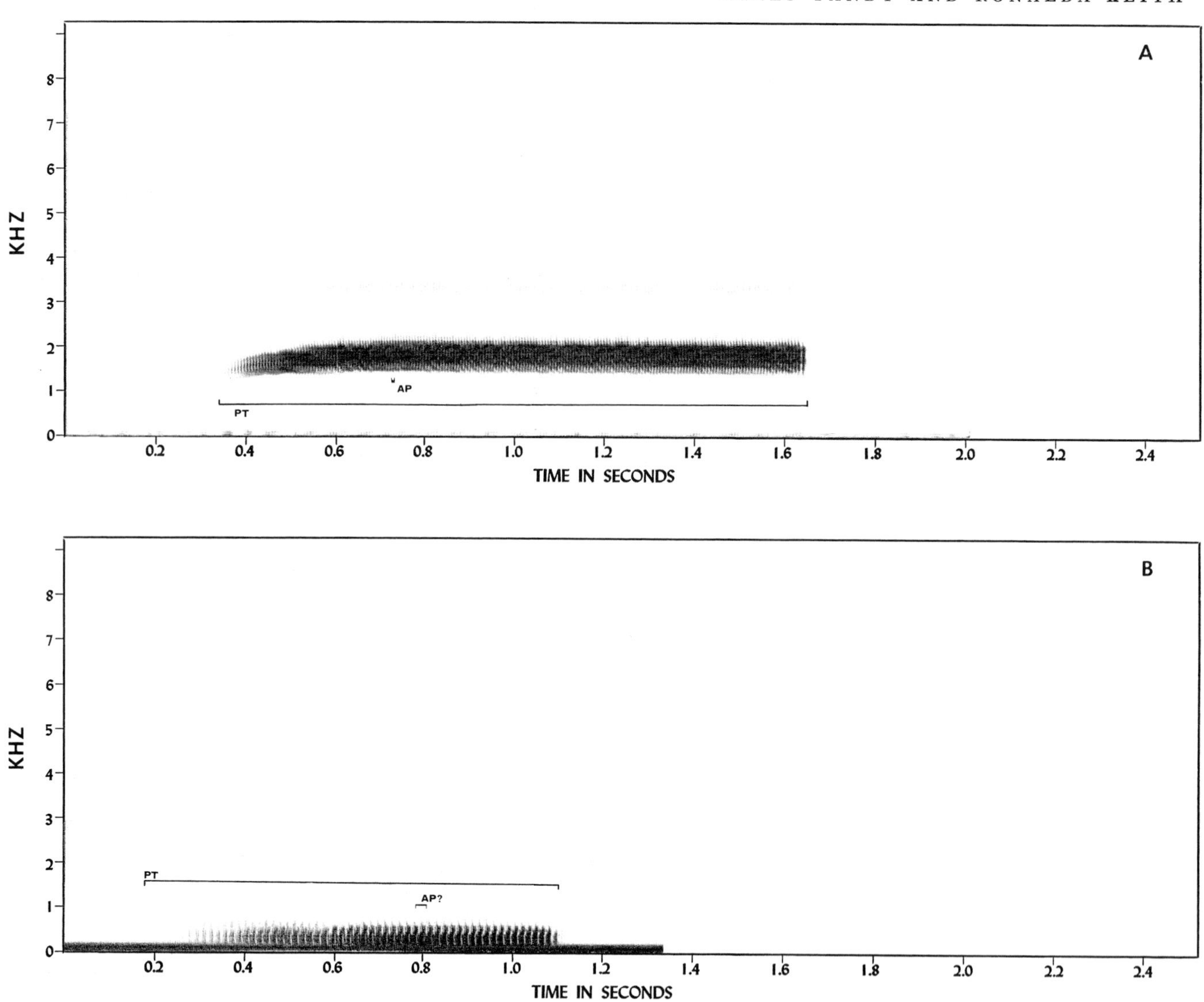

Fig. 9–4. Sound spectrograms of parts of the mating calls of two species of *Bufo* consisting of complex pulse trains. The first contains only active amplitude modulation. The mechanism of the origin of the pulses in the call of *B. carens* is not known. *(A) B. woodhousei*, near Weatherford, Oklahoma, U.S.A., May 13, 1957; University of Texas library of anuran sounds, station 144. Air temperature, 17.0° C; water temperature 25.0° C. (*B*) *B. carens,* near Iringa, Tanzania, by R. Keith, February 2, 1963. Effective filter band width 50 Hz. The slow rise time of the wave form of the pulses of *B. carens* suggests that they might be of active origin. The description by Wager (1965) of the vibration of the flanks of *B. carens* during inflation of the vocal sac suggests that this pulsation may be active.

active amplitude modulation (aam). It may give rise to pulses and simple pulse trains as in the call of *Bufo viridis.* Such active mechanisms may also produce complex pulse trains, as in the call of the *B. woodhousei,* or first-order sequences of complex pulse trains, as in the call of *B. perreti.* Thus, pulses are of either passive or active physiological origin. More complex temporal structures are of active origin. (See figures 9-2–9-4.)

With the exception of *B. viridis,* all African *Bufo* calls consist of either complex pulse trains or first-order sequences of such trains (see Table 9-1). The repetition rate of simple pulse trains within complex trains ranges from 0.059 to 16.0/second among Afri-

Table 9–1. The Temporal Structure of *Bufo* Mating Calls*

	Temporal Components and Their Mechanism of Production				
	p		pt	cpt	fcpt
Species	AAM	PAM	AAM	AAM	AAM
African species					
B. lönnbergi	—	+	+	+	—
B. gariepensis	—	+	+	+	+
B. bufo	+	+	+	+	+
B. carens	+(?)	—(?)	+	+	—
B. steindachneri	—	+	+	+	—
B. vittatus	—	+	+	+	—
B. sp. (Lake Rudolf)	—	+	+	+	—
B. brauni	—	+	+	+	—
B. garmani	—	+	+	+	—
B. sp. (Somali region)	—	+	+	+	—
B. kisoloensis	—	+	+	+	—
B. rangeri	—	+	+	+	—
B. regularis (E)	—	+	+	+	—
B. regularis (W)	—	+	+	+	—
B. kerinyagae	—	+	+	+	—
B. maculatus	—	+	+	+	—
B. perreti	—	+	+	+	+
B. pentoni	—	+	+	+	—
B. viridis	+	—	+	—	—
B. mauritanicus	+	+	+	+	+
Non-African species					
B. calamita	—	+	+	+	+(?)
B. melanostictus	—	+	+	+	—
B. valliceps	—	+	+	+	—
B. cognatus	—	+	+	+	—
B. speciosus	—	+	+	+	+
B. americanus	+	—	+	—	—
B. woodhousei	+	—	+	+	—
B. quercicus	+	—	+	—	—

* p = pulse
pt = simple pulse train
cpt = complex pulse train
fcpt = first-order sequence of complex pulse trains

can species. Most have rates between 0.2 and 4.0/second. What have been termed "calls" of members of the *regularis* "complex" (Keith 1968; see Figure 9-2) are simple pulse trains within complex trains.

Repetition of complex pulse trains to produce first-order sequences of such trains would be difficult to detect if the periodic rate were extremely slow. Attempts to measure it have not been made for many non-African species. The calls of some species, such as those of *B. cognatus* and *B. valliceps*, may actually be of such complexity, although it is not so indicated in table 9-1.

The mating calls of *B. perreti* (Fig. 9-3) and *B. mauritanicus* (Fig. 9-6 *A*, *B*) contain simple pulse trains of active origin within complex pulse trains, also of active origin, within first-order sequences of complex pulse trains, also of active origin, and appear to be structurally, and possibly mechanically, very similar to the simple pulse trains within each complex pulse train of the call of *B. marmoreus*. On the other hand, the simple pulse trains within the calls of *B. gariepensis* (Fig. 9-5*B*) and *B. regularis* (E) (Fig. 9-2*C*) appear to be temporally more similar to the complex pulse trains of the call of *B. perreti* than to the simple pulse trains of *B. perreti*.

Homologies of different active components between somewhat distantly related species are in several instances presently unclear. The call of *B. mauritanicus* exemplifies this. Are the 0.07-second-duration actively produced simple pulse trains of the *B. mauritanicus* call homologous to the actively produced simple pulse trains of similar duration within the call of *B. perreti*? The simple pulse trains of *B. perreti* apparently are the result of a qualitatively different timing mechanism from that producing the simple pulse trains of somewhat longer duration of *B. garmani* and *B. rangeri*. The call of *B. perreti* begins as a series of slowly repeated simple pulse trains, each containing passively produced pulses typical of species of the *regularis-maculatus-latifrons complex*[1] with which *B. perreti* shares the uniquely African chromosome number 20 (Bogart 1968). A second type of active amplitude modulation starts subsequently, breaking up each pulse train into 5 to 15 pulse trains—effectively converting each simple pulse train into a complex pulse train. Biochemical data of Low (Ch. 13) indicate fairly close similarity between *B. mauritanicus* and members of the *regularis complex* (max. p.a. = 82% with *B. regularis*, Vumba Mountains). *B. mauritanicus* also shows some genetic compatibility with members of this complex (Blair, Ch. 11). Thus, we might suspect that the 0.07-second-duration actively produced simple pulse trains of *B. mauritanicus* are comparable to the actively produced simple pulse trains of *B. perreti*.

Alternatively, are the simple pulse trains of *B.*

[1] In this paper *complex* refers to groups of species that are known to be closely related, such as the *regularis-maculatus-latifrons* complex, which are all African and have 2n chromosome number = 20, but which may not be a species group as defined by reproductive compatibility by Blair (Chap. 11).

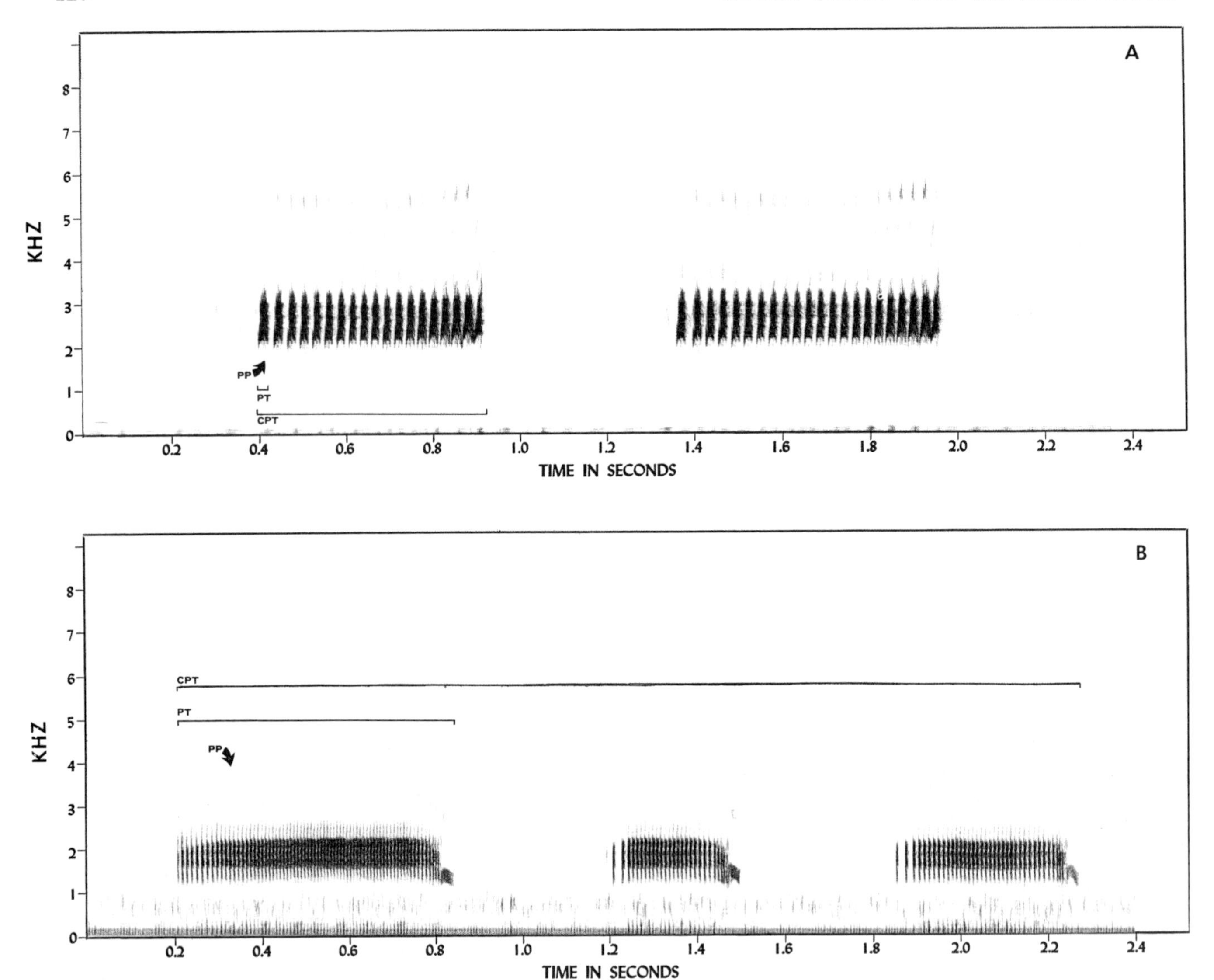

Fig. 9–5. Sound spectrograms of parts of *Bufo* mating calls consisting of first-order sequences of complex pulse trains containing active and passive amplitude modulation. (*A*) *B. speciosus*, 4 miles northeast of Fentress, Caldwell County, Texas, U.S.A., April 24, 1956; University of Texas library of anuran sounds, station 177. Air temperature, 17.0° C; water temperature 18.0° C. (*B*) *B. gariepensis*, near Vioolsdrif, Little Namaqualand, South Africa, by R. Keith, August 21, 1964. Air temperature, 6.1° C; water temperature, 8.9° C. Effective filter band width 50 Hz. In *B. gariepensis* each complex pulse train contains 2 to 4 simple pulse trains, whereas in *B. speciosus* each contains about 20 simple pulse trains.

mauritanicus comparable to simple pulse trains of *B. garmani* and *B. rangeri*, and do all these represent the same mechanism? In that case a different qualitative change would have occurred between the *B. garmani* type and the *B. mauritanicus* type — namely, the addition of an active temporal mechanism to repeat complex pulse trains periodically. A similar change has occurred between *B. cognatus* and *B. speciosus*. There is an almost complete continuum of simple pulse train durations and repetition rates within the *regularis* complex, ranging from those comparable to the slow rate of *B. kerinyagae* to somewhat faster ones, such as in *B. maculatus*, to fast ones, such as in *B. garmani*. The rate of *B. steindach-*

neri is even faster and approaches that of the Asian *B. melanostictus* — which has been shown through some lines of evidence (hybridization and osteology) to be fairly close to the *valliceps* and *marinus* groups of Central and South America. The latter groups are known to be phylogenetically close to the *guttatus* group of which *B. blombergi* is a member and with which *B. mauritanicus* shows a higher similarity of composition of parotoid secretions (96%) than with any other *Bufo* species (Low, Ch. 13). In *B. mauritanicus*, 2n = 22 as in *B. melanostictus*, *B. valliceps*, and *B. marinus*. The mating calls of *B. blombergi* and *B. mauritanicus* are also similar. Both species produce calls of two different types. One type consists of a first-order sequence of complex pulse trains, each complex pulse train containing simple pulse trains of passively produced pulses. In the other, irregular series of simple pulse trains of passively produced pulses and pulses of active origin lacking passive modulation are produced (see Fig. 9-6 *A–D*). Affinity between *B. mauritanicus* and this New World radiation seems apparent from many lines of evidence. Such evidence also indicates connection between these toads and the *regularis* complex, but the details of relationship of *B. mauritanicus* or the American species to the *regularis* complex are not clear. We have seen above that arguments can be presented suggesting a close connection with the *regularis* complex via *B. perreti* or a more distant one via *B. steindachneri* and the *B. melanostictus–valliceps–marinus* line. The confusion of homologous call elements among these species makes it difficult to clarify the problem by comparing call structure. Perhaps future physiological experimentation may shed some light on this.

There is less spectral variation among different species than is shown by the pulsatile components. In African species, as in all species of *Bufo* so far tested by W. F. Martin, a carrier frequency is generated by air passing over the anterior membranes of the larynx. Harmonics of this frequency, as well as harmonics of pulse repetition rates, are also produced. In the call of *B. sp.* (Fig. 9-7*B*), even-numbered harmonics of the pulse repetition rate are absent because of symmetry of pulses and interpulse intervals (see Watkins 1967). Often, frequencies are emphasized in three harmonically related regions (Fig. 9-7 *A* and *C*). These frequencies apparently represent the carrier frequency of the anterior membrane and its harmonics, and are referred to in descriptions of vocalizations in this paper as low-emphasized (low emp), mid-emphasized (mid emp), and high-emphasized (high emp) frequencies. One frequency is often dominant (dom), but this frequency is not always the resonant frequency of the vocal cords — an apparent difference between many African and non-African *Bufo* species. Not uncommonly, two harmonically related frequencies receive almost equal emphasis (Fig. 9-7*C*). In several species, a wide range of frequencies is emphasized (Fig. 9-7*B* and *C*).

Seasonal Breeding Patterns

All parts of Africa are characterized by rainy and dry seasons that occur at different times of the year in different regions. Only the extreme southwestern tip of the continent and the Mediterranean region are affected by marked thermal seasonality, although there is some thermal change correlated with the rainy and dry seasons even on the equator.

Many African toads, like most anurans, breed at the beginning of the rainy season. There are exceptions to this typical breeding pattern. Inger and Greenberg (1956) found that *B. fulginatus* females in the southeastern Democratic Republic of the Congo had a marked cyclic reproductive pattern with a peak in the rainy season, but that males did not show such a pattern (as indicated by development of secondary sex characters). At Idanre, western Nigeria, *B. regularis* (W) breeds in the typical pattern, with a sharp peak at the start of the rainy season. *B. maculatus* apparently breeds throughout the rainy season, also at Idanre, and *B. perreti* spawns midway through and at the height of the rains. Inger's data (1968) for *B. maculatus* and *B. regularis* (W) in the Parc National de la Garamba of the northeastern Democratic Republic of the Congo indicate patterns similar to those at Idanre. Winston (1955), however, recorded no definite breeding season in Nigeria for *B. "regularis."* This may be because of confusion with *B. maculatus*. *B. regularis* (W) at Ogbomosho, Nigeria, where there is normally a major and a minor rainy season each year, breeds primarily at the beginning of the major season, but some individuals breed at the start of the minor one. *B. garmani* has been observed breeding during the dry season (Keith 1968). *B. gariepensis* (= *B. angusticeps*?) at Cape Town, South Africa, breeds in winter during June, July, and August, when rains are most common there. *B. rosei* breeds in late August or early September, at the be-

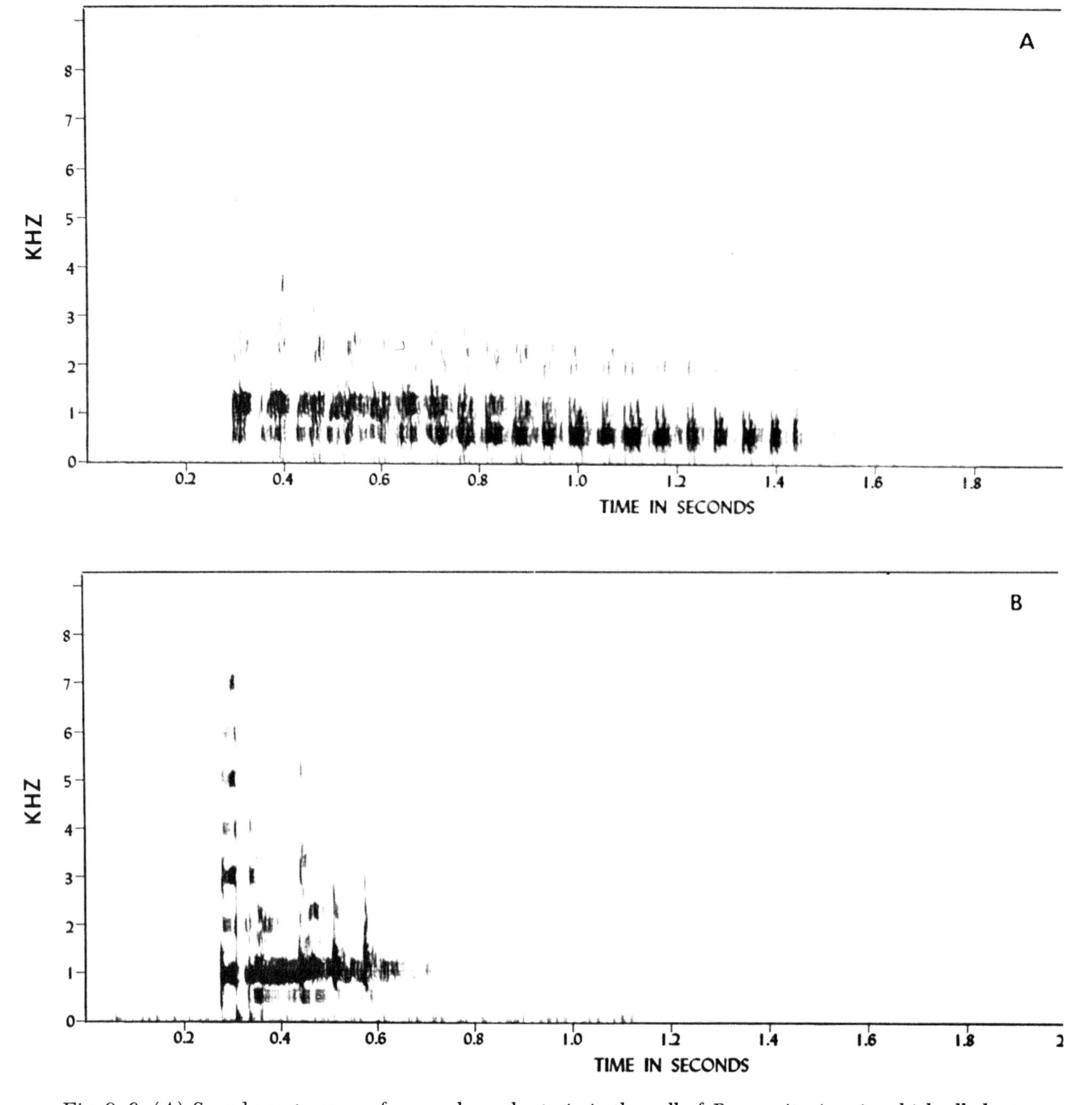

Fig. 9–6. (*A*) Sound spectrogram of a complex pulse train in the call of *B. mauritanicus* in which all short duration active amplitude modulations contain passive amplitude modulation. (*B*) A "quack" in the call of *B. mauritanicus* in which some active amplitude modulations contain little or no passive modulation. (*C*) A complex pulse train in the call of *B. blombergi* in which some active amplitude modulations contain passive modulation. (*D*) Two "hoots" in the call of *B. blombergi* in which long sections of active modulations lack

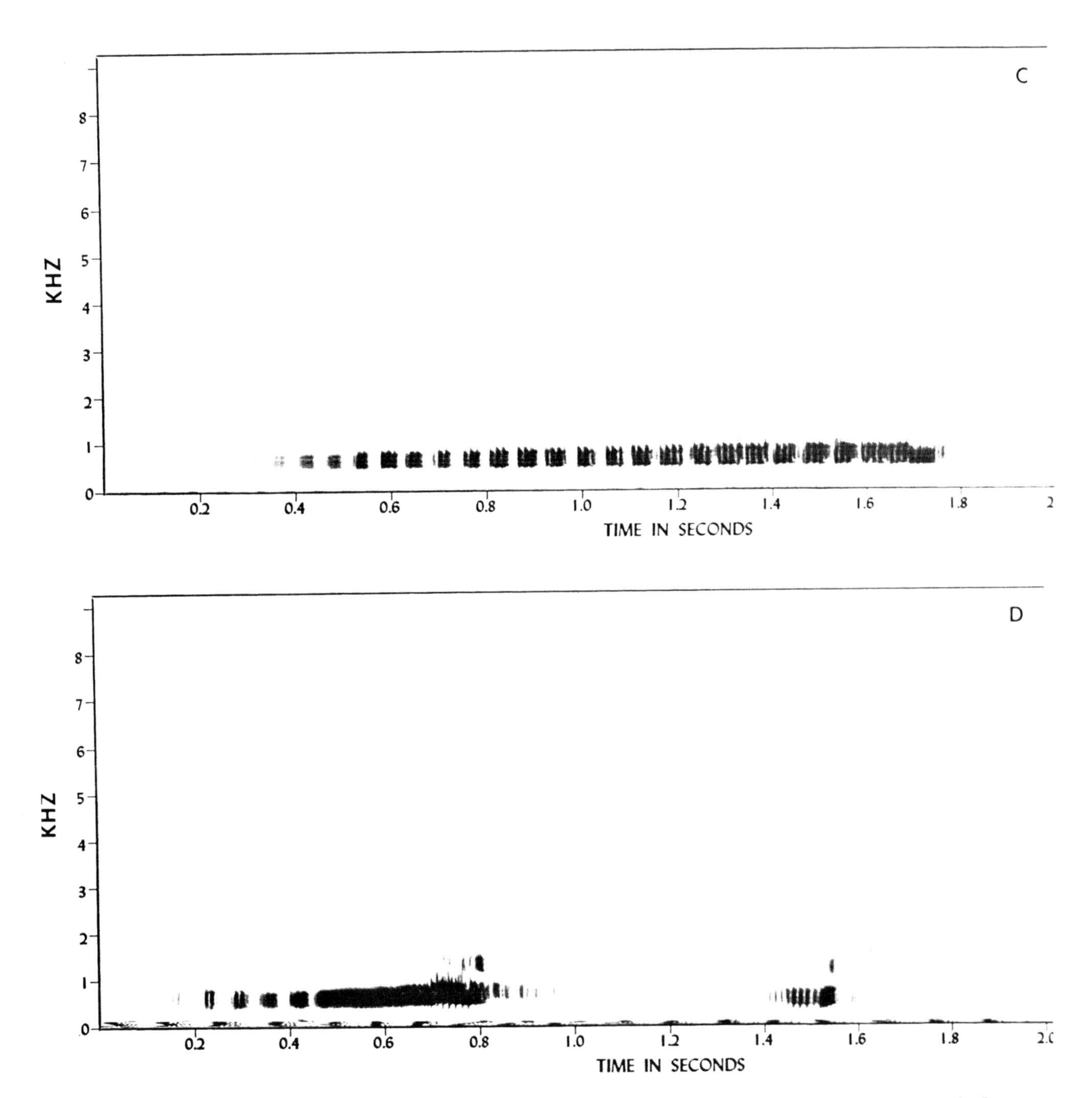

passive modulation. Locality for *B. mauritanicus* was given as "Morocco" by the dealer who supplied it. By W. F. Blair, February 7, 1961, laboratory at Austin, Texas. Air temperature 24.5° C. *B. blombergi* was obtained from Colombia from a dealer. By M. Tandy, in laboratory at Austin, Texas, April 7, 1969. Air temperature, 24.1° C; water temperature 22.5° C; Soil temperature 22.6 ° C.

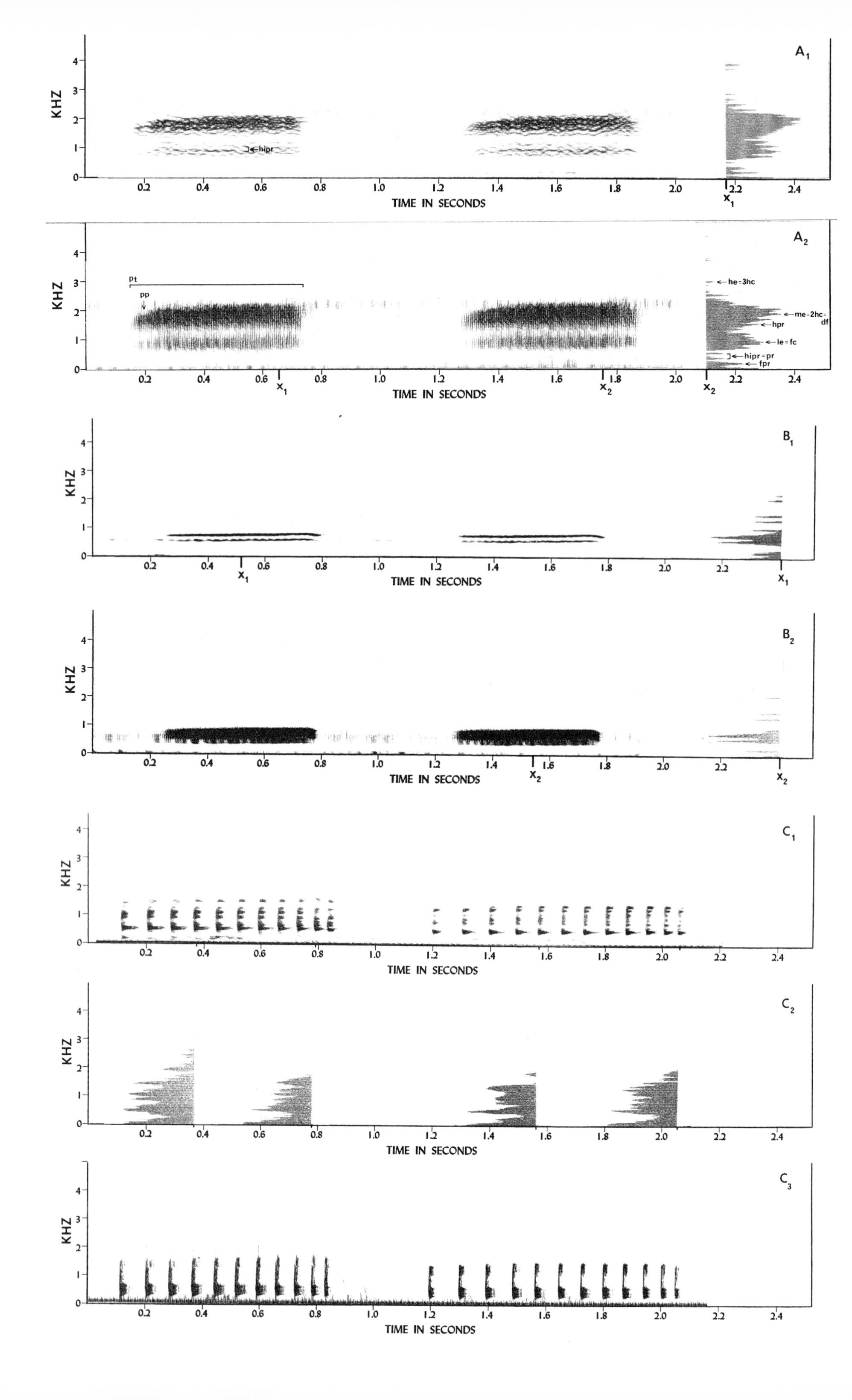
A1
KHZ
hipr
TIME IN SECONDS
X1
A2
pt
pp
he=3hc
me=2hc=df
hpr
le=fc
hipr=pr
fpr
X1
X2
B1
B2
C1
C2
C3

ginning of spring, on the summit of Table Mountain of the Cape Peninsula. The summit is well watered all year, but the winter rainy season of the Cape ends in late August. *B. regularis* (E) apparently starts breeding slightly later in spring in Pietermaritzburg, Natal, than *B. rangeri* (Poynton 1964). The population of *B. "regularis"* studied by Menzies (1963) in Sierra Leone showed a peak of breeding activity during the second month of the dry season. *B. regularis* (W) was recorded at Ikom, Nigeria, at the end of December 1964, which corresponds to the part of the major dry season during which Menzies observed breeding in Sierra Leone. It is not known whether the Ikom population exhibits cyclic breeding activity. A chorus of *B. regularis* (W) was recorded at Ibadan, Nigeria, in mid-January 1965, which is the third month of the four-month major dry season there. The population studied by Chapman and Chapman (1958) bred primarily during the first half of the six-month rainy season of the Rukwa Valley, but some toads bred during the dry season. These facts and Chapmans' developmental information suggest that at least two species were involved in their material.

Circadian Patterns

The pattern of activity of African toads is that typical for the genus. Most species breed at night. Males of vocal species usually begin calling at or near the breeding site at dusk and continue for several hours, sometimes until after dawn. Often there is little calling after one or two a.m. Internal and external factors apparently affect the calling behavior. Rain induces calling. During strong wind it often stops. Calling has been observed on many occasions to cease before dawn, which may be related to the gradual drop in air temperature during the night. Figure 9-8 shows that the temperatures of calling animals were highly correlated with the air temperatures at the time of calling and with the mean annual air temperatures for the localities but not with the water temperatures during calling. The regressions of air temperature with cloacal temperature and of mean annual temperature with cloacal temperature are not statistically different (Table 9-2). Neither of these regressions is significantly different from that of a 1:1 relationship between air and toad or mean annual and toad temperatures. The animals possibly cease calling when the body temperature cools to a

Table 9–2. Probabilities that Regressions Shown in Fig. 9–8 are Identical

Regression	Air-Cloacal Temperature	Mean Annual Air-Cloacal
Mean annual air-cloacal temperature	$0.10 > P > 0.05$	***
1:1	$P > 0.10$	$P > 0.10$

Figure 9–7 A–C. Sound spectrograms of parts of mating calls of three species of African *Bufo* illustrating spectral structure. (A_1&2) Two simple pulse trains of the call of *B. maculatus*, Orlu, Nigeria, by M. Tandy, August 17, 1965. Air temperature, 24.4°C. (A_1) Filter band width 300Hz. The zone of energy concentration around 1000Hz corresponds to the carrier frequency and various harmonics produced by the passive pulse repetition rate. The passive pulse repetition rate of about 160 pulses/second may be read directly from the harmonic intervals, which are apparent at this band width. This rate can also be read from the line that appears at its fundamental frequency of 160Hz. The zone of energy concentration at about 2000Hz corresponds to the second harmonic of the carrier frequency, and in this case it is the dominant frequency. (A_2) Filter band width 50Hz. At this band width, the passive pulses are visible as discrete pulses, and the passive pulse rate may be measured on the time scale. X_1 and X_2 mark the places where the spectral sections were made. Abbreviations: fpr=fundamental frequency of the passive pulse repetition rate; hipr=harmonic interval produced by the passive pulse repetition rate, which is equal to that rate (pr); hpr=an harmonic of the pulse repetition rate; le=low-emphasized frequency; me=mid-emphasized frequency; he=high-emphasized frequency; fc=the fundamental carrier frequency of the vocal-cord resonance; 2hc=second harmonic of the carrier frequency; 3hc=third harmonic of the carrier frequency; df=dominant frequency. (B_1&2) *B. sp.*, Voi, Kenya, by R. Keith, October 3, 1962. In the call of this species, the carrier frequency is the dominant frequency, and there is very little energy in the upper harmonics of the carrier. (B_1) Filter band width 300Hz. The passive pulse repetition rate of 170 pulses/second may be read from the harmonic interval of the spectrogram as 170Hz. (B_2) Filter band width 50Hz. Passive pulses are visible at this band width, and their rate may be measured on the time scale. (*C*) *B. regularis* (E), near the Loginya Swamp at Ol Tukai, Masai, Amboseli Game Reserve, Kenya, by M. Tandy, November 16, 1965. Air temperature 18.0° C; water temperature, 24.0° C; toad temperature (both), 19.5° C. One simple pulse train from the mating calls of each of two individual toads is shown with the narrow band spectrogram, the spectral sections, and the wide band spectrogram in temporal alignment. (C_1) Filter band width 300Hz. The carrier and at least its second and third harmonics are evident within each passive pulse. (C_2) Spectral sections. The carrier, its harmonics, and other spectral peaks are quite visible. In this case, the carrier is the dominant frequency, which is not always true for this species. Note the differences in intensities and distribution of spectral energy between each pp within each pulse train. (C_3) Filter band width 50Hz. Spectral structure is here less apparent, but the wave form of each passively modulated pulse is clear. All spectral sections were made at band width 300Hz.

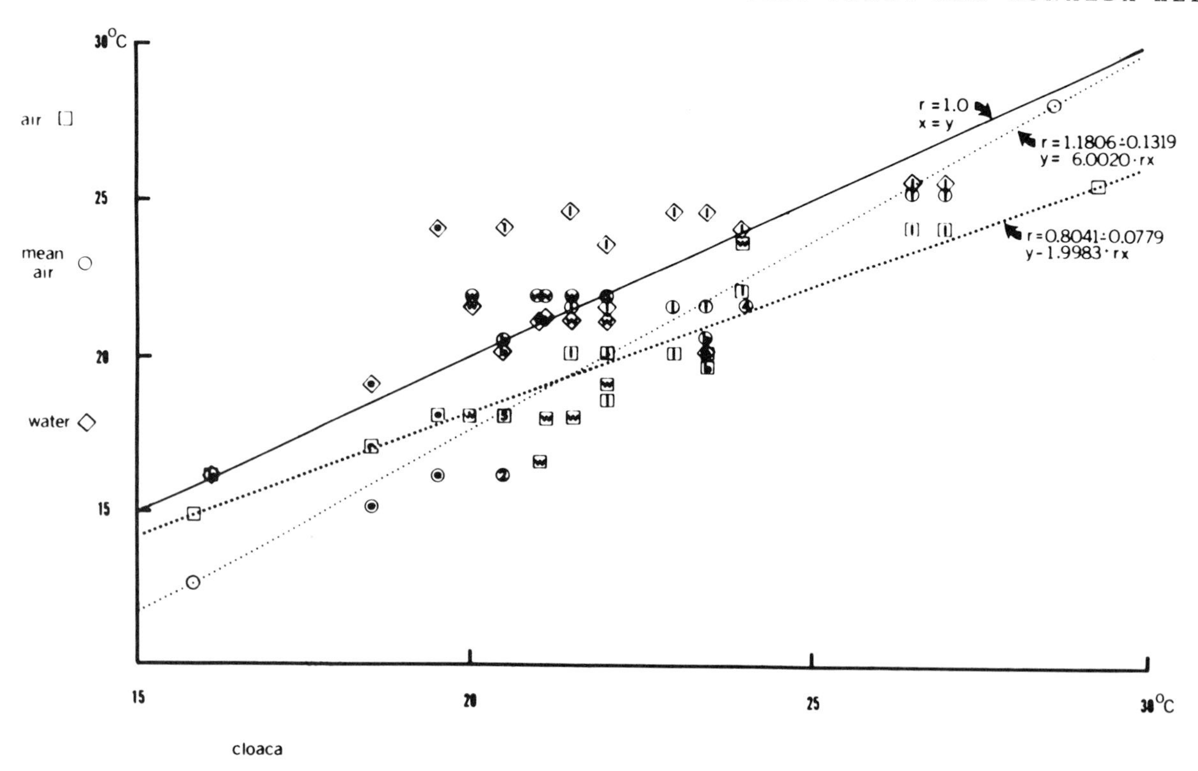

Fig. 9–8. The relationships between air temperature at the time of calling, mean annual air temperature, and water temperature at the time of calling to toad temperature. Regression lines and coefficients are given for a 1:1 relationship and for the significant regressions of air temperature and mean annual temperature to toad temperature. P for each of the regression coefficients of the latter being zero is less that 0.001. Abbreviations: b=*B. brauni*, e=*B. regularis* (E), k=*B. kerinyagae*, l=*B. maculatus*, w=*B. regularis* (W); 1, 2, 4, & 5 = *B. garmani* & *B. sp.*, 3 = *B. maculatus* and *B. regularis* (W). Each point on the graph represents one or more individuals. Total number of individuals used for each plot was 37.

certain level. However, we have no temperature data for noncalling animals.

B. rangeri was heard calling all day and night near Swellendam, South Africa, in early September 1965. Power (1927*a*) observed similar activity of *B. garmani* at Kimberly, South Africa, in November 1925. Power (1925) stated that *B. carens* breeds day and night just after rains. Chapman and Chapman (1958) reported that the *B. "regularis"* they studied called during the day at the height of breeding activity, but that this calling was less intense than nocturnal calling at the same time of year. It is not known whether such diurnal calling is typical. The earless *B. lönnbergi* (= *B. taitanus nyikae* ?) has been observed spawning during the day in small pools in grassy meadows on the Nyika Plateau in Malawi (Stewart 1967). The recording of *B. lönnbergi* at Nanyuki, Kenya, was made at night.

Reproductive Aggregation

The common anuran pattern of males congregating at ponds or other spawning sites during the breeding season is characteristic of many African *Bufo*. Somewhat exceptional are *B. brauni*, *B. gariepensis*, *B. sp.*, and *B. maculatus*, the males of which often space themselves several yards apart, but the antiphonic vocal response pattern (Capranica 1965) is still exhibited in these species. *B. gracilipes* apparently calls in small groups at small waterholes in high tropical forest (Schiøtz 1969). In many species — e.g., *B. regularis* (E), *B. regularis* (W), *B. carens*, *B. kisoloensis*, and *B. rangeri* — the aggregations are so dense that it is difficult to make tape recordings of individual animals.

Wager (1965) noted apparent segregation of choruses of two densely congregating species — *B. carens* and *B. regularis* (E) — at different breeding sites within close proximity of one another. The pattern was reestablished after individuals were artificially displaced from one site to the other. A. A. Martin (personal communication) has data on similar situations involving *B. rangeri* and *B. regularis* (E) on the high veld of South Africa.

The behavior of earless species is largely unknown, but, as mentioned above, aggregations of *B. lönnbergi* have been observed. Boulenger (1907) cited observations by Swynnerton of three male *B. anotis* in nuptial dress found in small water-filled holes at the foot of large trees in the Chirinda Forest. Fitz-Simons (1939) observed a group of five males in a water-filled cavity of a rotten log at the same locality. Pairs of *B. rosei* were observed spawning in a small shallow pool of rain water on the summit of Table Mountain on August 18, 1927, by Power and Rose (1929). On September 8, 1965, John Visser and Mills Tandy found several small shallow pools, about four inches deep, on Table Mountain near McLear's Beacon filled with thousands of *B. rosei* eggs. A search beneath damp decayed reeds overlying the muddy edges of these pools revealed an enormous aggregation of hundreds of adults, mostly males, literally packed among the reeds like sardines. Among one hundred individuals there were only five females, of which three were spent. The two other females threw their eggs that night in a collecting tin despite separation from the males. This aggregation of hundreds of individuals is the densest we have ever observed among anurans. *B. rosei* lacks a mating call and ear structures, so that the mechanism of aggregation is not acoustic, which enhances the suspicion that there is a chemical attraction related to the swollen pink inguinal area of breeding adults of the species, as suggested by Power and Rose (1929).

Selection of Call Site

In addition to differences in spacing by vocalizing males, there are species-specific differences in selection of calling sites. *B. carens* normally calls while floating in the water of a pond or small stream (Loveridge 1925; Power 1925; Wager 1965). *B. regularis* (E) and *B. regularis* (W) usually call while sitting in shallow water near the bank. *B. garmani, B. sp., B. gariepensis,* and *B. pentoni* (Schiøtz 1964) call from exposed sites near the bank. *B. rangeri* often calls from rocky outcrops in the middle of or alongside streams (Hewitt 1935; A. A. Martin [personal communication]; Poynton 1964). *B. brauni* and *B. maculatus* call from the bank, but well concealed among low vegetation. *B. perreti* calls from exposed sites on granite inselberg outcrops while sitting in small runnels on the rock surfaces that are also used for spawning.

Mate Selection

There have been very few observations of the actual selection of mates among African *Bufo*. Most pairs have been found in amplexus. Chapman and Chapman (1958) reported that their *B. "regularis"* males, which called while sitting in shallow water near the bank of a springfed pool, "sometimes" advanced out of the water toward females that approached them from the bank. Amplexus followed "at once." They did not state whether amplexus was effected in or out of the water. It presumably occurred wherever the male met the female. On the night of June 26, 1965, at Idanre, in western Nigeria, while attempting to induce a male *B. maculatus* to call from its hidden site among dense herbs surrounding a small stream by playing a recording he had just made of the voice of that individual, Tandy noticed another *B. maculatus* near the tape record er. This was a female. A few feet away, a male *B. perreti* had been calling from an exposed algal slick on the rock outcrop where the small stream emerged from the dense herbs. This male now sat silently. Tandy placed the female *B. maculatus* in a collecting bag and continued playing the *B. maculatus* recording and searched for the male, which he soon succeeded in capturing. He next noticed a pair of toads in amplexus near the recorder. The male was the *B. perreti* observed earlier — now in amplexus with a female *B. maculatus*. Both female *B. maculatus* were apparently attracted to the conspecific call coming from the tape recorder. The male *B. perreti* embraced one of them, possibly because of the female's proximity and visual availability to him. The difference in calling site of the males coupled with the conspecific responses of the females possibly keeps females out of the visual range of heterospecific males, thus functioning as a premating isolating mechanism. Conspecific pairings of these species have not been observed.

On the night of November 8, 1965, several pairings of *B. regularis* (E) were observed at a small pond near Mazeras, Kenya. Approximately ten *B. regularis* (E) males called while sitting near the bank with their front feet out of the water on the mud bank and their hind quarters in the water. Three male *B. maculatus* were also present and these called while sitting on the bank entirely out of the water. There was no low vegetational cover where they could conceal themselves. Two pairs were in amplexus in the pond when the senior author arrived, one each of *B. regularis* (E) and *B. maculatus.* The edges of the pond were surrounded on three sides by a vertical two-foot embankment. On the third side was a gradual slope down to the water's edge. Toads could leave and enter the pool only from the sloping side. A short time after Tandy arrived, a female *B. regularis* (E) approached the chorus fairly rapidly, making her way down the slope to the water's edge. The males took no apparent notice of her until she hopped into the water. At that time all the male *B. regularis* (E) ceased calling and lunged for the spot where the female had entered. The male *B. maculatus* merely stopped calling briefly. A male *B. regularis* (E) quickly amplexed the female while struggling with other males trying to do the same, and the pair then submerged. The other males would mount each other briefly but release when the other male uttered a release call. When the amplexed pair surfaced a short time later, three or four other males followed them and one usually clasped the properly amplexed male. This evoked a loud release call. Sometimes the second male would release, sometimes it retained its grip and also began giving release calls, and at other times it held on and kicked other males that were trying to amplex. After several submersings, the amplexed pair was left mostly alone, and the unpaired males resumed calling. The males appeared to be very sensitive after this activity and would occasionally lunge at another male when that male moved slightly.

Two more pairings similar to the first were observed on the same night. The males of *B. maculatus* never entered the water during this activity. They would stop calling briefly and then resume. Shortly after the second *B. regularis* (E) pairing was observed and as unpaired males were returning to calling positions along the bank, a male *B. regularis* (E) surfaced and put his feet on the bank near a male *B. maculatus.* The *B. maculatus* had been calling, but he stopped when the *B. regularis* (E) surfaced. He soon started to call but was attacked by the *B. regularis* (E) with a snap of the mouth. The *B. maculatus* withdrew a few inches, and later the two started calling, apparently taking little notice of one another.

Males of *B. regularis* (E) and *B. rangeri* have been observed entering the water and attempting to clasp objects that disturb the water surface. The observations at Mazeras suggest that this may be how the sexes of these species are brought together. Perhaps the female is attracted to the calls of the male from a distance, but the recognition of her presence by the males appears to be a response to her disturbance of the water surface. It appears that she does not select a particular male. This might not be true at a larger pond, where she might be able to enter the water closer to one male than to others, or she might approach a male on land. It was stated earlier that Chapman and Chapman (1958) reported a different pattern of female approach. However, the species identity of their material is uncertain. Their specimens of *B. "regularis"* were identified by Arthur Loveridge who at that time did not recognize many of the different species of 20-chromosome toads found in East Africa. Among the latter, B. *garmani, B. gutturalis, B. maculatus,* and *B. regularis* (E) might be expected to occur at Kafukola, Tanzania, Chapman's study site.

The behavior of the male *B. maculatus* at Mazeras suggests that they do not recognize the presence of females in the same manner as *B. regularis* (E). The behavior of the female *B. maculatus* at Idanre and the perfection of visual concealment of the males suggest that the females approach the males very closely before they pair. The fairly wide spacing of *B. maculatus* males also suggests that females of this species possibly choose an individual male.

Amplexus and Spawning

All African *Bufo* known exhibit thoracic amplexus, which is typical for the genus. Spawning patterns vary somewhat among different species. *B. lönnbergi* and *B. rosei* deposit their eggs unattached to debris on the bottoms of shallow pools about four inches deep (Power and Rose 1929; Stewart 1967). In these species, as in all other African *Bufo* observed, the eggs are connected in double gelatinous strings surrounded by outer transparent tubes of jelly. In the one set of *B. perreti* spawn observed, strings of eggs were fastened to a clump of grass growing from a rock crevice that lay directly in the path of water flowing over a granite outcrop. Although less than half the eggs were submerged, there was sufficient flow and turbulence to keep the eggs wet. The strings were not spread out but were all clumped together in a fairly small mass near the point of attachment to the grass.

B. carens deposits its eggs along the shore in shallow water in several parallel rows about four feet in length, also parallel to the bank, entwining them around stones and weeds. Up to twenty thousand eggs may be deposited in this manner (Power 1925, 1927*a*; Rose 1962; Chapman and Chapman 1958). Members of the *regularis* complex entwine their egg strings, in seemingly haphazard fashion, around submerged vegetation in shallow water less than one foot deep.

Ecological Limits and Geographic Distribution

No species of *Bufo* is distributed throughout the African continent. Most species are restricted to one or two major ecological zones, though not necessarily to zones of specific plant composition. Others are restricted to certain vegetational complexes, a fact particularly true of equatorial forest species. Some are endemic to very small regions. Several are restricted to the northwest (*B. mauritanicus* and *B. bufo*), the north (*B. viridis*), or northeast *B. dodsoni* and *B. blanfordi*), possibly reflecting their recent invasion from other continents. Temperate South Africa has several endemic forms. In the vast area of tropical Africa, altitude and rainfall play primary roles in determining distributions. Although some species exhibit wide tolerances to climate, all species whose distributions are reasonably well known occupy ecological zones that can be defined in terms of mean annual rainfall and temperature. Examples of this may be seen by comparing the distribution of *B. brauni, B. garmani, B. kisoloensis, B. rangeri*, and *B. superciliaris* shown in figure 9-12 with the maps of mean annual rainfall in figure 9-13, 9-14, and 9-15. *B. rangeri* is limited to areas of South Africa with mean annual rainfall greater than 53 cm. and usually mean annual temperature of 15-18°C. *B. kisoloensis* is found in montane areas of east-central Africa with rainfall greater than 125 cm/year and average temperature similar to that for *B. rangeri*. Such distributions are often correlated with natural vegetation zones. These natural vegetation zones do not always correspond to the specific types of vegetation growing in the areas today (Acocks 1953).

Ecologically Adaptive Morphology

Many desert, montane, and forest species have short legs, which are useful for burrowing forms. Short legs are obviously useful for desert species and species of regions with prolonged dry seasons. They are also well suited for the largely fossorial habits of montane species. The suitability of short legs for forest life is less apparent, but here also this adaptation may be related to burrowing.

Aquatic species such as *B. kisoloensis, B. lemairi*, and *B. rangeri* have unusually long legs and well-developed webbing suited to their habitat preferences.

Several species exhibit extreme ecological specializations. *B. perreti* is an example. The tadpole has a very long tail that enables it to "crawl" about half submerged on wet granite rock faces (Schiøtz 1964). The elongated appendages of the adults are also adaptive for the rupicolous habits of this toad. The tarsal spur of the desert-dwelling *B. pentoni* parallels such a development in the pelobatid spadefoots of North America. An elongate form and large oral sucker are adaptive for life in the rapid-flowing streams of Mount Cameroun and the Manengouba Mountains where *B. preussi* tadpoles live (Perret 1966). A dead-leaf color pattern is especially well developed in the forest species *B. brauni* and *B. superciliaris*. *B. dombensis* is colored remarkably like the lichen-covered rocks it inhabits.

The functions of some morphological specializations, such as the cranial fin fold of the larvae of *B. carens*, are not understood. Possibly it serves as an accessory respiratory organ. *B. carens* lives in dry areas and often breeds in stagnant pools. Power (1925) described the behavior of *B. carens* tadpoles in particularly stagnant water. They swam at the surface with this structure expanded.

General Comparative Morphology

Adults of most African species fall within a narrow size range. Forms showing great clinal variation in size (e.g., *B. gariepensis*) are exceptional. Variation in size within species is possibly related directly, or indirectly through its effects on the animals' vocalizations, to functions of premating isolation. There is usually sexual dimorphism in size, the female being the larger. This difference may be considerable — as in the case of *B. superciliaris* (♂♂ $\bar{x}$ 115 mm.; ♀♀ $\bar{x}$ 140 mm.) and *B. tuberosus* (♂♂ $\bar{x}$ 38 mm.; ♀♀ $\bar{x}$ 64 mm. snout-vent). In at least one species (*B. kerinyagae*) the males are about the same size as the females (♂♂ $\bar{x}$ 68.6 mm.; ♀♀ $\bar{x}$ 70.8 mm. snout-urostyle). The smallest African species is *B. micranotis*, in which the largest adult female known measures only 23 mm. The largest species is *B. superciliaris*. An enormous female from Oda, Ghana, in the Field Museum of Natural History collection, measures 168 mm. from snout to urostyle. This size is still short of the Colombian *B. blombergi*, the world's largest toad.

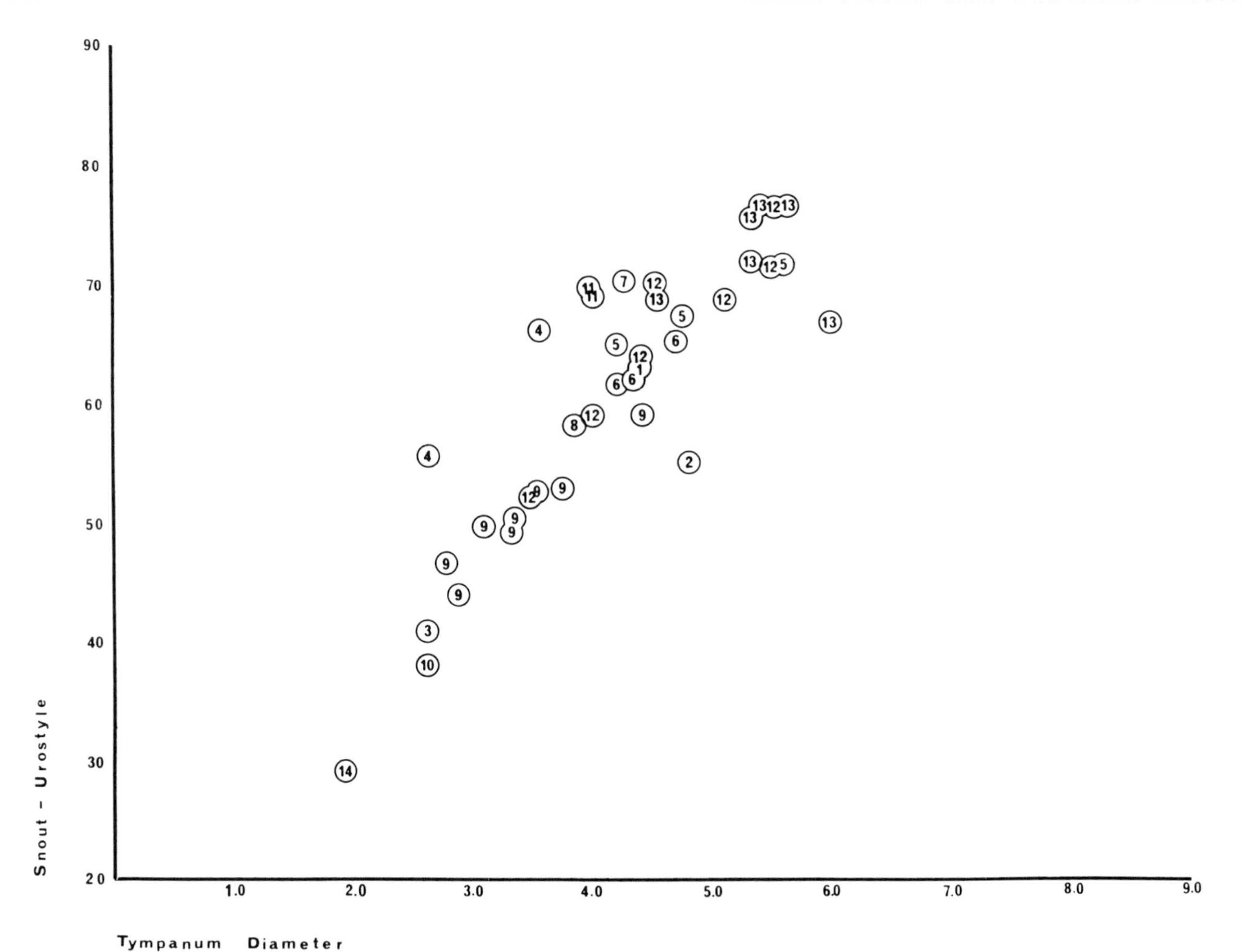

Fig. 9–9. The relationship between tympanum size and body size. Each point is a population mean of several individuals from a given locality. Abbreviations: 1 = *B. brauni,* Amani, Tanzania. 2 = *B. carens,* Iringa, Tanzania. 3=*B. sp.,* Loiengalani, Kenya. 4=*B. gariepensis,* Ladismith and Vioolsdrif, South Africa. 5=*B. garmani,* Namanga, and 01 Tukai, Kenya; Letaba, Kruger National Park, South Africa. 6=*B. sp.,* Mtito Andei, 01 Tukai, and Voi, Kenya. 7=*B. kerinyagae,* Nanyuki, Kenya. 8=*B. kisoloensis,* Kitale, Kenya. 9= *B. maculatus,* Lake Ejagham, Eyumojok, Mamfe Division, West Cameroon; Mazeras and Sigor, West Pokot, Kenya; Idanre and Orlu, Nigeria; Muheza, Tanzania; Entebbe, Uganda. 10 = *B. perreti,* Idanre, in western Nigeria. 11=*B. rangeri,* Port St. Johns, Transkei, and Swellendam, Cape Province, South Africa. 12=*B. regularis* (E), Mazeras, Nyeri, and 01 Tukai, Amboseli, Kenya; Lourenço Marques, Mozambique; Vumba Mountains, Rhodesia; Louis Trichardt, Transvaal and Port St. Johns, South Africa. 13=*B. regularis* (W), Ikom, Idanre, Ilesa, and Ogbomosho, Nigeria; Vumba Mountains, Rhodesia; Entebbe, and Kampala, Uganda. 14=*B. vittatus,* Buko Bay, near Entebbe, Uganda.

Most African species are cryptically colored. There is usually only slight sexual dimorphism in general color. The females often have patterns of greatest contrast. Exceptions are *B. kisoloensis* and *B. lönnbergi,* in which the males are bright greenish yellow and the females are patterned brown and black as in both sexes of most *B. regularis* (E) and *B. regularis* (W). This coloring may be related to the adaptive value of male advertisement during the breeding season versus concealment of the female, especially if the female makes the mating choice.

The parotoid glands are generally oblong. An exception is seen in the oval parotoids of *B. tuberosus.* In *B. regularis* (E), *B. regularis* (W), and *B. dom-*

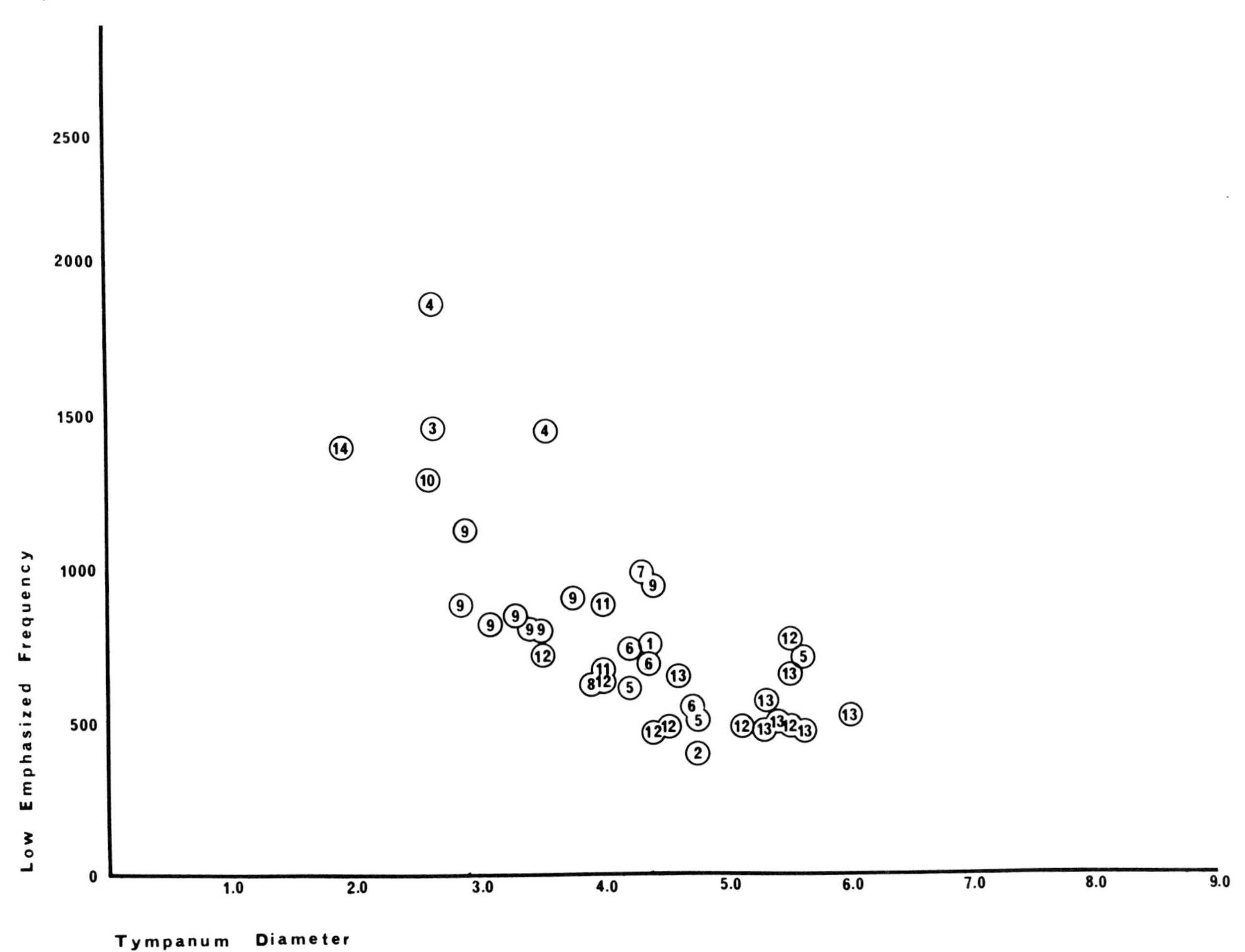

Fig. 9–10. The relationship between tympanum size and low-emphasized frequency of the mating call. (Symbols as in Fig. 9–9.)

bensis, they are somewhat kidney-shaped. In *B. blanfordi* the glands are indistinct. *B. carens* apparently lacks parotoids but has long dorsolateral glandular ridges that resemble parotoids. It is not known whether these are homologous to parotoid glands.

The size of the tympanum varies among different species. Among small species its size relative to the size of the animal is less variable than among large species (Fig. 9-9). The size of the tympanum appears to be correlated with the spectral structure of the mating call, particularly with the low-emphasized frequency (Fig. 9-10). The lesser variability in size of the tympanum and vocal mechanism (Martin, Chap. 15) and range of spectral components among small species are probably allometrically limited. Larger species are not so limited, and thus we find greater variability among them, which possibly reflects broad selective acoustical experimentation during their evolution.

Species among the large *taitanus* complex and *B. rosei* lack tympanic and other auditory structures. The breeding behavior of most of these taxa is largely unknown. As mentioned earlier, *B. lönnbergi* males emit a call in breeding aggregations very similar to mating calls of other species. The possibility of nonaural acoustic reception must be considered.

The males of all African *Bufo* for which mating calls are known develop median subgular vocal sacs. Many species that are acoustically unknown also have these structures. The bucal entrances to the vocal sac are situated laterally in the floor of the mouth. There may be a right or a left opening, or both, and the state of this character often varies considerably within species (Schmidt and Inger 1959; Keith

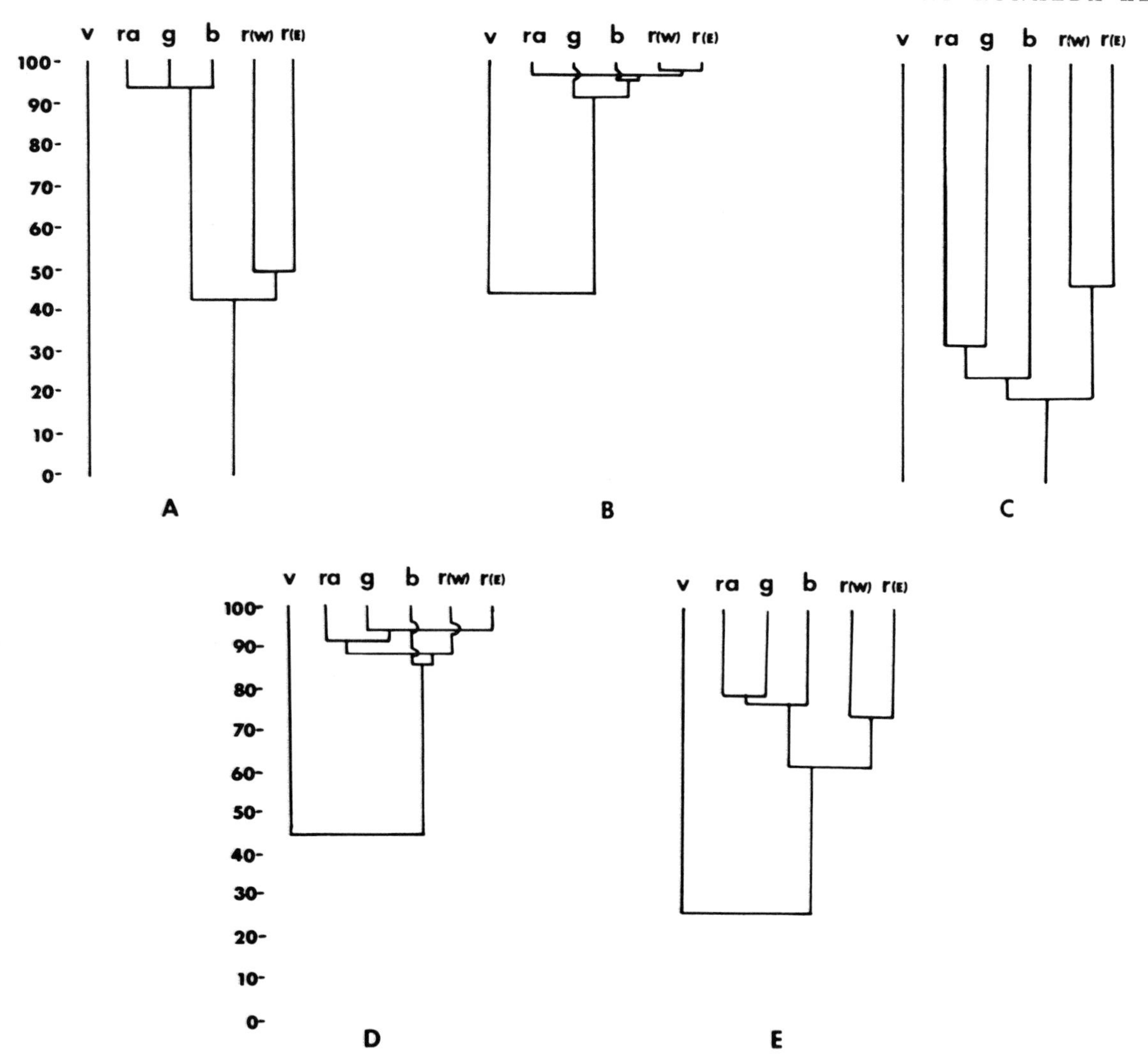

Fig. 9–11. Dendrograms of phenetic similarities of six African *Bufo* species based on four data systems and an average of them. (*A*) Passive pulse rate of mating call. (*B*) Parotoid-gland secretions. (*C*) Hemoglobins. (*D*) Hybridization experiments. (*E*) Average of first four.

1968). Some species lack vocal sacs and their openings, e.g., *B. superciliaris* and *B. fulginatus.* Males of most species with vocal sacs have the gular skin pigmented yellow, blackish green, or blue. Exceptions are *B. gariepensis, B. kerinyagae,* and *B. viridis.*

Males of all African species, with the possible exception of *B. superciliaris*, develop melanized nuptial asperities on at least the first finger during the breeding period.

There is considerable variation in the melanized dermal spines over much of the dorsal and lateral surfaces, which is correlated with breeding activity. Whether such spines increase in size and number or degenerate during the breeding season depends on the species and sex (Schmidt and Inger 1959; Inger 1968). Inger and Greenberg (1959) have suggested that such differences between sympatric *B. fulginatus* and *B. "regularis"* function as a premating isolating mechanism.

Evolutionary Relationships

Objectives

Although only fossil evidence or other historical documentation can give unequivocal answers to many phylogenetic questions, much can be learned from extant organisms. It is important to consider as many aspects of the forms in question as possible. Single characters reflect patterns similar to those based on many characters, but the validity of such patterns lies in their concordance with patterns based on many characters (Sturtevant 1942; Throckmorton 1968). The sheer magnitude of the mathematical tasks required for analyzing similarities of forms based on many characters has until recently retarded the development of techniques of multivariate analysis. Modern computers have, however, overcome this problem.

The combined efforts of several workers who have analyzed various systems within the genus *Bufo*, reported in this volume and elsewhere, have produced much data suitable for multivariate analysis, which also includes information on African species. As a basis for phylogenetic discussion, we have utilized these data to generate patterns of phenetic similarities (Sokal and Sneath 1963) and phylogenetic possibilities. Patterns of phenetic similarity per se are not considered to be the same as phylogenetic ones. Concordance among phenetic patterns based on different systems and patterns produced by combinations of different systems should provide the closest approximations to phylogenetic patterns.

Methods

Matrices of similarity values for operational taxonomic units (OTU; Sokal and Sneath 1963) were constructed from four different data systems (1. mating calls, 2. parotoid-gland secretions, 3. blood proteins, 4. hybridization experiments) and from a combination of these four systems. The derivation of these matrices differed slightly for the four systems as discussed below under each data system. The OTU's were of different species. This analysis deals with six African species for which data are available from the above four systems: *B. brauni, B. garmani, B. rangeri, B. regularis* (E), *B. regularis* (W), and *B. viridis*. The first five are members of the *regularis* group *sensu* Blair (Chap. 11). Matrices have been constructed based on data systems 1, 2, 3, and 4 listed above, and a combination matrix has been created by averaging these four matrices. These matrices and their clustering patterns, dendrograms generated by single linkage (Sokal and Sneath 1963), are illustrated in tables 9-3 to 9-7 and in figure 9-11. The matrices based on all four data systems were generated as follows:

1. *Mating call physical structure.* Only the passive pulse rate was used in this example. Similarities were calculated as ratios of slower to faster rates of each species pair.

2. *Parotoid-gland secretions.* These similarities are the weighted paired affinities shown in table 13-9 (Chap. 13). Paired affinity values used for *B. regularis* (E) and *B. regularis* (W) compared with other species were taken as those of the population which showed the highest paired affinity with the species in question.

Table 9–3. Affinities Based on the Passive Pulse Rate of the Mating Call

	b	g	ra	r(E)	r(W)	v
Bufo brauni						
B. garmani	.94					
B. rangeri	.57	.94				
B. regularis (E)	.21	.20	.12			
B. regularis (W)	.40	.40	.24	.49		
B. viridis	0	0	0	0	0	

Table 9–4. Affinities Based on Parotoid Secretions

	b	g	ra	r(E)	r(W)	v
Bufo brauni						
B. garmani	.89					
B. rangeri	.93	.82				
B. regularis (E)	.96	.93	.97			
B. regularis (W)	.96	.84	.93	.98		
B. viridis	.35	.34	.45	.44	.44	

Table 9–5. Affinities Based on Hemoglobins

	b	g	ra	r(E)	r(W)	v
Bufo brauni						
B. garmani	.25					
B. rangeri	.25	.33				
B. regularis (E)	.18	.09	.20			
B. regularis (W)	.15	.07	.15	.47		
B. viridis	0	0	0	0	0	

Table 9–6. Affinities Based on Data from Hybridization Experiments

	b	g	ra	r(E)	r(W)	v
Bufo brauni						
B. garmani	.82					
B. rangeri	.76	.92				
B. regularis (E)	.86	.94	.68			
B. regularis (W)	.74	.88	.88	.80		
B. viridis	.44	.44	.44	.44	.44	

Table 9–7. Average Affinities Based on Data from Four Systems

	b	g	ra	r(E)	r(W)	v
Bufo brauni						
B. garmani	.73					
B. rangeri	.63	.75				
B. regularis (E)	.55	.54	.49			
B. regularis (W)	.57	.55	.55	.69		
B. viridis	.20	.20	.22	.22	.22	

3. *Blood proteins.* This example includes only hemoglobins. The similarity values are paired affinities.

4. *Hybridization data.* This matrix was derived from table 11-7 (Chap. 11). Affinity values were calculated by scaling each reciprocal cross as follows and summing the value for each reciprocal:

Cross Result	Affinity Value
sterile (no development)	0.000
some cleavage	0.066
gastrula	0.132
neurula	0.200
larva	0.300
metamorphosis — to 100%	0.301–0.5

Each reciprocal had a maximum value of 0.5; each similarity value 1.0. Backcrosses or F_2's are not considered in this example. Where one reciprocal cross was lacking, the value of the other was doubled to obtain the similarity value.

Results

The matrices in tables 9-3 to 9-7 and the dendrograms of figure 9-11 show a number of common features. Members of the *regularis* group are markedly separated from B. *viridis. B. regularis* (E) and *B. regularis* (W) have higher affinities for each other than for other species in all systems except hybridization. The composite matrix and dendrogram also indicate this. *B. rangeri, B. garmani,* and *B. brauni* cluster closer together but are more distant from *B. regularis* (E) and *B. regularis* (W) in the dendrograms based on passive pulse rate, hemoglobin composition, and the composite. This dichotomy is not evident in the dendrograms for parotoid secretion and hybridization. It probably would have been shown in the dendrogram for hybridization if F_2 data had been included. An F_2 cross of hybrids between *B. brauni* and *B. garmani* produced larvae (Chap. 11). The matrices and dendrograms for parotoid secretion and hybridization are similar in that both indicate similar high levels of affinity among all the *regularis*-group species. In this way they vary from the passive pulse rate and hemoglobin data that indicate the intragroup dichotomy mentioned above. The dendrograms for passive pulse rate and hemoglobin are very similar in pattern, although they are divergent in actual affinity values. The pattern of the hemoglobin dendrogram is most similar to that of the composite.

Data on parotoid secretions for different populations of *B. garmani* and *B. rangeri* were not available. The *B. garmani* sample used includes specimens from Ol Tukai, Kenya, some of which are probably *B. sp.* The *B. rangeri* sample includes specimens from Port St. Johns, South Africa, many of which are now known to have been hybrids between *B. rangeri* and *B. regularis* (E).

Discussion

The agreement of all systems on the position of *B. viridis* in relation to members of the *regularis* group suggests that all these data systems may be equally effective in discriminating species groups. Members of more groups would have to be included to be certain of this.

Within the *regularis* group, the passive pulse rate of the mating call, the hemoglobin composition, and possibly hybridization best indicate intragroup relations. The patterns revealed by the first two of these data systems agree very closely with the composite pattern and with the intragroup pattern suggested by Bogart in chapter 10. This intragroup dichotomy does not seem to be an artifact, as will be discussed later. The effectiveness of parotoid data in revealing intragroup patterns is difficult to evaluate because of the mixed nature of the parotoid samples of *B. garmani* and *B. rangeri.*

Completion of analyses of other systems should help determine the validity of the patterns indicated in this roughly executed example. The concordance of results from data of very different types speaks for the value of a multivariate comparative approach.

Mechanisms of Speciation

Geographic and Ecological Isolating Factors

The ecological delimitation of extant African *Bufo* species was mentioned earlier. Changes in the distributions of ecological zones in geologic time and a comparison of these distributions with the relationships of certain species and their ecological limits may reveal mechanisms by which geographic and ecological isolation might have led to speciation.

There were marked ecological changes in Africa during the Pleistocene. The sequence of these changes and their magnitude in terms of rainfall and temperature are now known for several localities from several lines of evidence: (1.) Studies of marine terraces (Bourcart 1943; Bourcart, Choubert, and Marçais 1949; Lecointre 1952; Gigout 1957); (2.) Climate and river terraces (Choubert 1953; Bernard 1959; Wayland 1935; Cooke 1947, 1958, 1964; Clark 1950, 1959; Lowe 1952; Flint 1959); (3.) Aeolian deposits (Cooke 1958, 1964; Cahen 1954; Leakey 1949; Janmart 1953; Clark 1950; Bond 1946, 1957*a*); (4.) Cave deposits (Bosazza, Adie, and Brenner 1946; Peabody 1954; Brain 1958; Brain and Meester 1964; Robinson 1959; Piveteau 1957); (5.) Isotopic studies of marine shells (Emiliani 1955, 1958); (6.) Faunas and prehistory (Alimen 1955; Furon 1955, 1958, 1963; Leakey 1953; Wayland 1935; Arambourg 1943–1947, 1949, 1952, 1954; Arambourg and Hoffstetter 1954; Clark 1959, 1960; Castany 1954); (7.) Pollen analysis (van Zinderen Bakker 1953, 1956, 1957, 1959; Coetzee and van Zinderen Bakker 1952). The glacial stages of the Pleistocene in the north temperate zone were apparently accompanied by pluvial periods, with lower temperatures in equatorial regions. Interglacials had their counterparts of warmer interpluvials in tropical regions (Furon 1963; Leakey 1953; Simpson 1929–1930). In temperate zones, increased temperature associated with interglacials results in an increase in "free water" in lakes and rivers, whereas in tropical zones thermal increase is associated with a decrease in "free water." Glaciation in Africa during the Pleistocene was very minor and was confined to the Atlas Mountains of the northeast and to certain restricted montane regions in central Africa. There are glaciers on Mount Kenya, on the equator, today. A comparative chronology of Pleistocene climatic changes in Africa and Europe is shown in table 9-8. Seven major changes are evident among four pluvial (glacial) and four interpluvial (interglacial) periods. The present is considered to be part of the fourth interpluvial, which is now about ten thousand years old.

Table 9–8. Pleistocene Climatic Changes in Africa and Europe

	Africa	*Europe*
10,000 B.P. +	4th Interpluvial	4th Interglacial
	Gamblian Pluvial	Würm Glaciation
	3rd Interpluvial	3rd Interglacial
	Kanjeran Pluvial	Riss Glaciation
	2nd Interpluvial	2nd Interglacial
	Kamasian Pluvial	Mindel Glaciation
	1st Interpluvial	1st Interglacial
Villafranchian	Kageran Pluvial	Günz Glaciation

The magnitude of change in the mean air temperature during the Pleistocene is estimated to have been 5–7° C (Flint 1957; Emiliani 1955, 1958). Estimated rainfall for southern Africa during the Pleistocene was about 60 percent of present rainfall for interpluvials and 140 percent for pluvials (Bond 1957; Brain 1958). Although the quantity of precipitation has changed considerably, its general pattern is thought to have been fairly stable and not to have varied greatly from that of the present. Cooke (1964) constructed hypothetical vegetation maps for southern Africa for pluvial and nonpluvial conditions. Important changes were the great extension of forests during pluvial periods and their considerable reduction accompanied by marked expansion of dry savanna and desert during nonpluvials.

Brain and Meester (1964) constructed rainfall maps of zones receiving 30 inches/year for the present, for pluvial, and for nonpluvial conditions in order to illustrate how ancestral species of the shrew *Myosorex*, of primarily central African affinities, might have migrated to southern Africa during cool pluvial conditions. Such populations could have been isolated during nonpluvials by the formation of 30 inches/year rainfall "islands" in southern Africa. This could have initiated the speciation of the endemic southern African species *M. varius* (Smuts) and later *M. cafer* (Sundevall).

The distributions of species of the *B. regularis* group and their relationships suggest that their speciation was initiated in a similar manner. The incompletely known distributions of *B. brauni*, *B. garmani*, *B. kisoloensis*, *B. rangeri*, and *B. superciliaris* are shown in figure 9-12. The ranges of *B. regularis* (E) and *B. regularis* (W) are too imperfectly known to be mapped. The known localities for these species are shown in figure 9-25. Rainfall and temperature statistics of several localities for each of these species are shown in table 9-9. *B. brauni* and *B. superciliaris* are tropical rain-forest forms. The former occupies relict patches of rain forest in Tanzania. The latter is widely distributed throughout the equatorial forest from eastern Democratic Republic of the Congo to Nigeria. It is also known from restricted forest sites farther west in Ghana. *B. garmani* as well as *B. sp.* are species of the dry savanna. *B. garmani*

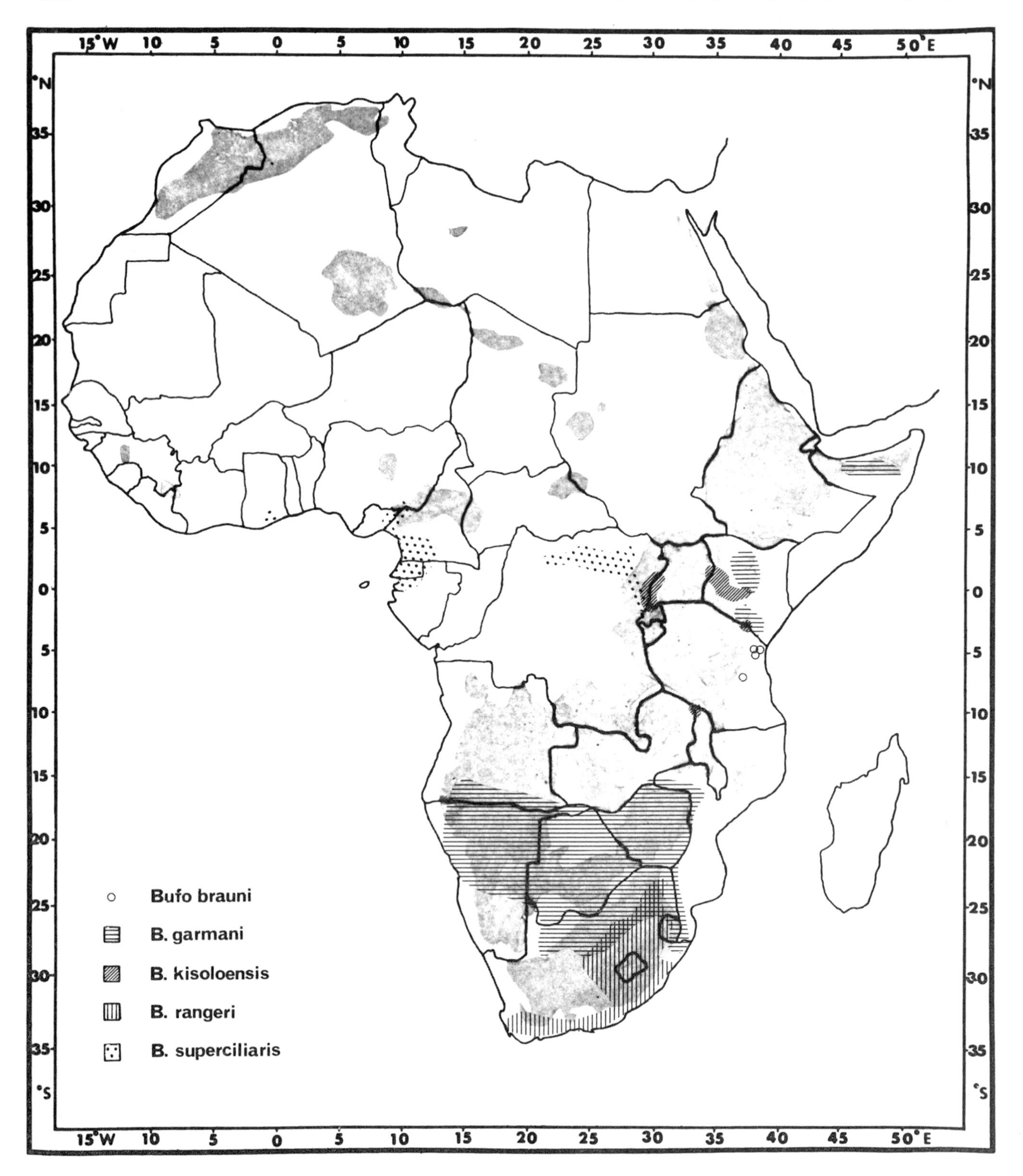

Fig. 9–12. Map of known distribution of *B. brauni*, *B. garmani*, *B. kisoloensis*, *B. rangeri*, and *B. superciliaris*. Land above 3000 meters is shaded gray. Base map modified from Thompson (1965), as in all distribution maps used in this chapter.

Table 9–9. Rainfall and Temperature Statistics for Eight Species of the *Regularis* Group*

	Mean Ann. Temp. °C		Mean Ann. Rainfall Cm.		N
	$\bar{x}$ + 2 se	Range	$\bar{x}$ + 2 se	Range	
B. regularis (E)	19.23 ±1.67	15.0-25.0	84.86 ±16.67	16.0-122.0	15
B. regularis (W)	23.79 ±1.36	19.0-26.5	146.21 ±36.53	7.0-260.1	13
B. brauni	20.50 —	— —	192.30 —	— —	1
B. superciliaris	22.83 ±1.69	20.0-36.0	194.88 ±41.33	158.0-259.5	6
B. garmani	19.27 ±1.46	15.0-25.0	58.23 ±8.70	29.5-99.5	18
B. sp.	21.94 ±2.77	16.0-26.0	59.71 ±17.79	29.0-100.0	8
B. kisoloensis	17.69 ±1.18	13.5-21.0	146.65 ±21.88	97.5-203.5	14
B. rangeri	16.66 ±1.05	14.0-20.0	79.60 ±26.81	16.0-168.6	11

*The geographic distributions of these species are shown in Figures 9–26 to 9–30 and 9–43 and 9–44. Climatic data from Jackson (1963), Wernstedt (1959), Great Britain Meteorological Office (1958), and Knoch and Schulze (1956). N = number of populations included in calculation of means and standard errors.

occurs in southern and eastern Africa. *B. sp.* is distributed in the Somali region of northeastern Africa and extends as far west as Chad. *B. kisoloensis* and *B. rangeri* are primarily cool, moist, highland forms. *B. kisoloensis* is known from areas above two thousand meters in east-central Africa. *B. rangeri* occupies the moist temperate highlands and the Cape temperate coastal lowlands of South Africa. *B. regularis* (E) and *B. regularis* (W) are primarily savanna forms, but they also occupy "farmbush" (Schiøtz 1963, 1967) localities in forest clearings. *B. regularis* (E) ranges at least from Uganda and Kenya to Natal. *B. regularis* (W) is known from Ghana to western Uganda and from Egypt. Both of these species, as well as *B. garmani, B. rangeri,* and *B. sp.*, can apparently thrive in areas of suboptimal rainfall if permanent water is available.

Fig. 9–13. Parts of Africa presently receiving 100 cm. or more of rainfall per year are shaded in black. Modified from Knoch and Schultze (1956).

The relationships of five of these species have already been discussed. Insufficient information was available to include *B. sp., B. kisoloensis,* and *B. superciliaris* in that analysis. Information from three of the four systems used in that analysis indicates highest affinity of *B. sp.* with *B. rangeri* (75%) and with *B. garmani* (54%). *B. sp.* also shares transferrins with these species (Guttman 1967). For those systems, only the passive pulse rate of the mating call of *B. kisoloensis* was available. It indicates highest affinity with *B. brauni* (91%), and next highest with *B. garmani* (85%). *B. kisoloensis* shares its one known transferrin with *B. garmani, B. sp.,* and *B. rangeri. B. kisoloensis* is morphologically most similar to *B. rangeri.* From the above systems, only parotoid data are available for *B. superciliaris.* These,

Fig. 9–14. Rainfall areas of 100 cm. or more postulated for a pluvial period of the Pleistocene. Modified from Knoch and Schultze (1956).

Fig. 9–15. Rainfall areas of 100 cm. or more postulated for a nonpluvial period of the Pleistocene. Modified from Knoch and Schultze (1956).

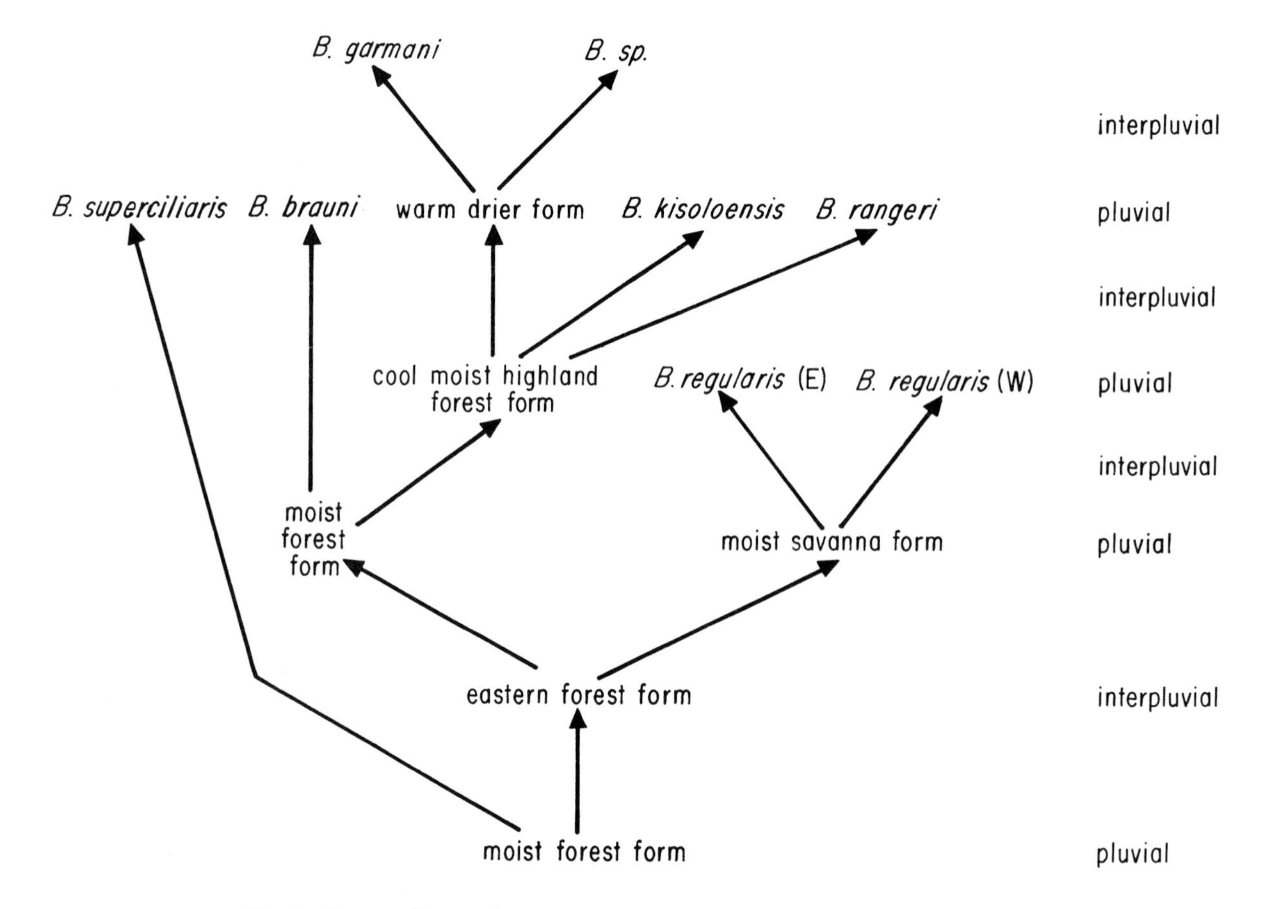

Fig. 9–16. Possible evolution of the *regularis* group during the Pleistocene.

however, indicate 98 percent affinity with *B. brauni* (Chap. 13). These two species are morphologically similar.

Hypothetical present, pluvial, and nonpluvial rainfall patterns for areas receiving more than 100 cm. (40 in.)/year for the entire African continent are shown in Figures 9-13, 9-14, and 9-15, respectively. These maps indicate a slightly greater change in rainfall — 50 percent of present (instead of 60%) for nonpluvials and 150 percent (instead of 140%) for pluvials — than seems to have actually occurred according to the data cited earlier. These percent changes were used because they were the only ones for which isohyet maps were available for the entire continent. The patterns produced by a 50 percent instead of a 60 percent change do not differ greatly. These maps illustrate the process of the merging and subsequent fragmentation of rainfall zones during pluvial-nonpluvial cycles, particularly in eastern and southern Africa.

When one compares these maps with the distributions of the *regularis*-group species, there are striking correlations. It is easy to visualize how these species could have been originally isolated by such a process of ecological change. A possible course of such events is illustrated in figure 9-16. The ancestral *regularis*-group species could have been a moist forest form. There is fossil evidence that forest extended over much of tropical and north Africa during the early Pleistocene (Meinertzhagen 1930; Loveridge 1933). An interpluvial could then have split its range into western isolates and one or more eastern isolates. Some eastern isolates would have been subjected to drier conditions and would have been selected for savanna conditions or have become extinct. Pluvial conditions might have enabled the eastern savanna forms to extend their ranges to suitable areas in the west. The subsequent nonpluvial would have repeated the process. Then, the savanna forms might have been split into western proto-*B. regularis* (W) and eastern proto-*B. regularis* (E) isolates separated by drier East African regions. But, their savanna habitats would have been extensive in the eastern and western segments, reducing the likelihood of further ecological isolation within the two savanna forms. Further isolation of savanna segments could have been possible, either by highland areas becoming surrounded by desert during nonpluvials or by lower rainfall areas being surrounded by forest during pluvials. Such isolation might account for the great polymorphism apparent within *B. regularis* (W) and especially *B. regularis* (E) (Chaps. 14 and 18). Forest forms could have been broken into eastern and western isolates, as before. A warm tropical forest form might have developed in the Usambara-Uluguru region, and cool moist-adapted forms could have evolved in eastern highland forest zones farther west. The subsequent pluvial would have extended the range of this highland forest zone further west. The subsequent pluvial would have extended the

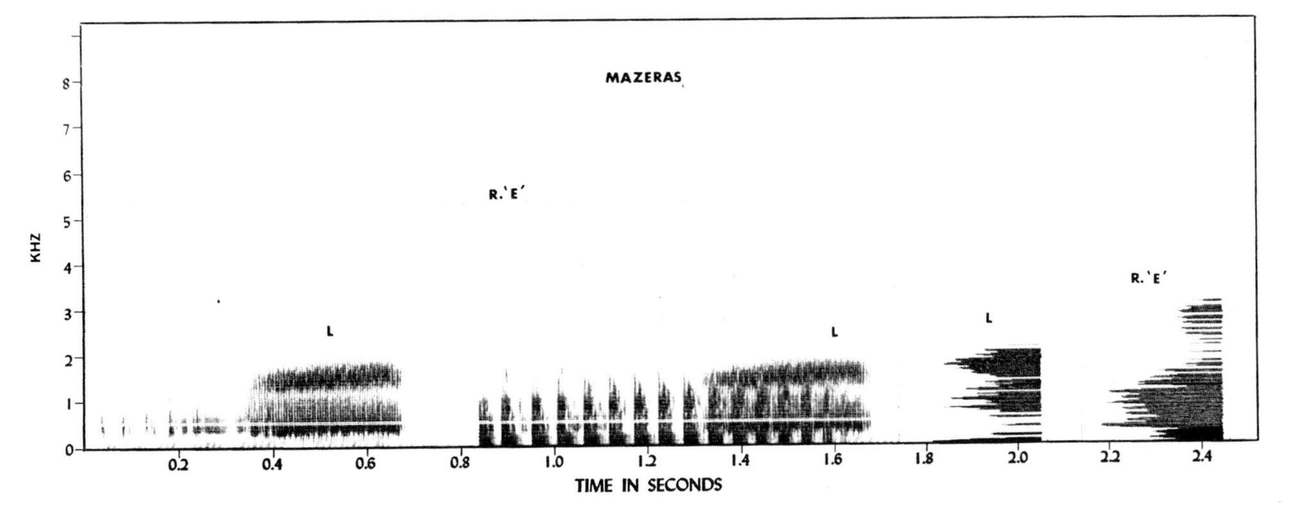

Fig. 9–17. Sound spectrogram of a chorus of *B. regularis* (E) and *B. maculatus* (-L), Mazeras, Kenya, by M. Tandy, November 8, 1965, 11:00 P.M. Air temperature, 24.0° C; water temperature, 25.5° C. Filter band width 50 Hz. Note how energy is distributed in different parts of the spectral range for each species, with each species having at least one of its dominant peaks at a frequency at which the call of the other has very little energy. Note also the differences in passive pulse rates.

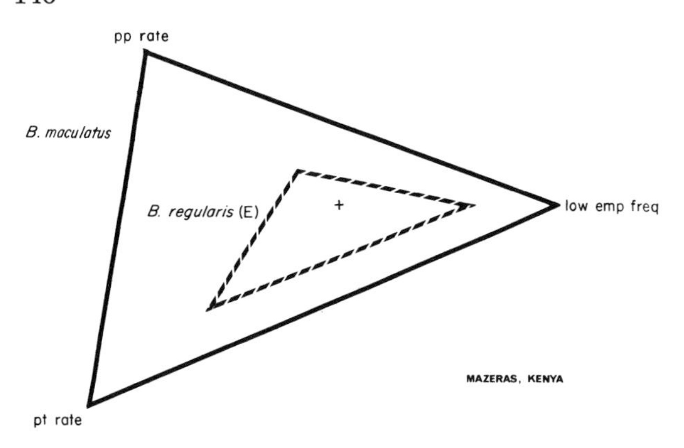

Fig. 9–18. Polygonal graph showing differences of three variables of the mating calls of *B. maculatus* and *B. regularis* (E) at Mazeras, Kenya.

range of this eastern forest form. The highland forms could have spread north and south along mountain chains and eventually to the temperate, moist south African region. The next interpluvial could have fragmented the highland form into several isolates. Cool moist forms (proto-*B. kisoloensis* and proto-*B. rangeri*) could have evolved from such isolates. Other isolates could have resulted in drier-adapted savanna forms, much as when the *B. regularis* (E)–*B. regularis* (W) stock split off. These savanna forms could have spread during the next pluvial and eventually given rise to *B. garmani* and *B. sp.* by isolation in the subsequent interpluvial.

The above is a hypothetical course of events based on what we know of the phylogeny and ecology of

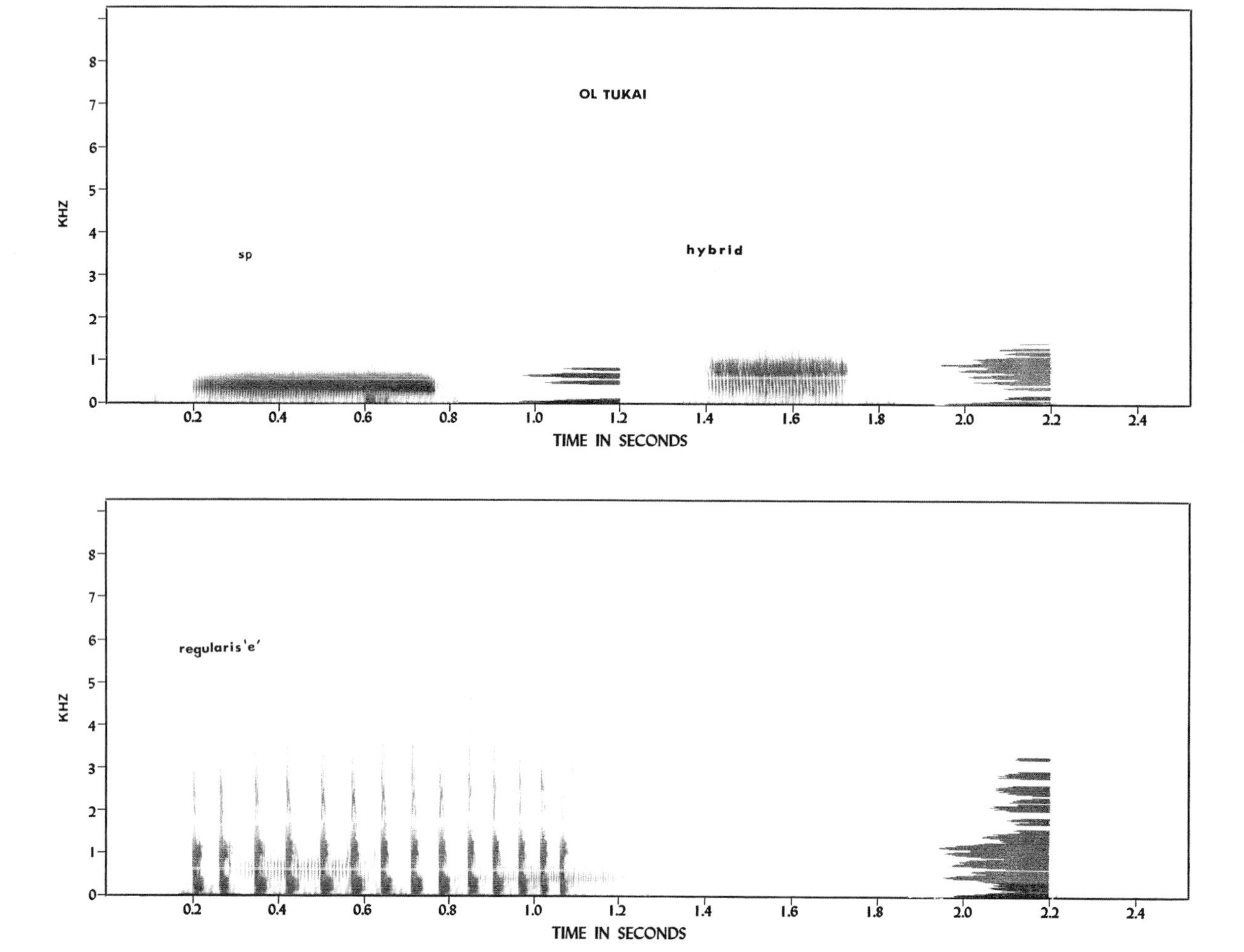

Fig. 9–19. Sound spectrograms of two sympatric species and hybrids at Ol Tukai, Kenya, by M. Tandy, November 16, 1965. Air temperature, 18.0° C; water temperature, 24.0° C; toad temperatures: hybrids and *B. sp.* – 20.5° C; *regularis* (E) – 19.5° C. Filter band width 50 Hz. Notice the different passive pulse rates for the three samples and the different distributions of spectral energy among them.

African *Bufo* and of climatic changes that occurred during the Pleistocene.

Premating Isolation Mechanisms in Sympatry

Many of the types of premating isolating mechanisms (Dobzhansky 1951) elaborated for Anura by A. P. Blair (1941, 1942), W. F. Blair (1955–1964), Littlejohn (1958–1965), Awbrey (1965, 1968), and others and reviewed by Bogert (1959) and Mecham (1961) have been observed among African *Bufo*. Five of these mechanisms (the mating call, size, microhabitat preference, breeding season, and calling site of the male) can act separately, but more commonly together, as effective barriers to hybridization between sympatric species. Several sympatric situations that have been observed in detail will illustrate this.

The chorus of *B. regularis* (E) and *B. maculatus* observed at Mazeras, Kenya, was described earlier. In this instance the mating calls and calling sites of the males appear to be of primary importance. The size difference may also be operative. Figures 9-17 and 9-18 and table 9-10 illustrate these differences.

In addition, the behavior of the female in approaching the male may be important. These two species are completely barred from gene exchange by the postmating isolating mechanism of hybrid inviability (Chap. 11).

A mixed chorus of *B. sp*, *B. regularis* (E), and their hybrids was observed at an artificial game watering tank near the Loginya Swamp at Ol Tukai,

Table 9–10. Physical Characteristics of the Mating Call and Body Size*

	B. maculatus		*B. regularis* (E)	
No. pp/pt	72.0	±2.8	14.6	±0.8
Low emp (Hz)	818.0	±25.0	483.0	±22.0
Mid emp (Hz)	1277.0	±66.0	873.0	±29.0
High emp (Hz)	1781.0	±59.0	1237.0	±43.0
Dominant (Hz)	1298.0	±151.0	533.0	±32.0
Pt dur (sec.)	0.393	±0.013	0.736	±0.045
Pp rate (/sec.)	183.9	±3.8	19.4	±0.4
Pt rate (/sec.)	1.208	±0.063	0.612	±0.021
N_i	8		7	
N_{pt}	27		21	
Snout-urostyle (in mm. ♂♂)	49.7	+1.0	68.7	+0.8

*For *Bufo maculatus* and *B. regularis* (E) at Mazeras, Kenya, November 8–9, 1965. Air temperature, 24.0°C; water temperature, 25.5 °C. The statistics are means and standard errors. N_i = number of individuals used to compute means; N_{pt} = total number of pt analyzed.

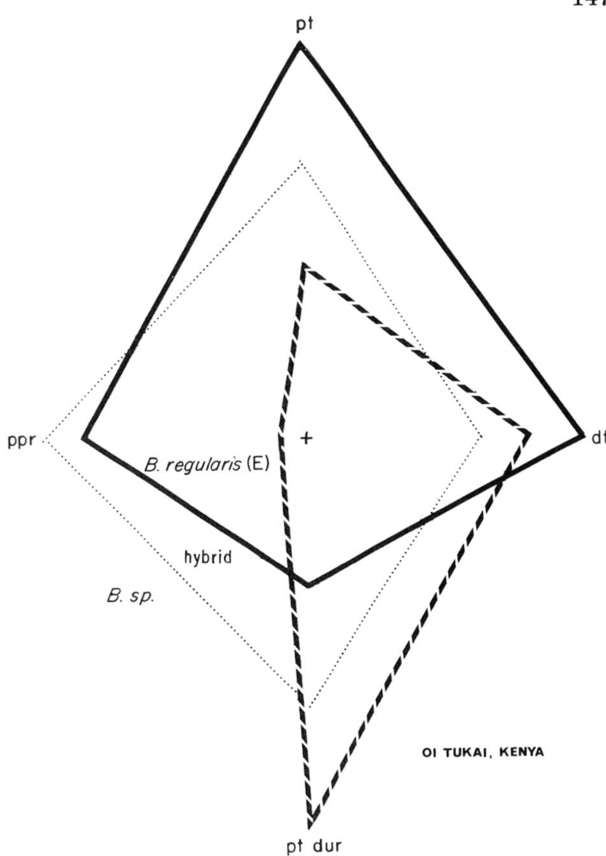

Fig. 9–20. Polygonal graph showing differences of four variables of the mating calls of three sympatric species at Ol Tukai, Kenya.

Masai Amboseli Game Reserve, Kenya, on the night of November 16, 1965. Both species as well as the hybrids were calling, although *B. regularis* (E) was most numerous. Large numbers of *Pyxicephalus delalandi* also called. All *Bufo* called either sitting in shallow water or on the bank. Of the five mechanisms mentioned above, only the isolating mechanism of mating call seems to have been operative here. Call and size statistics for this aggregation are shown in figures 9-19 and 9-20 and table 9-11. Three amplexed pairs of *B. regularis* (E) and a mixed pair (♀ *B. regularis* (E) x ♂ *B. sp.*) were taken in amplexus. A cross ♀ *B. sp.* x ♂ *B. regularis* (E) produced high percentage metamorphosis, while the reciprocal resulted in inviable larvae. Thus, at least the postmating mechanism of hybrid inviability seems to be operative here.

One locality with three sympatric species, Idanre, in western Nigeria, was visited several times during 1964–1965. *B. maculatus*, *B. perreti*, and *B. regularis*

Table 9–11. Physical Characteristics of the Mating Call and Body Size*

	hybrids		*B. sp.*		*B. regularis* (E)	
No. pp/pt	48.7	±2.3	89.9	±2.5	12.2	±0.4
Low emp	502.0	±32.0	548.0	±39.0	481.0	±12.0
Mid emp	870.0	±20.0	1210.0	±45.0	904.0	±19.0
High emp	1583.0	±132.0	1729.0	±57.0	1313.0	±46.0
Dominant	870.0	±20.0	548.0	±39.0	698.0	±115.0
Pt dur	0.363	±0.016	0.560	±0.024	0.811	±0.028
Pp rate	135.0	±3.4	160.6	±3.7	15.0	±0.3
Pt rate	1.218	±0.098	0.842	±0.063	0.519	±0.033
N_i	4		5		9	
N_{pt}	11		19		27	
Snout-urostyle (in mm. ♂♂)	67.5	±2.3	65.5	±1.6	76.5	±2.1

*For *B. sp.* and *B. regularis* (E) and their hybrids and at Ol Tukai, Masai Amboseli Game Reserve, Kenya. Statistics and codes as in Table 9–10.

Table 9–12. Physical Characteristics of the Mating Call*

	B. maculatus		*B. perreti*		*B. regularis* (W)	
No. pp/pt	102.4	±5.8	4.75	±.17	19.1	±0.6
No. pp/cpt	∞	—	72.7	±4.4	∞	—
No. pt/cpt	∞	—	12.27	±2.29	∞	—
Low emp	904.0	±29.0	1285.0	±56.0	558.0	±7.0
Mid emp	1496.0	±29.0	1847.0	±83.0	933.0	±41.0
High emp	2125.0	±325.0	2325.0	±62.0	1300.0	±18.0
Dominant	1484.0	±217.0	2008.0	±174.0	558.0	±7.0
Pt dur	0.610	±0.030	0.0500	±0.0	0.512	±0.019
Cpt dur	∞	—	0.728	±0.078	∞.	—
Pp rate	168.5	±1.5	109.3	±7.6	37.0	±0.6
Pt rate	0.690	±0.110	16.50	±1.99	0.764	±0.030
Cpt rate	?	—	0.698	±0.034	?	—
N_i	2		4		5	
N_{pt}	7		132		15	

*For *Bufo maculatus*, *B. perreti*, and *B. regularis* (W) at Idanre, western Nigeria. Statistics and codes as in Table 9–10.

all occur there, but they are markedly isolated from each other by a complex of premating mechanisms. Of the five mechanisms being considered, only temporal and size differences seem to be inoperative in the combination of *B. maculatus* x *B. perreti*. These differences are illustrated in figures 9-21 to 9-23 and table 9-12. All these species are completely isolated genetically (Chap. 11).

Sympatric associations among twenty-three African species have been investigated behaviorally. We now have a fairly good understanding of which premating and postmating isolating mechanisms are operative among them. All these species are sympatric with at least one other *Bufo* species. Among fifty sympatric situations, the percentages exhibiting the five premating isolating mechanisms are shown in table 9-13. The structure of the mating call, calling site, and microhabitat preference seem to be the most common mechanisms among these species. A complex of several mechanisms is usually operative. When there are apparently few and imperfect ones, such as at Ol Tukai, the species are thought to be very closely related and recently speciated. Because of the postmating mechanisms that exist between such species, there should be strong selective pressure to perfect the premating mechanisms. Isolating mechanisms seem to be very imperfect at Port St. Johns, Transkei, South Africa, where *B. regularis* (E) and *B. rangeri* have apparently formed a hybrid swarm, and introgression is occurring (Chap. 14). Experiments show

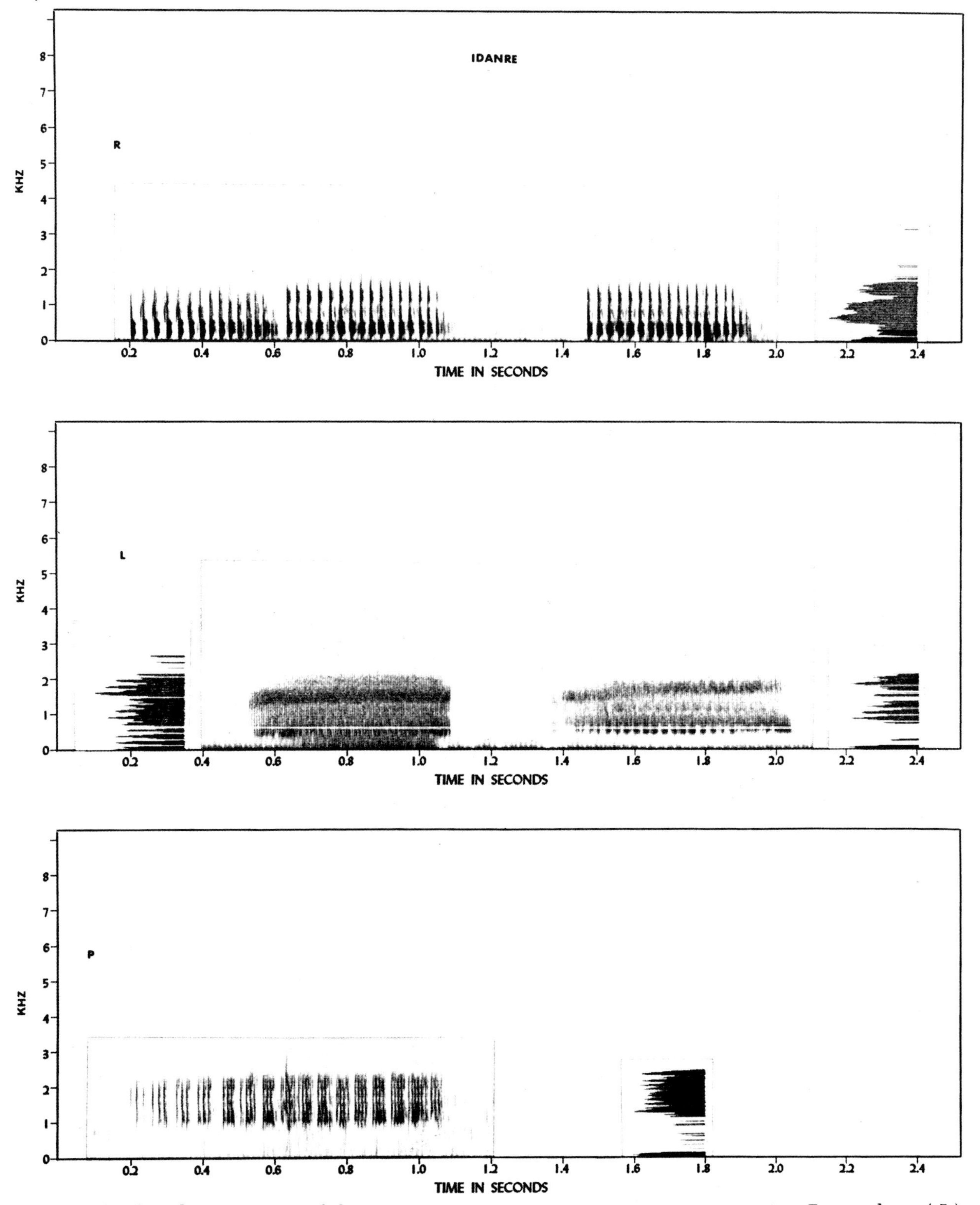

Fig. 9–21. Sound spectrograms of three sympatric species at Idanre, in western Nigeria – *B. maculatus* (-L), *B. perreti*, *B. regularis* (W), by M. Tandy. *B. maculatus* and *B. perreti*, June 26, 1965, 11:00 P.M.; air temperature, 23.3° C. *B. regularis* (W), February 28, 1965, 10:00 P.M.; air temperature, 24.8° C.

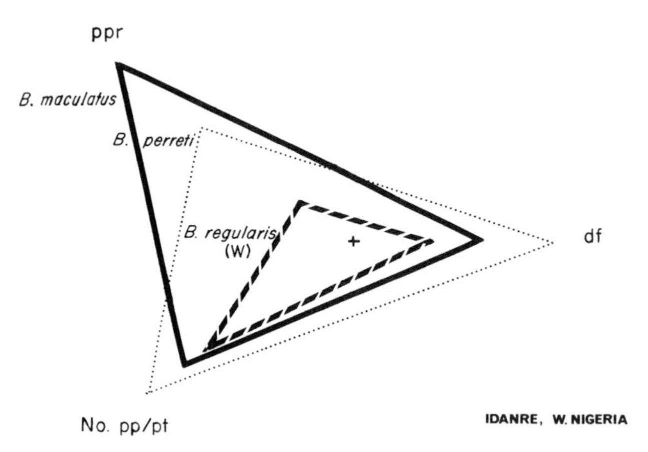

Fig. 9–22. Polygonal graph showing differences of three variables of three sympatric species at Idanre, western Nigeria. For *B. perreti*, the value for no. pp/pt is actually no. pp/cpt. (Same as Fig. 9–21).

Table 9–13. Premating Isolating Mechanisms in Sympatric Situations*

Premating Isolating Mechanism	% of Sympatric Situations Where Operative
Mating call	100%
Size	29%
Microhabitat preference	60%
Breeding seasons	46% (75% for temperate species; 25% for tropical species)
Male calling site	81%

*Percentages of 50 sympatric situations among 21 species of African *Bufo* in which various premating isolating mechanisms seem to be operative.

that, despite the apparent introgression, there is considerable genetic blockage between these two species, manifested mainly as hybrid sterility (Chap. 11) and chromosome-number polymorphism (Chap. 10). The pool in the stream where this hybrid swarm was found had no rock outcrops — the preferred calling site of *B. rangeri*. Here the *B. rangeri* called from the floating algal mats in the pool. *B. regularis* (E) called from near the bank. Three male *B. rangeri* were taken in amplexus at Swellendam, South Africa. A well-developed behavioral device, the male release call and vibrations (Fig. 9-24), had failed to function properly. If female *B. regularis* (E) entered the pool at Port St. Johns in the same manner as they were observed to do at Mazeras, Kenya, and if there were male *B. rangeri* nearby as sexually excited as those were at Swellendam, it is easy to imagine how hybrid pairs might have been formed. There appears to be reinforcement of size differences between these two species both at Port St. Johns and elsewhere in South Africa. At Port St. Johns *B. regularis* (E) is smaller than usual. However, this tends to lead to greater spectral similarity of the mating calls of *B. regularis* (E) and *B. rangeri* at this locality (see Fig. 9-28 and Table 9-14). The passive pulse rate of *B. regularis* (E) at this locality is highly variable (Fig. 9–27). Such a situation is evolutionarily dynamic and may well result in speciation if the Port St. Johns popula-

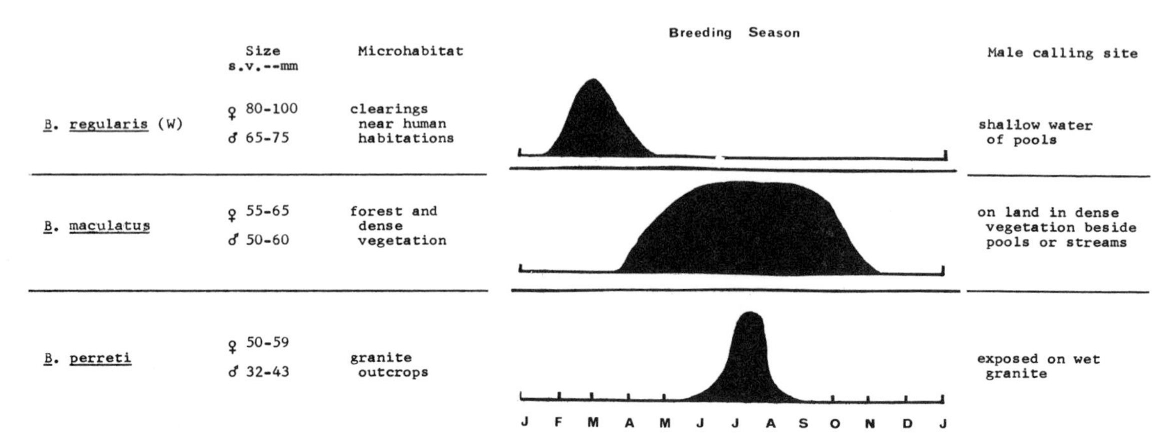

Fig. 9–23. Comparison of premating isolating mechanisms that seem to be operative among three sympatric species of African *Bufo* at Idanre, in western Nigeria. (Same as Fig. 9–21.)

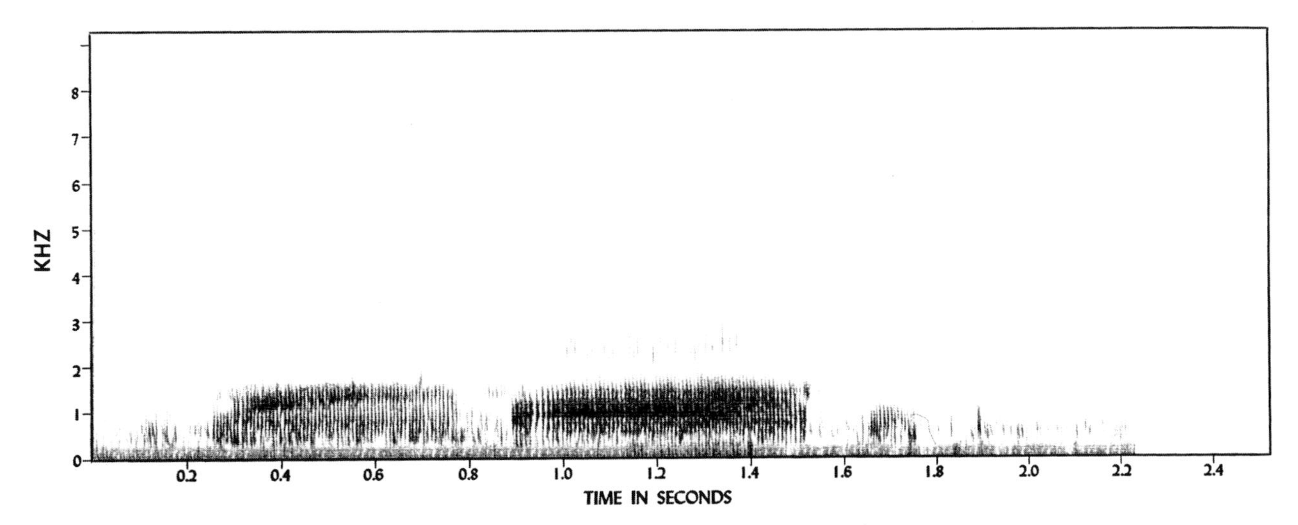

Fig. 9–24. Sound spectrograms of ineffective release calls given by three male *B. rangeri* in amplexus at Swellendam, South Africa, by M. Tandy, September 14, 1965, 11:00 P.M. Air temperature, 18.5° C.

tion of *B. regularis* (E) should become further ecologically isolated during the current interpluvial. The peculiarity of this population was recognized by Hewitt (1935), who suggested it might represent a separate subspecies.

Cryptic Species in the *Regularis* Group

The *regularis* group has an unusually large number of cryptic species, some of which are possibly yet to be discovered.

The following brief accounts are given for the three cryptic species now known within this group.

Bufo sp.

Bufo regularis gutturalis (non Power) Keith 1968.

General Morphology. This is a medium-sized 20-chromosome toad, externally very similar to *B. regularis* (E) and *B. maculatus* Hallowell. It exhibits scarlet femoral vermiculations characteristic of *B. garmani* Meek, *B. regularis* (E), and some populations of *B. maculatus*. Dorsal markings resemble those of *B. regularis* (E), *B. maculatus*, *B. regularis* (W), and hybrids of *B. regularis* with *B. garmani* and *B. rangeri*. Diagnostic external features are not yet apparent.

Ecological Limits and Geographic Distribution. This is a species of dry savanna. It is often sympatric with *B. garmani*. The only diagnostic feature known is the mating call. The species is acoustically documented from Ft. Lamy, Chad; Ghinda, Eritrea; Afgoi and Hoddur, Somalia; Kibwezi, Mtito Andei, Ol Tukai, and Voi in Kenya; Moroto, Uganda.

Physical Structure of the Mating Call. A sonogram of 2 simple pulse trains of the mating call is shown in figure 9-6B and 9-19. Statistics on the physical structure of the call are given in table 9-11. The call does not have features characteristic of hybrid mating calls; it does not seem "abnormal," and its structure is very consistent among different localities. Neither is its structure intermediate between those of the two species whose hybrids it might represent; *B. garmani* and *B. regularis* (E). It is distinctive and sympatric with at least one of these species at Ol Tukai, Kenya, and with both near Kibwezi, Kenya.

Call Site Selection by Males. *B. sp.* often calls from well-concealed sites among living or dead vegetation or rocks in, near, or up to seven meters away from a pond. But it also calls from exposed sites similar to the usual manner of *B. regularis* (E), *B. regularis* (W), and *B. garmani* in sympatric situations with all of these species. The not infrequent hybrids found in such localities may be partially the result of failure of perfection of different calling site preferences.

Systematic Comments. Because this form is sympatric with and distinctive from *B. garmani* and *B. regularis* (E), and because it exhibits genetic incompatibility with those species but not with conspecific specimens in laboratory crosses, this form merits specific status. Its possible origin has been discussed

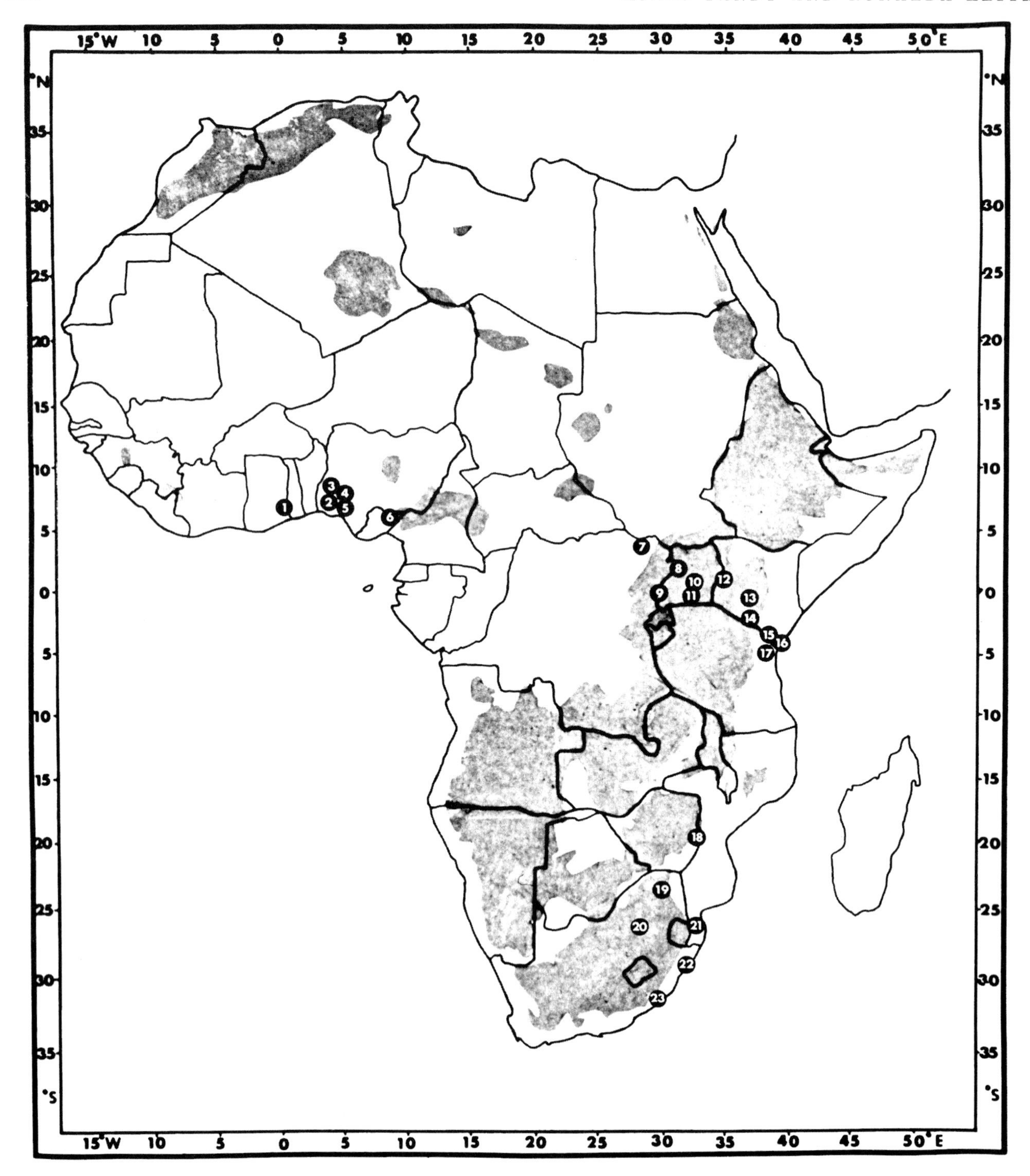

Fig. 9–25. Localities of *B. regularis* (E) and *B. regularis* (W) at which recordings were made by the authors. These numbers correspond to the population numbers of the diagrams showing the geographic variation of different variables of the mating calls of these species in Figs. 9–26 through 9–29 and to the population numbers listed in Table 9–14.

Table 9–14. Physical Characteristics of the Mating Call and Size*

	No.	pp/pt	low emp	mid emp	high emp	dom	pt dur	pp rate	pt rate	N_i	N_{pt}	Sno-Uro	N_p
B. regularis (W)													
Biakpa, Ghana	$\bar{x}$	21.0	400.0	0.0	1100.0	400.0	0.550	39.0	0.830	1	2	—	1
	σm	±0.0	±0.0	—	±0.0	±0.0	±0.0	±0.0	±0.0				
Ibadan, Nigeria	$\bar{x}$	19.2	490.0	835.0	1275.0	972.0	0.554	34.8	0.778	5	13	78.0	2
	σm	±1.9	±10.0	±33.0	±40.0	±198.0	±0.023	±2.5	±0.062				
Ogbomoso, Nigeria	$\bar{x}$	20.1	509.0	927.0	1191.0	868.0	0.680	28.6	0.646	5	20	66.8	3
	σm	±2.9	±13.0	±42.0	±55.0	±147.0	±0.074	±1.0	±0.048				
Ileṣa, Nigeria	$\bar{x}$	18.5	638.0	900.0	1313.0	688.0	0.500	36.0	1.110	1	4	68.8	4
	σm	±0.0	±0.0	±0.0	±0.0	±0.0	±0.0	±0.0	±0.0				
Idanre, Nigeria	$\bar{x}$	19.1	558.0	933.0	1300.0	558.0	0.512	37.0	0.764	5	15	72.0	5
	σm	±0.6	±7.0	±41.0	±18.0	±7.0	±0.019	±0.6	±0.030				
Ikom, Nigeria	$\bar{x}$	16.8	494.0	965.0	1305.0	865.0	0.450	35.5	0.905	2	9	76.7	6
	σm	±0.0	±6.0	±15.0	±5.0	±15.0	±0.0	±1.5	±0.025				
Dungu, R. D. du Congo	$\bar{x}$	19.5	680.0	1320.0	1810.0	1520.0	0.508	38.0	0.760	5	7	71.6	7
	σm	±2.3	±20.0	±63.0	±10.0	±195.0	±0.044	±1.6	±0.047				
Lake Albert, Uganda	$\bar{x}$	17.5	617.0	1067.0	1567.0	1567.0	0.570	29.5	0.685	2	6	—	8
	σm	±0.5	±34.0	±67.0	±34.0	±34.0	±0.030	±0.5	±0.015				
Kampala, Uganda	$\bar{x}$	19.7	473.0	858.0	1240.0	955.0	0.758	25.8	0.492	5	12	75.8	10
	σm	±2.7	±23.0	±38.0	±58.0	±150.0	±0.076	±2.1	±0.044				
Entebbe, Uganda	$\bar{x}$	24.8	463.0	842.0	1192.0	1098.0	0.845	29.0	0.550	2	7	76.8	11
	σm	±2.0	±38.0	±92.0	±9.0	±85.0	±0.045	±1.0	±0.110				
B. regularis (E)													
Queen Elizabeth Natl. Pk., Uganda	$\bar{x}$	12.1	640.0	997.0	1428.0	815.0	0.596	19.4	0.670	5	9	—	9
	σm	±0.6	±37.0	±38.0	±51.0	±169.0	±0.021	±0.8	±0.053				
Kitale, Kenya	$\bar{x}$	11.2	433.0	808.0	1239.0	1161.0	0.760	15.3	0.355	3	7	—	12
	σm	±0.4	±17.0	±46.0	±56.0	±122.0	±0.060	±0.7	±0.005				
Nyeri, Kenya	$\bar{x}$	12.4	471.0	876.0	1265.0	471.0	0.870	14.3	0.425	4	10	64.0	13
	σb	±0.5	±24.0	±15.0	±63.0	±24.0	±0.055	±1.0	±0.012				
Ol Tukai, Kenya	$\bar{x}$	12.2	481.0	904.0	1313.0	698.0	0.811	15.0	0.519	9	27	76.5	14
	σm	±0.4	±12.0	±19.0	±46.0	±115.0	±0.028	±0.3	±0.033				
Voi, Kenya	$\bar{x}$	9.0	550.0	938.0	1300.0	1300.0	0.480	18.0	0.520	2	4	—	15
	σm	±0.0	±0.0	±0.0	±0.0	±0.0	±0.0	±0.0	±0.0				
Mazeras, Kenya	$\bar{x}$	14.6	483.0	873.0	1237.0	533.0	0.736	19.4	0.612	7	21	68.7	16
	σm	±0.8	±22.0	±29.0	±43.0	±32.0	±0.045	±0.4	±0.021				
Lushoto, Tanzania	$\bar{x}$	12.3	469.0	840.0	1175.0	665.0	0.680	17.5	0.540	4	9	—	17
	σm	±0.9	±23.0	±38.0	±48.0	±197.0	±0.037	±0.5	±0.007				
Louis Trichardt, South Africa	$\bar{x}$	19.6	470.0	866.0	1301.0	551.0	1.061	18.6	0.449	8	25	70.2	19
	σm	±0.5	±21.0	±29.0	±51.0	±95.0	±0.038	±0.8	±0.030				
Rivonia, South Africa	$\bar{x}$	20.2	483.0	967.0	1392.0	1175.0	1.530	13.5	0.280	2	6	71.4	20
	σm	±1.2	±0.0	±34.0	±59.0	±142.0	±0.030	±0.5	±0.0				

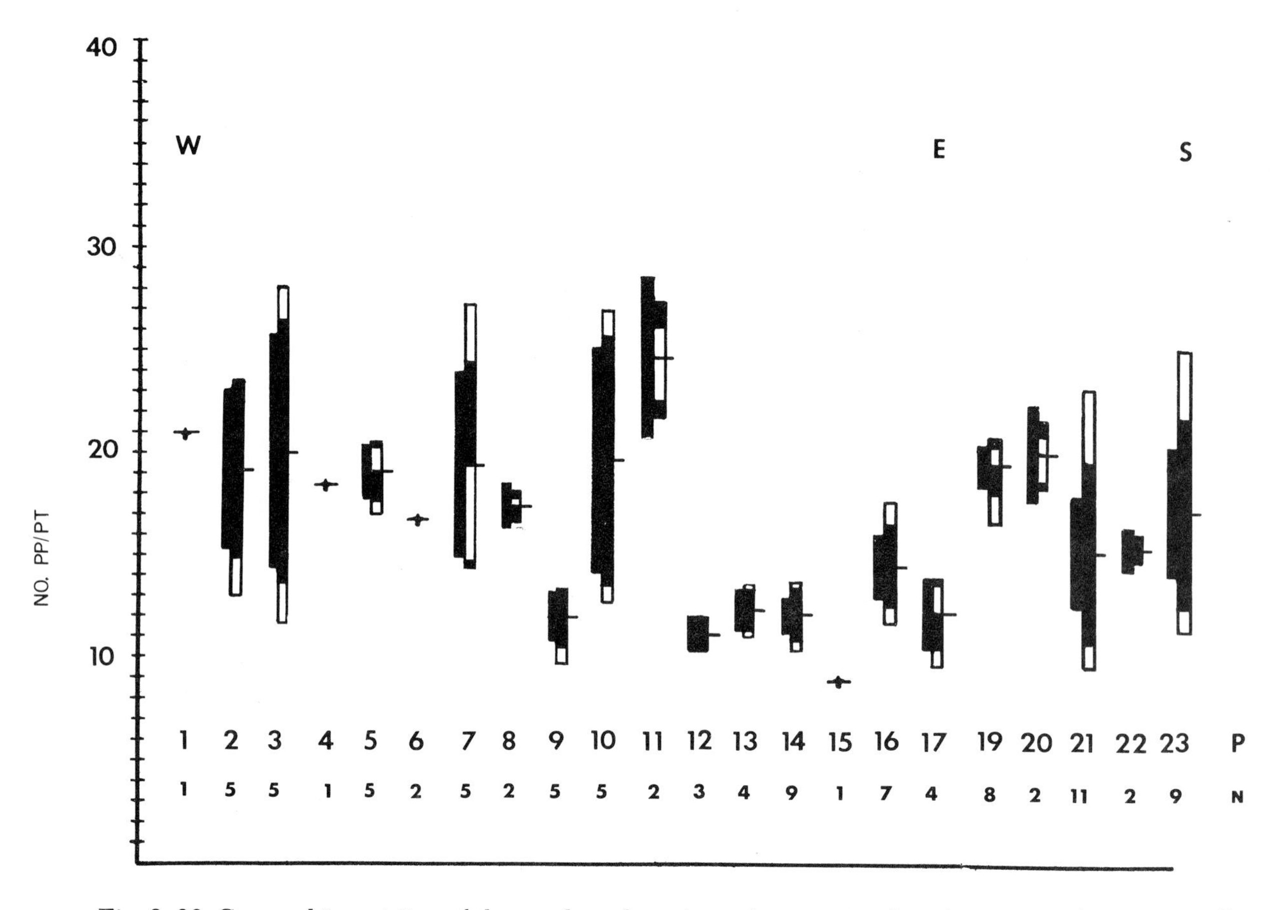

Fig. 9–26. Geographic variation of the number of passive pulses per simple pulse train in the mating calls of *B. regularis* (E) and *B. regularis* (W). 1–23 refer to population numbers for localities shown in Fig. 9–25. Populations 1–8 and 10–11 are *B. regularis* (W); populations 9 and 12–23 (except possibly 18) are *B. regularis* (E). N refers to the number of individuals recorded from each population. P=population numbers. Horizontal lines=means; each single black rectangle to the left indicates two standard errors each side of the mean; the black rectangle to the right indicates one standard deviation each side of the mean; hollow rectangles to the right indicate range. Notice the lesser variability of this call variable for *B. regularis* (E) populations closest to *B. regularis* (W). This may indicate reinforcement of call differences between proximal populations.

above. It resembles *B. maculatus* Hallowell in its call structure and external morphology, but the one cross of *B. maculatus* x *B. sp.* resulted in inviable larva (Chap 11). *B. maculatus* Hallowell (? = *B. pusillus* Mertens) may have been confused with this species.

Bufo regularis (E) in text

Bufo regularis gutturalis Power 1927. Trans. R. Soc. S. Afr. 14:416, Pl. 21. Lobatsi and Kuruman, Bechuanaland.[2]

General Morphology. This species is externally similar to *B. sp.*, described above, and *B. regularis* (W). It differs from the former in its more contrasting patterning and differently shaped parotoids. It differs from *B. regularis* (W) by the presence of some individual adult males with bright yellow throats rather than black ones as in almost every population of *B. regularis* (E). It further differs from *B. regularis* (W) in that most populations have many individuals with bright red femoral patches, as in *B. garmani* and *B. sp.* In general, *B. regularis* (E) is smaller than *B. regularis* (W).

Ecological Limits and Geographic Distribution. This is a savanna form common in moist savanna and in farmbush localities, including human habitations. It is known from Queen Elizabeth National Park in western Uganda to the Kenya coast and from there south to Natal. It has a considerable elevational range, up to at least two thousand meters in Kenya. Localities from which recordings are available are shown in figure 9-25.

Physical Structure of the Mating Call. Statistics for the physical variables of the call are shown for twelve train containing a variable number of simple pulse train repeated at a repetition rate of 0.45/second, populations in table 9-14. The call is a complex pulse each containing from eleven to twenty-six passive pulses with a rate about 16/second. The low-emphasized frequency (= carrier frequency) varies insignificantly from population to population, as shown in figure 9-28. Each passive pulse contains a wide array of frequencies approaching white noise, but narrowband spectrograms and sections reveal formants at the low emphasized frequency and at its second and third harmonics. The dominant formant is usually either the low emphasized frequency or its third harmonic. Simple pulse train duration varies considerably within individuals and among populations. The number of passive pulses per simple pulse train varies directly with the simple pulse train duration within individuals. There is geographic variation in both of these characters (Figs. 9-26, 9-27). The number of passive pulses per simple pulse train differs significantly from that of *B. regularis* (W) between geographically proximal populations of the two species, but not between very distant ones. Reinforcement between proximal populations is suggested. The most diagnostic feature of the call, which differs significantly from all *B. regularis* (W) populations, is the passive pulse rate (Fig. 9-27). This variable indicates proximal reinforcement.

Call Site Selection by Males. This has been discussed above. The *B. regularis* (E) prefers shallow water near the bank as does *B. regularis* (W). This preference may well serve as an effective premating isolating mechanism in sympatric situations with such species as *B. sp.* and *B. maculatus*, as elaborated above.

Systematic Discussion. Hybridization data (Blair, Chap. 11) indicate consistent one-way results – ♀ *B. regularis* (W) x ♂ *B. regularis* (E) – of larval inviability. The reciprocal cross yields high percentage metamorphosis of the F_1. These crosses refer to the same populations sampled acoustically.

Bufo regularis (W)

Bufo regularis Reuss 1834. Mus. Senck. (Abh.) 1:60 ("Egypt").

General Morphology. As discussed above, this species is very similar to *B. regularis* (E). It is larger than that species. The throats of breeding males are usually black. There are no red markings on the femur. White dorsal and lateral spotting is a fairly frequent character of this species, but such spotting also occurs in populations of *B. regularis* (E).

Ecological Limits and Geographic Distribution. This is a savanna and farmbush form ecologically very similar to *B. regularis* (E). It is known from Biakpa, Ghana, to Entebbe, Uganda, and probably occurs throughout the moist savanna region bordering the southern Sahara. It shows a similar elevational range to that of *B. regularis* (E), although most known localities are at lower elevations. Localities from which recordings are available are shown in figure 9-25.

Physical Structure of the Mating Call. Statistics for the physical variables of the call are shown for ten

[2] Lobatsi is in Botswana, while Kuruman is in the Transvaal of South Africa.

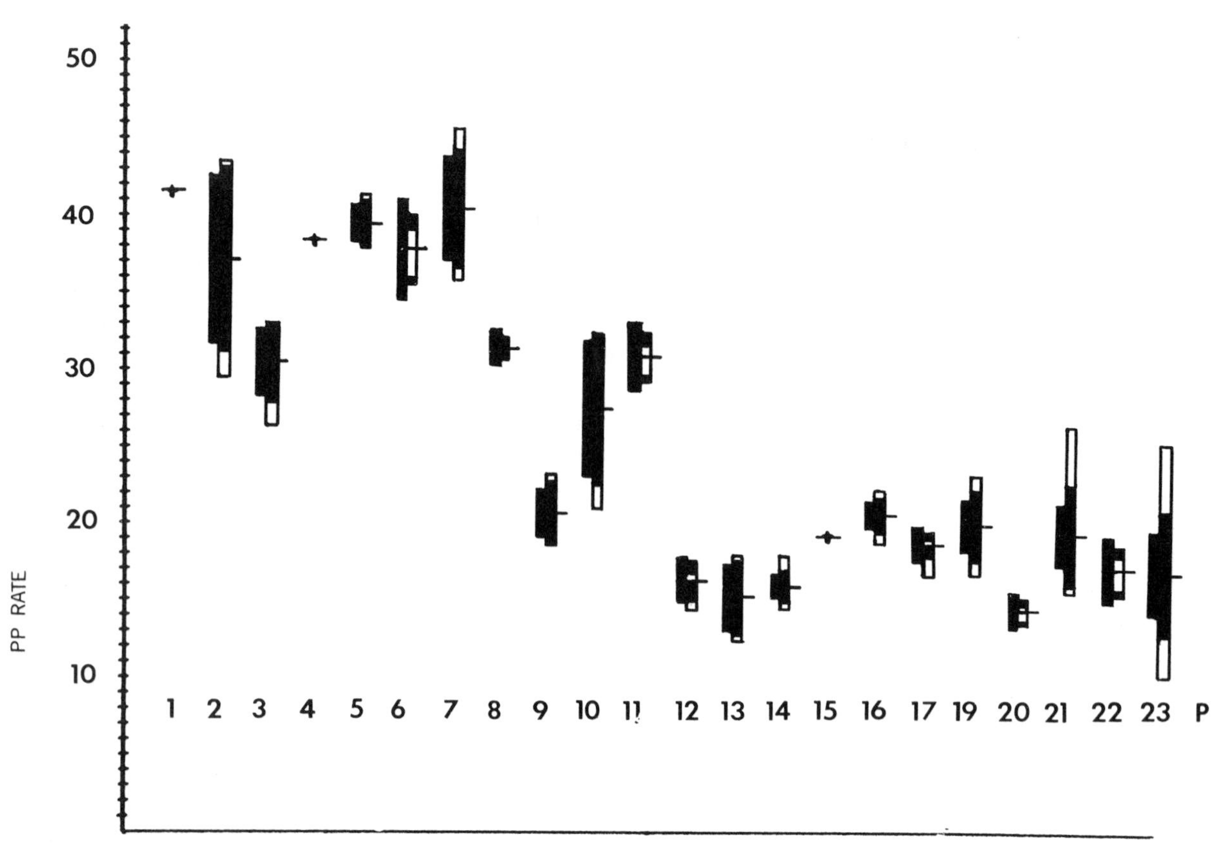

Fig. 9–27. Geographic variation of the passive pulse rate in the mating calls of *B. regularis* (E) and *B. regularis* (W). Note the consistent differences between proximal populations of the two species.

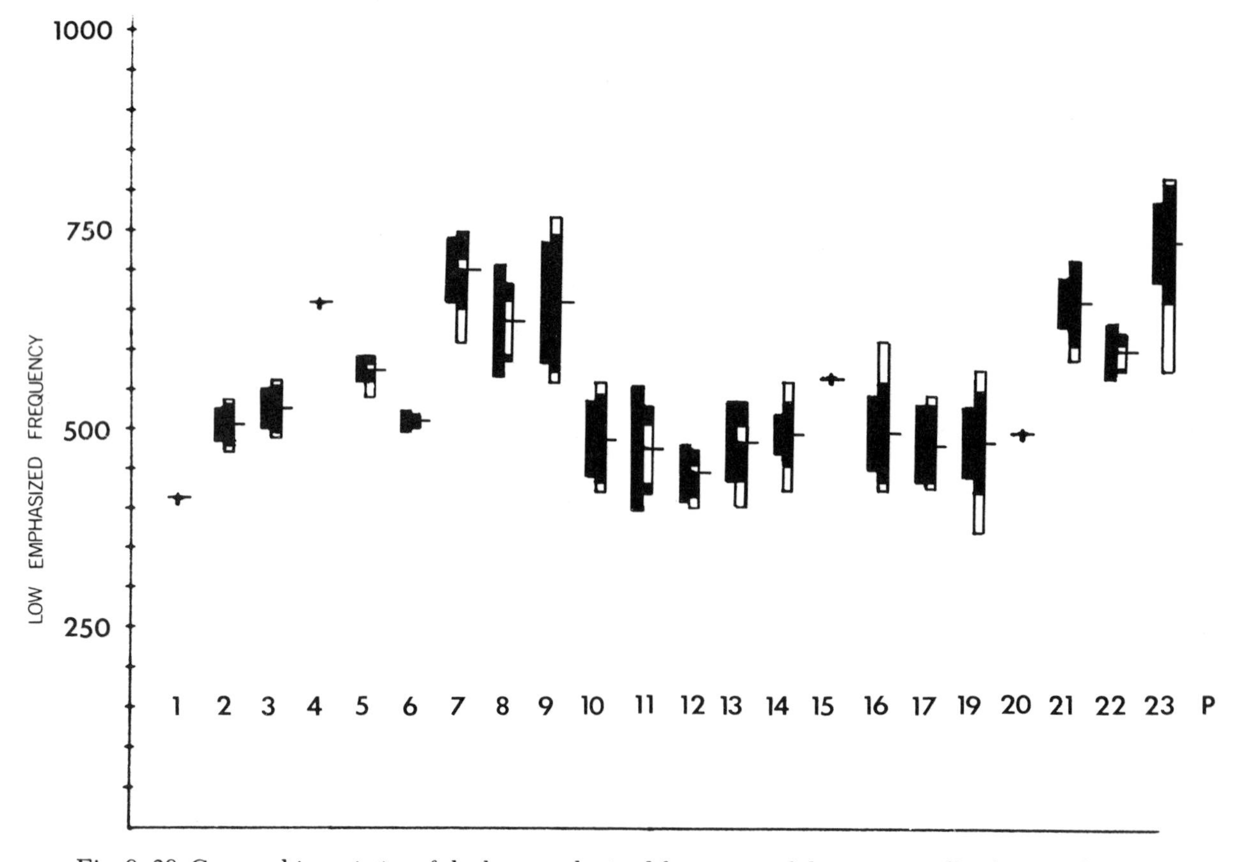

Fig. 9–28. Geographic variation of the low-emphasized frequency of the mating calls of *B. regularis* (E) and *B. regularis* (W). Note that the variation of this variable does not fit any obvious pattern.

populations in table 9-13. Sonograms are shown in figures 9-21 and 9-30. The call is a complex pulse train containing an indefinite number of simple pulse trains repeated at a rate of 0.71/second, each containing from twelve to twenty-eight passive pulses with a rate of about 33/second. The low-emphasized frequency varies insignificantly between populations (see Fig. 9-28). The spectral structure is very similar to that of *B. regularis* (E) decribed above. Geographic variation and comparison with *B. regularis* (E) is also described above under the latter species.

Systematic Discussion. See under *B. regularis* (E).

Call data not yet analyzed for Egyptian populations indicate that *B. regularis* (W) is the species common near Cairo. Positive hybridization results with other *B. regularis* (W) populations also indicate this.

Synopsis of African *Bufo*

Seven generic names have been used for African *Bufo*. Approximately ninety-two species and thirty-three subspecies of *Bufo* have been described or reported from Africa. The most recent literature recognizes about sixty species and fifteen subspecies (Scortecci 1936; Loveridge 1957; Poynton 1964; Schiøtz 1964*a*, 1964*b*; Perret 1966; Stewart 1967; and Marx 1968). We have worked with material of about fifty-seven of the species described and thirty-nine of those currently recognized. We believe that this material includes approximately fifty species — not all of them conforming to those currently recognized. The literature indicates that there are at least ten additional valid species on the continent. In many ways *B. carens* shows evidence of deserving separate generic status as was attributed to it by Smith — *Schismaderma carens*. See chapters 4 and 12, respectively, for accounts of the peculiar osteology and biochemistry of this species. If this status is valid, then two genera are represented in our material. Subspecies are not recognized, because there is no conceptual basis for this taxonomic category; populations referred to subspecific names cannot be delimited if the genetic continuity implied by the lack of specific status is real. Comparative descriptions are made of the variation among populations within several species, but no taxonomic status is attributed to more or less distinctive local populations. An attempt is made to explain this variation by differentially selective pressures acting on a common gene pool within a reproductively continuous species.

Criteria used for species recognition:

1. Genetic incompatibility in hybridization tests
2. Maintenance and especially enhancement of phenetic differences in sympatry
3. Level of phenetic divergence in cases where neither hybridization nor sympatrically distinctive evidence is available, particularly on the physical structure of the mating call (thought to be an important premating isolation mechanism), ecological preferences, and diagnostic characteristics likely to reflect marked genetic divergence.

A tentative list of taxa and suspected synonyms follows. Taxa utilized for morphological quantitative comparison are indicated with an asterisk(*). Those examined but not yet used in the above analysis are indicated with a plus sign (+). The species are grouped primarily according to morphological similarity but also according to genetic compatibility when that information is available. Species tested at least once for genetic compatibility (Chap. 11) are indicated with an "o." Names of taxa of which there is doubt about the listed status are followed by a question mark. Some of these taxa are known only from the literature. The taxa will be listed in the following manner:

Species complex or species group

  Species included

    Apparent synonymies

Bufo preussi group

Bufo preussi Matschie 1893 +

Atelopus africanus Werner 1898 +

Stenoglossa fulva Andersson 1903

Werneria fulva (Andersson) Poche 1903

This is a monotypic group represented by the single species *B. preussi*. It is distinguished by the lack of tympani and other ear structures and very slender habitus reminiscent of *Atelopus*. The tadpole with its elongate habitus and large oral sucker appears to be specialized for life in the swift-flowing mountain streams where it lives (Perret 1966; Inger 1959; Mertens 1940). The tadpole is similar to those of members of the genus *Heleophryne* of South Africa, which live in similar habitats (Hewitt 1935; van Dijk 1968). *B. preussi* is known from Mount Cameroun (950 m.) (Matschie 1893), from the Manengouba Mountains in the Federal Republic of Cameroun (Mertens 1940;

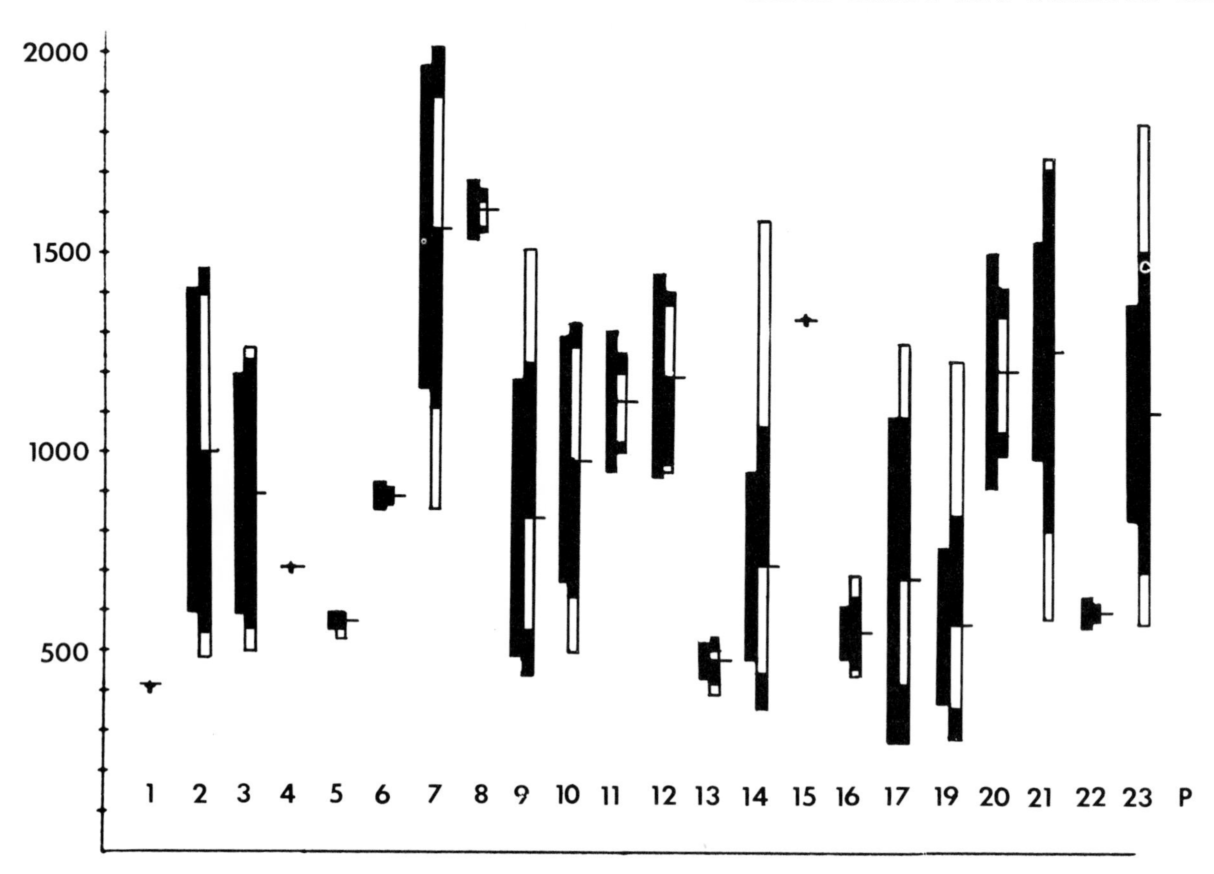

Fig. 9–29. Geographic variation of the dominant frequency of the mating calls of *B. regularis* (E) and *B. regularis* (W). Note the extreme intrapopulational variability of this variable and the lack of significant differences between populations. Compare the variability of the dominant frequency with the variability of the low emphasized frequency as shown in Fig. 9–28.

Perret 1966), and from Bismarckburg in the highlands of west-central Togo (Werner 1898).

Bufo taitanus complex

- *Bufo taitanus* Peters 1878 *
  - *Bufo katanganus* Loveridge 1932a *
  - *Bufo taitanus uzunguensis* Loveridge 1932a *
  - *Bufo taitanus beiranus* Loveridge 1932a +
  - *Bufo lindneri* Mertens 1955a *
- *Bufo anotis Boulenger* 1907 *
- *Bufo lönnbergi* Andersson 1911 *o
  - *Bufo mocquardi* Angel 1924 *
  - *Bufo lönnbergi nairobiensis* Loveridge 1932a *
  - *Bufo chappuisi* Roux 1936 ?+
  - *Bufo taitanus nyikae* Loveridge 1953 *
- *Bufo micranotis* Loveridge 1925 +
  - *Bufo micranotis rondoensis* Loveridge 1942
  - *Bufo osgoodi* Loveridge 1932a +
  - *Bufo jordani* Parker 1936a +
  - *Bufo melanopleura* Schmidt & Inger 1959 +

This large group consists of small and very small species lacking tympani and other ear structures. The digits often show reduction. In *Didynamipus sjöstedti* Andersson, an African bufonid apparently closely related to this complex, both the hands and feet have only four digits. The eggs of these species are very large for the size of the toad (2 mm. diameter) and have a dark animal and light vegetal pole. There are few eggs per female—twenty-two in a gravid *B. anotis* (Poynton 1964), which is the largest of these species. Members of this complex are mostly montane and forest forms distributed in eastern and central Africa from Ethiopia to Kenya, Tanzania,

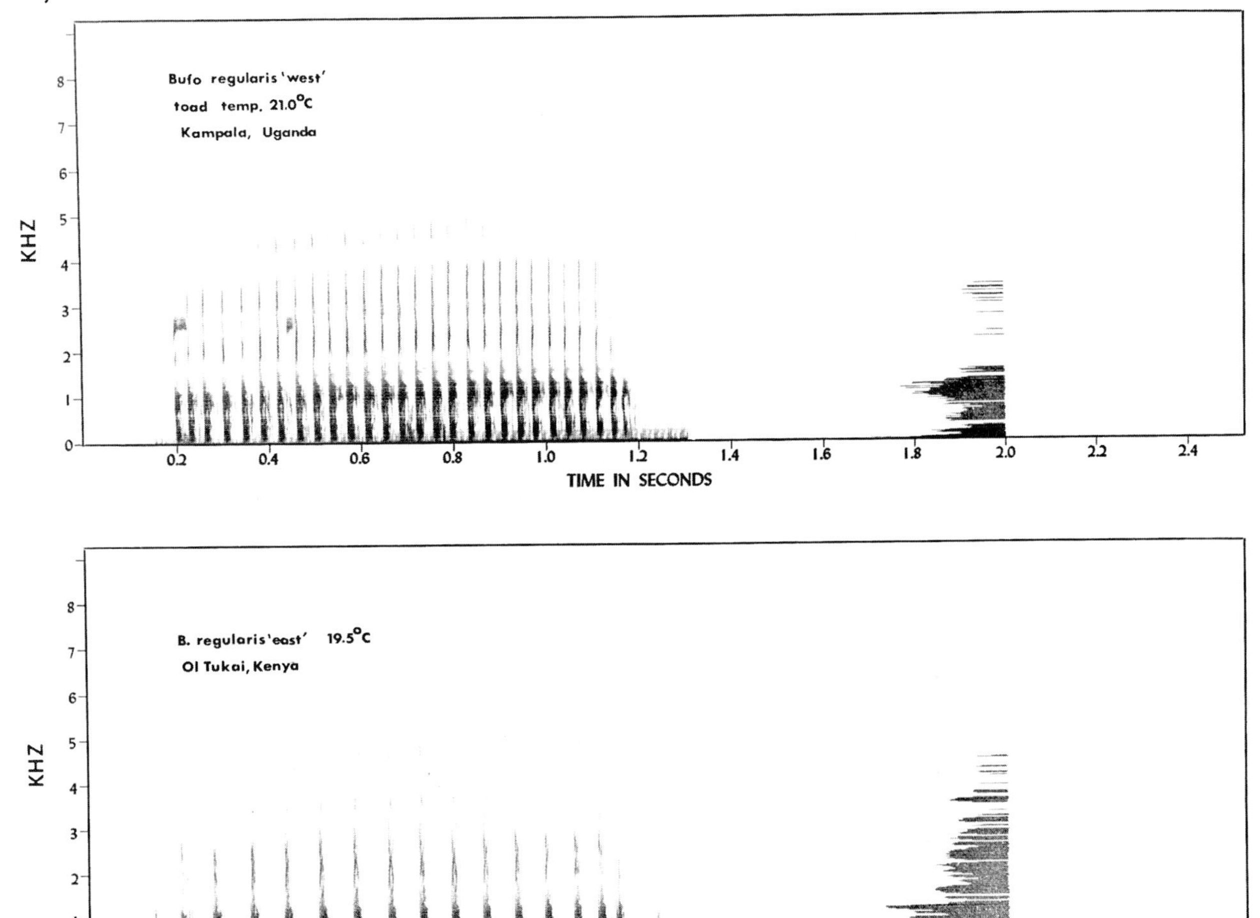

Fig. 9–30. Sound spectrograms of the mating calls of *B. regularis* (W) from Kampala, Uganda, and *B. regularis* (E) from 0l Tukai, Kenya. Note the difference in the passive pulse rates in these two calls recorded at a temperature difference of only 1.5° C. Kampala recording by M. Tandy, November 23, 1965. Air temperature, 16.5° C; water temperature, 21.0° C; toad temperature, 21.0° C. 0l Tukai recording by M. Tandy, November 16, 1965. Air temperature, 18.0° C; water temperature, 24.0° C; toad temperature, 19.5° C. Filter band width 50 Hz.

Malawi, Congo, Zambia to Namibia (South West Africa). *B. taitanus* has populations at lower elevations in Tanzania and Mozambique. *Didynamipus* is found on the island of Fernando Poo, Equatorial Guinea, as well as in mountains of mainland Cameroun.

Bufo ushoranus group

Bufo ushoranus Loveridge 1932a *

The one species concerned is seemingly related to the *B. taitanus* complex. This is a very small, earless species, which differs from members of the above complex in that its eggs are smaller and much more numerous per female than in any member of that complex, and the eggs are unpigmented.

Bufo rosei group

Bufo rosei Hewitt 1926a *o

This small species, which lacks a tympanum, resembles superficially members of the *B. taitanus* complex. Its detailed anatomy and distribution in relict populations on the summits of mountains of the Cape Peninsula and Cape Folded Mountains of South Africa suggest closer affinity to the *angusticeps* group than to the *taitanus* complex. The smooth skin and hypertrophied testes suggest such an affinity, as has been pointed out by Hewitt (1926*a*) and Poynton (1964).

Bufo angusticeps group

Bufo angusticeps Smith 1848?*
Bufo gariepensis Smith 1848 *o
Bufo tuberculosus Bocage 1896 +?
Bufo granti Boulenger 1903 +?
Bufo gariepensis nubicola Hewitt 1927 *?
Bufo nubicolus Hewitt, FitzSimons 1948 *?
Bufo gariepensis nubicolus Hewitt, Poynton 1964 *?
Bufo amatolica Hewitt 1925 *
Bufo angusticeps amatolica Hewitt 1925 *
Bufo tradouwi Hewitt 1926*b* +
Bufo inyangae Poynton 1963 *o?
Bufo gariepensis inyangae Poynton 1963 *o?

This is a group of small to medium-sized toads having smooth skins, hypertrophied testes with a peculiar biochemical effect on eggs of other species when squashed and used in artificial crosses (Chap. 11), and very short legs. These toads also have a distinctive, dark, reticulate dorsal color pattern, except in *B. tradouwi*, in which the color is more like that of *B. rosei*. The group is restricted to extreme southern Africa. The specific status of many forms is unclear.

Bufo blanfordi group

Bufo blanfordi Boulenger 1882 *
Bufo viridis somalacus Meek 1897 +
Bufo sibiliai Scortecci 1929 +?
Bufo chudeaui Chabanaud 1919?

These are small toads with acuminate snouts and inconspicuous parotoids. *B. blanfordii* is known from Eritrea and Somalia; *B. chudeaui* is known only from the type locality — Sahel de Nioro, Mali, on the southern edge of the Sahara in western Africa. The latter species was described from postmetamorphic material that cannot be identified as any known species.

Bufo bufo group (*sensu* Blair)

Bufo bufo Linné 1735 *o
Rana bufo Linné 1735
Bufo vulgaris Laurenti 1768 +
Bufo cinereus Schneider 1799+
Bufo bufo spinosus Daudin 1803+

B. bufo has been reported from Morocco where it occurs as an extension of its primarily Eurasian distribution.

Bufo carens group

Bufo carens Smith 1848 *o
Schismaderma carens Smith 1848 +
Schismaderma lateralis Smith 1848 +

Bufo carens lacks parotoids and shows other peculiar skeletal, biochemical, and developmental features, which makes uncertain its placement in the genus *Bufo* as discussed earlier.

Bufo funereus-steindachneri-vittatus complex

Bufo funereus Bocage 1866 *
Bufo benguelensis Boulenger 1882 (part) +?
Bufo buchneri Peters 1882 +?
Bufo decorsei Mocquard 1908 +?
Bufo berghei Laurent 1950 ?
Bufo steindachneri Pfeffer 1893*
Bufo incertus Scortecci 1933b +
Bufo vittatus Boulenger 1906*b* *
Bufo fuliginatus de Witte 1932 ? ?
Bufo funereus upembae Schmidt and Inger 1959 +
Bufo villiersi Angel 1940 +?
Bufo sp. A*?
Bufo sp. B

These are small to medium-sized toads lacking tarsal folds. The habitus is similar to that of *B. regularis* (E), but the skin is very spinose. The parotoid glands are unusually slender. *B. funereus* is a rain-forest toad. *B. sp.*, an undescribed species apparently belonging to this complex, is known only from Lake Rudolf and the Uaso Nyiro River northeast of Mount Kenya — areas of very low rainfall.

Bufo gracilipes group

Bufo gracilipes Boulenger 1899 *
Bufo benguelensis Boulenger 1882 (part) +
Bufo petiti Knoepffler 1967 +
Bufo urunguensis Loveridge 1932b *

These are very small rain-forest toads lacking tarsal folds and having small tympani. *B. gracilipes* is known from the Cameroun-Gabon area, while *B. urunguensis* is known only from a small relict rainforest area in the highlands of southwestern Tanzania.

Bufo vertebralis complex

Bufo vertebralis Smith 1848 *o
Bufo dombensis Bocage 1895a *
Bufo fenoulheti Hewitt & Methuen 1913 *o
Bufo fenoulheti obtusum Hewitt 1925 *
Bufo vertebralis albiventris Power 1927c*
Bufo fenoulheti rhodesianus Hewitt 1932 *
Bufo hoeschi Ahl 1934 *
Bufo fenoulheti damaranus Mertens 1954 *
Bufo vertebralis grindleyi Poynton 1963 *
Bufo lughensis Loveridge 1932a *o ?
Bufo gardoensis Scortecci 1933a +?
Bufo parkeri Loveridge 1932b *?

These are small to very small toads found in deserts and relatively dry highland areas of southern, eastern, and north-eastern Africa. They have small but distinct tympani, and at least two species have mating calls, although these are yet to be recorded.

Bufo regularis group

Bufo regularis Reuss 1834 *o
Grenouille ponctuée Geoffroy 1827
Bufo pantherinus Duméril & Bibron 1841 (part) ?
Rana mosaica Seetzen 1855 ?
Bufo nubicus Fitzinger MS ?
Bufo regularis (W) in text
Bufo garmani Meek 1897 *o
Bufo regularis poweri Hewitt 1935 *?
Bufo regularis humbensis Monard 1937 *
Bufo poweri Hewitt, Pienaar 1963 *
Bufo pseudogarmani Hulselmans 1969 +?
Bufo brauni Nieden 1910 *o
Bufo gutturalis Power 1927c *o
Bufo regularis gutturalis Power 1927c *o
Bufo regularis (non Reuss) Keith, 1968
Bufo regularis (E) in text
Bufo ngamiensis FitzSimons 1932 +?
Bufo regularis ngamiensis FitzSimons 1932 +?
Bufo kisoloensis Loveridge 1932a *o
Bufo regularis kisoloensis Loveridge 1932a *o
Bufo regularis marakwetensis Roux 1936
Bufo garmani (non Meek) Stewart 1967 +?
Bufo rangeri Hewitt 1935 *o
Bufo regularis rangeri Hewitt 1935 *o
Bufo sp. *o
Bufo regularis gutturalis (non Power) Keith 1968 *o

This is probably the largest and certainly the most intensively studied species group in Africa. There is great ecological diversification among the species. All are medium-sized to large toads. The group has some species in almost all parts of the continent inhabited by the genus, except possibly northwestern Africa. Those species sampled have a chromosome number of 2n = 20 (Chap. 10). Those species treated showed some or considerable reciprocal metamorphosis (Blair, Chap. 11). The mating calls of all these species are complex pulse trains containing simple trains of passively amplitude modulated pulses.

Bufo latifrons complex

Bufo latifrons Boulenger 1900 *
Bufo togoensis Ahl 1924a *?
Bufo cristiglans Inger & Menzies 1961 *?
Bufo pardalis Hewitt *o
Bufo regularis var. B Boulenger 1880
Bufo regularis pardalis Hewitt 1935 *o
Bufo camerunensis Parker 1936b *
Bufo polycercus (sic) (non Werner) Noble 1924
Bufo camerunensis poensis Parker 1936b
Bufo kerinyagae Keith 1968 *o

These are large toads with a distinctive pattern of dorsal markings consisting of paired dark bars, broad at the base and narrowly separated in the midline, which extend outward and transect the dorsal surfaces of the eyelids. These species are all found in high rainfall areas. The mating call of *B. kerinyagae* is of the *B. regularis* (E) type except for its high dominant frequency and long simple pulse train duration, the latter of which may exceed two seconds. *B. pardalis* is reported to have a call of similar duration (Hewitt 1937). *B. kerinyagae* has a chromosome number of 2n = 20. *B. pardalis* produced inviable hybrids in crosses with *B. maculatus*.

Bufo maculatus group

Bufo maculatus Hallowell 1854 *o
Bufo cinereus (non Schneider) Hallowell 1844 (part) *
Bufo maculatus Hallowell 1854 (part) *
Bufo guineensis Günther 1858 (part) +
Bufo guineensis Schlegel MS (part) +
Bufo spinosus Bocage 1867 (part) +
Bufo regularis spinosa Bocage, Boettger 1888 (part)
Bufo regularis pusillus Mertens 1937 *o
Bufo pusillus Mertens, Poynton 1964 *o
Bufo latifrons savannae Inger 1968 +

This group may be monotypic. *B. maculatus* is a

medium-sized toad known from high and very high rainfall tropical savanna and forest. The parotoid glands are often inconspicuous. The chromosome number is 2n = 20, but no reciprocal metamorphosis has been obtained in crosses with members of the *regularis* group (Chaps. 10 and 11). The mating call is like that of members of the *regularis* group. The status of *B. pusillus* is difficult to evaluate, especially since no crosses were made between it and *B. maculatus,* but there are no significant differences between the mating calls of populations which represent the two forms.

Bufo perreti group

Bufo perreti Schiøtz 1963 *o

The one species concerned is a specialized form endemic to granitic inselberg mountains near Idanre in western Nigeria. Both larvae and adults are highly modified morphologically for their ecological roles. The 2n chromosome number is 20, as in species of the *B. regularis-maculatus-latifrons* complex. Difficulties in interpreting the interesting structure of the *B. perreti* mating call phylogenetically have been mentioned. It may represent either an independent evolution of an active pulsatile mechanism or the retention of a primitive active mechanism from an ancestral line.

Bufo superciliaris group

Bufo superciliaris Boulenger 1887 *
Bufo laevissimus Werner 1897
Bufo chevalieri Mocquard 1908 +

This is a monotypic group which exhibits strong morphological and biochemical similarities (Low, Chap. 13) to members of the *regularis* complex. It is restricted to the high equatorial forest. It apparently has no mating call.

Bufo mauritanicus group

Bufo mauritanicus Schlegel 1841 *o
Bufo arabicus Gervais 1836
Bufo pantherinus Duméril & Bibron 1841 (part)

B. mauritanicus is found in the Mediterranean region from Morocco to Tunisia and as far south as Upper Volta (Lamotte, personal communication). This is a large species having a 2n chromosome number of 22. Biochemical data (Chap. 13) and possibly call structure suggest a distant relationship to members of the *B. regularis* complex.

Bufo tuberosus group

Bufo tuberosus Günther 1858 *
Bufo polycerus Werner 1897

This species does not even remotely resemble any other member of the genus in Africa. The development of its dermal tubercles is extreme, which gives the appearance of spines. In this respect and in the laterally displaced large ovoid parotoids, it somewhat resembles *B. asper* of Southeast Asia. It is confined to the high tropical forest of the Cameroun-Gabon-Congo region.

Bufo pentoni group

Bufo pentoni Anderson 1893 +

This is a medium-sized toad with peculiar tarsal spurs and short legs suited for burrowing. It is known from very dry savanna and desert on the southern borders of the Sahara from Upper Volta (Schiøtz 1964) to the Red Sea coast of Sudan, and from the southern Arabian Peninsula (Balletto and Cherchi, personal communication).

Bufo orientalis group (sensu Inger)

Bufo arabicus Rüppell 1827 +?
Bufo virdis arabicus Rüppel 1827 +?
Bufo dodsoni Boulenger 1895 +
Bufo orientalis Werner 1896? +
Bufo brevipalmata Ahl 1924b +
Bufo viridis orientalis Werner 1896 +
Bufo dhufarensis Parker 1931 +?

This group is described morphologically by Inger (Chap. 8). It is distributed in arid and desert regions of northeastern Africa and Arabia. It is acoustically unknown. It may represent either an independent evolution of an active pulsatile mechanism or the retention of a primitive active mechanism from an ancestral line.

Bufo lemairi group

Bufo lemairi Boulenger 1901b

Close relatives are not apparent for the one species concerned. The habitus is strikingly raniform. This is a medium-sized toad with a very acuminate snout, large tympani, and elongate parotoid glands. This toad is known from the central Congo highlands.

Bufo viridis group

Bufo viridis Laurenti 1768 *o
Bufo schreiberianus Laurenti 1768

Rana variabilis Pallas 1769
Rana bufina Müller 1776
Bufo variabilis Merrem 1820
Bufo boulengeri Lataste 1879 +
Bufo hemprichi Fitzinger MS

This primarily Eurasian group is found along the Mediterranean coast from Morocco to Egypt and beyond and in oases of the northern Sahara.

Origin of African *Bufo*

The origin of African *Bufo* is obscure. The primary evidence of relationship between African and other *Bufo* comes from the relationship between the *regularis* complex and *B. mauritanicus* and its relatives via the Asian (*melanostictus*)–Central American (*valliceps*)–South American (*guttatus*) line. It is not clear how close this relationship is. Species very similar to *B. regularis* are known from the Miocene of Morocco (Chap. 2). Africa and South America are not thought to have established land contact with Europe and North America, respectively, until the Oligocene or Miocene (Kurtén 1969). Thus, it is possible that the *regularis* group's progenitors invaded Africa from Eurasia in the mid-Cenozoic. But if that was the case, then why was there a huge radiation of this line in Africa but not in Asia? From where did the remainder of the large and diversified African radiation of bufonids come? Many of these species do not seem to have close relatives either in Africa or on other continents.

It has been suggested elsewhere in this volume that the genus *Bufo* probably evolved in South America and later spread to other parts of the earth.

Difficulties are encountered when one applies this interpretation to African *Bufo*. Several radiations are apparent within *Bufo* in Africa, and the *regularis* complex seems to be the most recent of these. But *regularis*-like toads were already present in Africa by the upper Miocene, although South America and Africa are thought to have only recently established land connections with North America and Eurasia by that time. Did *regularis* progenitors make the trans-Asian migration that rapidly and yet leave no close relatives in the oriental tropics? Where did the stocks come from that gave rise to the other African *Bufo* radiations? Investigation of amphibian fossils in Africa is in a very rudimentary stage, and even the better-explored North American and European geological formations have yielded only sparse evidence of anurans.

The existence of large disjunct leptodactylid radiations in South America and Australia and the isolated relict *Heleophryne* in South Africa represent one of the classic zoogeographical problems. But this picture may be illusory. Cytological evidence is bringing much of our present higher classification of Anura into serious question. There are genera in Africa and Asia that greatly resemble leptodactylids and ceratophrynids in their external features and their life histories but that are currently regarded as ranids or pelobatids on the basis of skeletal features. South America may not be unique in having large leptodactylid and bufonid radiations.

The vocal mechanisms and karyotypes of most African and Asian *Bufo* are unknown, not to mention those features of possible *Bufo* relatives in Africa and Asia – except for *Nectophryne*, which has a 2n chromosome number of 22 (Scheel 1970).

Few species of African and Asian *Bufo* have been subjected to the intensive investigation given the number of New World forms that have been studied. Intermediate and linking forms may yet be discovered for the Old World. Both of the major lines discussed above occur in the Old World. The *spinulosus* line is different in that it apparently penetrates Africa only peripherally.

Another explanation is possible for the similarities of the bufonid radiations of South America and Africa other than migration from one continent to the other. *Bufo* may have evolved prior to the break up of the southern landmass, Gondwanaland. Proto-Africa and South America were not effectively separated by barriers of land dispersal until well into the Cretaceous. Such a hypothesis more readily explains the large and apparently old bufonid radiations in Africa. This would also account for the present disjunct leptodactylid distribution. This explanation does not contradict the intercontinental dispersal patterns clearly apparent from the work presented in this volume but it does help to answer some of the enigmas posed by African *Bufo*.

Summary

Man's interest in African *Bufo* probably dates from the time of first contact between the two organisms. African literature is full of references to toads. Such sources of information have often been ignored by herpetologists, but they frequently contain interesting biological information about the Anura. European scientific study of African bufonids parallels that in other non-European parts of the world. There were early collections by travelers, next the intense competition of museum curators to describe species — an extension of European colonial competition during the nineteenth century — and later, museum-sponsored collecting expeditions of the twentieth century. The last fifty years have witnessed an increase of interest in African *Bufo* by evolutionarily oriented zoologists.

Little is known about breeding behavior of African *Bufo*. There is little information in older literature, and newer literature is scarce. Our work more than doubles the number of species that have been reported, but some of these are much less well known than others.

Mating calls of various species have temporal structures of 4 grades of complexity: pulses, pulse trains, complex pulse trains, and first-order sequences of complex pulse trains. Pulses are of passive or active origin. Homologies of active temporal components between African and non-African species are not clear. *B. mauritanicus* has been shown via many lines of evidence to be related to the 20-chromosome toads of Africa, but confusion of homologous active call elements prevents acoustics from showing how close this relationship is. Spectrally, calls of African species are similar to other *Bufo* except that several African species have dominant frequencies that are not the carrier frequency of the vocal cords.

Most African species breed in the typical anuran pattern at the beginning of rainy seasons. There are several exceptions.

Most African *Bufo* breed at night. Correlations of temperatures of vocalizing animals with air temperatures and mean annual air temperatures, plus the observation that toads often cease calling long before dawn while the night air temperature is dropping, suggest that calling may be related to circadian temperature changes. Diurnal calling has been observed for several species.

The males of most African *Bufo* form choruses during the breeding season. In some species, the males are spaced widely, but they still exhibit antiphonic response. Several species form very dense aggregations, and segregation of such choruses has been observed for three species. Males of earless species also form reproductive aggregations. That of *B. rosei* is one of the densest known among Anura.

Males exhibit species-specific patterns in their selection of calling sites, for instance, sitting in shallow water or floating in the water.

The few observations that have been made of the process of mate selection suggest that vocal species display species-specific differences in choosing mates. In all cases studied, females appear to respond to the mating calls of males from a distance. There are indications that females may choose particular individuals among species whose males call widely spaced or from hidden positions. Females of densely congregating species may not do this. Such females may create a stimulus, such as disturbing the surface of a pond by hopping into it, which causes males to respond by approaching the disturbed area and attempting to clasp the female. Such male-female behavior patterns appear to function as premating isolating mechanisms.

All African *Bufo* known exhibit thoracic amplexus. Eggs are laid in pairs of strings each surrounded by a jelly tube. Different spawning sites are used by different species, and the manner of depositing eggs differs among them. Most species lay their eggs in shallow water, but *B. perreti* deposits them only partly submerged in runnels on exposed surfaces of granite.

No species of *Bufo* is distributed throughout the African continent. Most species are restricted to one or two major ecological zones, though not necessarily to zones of specific plant composition. Most distributions can be defined in terms of mean annual temperature and rainfall. There are species endemic to small regions and nonendemic species that are restricted to northern Africa, apparently because of their recent invasion from other land masses. Rainfall and altitude play primary roles in determining distributions in tropical Africa.

Some African *Bufo* species exhibit morphological specializations specially adaptive for their ecological roles. These range from familiar adaptive phenomena, such as cryptic coloration and mechanically efficient limb proportions, to bizarre phenomena such as the pelobatidlike tarsal digging spur of *B. pentoni* and the large oral sucker of *B. preussi* tadpoles, which functions for clinging to rocks in fast-flowing streams.

A great size range is exhibited among African *Bufo*. The smallest species, *B. micranotis*, is one of the smallest species in the genus. *B. superciliaris*, the largest African species, does not reach the size of *B. blombergi*, *B. marinus*, or *B. paracnemis* — all of South America.

Both sexes of most species are cryptically colored.

Exceptions involve more brilliant nonpatterned coloring of males, possibly related to advertisement during the breeding season.

Most African species have parotoid glands of similar shape. Those of *B. tuberosus* are exceptional. The parotoid glands of *B. blanfordi* are inconspicuous, and *B. carens* seems to lack them entirely.

The size of the tympanum shows allometric correlation with body size, but such correlation is less marked among large species. Size of the tympanum is also correlated with the low emphasized frequency of the mating call. This correlation is as high among large as among small species. Selection for tympanum size to match mating call spectral structure independently of body size is indicated.

At least nine species lack tympani and other auditory structures. The significance of this is unknown because most of these species are unknown behaviorally. One, *B. lönnbergi*, emits a call in breeding aggregations very similar to the mating calls of other species.

The males of most species develop median subgular vocal sacs and associated gular pigmentation during the breeding period, accompanied by melanized nuptial asperities on at least the first fingers. The development of melanized dermal spines is also related to the breeding season and may be either enhanced or reduced during that time.

A simplified multivariate analysis of five members of the *regularis* group and *B. viridis* reveals concordance of phenetic patterns based on four different data systems: (1) mating call physical structure, (2) parotoid-gland secretions, (3) blood proteins, and (4) hybridization data. All these data systems clearly separate the *regularis*-group members from *B. viridis*. Call, blood proteins, and a combination of the four systems reveal a dichotomy within the *regularis* group that separates *B. brauni*, *B. garmani*, and *B. rangeri* from *B. regularis* (E) and *B. regularis* (W).

Climatic changes known to have occurred during the Pleistocene and the phylogeny within the *regularis* group as indicated by multivariate analysis suggest that speciation may have been initiated within this group by ecological isolation that resulted from changes in the extent of rainfall zones during the Pleistocene.

Many of the premating isolating mechanisms known for Anura occur among African *Bufo*. Five of these — the mating call, size, microhabitat preference, breeding season, and calling site of the male — can act separately, but more commonly together, as effective barriers to hybridization between sympatric species. Most sympatric species also exhibit postmating isolating mechanisms ranging from sterility to F_2 inviability, but introgression appears to be occurring in the Port St. Johns region of South Africa between *B. rangeri* and *B. regularis* (E) despite considerable hybrid sterility.

Three cryptic species are apparent within what has been regarded as *B. regularis* — *B. sp.*, *B. regularis* (E), and *B. regularis* (W) — on the basis of mating call analysis and hybridization experiments. Diagnostic external features are not yet apparent among these species.

There appear to be fifty or more species of *Bufo* inhabiting Africa today. These species are tentatively grouped into twenty-two species groups or species complexes based primarily on morphological criteria. Twelve of these groups are monotypic. Three groups are primarily Eurasian rather than African in distribution. Positive hybridization evidence is available for parts of only five of the eighteen strictly African groups — the four 20-chromosome groups plus *B. mauritanicus*. Limited negative hybridization evidence is available for parts of five additional groups.

There are at least sixteen species of 20-chromosome toads included in at least four species groups. The *regularis* group is the largest, as well as the largest of the genus in Africa, and includes at least nine species. The second largest grouping, which may include more than one species group, is the *taitanus* complex with about six species; next the *vertebralis* complex with about four species; and then the *angusticeps* group of approximately the same size. Two groups — the *rosei* group and the *ushoranus* group — contain single divergent species that seem to be closely related to the *angusticeps* group and the *taitanus* complex respectively. Seven groups — *preussi* group, *blanfordi* group, *carens* group, *gracilipes* group, *lemairi* group, *tuberosus* group, and the *pentoni* group — contain only one or two species and are apparently not closely related to any other *Bufo* species, African or otherwise.

In addition to this large radiation of *Bufo*, there are four other genera of bufonids in Africa including six species. *Didynamipus* appears to be related to the *taitanus* complex. *Nectophryne* and *Wolterstorffina* show similarities to *Bufo*. Both lay their eggs in strings (Perret 1966; Scheel 1969). *Nectophryne afra* has a 2n chromosome number of 22 (Scheel 1970). The bizarre ovoviviparous *Nectophrynoides* are well known.

REFERENCES

Acocks, J. P. H. 1953. Veld types of South Africa. Bot. Survey Mem. 28. Govt. Printer, Pretoria.

Ahl, E. 1924*a*. Neue Reptilien und Batrachier aus dem Zoologischen Museum. Arch. Naturgesch. (Berlin) 90:246–254.

———. 1924*b*. Über eine Froschsammlung aus Nordost-Afrika und Arabien. Mitteil. Zool. (Berlin) 11:1–12.

———. 1934. Über eine kleine Froschsammlung aus Deutsch-Südwestafrika. Zool. Anz. 107:333–336.

Alimen, H. 1955. Préhistorie de l'Afrique. Boubée, Paris.

Anderson, John. 1893. On a new species of *Zamenis* and a new species of *Bufo* from Egypt. Ann. Mag. Nat. Hist. 12:439–440.

Andersson, L. G. 1903. Neue Batrachier aus Kamerun, von den Herren Dr. Y. Sjöstedt und Dr. S. Junger gesammelt. Verh. Zool. Bot. Ges. (Vienna) 53:141–145.

———. 1911. Reptiles, batrachians and fishes collected by the Swedish Zoological Expedition to British East Africa, 1911. 2. Batrachians. Kungl. Svensk. Vetensk. Handl. 47:23–36.

Angel, F. 1924. Note préliminaire sur deux Batraciens nouveaux, des genres *Rappia* et *Bufo*, provenant d'Afrique Orientale anglaise (mission Alluaud et Jeannel, 1911–1912). Bull. Mus. Hist. Nat. (Paris) 1924:269–270.

———. 1940. Description de trois amphibiens nouveaux du Cameroun. Bull. Mus. Hist. Nat. Paris 12:238–243.

Arambourg, C. 1943–1947. Mission scientifique de l'Omo. Paris, Publ. Mus. Nat. Hist. Nat.

———. 1949. Présentations d'objects énigmatiques provenant du Villafranchian d'Algérie. Comptes Rendus Soc. Géol. Fr. 120–122.

———. 1952. Paléontologie des vertébres en Afrique du Nord française. C. R. XIXe Int. Géol. Congress (Algiers) 64 p.

———. 1954. L'hominien fossile de Ternifine (Algérie). C. R. Acad. Sci. 239:839–895.

———, and R. Hoffstetter. 1954. Découverte en Afrique du Nord de restes humains du Paléolithique inférieur. C. R. Acad. Sci. 239:72–74.

Awbrey, F. T. 1965. An experimental investigation of the effectiveness of anuran mating calls as isolating mechanisms. Ph.D. Thesis, Univ. Texas.

———. 1968. Call discrimination in female *Scaphiopus couchii* and *Scaphiopus hurterii*. Copeia 1968:420–423.

Barbour, T., and A. Loveridge. 1946. First supplement to typical reptiles and amphibians. Bull. Mus. Comp. Zool. (Cambridge) 96:57–214.

Beier, Ulli. 1966. The origin of life and death. St. Paul's Press Ltd., Malta.

Bernard, E. A. 1959. Les climats d'insolation des latitudes tropicales au Quaternaire. Bull. Acad. R. Sci. Col. 5:344–364.

Blair, A. P. 1941. Variation, isolation mechanisms, and hybridization in certain toads. Genetics 26:398–417.

———. 1942. Isolating mechanisms in a complex of four species of toads. Biol. Symposia. 6:235–249.

Blair, W. F. 1955*a*. Mating call and stage of speciation in the *Microhyla olivacea-M. carolinensis* complex. Evolution 9:469–480.

———. 1955*b*. Size difference as a possible isolation mechanism in *Microhyla*. Amer. Nat. 89:297–302.

———. 1955*c*. Differentiation of mating call in spadefoots, genus *Scaphiopus*. Texas J. Sci. 7:183–188.

———. 1956*a*. Call difference as an isolation mechanism in southwestern toads (genus *Bufo*). Texas J. Sci. 8:87–106.

———. 1956*b*. The mating calls of hybrid toads. Texas J. Sci. 8:350–355.

———. 1957. Structure of the call and relationships of *Bufo microscaphus* Cope. Copeia 1957:208–212.

———. 1958*a*. Distributional patterns of vertebrates in the southern United States in relation to past and present environments. p. 433–468. *In* C. L. Hubbs (ed.), Zoogeography. A.A.A.S. Publ. 51.

———. 1958*b*. Mating call in the speciation of anuran amphibians. Amer. Nat. 92:27–51.

———. 1961. Calling and spawning seasons in a mixed population of anurans. Ecology 42:99–110.

———. 1962. Non-morphological data in anuran classification. Syst. Zool. 11:72–84.

———. 1964. Isolating mechanisms and interspecies interactions in anuran amphibians. Quart. Rev. Biol. 39:334–344.

Bocage, J. V. B., du. 1866. Reptiles noveaux ou peu connus recueillis dans les possessions portugaises de l'Afrique occidentale, qui se trouvent au Muséum de Lisbonne. J. Sci. (Lisbon) 1:57–78.

———. 1867. Batraciens nouveaux de l'Afrique occidentale (Loanda et Benguella). Proc. Zool. Soc. (London) 843–846.

———. 1895*a*. Sur une espèce de crapaud à ajouter à la faune herpétologique d'Angola. J. Sci. (Lisbon) 4:51–53.

———. 1895*b*. Herpétologie d'Angola et du Congo.

Ouvrage publié sous les auspices du Ministère de la Marine et des Colonies. Lisbon. 8 vol.

———. 1896. Sur quelques reptiles et batraciens africains provenant du voyage de M. le Dr. Emil Holub. J. Sci. (Lisbon) 4:115–120.

Boettger, O. 1887. Zweiter Beitrag zur Herpetologie Südwest-und Süd-Afrikas. Ber. senckenburg. naturf. Gesch. 135–173.

———. 1888. Materialien zur Fauna des unteren Congo. II. Reptilien und Batrachier. Ber. senckenburg naturf. Gesch. 3–108.

Bogart, J. P. 1968. Chromosome number difference in the amphibian genus *Bufo:* The *Bufo regularis* species group. Evolution 22:42–45.

Bogert, C. M. 1959. Influence of sound on amphibians and reptiles, p. 166–260. *In* W. E. Lanyon & W. N. Tavolga (eds.), Animal sounds and communication. A.I.B.S. Publ. 7.

Bond, G. 1946. The Pleistocene succession near Bulawayo. Occas. Papers Nat. Mus. (S. Rhodesia) 12: 104–113.

———. 1957. Quaternary sands at the Victoria Falls. Prehistory, p. 115–122. *In* Proc. III Pan-Afr. Congr. Livingstone, 1955. Chatto and Windus, London.

Bosazza, V. L., R. Adie, and S. Benner. 1946. Man and the great Kalahari desert. Natal Univ. Coll. Sci. J. 5:1–9.

Boulenger, G. A. 1880. On the palaearctic and Aethiopian species of *Bufo*. Proc. Zool. Soc. (London) 1880: 545–574.

———. 1882. Catalogue of the Batrachia Salientia s. Ecaudata in the collection of the British Museum. 2nd ed. London.

———. 1887. A list of the reptiles and batrachians collected by Mr. H. H. Johnston on the Rio del Rey, Cameroons District, West Africa. Proc. Zool. Soc. (London) 1887:564–565.

———. 1895. An account of the reptiles and batrachians collected by Dr. A. Donaldson Smith in western Somaliland and the Galla Country. Proc. Zool. Soc. (London) 1895:530–540.

———. 1899. Descriptions of new Batrachians in the collection of the British Museum (Natural History). Ann. Mag. Nat. Hist. 3:273–277.

———. 1900. A list of the batrachians and reptiles of the Gaboon (French Congo) with descriptions of new Genera and species. Proc. Zool. Soc. (London) 1900: 433–456.

———. 1901*a*. Matériaux pour la Faune du Congo. Batraciens et reptiles nouveaux. Ann. Mus. Congo (Zool.) 2(1):1–14.

———. 1901*b*. Batraciens nouveaux. Ann. Mus. Congo (Zool.) 2(1):1–14.

———. 1903. On a collection of batrachians and reptiles from the interior of Cape Colony. Ann. Mag. Nat. Hist. 12:215–217.

———. 1905*a*. A list of batrachians and reptiles collected by Dr. W. J. Ansorge in Angola, with descriptions of new species. Ann. Mag. Nat. Hist. 16:105–115.

———. 1905*b*. On a collection of batrachians and reptiles made in South Africa by Mr. C. H. B. Grant, and presented to the British Museum by Mr. C. D. Rudd. Proc. Zool. Soc. (London) 1905:248–255.

———. 1906*a*. Report on the batrachians collected by the late L. Fea in West Africa. Ann. Mus. Civ. St. Nat. (Genoa). Ser. 3, 2:157–172.

———. 1906*b*. Additions to the herpetology of British East Africa. Proc. Zool. Soc. (London) 1906:570–573.

———. 1907. Description of a new toad and a new amphisbaenid from Mashonaland. Ann. Mag. Nat. Hist. 20:47–49.

———. 1910. A revised list of the South African reptiles and batrachians, with synoptic tables, with special reference to the specimens in the South African Museum, and descriptions of new species. Ann. S. Afr. Mus. 5:455–543.

Bourcart, J. 1943. La géologie du Quaternaire au Maroc. Rev. Sci. 311–336.

———, and J. Marçais. 1949. Sur la stratigraphie du Quaternaire cotier à Rabat. C.R. Ac. Sci. 228:108–109.

Brain, C. K. 1958. The Transvaal ape-man-bearing cave deposits. Mem. Transvaal Mus., no. 11.

———, and J. Meester. 1964. Past climatic changes as biological isolating mechanisms in Southern Africa, p. 332–341. *In* D. H. S. Davis (ed.), Ecological studies in Southern Africa. Monogr. Biol. 14.

Broughton, W. B. 1963. Method in bio-acoustic terminology, p. 3–24. *In* R. G. Busnel, Acoustic behavior of animals. Amsterdam.

Cahen, L. 1954. Géologie du Congo Belge. Liège.

Capranica, R. R. 1965. The evoked vocal response of the bullfrog. Monograph 33. M. I. T. Press, Cambridge, Mass.

Castany, G. 1954. Le niveau à Strombes de Tunisie: sa place dans la chronologie préhistorique et la paléogéographie du Quaternaire. C. R. Soc. Géol. Fr. 55–56.

Chabanaud, Paul. 1919. Description d'une espèce nouvelle de Batracien du Sénégal. Bull. Mus. Hist. Nat. (Paris) 1919:454–455.

Chapman, B. M., and R. F. Chapman. 1958. A field study of a population of leopard toads (*Bufo regularis regularis*). J. Anim. Ecol. 27:265–286.

Choubert, G. 1953. Les rapportes entre les formations marines et continentales quaternaires. Actes IVe Congr. INQUA 2:576–590.

Clark, J. D. 1950. The stone age cultures of Northern Rhodesia. S. Afr. Archaeol. Society, Cape Town.

———. 1959. The prehistory of southern Africa. Penguin Books, Harmondsworth.

———. 1960. Human ecology during Pleistocene and later times in Africa south of the Sahara. Current Anthrop. 1:307–324.

Coetzee, J. A., and E. M. van Zinderen Bakker. 1952. Pollen spectrum of the southern middleveld of the Orange Free State. S. Afr. J. Sci. 48:275–281.

Cooke, H. B. S. 1947. The development of the Vaal river and its deposits. Trans. Geol. Soc. S. Afr. 49: 243–259.

———. 1958. Observations relating to Quaternary environments in east and southern Africa. Trans. Geol. Soc. S. Afr. Annexure to Vol. 60. 73 p.

———. 1965. The Pleistocene environment in southern Africa, p. 1–24. *In*, D. H. S. Davis (ed.), Ecological studies in southern Africa. Monogr. Biol. 14.

Daudin, F. M. 1803 Histoire naturelle des rainettes, des grenouilles et des crapauds. Paris.

Davis, L. I. 1964. Biological acoustics and the use of the sound spectrograph. Southwestern Naturalist 9:118–145.

De Villiers, C. 1929*a*. Some observations on the breeding habits of Anura of the Stellenbosch flats, in particular of *Cacosternum capense* and *Bufo angusticeps*. Ann. Transvaal Mus. 13:123–141.

———. 1929*b*. Some features of the early development of *Breviceps*. Ann. Transvaal Mus. 13:142–151.

Dobzhansky, T. 1951. Genetics and the origin of species. New York, Columbia Univ. Press.

Duméril, A., and G. Bibron. 1841. Erpétologie générale. 8:687.

Emiliani, C. 1955. Pleistocene temperatures. J. Geol. 63:538–578.

———. 1958. Ancient temperatures. Sci. Amer. 198:2–11.

Espinal T., Luis Sigfredo. 1968. Vision ecologica del departamento del valle del cauca. Universidad del Valle, Colombia.

Fischer, J. V. 1883. On the habits of *Bufo mauritanicus*, Schleg. in captivity. Naturwiss. Beobachter (Frankfurt a. M.) 24:43–45.

FitzSimons, V. 1932. Preliminary descriptions of new forms of South African Reptilia and Amphibia, from the Vernay-Lang Kalahari Expedition, 1930. Ann. Transvaal Mus. 15:35–40.

———. 1935. Scientific results of the Vernay-Lang Kalahari Expedition, March to September, 1930. Reptilia and Amphibia. Ann. Transvaal Mus. 16:295–397.

———. 1937. Notes on the reptiles and amphibians collected and described from South Africa by Andrew Smith. Ann. Transvaal Mus. 17:259–274.

———. 1939. An account of the reptiles and amphibians collected on an expedition to southeastern Rhodesia during December 1937 and January 1938. Ann. Transvaal Mus. 20:17–46.

———. 1948. Notes on some reptiles and amphibians from the Drakensberg, together with a description of a new *Platysaurus* from northern Natal. Ann. Transvaal Mus. 21: 73–80.

Flake, R. H., and B. L. Turner. 1968. Numerical classification for taxonomic problems. J. Theor. Biol. 20: 260–270.

Flake, R. H., E. von Rudloff, and B. L. Turner. 1969. Quantitative study of clinal variation in *Juniperus virginiana* using terpenoid data. Bull. Nat. Acad. Sci. (*in press*).

Flint, R. F. 1957. Glacial and Pleistocene geology. New York, John Wiley and Sons.

———. 1959. Pleistocene climates in eastern and southern Africa. Bull. Geol. Soc. Amer. 70:343–374.

Fullard, H. (ed.). 1965. Philips' modern college atlas for Africa. Geo. Philip and Daughters, Ltd. London.

Furon, R. 1955. Notules de voyage sur le Quaternaire de Tunisie. Bull. Mus. Nat. Hist. Nat. 27:262–265.

———. 1958. Manuel de préhistoire générale. Paris, Payot.

———. 1963. Geology of Africa. Oliver & Boyd Ltd., Edinburgh.

Geoffroy Saint Hilare, Étienne, and Isidore Geoffroy Saint Hilare. 1827. Description des reptiles qui se trouvent en Égypte. Commission d'Égypte. Description de l'Égypte, &c. Histoire Naturelle. Vol. 1, part 1.

Gervais, P. 1836. Enumération de quelques espèces de reptiles provenant de Barbaries. Ann. Sci. Nat. 6:308–313.

Gigout, M. 1957. L'Oulijien dans le cadre du Tyrrhenien. Bull. Soc. Géol. Fr. 7:385–400.

Great Britain Meteorological Office. 1958. Tables of temperature, relative humidity, and precipitation for the world. 4. Her Majesty's Stationery Office, London.

Guibé, J. 1949. Catalogue des types d'amphibiens de Musée d'Histoire Naturelle.

———, and M. Lamotte. 1958. La réserve naturelle intégrale du Mont Nimba. XII. Batraciens. Mem. Inst. Franç. Afr. Noire 53:241–273.

Günther, A. 1858. Catalogue of the Batrachia Salientia in the collection of the British Museum. London.

Guttman, Sheldon I. 1967. Evolution of blood proteins within the cosmopolitan toad genus *Bufo*. Ph.D. thesis, Univ. Texas.

Hallowell, E. 1844. Descriptions of new species of African reptiles. Proc. Acad. Nat. Sci. (Philadelphia) 2: 169–172.

Hewitt, John. 1909. Description of a new frog belonging to the genus *Heleophryne* and a note on the systematic position of the genus. Ann. Transvaal Mus. 2:45–46.

———. 1911. A key to the species of the South African Batrachia with some notes on the specific characters and a synopsis of the known facts of their distribution. Rec. Albany Mus. 2:189–288.

———. 1912. Notes on the specific characters and distribution of some South African Ophidia and Batrachia. Rec. Albany Mus. 2:264–281.

———. 1913. Notes on the distribution and characters of reptiles and amphibians in South Africa, considered in relation to the problem of discontinuity between closely allied species. S. Afr. J. Sci. 10:238–253.

———. 1919. *Anhydrophryne rattrayi*, a remarkable new frog from Cape Colony. Rec. Albany Mus. 3:182–189.

———. 1925. On some new species of reptiles and amphibians from South Africa. Rec. Albany Mus. 3:343–368.

———. 1926*a*. Description of new and little-known lizards and batrachians from South Africa. Ann. S. Afr. Mus. 20:413–431.

———. 1926*b*. Some new or little-known reptiles and batrachians from South Africa. Ann. S. Afr. Mus. 20: 473–490.

———. 1927. Further descriptions of reptiles and batrachians from South Africa. Rec. Albany Mus. 3: 371–415.

———. 1932. Some new species and subspecies of South African batrachians and lizards. Ann. Natal Mus. 7:105–128.

———. 1935. Some new forms of batrachians and reptiles from South Africa. Rec. Albany Mus. 4:282–294.

———. 1937. A guide to the vertebrate fauna of the eastern Cape Province, South Africa, Part II. Albany Mus., no. 85.

———, and P. A. Methuen. 1913. Descriptions of some new batrachia and lacertilia from South Africa. Trans. R. Soc. S. Afr. 3:107–111.

Hulselmans, J. L. J. 1969. A new species of *Bufo* from South-West Africa. Rev. Zool. Bot. Afr. 79:393–402.

———. 1970. Preliminary notes on African Bufonidae. Rev. Zool. Bot. Afr. 81:149–155.

Inger, R. F. 1968. Exploration du Parc National de la Garamba, Mission H. De Saeger. Institut des Parcs Nationaux du République Démocratique du Congo. Fascicule 52, Kinshasa.

———, and B. Greenberg. 1956. Morphology and seasonal development of sex characters in two sympatric African toads. J. Morph. 99:549–574.

———, and J. I. Menzies. 1961. A new species of toad (*Bufo*) from Sierra Leone. Fieldiana Zoology 39:589–594.

Jackson, S. P. 1963. C.C.T.A. Climatological Atlas of Africa, Pretoria.

Janmart, J. 1953. The Kalahari sands of the Lunda (N. E. Angola), their earlier redistributions and the Sangoan culture. Publ. Cult. Cia Diamant. Angola 20.

Keith, Ronalda. 1968. A new species of *Bufo* from Africa, with comments on the toads of the *Bufo regularis* complex. Amer. Mus. Novitates, no. 2345, pp. 1–22.

Kendrew, W. G. 1961. The climates of the continents. 5th ed. Clarendon Press, Oxford.

Knoch, K., and A. Schulze. 1956. Precipitation, temperature, and sultriness in Africa. Falk-Verlag, Hamburg.

Knoepffler, L.-Ph. 1967. *Bufo petiti* n. sp. Crapaud nain de la foret gabonaise. Biologica Gabonica 3:249-252.

Kurtén, Björn. 1969. Continental drift and evolution. Sci. Amer. 220:54–64.

Lataste, F. 1879. Remarks on the genus *Bufo*, with indication of *B. boulengeri*, *Sp. n.* Rev. Intern. Sci. Biol. 3:436–438.

Laurent, R. F. 1950. Diagnoses preliminaires de treize batraciens nouveaux d'Afrique centrale. Rev. Zool. Bot. Afr. 44:1–18.

———. 1952. *Bufo kisoloensis* Loveridge and *Chamaeleo ituriensis* Schmidt revived. Herpetologica 8:53–55.

Laurenti, Joseph Nicolas. 1768. Specimen medicum, exhibens synopsin Reptilium emendatam cum experimentis circa venena et antidota Reptilium Austriacorum. Vienna.

Leakey, L. S. B. 1949. Tentative study of the Pleistocene climatic changes and stone-age cultural sequence in N. E. Angola. Publ. Cult. Cia Diamant, Angola 4.

———. 1953. Adam's ancestors. Harper & Brothers, New York.

Lecointre, G. 1952. Recherches sur le Néogène et le Quaternaire marins de la côte atlantique du Maroc. Notes et Mém. Serv. Géol. Maroc (Rabat), no. 99.

Linné, C. 1735. Systema naturae, with an introduction and a first English translation of the "Observationes."

Littlejohn, Murray J. 1958. Mating behavior in the tree frog *Hyla versicolor*. Copeia 1956:222–223.

———. 1959. Call differentiation in a complex of seven species of *Crinia* (Anura, Leptodactylidae). Evolution 13:452–468.

———. 1965. Premating isolation in the *Hyla ewingi* complex (Anura: Hylidae). Evolution 19:234–243.

Liu, C. C. 1935. Types of vocal sac in the Salientia. Proc. Boston Soc. Nat. Hist. 41:19–40.

Loveridge, A. 1925. Notes on East African batrachians collected 1920–1923, with the description of four new species. Proc. Zool. Soc. (London) 1925:763–791.

———. 1929. East African reptiles and amphibians in the United States National Museum. U.S. Nat. Mus. Bull. 15:1–135.

———. 1932*a*. Eight new toads of the genus *Bufo* from East and Central Africa. Occas. Papers Boston Soc. Nat. Hist. 8:43–53.

———. 1932*b*. New reptiles and amphibians from Tanganyika Territory and Kenya Colony. Bull. Mus. Comp. Zool. 72:373–387.

———. 1933. Reports on the scientific results of an

expedition to the southwestern highlands of Tanganyika Territory. Bull. Mus. Comp. Zool. 75:1–43.

———. 1936. African reptiles and amphibians in Field Museum of Natural History. Field Mus. Nat. Hist. (Zool.) 22:3–11.

———. 1942. Scientific results of a fourth expedition to forested areas of east and central Africa. Bull. Mus. Comp. Zool. 91:377–436.

———. 1953. Zoological results of a fifth expedition to East Africa. IV. Amphibians from Nyasaland and Tete. Bull. Mus. Comp. Zool. 110:323–406.

———. 1957. Check list of the reptiles and amphibians of East Africa (Uganda; Kenya; Tanganyika; Zanzibar). Bull. Mus. Comp. Zool. 117 (2):153–362.

Lowe, C. van Riet. 1952. The Vaal river chronology: An up to date summary. S. Afr. Archaeol. Bull. 7:1–15.

Martin, W. F. 1967. Mechanism and evolution of sound production in the toad genus *Bufo*. Masters Thesis, Univ. Texas.

Marx, Hymen. 1958. Catalogue of type specimens of reptiles and amphibians in Chicago Natural History Museum. Fieldiana Zool. 36:407–497.

———. 1968. Checklist of the reptiles and amphibians of Egypt. Special Pub. U.S. Naval Med. Res. Unit No. 3, Cairo.

Matschie, Paul. 1893. Einige anscheinend neue "Reptilien und Amphibien aus West-Afrika." Sitzber. Ges. naturk. Freunde Berlin. 170–175.

Mecham, J. S. 1961. Isolating mechanisms in anuran amphibians, pp. 24–61. *In* W. F. Blair (ed.), Vertebrate speciation. Univ. Texas Press, Austin, 642 p.

Meek, S. E. 1897. List of fishes and reptiles obtained by the Field Columbian Museum East African Expedition to Somaliland in 1896. Publ. Field Mus. (Zool.) 1(8): 165–183.

Menzies, J. I. 1963. The climate of Bo, Sierra Leone, and the breeding behavior of the toad, *Bufo regularis*. J. W. Afr. Sci. Assoc. 8:60–73.

Merrem, B. 1820. Versuch eines Systems der Amphibien. Johann Christian Krieger, Marburg.

Mertens, R. 1937. Reptilien und Amphibien aus dem südlichen Inner-Afrika. Abh. senckenb. naturk. Ges. 435:1–23.

———. 1938. Herpetologische Ergebnisse einer Reise nach Kamerun. Abh. Senckenb. naturf. Ges. 442:1–52.

———. 1939. Über das Höhenvorkommen der Froschlurche am Grossen Kamerun-Berge. Abh. Ber. Mus. Naturk. Magdeburg 7:121–128.

———. 1940. Amphibien aus Kamerun, gesammelt von M. Köhler und Dr. H. Graf. Senckenbergiana 22:103–135.

———. 1954. Eine neue Kröte aus Südwestafrika. Senckenb. Biol. 35:1–2.

———. 1955*a*. Amphibien und Reptilien aus Ostafrika. Jahrb. Ver. vaterl. Naturf. (Württemberg) 110:47–61.

———. 1955*b*. Die Amphibien und Reptilien Südwestafrikas. Abh. senckenb. naturf. Ges. 490:1–171.

———, & Heinz Wermuth. 1960. Die Amphibien und Reptilien Europas. Frankfurt am Main, Kramer.

Mocquard, F. 1908. Description de quelques Reptiles et d'un Batracien nouveaux de la collection du Muséum. Bull. Mus. Hist. Nat. (Paris) 1908:259–262.

Monard, A. 1937. Contribution à la batrachologie d' Angola. Bull. Soc. Neuchat. Sci. Nat. 62:5–59.

———. 1938. Contribution à la batrachologie d'Angola. Arq. Mus. Bocage 9:52–118.

Müller, O. F. 1776. Zoologiae Danicae prodromus, seu Animalium Daniae et Norvegiae indigenarum characteres, nomina, et synomina imprimis popularum.

Nieden, Fritz. 1910. Verzeichnis der bei Amani in Deutschostafrika vorkommenden Reptilien und Amphibien. Sitzber. Ges. naturf. Freunde (Berlin) 441–452.

———. 1923. Anura I. Das Tierreich. Berlin and Leipzig.

———. 1926. Anura II. Das Tierreich. Berlin and Leipzig.

Nikolski, A. 1903. On three new species of reptiles, collected by Mr. N. Zarudny in Eastern Persia in 1901. Annuaire Mus. St. Petersb. 8:95–98.

Noble, G. K. 1924. Contributions to the herpetology of the Belgian Congo based on the collection of the American Mus-Congo Expedition, 1909–1915. Bull. Amer. Mus. Nat. Hist. 49:147–347.

Pallas, P. S. 1769. Spicilegia Zoologica (quibus novae . . . et obscurae animalium species . . . illustrantur). Berlin.

———. 1771–1776. Reise durch verschiedene Provinzen des Russischen Reichs (1768–1774). St. Petersburg.

Parker, H. W. 1931. Some reptiles and amphibians from S. E. Arabia. Ann. Mag. Nat. Hist. Ser. 10, 8:514–522.

———. 1932. Two collections of reptiles and amphibians from British Somaliland. Proc. Zool. Soc. (London) 1932:335–367.

———. 1936*a*. Dr. Karl Jordan's expedition to southwest Africa and Angola: Herpetological collections. Nov. Zool. 40:115–146.

———. 1936*b*. The amphibians of the Mamfe Division, Cameroons. I. Zoogeography and systematics. Proc. Zool. Soc. (London) 1936:135–163.

Pasteur, G., and J. Bons. 1959. Les batraciens du Maroc. Travaux de l'Institut Scientifique Cherifien, Sér. Zool. 17. Rabat.

Peabody, F. E. 1954. Travertines and cave deposits of the Kaap escarpment of South Africa, and the type locality of *Australopithecus africanus* Dart. Bull. Geol. Soc. Amer. 65:671–706.

Perret, J. L. 1966. Les amphibiens du Cameroun. Zool. Jb. Syst. 93:289–464.

———, and R. Mertens. 1957. Étude d'une collection herpétologique faite au Cameroun de 1952 à 1955. Bull. Inst. Franç. Afr. Noire. 19:548–601.

Peters, W. 1878. Über die von Hrn. J. M. Hildebrandt während seiner letzten ostafrikanischen Reise gesammelten Säugethiere und Amphibien. Monatsber. Akad. Wiss. (Berlin) 194–209.

———. 1881. Amphibien der Expedition nach Kufra. Leipzig, Brockhaus.

———. 1882. Ueber drei neue Batrachier (*Amblystoma krausei*; *Nyctibatrachus sinensis*; *Bufo buchneri*). Sitzber. Ges. naturk. Freunde (Berlin) 145–148.

Pfeffer, G. 1893. Ostafrikanische Reptilien und Amphibien gesammelt von Herrn Dr. F. Stuhlmann. Jahrb. Hamburg. Wiss. Anst. 10(1892):71–105.

Pienaar, U. de V. 1963. The zoogeography and systematic list of the Amphibia in the Kruger National Park. Koedoe 6:76–82.

Piveteau, J. 1957. Traité de paléontologie, Vol. 7. Paris.

Poche, F. 1903. Einige nothwendige Aenderungen in der herpetologischen Nomenclatur. Zool. Anz. 26:698–703.

Power, H. J. 1925. Notes on the habits and life-histories of certain little-known Anura, with descriptions of the tadpoles. Trans. R. Soc. S. Afr. 13:107–117.

———. 1927*a*. Notes on the habits and life histories of South African Anura with descriptions of the tadpoles. Trans. R. Soc. S. Afr. 14:237–247.

———. 1927*b*. Some tadpoles from Griqualand West. Trans. R. Soc. S. Afr. 14:249–254.

———. 1927*c*. On the herpetological fauna of the Lobatsi-Linokana area. 1. Lobatsi. Trans. R. Soc. S. Afr. 14:405–422.

———. 1932. On the herpetological fauna of the Lobatsi-Linokana area. 2. Linokana. Trans. R. Soc. S. Afr. 20: 39–50.

———, and W. Rose. 1929. Notes on the habits and life-histories of some Cape Peninsula Anura. Trans. R. Soc. S. Afr. 17:109–115.

Poynton, J. C. 1963. Descriptions of southern African amphibians. Ann. Natal Mus. 15:319–332.

———. 1964. The amphibia of southern Africa. Ann. Natal Mus. 17:1–334.

Reuss, Adolph. 1834. Zoologische Miscellen, Reptilien. Saurier. Batrachier. Mus. Senckenbergianum. Abhandl. Gebiete beschreibender Naturgesch. 1:27–62.

Robinson, J. T. 1959. The Sterkfontein tool-maker. Leech 28:94–100.

Rodenwaldt, E., and H. J. Jusatz. 1965. World maps of climatology. Springer Verlag, New York.

Romer, J. D. 1952. Racial variation in the common African toad. Nigerian Field 17:82–83.

Rose, W. 1962. The reptiles and amphibians of southern Africa. Standard Press Ltd., Cape Town.

Roux, J. 1906. Synopsis of the toads of the genus *Nectophryne* B. & P., with special remarks on some known species and description of a new species from German East Africa. Proc. Zool. Soc. 1906:58–65.

———. 1935. Mission Scientifique de l'Omo. 3, Zool., 180.

———. 1936. Mission Scientifique de l'Omo. Reptilia et amphibia. Mém. Mus. Hist. Nat. 4:157–190.

Rüppell, Edward. 1827. Atlas zu der Reise im Nördlischen Afrika. Frankfurt am Main.

Sanderson, I. T. 1936. The amphibians of the Mamfe Division, Cameroons. II. Ecology of the frogs. Proc. Zool. Soc. (London) 1936:165–208.

Savory, H. J. 1963. Junior biology, science for tropical secondary schools. London, Th. Nelson and Sons.

Scheel, J. J. 1970. Notes on the biology of the African tree-toad *Nectophryne afra* Bucholz and Peters, 1875 (Bufonidae, Anura) from Fernando Zoo. Rev. Zool. Bot. Afr. 81:225–236.

Schiøtz, Arne. 1963. The amphibians of Nigeria. Vidensk. Medd. dansk. naturh. Foren. (Copenhagen) 125: 1–92.

———. 1964*a*. The voices of some West African amphibians. Vidensk. Medd. dansk. naturh. Foren. Copenhagen) 127:35–83.

———. 1964*b*. A preliminary list of amphibians collected in Ghana. Vidensk. Medd. dansk naturh. Foren. (Copenhagen) 127:1–17.

———. 1966. On a collection of Amphibia from Nigeria. Vidensk. Medd. dansk naturh. Foren. (Copenhagen) 129:43–48.

———. 1967. The treefrogs (*Rhacophoridae*) of West Africa. Spolia zoologica Musei hauniensis, vol. 25.

———. 1969. The Amphibia of West Africa: A review. The Nigerian Field 34:4–17.

Schlegel, Hermann. 1841. Bemerkungen über die in der Regentschaft Algier gesammelten Amphibien, pp. 106–139. *In* Moritz Wagner, Reisen in der Regentschaft Algier in den Jahren 1836, 1837, und 1838. Leipzig.

Schmidt, K. P. 1936. The amphibians of the Pulitzer Angola Expedition. Ann. Carnegie Mus. 25:127–133.

———, and R. F. Inger. 1959. Exploration du Parc National de l'Upemba. Inst. Parcs Nationaux Congo Belge. Fascicule 56. Brussels.

Schneider, H. 1966. Die Paarungsrufe einheimischer Froschlurche (Discoglossidae, Pelobatidae, Bufonidae, Hylidae). Z. Morph. Okol. Tiere 57:119–135.

Schneider, J. G. 1799. Historiae amphibiorum naturalis et literariae Fasciculus primus (secundus). 2 pt. Jena.

Schwarzbach, Martin. 1963. Climates of the past. Van Nostrand, London.

Scortecci, G. 1929. Contributo alla conoscenza degli anfibi del'Eritrea. Atti. Soc. Ital. Milano 68:174–192.

———. 1932. Nuove specie di anfibi e rettili della Somalia Italiana. Atti. Soc. Ital. Milano 71:264–269.

———. 1933. Anfibi della Somalia Italiana. Atti. Soc. Ital. Milano 72:5–69.

———. 1936. Gli anfibi della Tripolitania. Atti. Soc. Ital. Milano 75:129–226.

———. 1945. Sahara. Ulrico Hoepli, Milan.

Seetzen, U. J. 1855. Reisen durch Syrien, Palastina, Phönicien, die Transjordan-Länder, Arabia Petraea und Unter-Aegypten. Berlin. 3 vol.

Simpson, C. G. 1929–1930. The climate during the Pleistocene period. Proc. R. Soc. (Edinburgh).

Smith, A. 1848. Illustrations of the zoology of South Africa, Reptilia. Smith, Elder & Co., London.

Sokal, R. R., and P. H. A. Sneath. 1963. Principles of numerical taxonomy. W. H. Freeman, San Francisco & London.

Stewart, M. M. 1967. Amphibians of Malawi. State Univ. of New York Press.

Sturtevant, A. H. 1942. The classification of the genus *Drosophila*, with descriptions of nine new species. Univ. Texas Publ. (4213):7–51.

Sutton, J. R. 1922. A contribution to the study of the rainfall map of South Africa. Trans. R. Soc. S. Afr. 10: 367–414.

Thompson, B. W. 1965. The climate of Africa. Oxford Univ. Press, Nairobi and New York.

Throckmorton, L. H. 1968. Concordance and discordance of taxonomic characters in *Drosophila* classification. Syst. Zool. 17:355–387.

Tihen, J. A. 1960. Two new genera of African bufonids, with remarks on the phylogeny of related genera. Copeia:225–233.

van Dijk, D. E. 1966. Systematic and field keys to the families, genera and described species of southern African anuran tadpoles. Ann. Natal Mus. 18:231–286.

van Zinderen Bakker, E. M. 1953. South African pollen grains and spores. I. Balkema, Cape Town.

———. 1956. South African pollen grains and spores. II. Balkema, Cape Town.

———. 1957. A pollen analytical investigation of the Floresbad deposits. (South Africa). Prehistory, pp. 56–67. *In* Proc. III Pan Afr. Congr. Livingstone, 1955. Chatto and Windus, London.

———. 1959. South African pollen grains and spores. III. Balkema, Cape Town.

Verheyen, R. 1960. Note on the altitudinal range of the amphibians collected in the National Upemba-Park (Belgian Congo). Rev. Zool. Bot. Afr. 61:82–86.

Wager, Vincent A. 1965. The frogs of South Africa. Purnell & Sons, Ltd., Capetown & Johannesburg.

Watkins, W. A. 1967. The harmonic interval fact or artifact in spectral analysis of pulse trains. Marine Bio-acoustics 2:15–43.

Wayland, E. 1935. The M. Horizon. A result of climatic oscillation in the second pluvial period. Bull. Geol. Surv. Uganda 2:69–76.

Werner, F. 1896. Über eine Sammlung von Reptilien aus Persien, Mesopotamia und Arabien. Verh. Zool. Bot. Ges. 45:13–20.

———. 1897. Ueber einige neue oder seltene Reptilien und Frosche der zoologìchen Sammlung des Staates in München. Sitzber. Akad. Wiss. (Munich) 27:203–220.

———. 1898. Ueber die Reptilien und Batrachier aus Togoland, Kamerun und Tunis, aus dem Kgl. Museum für Naturkunde in Berlin. II. Verh. Zool. Bot. Ges. (Vienna) 48:191–213.

Wernstedt, F. L. 1959. World climatic data — Africa. Dept. of Geography, Penn. State Univ.

Winston, R. M. 1955. Identification and ecology of *Bufo regularis*. Copeia 1955:293–302.

Witte, G. -F. de. 1930. Liste des batraciens du Congo Belge. I. Rev. Zool. Bot. Afr. 19:232–274.

———. 1932. Description d'un batracien nouveau du Katanga. Rev. Zool. Bot. Afr. 22:1.

———. 1933. Reptiles récoltés au Congo Belge par le Dr. H. Schouteden et par M. G. -F. de Witte. Ann. Mus. Congo (Zool.) 2:55–100.

———. 1934. Batraciens récoltés au Congo Belge par le Dr. H. Schouteden et par M. G. F. de Witte. Ann. Mus. Congo (Zool.) 3:153–188.

———. 1941. Batraciens et reptiles. Explor. Parc Nat. Albert Miss. de Witte 33:1–261.

10. Karyotypes

JAMES P. BOGART

Louisiana Tech University
Ruston, Louisiana

Introduction

My research has had two main aims: (1) to examine the karyotypes of many species in the genus *Bufo,* in the hope that an analysis of these karyotypes would be phylogenetically significant, and (2) to examine the chromosomes of hybrid tadpoles (Chap. 11). Abnormalities in amphibian hybrids have been well documented (Moore 1955; Fankhauser 1945; Briggs and King 1959; Kawamura 1950; and others). It is essential, therefore, that the chromosomal status of hybrids be ascertained in crosses of divergent species if the percentage of hatched tadpoles is low or if the tadpoles appear abnormal. The chromosomes of hybrids can also be analyzed to determine whether any alterations are correlated with particular hybrid combinations. Navashin (1934) indicated that hybrids sometimes contain chromosomes that are unlike the chromosomes of either parent. A third aim was to develop a method of analysis that took into account the variabilities of chromosomes resulting from natural causes and from procedures of slide preparation.

Methods and Materials

Preparation of Chromosomes for Study

Somatic chromosomes were obtained from tadpole tail epithelial cells or the stratified squamous epithelium overlying Bowman's membrane of the cornea. The latter procedure included the following steps:

1. Eyes were dissected from freshly pithed toads and were placed in a .008 percent colchicine solution for two hours.
2. The cornea was then fixed in acetic acid vapor by suspending the eye over glacial acetic acid for one minute.
3. The eye was then placed in distilled water and the corneal epithelium scraped off with a small scalpel.
4. The cells were stained for one minute with aceto-orcein consisting of 3 percent synthetic orcein in 70 percent acetic acid. The cells were then squashed.

Material for chromosomal studies could be obtained by taking a small snip from the tail of a tadpole without killing it. This was important because few tadpoles were viable in many of the crosses. Tail tips were removed from actively growing tadpoles, placed in a .008 percent solution of colchicine for two hours, stained, and squashed. After the tadpole was sampled, another snip could be taken in approximately five days if the first yielded poor results.

Meiotic stages were examined from fresh testes,

and testes that were prefixed in a modified Carnoy solution were also stained and squashed. The method used was the same as the one employed by Duellman and Cole (1965).

Sampling Methods

The number of toads sampled in each species and the number of tadpoles sampled from each cross varied and was determined by the number available as well as by time limitations. Sometimes, very few cells could be obtained that were suitable for analysis even though the sample size was very large. For example, *Bufo quercicus* is a very small toad, and the eyes of this species present operational difficulties. I was able to obtain suitable cells in only two of twelve specimens examined. Conversely, in some species, ample material could be obtained from a single eye or from a single tail tip. At least twenty cells from two individuals of each species or hybrid combination were studied to be certain of the number of chromosomes. Only well-spread chromosome complements were used for percentage and centromeric-ratio analysis. At no time did I find any divergence in the number of chromosomes within an individual or a species.

Two or more tadpoles, depending on the number of crosses in progress at the time, were selected at random and sampled. If the sampled tadpoles were polyploids, other tadpoles in the same cross were sampled until an approximation of the numbers of polyploids to diploids could be made. Abnormal tadpoles were sampled separately unless the entire cross resulted in phenotypically abnormal tadpoles. Even in a control cross there was often a small percentage of abnormal tadpoles. Individuals showing such abnormalities as pigment deficiencies, size discrepancies, or the failure of a few hybrids to metamorphose were examined for chromosomal abnormalities whenever possible.

Measurement of Chromosomes and Constriction Ratios

Chromosomes were photographed at a magnification of 1,300X, and measurements were taken using rear projection at a magnification of 17,160X. Idiograms were constructed of each species, using measurements of each chromatid arm of four homologous chromosomes in the two best cells that could be obtained. Thus, each arm in the idiogram represents the average measurement of eight arms. To minimize differential contraction between cells, the longest chromosome of each idiogram was assigned an arbitrary value of 100 percent and the other chromosomes in the idiogram were assigned percentage values relative to the longest chromosome. Chromosomes more than 50 percent of the measurement of the longest chromosome are considered large, those from 40 percent to 50 percent intermediate, and those less than 40 percent, small. The centromeric ratio was determined by dividing the long arm by the short arm following Levan et al. (1964). Secondary constriction ratios were determined as if the secondary constriction were the centromere and ignoring the position of the centromere. The length of the arms composing the idiogram were determined using the following formulae

$$\text{short arm} = \frac{\text{percentage length}}{1 + \text{centromeric ratio}}$$

$$\text{long arm} = \text{percentage length} - \text{short arm}$$

Karyotypes

Karyotype is defined as the basic chromosome set of a species (Swanson 1957). Since only complete, well-spread cells were used for chromosome analysis, the linear arrangement of the chromosomes from each species is referred to as the karyotype for that species. A question of terminology arises about what to call the chromosome set of a hybrid. To avoid superfluous terms, I will refer to the chromosome set of a hybrid as the *hybrid karyotype* and, as such, it represents a combination of chromosomes of the two parental species.

Hybrid chromosomes could often be referred to one parent or the other by comparing the parental and hybrid karyotypes. Even when the chromosomes could not be definitely referred to one parent, the karyotype number could be inferred from the size of the chromosomes and from their centromeric ratios. Since comparisons are made between the chromosomes with common karyotype numbers, it was not critical to define absolutely the parent for each hybrid chromosome. What is very important, however, is to ensure that the two chromosomes sharing a hybrid karyotype number are from different parents.

Action of Colchicine

Pernice, in 1889, described the action of colchicine on mitosis, but it was not until 1938, when Levan studied colchicine's effect on mitosis in *Allium*, that the potential of this chemical was fully realized. "C-metaphase" or the metaphase that develops as a reaction to colchicine is different from the normal metaphase (Kihlman 1966). "C-pairs" accumulate over a period of time, thus increasing the chance of obtaining identifiable chromosomes in a given tissue. All the chromosomes studied are in the same C-metaphase stage, making comparisons and homologies more realistic than comparisons of different stages between different cells. The only disadvantage of using colchicine is contraction of chromosomes in C-metaphase, which is probably due to increased coiling (see Swanson 1957; White 1961), this being a function of both the concentration of colchicine (Eigsti and Dustin 1955) and the length of exposure to the chemical. A standard concentration of colchicine was used to standardize the first of the above-mentioned variables. Even though C-pairs in different cells in the squashed tissue were at slightly different stages of contraction, the C-pairs within a cell all appeared to be at the same stage of contraction. To minimize some of these errors, cells with extremely uncontracted or extremely contracted chromosomes were not used for analysis, and a percentage analytical approach circumvented some of the between-cell variation.

The question of polyploidy arises. Are the observed polyploid cells the result of the colchicine treatment? Polyploid plant cells are produced quite readily in plants, but animal cells seldom become polyploid (Eigsti and Dustin 1955). Since the tissue is subjected to colchicine for only a short time (2 hours) and since all the cells in an individual are polyploid, if any are, it is extremely unlikely that colchicine causes the observed polyploid or aneuploid cell.

Percentage Lengths and Centromeric Ratios

The karyotypes, percentage lengths, centromeric ratios, and idiograms for the karyotypes of the various species are presented in Appendixes D and E.

Indispensable, and prerequisite, to any analysis of karyotypes is to determine the variation that results from the squash technique, errors in measurement, and the natural variation of chromosome lengths. I used triploid hybrids to determine the intracell error involved in percentage lengths and centromeric ratios. When a triploid hybrid is examined and the chromosomes arranged according to size and centromeric position, there should be two identical chromosomes in each set with a common hybrid karyotype number. Thus, the most similar chromosome pair of each karyotype number should vary only according to squash technique, errors in measurement, and the natural variation of chromosome lengths.

The percentage lengths and centromeric ratios were calculated for 873 chromosomes, which represented the chromosome complements of 28 triploid hybrids. The most similar pair of chromosomes were chosen from each set of three chromosomes sharing a common hybrid karyotype number. This yielded 291 pairs of chromosomes that should be genetically identical. The percentage length between these "identical" chromosomes was found to vary from 0 to 10 percent, and the centromeric ratios from 0 to 1.4. The number of chromosome pairs in the various centromeric-ratio differences or percentage-length differences are listed in table 10-1.

Table 10–1. Variation Found Between the Two Most Similar Chromosomes of Each Hybrid Karyotype Number of Twenty-Eight Triploid Hybrids

Ratio Differences	Number of Chromosomes	Percentage Differences	Number of Chromosomes
0	94	0	96
.1	58	0-1	11
.2	68	1-2	74
.3	34	2-3	46
.4	20	3-4	13
.5	10	4-5	22
.6	2	5-6	12
.8	1	6-7	2
.9	2	7-8	6
1.3	1	8-9	5
1.4	1	9-10	4
Total	291		291

This enumeration demonstrates that a large range of error must be considered in the mathematical analysis of the chromosomes. From these data I arbitrarily chose a centromeric ratio error of .5 and a percentage-length error of 10 to be within the range of natural variation. A ratio variation of .5 encompassed variation in 97.6 percent of the 291 chromosome pairs. A percentage difference of 10 encompassed variation in all the chromosome pairs. If a pair of chromosomes in the hybrid karyotype having the same karyotype number has a centromeric-ratio variation less than .5 or percentage-length variation less than 10, these chromosomes cannot be assumed to be different because of intracell variation in centromeric ratio and percentage length. However, if a pair of chromosomes in a hybrid karyotype sharing the same karyotype number has a centromeric-ratio difference

Table 10–2. Comparison of Significantly Different Chromosomes Between Closely Related Species

	B. americanus	B. houstonensis	B. microscaphus	B. terrestris	B. hemiophrys	B. woodhousei
B. americanus	0					
B. houstonensis	3	0				
B. microscaphus	1	1	0			
B. terrestris	5	3	2	0		
B. hemiophrys	2	2	1	2	0	
B. woodhousei	3	2	1	1	0	0

Table 10–3. Comparison of Significantly Different Chromosomes of a Pair of Species from Each of Four Continents

		B. americanus	B. valliceps	B. calamita	B. bufo	B. arenarum	B. spinulosus	B. lönnbergi	B. mauritanicus
North America	B. americanus	0							
	B. valliceps	3	0						
Europe	B. calamita	1	3	0					
	B. bufo	3	1	2	0				
South America	B. arenarum	3	2	1	1	0			
	B. spinulosus	3	6	3	2	3	0		
Africa	B. lönnbergi	3	3	4	2	2	2	0	
	B. mauritanicus	2	1	2	4	1	3	3	0

greater than .5 or a percentage-length difference greater than 10, the difference is probably not due to intracell error and should represent real variations between the two parental karyotypes. Chromosome pairs in hybrid karyotypes that have a centromeric-ratio variation of more than 10 are considered to be significantly different chromosomes.

To test the usefulness of centromeric ratio and percentage-length analysis between the various species, the number of significantly different chromosomes between the karyotypes of closely related species was compared with the number of significantly different chromosomes between species that are not considered to be closely related. The same error factor previously defined for hybrid karyotypes was employed in order to account for intracellular variation.

The *americanus* species group is considered a natural assemblage of closely related species as evidenced by compatibility experiments (Blair 1959, 1961, 1963), general morphology (Baldauf 1959), osteology (Tihen 1962), and parotoid-gland secretions (Low 1967). Speciation in this group may have resulted from Pleistocene glaciation (Blair 1963). Six species of the *americanus* group were compared in table 10-2. Six distantly related species representing six different species groupings from four continents were also compared (Table 10-3). If the number of significantly different chromosomes according to percentage lengths and centomeric ratios had any evolutionary significance, it would be expected that table 10-2 would contain much lower numbers than table 10-3. From the data in table 10-2 and table 10-3 it appears that *B. americanus* is more closely related to *B. calamita* and *B. mauritanicus* than it is to *B. houstonensis*, *B. hemiophrys*, or *B. woodhousei*. These results are biogeographically difficult, if not impossible, to explain. Furthermore, there are inconsistencies in the data. For example, it appears (Table 10-3) that *B. valliceps* is closely related to *B. bufo* and *B. mauritanicus*, since there is only one chromosome that is significantly different. It might be expected, on the basis of this evidence, that *B. bufo* and *B. mauritanicus* would be chromosomally quite similar. However, there are four significantly different chromosomes between *B. bufo* and *B. mauritanicus*.

Even though there are some phylogenetically realistic similarities, they are not consistent enough to be of evolutionary value. It may be that karyotype evolution has not paralleled species evolution in the genus *Bufo*, or the apparent inconsistencies may be the result of the large error that has to be considered.

All of the 22-chromosome *Bufo* species that were studied have similar karyotypes. There are five large pairs, one intermediate pair, and five smaller pairs. Most of the chromosomes are metacentric, but there is usually one large and one small submetacentric pair. Slight interspecific discrepancies from these generalizations do occur. The discrepancies, however, do not seem to reflect sequences of past chromosomal polymorphisms whereby one sequence reached fixation in one evolutionary lineage, the other sequence in a second line of evolution. Many species of *Bufo* are interspecifically compatible, and artificially produced hybrid combinations are often viable (Blair 1959, 1961, 1962, 1963*a*, 1963*b*, 1964, 1966; *see* Chap. 11). This degree of compatibility does not occur among animal groups that have variable chromosome karyotypes. Rates of karyotype differentiation have been shown to differ among groups of animals (Patterson and Stone 1952; White 1954). The ability to produce hybrids indicates that *Bufo* species have conservative karyotypes and that the rate of karyotype differentiation has been rather slow.

Evolutionarily significant changes could have occurred in *Bufo* karyotypes that cannot be detected using percentage lengths and centromeric ratios. Single or even multiple gene mutations at specific loci would not be detected. Slight deletions, duplications, or both occurring simultaneously would also not be detected. Translocations involving similar amounts of chromatin would not alter the appearance of the karyotype. Neither paracentric inversions nor pericentric inversions involving the same amount of chromatin on either side of the centromere would be detected in mitotic cells. Meiotic squashes revealed no evidence of inversions. Even if pericentric inversions or translocations that involved different amounts of the chromatin occurred, the chromosomes involved may just switch positions on the percentage-length scale, revealing no difference in the analysis of the karyotype.

To determine the cause of the error that was established by comparing homologous chromosomes in triploids, several factors must be considered. Although there is some controversy on the subject, most authors agree that chromosomes are made up of microfibrillar units 250Å (Wolfe and John 1965) or 100–129 Å (Ris and Chandler 1963) in diameter Contraction of the chromosome is due to coiling of the microfibrillar units, as postulated by Ris (1945) and verified by several authors. As the chromosomes progress through the division stages, they become more contracted until metaphase is reached. Different chromosomes, or parts of chromosomes, have been shown to exhibit differing degrees of contraction within the same cell. Coleman (1943) demonstrated that the X-chromosome of male grasshoppers is in a much more condensed state than the autosomes. Brown (1949) studied contraction of chromosomes in the tomato and found differential contraction to be a function of the stage of meiotic division. There is some evidence of differential contraction in *Bufo*. Invariably, in this study, submetacentric chromosomes had a lower centromeric ratio if the chromosomes were very condensed; less condensed submetacentric chromosomes had higher centromeric ratios. This could indicate that the smaller arms condense more rapidly than the longer arms, since condensation increases with time as the result of colchicine treatment (Eigsti and Dustin 1955). Submetacentric chromosomes observed in mitosis were almost all metacentric in meiotic metaphase; meiotic metaphase chromosomes are much more contracted than mitotic C-metaphase chromosomes. Secondary constrictions were also affected and were not evident in very condensed mitotic chromosomes or meiotic metaphase chromosomes. Since the heterochromatic regions are in a less contracted state, variations in chromatid length might be expected as the result of a different ratio of heterochromatic to euchromatic regions between homologous chromosomes even in the same cell. It is impossible to identify all the heterochromatic regions that probably occur throughout the length of any chromatid, since only the major regions are visually evident. Differential contraction may be responsible for much of the chromosomal size variation.

Even though the chromosomes were magnified considerably in order to reduce inaccuracies in the measurements, the "map measurer" could only be considered accurate to .5 cm. of the highly magnified chromosome. This would correspond to about one micron in actual chromosome length. A difference of this magnitude is substantial in the smallest chromosome arms. Differences of a small magnitude could often switch a chromosome's karyotype position, because nonhomologous chromosomes often had similar sizes.

In summary, centromeric-ratio analysis and percentage-length determinations are not valid criteria for outlining species evolution in the genus *Bufo*. An intracellular error factor of .5 for the centromeric ratio and 10 for the percentage length was arbitrarily determined from the analysis of homologous chromosomes in triploid hybrids. This error is probably the result of differential contraction of the chromosomes and the actual error in measurement. Since the karyotypes of many *Bufo* species are very similar, any phylogenetically realistic differences in centromeric ratio or percentage length is probably masked by intracellular variation of chromosomes using the present technique.

Chromosomal Characters

Secondary Constrictions

Heitz, in 1931, was the first person to associate secondary constrictions with nucleolar formation. More recently, Sirlin (1960), using electron micrographs, found nucleolar RNA in intimate association with the secondary constrictions in chironomid nuclei. Secondary constrictions have been found in most species that have been studied, and these heteropycnotic regions have been associated with the formation of the nucleolus in many cases. McClintock (1934) utilized X-rays in Maize to fracture the chromosomes. If a fracture separated the nucleolar organizer into unequal segments, both portions produced nucleoli and the larger segment produced a larger nucleolus than did the smaller segment. When only one segment was present, it still produced a full-sized nucleolus. The rate of nucleolus production from the material present was critical.

In *Bufo*, secondary constrictions were found in most species. In a few species, the chromosomes were very contracted and the secondary constrictions were probably not distinguishable. Often there were several secondary constrictions in a single karyotype. In F_1 hybrids, some secondary constrictions appear to dominate the hybrid karyotype. The secondary constriction associated with chromosome 1 of *B. woodhousei* or *B. speciosus* always appeared in any hybrid combination involving one of these species. The secondary constriction is often exaggerated. The secondary constriction on chromosome 7 of *B. marinus*, however, did not appear at all in any hybrid combinations with African species. Secondary constrictions sometimes appeared in the hybrid karyotype that were not discovered in the parent species. The explanation for this might be found in the "strength" of the nucleolus organizer (Navashin 1934). In F_1 hybrids of *Crepis*, Navashin found one species to have a "stronger" nucleolar organizer, which would suppress the nucleolar activity of the other species in the hybrid combination. A secondary constriction might not even be apparent in a "suppressed" state.

Secondary constrictions are areas subject to bending and variance in area (Resende 1940; Therman-Suomaleinen 1949). In *Bufo*, breaks in chromosomes were rarely encountered but, when they did occur, they were at the secondary constriction. In some species of *Bufo*, secondary constrictions appeared to be on chromosomes having the same karyotype number and to have similar constriction ratios. Various dichotomies in the evolution of the karyotype in the genus *Bufo* may be inferred through a study of similar secondary constrictions among different species of *Bufo*. It cannot be determined that a particular constriction is definitely homologous between two species, but it is evident that highly compatible species have similar secondary constriction patterns. If a secondary constriction occurred on the same chromosome, in a similar position, in two or more species, the secondary constriction was designated with an alphabetical letter in order to prevent repetitive discussion.

The following classification of secondary constrictions was based on the species that had secondary constrictions on chromosomes having the same karyotype number and similar secondary constriction ratios.

Chromosome 1

A (secondary constriction): a metacentric constriction on the long arm. Since many species have a metacentric first chromosome, this constriction may appear to be on the short arm. This constriction can be confused with the D secondary constriction if chromosomes 1 and 2 are similar in size. The A constriction has been found in *B. crucifer*, *B. haematiticus*, *B. canaliferus*, *B. cognatus*, *B. compactilis*, *B. hemiophrys*, *B. luetkeni*, *B. microscaphus*, *B. occidentalis*, *B. punctatus*, *B. retiformis*, *B. speciosus*, *B. terrestris*, *B. valliceps*, *B. woodhousei*, *B. bufo*, *B. calamita*, *B. melanostictus*, and *B. viridis*. The distribution of the various species of *Bufo* is shown in Appendix F.

B (secondary constriction): a submetacentric secondary constriction on the long arm. Also, this constriction may appear to be on the short arm since chromosome 1 is often metacentric. It has been found in *B. arenarum*, *B. crucifer*, *B. bocourti*, *B. granulosus*, *B. holdridgei*, *B. alvarius*, *B. kelloggi*, *B. luetkeni*, *B. marmoreus*, *B. nelsoni*, *B. quercicus*, *B. retiformis*, and *B. stomaticus*.

C (secondary constriction): a subtelocentric secondary constriction on the long arm of chromosome 1. It has been found in *B. canaliferus*, *B. americanus*, *B. houstonensis*, *B. nelsoni*, and *B. valliceps*.

Chromosome 2

D (secondary constriction): a metacentric secondary

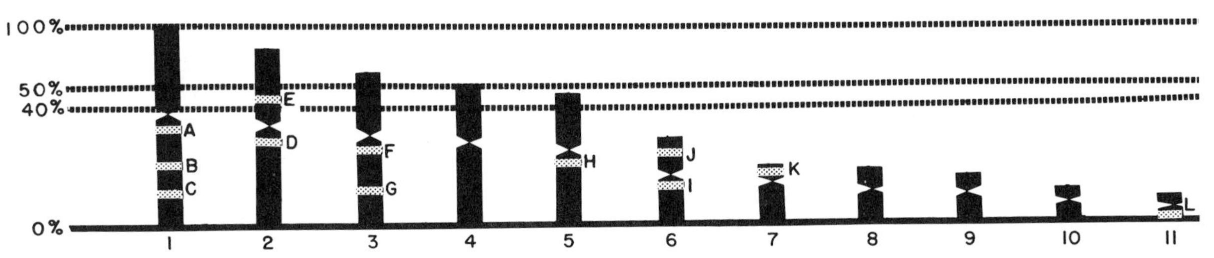

Fig. 10–1. Schematic ideogram of species in the genus *Bufo*. The scale 0% to 100% represents relative percentage lengths of the longest chromosome in the karyotype. *A* to *L* represents the positions of secondary constrictions found to be present in various species of *Bufo*.

constriction. Since chromosome 2 is often metacentric, this constriction may appear to occur on the short arm. Chromosomes 1 and 2 are often of a very similar length, so that it is often difficult to distinguish this constriction from the A secondary constriction. It has been found in *B. crucifer*, *B. americanus*, *B. canorus*, *B. hemiophrys*, *B. houstonensis*, *B. microscaphus*, *B. punctatus*, *B. retiformis*, and *B. viridis*.

E (secondary constriction): a submetacentric secondary constriction on the short arm. It has been found in *B. spinulosus*, *B. canaliferus*, *B. alvarius*, *B. retiformis*, and *B. stomaticus*.

Chromosome 3

F (secondary constriction): a metacentric secondary constriction on the long arm. It has been found in *B. spinulosus*, *B. alvarius*, *B. canaliferus*, *B. occidentalis*, *B. retiformis*, *B. valliceps*, *B. bufo*, *B. stomaticus*, and *B. viridis*.

G (secondary constriction): a subtelocentric secondary constriction found on the long arm. It has been found in *B. kelloggi*, *B. nelsoni*, *B. retiformis*, and *B. valliceps*.

Chromosome 5

H (secondary constriction): a metacentric secondary constriction found on the long arm. It has been found in *B. kelloggi*, *B. nelsoni*, *B. retiformis*, and *B. valliceps*.

Chromosome 6

I (secondary constriction): a metacentric secondary constriction on the long arm. It has been found in *B. americanus*, *B. canorus*, and *B. houstonensis*.

J (secondary constriction): a subtelocentric secondary constriction found on the short arm. It has been found in *B. crucifer*, *B. alvarius*, *B. marmoreus*, *B. retiformis*, and *B. stomaticus*.

Chromosome 7

K (secondary constriction): a subtelocentric (rarely submetacentric) secondary constriction on the short arm. This secondary constriction often produces a satellite. It has been found in *B. atacamensis*, *B. arenarum*, *B. crucifer*, *B. marinus*, *B. paracnemis*, *B. poeppigi*, *B. spinulosus*, *B. variegatus*, *B. coniferus*, *B. granulosus*, *B. haematiticus*, *B. valliceps*, and *B. stomaticus*.

Chromosome 11

L (secondary constriction): a telocentric constriction on the long arm. It has been found in *B. chilensis*, *B. calamita*, *B. melanostictus*, *B. mauritanicus*, and *B. flavolineatus*.

R (secondary constriction): a metacentric secondary constriction on the long arm of chromosome six. This constriction has been found in the 20-chromosome *regularis* species group and is probably unique to the 20-chromosome toads.

Figure 10-1 shows a hypothetical *Bufo* idiogram that contains all these secondary constrictions as they have been classified.

In some species of Bufo constrictions appear that cannot be related to two or more species. Also, it is often difficult to distinguish between some of the constrictions if two nonhomologous chromosomes are similar in size and there are no reference constrictions. For example, chromosomes 1 and 2 are often very similar in length in many *Bufo* species. If a metacentric secondary constriction is apparent on one of these chromosomes, it could be an A or a D secondary constriction. But, if B, C, or E secondary constrictions can be identified, it is possible to distinguish between chromosomes 1 and 2. Sometimes secondary constrictions may be identified in hybrid karyotypes that are not found in the species karyotype and these may be referred to one parent or the other.

The karyotypes of many specimens of *B. regularis* from various parts of Africa were examined. The positions of the secondary constrictions vary in this species or "group of species." The only constriction that all specimens seem to have in common is the R secondary constriction. This was evident in corneal chromosomes of specimens from El Mahalla el Kubra, Egypt; Umtali, Rhodesia; and the Vumba Mountains, Rhodesia. Artificial cross combinations indicate that this constriction is also present in specimens from Ol Tukai, Kenya; Ghana; and Port St. Johns, South Africa. No secondary constrictions have yet been found in the karyotype of specimens from Lourenço Marques, Mozambique. El Mahalla el Kubra, Egypt, and Umtali, Rhodesia, are the only two localities from which *B. regularis* specimens had metacentric secondary constriction on the long arm of chromosome 2. A Vumba Mountains, Rhodesia, specimen demonstrated a submetacentric secondary constriction on the long arm of chromosome 2. *B. regularis* of Mazeras, Kenya, have a submetacentric secondary constriction on the long arm of chromosome 4, which may also be present in Nyeri, Kenya, *B. regularis*, as evidenced in hybrid combinations. A metacentric secondary constriction is found on the long arm of chromosome 4 in a specimen from Vumba Mountains, Rhodesia. A submetacentric secondary constriction is present on the short arm of chromosome 5 in *B. regularis* from Ol Tukai, Kenya, which is not found in specimens of *B. regularis* from other localities. Other unique constrictions confined to particular localities include: a subtelocentric secondary constriction on the long arm of chromosome 7 in *B. regularis* from Nyeri, Kenya; a telocentric secondary constriction on the long arm of chromosome 1, a telocentric secondary constriction on the short arm of chromosome 2, and a metacentric secondary constriction on the long arm of chromosome 3 in *B. regularis* from Port St. Johns; and a submetacentric secondary constriction on the long arm of chromosome 10 in *B. regularis* from El Mahalla el Kubra, Egypt.

The R secondary constriction found on chromosome 6 in *B. regularis* is the same as the I constriction, but these constrictions are probably not homologous. To avoid confusion, I prefer to treat the 20-chromosome African toads as a separate section in constriction classification.

A summary of the secondary constrictions found to be present in the karyotypes or hybrid karyotypes of species of *Bufo* is presented in Appendix F.

Some of the secondary constrictions have also been discovered by other investigators. Bianchi and Laguens (1964) described the secondary constriction on the seventh chromosome in *B. arenarum* (the K constriction). Morescalchi and Garguilo (1968) illustrated many secondary constrictions in *Bufo*: a submetacentric secondary constriction on the short arm of chromosome 1 in the karyotype of *B. hemiophrys* (probably the A constriction); a secondary constriction on the short arm of chromosome 7 in the karyotype of *B. arenarum*, *B. marinus*, *B. paracnemis*, and *B. spinulosus* (the K constriction); secondary constriction of chromosome 11 in the karyotypes of *B. calamita*, *B. melanostictus*, and *B. mauritanicus* (the L constriction); a secondary constriction on the short arm of chromosome 2 for *B. garmani*; and a secondary constriction on the long arm of chromosome 6 in *B. rangeri* (the R constriction.) A few constrictions, however, were illustrated by Morescalchi and Garguilo (1968) that I did not discover in the same species. They illustrate secondary constrictions present on the long arm of chromosome 6 in *B. bufo* and *B. viridis*, a secondary constriction on the long arm of chromosome 8 in *B. viridis*, a secondary constriction on the short arm of chromosome 11 in *B. cognatus*, and a secondary constriction on the short arm of chromosome 2 in *B. regularis*. These few discrepancies could be the result of the different populations sampled, the different technique employed, or the different tissue that was used.

Cole et al. (1968) studied the karyotypes of *B. alvarius*, *B. cognatus*, *B. microscaphus*, *B. punctatus*, *B. retiformis*, *B. valliceps*, *B. woodhousei*, and *B. marinus*. They recognized a distinct secondary constriction near the centromere of chromosome 1 in the karyotypes of all species except *B. marinus*, which has a secondary constriction on the short arm of chromosome 7 (the K constriction). Although they only present figures for *B. microscaphus*, *B. retiformis*, and *B. marinus*, I believe that they failed to distinguish many secondary constrictions that are present in some of the species' karyotypes, and they failed to distinguish between the A and D secondary constrictions.

Chromosome Numbers

Chromosome number in *Bufo* has been discussed previously (Bogart 1968). Essentially, all species of *Bufo* examined have 22 chromosomes, except in Africa. In Africa, the *B. regularis* species group has only 20 chromosomes (2n), but there are other species in Africa that have 22 chromosomes. White (1961)

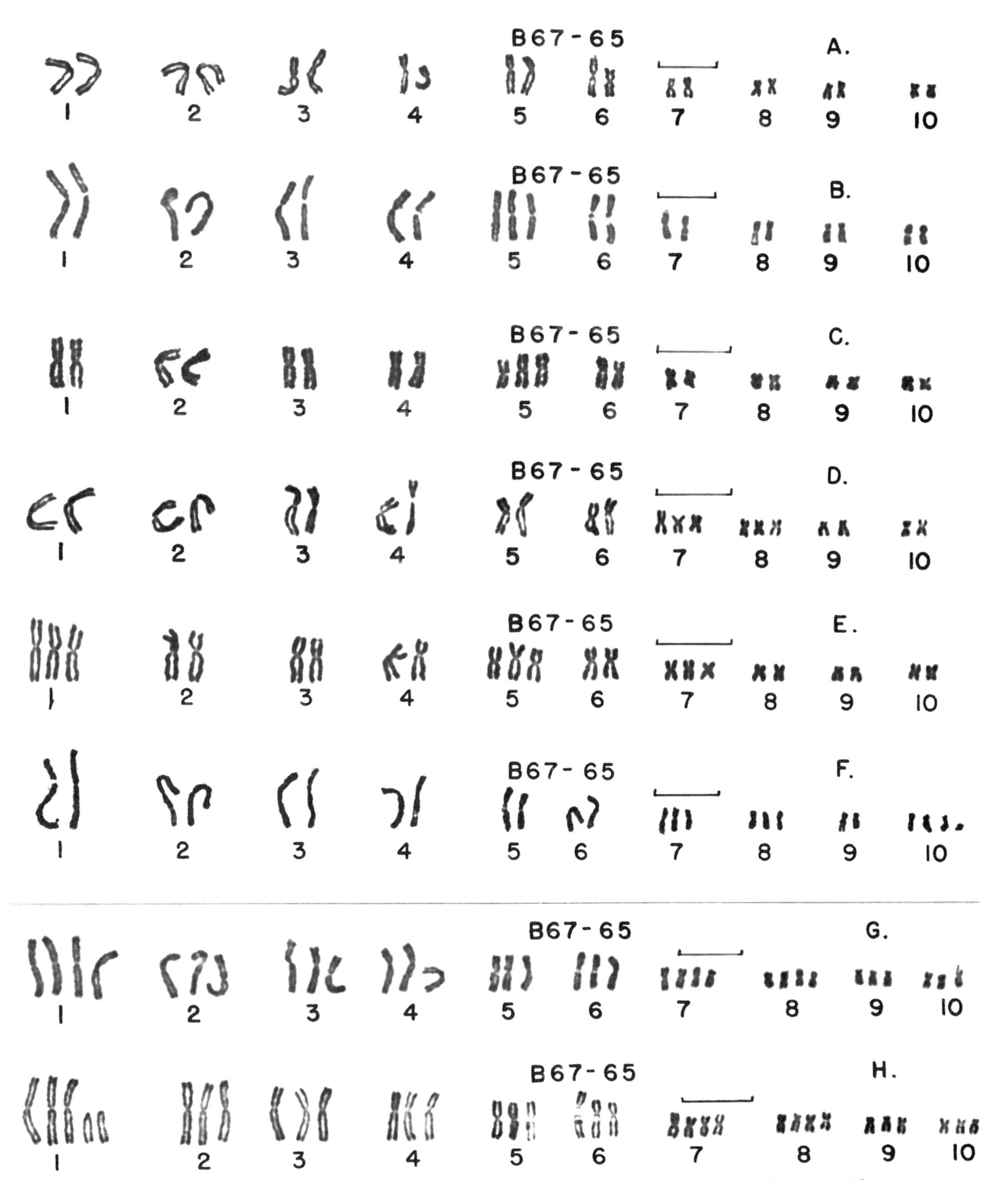

Fig. 10–2. Hybrid karyotypes from tadpoles produced in the cross B67-65 of a female *Bufo regularis* from Ghana and a triploid male hybrid from cross B66-93 (*B. regularis* female from El Mahalla, Egypt, and a male *B. regularis* from the Vumba Mountains in Rhodesia).

stated that there was little evidence that evolutionary changes in the chromosome number of animals have occurred through duplication or deletions of whole chromosomes, but it is frequently found that a pair of metacentric chromosomes in one species is represented by two pairs of acrocentrics in a related species. Schrader and Hughes-Schrader (1958) suggested that complete chromosomes could be duplicated by "chromatid autonomy," but there would be a size reduction in the resultant chromosomes because of the duplication caused by a fission of DNA.

All available evidence points to a 22-chromosome ancestor for *Bufo*. *Atelopus, Odontophrynus, Pleurodema, Proceratophrys,* and *Leptodactylus* are considered allied to *Bufo* (Noble 1931). I have examined four species of *Atelopus*, one of *Pleurodema*, five of *Leptodactylus*, one of *Proceratophrys*, and two of *Odontophrynus*. All these genera have 22 chromosomes. *Odontophrynus americanus* is a tetraploid anuran but had a 22-chromosome ancestor (Bogart 1967). Two species of fairly specialized *Eleutherodactylus* and species of *Dendrobates* have 20 chromosomes, but the karyotypes are very different from those of *Bufo*. Since 20- and 22-chromosome species of *Bufo* produce viable hybrids (Chap. 11), the number dichotomy must have occurred recently enough to exclude convergent evolution from two very primitive ancestors.

No acrocentric chromosomes have been found in any of the 22-chromosome *Bufo*, and it is unlikely that centric fusion of acrocentric chromosomes gave rise to the 20-chromosome group of toads. Morescalchi (1968) studied the chromosomes in two female individuals of *Leiopelma hochstetteri*, a primitive anuran species that lives in New Zealand. He found one of the specimens had 34 chromosomes and the other specimen had 23 chromosomes (2n). One specimen apparently lacked 11 "microchromosomes." He assumed that the microchromosomes may be genetically inert, such as the supernumerary chromosomes (or B-chromosomes) of various plants and of certain animals, and that these chromosomes could be lost with no apparent effect on the gene system of the species involved. Morescalchi (1968) speculated that the same phenomena may be responsible for the chromosome-number difference in the genus *Bufo*. There is no evidence available that any *Bufo* chromosome is entirely heterochromatic and it seems rather unrealistic to assume that the DNA contained in an entire chromosome is so superfluous that it could simply disappear with no apparent effect on the species involved.

Different chromosome numbers were encountered in the backcross B67-65 of a female *B. regularis* from El Mahalla el Kubra, Egypt, and an F_1 male *B. regularis* from El Mahalla el Kubra female x male *B. regularis* from Vumba Mountains, Rhodesia. Tadpoles sampled from this backcross had 20, 21, 22, 23, 24, and 33 chromosomes (Fig. 10-2). The male used in this cross was a somatic triploid. The tadpoles appeared fairly normal phenotypically even though their chromosome number was altered. Only a small proportion (5.8%) of the fertilized eggs metamorphosed (Chap. 11). The metamorphosed individuals were not examined for chromosomes. It is evident that the triploid male used in the backcross was not sterile and that he produced viable sperm. Some viable sperm must have contained 10, 11, 12, 13, 14, and 23 chromosomes and, possibly, other numbers.

Triploids are often produced in the laboratory between closely related species of anurans and even in control crosses. No triploid anurans have been discovered in nature, although Fankhauser (1945) has discovered triploid salamanders under natural conditions. If fertile triploid anurans were able to survive in a natural population, a polymorphism for chromosome numbers might be established. Fixation of various numbers could occur in isolated groups and various numbers be maintained. Essentially, this would be a mechanism for a species' experimentation with chromosome-number differences. It is obvious that lethal combinations and sterility would result from abnormal chromosome numbers, but surprisingly many abnormalities were found to exist in viable hybrid tadpoles.

Invariably, in 21-chromosome hybrid tadpoles resulting from crosses of 20- and 22-chromosome species, the smallest chromosome in the hybrid karyotype is from the 22-chromosome parent and there is no "homolog" for this smallest chromosome.

It is possible that a fusion of two chromosomes resulted in the chromosome-number dichotomy in *Bufo*, and chromosome 11 of a 22-chromosome species was probably involved in the chromosome-number reduction. It is difficult to determine exactly how the chromosomes combined or which chromosomes were involved in the combination. No information was obtained from testis squashes of adult hybrids. Secondary constrictions could be involved in the chromosomal fusion. Most of the 22-chromosome toads that produced viable hybrids with the 20-chromosome species have a K constriction or an L constriction (Chap. 11). If the satellites produced by these constrictions were lost, and the two chromosomes fused, a chromosome very similar to chromosome 6 of the 20-chromosome species could be produced. Future radiation experiments might substantiate this speculation.

Size Differences

Differences between the number 1 chromosomes in the hybrid karyotype were sometimes greater than the arbitrarily established error of 10 percent which could indicate that the longest chromosome has lost or gained chromatin from another chromosome or that the entire karyotype of one species might be larger than that of the other in the hybrid combination. Since there appears to be a general correspondence between chromosome size and DNA value (Hughes-Schrader and Schrader 1956; Mirsky and Ris 1951; Ullerich 1966), some species of *Bufo* may have a difference in absolute DNA value. Manton (1952) found that the amount of DNA in ferns was related to phylogeny. Primitive Osmundaceae had the largest chromosomes, and the specialized Salviniaceae, the smallest. The absolute value of DNA may have phylogenetic implications in some other groups (Mirsky and Ris 1951; Stebbins 1950). Obviously, chromosome size is not universally useful as a phylogenetically sound criterion, or we might group the chromosomally large Amphibia with the Orthoptera, thus distinct from the lizards and the dicotyledonous plants, which have small chromosomes.

Comparing the various hybrid combinations, the size differences that may reflect DNA value differences did not appear to have any phylogenetic significance. Chromosome 1 in *B. spinulosus* was 27.8 percent longer than chromosome 1 in hybrid combinations with *B. luetkeni*. Chromosome 1 in *B. luetkeni* was 20.4 percent longer than chromosome 1 in some hybrid combinations with *B. speciosus*. Chromosome 1 in *B. speciosus* was 29.1 percent longer than chromosome 1 in some hybrid combinations with *B. canaliferus*. Chromosome 1 in *B. canaliferus* was 16.7 percent longer than chromosome 1 in hybrid combinations with *B. arenarum*. This implies that the chromosome 1 in *B. arenarum* should be much shorter than the longest chromosome in *B. spinulosus*. In the hybrid combination *B. spinulosus* x *B. arenarum*, the longest chromosome of *B. spinulosus* is only 11.7 percent longer than chromosome 1 of *B. arenarum*, and this difference is only slightly greater than the arbitrarily established error.

Ullerich (1966) found that *B. bufo* had a greater amount of DNA and larger chromosomes than *B. viridis*. However, I found *B. viridis* to have a slightly larger chromosome 1 in hybrid combinations with *B. bufo*. The percentage difference was 11.6, which is only slightly greater than the 10 percent error. Apart from the difference noted in chromosome 1, the hybrid karyotype was only significantly different in size in chromosome 5. Chromosome 5 of *B. viridis* was 11.1 percent longer than chromosome 5 of *B. bufo*. If the size differences reflected phylogenetically significant DNA values, greater size difference would be expected to occur between species that do not produce diploid hybrids. Hybrid combinations involving a *B. viridis* female and a *B. debilis* male produced only triploids. Parental chromosomes are accurately separable in triploids and there was no significant size difference in chromosome 1 of the two parents.

It seems that if the DNA content in species of *Bufo* is not constant, the apparent differences may be caused by experimental error or minor fluctuations that are not phylogenetically significant.

Haploids and Polyploids

When interspecific hybrid combinations are produced and analyzed, it is important to determine the chromosome complement in the resulting hybrids. Little significance can be applied to compatibility experiments if it is proven that both parents have not contributed equally to the progeny. Cleavage rates and the development to gastrulation in anurans appear to be controlled entirely by the egg cytoplasm: the nuclei make no specific contribution to pregastral development (Briggs and King 1959; Moore 1955).

Gynogenetic and androgenetic haploid amphibians have been experimentally produced using X-radiation, cold treatment, microdissections of nuclei, mechanical stimulation of unfertilized eggs, and foreign sperm (Moore 1950, 1955; Briggs and King 1959; Fankhauser 1945; Rollason 1949; Rugh 1939). Some haploid larvae have appeared spontaneously without any treatment (Fankhauser 1945). Haploid embryos often hatch but rarely metamorphose. Briggs (1949), however, induced androgenetic haploids in "smaller than normal ova," and some of these haploids were able to feed and survived as long as nine months. Fankhauser (1945) maintained a haploid *Triturus* (= *Notopthalmus*) that lived up to the conclusion of metamorphosis. Many anuran interspecific hybrid combinations do not progress past the stages through which haploids might ordinarily progress. Also, abnormal diploid interspecific hybrids may demonstrate aspects of the "haploid syndrome" of Moore (1955) because of genetical factors resulting from incompatible genes rather than actual haploidy. Thus, it is important to determine the chromosomal status of hybrids in compatibility experiments.

I have examined cytologically more than one thousand tadpoles from various crosses of many species of *Bufo* and I have never found a haploid individual. It would appear that haploids are rarely, if at all, produced in *Bufo* as the result of hybridization or escape detection if their development was halted at a prehatching stage.

A small percentage of polyploid amphibians occurs in natural populations (Fankhauser 1945). Cold shock and foreign sperm have also induced triploidy (Kawamura 1952; Muto 1952). Muto (1951) found a chromosomal mosaic in *Bufo*. The mosaic was triploid on one side of the embryo and haploid on the other side. This mosaic reached adult stage and possessed normal adult structures. Diploid eggs have been reported occasionally in *B. calamita* (Bataillon and Tchou-Su 1929). Polyspermy is another way of producing polyploid hybrids. Brachet (1954) indicated that polyspermy occurred if the sperm concentration was increased in anurans or if the eggs were not in optimum condition. Polyspermy was identified by irregular or "baroque" cleavages in Brachet's experiments. My results do not support polyspermy in *Bufo* for reasons that will be discussed.

Polyploids were encountered in many cross combinations that were studied. Triploids were found most commonly, but some pentaploids were also discovered. No tetraploid cells, tetraploid individuals, or mosaics were found. Triploids and pentaploids occurred even in control crosses. Invariably, where it could be determined, the female was responsible for the duplicated set(s) of chromosomes in triploids and pentaploids. Secondary constrictions on the chromosomes were often evident to indicate that the female chromosome complement doubled in triploid hybrids. A pentaploid tadpole was produced in the cross ♀ *B. rangeri* x ♂ *B. valliceps*. This tadpole had 51 chromosomes. This number can be explained only by assuming that the 20-chromosome female *B. rangeri* contributed four sets of chromosomes and the *B. valliceps* male contributed one set. A 20-chromosome female *B. garmani* and a 22-chromosome male *B. marinus* produced a triploid hybrid that had 31 chromosomes. Constrictions on two of the three largest chromosomes indicated the homologous number 1 chromosomes of *B. garmani*. A metamorphosed adult resulted from a cross of a *B. garmani* female and a *B. occidentalis* male. This was a triploid and had 31 chromosomes. Again, the female must have contributed two sets of chromosomes. A ♀ *B. regularis* x ♂ *B. perplexus* cross resulted in a pentaploid hybrid with 51 chromosomes. A ♀ *B. regularis* x ♂ *B. mauritanicus* cross also resulted in a metamorphosed individual with 31 chromosomes. Constrictions on two of the number 6 chromosomes identified the females as contributing the extra set of chromosomes. In other instances, secondary constrictions in the triploid or pentaploid hybrid karyotype apparently prove that the female donated the extra set(s) of chromosomes. Polyploids were often found in more than one cross of the same female.

Only two tadpoles developed in a cross of a ♀ *B. viridis* x ♂ *B. valliceps*. They were phenotypically normal and were found to be diploid. This result was unexpected. Ordinarily, if a small percentage of the fertilized eggs produces viable tadpoles, the tadpoles are polyploid. There were no constrictions in the karyotype that were referable to *B. valliceps* and there were no significantly different chromosomes in the hybrid karyotype. Only one tadpole developed in another cross of the same ♀ *B. viridis* x ♂ *B. lughensis*. The tadpole from the latter cross was also diploid and its hybrid karyotype was very similar to that found in the former cross. The obvious conclusion is that these two crosses resulted in gynogenetic diploid tadpoles. Since the female used in these two crosses is the same *B. viridis* female that was used for three other crosses that had triploid tadpoles, a small percentage of the eggs was probably diploid, and the development of a few diploid eggs initiated by foreign sperm produced the gynogenetic diploids. Since these were the only gynogenetic tadpoles discovered in any of the crosses sampled, it seems to be a rare phenomenon in *Bufo*. A gynogenetic diploid was produced by crossing a *Rana pipiens* female with a *B. valliceps* male by Lawrence Licht (personal communication). The one tadpole that developed had 26 chromosomes, which is the diploid number for *Rana pipiens.*

Tchou-Su (1936) has described diploid eggs in *Bufo* that arose by suppressing the second maturation division in the oocyte. "When the ova are shed normally, or obtained experimentally from the uterus, the first polar body has been given off and the nucleus is in the metaphase of the second meiotic division" (Moore 1955). Fertilization of a diploid egg would result in the formation of a triploid. Pentaploids could result from an abnormal suppression of the first maturation division in the oocyte producing tetraploid eggs. Fertilization of a tetraploid egg would result in the formation of a pentaploid.

Polyploids seem more viable than diploids in hybrid combinations of distantly related species. In crosses of closely related species, the polyploid tadpoles are not more viable than the diploids and are often less viable. The pentaploid tadpole discovered in a *B. regularis* control cross was much smaller in size than the diploid tadpoles and died before metamorphosis. Triploids and diploids did not appear to be phenotypically different in most of the crosses.

In hybridization experiments, no defensible conclusions about the relations of the parents can be drawn if polyploids and diploids occur together in the same cross.

The results would indicate the following points about haploids and polyploids as the result of hybridization in the genus *Bufo.*

1. Haploids are rarely, if at all, produced.
2. A small percentage of eggs produced by female *Bufo* species are diploid or tetraploid, which is probably the result of suppressed maturation divisions in some oocytes.
3. Triploid and pentaploid hybrids result from fertilization of diploid and tetraploid eggs. Unfertilized diploid eggs may occasionally be stimulated (by foreign sperm) to produce gynogenetic diploid tadpoles.
4. Tetraploids are rarely, if at all, produced.
5. If polyspermy does occur, it must be very rare.
6. Triploid and diploid tadpoles do not appear to be phenotypically distinct.
7. Gynogenetic diploid and polyploid tadpoles are capable of metamorphosing.
8. Polyploid and gynogenetic diploid or "false hybrids" should be identified and rejected from an analysis of parental relationships, especially if the percentage of developing hybrids is much reduced from the percentage of fertilized eggs.

Chromosomal Evolution

We attempt here to apply chromosomal evidence in outlining the paths of evolution in the genus *Bufo.* Secondary constrictions, chromosome numbers, and the production of normal diploid hybrids are used as evolutionary criteria.

Prerequisite to a treatment of karyotype evolution there should be an attempt to determine a primitive karyotype. It is unlikely that any contemporary species still possesses a karyotype that has not morphologically changed at all since the first toad, but there might be some species with chromosomal characters typical of an early "*Protobufo.*" As indicated earlier, the primitive karyotype probably had 22 chromosomes. Since most of the species in various parts of the world have five large pairs, one intermediate pair, and five smaller pairs of chromosomes, this situation was probably true of the early *Bufo.* Any deviation from the above may be considered to be secondarily evolved.

The two most likely secondary constrictions to have been included in a primitive karyotype are the A and K constrictions. The A constriction is present in twenty species, most of which occur in North America. The K constriction occurs in thirteen species, most of which are in South America. Both constrictions occur in species of three major continental landmasses. African species form diploid hybrids with *B. arenarum, B. marinus, B. occidentalis, B. granulosus,* and *B. valliceps* (Chap. 11). All these species have a K constriction or an A constriction or both. *B. arenarum* is the only species that produced diploid hybrids with all the following 20-chromosome toads: *B. brauni, B. rangeri, B. garmani,* and *B. regularis. B. arenarum* has a K constriction. Of the 22-chromosome toads in Africa, *B. lönnbergi* has an A constriction and *B. mauritanicus* has an L constriction. A K constriction was not found in any African species examined, unless chromosome 6 of a 20-chromosome species was produced by a fusion of chromosome 7, which contains a K constriction, with the smallest chromosome of an early 22-chromosome species.

B. marinus, B. atacamensis, B. paracnemis, and *B. variegatus* could have arisen from a primitive ancestor possessing only a K constriction. The only other constriction evident is a submetacentric constriction on the short arm of chromosome 5 in *B. marinus.* This constriction has not been found in any other species and might have evolved recently. The appearance of the A constriction probably represents an early dichotomy from the line giving rise to the above species. *B. poeppigi, B. crucifer, B. haematiticus, B. valliceps,* and *B. stomaticus* can all be related to an ancestor that had both a K constriction and an A constriction.

B. poeppigi could not have diverged very much from *B. marinus,* since these species produce normal diploid hybrids. *B. crucifer* has probably not diverged greatly and also produces small numbers of normal diploids with *B. marinus. B. crucifer* does, however, possess constrictions that indicate important dichotomies that have led to other divergent lines of evolution. An early dichotomy on the *B. crucifer* line is represented by the formation of a B constriction, which was probably also responsible for the evolution of the line leading to *B. arenarum.* This last species must have secondarily lost the A constriction. *B. arenarum* still has a fairly unspecialized karyotype, since diploid hybrids involving *B. arenarum* may be produced with some species in all continents where *Bufo* occurs. *B. granulosus* may have been derived similarly, since the constrictions are the same in the two, and they produce diploid hybrids with each other. J and D constrictions, evident in *B. crucifer,* were probably the result of later dichotomies.

A major dichotomy must have taken place close to the line that gave rise to *B. crucifer. B. valliceps* maintains many of the constrictions that characterize this "major radiation." The K and A constrictions relate *B. valliceps* to a primitive ancestor. *B. valliceps* also possesses, however, an F constriction, a G constriction, an H constriction, and an E constriction. Other species have combinations of these constrictions. *B. valliceps* produces diploid hybrids with *B. regularis. B. occidentalis* is the only other North American species that produces diploid hybrids with the African species. This indicates that *B. valliceps* still maintains a fairly unspecialized karyotype and provides further evidence for an early dichotomy of the line that led to *B. valliceps.* The appearance of new secondary constrictions and the elimination of old secondary constrictions must have taken place many times. Some species maintained various constrictions and provide a measure of relationship. For example, G and F constrictions must have developed early in the "major radiation" that led to *B. valliceps.*

B. stomaticus also has an A and a K constriction, which relates this species to the "major radiation." E, F, and G constrictions are also present. *B. valliceps* and *B. stomaticus* derived apparently from the same line of evolution but have diverged in different directions, for they do not produce diploid hybrids. Since *B. stomaticus* also contains a B and an F constriction, the dichotomy for the "major radiation" must have taken place close to the line giving rise to *B. crucifer. B. valliceps* has secondarily lost the J and B constrictions.

B. haematiticus still maintains the A, K, and G constrictions. This species probably broke off very early from the same "major radiation" that gave rise to the large North American and Eurasian radiations. It must be assumed that this species lost the B, J, and F constrictions. Perhaps the F constriction had not yet appeared before this species' early dichotomy. *B. haematiticus* probably diverged considerably after its severance, which is reflected by morphological difference and the incapability of forming diploids with any of the species where crosses have been attempted. *B. coniferus* still maintains the K constriction but produces diploids only with *B. valliceps.* This species probably broke off the *B. valliceps* lineage early and has lost most of its constrictions.

B. spinulosus has constrictions indicating that it was derived from an ancestor on the main trunk of the "major radiation" that led to the Eurasian and North American radiations. E, F, K, and L constrictions are evident in the karyotype of *B. spinulosus.* Diploids are produced in hybrid combinations between *B. spinulosus* and five species in North America, *B. viridis* in Eurasia, and *B. arenarum* and *B. flavolineatus* in South America. The K constrictions are quite different in *B. flavolineatus* and *B. spinulosus,* and it must be assumed that the K constriction of *B. flavolineatus* was derived from the *B. spinulosus* K constriction. *B. spinulosus* probably broke off the "major radiation" at a point close to the North American radiation and close to the dichotomy of the Eurasian radiation. This would explain the compatibility with the North and Central American *Bufo* and with *B. viridis.* The L constriction is present in two Asian species (*B. melanostictus* and *B. calamita*) and the South American *B. chilensis.* If this constriction is homologous in the three species, *B. chilensis* is probably quite closely related to a Eurasion radiation.

Of all the Eurasian species studied, *B. stomaticus*

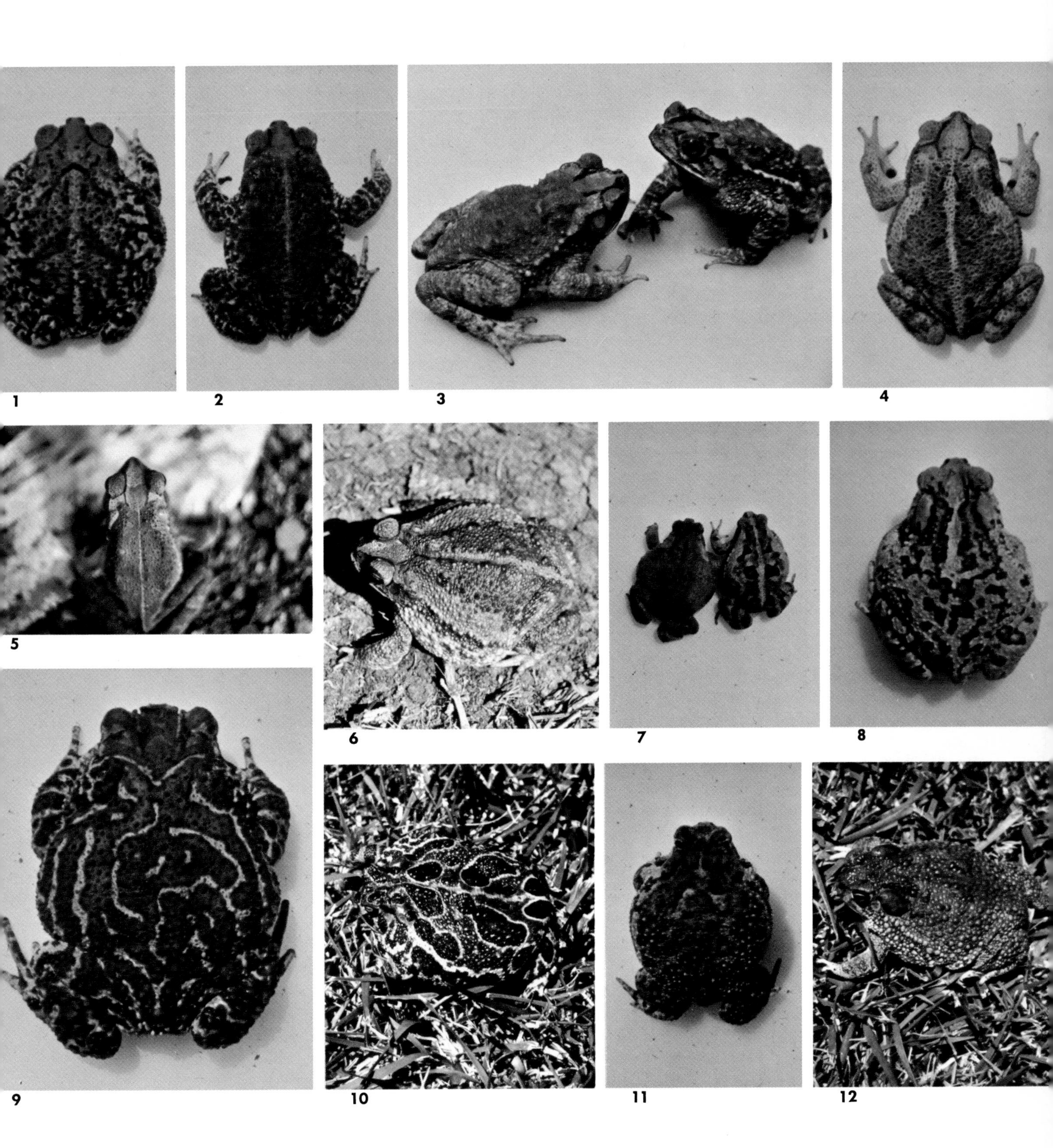

III. Toads of the broad-skulled Central and North American *valliceps* group and relatives: (1) *B. gemmifer*, (2) *B. mazatlanensis*, (3) *B. coniferus*, (4) *B. valliceps*, (5) *B. melanochloris*, (6) *B. luetkeni*, (7) *B. canaliferus*, (8) *B. occidentalis*, and (9) *B. peltacephalus*; the three species (10) *B. cognatus*, (11) *B. compactilis*, and (12) *B. speciosus* of the *cognatus* group.

has the largest number of secondary constrictions. Many of the constrictions indicate a common ancestor in the "major radiation" stemming from an ancestor of *B. crucifer*. K, B, and J constrictions are evident in this line. Later, F and G constrictions arose. *B. stomaticus* must have secondarily lost the D and A constrictions that were also present in this line. The E constriction probably evolved in the New World, since *B. alvarius, B. retiformis, B. canaliferus,* and *B. spinulosus* still maintain this constriction. Thus, the ancestor of *B. stomaticus* was possibly in the New World after a great number of dichotomies had occurred. The arrival of *B. stomaticus* in Eurasia is recent in the history of the genus, which would explain its ability to produce diploid hybrids with many North and Central American species. One rather puzzling inconsistency is the ability of *B. stomaticus* to form diploid hybrids with *B. atacamensis*. *B. atacamensis* is the only South American species shown to produce diploid hybrids with *B. stomaticus*. This inconsistency could be circumvented by assuming that *B. atacamensis* has secondarily lost all constrictions but the K constriction and was derived close to the dichotomy that gave rise to the Eurasian species and not directly from a more primitive ancestor.

B. viridis maintains D and A constrictions, which have evidently been lost in *B. stomaticus*. The F constriction is found in *B. stomaticus, B. viridis,* and *B. bufo* in Eurasia. It is also found in *B. retiformis, B. occidentalis, B. valliceps, B. canaliferus,* and *B. spinulosus*. *B. stomaticus* probably had a common ancestor that broke off the major radiation close to the dichotomy that gave rise to *B. spinulosus* (and perhaps *B. atacamensis*). Shortly after their origin, the two species diverged. The line that gave rise to *B. viridis* maintained the D, A, and F constrictions and lost the J, K, G, B, and E constrictions. The line that produced *B. stomaticus* maintained the K, B, E, F, and G constrictions but lost the D and A constrictions.

B. bufo must have been derived from the line giving rise to *B. viridis* after its severance from the line producing *B. stomaticus*. The F and either the A or D constrictions are still maintained in *B. bufo*.

B. melanostictus has only the A and L constrictions. *B. melanostictus* produces diploid hybrids with *B. alvarius, B. boreas, B. valliceps, B. arenarum,* and *B. marinus*. The hybrid combinations with *B. arenarum* and *B. marinus* suggest that *B. melanostictus* did not follow the same evolutionary path that was taken by *B. viridis, B. bufo,* and *B. stomaticus*. *B. chilensis* is the only South American toad that has demonstrated the L constriction. The most plausible explanation would be that the L constriction appeared early in the evolution of *Bufo* and has been maintained in only a very few species lines. *B. melanostictus* probably had a common ancestor with *B. chilensis*, but *B. melanostictus* separated from the line leading to *B. chilensis* at an early date, which would explain the L constriction in both species and the different compatibilities with *B. marinus*. Perhaps the evolution of *B. melanostictus* took place at an earlier period than that of *B. stomaticus, B. bufo,* and *B. viridis*. *B. melanostictus* might even be the product of an earlier radiation into Eurasia.

B. calamita has either an A or a D constriction and it also has an L constriction. On the basis of secondary constrictions, *B. calamita* would be related to *B. melanostictus*, but *B. calamita* produces diploids in hybrid combinations with *B. viridis*. *B. calamita* is apparently a derivative of the *B. viridis* line and has obtained the L constriction from a common ancestor with *B. chilensis*. Thus, the evolution of *B. calamita* might have taken place at a later time than the evolution of *B. melanostictus*.

B. canaliferus has A, C, and E constrictions. The A constriction and the E constriction were present in the lineage of toads prior to the Eurasian radiation. Other species possessing the E constriction are *B. spinulosus, B. alvarius, B. retiformis,* and *B. stomaticus*. These data suggest that *B. canaliferus* had an ancestor close to the dichotomy that produced *B. spinulosus* and the *B. stomaticus* line. The C constriction arose for the first time on the line giving rise to *B. canaliferus*.

B. retiformis has constrictions in common with *B. bufo* and *B. marmoreus* that are found in no other species examined. This species has maintained many of the constrictions evident in the "major radiation" and probably represents an offshoot of the line that gave rise to the Eurasian radiation.

B. marmoreus and *B. perplexus* produce diploid hybrids with *B. stomaticus*. The three constrictions evident in *B. marmoreus* are also present in *B. retiformis*. *B. marmoreus* and *B. perplexus* were probably derived like *B. retiformis*.

All the constrictions evident in the karyotype of *B. alvarius* were present early in the "major radiation," and *B. alvarius* could be derived at a number of early dichotomies. Since *B. alvarius* produces diploid hybrids with both *B. melanostictus* and *B. vi-*

ridis, the dichotomy must have taken place close to the origin of the *B. spinulosus* lineage. *B. alvarius* possesses two constrictions that cannot be traced to any other species, which suggests that *B. alvarius* diverged early and evolved these two constrictions in its own line. The karyotype, however, must have remained fairly unspecialized, since *B. alvarius* produces diploid larvae with both *B. melanostictus* and *B. viridis*.

B. occidentalis has either an A or a D constriction and an F constriction. Since this is one of the two North American species that produce diploid hybrids with the African 20-chromosome species, the dichotomy for this species probably occurred early. Both *B. occidentalis* and *B. alvarius* have an F constriction and they may have had a common ancestor.

If the C constriction arose in the line giving rise to *B. canaliferus*, then *B. americanus, B. nelsoni,* and *B. valliceps* may all be related and be derived from the same line of evolution. *B. valliceps* itself has a long history of unspecialization in its karyotype, which can be traced to the first "major dichotomy" in South America and to recent dichotomies in North America. This unspecialization is also reflected in its ability to produce diploid hybrids with at least one species in every continent where *Bufo* occurs. The B constriction was probably involved in the "major radiation," for it was present in the line giving rise to *B. crucifer*. Most of the lines that broke from the "major radiation" lost the B constriction, but it was probably maintained in the Eurasian radiation from which the line giving rise to *B. canaliferus* and *B. retiformis* was evolved.

B. americanus and *B. houstonensis* still maintain a C and a D constriction. These two species also possess an I constriction. *B. americanus* and *B. houstonensis* are probably closely related. *B. microscaphus* has both a D and an A constriction. Since *B. microscaphus* has an A constriction, it probably diverged before the A constriction was lost in *B. americanus* and *B. houstonensis*. *B. terrestris* also has a D and an A constriction, which would relate this species to *B. microscaphus*. In hybrid combinations, *B. terrestris* produces diploids with *B. quercicus, B. asper, B. woodhousei,* and *B. hemiophrys*. Since *B. terrestris* produces diploid hybrids with an Asian species, it probably broke off the line leading to *B. americanus* early and is closely related to *B. woodhousei*.

B. terrestris produced diploid hybrids with *B. asper*, which may indicate that *B. asper* is a fairly recent Asian invader. Unfortunately the chromosomes obtained from *B. asper* were very condensed and no constrictions were evidenced for this species. The hybrid karyotype suggests that *B. asper* has at least an A constriction.

B. hemiophrys has been found to have both D and A constrictions. This species also produces diploid hybrids with a Eurasian species (*B. stomaticus*); thus, *B. hemiophrys* probably also arose from the line leading to *B. americanus* at a time close to the point of origin of *B. terrestris*. *B. woodhousei* produces diploids with *B. stomaticus*. From the variety of species that are compatible with *B. woodhousei*, it appears that *B. woodhousei* broke off the line leading to *B. americanus* at an even earlier time than *B. hemiophrys* or *B. terrestris*. There is no D constriction but there is an A constriction, and this would indicate that *B. woodhousei* may be close to the dichotomy that produced *B. valliceps*.

B. luetkeni has shown only an A constriction. Hybrid combinations indicate that *B. luetkeni* produces diploids with species from the line giving rise to *B. americanus*. Since diploids are also produced with *B. spinulosus, B. luetkeni* must be derived from an even earlier dichotomy than *B. woodhousei*.

B. cognatus and *B. speciosus* have only an A constriction, which can be referred to many species. Hybrid combinations producing fertile males (Chap. 11) reveal that these two species are more closely related to one another than either is to any of the other species with which they have been hybridized. Both species produce diploid hybrids with a variety of species representing different lines of evolution. *B. speciosus* produces diploid hybrids with *B. granulosus, B. arenarum,* and *B. stomaticus*. The karyotypes of *B. speciosus* and *B. cognatus* must be quite unspecialized, even though the majority of the constrictions have been lost. *B. compactilis* was considered conspecific with *B. speciosus* until their distinctness was recognized on the basis of call differentiation (Bogert 1960). I have not obtained hybrid material from *B. compactilis*. The karyotype has an A constriction and a constriction on chromosome 11 that might be similar to one found in *B. debilis* and *B. kelloggi*. This could indicate that *B. cognatus, B. compactilis,* and *B. speciosus* are derived from an ancestor common to *B. debilis* and *B. kelloggi*.

B. kelloggi has a B constriction and an H constriction. This combination has been found only in *B. retiformis, B. nelsoni,* and *B. kelloggi*. *B. kelloggi* is probably closely related to *B. debilis* since they have a similar constriction on chromosome 8 or 9. Diploid hybrids are produced when *B. kelloggi* is crossed with *B. arenarum* and *B. viridis*. The karyotype of *B. retiformis* has a generalized pattern of secondary constriction. It could be that *B. kelloggi, B. debilis,* and *B. nelsoni* were derived from the line leading to *B. retiformis* and that *B. kelloggi* and *B. debilis* were derived from a single dichotomy of this line. The presence of a D constriction in *B. nelsoni* might indicate that the *B. americanus* line was an early offshoot of the *B. retiformis* line.

B. punctatus produces diploid hybrids with *B. debilis, B. kelloggi,* and *B. nelsoni*. A and D constrictions are present in *B. punctatus*. The secondary constrictions indicate a position on the line leading to *B. americanus* for the dichotomy of the line giving rise to *B. punctatus*. The line giving rise to *B. americanus* probably broke off the line giving rise to *B. retiformis,* which in turn may have broken from the line producing *B. canaliferus*.

B. quercicus has an A or a D constriction and may have a B constriction. Diploid hybrids have been produced with *B. woodhousei, B. terrestris,* and *B. americanus*. The satellite on the long arm of chromosome 3 in *B. quercicus* is probably not homologous to that found in *B. granulosus*. The most likely explanation for the evolution of *B. quercicus* is an early dichotomy from the *B. retiformis* line close to the juncture of the line giving rise to *B. americanus. B. quercicus* evolved independently and this would explain its morphological divergence from the ancestral lines.

B. nelsoni appears to have C and H constrictions. This species may also have a B constriction. Since the C constriction is present, *B. nelsoni* may be derived from the line that gave rise to *B. americanus*. The I constriction probably arose on the same line that produced the C constriction. *B. nelsoni* probably has an ancestor in common with *B. americanus, B. houstonensis,* and *B. canorus. B. canorus* produces diploid hybrids with *B. stomaticus*. This may indicate that *B. canorus* was derived from the radiation that led to *B. stomaticus* and also that *B. canorus* was an early derivative of the *B. canaliferus* line. The I constriction probably appeared early in this line. Little information can be gained from the constriction evident in the karyotype of *B. boreas*, but from hybridization evidence (Chap. 11) *B. boreas* is closely related to *B. canorus*.

B. bocourti maintains a K constriction and a B constriction. Diploids have been produced with *B. speciosus, B. canorus,* and *B. woodhousei*. Hybrids have been produced with *B. viridis* (Chap. 11). *B. bocourti* probably arose from the line leading to the Eurasian radiation close to the North American radiation and has maintained a K constriction.

B. holdridgei has demonstrated only a B constriction. No hybrid evidence is available for this species. *B. holdridgei* could be derived from a number of different lines.

A male *Melanophryniscus stelzneri* was crossed with a female *B. woodhousei*. Some tadpoles developed and these were determined to be diploid. The hybrid karyotype indicated that the progeny were not gynogenetic. Some of the tadpoles, however, had cells containing broken chromosomes and none of the tadpoles developed to a very large size. I have not been able to obtain a parental karyotype for *Melanophryniscus stelzneri*, but, from the hybrid karyotype, I judge that this species probably has 22 chromosomes. Recently, Morescalchi (1968) presented the karyotype of *Atelopus* (= *Melanophryniscus*) *stelzneri*. He stated that the karyotype was very similar to *Bufo*, which would confirm the hypothesis deriving the Atelopodidae from *Bufo*. He found a heteromorphic zone to occur in the long arm of chromosome 11.

If the constriction identified by Morescalchi in *Melanophryniscus stelzneri* is the L constriction, then this constriction may prove to be an "older" constriction then the K constriction. A constriction similar to the L constriction is found in *Odontophrynus americanus* (Bogart 1967), but it is impossible, at the present time, to prove any constrictions are actually homologous.

B. mauritanicus has an L constriction on chromosome 10 or 11. If this constriction is homologous to that found in *B. chilensis, B. melanostictus,* and *B. calamita*, then the ancestor(s) that entered Africa might have possessed both an L constriction and a K constriction. *B. mauritanicus* has lost the K constriction. This species could have broken away and diverged from the ancestral line before the formation of the 20-chromosome species, which would explain the formation of diploid hybrids between *B. mauritanicus* and *B. regularis. B. mauritanicus* is the only 22-chromosome species in Africa among those tested that produces diploid hybrids with 20-chromosome species (Chap. 11).

B. melanostictus and *B. calamita* may have had the same New World ancestor as the African radiation of 20-chromosome toads. The divergence of the 20-chromosome ancestor possibly took place before the

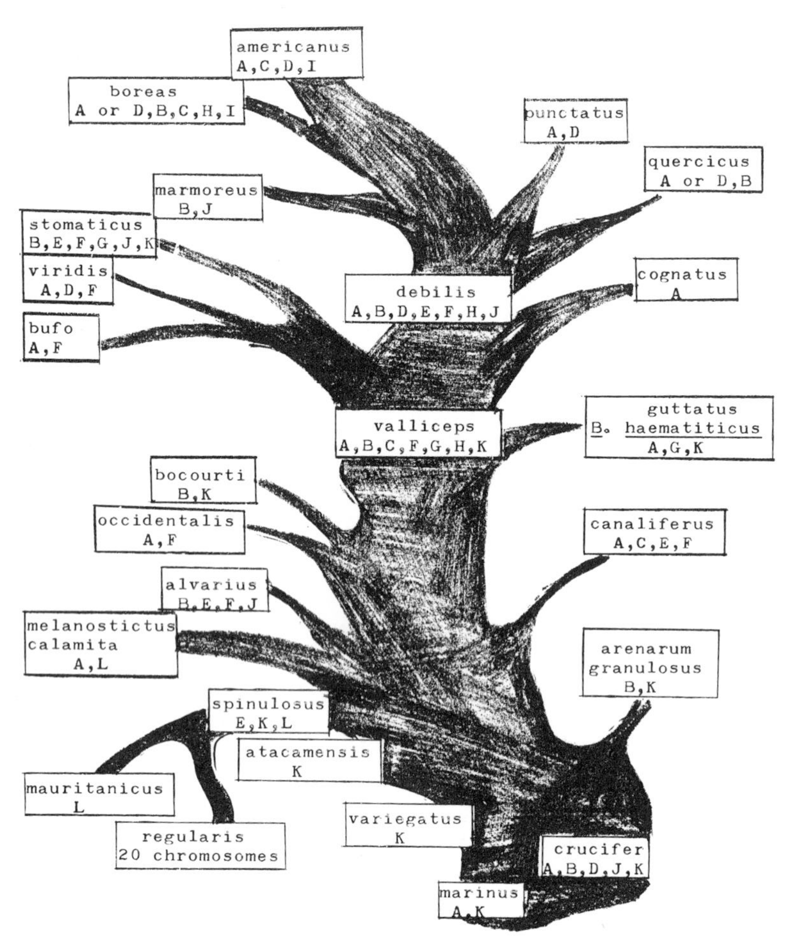

Fig. 10–3. Dendrogram of the hypothetical relationships of *Bufo* species groups as indicated by the chromosomal analysis. Secondary constrictions found to be present in the species groups are presented.

formation of the lines leading to *B. chilensis, B. melanostictus,* and *B. calamita,* causing some of the incompatibility in hybrid combinations between the 20-chromosome species and the same species possessing the L constriction (Chap. 11).

B. regularis specimens from different localities seem to have differences in their karyotypes. *B. regularis* is apparently a very polymorphic species; otherwise *B. regularis* is the name applied to what is in reality more than one species. Some populations, when crossed, did not form diploids at all: *B. regularis* from Ogbomosho, Nigeria crossed with *B. kerinyagae* of the *regularis* group from Mazeras, Kenya produced only triploids, as did a Ghana female crossed with a *B. regularis* (E) from Ol Tukai, Kenya. Thus, some of the populations must represent species, because they are genetically isolated. Secondary constrictions are not of much value in separating the populations of *B. regularis.* Some of the populations have unique constrictions, and there are not evident, major dichotomies in the presence or absence of constrictions.

B. garmani also produces diploid hybrids with some of the American species, which indicates that *B. garmani* still has a fairly unspecialized karyotype and is closely related to *B. regularis. B. garmani* has many secondary constrictions. Since all the 20-chromosome toads have an R constriction, this constriction is not useful for determining relationships between the various 20-chromosome species. *B. garmani* has, on the short arm of chromosome 3, a subtelocentric secondary constriction, which is similar to one found in *B. brauni* and *B. maculatus.* A submetacentric secondary constriction on the short arm of chromosome 2 is found in *B. gutturalis, B. garmani,* and *B. rangeri. B. garmani* probably broke away from an early 20-chromosome ancestor close to the dichotomy of *B. regularis.* Many of the constrictions present in the karyotype of *B. garmani* indicate that *B. brauni, B. maculatus, B. gutturalis,* and *B. rangeri* are more closely related to *B. garmani* than to *B. regularis.*

B. rangeri has a constriction on the short arm of chromosome 2, which relates this species to *B. garmani* and *B. gutturalis* and indicates that *B. rangeri* diverged at an early time from the line leading to *B. garmani* by the secondary constriction on the short arm of chromosome 2. This constriction might indicate that *B. gutturalis* diverged farther from *B. regularis* on the line that led to *B. garmani* or *B. rangeri.*

There is no constriction on the short arm of chromosome 2 in *B. brauni,* but a subtelocentric secondary constriction is present on the short arm of chromosome 3, which is similar to *B. garmani* and to *B. maculatus. B. maculatus* produces diploid hybrids with *B. garmani, B. brauni,* and *B. regularis.* The subtelocentric secondary constriction on the short arm of chromosome 3 may relate *B. maculatus* to *B. garmani* and *B. brauni. B. maculatus* might have been an offshoot from the line giving rise to *B. brauni.*

Although no karyotype is available for *B. perreti,* hybrid combinations reveal that *B. perreti* produces diploid hybrids in crosses with *B. garmani* and *B. regularis. B. perreti* might have diverged from the line giving rise to *B. garmani.*

No relationships can be determined for *B. gariepensis* or *B. lönnbergi.* These 22-chromosome African toads did not produce any viable hybrids, although

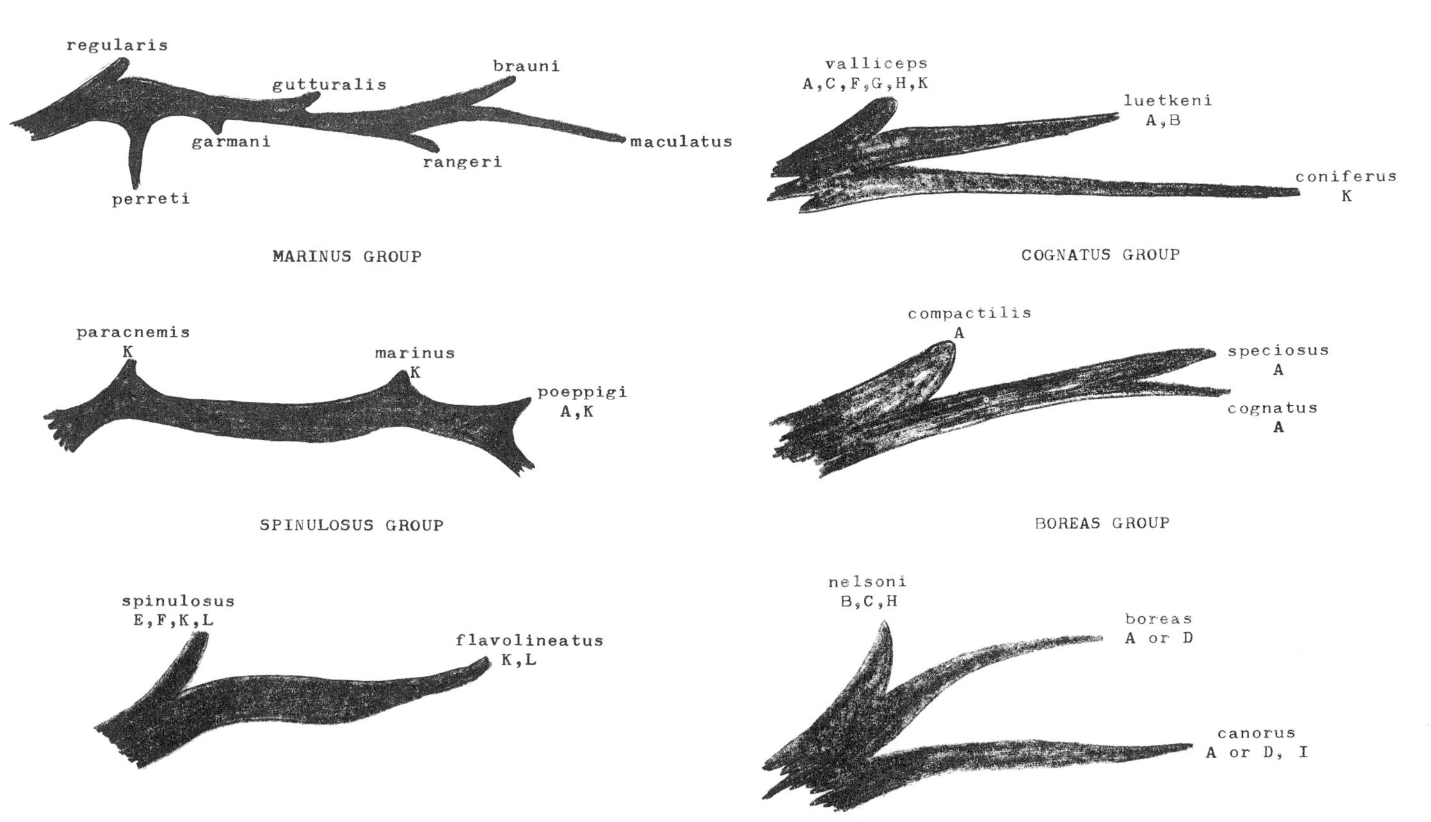

Fig. 10–4. Possible dichotomies within the *regularis*, *marinus*, and *spinulosus* species groups based on chromosomal evidence.

Fig. 10–5. Possible dichotomies within the *valliceps*, *cognatus*, and *boreas* species groups based on chromosomal evidence.

the A secondary constriction present in the karyotype of *B. lönnbergi* may be traced to many other 22-chromosome species. Since these species are probably not closely related to any 20-chromosome toads, they might be related to an early invader that entered Africa. *B. viridis* also occurs in Africa. Only the A constriction evident in the karyotype of *B. viridis* can be related to any African species. Diploids were not produced when *B. viridis* was crossed with many of the African species. *B. viridis* is probably a very recent invader into Africa and is not closely related to any of the Old World species in Africa today.

From the various relationships evident in this synthesis, a dendrogram of relationships was drawn (Fig. 10-3). This dendrogram may reflect the evolutionary sequence that led to the formation of the present-day species from ancestral species. The species names used in figure 10-3 refer to species groups (Chap. 11). If more than one species were examined in a species group, the relationships of the species within the group were attempted (Figs. 10-4, 10-5, 10-6). The letters beneath each species or species group refer to the secondary constrictions (see Appendix F) present according to the classification in the section on secondary constrictions.

Differences in secondary constriction patterns were often found between karyotypes of species considered to be in the same species groups (Chap. 11). It must be assumed that, within the same species group, some karyotypes retain basic constrictions, while the karyotypes in other species are more specialized and have lost many constrictions. *B. valliceps*, of the *valliceps* group, has the largest number of constrictions evident in the species that have been examined in that group. Some of the secondary constrictions evident in the karyotype of *B. valliceps* are found in many distantly related species. Apparently, *B. valliceps* has maintained a very unspecialized

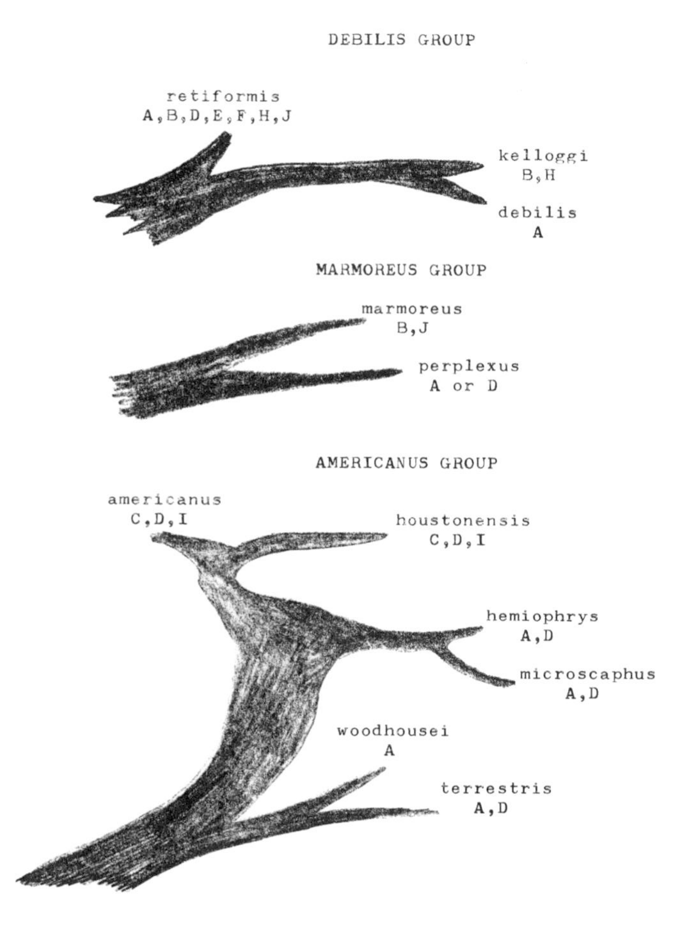

Fig. 10–6. Possible dichotomies within the *debilis, marmoreus*, and *americanus* species groups, based on chromosomal evidence.

karyotype that could be similar to the ancestral species for that group. By similar reasoning, *B. retiformis* (*debilis* group), *B. chilensis* (*spinulosus* group), and *B. poeppigi* (*marinus* group) could be considered most similar to the ancestral species for each of their species groupings.

Biogeographical Implications

The first members of the genus *Bufo* probably arose from a 22-chromosome ancestor in South America. Thus, South America was the site of the first radiation of the genus, and subsequent radiations into other continents were derived from the ancestral South American radiation. Some *Bufo* species were derived in South America and confined to that continent, but there was a tendency for northward dispersal out of South America. A connection for *Bufo* between South and Central America must have existed for a long time or at more than one time.

Central America was probably a corridor through which ancestral *Bufo* passed in northward dispersion. Many species may have arisen in Central America in adaptation to new habitats. Some of the species that arose may have become isolated in Central America because of narrow ecological tolerances; other species could have arisen and later radiated northward.

There are many species of *Bufo* in North America and most of these species probably originated in this continent. Major climatic changes that have occurred in the past may have caused explosive speciation at a comparatively recent date. All the species examined that occur in Eurasia can be related to American ancestry. The most likely explanation is migration from North America to Eurasia by a land bridge across the Bering Strait.

B. melanostictus may be related to an early ancestor that arose in South America. *B. viridis, B. bufo*, and *B. stomaticus* may be related to an ancestor in North America, and *B. calamita* may be related to an intermediate ancestor between the one giving rise to *B. melanostictus* and the one giving rise to *B. viridis, B. bufo*, and *B. stomaticus*. Therefore, there could have been more than one and probably at least three radiations into Asia, which would indicate that the Bering Strait was no barrier for *Bufo* over a long period of time or that the barrier was broken and reformed several times. The few species in Europe and northern Asia are karyotypically as closely related to North American species as they are to each other. Only a very few species from Southeast Asia were available for this study. If *B. melanostictus* was the result of a very early invader, speciation should have occurred in Asia from this evolutionary line, thus accounting for many of the species and species groups outlined by Inger (Chap. 8) that occur in South Asia.

The species diversification in Africa is great and may be comparable to the diversification in North and Central America. If the rate of speciation was fairly constant, the ancestor(s) of the African species must have arisen at about the same time as the ancestor(s) of the Central and North American species. The karyotypes of *B. melanostictus, B. mauritanicus*, and *B. spinulosus* may be related to a common ancestral line that possessed the L constriction. This one evolutionary line may have crossed the Bering land

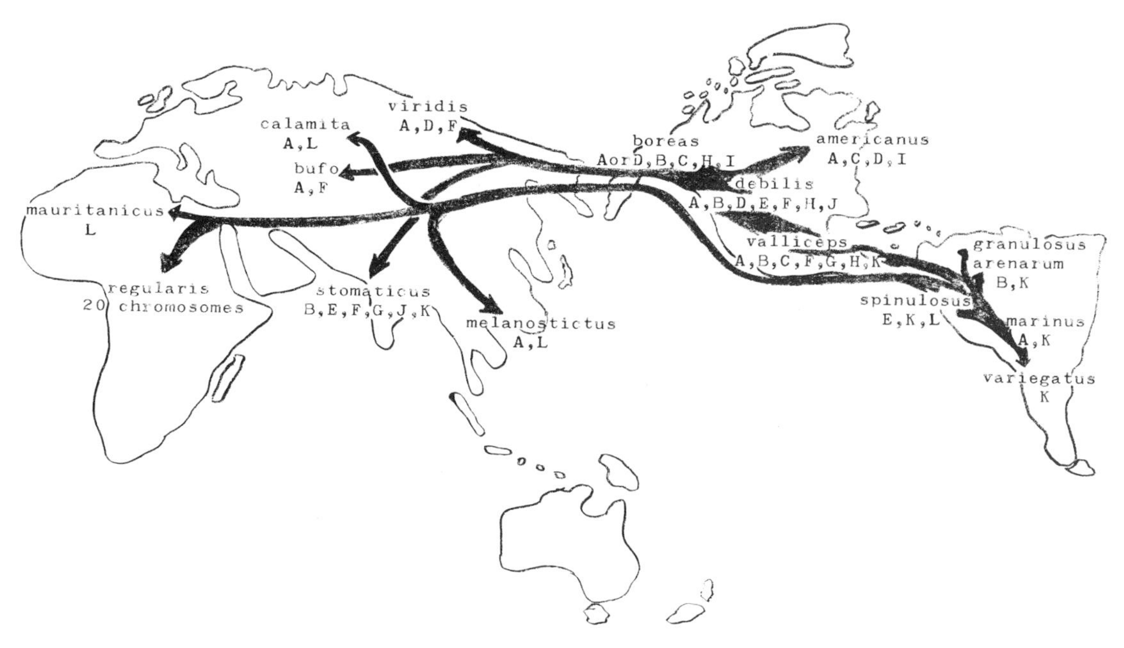

Fig. 10–7. Possible dispersal pattern involved in the evolution of species in the genus *Bufo,* based on chromosomal evidence. The secondary constrictions are indicated for the various species groups.

bridge, and a dichotomy could have occurred early, after its penetration into Asia. The two lines, however, could have radiated independently into Asia at different times. An examination of other species in Southeast Asia may help to solve this problem. In any case, one line entered Africa and was probably isolated in that continent.

B. mauritanicus was perhaps an early evolved species. A 20-chromosome ancestor evolved in Africa, which led to the formation of the widely spread and diversified 20-chromosome species in Africa. The appearance of *B. viridis* in North Africa is probably a recent event. The other 22-chromosome species in Africa may be the result of an early dichotomy of the 22-chromosome ancestor entering Africa. When more 22-chromosome African species are examined, their relationships will be more easily defined.

The theory presented here (Fig. 10-7) is in direct opposition to that presented by Darlington (1957). Darlington believes that *Bufo* originated in Africa and radiated from the Old World tropics. His main argument stems from the existence of two bufonid genera in Africa (*Werneria* and *Nectophrynoides*). Darlington also says, however, that the bufonids probably arose from leptodactylids. He does not find it illogical that leptodactylids do not occur in tropical Africa, for, when speaking of the origin of the leptodactylids, he says, "... true bufonids have apparently been derived from leptodactylids in the main part of the Old World Tropics, which indicates the former presence of leptodactylids there." I think it is much more logical to assume that *Bufo* arose in South America where the leptodactylid fauna is large and diversified. Some of the genera in the Leptodactylidae in South America have karyotypes similar to *Bufo*. The genus *Werneria* is known only from the type and *Nectophrynoides* is viviparous (Noble 1931). The chromosomes of some of the non-*Bufo* bufonid species have been obtained (Fig. 10-8). *Pedostibes hosei* from Borneo has 22 chromosomes consisting of six large and four smaller pairs (Fig. 10-8A). Only one female specimen was studied, but it is evident that this species has a secondary constriction on chromosome 11 that is very similar to the L constriction of some *Bufo* species. *Melanophryniscus rubriventris* from Argentina (Fig. 10-8B) also has 22 chromosomes, but the secondary constriction appears to be on chromosome 10 and not chromosome 11. *Melanophryniscus stelzneri,* however, does have a constriction on chromosome 11 (Morescalchi

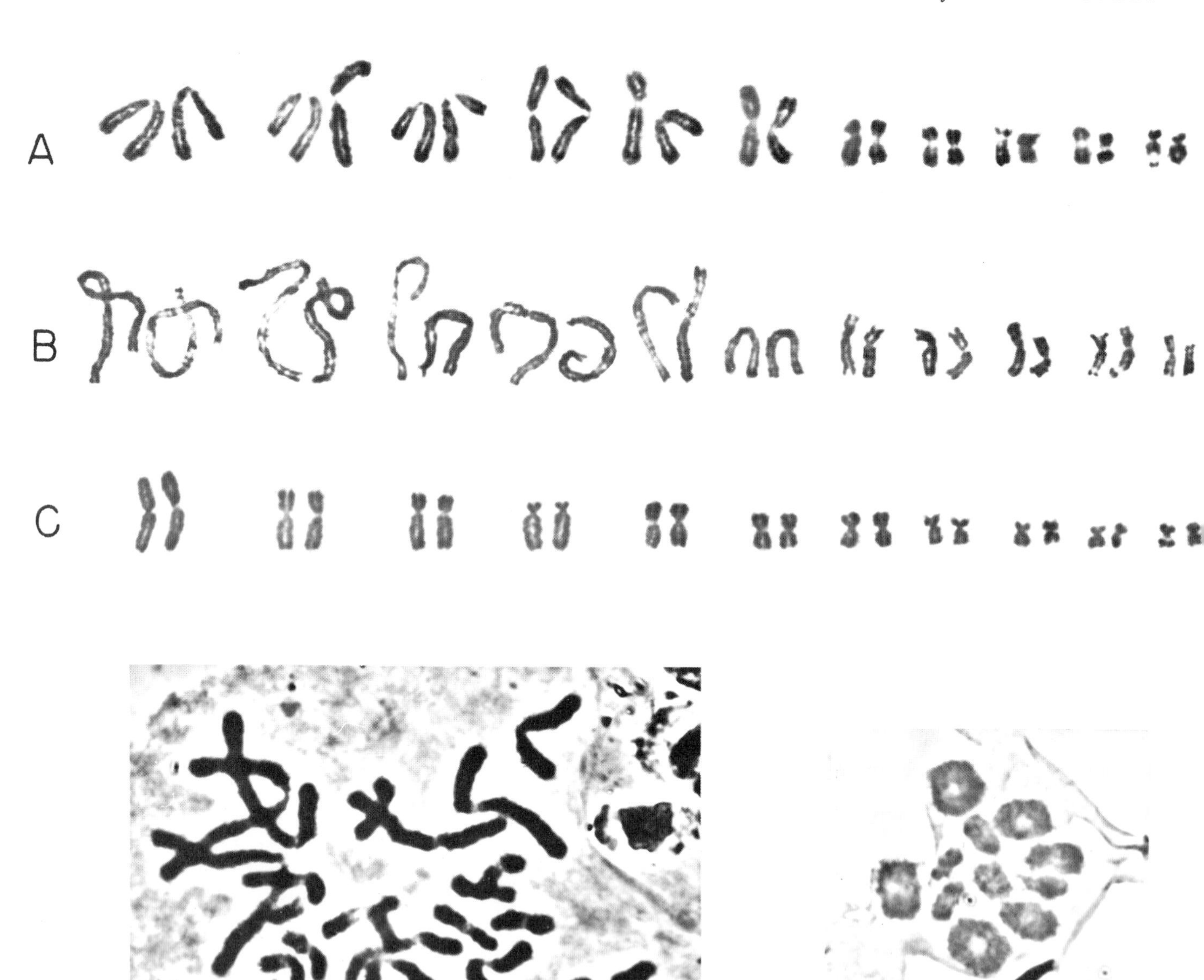

Fig. 10–8. Chromosomes of *Bufo* relatives. *A* is the karyotype of *Pedostibes hosi* from Borneo. *B* is a karyotype of *Melanophryniscus rubriventris* from Argentina. *C* is the karyotype of *Odontophrynus occidentalis* from Argentina. *D* is a mitotic anaphase figure from a testicular squash of *Nectophryne afra*, and *E* is a meiotic metaphase figure from *Nectophrynoides tornieri.* (*D* and *E* courtesy of Colonel J. J. Scheel.)

and Gargiulo 1968). *Odontophrynus occidentalis* is considered a leptodactylid species, but this species has a karyotype that is very similar to many *Bufo* species (Fig. 10-8C) and possesses an L constriction.

Recently, Colonel J. J. Scheel (personal communication) obtained chromosomes from *Nectophryne afra* and *Nectophrynoides torneri*. Both these African species have 22 chromosomes (Fig. 10-8D and E). *Nectophryne afra* possibly has the L constriction (Fig. 10-8D, arrow). The meiotic configurations obtained by Colonel Scheel cannot be satisfactorily compared with the C-metaphase configurations obtained in this study. However, this new evidence does support the hypothesis that *Bufo* species were derived from a 22-chromosome ancestor. Even if these genera do prove to be closely related to *Bufo*, they are extremely specialized and not typical ancestral forms of *Bufo*. Noble (1931) admits that the bufonidae (his bufoninae) probably represent an unnatural group of toads showing closest affinities to the criniinae. A karyotypic study of these genera should prove illuminating.

Summary

The chromosomes from 50 species of *Bufo* and 175 hybrid combinations were examined and compared. Karyotypes were analyzed mathematically. Some African species were found to have a diploid complement of 20 chromosomes, but the majority of *Bufo* species have 22 chromosomes, which can be grouped into five large pairs, one intermediate pair, and five smaller pairs. Because of the large error that must be considered, as well as the similarity of *Bufo* chromosomes, percentage lengths and centromeric ratios were not found to be valid criteria for phylogenetic groupings of the karyotypes. The absolute size of the chromosomes did not vary interspecifically or intraspecifically very much outside the arbitrarily determined range of error, which probably reflects a fairly constant level of DNA in all the species that were examined.

The primitive, ancestral, chromosome number in the genus *Bufo* is believed to be 22. An ancestral 20-chromosome species was probably derived from a 22-chromosome species by fusion. The smallest chromosome was probably involved in the fusion.

Polyploidy was frequently encountered in hybrid tadpoles. When the chromosomes could be distinguished, it was found that the female parent contributed the extra set(s) of chromosomes to a polyploid toad. There was never any evidence for polyspermy. Gynogenetic diploid tadpoles were discovered in only two hybrid combinations. Gynogenetic, triploid, and pentaploid hybrid tadpoles are phenotypically very similar to diploid hybrid tadpoles and are capable of metamorphosing. No tetraploid or mosaic tadpoles were found. A triploid male was found to be fertile and to produce sperm that contained varying numbers of chromosomes.

Secondary constrictions, chromosome number, and the species that produced normal diploid hybrids were used to outline evolution in the genus *Bufo*. The primary findings were as follows.

1. The genus *Bufo* probably arose in South America from a leptodactylid ancestor possessing 22 chromosomes.
2. Ancestral species dispersed from South America into Central America, North America, Eurasia, and Africa.
3. Invasions of ancestral species into Eurasia must have taken place once and perhaps three times or more. A very early ancestor was probably responsible for the African radiation and perhaps for *B. melanostictus*. A more recent invasion probably gave rise to *B. bufo*, *B. viridis*, *B. stomaticus*, and perhaps *B. calamita*. The last-named species, however, may be the result of a third invasion that was intermediate in time between the other two invasions.
4. Toads were subsequently isolated in Africa until very recently. All the 20-chromosome toads form a cohesive unit related to one 20-chromosome ancestor. The number change must have occurred in Africa.

REFERENCES

Baldauf, R. J. 1959. Morphological criteria and their use in showing bufonid phylogeny. J. Morph. 104:527–560.

Bataillon, E., and Tchou-Su. 1929. Analyse de la fecondation par l'hybridation et la polyspermie physiologique. Arch. Entw. Mech. Org. 115:779–824.

Bianchi, N. O., and R. Laguens. 1964. Somatic chromosomes of *Bufo arenarum*. Cytologia 29:151–154.

Blair, W. F. 1959. Genetic compatibility and species groups in U.S. toads (*Bufo*). Texas J. Sci. 11:427–453.

———. 1961. Further evidence bearing on intergroup and intragroup compatibility in toads (genus *Bufo*). Texas J. Sci. 13:163–175.

———. 1962. Non-morphological data in amphibian classification. Syst. Zool. 11:72–84.

———. 1963*a*. Evolutionary relationships of North American toads of the genus *Bufo*: A progress report. Evolution 17:1–16.

———. 1963*b*. Intragroup genetic compatibility in the *Bufo americanus* species group of toads. Texas J. Sci. 15:15–34.

———. 1964. Evidence bearing on the relationships of the *Bufo boreas* group of toads. Texas J. Sci. 16:181–192.

———. 1966. Genetic compatibility in the *Bufo valliceps* and closely related groups of toads. Texas. J. Sci. 18:333–351.

Bogart, J. P. 1967. Chromosomes of the South American amphibian family Ceratophridae with a reconsideration of the taxonomic status of *Odontophrynus americanus*. Canad. J. Genet. Cytol. 9:531–542.

———. 1968. Chromosome number difference in the amphibian genus *Bufo*: The *Bufo regularis* species group. Evolution 22:42–45.

Bogert, C. M. 1960. The influence of sound on the behavior of amphibians and reptiles, p. 137–320. *In*, W. E. Lanyon and W. N. Tavolga (eds.), Animal sounds and communication. A.I.B.S. Publ. 7.

Brachet, J. 1954. Nouvelles observations sur les hybrides letaux entre batraciens et entre echinoderms. Arch. Biol. 65:1–72.

Briggs, R. W. 1949. The influence of egg volume on the development of haploid and diploid embryos of the frog *Rana pipiens*. J. Exp. Zool. 111:255–294.

———, and T. J. King. 1959. Nucleocytoplasmic interactions in eggs and embryos, p. 538–617. *In* J. Brachet and A. E. Mirsky (eds.), The cell. I. Academic Press, New York.

Brown, S. W. 1949. The structure and meiotic behavior of the differentiated chromosomes of tomato. Genetics 34:437–461.

Cole, C. J., C. H. Lowe, and J. W. Wright. 1968. Karyotypes of eight species of toads (genus *Bufo*) in North America. Copeia 1968:96–100.

Coleman, L. C. 1943. Chromosome structure in the Acrididae with special reference to the X-chromosome. Genetics 28:2–8.

Darlington, P. J. 1957. Zoogeography: The geological distribution of animals. John Wiley and Sons, Inc., New York.

Duellman, W. E., and C. J. Cole. 1965. Studies of chromosomes of some anuran amphibians (Hylidae and Centrolenidae). Syst. Zool. 14:139–143.

Eigsti, O. J., and P. Dustin, Jr. 1955. Colchicine. Iowa State College Press, Ames, Iowa.

Fankhauser, G. 1945. The effects of change in chromosome number on amphibian development. Quart. Rev. Biol. 20:20–78.

Hughes-Schrader, S., and F. Schrader. 1956. Polyteny as a factor in the chromosomal evolution of the Pentatomini (Hemiptera). Chromosoma 8:135–145.

Kawamura, T. 1950. Studies on hybridization in amphibians. II. Interspecific hybrids in red-colored frogs. J. Sci. Hiroshima Univ. Ser. B, Div. 1, 11:61–70.

———. 1952. Triploid hybrids of *Rana japonica* Gunther female x *Rana temporaria ornativentria* Werner male. J. Sci. Hiroshima Univ. Ser. B, Div. 1, 12:39–46.

Kihlman, B. A. 1966. Action of chemicals on dividing cells. Prentice-Hall, Inc., Englewood Cliffs, New Jersey.

Levan, A., K. Fredga, and A. A. Sanberg. 1964. Nomenclature for centromeric positions on chromosomes. Hereditas 52:201–220.

Low, Bobbi S. 1967. Evolution in the genus *Bufo*: Evidence from parotoid secretions. Ph.D. Thesis, Univ. Texas.

Manton, I. 1950. Problems of cytology and evolution in the *Pteridophyta*. London, Cambridge University Press.

McClintock, B. 1934. The relation of a particular chromosomal element to the development of the nucleoli in *Zea Mays*. Z. Zellforsch. Mikr. Anat. 21:294–328.

Mirsky, A. E., and H. Ris. 1951. The desozyribonucleic acid content of animal cells and its evolutionary significance. J. Gen. Physiol. 34:451–462.

Moore, J. A. 1950. Further studies on *Rana pipiens* racial hybrids. Amer. Nat. 84:247–254.

———. 1955. Abnormal combinations of nuclear and

cytoplasmic systems in frogs and toads, p. 139–182. *In*, Advances in genetics, edited by M. Demerec. Academic Press, Inc., New York, New York.

Morescalchi, A. 1968. The karyotypes of two specimens of *Leiopelma hochstetteri* Fitz. (Amphibia Salientia). Caryologia 21:37–46.

———, and G. Gargiulo. 1968. Su alcune relazioni cariologiche del genero *Bufo* (Amphibia Salientia). Rend. Acc. Sc. Fis. Mat. (Naples) 35: 117–120.

Muto, Y. 1951. Haploid-triploid mosaic toads induced by heat treatment of the unfertilized eggs. J. Sci. Hiroshima Univ. Ser. B, Div. 1, 12:39–46.

———. 1952. Production of triploid toads, *Bufo vulgaris formosus* (Boulenger), by a temperature shock on fertilized eggs. J. Sci. Hiroshima Univ. Ser. B, Div. 1, 13:163–171.

Navashin, M. 1934. Chromosome alterations caused by hybridisation and their bearing upon certain central genetic problems. Cytologia 5:169.

Noble, G. K. 1931. The biology of the Amphibia. McGraw-Hill (Dover Publ. Inc.), New York.

Patterson, J. T., and W. S. Stone. 1952. Evolution in the genus *Drosophila*. Macmillan, New York.

Resende, F. 1940. Über die Chromosomenstruktur in der Mitose der Würzelspitzen. II. SAT-Differenzierungen, Spiralbau, and Chromonemata. Chromosoma 1:486–520.

Ris, H. 1945. The structure of meiotic chromosomes in the grasshopper and its bearing on the nature of "chromomeres" and "lampbrush chromosomes." Biol. Bull. 89:242–257.

———, and B. L. Chandler. 1963. The ultrastructure of genetic systems in prokaryotes and eukaryotes. Cold Spring Harb. Symp. Quart. Biol. 28:1–16.

Rollason, G. S. 1949. X-radiation of eggs of *Rana pipiens* at various maturation stages. Biol. Bull. 97:169–186.

Rugh, R. 1939. Developmental effects resulting from exposure to X-rays. I. Effect on the embryo of irradiation of frog sperm. Proc. Amer. Phil. Soc. 81:447–471.

Schrader, F., and S. Hughes-Schrader. 1958. Chromatid autonomy in *Banasa* (Hemiptera: Pentatomidae). Chromosoma 9:193–215.

Sirlin, J. L. 1960. The nucleolus problem. Nature 186: 275–277.

Stebbins, G. L., Jr. 1950. Variation and evolution in plants. Columbia Univ. Press, New York.

Swanson, C. P. 1957. Cytology and cytogenetics. Prentice Hall, Inc., Englewood Cliffs, N.J.

Tchou-Su. 1936. L'hybridation chez les anoures de Canton (Chine). C. R. Acad. Sci. 202:242–244.

Therman-Suomaleinen, E. 1949. Investigations on secondary constrictions in Polygonatum. Hereditas 35: 86–108.

Tihen, J. A. 1962. Osteological observations on New World *Bufo*. Amer. Midland Nat. 67:157–183.

Ullerich, F. H. 1966. Karyotype und DNS-Gehalt von *Bufo bufo, B. viridis, B. bufo* x *B. viridis* und *B. calamita* (Amphibia, Anura). Chromosoma 18:316–342.

White, M. J. D. 1954. Animal cytology and evolution. 2nd ed. London, Cambridge University Press.

———. 1961. The chromosomes. John Wiley and Sons, Inc., New York. 188 p.

Wolfe, S. L., and B. John. 1965. The organization and ultrastructure of male meiotic chromosomes in *Oncopeltus fasciatus*. Chromosoma 17:85–103.

11. Evidence from Hybridization

W. FRANK BLAIR

The University of Texas at Austin
Austin, Texas

Introduction

Artificial hybridization experiments provide a powerful tool for evaluating degree of affinity of species. This kind of work with the genus *Bufo* was pioneered by A. P. Blair (1941), and hundreds of hybridization experiments have been performed during the past sixteen years in my own laboratory. The results of these experiments now provide a large body of data that contribute to knowledge of how closely species are related and hence place of origin and direction and distance of the spread of the various differentiates within the genus. Relationships of the North and Middle American species were discussed in a preliminary paper by W. F. Blair (1963*a*). Most conclusions from that report remain unchanged, but we now have an enormously greater understanding of this group of species as well as of the genus on a world-wide basis. Detailed results of hybridization experiments within the *boreas* group (W. F. Blair 1964*a*), the large *americanus* group (W. F. Blair 1963*b*), and the large *valliceps* group and related species (W. F. Blair 1966) have been published.

Methods and Rationale

The methods of hybridization have been previously described (W. F. Blair 1959). The most pertinent point to emphasize is that almost all the crosses were made in vitro, which means that premating isolating mechanisms were not functioning to prevent the interspecific hybridizations. Eggs were squeezed into a sperm suspension from crushed testes. A single large female could thus be crossed with a large number of males. Attempts were made to raise F_1 hybrids to sexual maturity and to test their fertility by backcrossing to one or both of the parental species. All F_1 hybrids were not tested because space limitations did not permit raising large numbers of toads to sexual maturity.

The results of the crosses range all of the way from essentially complete interfertility of both male and female F_1 hybrids to failure of the zygote to develop beyond gastrulation. In two groups of species, there is a special situation in that some interaction between egg and testes contents (with the testes greatly hyperthrophied) prevents fertilization or restricts development to early cleavage stages, which are often abnormal. Aside from these two groups, fertilization ordinarily occurs even in the most distantly related species of *Bufo*. The results form very nearly a continuum, but some levels do predominate, as follows:

1. Failure at gastrulation, with or without exogastrulation.
2. Failure as a "monster" at neurula or tailbud stages.
3. Failure as abnormal larvae (microcephalic, edematous, or "dumpy"), which hatch but never feed.
4. Failure as seemingly normal larvae, which feed and grow, but after several days apparently lack something essential for further growth and die. This phenomenon is mostly restricted to the large African *regularis* group.
5. Inability to metamorphose.
6. Metamorphosis of abnormal individuals that remain dwarfed, are pale colored, or lack one or both eyes or one or more limbs.
7. Metamorphosis of vigorous, sometimes heterotic, but sterile hybrids. The gonads are sometimes undeveloped, abnormal, or apparently normal. There is sometimes a well-developed testis as well as a Bidder's organ containing pigmented eggs.
8. Adult hybrid males produce large numbers of abnormal sperm, which fertilize eggs to produce an inviable backcross generation.
9. Low percentage of fertile F_1 of one or the other sex, which can produce a few viable backcross individuals.
10. High percentage of fertile F_1 of both sexes, which can produce many viable backcross individuals.

It is not possible to assess the degree of genotypic difference of the parents that is associated with the various levels of genetic incompatibility. However, there can be little argument with the proposition that there is a general agreement between the level of genetic differentiation and the level of compatibility between species. Furthermore, while it is theoretically possible that a mutation at a single locus could result in genetic incompatibility, the graded response in essentially all crosses indicates otherwise and implies that polygenic systems are involved.

Most of the hybrids produced in the last four years have been checked for karyotype by James Bogart (Chap. 10). In some instances the viable hybrids between distantly related species have proved to be triploids or pentaploids, and in two instances gynogenetic diploids. In no instance has a haploid larva been identified. High genetic compatibility, that is, fertility of both male and female F_1 hybrids has been found only in crosses between species that, according to other evidence, are members of the same species group. On the other hand, it is not uncommon for male and female gametes to behave very differently in crosses. For example, females of the *americanus* group can produce hybrids with males of a great many species, and it is the exception when the F_1 hybrids are inviable. On the other hand, the males of this species group are highly incompatible with the females of most other species outside their own group.

South American Radiation

South America is taken as the starting point in this discussion because, as will be evident below, the weight of the available evidence points more strongly to South America, or at least to the American tropics, as the probable center of origin of the genus than to any other area of the world. One notable point about the South American radiation is that it involves two morphological and ecological extremes. One of these is represented by *B. marinus* and related species, which are primarily lowland in distribution, with *B. marinus* occupying the Amazon basin and other tropical areas and with the related species spreading into drier lowland areas. The skull in this morphological type is broad in the frontoparietal region (Fig. 4-1). The roof of the skull is heavily ornamented with dermal bone, and there are well-developed supraorbital crests. The occipital canals are closed over by the development of the dermal ornamentation. The other type is represented by *B. spinulosus* and closely related species and has its distribution along the cordillera of western South America, where it ranges up to high elevations and cold temperatures. In this morphological type, the skull is thin-roofed, and the frontoparietal region is narrow, with more or less parallel sides. The occipital canal is open.

Hybridization within the *Spinulosus* Group

Crosses have been made in four combinations among the allopatric members of the *spinulosus* group (♀ *spinulosus* x ♂ *chilensis* and the reciprocal; ♀ *flavolineatus* x ♂ *spinulosus*; ♀ *flavolineatus* x ♂ *chilensis*). A low percentage of metamorphosis (0.1% to 8.5%) occurred in all these crosses, but the implied incompatibility could be spurious, since the metamorphosis in the controls was also low.

Hybridization within the *Marinus* Group

Ten hybrid combinations have been produced

among five members of the *marinus group*. There was moderate percentage metamorphosis (15.0 and 27.7% respectively) in crosses between female *B. marinus* and males of *B. paracnemis* and *B. ictericus*. James Bogart found *B. marinus* and *B. poeppigi* existing sympatrically at El Bosque Nacional de Iparia in the upper Amazonas of Peru. Low percentage metamorphosis (3.2%) occurred in a cross between a female *B. poeppigi* from Iparia and a male *B. marinus* from Guatemala. Moderate percentage metamorphosis (40.8%) occurred in a cross between a female *B. poeppigi* from Iparia and a male *B. marinus* from Iparia, and both low and high (16.4% and 82.7%) in crosses with males from two localities in Costa Rica. However, there was 100 percent metamorphosis in subsamples of these three crosses, and the control showed similar variance, so that disease is a likely cause of the differences. In a cross between a female *B. paracnemis* and male *B. ictericus*, the percentage metamorphosis was high (64.3%). In crosses between a female *B. ictericus* and male *B. marinus* and *B. paracnemis*, the percentage metamorphosis was low (2.2 and 8.8% respectively).

B. arenarum is an evolutionarily significant member of this group because of its compatibility with members of other South American species groups as well as with members of this group. In three crosses of ♀ *arenarum* x ♂ *marinus* percentage metamorphosis ranged from zero (failure at gastrulation, with exogastrulation) to 23.1 and 38.5 percent. In one cross of ♀ *arenarum* x ♂ *poeppigi* (the latter from Iquitos, Peru) there was much exogastrulation, and only 0.6 percent metamorphosed. In two crosses of ♀ *marinus* x ♂ *arenarum*, one lot failed as larvae after many had hatched, and in the other there was 2.3 percent metamorphosis. There was low percentage metamorphosis (12.4%) in one cross of ♀ *poeppigi* x ♂ *arenarum*. Crosses between female *B. ictericus* and *B. paracnemis* and male *B. arenarum* gave vastly different results. In the cross of ♀ *ictericus* x ♂ *arenarum*, there was 83.6 percent metamorphosis. In the cross of ♀ *paracnemis* x ♂ *arenarum*, there was 65.0 percent metamorphosis, and one hybrid male was tested by backcrossing to a *B. arenarum* female and was highly fertile. One female hybrid sacrificed at the age of twenty-five months had pigmented eggs in her ovaries. These results obviously imply greater genetic similarity among *B. ictericus*, *B. paracnemis*, and *B. arenarum* than there is between any one of them and *B. marinus* or *B. poeppigi*. However, in the cross of ♀ *arenarum* x ♂ *paracnemis* there was high prehatchery mortality, and only 0.6 percent metamorphosed.

Intergroup Hybridization among South American Toads

One remarkable feature of the South American *Bufo* is that, in genetic compatibility, there is a connection by way of *B. arenarum* between the extremes of skull types represented by the large thick-skulled toads of the *marinus* group and the small or thin-skulled toads of the *spinulosus* group. Three hybrid combinations have been made between female *B. arenarum* and males of the *B. spinulosus* group. In three crosses of ♀ *arenarum* x ♂ *chilensis*, one set failed to develop beyond the larval stage; percentage metamorphosis was only 0.8 percent in the second and 7.6 percent in the third. In a cross of ♀ *arenarum* x ♂ *flavolineatus*, the percentage metamorphosis was 28.1 percent. In a cross of ♀ *arenarum* x ♂ *flavolineatus*, the percentage metamorphosis in a very small sample was 66.7 percent.

Two crosses have been made between males of *B. arenarum* and females of the *spinulosus* group. In a cross of ♀ *chilensis* x ♂ *arenarum*, there was only 4.5 percent metamorphosis. In a cross of ♀ *spinulosus* x ♂ *arenarum*, there was 6.1 percent metamorphosis. Thus, there was metamorphosis in both reciprocals of crosses between *B. arenarum* of the *marinus* group and members of the *spinulosus* group, although the percentage of metamorphosis was consistently low. Contrastingly, there was variable percentage metamorphosis in crosses between females of *B. marinus* and males of the *spinulosus* group and no metamorphosis in reciprocal crosses. In a cross of ♀ *poeppigi* x ♂ *spinulosus*, development failed at the larval stage. In a cross of ♀ *marinus* x ♂ *spinulosus*, there was low percentage hatching, and only 0.2 percent metamorphosis; many other larvae grew to large size, but were unable to metamorphose. Hybrids produced in crosses of female *B. chilensis*, *B. atacamensis*, and *B. spinulosus* with males of *B. marinus* developed only to the gastrula stage.

There is little compatibility between *B. crucifer*, which externally shows considerable resemblance to *B. valliceps* of North America, and the other South American species with which it has been tested. There is reciprocal incompatibility with the *spinulosus* group. In a cross of ♀ *crucifer* x ♂ *flavolineatus*, development stopped at gastrula or neurula, and in a cross of ♀ *atacamensis* x ♂ *crucifer* development stopped at gastrulation.

In the only cross between *B. crucifer* and a member of the *granulosus* group (♀ *fernandezae* x ♂ *crucifer*) there was extensive exogastrulation and failure at that stage.

There is reciprocal incompatibility with *B. arenarum* of the *marinus* group. In a cross of ♀ *crucifer* x ♂ *arenarum,* abnormal larvae hatched but died without feeding, while in a cross of ♀ *arenarum* x ♂ *crucifer,* development failed at gastrulation. There is, however, low reciprocal compatibility between *B. crucifer* and other members of the *marinus* group. In two crosses of ♀ *crucifer* x ♂ *marinus,* 0.5 percent of the hybrids metamorphosed in one, and 3.5 percent in the other. James Bogart found both 2n and 3n chromosome counts in the latter group. In three crosses of ♀ *poeppigi* x ♂ *crucifer,* a low percentage hatched in one, and two of the three larvae that hatched did metamorphose.

There are few crosses between *B. typhonius* and other South American species, so that the hybridization data contribute little to the knowledge of relations of this widely distributed species. In a cross of ♀ *B. typhonius* (from Panama) x ♂ *B. marinus,* most of the hybrids died before hatching; one abnormal larva hatched but died without feeding. In two crosses of ♀ *B. fernandezae* x ♂ *B. typhonius,* development ceased at gastrulation.

The *granulosus* group is a widely distributed complex of poorly known species that has been treated as a single species by Gallardo (1965), who used only morphological characters and a typological concept. His subspecies names are used here as though they were species, since there is evidence that at least some of them are good biological species (Chap 6). No compatibility has been found between toads of this group and *B. marinus.* In a cross of ♀ *B. fernandezae* x ♂ *B. marinus* and one of ♀ *B. humboldti* x ♂ *B. poeppigi* there was failure at gastrulation, and in a cross of ♀ *B. marinus* x ♂ *B. pygmaeus* and in one of ♀ *B. poeppigi* x ♂ *B. humboldti,* there was failure at the stage of abnormal larvae, which failed to feed and died at that stage.

There was percentage metamorphosis at 27.8 percent in a cross of ♀ *B. humboldti* x ♂ *B. spinulosus,* but the hybrids failed to develop beyond the stage of abnormal neurulae in a cross of ♀ *B. chilensis* x ♂ *B. humboldti.* In a cross of ♀ *B. pygmaeus* (from Argentina) x ♂ *B. arenarum* there was 4.4 percent metamorphosis, and in a cross of ♀ *B. fernandezae* x ♂ *B. arenarum* there was 50.0 percent metamorphosis. In reciprocal group crosses, there was 50.9 percent metamorphosis in a cross of ♀ *B. arenarum* x ♂ *B. humboldti* and 22.1 percent metamorphosis in a cross of ♀ *B. arenarum* x ♂ *B. fernandezae.* The incompatibility in crosses between female *B. fernandezae* and males of *B. crucifer, B. typhonius,* and of *B. ictericus* and *B. poeppigi* of the *marinus* group already has been mentioned.

The *guttatus* group of South American toads is represented in our work only by the large *B. blombergi* of Colombia and the small *B. haematiticus* of Central America. Relationships of these toads with other South American species are obscure, because only males have been available for crossing and because the hypertrophied testes of these toads contain some compound that inhibits fertilization of foreign eggs.

Males of the toad that has been called *Melanophryniscus stelzneri* were used in attempted in vitro crosses with females of *B. spinulosus, B. bocourti,* and *B. regularis,* but no evidence of cleavage was detected. In a cross with *B. woodhousei,* however, 77.4 percent of the eggs cleaved, and 83.3 percent of the fertilized eggs hatched into abnormal larvae, which died without feeding. These latter results cast doubt on the merit of recognizing *Melanphryniscus* as a separate genus.

Main Features of the South American Radiation

The salient features of the South American radiation of the genus *Bufo* with respect to genetic compatibility may be summed up as follows:

1. The broad-skulled toads of the *marinus* group (*B. marinus, B. paracnemis, B. poeppigi, B. ictericus,* and *B. arenarum*) demonstrate metamorphosis in reciprocal intragroup crosses, and the male hybrid has been shown to be fertile in the cross of ♀ *B. paracnemis* x ♂ *B. arenarum.*
2. The narrow-skulled toads of the *spinulosus* group exist as isolated populations that are probably allopatric species. There was low percentage metamorphosis in intragroup crosses, but the results are inconclusive because controls were also affected.
3. *B. arenarum* is an important connecting link between the lines of broad-skulled and narrow-skulled toads in genetic compatibility.
4. *B. crucifer* shows high genetic incompatibility with most other South American toads with which it has been tested, but the evidence places it closest to the *marinus* group.
5. The name *B. granulosus* applies to a complex of presently poorly understood species (Chap. 6).

Table 11–1. Results of Crosses within the *Valliceps* Group

♀	♂						
	valliceps	*cavifrons*	*ibarrai*	*mazatlanensis*	*gemmifer*	*luetkeni*	*coniferus*
valliceps		71	69	96	28	90	81
cavifrons							
ibarrai							
mazatlanensis	84					33	N
gemmifer							
luetkeni	85		L	42	43		G
coniferus							

Best results for the interspecies combination: G = failed at gastrulation; N = at least some to neurula; L = at least some to larva. Numbers indicate percentages of fertilized eggs that resulted in metamorphosed individuals.

Intragroup crosses have not been attempted. In crosses with other groups, there is no evidence of differences in the compatibility of these different entities of the *granulosus* group with members of other groups.

6. The relations of *B. typhonius* are still uncertain.
7. *B. blombergi* and *B. haematiticus* of the poorly known *guttatus* group of northern South America have hypertrophied testes that contain a presently unknown substance that adversely affects the eggs in attempted in vitro crosses, hence the compatibility with other South American species is presently undetermined.
8. The toad that has been known as *Melanophryniscus stelzneri* seems evolutionarily no more distant from *Bufo* than some currently recognized members of the genus are from one another.

Central and North American Radiation

There is greater diversity among the North and Central American *Bufo* than there is among the South American representatives of the genus. By contrast with the five species groups (6 if *Melanophryniscus* is included in *Bufo*) in South America, there is great genetic diversity in North America where fifteen species groups are recognizable. Seven of these have two or more species, and eight are monotypic. The greater diversification of the genus in North and Central America over South America may be a reflection of the greater environmental diversity of the northern continent and also of the greater climatic fluctuations in the northern continent during the Pliocene and Pleistocene.

Hybridization within the *Valliceps* Group

Genetic compatibility within the large *valliceps* group has been discussed by W. F. Blair (1966). Since that time, additional hybridizations have been performed. With the exception of *B. melanochloris* of Costa Rica, all eight species have been crossed with one or more different species of the group. Fourteen hybrid combinations have been produced in this species group. Female *B. valliceps* have been crossed with males of all the other species except *B. melanochloris.* There is no evident difference in the degree of viability of hybrids from the cross of female *B. valliceps* with males of the six species of the *valliceps* group with which it has been crossed (Table 11-1; Appendix H).

The percentage metamorphosis was high in all crosses except with *B. gemmifer,* where it was 28 percent. Crosses between female *B. mazatlanensis* and males of *B. valliceps, B. luetkeni,* and *B. coniferus* gave quite different results, for there was very high percentage metamorphosis in the cross with *B. valliceps* and moderate in the cross with *B. luetkeni* but failure of the hybrids to develop beyond the neurula stage in two attempted crosses with *B. coniferus.* Similarly, the hybrids involving female *B. luetkeni* and males of other species of the group showed variable viability, which was high when the male parent was a *B. valliceps,* moderate when it was *B. mazatlanensis* or *B. gemmifer,* but zero when it was *B. ibarrai;* or *B. coniferus* (Table 11-1).

The difference in viability of hybrids between *B. valliceps* and *B. coniferus* from that of hybrids between *B. mazatlanensis* or *B. luetkeni* and *B. coniferus* is consistent with earlier evidence (W. F. Blair 1966) indicating a dichotomy between a western group and an eastern group within the *valliceps* group. The evidence presented here also indicates a genetic difference of *B. coniferus* from other members of the group that is greater than that among the six other members that have been tested.

Although there is high percentage survival of F_1 hybrids among the species of the *valliceps* group

Table 11–2. Summary of the Results of Backcrossing Intragroup F_1 Hybrids in the *Valliceps* Group

Hybrid Combinations	No. F_1 ♂♂ Tested	No. F_1 ♂♂ with Sperm	No. with Offspring to Met.	% Fert.	% Fert. Hat.	% Fert. Met.
♀ *valliceps* x ♂ *mazatlanensis*	3	3	2	6-93-24	75-4-45	0-0.1-19
♀ *mazatlanensis* x ♂ *valliceps*	3	2	1	0-1-30	0-50-36	0-0-0.2
♀ *valliceps* x ♂ *luetkeni*	2	2	2	12-68	18-20	15-2
♀ *luetkeni* x ♂ *valliceps*	6	4	2	0-0-3-5-21-52	0-0-50-0-50-0	33-<25
♀ *valliceps* x ♂ *ibarrai*	2	2	1	(15-30)(3-8)		3
♀ *valliceps* x ♂ *gemmifer*	2	2	1	few–many	?	<46
♀ *mazatlanensis* x ♂ *luetkeni*	1	1	1	98	100	97
♀ *luetkeni* x ♂ *mazatlanensis*	3	3	2	2-8-11	0-20-60	0-20-37

that have been tested, the results of backcrossing male hybrids to the parental species indicate considerable genetic differentiation among the species of this group. The results of these backcrosses are summarized in table 11-2. From these results, it is clear that there is much reduction of fertility in most of the hybrids, as pointed out earlier (W. F. Blair 1966). Only in the combination ♀ *B. mazatlanensis* x ♂ *B. luetkeni* is there high fertility.

Hybridization within the *Boreas* Group

Reciprocal crosses have been made between *B. boreas* and the relictual *B. canorus*, and a female *B. boreas* has been crossed with a male of the allopatric isolate *B. exsul*. W. F. Blair (1964) reported high percentage metamorphosis in two crosses of ♀ *B. boreas* x ♂ *B. canorus*, and one of the male hybrids was subsequently backcrossed to a female *B. boreas*. Survival to metamorphosis in the backcross generation was 32.3 percent, which compares favorably with 45.0 percent in the control. Survival through metamorphosis was 72.2 percent in a cross of ♀ *B. boreas* x ♂ *B. exsul*.

Hybridization within the *Cognatus* Group

Three crosses of ♀ *B. cognatus* x ♂ *B. speciosus* have been made, and metamorphosis has ranged from 5.4 to 68.0 percent. One hybrid male was backcrossed to *B. cognatus*, and another to *B. speciosus*. In the backcross to ♀ *B. speciosus*, only 25.6 percent of the eggs were fertilized (higher than in the control), and 38.7 percent of the fertilized eggs resulted in metamorphosed toads. In the backcross to *B. cognatus*, 73.7 percent of the eggs were fertilized, and 32.8 of the fertilized eggs resulted in metamorphosed toads. However, 8.0 percent of the larvae were unable to metamorphose. The reciprocal cross has been made six times, and percentage metamorphosis ranged from 2.3 to 37.7 percent. In a cross of ♀ *B. cognatus* x ♂ *B. compactilis*, percentage metamorphosis was 25.1 and in a cross of ♀ *B. compactilis* x ♂ *B. speciosus*, 4.6 percent of the hatched larvae metamorphosed. Possibly, the low percentages for these crosses do not indicate genetic differences, since these are difficult toads to raise through the larval stage in the laboratory.

Hybridization within the *Marmoreus* Group

The two species of the *marmoreus* group have been reciprocally hybridized, and there was high percentage metamorphosis in both reciprocals (Appendix H). An F_1 male of the combination ♀ *B.*

Table 11–3. Highest Percentage Metamorphosis among Crosses of Species of the *Americanus* Group

♀	♂					
	americanus	*houstonensis*	*terrestris*	*hemiophrys*	*microscaphus*	*woodhousei*
americanus		69	50	65	75	47
houstonensis	80		69		27	86
terrestris	90	95		95	21	19
hemiophrys	16	24	L		5	60
microscaphus	36		52	L		37
woodhousei	57	81	90	31	34	

marmoreus x ♂ *B. perplexus* was backcrossed to a female *B. marmoreus* and proved fertile, although only 2 of 165 eggs were fertilized, and only one of these survived through metamorphosis. These species are interfertile despite their difference in size, the presence of sexual dichromatism in *B. marmoreus*, which is absent in *B. perplexus*, and the laying of single, adhesive eggs by *B. marmoreus* and normal, stranded eggs by *B. perplexus*.

Hybridization within the *Debilis* Group

Only one intragroup cross has been made among the three species of the *debilis* group. In the cross of ♀ *B. kelloggi* x ♂ *B. debilis*, 10.8 percent of the fertilized eggs resulted in metamorphosed toads.

Hybridization within the *Americanus* Group

Laboratory hybridization and intragroup fertility within the *americanus* group was first described by A. P. Blair (1941), who later (1955) reported hy-

Table 11–4. Results of Hybridization between Species Groups of North and Central American *Bufo**

♀	♂												
	valliceps	*canaliferus*	*occidentalis*	*coccifer*	*bocourti*	*boreas*	*cognatus*	*alvarius*	*marmoreus*	*punctatus*	*debilis*	*americanus*	*quercicus*
valliceps (8)		**96**	**17**	L	4	L	S♂ **92**	**22**	S♂ **68**	L	L	F♂ **3**	L
canaliferus (1)	**26**			**3**	L	N	**25**	G	53	N	1	G	L
occidentalis (1)	S♂ **27**	L		N		L	L	L	L	N	N	N	G
coccifer (2)	21	L			L	2	L	G	3	33	L	<**.1**	
bocourti (1)	G			G		G	L		L	G		G	
boreas (4)	59	L	G	G	13		**66**	**83**	63	**67**	7	**7**	
cognatus (3)	**73**	**32**	15	G	33	**100**		**25**	**91**	L	10	L	L
alvarius (1)	**14**			G	L	**9**	**81**		**82**	L	L	N	N
marmoreus (2)	S♂ **93**	L	G	G	S♂ 49	G	**36**	**87**		L	7	N	L
punctatus (1)	S♂ 31	L	G	G	79	**47**	88	4	100		28	L	N
debilis (3)	G					G	G	G		G		N	
americanus (6)	S♂ **72**	25	L	**29**	61	**80**	S♂ 19	80	66	54	8		25
quercicus (1)													

*Number following female parent is the number of species in that group. Best performance for the intergroup combination: *G* = failed at gastrulation; *N* = at least some to neurula before failure; *L* = at least some to larva before failure. Numbers indicate percentage of fertilized eggs that resulted in metamorphosed individuals. Numbers in boldface indicate crosses in which at least some hybrids metamorphosed in both reciprocals.

bridization of *B. americanus* and *B. microscaphus*. The existence of interfertility at somewhat reduced levels among various species combinations within the group was reported by W. F. Blair (1963). The species *B. woodhousei,* which occurs sympatrically with several of the other species and often forms sympatric hybrids with them, shows greater incompatibility with the other species than they do with one another.

Additional combinations within this group have been effected since my 1963 publication, so that now all but three of the thirty possible combinations among the six species have been produced (Table 11-3). The one instance in which there was no metamorphosis (♀ *B. microscaphus* x ♂ *B. hemiophrys*) is probably without significance since the eggs were small and apparently immature, and only three were fertilized.

Fertility in the F_1 generation, as reported by W. F. Blair (1963), is considerably higher than in intragroup crosses in the *valliceps* group, but there is evidence of some reduction in fertility in both backcross and F_2 generations. Unlike the situation in the *valliceps* group, the hybrid females in the *americanus* group could be induced to ovulate, and they were fertile to some degree in all combinations tested.

Intergroup Hybridization among North American Toads

The results of intergroup crosses among the *Bufo* of North and Central America are summarized in table 11-4. The main elements of evolutionary relationships of the North American *Bufo* are best seen in figure 11-2, where groups that show metamorphosis of reciprocal hybrids are connected by lines.

The *Valliceps* Group. The thin-skulled type of toad is represented by such groups as *bocourti* of Guatemala and *boreas, alvarius,* and *americanus* of North America. Other groups are intermediate, as is also the case in South America. It can be seen from table 11-4 that toads of the *valliceps* group and certain other thick-skulled toads undergo metamorphosis in both reciprocals of intergroup crosses. There was high percentage metamorphosis in the cross of ♀ *B. valliceps* group x ♂ *B. canaliferus* and relatively high in the reciprocal (W. F. Blair 1966). Females of *B. valliceps, B. luetkeni,* and *B. mazatlanensis* have been crossed with male *B. canaliferus*. Metamorphosis of the hybrids ranged from 3.2 percent in the combination ♀ *B. luetkeni* x ♂ *B. canaliferus* to 96.1 percent in the combination of ♀ *B. mazatlanensis* x ♂ *B. canaliferus*. Two F_1 hybrid males from the combination ♀ *B. valliceps* x ♂ *B. canaliferus* were backcrossed to *B. valliceps*, but, although there was moderate (22.7%) to high (73.6%) fertilization, none of the backcross generation survived beyond gastrulation.

In four crosses of ♀ *B. canaliferus* x ♂ *B. valliceps* percentage metamorphosis ranged from 14.5 to 25.8 percent. In a cross of ♀ *B. canaliferus* x ♂ *B. gemmifer*, percentage metamorphosis was 21.2 percent. In a cross of ♀ *B. canaliferus* x ♂ *B. mazatlanensis,* there was no count of prehatching mortality, but all of the seventy-two larvae that hatched did continue through metamorphosis. In a cross of ♀ *B. canaliferus* x ♂ *B. luetkeni,* only 9.1 percent of the fertilized eggs resulted in metamorphosed hybrids. In all the hybrids with *B. canaliferus* as the female parent in which prehatching mortality was recorded, it was high.

When two males of the *B. valliceps* x *B. canaliferus* hybrid combination were tested by backcross to *B. valliceps* females, one male fertilized 73.4 percent and the other 20.2 percent of the eggs, but cleavage was abnormal and the embryos failed at gastrulation. One hybrid male of the reciprocal combination, which was also tested by backcrossing to a *B. valliceps* female, gave the same results (W. F. Blair 1966), with 73.5 percent fertilization but failure at gastrulation.

Exchange of genes between the toads of the *valliceps* group and *B. occidentalis* is also genetically blocked. Low percentage metamorphosis occurred in both reciprocals of the combination ♀ *B. valliceps* x ♂ *B. occidentalis,* with 17.4 and 26.9 percent, respectively, surviving through that stage. However, in the two crosses of ♀ *B. occidentalis* x ♂ *B. valliceps,* the hybrids failed as larvae. In crosses of ♀ *B. occidentalis* x ♂ *B. ibarri, B. mazatlanensis, B. luetkeni,* and *B. coniferus,* the hybrids died as larvae without feeding. Two hybrid males of the combination ♀ *B. occidentalis* x ♂ *B. valliceps* contained no identifiable testes. The results of these crosses suggest lesser relationship between *B. occidentalis* and the *valliceps* group than between that group and *B. canaliferus.*

The previously mentioned dichotomy in the *valliceps group* is reflected in the results of crosses between members of that group and the two members of the *marmoreus* group. In three crosses of ♀ *B. marmoreus* and two of *B. perplexus* with ♂ *B. valliceps,* the hybrids failed at stages from gastrula to nonfeeding larva. Hybrids between ♀ *B. perplexus* and ♂ *B. coniferus* failed as gastrulae. By contrast, there was high percentage metamorphosis of hybrids between ♀ *B. marmoreus* and ♂ *B. mazatlanensis* (77.9%) and between ♀ *B. perplexus* and ♂ *B. luetkeni* (93.1%) and ♂ *B. ibarrai* (58.0%).

In the reciprocal crosses between the *valliceps* and *marmoreus* groups, there was low (1.5%) metamorphosis in a cross of ♀ *B. valliceps* x ♂ *B. marmoreus*; there was low (0.9%) metamorphosis in the combination of ♀ *B. valliceps* x ♂ *B. perplexus* and failure as larvae in a replication of this cross. Conversely, there was high percentage metamorphosis (68.3%) in a cross of ♀ *B. mazatlanensis* x ♂ *B. marmoreus*. There also was high percentage metamorphosis in crosses of ♀ *B. mazatlanensis* and *B. luetkeni* x ♂ *B. marmoreus* and *B. perplexus*. These results also add to the evidence of dichotomy in the *valliceps* group.

Two F_1 hybrid males of the combination ♀ *B. marmoreus* x ♂ *B. mazatlanensis* were backcrossed to a ♀ *B. marmoreus* and to a ♀ *B. mazatlanensis* respectively. In the first instance, no eggs were fertilized; in the second, 53.6 percent were fertilized, but the embryos failed at gastrulation. Two hybrid males of the combination ♀ *B. perplexus* x ♂ *B. mazatlanensis* were backcrossed to two female *B. perplexus*, but no eggs were fertilized. Two hybrid males of the combination ♀ *B. mazatlanensis* x ♂ *B. marmoreus* were backcrossed respectively to a female *B. marmoreus* and a female *B. mazatlanensis*. In the first, there was no fertilization, but in the second 47.5 percent of the eggs were fertilized but failed at gastrulation. In one attempted backcross of one male F_1 hybrid of ♀ *B. luetkeni* x ♂ *B. marmoreus* to *B. luetkeni*, there was no fertilization. The results of all of these backcrosses imply that there is little likelihood of gene exchange between the two species groups.

There is compatibility in reciprocal crosses between members of the *valliceps* and *cognatus* group. As in previously mentioned crosses involving the *valliceps* group, there is a dichotomy in compatibility, with the highest percentage metamorphosis in both reciprocals occurring when *B. mazatlanensis* or *B. luetkeni* is crossed with a member of the *cognatus* group. In four replications of the cross ♀ *B. valliceps* x ♂ *B. cognatus*, two sets failed as gastrulae or larvae and the highest percentage metamorphosis in the two others was 28.5 percent. In three crosses of ♀ *B. valliceps* x ♂ *B. speciosus*, one set failed as larvae, one showed low (8.7%) metamorphosis of the hatched larvae, while the other showed high (61.8%) metamorphosis of hatched larvae. In one cross of ♀ *B. valliceps* x ♂ *B. compactilis*, there was only 18.8 percent metamorphosis.

Hybrids of the combination ♀ *B. mazatlanensis* x ♂ *B. speciosus* showed high (92.0%) metamorphosis, while hybrids of the combination ♀ *B. mazatlanensis* x ♂ *B. cognatus* showed relatively high (42.2%) metamorphosis in one set and failed as gastrulae in the other. Hybrids from two crosses of ♀ *B. luetkeni* x ♂ *B. speciosus* metamorphosed at the rate of 16.5 and 84.0 percent. Hybrids from the combination ♀ *B. luetkeni* x ♂ *B. cognatus* showed high (56.6%) metamorphosis. One set of ♀ *B. luetkeni* x ♂ *B. compactilis* hybrids showed high (48.1%) metamorphosis.

In three crosses of ♀ *B. cognatus* x ♂ *B. valliceps*, two sets failed as larvae, and percentage metamorphosis was low (15.8% of hatched larvae) in the third. In an equal number of crosses of ♀ *B. speciosus* x ♂ *B. valliceps*, one set failed as larvae, and the others showed only low (2.8 and 5.4%) metamorphosis. In two crosses of ♀ *B. cognatus* x ♂ *B. luetkeni*, one showed high (72.6%) metamorphosis, but the other failed to be counted. In crosses of ♀ *B. speciosus* x ♂ *B. luetkeni* and *B. ibarrai* metamorphosis was 26.0 and 18.4 percent respectively.

In one cross of ♀ *B. speciosus* x ♂ *B. mazatlanensis*, 26.8 percent of the fertilized eggs resulted in metamorphosed toads. The rather divergent genotype of *B. coniferus* among members of the *valliceps* group is reflected in the incompatibility between males of this species and females of the *cognatus* group. In two crosses of ♀ *B. speciosus* x ♂ *B. coniferus*, one set failed as exogastrulated gastrulae, and in the other, one abnormal larva hatched but died without feeding. Hybrids of ♀ *B. cognatus* x ♂ *B. coniferus* also died as exogastrulae.

In an attempted backcross of a male of the combination of ♀ *B. mazatlanensis* x ♂ *B. speciosus*, 24.4 percent of the eggs were fertilized, but they failed to survive beyond gastrulation. In the attempted backcross of one hybrid male of the combination ♀ *B. valliceps* x ♂ *B. cognatus* to two female *B. valliceps* and of one male of the combination ♀ *B. luetkeni* x ♂ *B. cognatus*, there was no fertilization, and these male hybrids were presumed sterile.

There is low genetic compatibility between the *valliceps* group and *B. alvarius*. In three crosses of ♀ *B. valliceps* x ♂ *B. alvarius*, one set failed at gastrulation, one showed 3.4 percent metamorphosis, and one showed 21.7 percent metamorphosis. In one cross of ♀ *B. mazatlanensis* x ♂ *B. alvarius*, the hybrid embryos failed as larvae. In the reciprocal intergroup cross, hybrids from two crosses of ♀ *B. alvarius* x ♂ *B. valliceps* failed at the larval stage, and those from one cross of ♀ *B. alvarius* x ♂ *B. ibarrai* failed as gastrulae. However, 14.3 percent of the hybrids of ♀ *B. alvarius* x ♂ *B. luetkeni* metamorphosed.

The *americanus* group is the only other North American species group with which metamorphosis is known in both reciprocals of crosses with the *valliceps* group. As is true in many crosses involving these females, there is at least some metamorphosis of hybrids between *americanus* group females and males of the *valliceps* group. This was first demonstrated by A. P. Blair (1941) in a cross of ♀ *B. terrestris* x ♂ *B. valliceps*, and it was repeated by Volpe (1959). Thornton (1955) crossed ♀ *B. woodhousei* x ♂ *B. valliceps* and demonstrated sterility of the intersexual malelike hybrids. Volpe (1956) repeated the cross, but not the attempted backcross, and showed that there is metamorphosis in crosses of ♀ *B. americanus* x ♂ *B. valliceps*. Kennedy (1962) showed that there is metamorphosis in a cross of ♀ *B. houstonensis* x ♂ *B. valliceps*.

We have crossed females of *B. terrestris* with males of *B. valliceps*, *B. mazatlanensis*, and *B. luetkeni*. We have crossed a female *B. americanus* with a male *B. gemmifer*. We have crossed a female *B. microscaphus* with a male *B. valliceps*, and females of *B. hemiophrys* with males of *B. valliceps* and *B. mazatlanensis* (Appendix H). Metamorphosis occurred in all combinations of crosses, with the highest percentage (72.2%) in the cross of ♀ *B. hemiophrys* x ♂ *B. mazatlanensis*, and the lowest (3.1%) in a cross of ♀ *B. hemiophrys* x ♂ *B. valliceps*.

There is high incompatibility between females of the *valliceps* group and males of the *americanus* group. We have made twenty-seven crosses in sixteen combinations between females of the *valliceps* group and males of the *americanus* group, and Thornton (1955) and Volpe (1956, 1959) have reported others. In all these crosses (Appendix H), the embryos have failed at gastrulation, neurulation, or as abnormal nonfeeding larvae except in one combination. In one cross of ♀ *B. luetkeni* x ♂ *B. microscaphus*, only three (5.2%) of the fertilized eggs hatched, but two of these larvae metamorphosed. These were not checked for the possibility that they were polyploids, but one was tested by backcrossing to a female *B. luetkeni*. There was 54.5 percent fertilization, and 30.5 percent of the backcross larvae metamorphosed.

Males of the *valliceps* group show compatibility to the extent that they produce metamorphosed hybrids with members of three other North American groups, but females of the *valliceps* group are genetically incompatible with the males of these same groups. In ten crosses of female *B. valliceps*, *B. mazatlanensis*, and *B. luetkeni* with male *B. boreas*, *B. canorus*, and *B. exsul* of the *boreas* group, the embryos failed between gastrula and larva (Appendix H). Metamorphosis occurred in four reciprocal intergroup crosses involving female *B. boreas* and males of *B. valliceps* (2.4 and 21.7%), *B. ibarrai* (13.0%), and *B. mazatlanensis* (58.9%).

In fourteen crosses between female *B. punctatus* and six members of the *valliceps* group, metamorphosis occurred in two combinations; one of two sets involving ♀ *B. punctatus* x ♂ *B. gemmifer* failed at or before the larval stage, but in the other there was 0.4 percent metamorphosis. In a cross of ♀ *B. punctatus* x ♂ *B. ibarrai*, there was 30.7 percent metamorphosis. In crosses of female *B. punctatus* with males of *B. valliceps*, *B. luetkeni*, *B. mazatlanensis*, and *B. coniferus*, the embryos failed at stages ranging from gastrula to larva. Four reciprocal intergroup crosses involved male *B. punctatus* and females of *B. valliceps*, *B. luetkeni*, and *B. mazatlanensis*, and all embryos failed from gastrula to early larva.

The third group with which the *valliceps* group shows one-way compatibility is the *coccifer* group. In five intergroup crosses, the embryos reached only the neurula stage in the cross of a female *B. coccifer* to a *B. coniferus* male, but a few metamorphosed in crosses to *B. valliceps* (9.3 and 20.8%), *B. luetkeni* (10.2%), and *B. ibarrai* (0.3%) males. The reciprocal intergroup cross was attempted a total of nine times with *B. valliceps*, *B. luetkeni*, and *B. mazatlanensis* females, but the hybrid embryos failed at stages from gastrula to larva.

In attempted crosses between toads of the *valliceps* group and *B. bocourti*, the relationship discussed above is reversed, for the only compatibility is found where the female parent belongs to the *valliceps* group. Four crosses were attempted between *B. bocourti* males and four females of *B. valliceps*, one female *B. mazatlanensis* and one female *B. luetkeni*. In five of the six, the hybrid embryos failed as larvae, but in one of the four crosses involving a *B. valliceps* female there was metamorphosis of 4.5 percent of the hybrids, which were not examined for evidence of polyploidy. In the reciprocal combination, three attempts to cross female *B. bocourti* with *B. valliceps* males resulted in failure at gastrulation, with exogastrulation.

Members of the *valliceps* group show no compatibility in reciprocal crosses with the members of the *debilis* group. Four attempted crosses of a female *B. debilis* and one of *B. kelloggi* with male *B. valliceps* ended in failure at gastrulation. Four attempted crosses of male *B. debilis* with female *B. valliceps,* one with *B. mazatlanensis,* and one with *B. luetkeni* all ended at the early larval stage.

The *quercicus* group, with its single species, shows no compatibility with the *valliceps* group in reciprocal crosses. In two attempted crosses of female *B. valliceps* and one of female *B. luetkeni* with male *B. quercicus,* the embryos failed as larvae. In the reciprocal intergroup cross with male *B. valliceps,* the hybrids failed at gastrulation.

The *Marmoreus* Group. The two species of the *marmoreus* group, in addition to being highly crossable with toads of the *valliceps* group, can also cross in both reciprocals with toads of the *cognatus* group and with *B. alvarius.* Fourteen crosses between members of these two groups include all possible intergroup combinations (Appendix H). Hybrids of ♀ *B. perplexus* x ♂ *B. cognatus* or *B. speciosus* seem to be more viable than those in which the female parent is a *B. marmoreus.* Hybrids between *cognatus*-group females and males of the two species of the *marmoreus* group show high percentage metamorphosis regardless of whether the male parent is *B. marmoreus* or *B. perplexus.* Five hybrid males of the combination ♀ *B. speciosus* x ♂ *B. marmoreus* were tested in attempted backcrosses to a female *B. speciosus,* but no eggs were fertilized, and the males were presumed sterile. One hybrid male of the combination ♀ *B. cognatus* x ♂ *B. perplexus* was tested in attempted backcrosses to two female *B. perplexus,* but no eggs were fertilized. Both gonads in this male were abnormal, and the Bidder's organ was about four times as large as the testis.

In reciprocal crosses between the two members of the *marmoreus* group and *B. alvarius,* percentage metamorphosis was usually high. Hybrids of the combination ♀ *B. alvarius* x ♂ *B. marmoreus* showed 52.2 and 56.7 percent metamorphosis in two crosses, and hybrids between female *B. alvarius* and male *B. perplexus* showed 23.5 and 81.7 percent metamorphosis in two crosses. Reciprocals of this hybrid also metamorphosed in large numbers (87.4%), but in those of the combination ♀ *B. marmoreus* x ♂ *B. alvarius* there was heavy prehatching mortality, and the few larvae that hatched failed to survive.

In crosses with five species groups of North American *Bufo,* there is metamorphosis when the male is a member of the *marmoreus* group and the female belongs to one of five other groups, but there is premetamorphic failure when the female belongs to the *marmoreus* group. The five groups are (1) *punctatus* (33.4–94.1% metamorphosis in 3 crosses); (2) *boreas* (11.1% in one cross to *B. canorus* to 58.6–63.4% in two crosses with *B. boreas*); (3) *americanus* (larval failure to 66.3% metamorphosis in seven crosses of male *B. marmoreus* with five members, *B. americanus, B. terrestris, B. woodhousei, B. hemiophrys,* and *B. microscaphus,* and larval failure to 11.8% metamorphosis in crosses of three species, *B. woodhousei, B. terrestris,* and *B. hemiophrys,* with male *B. perplexus*); (4) *canaliferus* (52.9% metamorphosis of hatched larvae when the male was *B. marmoreus* and 35.9–44.6% in two crosses when the male was *B. perplexus*); (5) *coccifer* (low, 1.9 and 3.0% metamorphosis in crosses with male *B. marmoreus* and *B. perplexus,* respectively).

In reciprocal intergroup crosses, two attempted crosses of female *B. marmoreus* and two of *B. perplexus* with male *B. punctatus* failed at the larval stage. Two attempted crosses of each and one of ♀ *B. marmoreus* x ♂ *B. canorus* failed as gastrulae or abnormal nonfeeding larvae.

With the *americanus* group there was failure at gastrula or neurula in seven crosses of female *B. marmoreus* with *B. woodhousei, B. americanus, B. terrestris, B. microscaphus,* and *B. hemiophrys.* There was failure at gastrulation in five crosses between female *B. perplexus* and males of *B. woodhousei, B. americanus, B. terrestris, B. hemiophrys,* and *B. microscaphus* of the *americanus* group.

In one attempted cross of ♀ *B. marmoreus* x ♂ *B. canaliferus,* there was failure as larvae, and in three in which the female was a *B. perplexus* there was failure at gastrulation. In one intergroup cross involving a male *B. marmoreus,* 52.9 percent of the hatched larvae metamorphosed, and in two involving male *B. perplexus* 35.9 and 44.6 percent of the fertilized eggs resulted in metamorphosed toads.

A cross was attempted between females of *B. marmoreus* and *B. perplexus* and males of *B. coccifer,* but in both instances there was failure at gastrulation. The reciprocal intergroup cross with male *B. marmoreus* gave 1.9 percent metamorphosis, and with *B. perplexus,* 3.0 percent.

In two intergroup combinations (*debilis* group and *bocourti* group) the only compatibility was found when the female parent was of the *marmoreus* group. In one attempted cross between a female *B. marmoreus* and a male *B. debilis,* and one in which the female was a *B. perplexus,* there was failure as abnormal larva, but in a replication of the latter cross 6.9 percent metamorphosed. In one attempted cross between a female *B. marmoreus* and a male *B. kelloggi* and two between females of *B. marmoreus* and male *B. retiformis,* the embryos failed at stages from gastrula to abnormal larva. It has not been possible to make the reciprocal cross.

In three crosses of ♀ *B. marmoreus* x ♂ *B. bocourti* there was metamorphosis of up to 48.4 percent, and in a cross involving a female *B. perplexus* there was 76.3 percent metamorphosis, but 13.6 of the tadpoles were unable to metamorphose. A male of each hybrid combination was tested by attempted backcross, but none fertilized any eggs.

High incompatibility with *B. occidentalis* is indicated. In one attempted cross of ♀ *B. marmoreus* x ♂ *B. occidentalis* and in two reciprocal intergroup crosses involving males of *B. marmoreus* and two involving males of *B. perplexus,* development failed at stages from gastrula to abnormal larva.

Compatibility between the *marmoreus* and *quercicus* groups was tested only by two attempted crosses in the combination ♀ *B. perplexus* x ♂ *B. quercicus* and one in the combination ♀ *B. marmoreus* x ♂ *B. quercicus.* Hybrids of the first combination reached abnormal larva, but in the second they failed at gastrulation. There was no opportunity to attempt the reciprocal intergroup cross.

The *Alvarius* Group. The hybridization data suggest that *B. alvarius* is most closely related to the *cognatus* group and somewhat less closely to the *boreas group.* In two crosses of female *B. alvarius* with male *B. cognatus* and one with male *B. speciosus,* metamorphosis was 50.0, 68.7, and 81.1 percent respectively. In three reciprocal intragroup crosses, there was failure as larvae in one, 25.2 percent metamorphosis in another involving female *B. cognatus,* and 25.2 percent in one involving a female *B. speciosus.*

In four attempted crosses of ♀ *B. alvarius* x ♂ *B. boreas,* there was failure at the larval stage. In the combination ♀ *B. alvarius* x ♂ *B. nelsoni,* one attempt failed at the larval stage, but 9.0 percent of the hybrids metamorphosed in a second attempt. In three reciprocal intergroup crosses (♀ *B. boreas* x ♂ *B. alvarius*) percentage metamorphosis was 52.7, 72.7, and 83.3 percent. One of these hybrids was tested and failed to fertilize eggs in an attempted backcross.

In crosses between *B. alvarius* and members of two groups (*americanus, punctatus*) there is metamorphosis of hybrids when *B. alvarius* is the male parent but not when it is the female parent. In crosses of female *B. alvarius* with males of five *americanus* group species (*B. americanus, B. terrestris, B. hemiophrys, B. microscaphus, B. woodhousei*) the hybrids failed as gastrulae or neurulae. In five reciprocal intergroup crosses involving females of *B. americanus* and *B. woodhousei,* the percentage metamorphosis of the hybrids ranged up to 80.0 percent.

In crosses of *B. alvarius* with *B. punctatus,* the hybrids failed as larvae when the *B. alvarius* was the female parent, but 4.2 percent metamorphosed in one of two crosses in which *B. alvarius* was the male parent. No compatibility between *B. alvarius* and the *debilis* group is indicated by the limited data. In a cross of ♀ *B. alvarius* x ♂ *B. debilis,* inviable larval monstrosities were formed; in the reciprocal cross, development stopped with gastrulation. In a cross of a female *B. alvarius* with a male *B. bocourti* inviable larvae were formed, and with male *B. coccifer* development ceased at gastrulation with exogastrulation. In two crosses of ♀ *B. occidentalis* x ♂ *B. alvarius,* the hybrids failed as larvae. Reciprocal crosses have not been attempted for any of these last three intergroup combinations.

In the only test of compatibility of *B. alvarius* with *B. canaliferus,* the hybrids failed at gastrulation in a cross of ♀ *B. canaliferus* x ♂ *B. alvarius.*

The *Cognatus* Group. In addition to its compatibility with the *marmoreus* group and with *B. alvarius,* the *cognatus* group shows variable reciprocal compatibility with the *boreas* group and lesser reciprocal compatibility with *B. canaliferus.* In four crosses of female *B. boreas* – two with *B. cognatus,* one with *B. compactilis,* and one with *B. speciosus* – there was percentage metamorphosis ranging from 1.3 percent of hatched larvae to 66.5 percent of fertilized eggs. However, in three crosses of ♀ *B. canorus* x ♂ *B. cognatus* and one of ♀ *B. canorus* x ♂ *B. speciosus* the hybrids failed at stages from gastrula to larva.

In eleven reciprocal intergroup crosses involving

three crosses of ♀ *B. cognatus* x ♂ *B. boreas* and a total of seven crosses of females of *B. speciosus* with *B. boreas, B. exsul, B. nelsoni,* and *B. canorus,* the hybrids failed as larvae in three and as gastrulae in one (where only 2% fertilization indicated poor eggs). In the others, metamorphosis of hybrids ranged up to 75.0 percent.

In two crosses of ♀ *B. cognatus* x ♂ *B. canaliferus* the embryos failed at neurula or earlier in one and showed 10.0 percent metamorphosis in the other. In three crosses, with female *B. speciosus* as parents, the hybrids failed as gastrulae in one and showed 16.6 and 31.7 percent metamorphosis in the others. In four reciprocal intergroup crosses, two involving *B. cognatus* showed failure as larvae and metamorphosis of 0.2 percent, while the two involving *B. speciosus* males showed 1.2 and 17.3 percent, respectively.

In intergroup combinations of the *cognatus* group and members of the *debilis* group and *B. bocourti* and *B. occidentalis,* the only metamorphosis occurred when the female parent was a member of the *cognatus* group. In two crosses of ♀ *B. cognatus* x ♂ *B. debilis,* 75.0 percent of the hatched larvae metamorphosed in one, and 9.7 percent of the fertilized eggs resulted in metamorphosed toads in the other. However, in five crosses between female *B. speciosus* and males of *B. debilis, B. kelloggi,* and *B. retiformis* development ceased at stages from gastrula to larva. In one reciprocal intergroup cross (♀ *B. debilis* x ♂ *B. speciosus*), the hybrids failed at gastrulation.

In three crosses of ♀ *B. cognatus* x ♂ *B. bocourti* metamorphosis ranged from 5.5 to 33.3 percent, and in two crosses in which the female parent was a *B. speciosus* it ranged from 1.5 to 11.3 percent. However, some individuals were unable to metamorphose in both combinations, some of the metamorphosed hybrids were pale in color, and in two instances some lacked forelegs. In one reciprocal intergroup cross (♀ *B. bocourti* x ♂ *B. cognatus*) the hybrids failed as small, slender larvae or at earlier stages.

In one cross of ♀ *B. speciosus* x ♂ *B. occidentalis* 12.9 percent of the fertilized eggs resulted in metamorphosed toads. In four reciprocal intergroup crosses (3 with *B. cognatus* males, one with *B. speciosus*) the hybrids failed as abnormal, nonfeeding larvae.

The situation on compatibility is directly reversed in intergroup crosses of the *cognatus* group with the *americanus* group and with *B. punctatus.* In fourteen crosses of female *B. cognatus* or *B. speciosus* with six members of the *americanus* group the embryos were lost as abnormal larvae or failed at earlier stages. In seventeen reciprocal intergroup crosses involving females of all *americanus*-group species except *B. houstonensis* and males of *B. cognatus, B. speciosus,* and *B. compactilis,* only the cross of ♀ *B. hemiophrys* x ♂ *B. speciosus* failed to metamorphose. The hybrids from this cross failed as small, abnormal larvae. In the other crosses, the percentage metamorphosis ranged from 2.3 to 78.6 percent.

Hybrid males that were tested by backcrossing proved sterile. One of two hybrids from the combination ♀ *B. americanus* x ♂ *B. cognatus* produced one abnormal, inviable larva in a natural backcross to a ♀ *B. terrestris,* but another failed to fertilize any eggs in a comparable backcross. In the backcross of the three hybrid males of the combination ♀ *B. woodhousei* x ♂ *B. cognatus* to *B. woodhousei,* two sets showed abnormal cleavage but no survival to gastrula, while the third showed no cleavage. In attempted backcrosses of one male of the combination ♀ *B. woodhousei* x ♂ *B. compactilis* to *B. woodhousei,* no eggs were fertilized.

In seven crosses between female *B. punctatus* and males of the *cognatus* group, there was 27.0 to 88.0 percent metamorphosis in the 6 crosses involving male *B. cognatus* or *B. speciosus.* In one cross of *B. compactilis,* only 3 of 112 cleaved, and these stopped at gastrulation. These last results are probably insignificant, because the low percentage fertilization reflects stale eggs. In 2 reciprocal intergroup crosses of female B. *cognatus* and one of *B. speciosus,* the hybrids failed at stages from gastrula to abnormal larva.

There was no compatibility in either reciprocal in crosses between the *cognatus* group and *B. coccifer.* The hybrids failed at gastrulation in one cross of a female *cognatus* and one of a *speciosus* with male *B. coccifer.* In two reciprocal intergroup crosses of ♀ *B. coccifer* x ♂ *B. cognatus,* the hybrids failed as nonfeeding larvae.

In two crosses of ♀ *B. speciosus* x ♂ *B. quercicus,* one set of hybrids failed as larvae, and in the other there was 14.5 percent metamorphosis. In the reciprocal cross the hybrids failed at gastrulation.

The *Boreas* Group. The *boreas* group, in addition to the previously mentioned reciprocal metamorphosis in crosses with the *cognatus* group and with *B. alvarius,* also shows high percentage metamorphosis in reciprocal crosses with *B. punctatus* and quite variable rates in reciprocal crosses with toads of the *americanus* group. In two crosses of ♀ *B. boreas* x ♂ *B. punctatus,* there was 20.5 and 66.7 percent metamorphosis respectively. In four reciprocal intergroup crosses involving males of *B. boreas, B. exsul,* and *B. nelsoni,* there was 45.7, 35.0, and, in two, 8.0 and 47.2 percent metamorphosis respectively. In the ♀ *B. punctatus* x ♂ *B. nelsoni* cross with only 8.0 percent metamorphosis, nine of the eleven individuals that metamorphosed did so without developing even vestiges of forelegs.

The crosses between toads of the *boreas* group and those of the *americanus* group yielded quite different results. In five crosses between females of four species (*B. woodhousei, B. terrestris, B. microscaphus,* and *B. hemiophrys*) and males of *B. boreas,* percentage metamorphosis ranged from 4.8 to 72.2 percent, and in one cross of ♀ *B. woodhousei* x ♂ *B. nelsoni,* it was 11.6 percent. Much lower compatibility is indicated for the reciprocal cross. In eight crosses between female *B. boreas* and all *americanus*-group males except *B. houstonensis,* six sets failed as larvae. In one of two in which the male parent was a *B. americanus,* there was 0.3 percent metamorphosis, and in one in which the male parent was a *B. terrestris,* there was 6.9 percent metamorphosis.

In two intergroup combinations, the hybrids metamorphose when the female parent is of the *boreas* group. There was 3.1 percent metamorphosis in a cross of ♀ *B. boreas* x ♂ *B. bocourti*; there was failure at gastrulation in one and 12.9 percent metamorphosis in the other of two crosses of ♀ *B. canorus* x ♂ *B. bocourti.* In two reciprocal intergroup crosses involving males of *B. boreas,* there was failure at gastrulation.

In three crosses of ♀ *B. boreas* x ♂ *B. debilis,* the hybrids failed as larvae in one, but 6.7 percent metamorphosed in one and 6.9 percent in the other. In one cross of ♀ *B. canorus* x ♂ *B. retiformis,* none metamorphosed, but one larva lived for thirty days and had only rudimentary hind legs when it died. The reciprocal intergroup cross has not been made.

In intergroup crosses of female *B. boreas* with male *B. coccifer,* 1.6 percent of the hybrids hatched in crosses in which the female parent was a *B. boreas.* In the reciprocal cross, the hybrids failed as gastrulae.

In reciprocal crosses of *B. boreas* and *B. canaliferus,* the hybrids in both reciprocals failed at stages from gastrula to abnormal larva. Similar results were obtained in reciprocal crosses between the *boreas* group (*B. boreas* and *B. canorus*) and *B. occidentalis.*

The *Punctatus* Group. *B. punctatus* shows metamorphosis of hybrids in both reciprocals of only the previously mentioned cross with the *boreas* group. In four intergroup combinations, including previously mentioned combinations with members of the *marmoreus* and *valliceps* group, as well as members of the *debilis* group and *B. bocourti,* metamorphosis of the hybrids occurred only when the female parent was a *B. punctatus.* In two replications of the cross of ♀ *B. punctatus* x ♂ *B. bocourti,* there was metamorphosis of 57.1 and 79.0 percent of the hybrids. One hybrid male that was tested by attempted backcross to a female *B. punctatus* had very small testes and fertilized no eggs. The reciprocal intergroup cross was attempted only once and resulted in failure at gastrulation. When female *B. punctatus* were crossed with males of the *debilis* group, metamorphosis was 25.6 percent with *B. debilis,* 28.4 percent with *B. kelloggi,* and 26.5 percent with *B. retiformis.* In the *B. debilis* combination, two other replications ended at gastrula or larva.

In addition to two previously mentioned intergroup combinations (*B. punctatus* x *cognatus* group; *B. punctatus* x *B. alvarius*) in which metamorphosis of the hybrid generation occurred only when the male parent was a *B. punctatus,* two other intergroup combinations showed similar results. In six attempted crosses between females of *B. punctatus* and males of all six species of the *americanus* group except *B. houstonensis,* development failed at gastrula to abnormal larva. In the reciprocal intergroup cross, however, metamorphosis ranged from 5.3 percent (*B. microscaphus*), 24.4 percent (*B. americanus*), to 35.3 percent (*B. hemiophrys*), to 46.1 and 53.8 percent (*B. woodhousei*).

In a cross of ♀ *B. coccifer* x ♂ *B. punctatus,* there was 33.3 percent metamorphosis, but in the reciprocal cross the hybrids failed as gastrulae.

B. punctatus shows incompatibility in both reciprocals of crosses with two North American species groups. In two attempted crosses of ♀ *B. punctatus* x ♂ *B. occidentalis* and in three replications of the attempted reciprocal cross, the hybrids failed as gastrulae or tailbud monsters.

In two attempted crosses of ♀ *B. punctatus* x ♂ *B. quercicus,* the hybrids failed at gastrulation, but in a third cross 4.5 percent of the 2n hybrids metamorphosed. In the reciprocal cross, the hybrids failed at gastrulation.

The *Americanus* Group. In addition to showing metamorphosis in both reciprocals in crosses previously mentioned with the *valliceps* and *boreas* groups, the *americanus* group also produces metamorphosing hybrids in reciprocal crosses with *B. coccifer*. The hybrids failed at gastrulation in a cross of a female *B. terrestris* and of a female *B. americanus* with a male *B. coccifer*, and they failed as dumpy, abnormal larvae in a cross of ♀ *B. hemiophrys* x ♂ *B. coccifer*. However, there was 29.3 percent metamorphosis in a cross of ♀ *B. woodhousei* x ♂ *B. coccifer*. In the reciprocal intergroup cross, crosses of female *B. coccifer* with males of *B. woodhousei*, *B. microscaphus*, *B. hemiophrys*, and *B. americanus* resulted in failure of the F_1 hybrids at gastrula to larva. However, there was 0.2 percent metamorphosis in one cross of ♀ *B. coccifer* x ♂ *B. terrestris* and failure at the larval stage in a second attempted cross.

The *americanus* group is notable for the ability of its females to produce viable hybrids in intergroup crosses and for the inability of its males to do so. In addition to the previously mentioned cases in which females of the *americanus* group produce hybrids with males of *B. punctatus*, the *cognatus* group, *B. alvarius*, and the *marmoreus* group, there is also metamorphosis when females of the *americanus* group are crossed with males of the *debilis* group, *B. bocourti*, *B. canaliferus*, and *B. quercicus*. In fourteen replications of reciprocal crosses of males of the six species of the *americanus* group with females of the *cognatus* group, there was hatching in four, but no metamorphosis. The same was true in one of three replications of the cross of female *B. quercicus* with males (*B. americanus* and *B. woodhousei*) of the *americanus* group. In all other cases the reciprocal hybrid failed before hatching.

Four crosses have been made between females of the *americanus* group and male *B. bocourti*. There was metamorphosis of 60.9 percent of the hybrids with *B. americanus*, 7.3 and 2.3 percent with two *B. woodhousei*, and 5.5 percent with a female *B. terrestris*. A hybrid male of the combination of ♀ *B. woodhousei* x ♂ *B. bocourti* was fertile and a backcross to *B. woodhousei* was accomplished. A male of the backcross generation was a somatic triploid but produced normal, haploid sperm (see Chap. 10) and fertilized eggs in a second backcross.

In reciprocal intergroup crosses, female *B. bocourti* has been crossed with male *B. woodhousei* (3 replications) and *B. terrestris* (once). The hybrid embryos failed at gastrulation, with exogastrulation.

Three females of the *americanus* group have been crossed with male *B. canaliferus*. With *B. terrestris*, there was 25.0 percent metamorphosis of the hybrids, with *B. woodhousei* there was 10.1 percent, but with *B. hemiophrys* the hybrids showed exogastrulation and failed at stages from gastrula to abnormal larva. The reciprocal intergroup cross has been attempted five times (*B. americanus*, twice; *B. terrestris*; *B. microscaphus*; *B. woodhousei*) and has always resulted in failure at gastrulation.

Compatibility is low between females of the *americanus* group and male *B. debilis*. There was only 8.3 percent metamorphosis with *B. hemiophrys*, 0.3 percent with *B. woodhousei*, and 3.2 percent of hatched larvae with *B. terrestris*. With females of *B. americanus* and *B. microscaphus*, the hybrids failed as larvae.

The reciprocal intergroup cross has been attempted eight times, with three crosses of female *B. debilis* with male *B. woodhousei*, two with *B. americanus* and one with *B. hemiophrys*, and two crosses of female *B. retiformis* with male *B. woodhousei*. All crosses ended at gastrulation or neurulation.

Four crosses of three *americanus*-group females with *B. quercicus* have been effected. There was failure at or before larva in one and 0.4 percent metamorphosis in the other of two crosses with *B. woodhousei*, 5.9 percent with *B. microscaphus*, and 25.0 percent with *B. terrestris*. In the reciprocal intergroup cross a few diploid larvae hatched but failed to metamorphose.

The available evidence points to reciprocal intergroup incompatibility between the *americanus* group and *B. occidentalis*. Hybrids from one cross of female *B. woodhousei* and one of female *B. hemiophrys* with males of *B. occidentalis* failed as larvae. In one instance a single larva survived but was unable to metamorphose.

The reciprocal intergroup cross has been attempted with five males (*B. americanus*, *B. hemiophrys*, *B. terrestris*, *B. microscaphus*, and *B. woodhousei*). All hybrids from these five crosses failed as gastrulae or neurulae.

The *Coccifer* Group. The data from hybridization do not give here the same clear indication of relationship that they do for most of the other New

World species. Aside from the previously mentioned low level of metamorphosis of reciprocal hybrids with the *americanus* group, in which the possibility of polyploidy has not been eliminated, *B. coccifer* has been shown to be unable to produce viable hybrids in both reciprocal crosses with other species group.

As previously mentioned, there is relatively low percentage metamorphosis of hybrids between female *B. coccifer* and males of the *valliceps* group, *marmoreus* group, and *B. punctatus.*

On the other hand, there is the previously mentioned low percentage metamorphosis when a male *B. coccifer* was crossed with a female *B. boreas.* There is comparable evidence for *B. canaliferus.* One cross of ♀ *B. coccifer* x ♂ *B. canaliferus* resulted in hybrids that failed as larvae. In three reciprocal crosses, two failed in the larval stage and the other showed 3.5 percent metamorphosis.

Inviability of the hybrids in both reciprocals of the cross between *B. coccifer* and the *cognatus* group has been previously mentioned. A similar situation exists with *B. bocourti,* in one cross of ♀ *B. coccifer* x ♂ *B. bocourti* the hybrids failed as larvae; in three replications of the reciprocal cross they failed as gastrulae.

In the previously mentioned cross of ♀ *B. alvarius* x ♂ *B. coccifer,* the embryos failed as gastrulae, but no reciprocal cross was made. In a cross of ♀ *B. coccifer* x ♂ *B. debilis* the hybrids failed as larvae, but no reciprocal cross has been made.

In two attempted crosses of ♀ *B. occidentalis* x ♂ *B. coccifer* the hybrids failed as gastrulae or monsters at the neurula and post neurula stages, but no reciprocal cross has been made. No crosses of *B. coccifer* with *B. quercicus* have been made.

The *Canaliferus* Group. There is metamorphosis in reciprocal crosses of *B. canaliferus* only in the previously mentioned crosses with toads of the *valliceps* and *cognatus groups.* As mentioned earlier, there is metamorphosis of hybrid larvae in the cross of ♀ *B. canaliferus* x ♂ *B. coccifer* but failure in the reciprocal cross. There is metamorphosis of hybrid larvae in crosses of male *B. canaliferus* with females of the *americanus* and *marmoreus* groups but failure in the reciprocal crosses. There is reciprocal intergroup incompatibility with the *boreas* and *punctatus* groups. Hybrids between female *B. canaliferus* and male *B. alvarius* fail as gastrulae, but the reciprocal cross has not been made.

In intergroup crosses of female *B. canaliferus* (with ♂ *B. bocourti* and ♂ *B. debilis*) the hybrids failed as abnormal, nonfeeding larvae, but no reciprocal crosses were made. Four crosses of female *B. canaliferus* with males of three species of the *debilis* group have been attempted. In a cross with *B. debilis* 0.7 percent of the hatched larvae metamorphosed, but in one cross with *B. kelloggi* and two with *B. retiformis* the hybrids failed at stages from gastrula to neurula. There is no reciprocal cross. In one cross of ♀ *B. occidentalis* x ♂ *B. canaliferus* the hybrids failed as larvae. No reciprocal cross has been made.

The *Occidentalis* Group. *B. occidentalis* shows metamorphosis in both reciprocals of the cross only in the earlier mentioned cross with *B. valliceps.* There is metamorphosis in the previously mentioned cross of ♀ *B. cognatus* x ♂ *B. occidentalis,* but the reciprocal hybrids are inviable.

There is incompatibility in both reciprocals of crosses of ♀ *occidentalis* x ♂ *marmoreus* group, *boreas* group, *americanus* group, and *B. punctatus.* As previously mentioned, crosses of ♀ *B. occidentalis* x ♂ *B. coccifer, B. canaliferus, B. alvarius,* and *B. debilis* resulted in hybrid larvae that died at stages from gastrula to larva, but no reciprocal crosses were made. In one additional cross of ♀ *B. occidentalis* x ♂ *B. quercicus* there was failure at gastrulation. The reciprocal cross has not been made. No crosses have been made between *B. occidentalis* and *B. canaliferus* and *B. bocourti.*

The *Bocourti* Group. *B. bocourti* forms viable hybrids in reciprocal crosses with none of the other North American groups with which it has been tested. As previously mentioned, there is metamorphosis of the F_1 hybrid when the male *B. bocourti* is crossed with females of the *valliceps, marmoreus, cognatus, boreas, punctatus,* and *americanus* groups. The reciprocal hybrids are inviable. There is failure at the larval stage in crosses of male *B. bocourti* with *B. canaliferus* and *B. alvarius.* No crosses have been attempted with *B. occidentalis,* with *B. quercicus,* or with members of the *debilis* group.

The *Debilis* Group. The *debilis* group has been shown to produce metamorphosing hybrids in reciprocal crosses with no other group. There is metamorphosis in crosses between females of *B. americanus, B. cognatus, B. punctatus, B. boreas,* and *B. marmoreus* and males of the *debilis* group. There was no metamorphosis in earlier mentioned crosses between females of *B. coccifer, B. canaliferus,* and *B. occidentalis* and male *B. debilis,* with the hybrids failing as gastrulae or larval monsters. No crosses with *B. quercicus* have been made. As mentioned earlier, there is incompatibility in both reciprocals of crosses between the *debilis* group and the *valliceps* and *alvarius* groups.

The *Quercicus* Group. The hybridization data do not indicate clearly the evolutionary position of the tiny *B. quercicus*. This is true in part because it has not yet been possible to cross females of this species with representatives of more than a few other groups. Crosses of males of this group do give some clues. The only group with which the males of *B. quercicus* have been shown to be able to form viable F_1 hybrids are the *americanus* and *cognatus* groups, there being relatively low percentage metamorphosis in both instances. Failure of the hybrids comes in the larval stage in crosses with the *valliceps* group and with *B. canaliferus*. Failure comes at gastrula in crosses with the *marmoreus* group and with *B. punctatus*, *B. alvarius*, and *B. occidentalis*. Hybrids between female *B. quercicus* and males of the *cognatus* and *valliceps* groups failed at gastrulation. In the crosses with males of two species of the *americanus* group there was failure at gastrulation with *B. woodhousei*; although a few apparently normal diploid larvae with *B. americanus* hatched, these died as larvae.

Main Features of the North American Radiation

The main features of the North and Central American radiation of *Bufo* may be summed up as follows:

1. There is great diversity, as indicated by the eleven species groups discussed above plus such representatives of mainly South American groups as *B. marinus* (*marinus* group), *B. haematiticus* (*guttatus* group), *B. holdridgei* (possibly *guttatus* group), and *B. periglenes* of presently unknown affinity.

2. The number of species in North and Central America is roughly 50 percent greater than the number in South America. Both the greater number of species and species groups in North America seem attributable to the greater area, to the greater ecological diversity, and to the more pronounced effects of Pleistocene and probably earlier climatic change of the northern continent over the southern.

3. The *valliceps* group is a major tropical and subtropical element in the North American radiation. *B. canaliferus* appears to be an offshoot of this stock that has evolved to small size and that remains capable of producing viable but sterile offspring with the species of the parent group. *B. occidentalis* is seemingly an offshoot of the *valliceps* group that has invaded the Mexican Plateau.

4. A series of species groups in Mexico and the western United States is interconnected by the ability to form viable F_1 hybrids in both reciprocals of intergroup crosses. These form a sequence from the *valliceps* group to the *marmoreus* group and *B. alvarius* and from the *valliceps* group to the *cognatus* to the *boreas* group to *B. punctatus*.

5. The *americanus* group, with high intragroup genetic compatibility, is set off markedly from the other groups by its high incompatibility with them.

6. The *coccifer* group does not seem close to any other group and possibly represents an early separation from the part of the *Bufo* radiation that led to the *americanus* group. *B. quercicus* may have had a similar but separate history.

7. The *debilis* group represents a rather specialized group of small toads with closest affinity to the complex of western U.S. species groups represented by the *cognatus*, *boreas*, and *punctatus* groups. *B. bocourti* of Guatemala is an isolate with apparently similar but quite independent affinities.

Eurasian Radiation

The Eurasian radiation is more poorly represented among the species that we have had available for hybridization than that of any other landmass. However, the available species are of great interest for the spread of the genus around the world.

Three of the species are tropical in distribution. These are *B. stomaticus*, *B. melanostictus*, and *B. asper*. All are common toads of southeastern Asia.

The *Asper* Group. There is no evidence from the hybridization data to indicate close relationship of this broad-skulled toad to any other Eurasian species. One cross of ♀ *B. asper* x ♂ *B. melanostictus* resulted in failure at gastrulation. One cross of ♀ *B. asper* x ♂ *B. bufo* had a similar result. One reciprocal intergroup cross of ♀ *B. melanostictus* x ♂ *B. asper* resulted in 0.9 percent metamorphosis, but the only hybrid examined cytologically was a triploid (Chap. 10). Two crosses of ♀ *B. stomaticus* x ♂ *B. asper* resulted in failure at gastrulation in one instance and metamorphosis of 5.2 percent in the other. However, the one hybrid sampled cytologically was a triploid. There is, thus, no evidence of close relationship to any other Eurasian group with which we have been able to work.

The *Melanostictus* Group. *B. melanostictus*, a broad-skulled species of southeastern Asia, shows no evidence of close affinity with any other Eurasian species available to us. In two crosses of ♀ *B. viridis* x ♂ *B. melanostictus* and one of ♀ *B. stomaticus* x ♂ *B. melanostictus* the hybrids failed as gastrulae or neurulae.

The *Stomaticus* Group. *B. stomaticus* shows little affinity with other Eurasian species. In crosses of ♀ *B. stomaticus* x ♂ *B. viridis*, *B. bufo*, and *B. calamita*, the hybrids failed as larvae. In a cross of ♀ *B. stomaticus* x ♂ *B. melanostictus* there was failure at gastrulation. In two crosses of ♀ *B. stomaticus* x ♂ *B. asper*, as previously mentioned, there was failure as gastrulae in one cross and 5.2 percent metamorphosis in the other.

In a reciprocal intergroup cross, there was high (68.9%) metamorphosis in a cross of ♀ *B. viridis* x ♂ *B. stomaticus*. In a second reciprocal intergroup cross, there was failure at stages from gastrula to abnormal neurulae in two crosses of ♀ *B. melanostictus* x ♂ *B. stomaticus*.

The *Viridis* Group. *B. viridis* is the only one of the three temperate-zone narrow-skulled Eurasian species with which we have been able to make an adequate number of intergroup crosses. Females of *B. viridis* have been crossed at least once with all of the five other Eurasian species we have had available to us.

In two crosses of ♀ *B. viridis* x ♂ *B. calamita* there was failure at larva, with exogastrulation, in one and 5.1 percent metamorphosis in the other. In a pair of similar crosses with a male *B. bufo*, there was failure at the larval stage in one and 6.1 percent metamorphosis in the other. The highest percentage metamorphosis (68.9%) occurred in a cross with a male *B. stomaticus*. However, the reciprocal hybrids, in one cross, failed as larvae. Flindt and Hemmer (1967) reported metamorphosis in crosses between female *B. viridis* and male *B. calamita*, but the reciprocal hybrids died as larvae.

Reciprocal hybrids between *B. viridis* and *B. melanostictus* are inviable. In two replications with female *B. viridis* the hybrids failed at gastrulation, and in one reciprocal cross they failed at gastrulation or very abnormal neurulae.

In one cross of ♀ *B. viridis* x ♂ *B. asper*, the hybrids failed as exogastrulated gastrulae.

The *Bufo* Group. Males of *B. bufo*, as mentioned earlier, have been crossed with females of four Eurasian species, *B. viridis*, *B. stomaticus*, *B. melanostictus*, and *B. asper*. Only with the first of these is there even low compatibility (6.1% metamorphosis). In the others, the hybrids failed as larvae (*stomaticus*) or gastrulae.

The *Calamita* Group. *B. calamita* males have been crossed with other Eurasian species only in the three crosses, mentioned earlier, with *B. viridis* (larval failure and 5.1% metamorphosis) and *B. stomaticus* (larval failure).

Some Features of the Eurasian Relations

Difficulties of obtaining Asian material makes it impossible to treat the evolutionary relations of the Eurasian species in the same way that we can treat them for the *Bufo* of other continents. However, some important clues toward understanding worldwide relations of the genus are provided in the available evidence.

1. The three thin-skulled species, *B. bufo*, *B. viridis*, and *B. calamita*, the only species of *Bufo* to reach Europe, appear widely divergent from one another on the basis of the hybridization data, and each would fit our concept of a separate species group.
2. Thin-skulled *B. stomaticus* of southeastern Asia shows its greatest affinity with the other thin-skulled species, particularly with *B. viridis* rather than with the two broad-skulled southeastern Asian species with which it has been tested.
3. It seems very significant that, while the sample of six Eurasian species is small, these six species are quite divergent from one another and show no evidence of the annectant forms that are such an important feature of the North American and South American toads.

African Radiation

Africa shows greater diversity of *Bufo* than South America but less than the North American continent. We have had representation of at least some individuals of ten of the probable species groups. The African toads divide naturally on the basis of chromosome number (Chap. 10). The numerous broad-skulled species of the *regularis* group as well as *B. latifrons* and *B. perreti* have a 2n of 20; all other species groups examined share the 2n of 22 with all other *Bufo* for which the chromosome number has been determined.

The 22-chromosome toads for which we have hybridization data include the broad-skulled *B. mauritanicus* and the narrow-skulled *B. lönnbergi*, *B. lughensis*, *B. vertebralis*, *B. carens*, and the three

Table 11–5. Stage Reached or Percentage Metamorphosis in Crosses between Eastern and Western Cryptic Species Included under the Name *"Regularis"*

♀	♂																
	Western							Eastern									
	El Mahalla	Kampala	Entebbe	Cape coast	Ibadan	Ogbomosho	Giza	Port St. Johns	Mazeras	Umtali	Nyeri	Ol Tukai	Vumba Mts.	Kabete	Louis Trichardt	Pietermaritzburg	Lourenço Marques
Western																	
El Mahalla	89.3	85.3				36.0		0.5	L	L	2.8	L	2.6				1.0
Kampala			92.9			89.1		5.8							L	L	
Entebbe																	
Cape coast				35.7								1.0					
Ibadan														5.8			
Ogbomosho																	
Maadi							38.2										
Eastern																	
Port St. Johns			16.5					11.7			16.1		27.3			2.1	
Mazeras						91.8		G			40.1						
Umtali						58.7									21.7		
Nyeri																	

species, *B. gariepensis*, *B. inyange*, and *B. rosei* of the *gariepensis* group.

Hybridization within the *Regularis* Group

The *regularis* group is the most widely distributed and most important group of African *Bufo*. The exact number of species is still in question, but there is little doubt that this is one of the largest species groups of the genus *Bufo*. Nine species have been available for hybridization, some of them from several widely separated localities. Others remain to be studied.

One of the most interesting results has been the revelation of the existence of cryptic species that replace one another allopatrically and that have masqueraded under the name *B. regularis*. The break between the two comes at Lake Victoria and the Rift Valley. The results of crosses between these two are shown in table 11-5. There is a high degree of incompatibility between the western females and eastern males. Twenty-two crosses were attempted between the western females and eastern males. In no instance did more than 5.8 percent of the hybrids metamorphose, and in ten of the twenty-two crosses the hybrids did not survive beyond the larval stage. By contrast, there is high compatibility in crosses between eastern females and western males and in crosses of western females and western males from different localities or between eastern females and eastern males. The only exceptions involved toads from the isolated Port St. Johns, South Africa, population, which appears to be a hybrid swarm (Chap. 10).

Even though very few hybrids between western females and eastern males were viable, backcrosses of two of the hybrid males to western females resulted in some metamorphosis (Table 11-6). Both males fertilized a relatively high percentage of the eggs and a high percentage of the larvae hatched. However, both had high mortality as larvae, with only 12.3 and 7.6 percent of the fertilized eggs resulting in metamorphosed larvae. Cytological evidence (Chap. 10) indicates strong disharmony between the genotypes of these cryptic species. One female from a cross between stocks from two western localities (El Mahalla el Kubra, Egypt, and Kampala, Uganda) was backcrossed to a male from the Cape coast. There was relatively high percentage fertilization (59.2%) and hatching (59.4%), but the offspring died as larvae. In another instance a female from the cross of a

Table 11–6. Results of Backcrosses in the *Regularis* Group

Stock No.	Hybrid	Species to Which Backcrossed	Sex	% Fert.	% Hatch	No. Larvae	% Met.
66-95	♀ *reg.* (W) El Mahalla x ♂ *reg.* (E) Lourenço Marques	*reg.* (W) El Mahalla	♂	87.4	100.0	203	12.3
66-93	♀ *reg.* (W) El Mahalla x ♂ *reg.* (E) Vumba Mts.	*reg.* (W) El Mahalla	♂	55.6	77.0	157	7.6
66-81	♀ *reg.* (W) El Mahalla x ♂ *reg.* (W) Kampala	*reg.* (W) Cape coast	♀	59.2	59.4	19	—
65-437	♀ *garmani* x ♂ *reg.* (W) Ogbomosho	*reg.* (W) Ogbomosho	♂	—	—	—	—
66-123	♀ *garmani* x ♂ *reg.* (W) Kampala	*reg.* (W) F_1 El Mahalla-Kampala	♂	—	—	—	—
66-108	♀ *garmani* x ♂ *reg.* (W) El Mahalla	*reg.* (W) F_1 El Mahalla-Kampala	♂	—	—	—	—
66-108	♀ *garmani* x ♂ *reg.* (W) El Mahalla	*reg.* (W) F_1 El Mahalla-Kampala	♂	—	—	—	—
66-108	♀ *garmani* x ♂ *reg.* (W) El Mahalla	*reg.* (W) F_1 El Mahalla-Kampala	♂	—	—	—	—
66-97	♀ *reg.* (W) El Mahalla x ♂ *garmani*	*reg.* (W) F_1 El Mahalla-Kampala	♂	no gonad	—	—	—
66-97	♀ *reg.* (W) El Mahalla x ♂ *garmani*	*reg.* (W) F_1 El Mahalla-Kampala	♂	—	—	—	—
65-429	♀ *garmani* x ♂ *reg.* (E) Pietermaritzburg	*reg.* (W) El Mahalla	♂	36.7	—	2	—
65-429	♀ *garmani* x ♂ *reg.* (E) Pietermaritzburg	*reg.* (W) El Mahalla	♂	2.0	—	—	—
66-125	♀ *garmani* x ♂ *reg.* (E) Vumba Mts.	*reg.* (W) El Mahalla	♂	—	—	—	—
65-379	♀ *rangeri* x ♂ *reg.* (W) Ogbomosho	*reg.* (W) El Mahalla	♂	10.2	—	—	—
65-456	♀ *brauni* x ♂ *reg.* (W) Ogbomosho	*reg.* (W) F_1 El Mahalla-Kampala	♂	—	—	—	—
65-456	♀ *brauni* x ♂ *reg.* (W) Ogbomosho	*reg.* (W) El Mahalla	♂	16.4	5.5	20	10.0
65-517	♀ *brauni* x ♂ *reg.* (E) Nyeri	F_1 (*brauni* x *garmani*)	♂	—	—	—	—
65-475	♀ *brauni* x ♂ *garmani*	*garmani* (67-49)	♂	37.5	84.6	85	—

Table 11–7. Best Results in Crosses among the African 20-Chromosome Toads

♀	♂										
	regularis (W)	*regularis* (E)	*sp.*	*garmani*	*rangeri*	*kisoloensis*	*brauni*	*kerinyagae*	*pardalis*	*maculatus*	*perreti*
regularis (W)		5.8	32.3	58.0	55.4	L	13.1	3.6		L	
regularis (E)	91.8		L	74.2	23.3		56.2			L	0.4
sp.	36.5	64.8									
garmani	80.2	81.0	L		81.6		34.2	70.1		L	1.8
rangeri	84.5									L	L
kisoloensis											
brauni	55.6 F♂	77.4 S♂	G	77.8	37.5					L	
kerinyagae					17.8					L	
pardalis											
maculatus	L	L	L	L	L		L	L	L		L
perreti											
steindachneri	L									L	

female from El Mahalla el Kubra, Egypt, with a male from Kampala, Uganda, was crossed with a male from the cross of a female from Kampala with a male from Ogbomosho, Nigeria. A fairly high percentage (58.8%) of the eggs were fertilized but only 32.1% of the fertilized eggs hatched and these died as larvae. These results from the crossing of individuals from different localities within the range of what we are calling the western species suggest that the relationships within what has been called *B. regularis* are even more complex than we thought. In other words, the "western species" may involve additional cryptic species.

B. garmani is a relatively distinct form that occurs sympatrically with the species that appear under the name *B. regularis*. It is of chief interest within the *regularis* group because of its high compatibility with other species. Females of *B. garmani* have been crossed with males of the eastern *B. regularis* type from eight localities (Table 11-7; Appendix H), and survival of the F_1 hybrids was very high (65.2 to 84.0%). One hybrid male from the combination ♀ *B. garmani* x ♂ *B. regularis* (E) from the Vumba Mountains, Rhodesia, and two from the combination ♀ *B. garmani* x ♂ *B. regularis* (E) from Pietermaritzburg, South Africa, were crossed with a female *B. regularis* (W) in the absence of any opportunity to backcross to the parental species. The male from the first combination fertilized no eggs, while one from the second fertilized 2.0 percent of the eggs, with subsequent failure at gastrulation. The second of these males fertilized 36.7 percent of the eggs, but the hybrids failed as larvae or earlier.

Males of *B. garmani* have been crossed with females of *B. regularis* (E) from two localities (Appendix H). Metamorphosis reached 28.1 percent in one set and 74.2 percent in the other. The ability of *B. garmani* to produce large numbers of F_1 hybrids in both reciprocals of crosses with *B. regularis* (E)[1] is self-evident, but the lack of the proper backcross leaves the question of possible introgression in doubt.

Females of *B. garmani* have been crossed with males of *B. regularis* (W) from three widely separated populations: Kampala, Uganda; Ogbomosho, Nigeria; and El Mahalla el Kubra, Egypt. The F_1 hybrids were highly viable, with 80.2 to 90.4 percent metamorphosing. However, four of the hybrids were shown to be sterile when they fertilized no eggs in attempted backcrosses.

In one reciprocal cross, there was 58 percent metamorphosis. Two hybrids that showed male secondary sexual characters were found to be without identifiable gonads. The evidence thus points to sterility of males at least in both reciprocals of the cross between *B. garmani* and *B. regularis* (W).

The cross of ♀ *B. garmani* x ♂ *B. rangeri* yields highly viable hybrids (81.6% metamorphosis). The

[1] The East African species will be referred to as *B. regularis* (E) and the western species as *B. regularis* (W) in this book.

IV. The North American *americanus* group of narrow-skulled toads, including: (1) *B. houstonensis*, (2) *B. terrestris*, (3) *B. americanus*, (4) *B. hemiophrys*, (5) *B. microscaphus*, and (6) *B. woodhousei*; (7) the related *B. coccifer*; the two species (8) *B. marmoreus* and (9) *B. perplexus* of the *marmoreus* group; (10) *B. punctatus*; the two species (11) *B. debilis* and (12) *B. retiformis* of the *debilis* group; and (13) *B. quercicus*.

hybrids from ♀ *B. garmani* x ♂ *B. brauni* showed 34 percent metamorphosis, and in two reciprocal crosses 61.1 and 77.8 percent of the hybrids metamorphosed. One of the latter hybrid males fertilized 37.5 percent of the eggs in a backcross to *B. garmani,* and 84.6 percent of the embryos hatched. However, these failed as larvae. An F_2 generation gave similar results, with 52.3 percent fertilization, 88.2 percent hatch, and failure as larvae.

One cross was made between a female *B. garmani* and a male *Bufo* sp. There was less than 0.05 percent fertilization, and the small stock of hybrids died as larvae.

High viability of hybrids was evident in a cross of ♀ *B. garmani* x ♂ *B. kerinyagae,* with 70.1 percent of the fertilized eggs resulting in metamorphosed toads.

B. rangeri, which occurs sympatrically with *B. regularis* (E), showed relatively low compatibility in the cross of males with female *B. regularis* (E) from Umtali, Rhodesia, and Mazeras, Kenya. In one set only 23.3 percent of the hybrids metamorphosed, and in the other only 13.2 percent did so.

There is also evidence of incompatibility in hybrids between male *B. rangeri* and females of *B. regularis* (W). There was metamorphosis of 55.4 percent of the hybrids when the female parent was from Kampala, Uganda, but there was failure at the larval stage with a female from Ibadan, Nigeria, and at gastrulation with a female from El Mahalla el Kubra, Egypt. In the reciprocal cross, however, which was made twice with males of *B. regularis* (W) from Ogbomosho, Nigeria, 63.3 and 84.5 percent of the hybrids metamorphosed. In one backcross to *B. regularis* (W) from El Mahalla el Kubra, the hybrid male fertilized 10.2 percent of the eggs, but none of the backcross hybrids hatched.

Viability of the F_1 hybrids is only moderate in crosses of male *B. rangeri* and female *B. brauni,* with 28.7 percent metamorphosis in one set and 37.5 percent in another. It is also low in a cross of male *B. rangeri* with a female *B. kerinyagae,* with only 17.8 percent metamorphosis.

In addition to the crosses mentioned above, there has been one cross of female *B. brauni* with a male *Bufo* sp. from Ol Tukai, Kenya. There was only 1.9 percent fertilization, and the embryos failed as gastrulae.

In a cross of ♀ *B. brauni* x ♂ *B. regularis* (W) from Ogbomosho, Nigeria, there was 55.6 percent metamorphosis. Two of the hybrid males were tested in backcrosses to *B. regularis* (W). One fertilized no eggs, but the other fertilized 16.4 percent, and 5.5 percent of the embryos hatched. Ten percent of the backcross hybrids metamorphosed, but all those examined were triploids (Chap. 10). In the reciprocal species cross, involving a male from El Mahalla el Kubra, only 13.1 percent of the F_1 hybrids metamorphosed.

In a cross of ♀ *B. brauni* x ♂ *B. regularis* (E) from Nyeri, Kenya, there was 74.7 percent metamorphosis. One hybrid male was tested for fertility in a cross of an F_1 (*B. brauni* x *B. garmani*) hybrid. A few eggs were fertilized and eight larvae hatched but failed to metamorphose. In the reciprocal cross, 56.2 percent of the fertilized eggs resulted in metamorphosed hybrids.

In a cross between a female *B. regularis* (W) from Uganda and a male *B. kisoloensis* from Kenya there was high (91.5%) fertilization and high (81.5%) hatch of the hybrid larvae, but these remained small and were never able to metamorphose.

Genetic compatibility among members of the *regularis* group may be summed by the statement that highly viable but generally sterile hybrids result from interspecific matings within the group. This is a situation under which strong pressures for the perfection of systems of premating isolating mechanisms would be expected.

The *Maculatus* Group. *B. maculatus* is widely sympatric with the cryptic species that have been confused under the name *B. regularis* and has been confused with those species in spite of its much smaller size, different male throat color, and different mating call.

All evidence from hybridization points to complete genetic incompatibility between *B. maculatus* and the other 20-chromosome toads. Female *B. maculatus* have been crossed with male *B. regularis* (E) from three localities, with male *B. regularis* (W) from two localities, and with males of *B. garmani, B. rangeri, B. brauni, B. pardalis,* and *B. kerinyagae* of the *regularis* group and with *B. perreti.* In every instance at least some of the embryos reached the larval stage but failed to proceed to metamorphosis.

Reciprocally, male *B. maculatus* were crossed with females of *B. regularis* (E) from three localities, *B. regularis* (W) from one locality, and with *B. garmani, B. rangeri,* and *B. brauni* (twice). The results were the same as in the reciprocal crosses, except that in one cross (with *B. regularis* (E) from Mazeras, Kenya) the embryos failed as gastrulae. From the results of these crosses it seems unlikely that any introgression occurs in nature between *B. maculatus* and sympatric species of 20-chromosome toads.

The *Perreti* Group. Males of the specialized *B. perreti* have been crossed with females of four other species of 20-chromosome toads. The embryos failed as larvae in crosses with *B. maculatus* and *B. rangeri.* However, 0.4 percent of the hybrids metamorphosed in a cross with *B. regularis* (E) from Umtali, Rhodesia, and 1.8 percent metamorphosed in a cross with *B. garmani.* A female *B. perreti* was crossed with two male *B. regularis* (W) from Giza and Maadi, Egypt, and with two male *B. maculatus* from Monrovia and Lamco, Liberia. The hybrids failed as exogastrulated neuralae or earlier in all crosses, while all fertilized eggs hatched in the control. Thus, the morphologically quite differentiated *B. perreti* retains only slightly greater compatibility with toads of the *regularis* group than does the morphologically little differentiated *B. maculatus.*

B. steindachneri. This 20-chromosome species is regarded by Tandy (Chap. 9) as a member of a "*Bufo funereus–steindachneri–vittatus* complex". In one cross of a female *B. steindachneri* with a male *B. regularis* (W) and one with a male *B. maculatus,* the hybrids failed as larvae.

Compatibility among African Groups

With one seemingly significant exception the 20-chromosome toads have shown incompatibility in all attempted crosses with the 22-chromosome African toads. In one cross of ♀ *B. maculatus* x ♂ *B. mauritanicus* there was <0.1% metamorphosis, and in two crosses of ♀ *regularis* (W) x ♂ *B. mauritanicus* there was 3.3 and 14.2 percent metamorphosis.

Females of *B. regularis* (E), *B. garmani, B. rangeri,* and *B. brauni* were crossed a total of five times with males of *B. carens,* and the hybrid embryos consistently failed at stages from gastrula to grotesque neurula. Reciprocally, males of *B. regularis* (W), *B. regularis* (E), *B. rangeri, B. maculatus,* and *B. perreti* were crossed a total of seven times with females of *B. carens,* and the resultant hybrids also failed as gastrulae or abnormal neurulae.

One female *B. regularis* (W) was crossed with a male *B. vertebralis,* and the hybrids failed at gastrulation. One cross to *B. lönnbergi* was made with a female *B. regularis* (W), five with *B. regularis* (E), and three with *B. brauni.* In the first two combinations the embyros failed as grossly abnormal neurulae, and in the third at gastrulation. A cross was made reciprocally between *B. regularis* (W) and *B. lughensis,* and in both combinations the hybrids failed at gastrulation.

Attempted crosses between females of the 20-chromosome toads and males of the *gariepensis* group failed to show evidence of cleavage or failed prior to gastrulation, presumably as the result of an inhibitory substance in the greatly hypertrophied testes. One cross was attempted between female *B. regularis* (E) and male *B. gariepensis,* one between female *B. regularis* (E) and two between female *B. brauni* and males of *B. inyangae.* One each was attempted between females of *B. regularis* (E), *B. rangeri,* and *B. brauni* and males of *B. rosei.* Gravid females of the *gariepensis*-group species have not been available.

The 22-Chromosome Species

Both broad-skulled and narrow-skulled species are among the African toads with the presumably primitive chromosomal number of $2n = 22$ (Chap. 4). *B. mauritanicus* is of the broad-skulled type. *B. carens, B. vertebralis, B. luqhensis, B. lönnbergi,* and the South African toads of the *gariepensis* group are of the narrow-skulled type. No chromosomal count is available for the broad-skulled *B. superciliaris.*

The *Mauritanicus* Group. *B. mauritanicus* of Morocco has no near relative among the other 22-chromosome toads of Africa. Aside from the earlier mentioned crosses of male *B. mauritanicus* with females of the *regularis* and *maculatus* groups, there has not been opportunity to cross this species with any other African toad.

The *Superciliaris* Group. This large broad-skulled toad of western Africa shows external similarity with the large broad-skulled *B. blombergi* of Colombia, and, as in that species, the testes are hypertrophied (Noble 1924). We have had only a few live females of *B. superciliaris,* and no crosses have been made.

The *Vertebralis* Group. Aside from the earlier mentioned cross with the *regularis* group, the only attempted cross of *B. vertebralis* with another African species involved a female *B. vertebralis* and a male *B. gariepensis.* Development of the hybrids ceased at or before gastrulation, presumably as a result of an inhibitory compound in the testes. One cross between a female *B. luqhensis* and a male *B. gariepensis* yielded similar results.

The *Lönnbergi* Group. No crosses between *B. lönnbergi* and other African toads were made, other than the previously mentioned ones with toads of the *regularis* group.

Table 11–8. Summary of the Number of Intergroup Crosses of *Bufo*

♀ Species Group	Number Species	Groups	♂ South America No.	South America No. Spec.	South America No. Spec. Gr.	North America No.	North America No. Spec.	North America No. Spec. Gr.	Eurasia No.	Eurasia No. Spec.	Eurasia No. Spec. Gr.	Africa No.	Africa No. Spec.	Africa No. Spec. Gr.
spinulosus	4		8	5	3	49	21	11	8	5	5	5	3	2
granulosus	3		10	7	4	21	13	10	8	4	4	1	1	1
marinus	5		19	8	4	85	25	13	15	6	6	22	9	5
crucifer	1		5	4	3	17	14	19	4	4	4	—	—	—
typhonius	1		3	2	1	7	6	4	—	—	—	1	1	1
South America	14	5	45	14	5	179	25	13	35	6	6	28	9	5
bocourti	1		6	4	4	15	8	7	6	6	6	2	2	1
boreas	2		10	8	4	38	21	10	13	6	6	3	3	3
marmoreus	2		15	5	4	71	15	12	9	6	6	7	5	5
punctatus	1		9	6	4	52	28	12	11	5	5	4	3	2
cognatus	3		12	7	4	74	27	13	8	6	6	8	6	5
debilis	3		1	1	1	17	11	8	—	—	—	—	—	—
alvarius	1		6	5	4	31	19	10	4	3	3	3	3	3
valliceps	7		33	12	6	93	24	13	20	5	5	11	6	4
occidentalis	1		7	6	5	35	23	10	3	3	3	1	1	1
canaliferus	1		8	3	2	37	23	11	4	4	4	1	1	1
coccifer	1		5	4	4	22	17	9	6	5	5	1	1	1
americanus	6		32	12	6	103	20	12	43	6	6	15	6	2
quercicus	1		1	1	1	7	6	4	—	—	—	—	—	—
North America	30	13	145	18	8	595	31	12	127	6	6	56	11	6
viridis	1		13	7	5	35	21	13	9	5	5	14	9	6
calamita	1		—	—	—	1	1	1	—	—	—	—	—	—
stomaticus	1		7	6	4	24	18	12	6	5	5	1	1	1
asper	1		3	3	3	16	16	10	2	2	2	—	—	—
melanostictus	1		9	5	3	34	19	12	5	4	4	1	1	1
Eurasia	5	5	32	9	5	110	26	14	22	6	6	16	9	5
vertebralis	1		2	2	1	9	7	7	3	2	2	2	2	2
lughensis	1		1	1	1	2	2	2	—	—	—	2	2	2
carens	1		6	5	4	29	15	11	9	6	6	9	6	4
regularis	7		34	9	6	62	23	15	18	6	6	33	8	6
maculatus	1		13	8	4	32	21	13	5	3	3	6	8	3
steindachneri	1		1	1	1	2	2	2	—	—	—	2	2	2
perreti	1		—	—	—	—	—	—	—	—	—	4	2	2
Africa	11	7	56	12	6	128	27	14	32	6	6	52	14	7

The *Carens* Group. Two crosses were attempted between female *B. carens* and male *B. gariepensis.* There was no cleavage, presumably as a result of chemical inhibition.

Main Features of the African *Bufo*

1. The only known deviation of *Bufo* from the chromosomal number of 2n = 22 is found in Africa, where toads with 2n = 20 have undergone extensive speciation.

2. Genetic incompatibility at the level of sterility of F_1 hybrids characterizes the species of the *regularis* group, and two 20-chromosome species, *B. maculatus* and *B. perreti,* have developed essentially complete genetic isolation from the other 20-chromosome toads.

Table 11–9. Best Results of Crosses between South American Species Groups of *Bufo* and Species Groups from other Continents

South American / Other	North American														Eurasian						African							
	bocourti	*boreas*	*alvarius*	*marmoreus*	*punctatus*	*debilis*	*periglenes*	*quercicus*	*valliceps*	*canaliferus*	*occidentalis*	*coccifer*	*cognatus*	*americanus*	*asper*	*melanostictus*	*stomaticus*	*viridis*	*calamita*	*bufo*	*mauritanicus*	*regularis*	*gariepensis*	*lönnbergi*	*vertebralis*	*carens*	*maculatus*	*steindachneri*
spinulosus ♀	1.6	L	L	30.9	N	L	L		3.7	G		G	12.5	N	G	G	10.0	L	N	L		N					N	
spinulosus ♂	N	59.7	G	6.2	3.1				72.8	50.0	3.0	G	L	1.3	G	8.3	5.5	13.6				L				G		
granulosus ♀	N	L	G		N	G		G	1.9			N	N	N			L	L	L			L						
granulosus ♂	G	17.8		9.1	N				44.3			4.0	4.2	5.7				8.8				1.8						
typhonius ♀	N			G					L					G								G						
typhonius ♂									G					G														
arenarum + *ictericus* ♀	N	L	21.4	2.7	N	1.7		L	52.6	L	L	N	L	G	L	G	L	N		G	G	G						
arenarum + *ictericus* ♂	N	0.9	0.6	L	N				85.3	44.4	12.5		20.6	53.7	N	43.7	G	N				82.5		N	N	N	L	
other *marinus* sp. ♀	G	G	4.2	L	G	G		N	L	L		N	L	G	N	G		N	G	N	L	G	O				N	N
other *marinus* sp. ♂	G	G	G	G	G	G			L	0.7	N	N	N	G	G	81.9	G	L				56.9			N	G	0.8	
crucifer ♀	G	G		N		G			L	L		N	G	G	1.2	G		N		G							L	
crucifer ♂									N				G	G		22.2		G				low				G	low	
guttatus ♂			G	G	G				G		G		G	G				G				0				G		

3. Both broad-skulled and narrow-skulled lines of toads are represented in Africa, but intermediate types are lacking, unlike the situation in the New World. Similarly, there is an incompatibility gap between the African narrow-skull toads and broad-skull toads.

4. Chromosomal numbers of 2n = 20 and 2n = 22 are found among the African broad-skulled toads, but all known species of narrow-skulled toads have 2n = 22.

5. Some substance, presently unidentified, is contained in the hypertrophied testes of the toads of the narrow-skulled *gariepensis* group, which inhibits fertilization and development. Hypertrophy of the testes in the broad-skulled *B. superciliaris* suggests a second and independent origin of this mechanism.

Results of Intercontinental Crosses

Many crosses have been made between toads of the different continents in an effort to learn the evolutionary history and the routes of dispersal of this genus. Results of the individual crosses are shown in Appendix I. Distribution of these crosses according to continents and number of species and species groups are shown in table 11-8.

Turning first to the crosses of the South American toads with those of other continents, the best results for each combination are shown in Table 11-9. Metamorphosis of the F_1 hybrid in both reciprocals of the cross occurs when toads of the *spinulosus* group are crossed with toads of the *marmoreus* and *valliceps* groups of North America and *B. stomaticus* of southeastern Asia. There is low percentage metamorphosis of F_1 hybrids between females of the *spinulosus* group and males of *B. bocourti* and of the *cognatus* group (1.6% and 12.5% respectively), but there is inviability of the reciprocal hybrids.

The F_1 hybrids from crosses of *spinulosus*-group males and females of the *boreas, marmoreus,* and *americanus* groups and with *B. punctatus, B. canaliferus, B. occidentalis,* and *B. viridis* metamorphose at rates ranging from very low (1.3% with *americanus* group) to high (59.7% with the *boreas* group). Hybrids between female *B. americanus* and male *B. variegatus* failed as larvae with a normal 2n chromosomal set. There was 8.3 percent metamorphosis in a cross with *B. melanostictus* of southeastern Asia, but the individuals examined were triploids (Chap 10). All other attempted combinations resulted in failure of the hybrids prior to metamorphosis. There is, in these results, a definite pattern of compatibility between narrow-skulled species of South America, North America, and Eurasia, as will be discussed later.

Toads of the *granulosus* group show genetic compatibility in both reciprocals of the cross with only one extracontinental group, the North American *valliceps* group. Even here, the survival of F_1 hybrids to metamorphosis is very low when the female parent is of the *granulosus* group. Females of the *granulosus* group show no compatibility in other intercontinental crosses.

Hybrids fathered by males of the *granulosus* group underwent relatively low percentage metamorphosis in crosses of those males with females of the *boreas, marmoreus, cognatus,* and *americanus* groups and with *B. coccifer, B. viridis,* and *B. garmani* (1.8%).

In the few tests that have been made of *B. typhonius,* no compatibility has been shown in any cross.

The *marinus* group, with its considerable amount of intragroup variation in compatibility, is divisible into two groups according to the performance in intercontinental crosses. *B. arenarum* and *B. ictericus* form one group; *B. marinus, B. poeppigi,* and *B. paracnemis* form the other. The contrasts are most obvious in the high (85.3% and 52.2%) percentage metamorphosis of hybrids between *B. arenarum* in both reciprocals of the cross with the *valliceps* group of southern North America and the somewhat lower (37.5 and 14.3% respectively) metamorphosis in reciprocal crosses of *B. ictericus* and toads of the *valliceps* group. This contrasts with failure of both reciprocal hybrids as larvae when other members of the *marinus* group are crossed with toads of the *valliceps* group. Furthermore, there is high (up to 53.7%) metamorphosis when a female *americanus*-group toad is crossed with a *B. ictericus* male and, in one cross, 26.7 percent when a female *B. woodhousei* was crossed with a *B. arenarum* male. In the latter cross, three of eleven hybrid males proved to be fertile at low levels and produced a backcross generation with females of *B. woodhousei.* By contrast, there is failure at gastrulation in both reciprocals of crosses between the other members of the *marinus* group and members of the *americanus* group.

The *B. arenarum–B. ictericus* section of the marinus group shows generally greater compatibility with additional North American groups than do the three other species. There was metamorphosis (21.4 and 0.6%) in reciprocal crosses of *B. arenarum* and *B. alvarius,* but the hybrids were not examined for ploidy. When females of the *arenarum–ictericus* complex were crossed with North American males, there was metamorphosis in crosses with *B. canaliferus* (44.4%), *B. occidentalis* (12.5%), *cognatus* group (20.6%), *boreas* group (0.9%), *B. coccifer* (0.5%), and the previously mentioned *americanus* group (53.7%),

The other species of the *marinus* group show great incompatibility with the North American toads, for only in the combination of ♀ *B. poeppigi* x ♂ *B. alvarius* was there even low metamorphosis (4.2%) and in the combination of ♀ *B. canaliferus* x ♂ *B. marinus* (0.7%).

Both of the subgroups of the *marinus* group showed relatively high compatibility in crosses of their males with females of the Eurasian broad-skulled *B. melanostictus,* with highs of 43.7 and 81.9 percent respectively. The same was true in the crosses of these *marinus*-group toads with the 20-chromosome, broad-skulled *regularis* group. Here the highest survival to metamorphosis was 82.5 percent for the hybrids from the *B. arenarum–B. ictericus* complex and 56.9 percent for those from the other. With the exception of the survival to metamorphosis of few (0.8%) hybrids from the combination of a female 20-chromosome *B. maculatus* and male *B. marinus,* there was no evidence of compatibility between toads of the *marinus* group and other Eurasian or African groups.

B. crucifer is notable for its high degree of incompatibility with toads of other continents as well as those of South America. Hybrids between female *B. crucifer* and male *B. asper* metamorphosed at a low rate (1.2%), but all examined were triploids (Chap. 10). Crosses of male *B. crucifer* with members of the *regularis* group and with *B. maculatus* and *B. melanostictus* resulted in metamorphosed toads, but all individuals examined were triploids.

Six combinations of North American with Eurasian *Bufo* have demonstrated some degree of metamorphosis in both reciprocals of the crosses (Table 11-10). One of these (*B. valliceps* x *B. viridis*) is probably without evolutionary significance, for the hybrid with a female *B. valliceps* parent was a triploid and the reciprocal hybrid was a gynogenetic diploid (Chap. 10). Three western, closely related species (*B. boreas, B. alvarius,* and *B. cognatus*) produced some metamorphosing hybrids in reciprocal crosses with *B. viridis.* Hybrids between females of the *cognatus* group and *B. calamita* metamorphosed at a fairly high rate (32.7%), but the reciprocal cross has not been made. Females of another close relative of these western species (*B. punctatus*) showed high compatibility with *B. viridis,* but the reciprocal hybrids failed as larvae.

The highest viability (73.9 and 66.7% metamorphosis, respectively) of F_1 hybrids between North Amercan and Eurasian species was found in the offspring of the combination of *marmoreus* group x *B. stomaticus. B. punctatus,* a near relative of the *marmoreus* group, showed low compatibility (13.0 and 9.2% metamorphosis) in reciprocal crosses with *B. stomaticus.* Another western species group with affinities close to these groups of species, the *debilis* group, showed low (8.1%) metamorphosis of hybrids between the *debilis*-group male and *B. viridis* female.

There was low percentage metamorphosis of hybrids between females of the *boreas* group and males of the Eurasian narrow-skulled *B. calamita* (5.8%) and *B. bufo* (5.4%). There was low percentage metamorphosis of diploid hybrids between male narrow-skulled *boreas* group toads of North America and female broad-skulled *B. melanostictus* of Asia.

There was moderate (29.7%) metamorphosis of hybrids between males of the broad-skulled *valliceps* group of North America and females of the broad-skulled *B. melanostictus* of Asia.

The six species of the *americanus* group provide a useful tool for assay of relationship. Females of this group are compatible with males of many species, while the males are incompatible with the females of most other species. Thus, negative results of crosses involving females of this group are suggestive of distant relationship, while positive results in crosses involving males of the group imply close relationship. The only intercontinental crosses involving females of the *americanus* group in which there was no metamorphosis were ones with broad-skulled toads: *B. crucifer, B. typhonius,* and *B. marinus, B. poeppigi,* and *B. paracnemis* of the *marinus* group, and *B. blombergi* of the *guttatus* group in South America; *B. melanostictus* of southern Asia; and the *regularis* group and *B. mauritanicus* of Africa. The hybrids did metamorphose, however, in crosses of *americanus*-group females with males of the broad-skulled *B. arenarum* (26.7%) and *B. ictericus* (53.7%). Three of 11 F_1 hybrids from the cross with *arenarum* proved fertile when backcrossed.

No hybrids reached metamorphosis in any intercontinental cross with an *americanus*-group male.

The low compatibility between the Eurasian and African toads (Table 11-11) contrasts strongly with that between the toads of the other continents (Ta-

Table 11–10. Best Results of Crosses between North American Species Groups of *Bufo* and Species Groups from other Continents

	South American							Eurasian						African							
	spinulosus	*granulosus*	*typhonius*	*arenarum + ictericus*	*marinus*	*crucifer*	*guttatus*	*asper*	*melanostictus*	*stomaticus*	*viridis*	*calamita*	*bufo*	*mauritanicus*	*regularis*	*maculatus*	*perreti*	*vertebralis*	*carens*	*gariepensis*	*steindachneri*
bocourti ♀	N	G		N	G			G	G-L	G	G	G	G		G						
bocourti ♂	4.1	N	N	N	G	G		G	L	L	8.0				N				G		
boreas ♀	59.7	17.8		0.9	G			G	G	4.0	14.8	5.8	5.4		L	G	L				
boreas ♂	L	L		L	G	G		G	2.5	L	1.9				N	L		L	G		
marmoreus ♀	1.1	9.1		L	G		G	G	G	73.9	L	N	G	G	G	G	G			G	
marmoreus ♂	30.9			12.1	N			G	L	66.7	L				L	N		L	G		
cognatus ♀	L	4.2		20.6	N	G	G	G	G	2.3	55.9	32.7	L	G	N	G	G			G	
cognatus ♂	12.5	N		L	N	G		G	L	N	22.0				16.6	N		L	N		
punctatus ♀	3.1	N		N	G	G		G		13.0	84.0	L	L		N		G				
punctatus ♂	N	G		L	G			G	N	9.2	L				N	L		L	G		
debilis ♀					G																
debilis ♂	L	G		1.7	G	G			N	L	8.1				1.8	N		L	N		
alvarius ♀	G			0.6	G		G				6.3	L	G		G	G	G				
alvarius ♂	<100.0	G		21.4	4.2			G	L	L	24.2				L	L					
valliceps ♀	72.8	44.3	G	85.3	L	N	N	G	G		2.4	L	L		G	G	G		G		
valliceps ♂	3.7	1.9	L	52.6	L	L		G	29.7	L	1.6				1.8	L		G	G		G
occidentalis ♀	3.0	L		50.0	N	G					L	G	G		G						
occidentalis ♂				L											72.8						
canaliferus ♀	50.0			43.8	0.7			L	G		L		G		G						
canaliferus ♂	G			L		L		N		N					1.4	L			L		
coccifer ♀	G	4.0		0.5	N			N	G	L	L		G		G						
coccifer ♂	G	N		N	N	N		G	L	G	G				N	L			G		
americanus ♀	1.3	5.7	G	26.7	G	N	N	25.1	N	52.9	48.1	52.5	14.8	G	N						
americanus ♂	N	L	G	G	G	G		G	N	G	N	G			N	N		G	G		G
quercicus ♂		G		N	G				L	L	L				L	G			G		
periglenes ♂	L										G										

bles 11-9 and 11-10). In sixteen crosses involving females of five species and five species groups of Eurasian *Bufo* and in thirty-two crosses involving females of nine species and five species groups of African *Bufo*, metamorphosis occurred in only three combinations, none of which is evolutionarily significant. Somewhat fewer than 40.0 percent of the hybrids metamorphosed in a cross between a female *B. viridis* and a male *B. mauritanicus*, but all hybrids examined were pentaploids. In a cross of female *B. viridis* with a male of the *vertebralis*-group (*B. lughensis*), 7.1 percent of the offspring metamorphosed, but these proved to be gynogenetic diploids (Chap. 10). There was 0.8 percent metamorphosis of hybrids between a female *B. regularis* (E) and a male *B. asper*, but all individuals examined were triploids.

Bearing of Hybridization Data on Evolutionary History

The results of the hybridization experiments permit formulation of a series of generalizations on the

Table 11–11. Best Results of Crosses between Eurasian Species Groups of *Bufo* and Species Groups from other Continents

Other / Eurasian	South American					North American												African							
	spinulosus	*granulosus*	*arenarum*	*marinus*	*crucifer*	*bocourti*	*boreas*	*marmoreus*	*cognatus*	*punctatus*	*debilis*	*alvarius*	*valliceps*	*occidentalis*	*canaliferus*	*coccifer*	*americanus*	*quercicus*	*periglenes*	*mauritanicus*	*regularis*	*maculatus*	*vertebralis*	*carens*	*gariepensis*
asper ♀	G		N	G		G	G	G	G	G		G	G		N	G	G								
asper ♂	G		L	N	1.2	G	G	G	G	G			G		L	N	25.1				G			G	
melanostictus ♀	8.3		43.7	81	22.2	L	2.5	L	L	N	N	L	29.7			L	N	L			G				
melanostictus ♂	G		G	G	G	L	G	G	G	G			G		G	G	N				L			G	
stomaticus ♀	5.5		G	G		L	L	66.7	N	9.2	L	L	L		N	G	G	L			G				
stomaticus ♂	10.0	L	L			L	4.0	73.9	2.3	13.0						L	52.9				N			G	
viridis ♀	13.6	8.8	N	L	G	8.0	1.9	L	22.0	L	8.1	24.2	1.6			G	N	L	G	<40.0	G	G	<7.1		G
viridis ♂	L	L	N	N	N	G	14.8	L	55.9	84.0		6.3	2.4	L	L	L	48.1				L	N		N	
calamita ♀																	G								
calamita ♂	N	L		G		G	5.8	N	32.7	L		L	L	G			52.5				N	G		G	
bufo ♂	L		G	N	G	G	5.4	G	L	L		G	L	G	G	G	14.8				L	G	N	N	

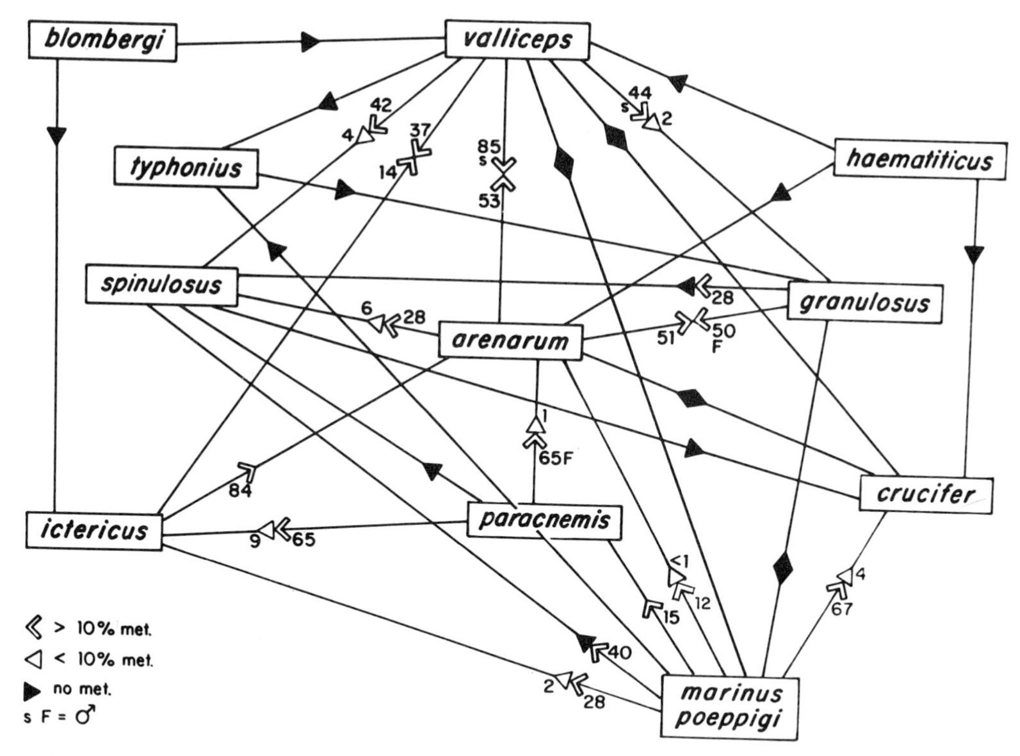

Fig. 11-1. Results of crosses among South American *Bufo* and with the North American *valliceps* group. Numbers indicate percentages of fertilized eggs that resulted in metamorphosed toads. *S*=sterile, *F*=fertile male F_1 hybrid.

evolutionary history of the genus to be formulated that can be tested by other kinds of evidence. This will be done in chapter 18.

The first of these generalizations, and one that will be further documented in chapter 18, is that the genus had its origin in the New World and subsequently spread to its present, nearly world-wide distribution. Some of the strongest evidence for this generalization from the hybridization data comes from the existence in South America and in southern North America of intermediate types, both in morphology and in crossability, that connect the broad-skulled and narrow-skulled lines of *Bufo*. In South America, these intermediate forms include such species as *B. arenarum*, *B. icterícus*, and members of the *granulosus* group. In the *spinulosus* group of mainly narrow-skulled toads, two species show the broad-skull condition (Chap. 4). In North America, they include such species as members of the *valliceps* group, *B. occidentalis*, *B. canaliferus*, the *marmoreus* group, the *cognatus* group, and *B. alvarius*. No such situation has been found in Eurasia or Africa, where the evidence points to sharp distinction between narrow-skulled and broad-skulled lines. The latter situation in Eurasia and Africa would be expected if separate evolutionary lines of *Bufo* had dispersed there from their area of origin in the New World.

A second major generalization is that the *Bufo* of North and Central America have evolved as a single radiation rather than as separate continental groups. This is graphically evident from Figure 11-1, which shows lines of genetic compatibility. The most striking feature is the high compatibility of reciprocal crosses between the *valliceps* group of southern North America and *B. arenarum* of southern South America. In addition there is the evidence of metamorphosis in both reciprocals of crosses between the *granulosus* and *valliceps* groups, *B. icterícus* and the *valliceps* group, the *spinulosus* and *valliceps* groups, and the *spinulosus* and *marmoreus* groups. The implication is that through the Tertiary the interchange of *Bufo* between North and South America was such that the evolution of the genus proceeded as though North and South America were a single landmass.

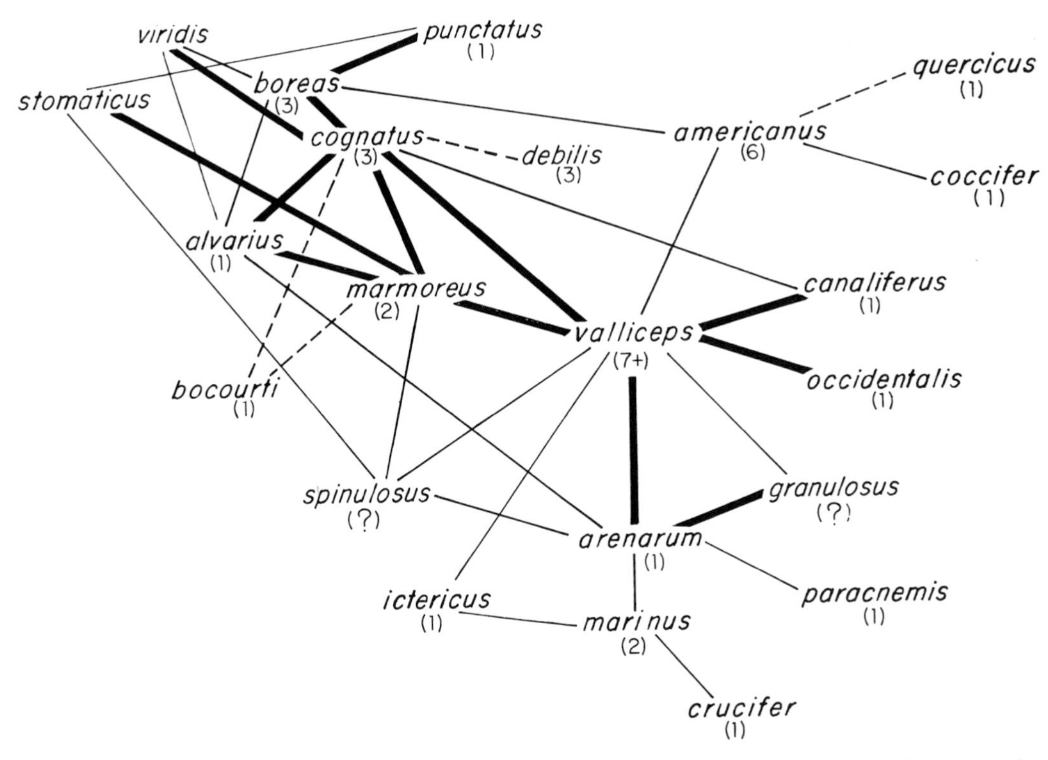

Fig. 11–2. All species groups of *Bufo* between which metamorphosis of reciprocal hybrids has been demonstrated. Heavy lines connect groups in which there was more than 10% metamorphosis in both reciprocals. Light lines connect groups for which one or both reciprocals metamorphosed at less than 10%. Broken lines indicate best estimate of relationship where there was no metamorphosis of reciprocal hybrids.

A third major generalization is that there has been dispersal of a line of narrow-skulled toads northward from South America, through western North America, across a Tertiary land bridge, bridging the Bering Strait to Asia, followed by movement across Eurasia to North Africa. Evidence for this is seen in Figure 11-2. Metamorphosis in reciprocal crosses occurs when *B. viridis* is crossed with the *boreas* and *cognatus* groups and with *B. alvarius* of western North America. There is metamorphosis in crosses between female *B. viridis* and males of *B. bocourti* of the Guatemalan highlands. There is metamorphosis in crosses of ♀ *B. viridis* x ♂ *B. spinulosus* group and with males of the *B. granulosus* group.

All this implies that narrow-skulled toads moved northward through the western American highlands, with which they are still associated, and crossed the Bering land bridge at a time in the Tertiary that was favorable for a group of cold-adapted toads. Conditions under which this could have happened existed as recently as 4 to 10 million years ago in the Pliocene (see Hopkins 1967), but we cannot rule out the possibility of an earlier crossing, possibly in the early Miocene.

The fourth major generalization is that there was a movement northward in the Tertiary of an intermediate line, represented today in southern Mexico by the *marmoreus* group and in southeastern Asia by *B. stomaticus.* The high percentage metamorphosis of reciprocal hybrids between these groups (Table 11-10) is most impressive. Because of the adaptation of these toads to tropical environments, the crossing of this evolutionary complex to Asia must have come at a time different from that of the cold-adapted, narrow-skulled line. Elimination of these toads from temperate Asia is probably attributable to Pleistocene glacial-stage climates that forced warmth-adapted species southward against east-west mountain chains and hence provoked extensive extinctions (see Szarski 1960).

A fifth generalization is that the broad-skulled line of toads followed this same route northward and

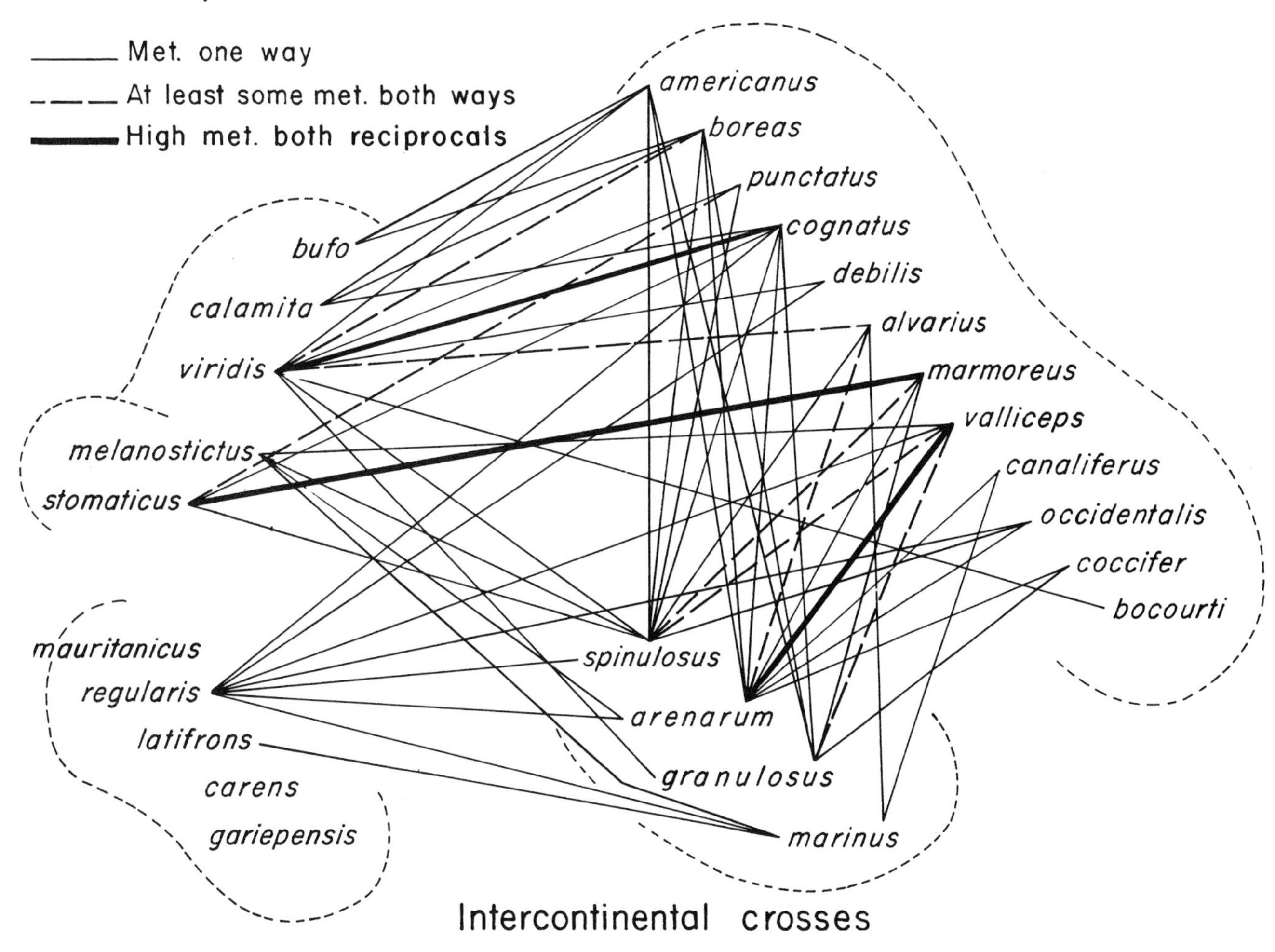

Fig. 11–3. Results of intercontinental crosses in which there was metamorphosis of one or both of the reciprocal hybrids. Results are omitted where only polyploid or gynogenetic hybrids were found. *(Latifrons = maculatus.)*

across the Bering land bridge to Asia and then on into Africa. The evidence for this generalization is seen in table 11-11 and is graphically shown in figure 11-4, which shows the results of hybridization between *B. arenarum* and toads of other continents. *B. melanostictus* of southeastern Asia is a remnant of this evolutionary line of broad-skulled toads. It has differentiated so much that males of *B. melanostictus* show no compatibility with the broad-skulled, or with any other, toads of the New World or of Africa. However, female *B. melanostictus* remain highly compatible with male *B. marinus* (Table 11-12) and show lesser compatibility with other broad-skulled, New World species, such as *B. arenarum, B coniferus,* and *B. valliceps.*

The invading broad-skulled toads evidently crossed into Africa and evolved into *B. mauritanicus,* with a normal chromosomal complement, and, after a chromosomal fusion (Chap. 10), which reduced the count to 2n = 20, evolved into the large *regularis* group and the derived *B. maculatus* and *B. perreti.*

In spite of this chromosomal change, the females of at least some species in the *regularis* group are highly compatible with some of the New World, broad-skulled toads (Table 11-12). The species in the *regularis* group differ considerably in their compatibility with the New World toads. Female *B. garmani* produce remarkably high percentages of viable, metamorphosing hybrids in crosses with *B. arenarum, B. occidentalis,* and *B. marinus.* A female F_1 hybrid between ♀ *B. garmani* and ♂ *B. occidentalis* had pigmented eggs in the ovary. However, male

Table 11–12. Stage Reached or Percentage Metamorphosis*

♀	♀							
	melanostictus	*garmani*	*rangeri*	*regularis* (W)	*regularis* (E)	*kerinyagae*	*brauni*	Best of *Regularis* group
marinus	81.9	56.9, 51.8	23.5	L, 0.5, L	1.1, 0.5, L	36.6	L, G, L	56.9
poeppigi	—	—	—	L	—	14.9	—	14.9
paracnemis	—	—	—	L	—	—	—	L
arenarum	4.8, 43.7	82.5	13.8	—	0.7	—	N, G, N	82.5
occidentalis	—	72.8**	—	—	—	—	—	72.8
valliceps	11.2, 12.8, 29.7	L	1.8 (5n)	L, L	L, L	14.0	L	14.0
coniferus	8.5	—	—	G, L	L	—	—	L
luetkeni	L	—	—	G	L	—	—	L
canaliferus	L	—	—	—	1.4	—	—	1.4
regularis gp.	G	—	—	—	—	—	—	G
melanostictus	—	—	—	G, N	—	—	—	N

*Crosses of female *B. melanostictus* and females of the *regularis* group with males of New World broad-skulled toads.

**Pigmented eggs in ovary.

hybrids between female *B. garmani* and male *B. marinus* and *B. arenarum* failed to fertilize any eggs in attempted backcrosses to female *B. marinus* and *B. arenarum* respectively. *B. brauni* is the only one of the six members of the *regularis* group tested that showed no compatibility in producing metamorphosed hybrids with the New World broad-skulled toads. The other species tested (*B. rangeri*, *B. regularis* (W), *B. regularis* (E), and *B. kerinyagae*) were intermediate in this respect between the extremes represented by *B. garmani* on the one hand and *B. brauni* on the other.

In both *B. melanostictus* and in the *regularis* group it is the female that has retained genetic compatibility with the New World broad-skulled toads. Nevertheless, these African and Asian broad-skulls have differentiated from one another so that they are unable to produce metamorphosed hybrids in either reciprocal of the intercontinental cross. As with the other crossings of the Bering land bridge, it is not possible to date such crossing with any degree of accuracy. The ecological adaptations of these toads are such that they could have crossed at the same time as the ancestors of *B. stomaticus*, that is, possibly in the early Miocene. If this is so, however, one must postulate a more rapid rate of evolution for the broad-skulled line than for the intermediate type represented by *B. stomaticus*. An alternative is to postulate an earlier spread of the broad-skulls, and this may well be correct.

A final broad generalization is that, despite occasional interruptions of their land connections, the continents of South and North America and the Eurasian landmass have shared largely their toad faunas. Africa, however, while receiving an input of toads from these other continents, has had its toad fauna evolve independently of these other continents. Among forty-eight attempted crosses between thirteen species and seven species groups of African toads and six species and six species groups of Eurasian toads, the only metamorphosis of hybrids, as detailed earlier, is explained by polyploid hybrids or gynogenetic diploidy. *B. viridis* does represent a present-day extension of the narrow-skulled line into Af-

Table 11-13. Data for Intergroup Crosses in which Fertility was Demonstrated by Backcrosses

Cross No.	Cross Combination	% Fert.	% Hatch	% Met.	% Fert. Met.	Backcross No.	Backcross Combination	% Fert.	% Hatch	% Met.	% Fert. Met.
64-9	*woodhousei* x *bocourti*	100.0	54.6	13.5	7.4	65-7	♀ *woodhousei* x F_1 ♂	11.4	100.0	18.4	18.4
65-7	*woodhousei* x F_1 ♂	11.4	100.0	18.4	18.4	66-194	♀ *woodhousei* x B.C. ♂	64.4	100.0	23.7	23.7
63-8	*woodhousei* x *arenarum*	100.0	61.6	43.7	26.7	64-3	♀ *woodhousei* x F_1 ♂	5.0	100.0	100.0	100.0
63-8	*woodhousei* x *arenarum*	100.0	61.6	43.7	26.7	64-23	♀ *woodhousei* x F_1 ♂	2.8	100.0	66.7	66.7
63-8	*woodhousei* x *arenarum*	100.0	61.6	43.7	26.7	64-24	♀ *woodhousei* x F_1 ♂	4.8	20.0	83.3	16.7
63-66	*hemiophrys* x *valliceps*	100.0	84.0	3.8	3.1	64-119	♀ *valliceps* x F_1 ♂	3.7	100.0	22.2	22.2
64-214	*pygmaeus* x *arenarum*	73.3	13.6	32.1	4.4	65-79	♀ *arenarum* x F_1 ♂	82.3	58.7	5.8	3.4
64-162	*luetkeni* x *microscaphus*	35.8	5.3	66.7	3.5	65-129	♀ *luetkeni* x F_1 ♂	22.9	54.5	56.0	30.5

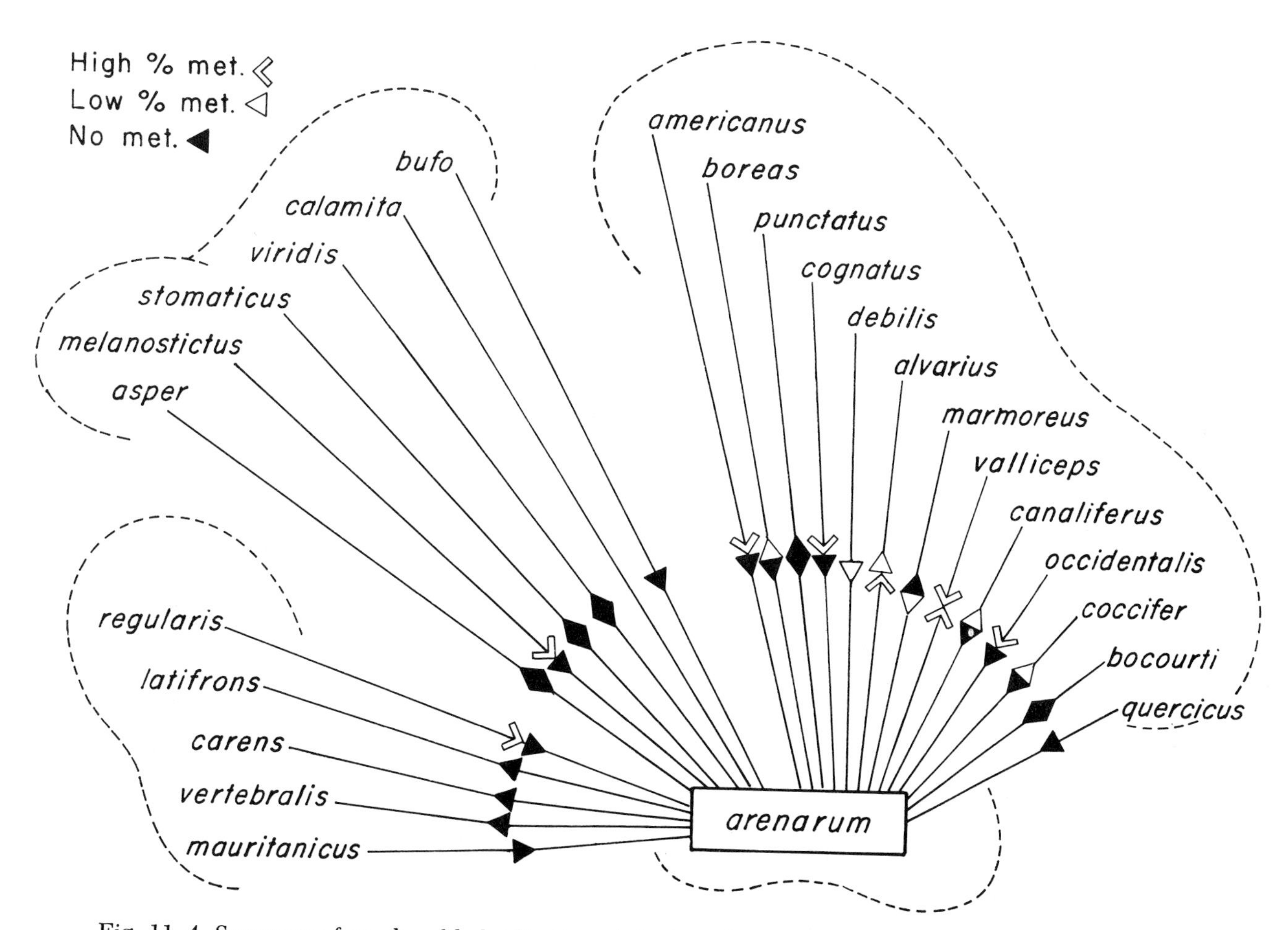

Fig. 11–4. Summary of results of hybridization of the broad-skulled *B. arenarum* of South America with species groups from other continents. High and low percentage metamorphosis as in Fig. 11–1. (*Latifrons* = *maculatus*.)

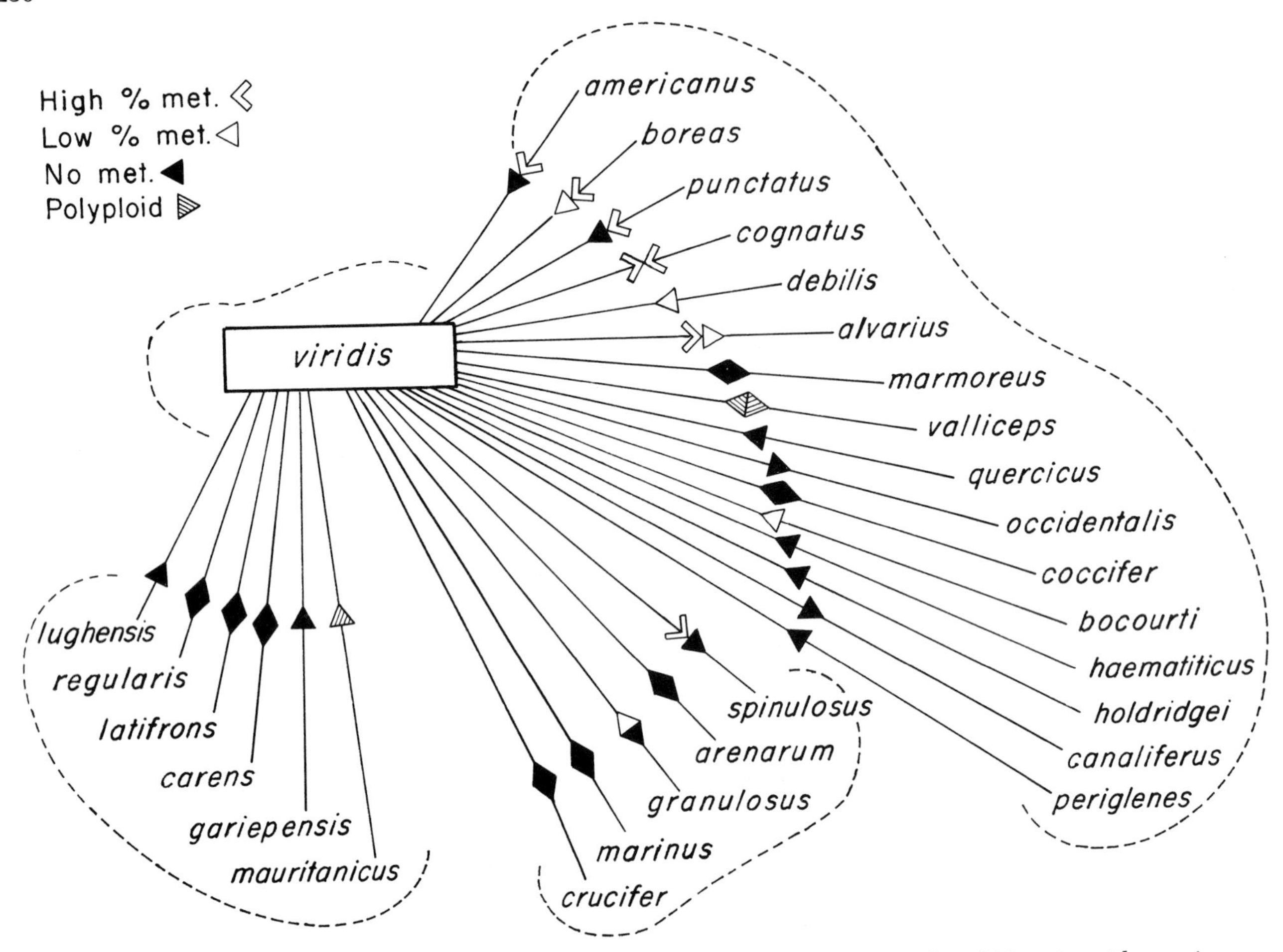

Fig. 11–5. Summary of results of hybridization of the narrow-skulled *B. viridis* of Eurasia with species groups from other continents. High and low percentage metamorphosis as in Fig. 11–1. (*Latifrons* = *maculatus*.)

rica, but its distribution, restricted to the far north of Africa, suggests very recent arrival on that continent.

Significance of Fertility

As demonstrated earlier and in earlier publications (A. P. Blair 1942; W. F. Blair 1962, 1964*b*), acquisition of premating isolating mechanisms generally precedes the acquisition of the genetic barriers of incompatibility and interspecies sterility in *Bufo*. Hence, members of the same species group tend to retain the ability for gene exchange, while at the same time their adaptive genotypes are protected by a complex of premating isolating mechanisms. In addition to this fertility in intragroup crosses, a few of the numerous backcrosses of intergroup hybrids that have been made have revealed fertility of some intergroup hybrids. Three of these involved female parents from the *americanus* group (Table 11-13). In the cross of ♀ *B. woodhousei* x ♂ *B. bocourti*, both narrow-skulled toads but rather distantly related on other evidence, 7.4 percent of the fertilized eggs resulted in metamorphosed hybrids. One hybrid male was backcrossed to a *B. woodhousei* female and fertilized 11.4 percent of her eggs. Of the fertilized eggs, 18.4 percent resulted in metamorphosed backcross hybrids. One of these was backcrossed a second time to a female *B. woodhousei*, at which time he was found to be a somatic triploid (Chap. 10). Nevertheless, normal haploid sperm were produced, and there was metamorphosis of a second backcross generation.

As mentioned earlier, three of eleven male F_1 hybrids between a narrow-skulled *B. woodhousei* female and a broad-skulled *B. arenarum* male fertil-

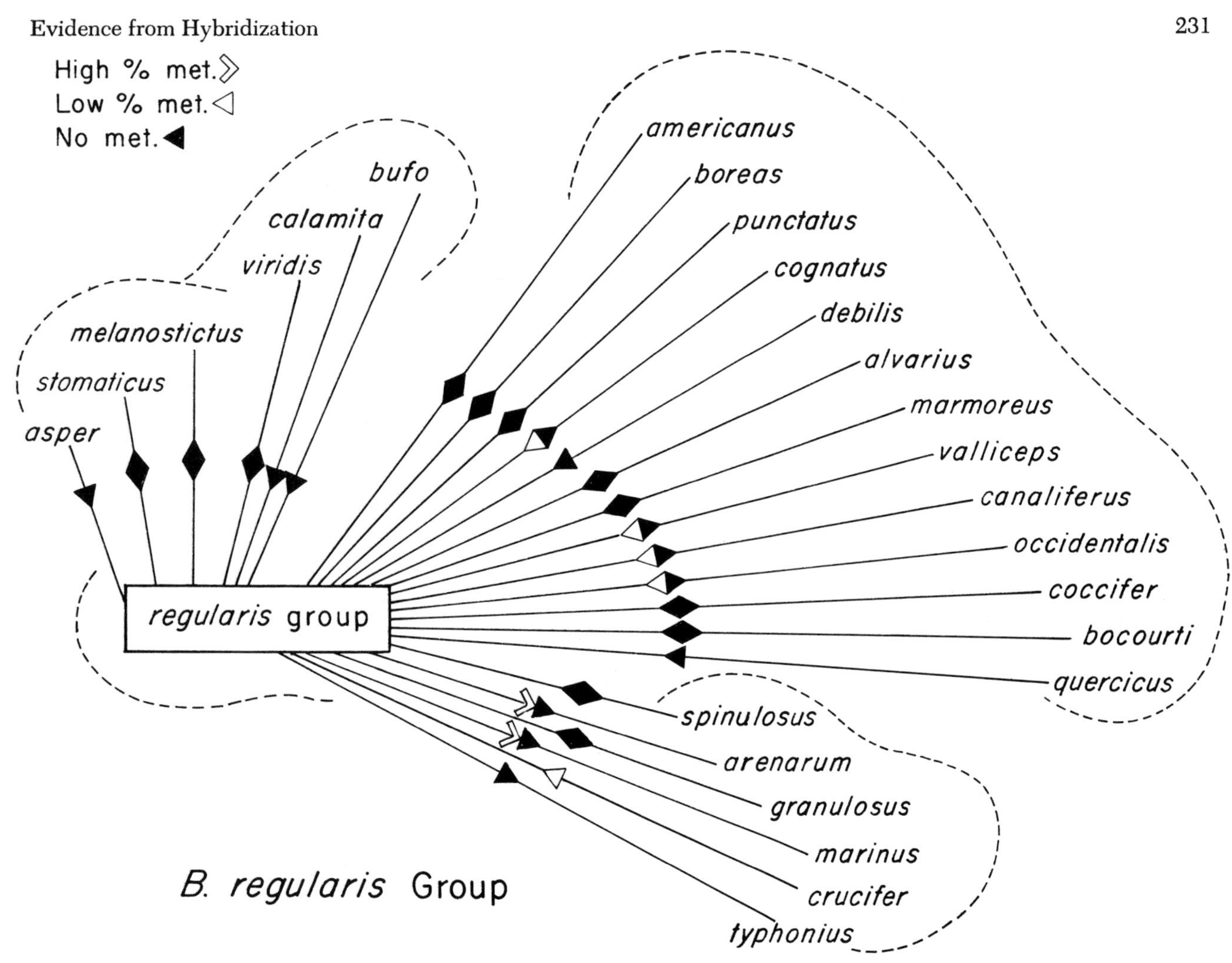

Fig. 11–6. Summary of results of hybridization of the narrow-skulled *regularis* group of Africa with species groups from other continents. High and low percentage metamorphosis as in Fig. 11–1.

ized small fractions of the eggs (2.8 to 5.0%) in backcrosses to *B. woodhousei*. The few backcross hybrids metamorphosed at a high rate (16.7% to 100%, where only one egg was fertilized). The third combination involved a ♀ *B. hemiophrys* x ♂ *B. valliceps* cross. Here only 3.1 percent of the F_1 hybrids metamorphosed. In a backcross to a *B. valliceps* female, one of the hybrid males fertilized only 3.7 percent of her eggs, but 22.2 percent of the backcross hybrids metamorphosed.

In one instance, a fertile hybrid resulted from the cross of a *valliceps*-group female (*B. luetkeni*) with an *americanus*-group male (*B. microscaphus*). Survival of the F_1 hybrid to metamorphosis was unusual, for in twenty-six other intergroup crosses in sixteen combinations no other F_1 hybrid did metamorphose. Nevertheless, 3.5 percent of the F_1 hybrids metamorphosed, and one of these was backcrossed to a female *B. luetkeni*. There was fertilization of 22.9 percent of the eggs, and 30.5 percent of the backcross hybrids metamorphosed.

In a cross of a female *B. pygmaeus* of the *granulosus* group with a male *B. arenarum* of the *marinus* group, only 4.4 percent of the F_1 hybrids metamorphosed. In a backcross to a *B. arenarum* female, one of these hybrid males fertilized 82.3 percent of the eggs. However, prehatching and posthatching mortality combined to reduce the percentage metamorphosis of the backcross hybrids to 3.4 percent.

The great majority of the F_1 hybrid males from intergroup crosses simply failed to fertilize any eggs in attempted backcrosses. The testes of such toads were usually, but not always, small, abnormal, or sometimes unidentifiable and presumably absent.

In a few instances, the hybrid males fertilized eggs, but the larvae failed to metamorphose. In all these cases, the species groups were relatively close to one another on other evidence. Male hybrids between female *B. perplexus* of the *marmoreus* group and male *B. luetkeni* of the *valliceps* group fertilized eggs in backcrosses to females of both parental species, but the backcross hybrids failed to metamorphose. A hybrid male from the combination ♀ *B. woodhousei* x ♂ *B. punctatus* fertilized one of 442 eggs in a backcross to a female *B. woodhousei.* This backcross hybrid survived twenty-seven days as a larva but could not metamorphose.

Three F_1 hybrid males from the combination ♀ *B. valliceps* x ♂ *B. arenarum* fertilized 3 of about 2,000 eggs, 7 of 108 eggs, and one of a large, uncounted sample, but all the backcross hybrids died as larvae.

Exact relation between genotypic differences and successful development of an individual or the fertility of that hybrid individual are, of course, unknown. The few fertile hybrids from crosses between species that, on other evidence, seem relatively distant from each other probably have little bearing on degree of affinity. Only one of these was examined for the presence of ploidy, and it was a triploid. The development of the hybrids to the larval stage does fit the pattern of relationship as indicated from other evidence and is apparently significant.

REFERENCES

Blair, A. P. 1941. Variation, isolation mechanisms, and hybridization in certain toads. Genetics 26:398–417.

———. 1942. Isolating mechanisms in a complex of four species of toads. Biol. Symposia 6:235–249.

———. 1955. Distribution, variation, and hybridization in a relict toad (*Bufo microscaphus*) in southwestern Utah, Amer. Mus. Novitates, no. 1722, pp. 1–38.

Blair, W. F. 1959. Genetic compatibility and species groups in U. S. toads (*Bufo*). Texas J. Sci. 11:427–453.

———. 1962. Non-morphological data in anuran classification. Syst. Zool. 11:72–84.

———. 1963*a*. Evolutionary relationships of North American toads of the genus *Bufo*: A progress report. Evolution 17:1–16.

———. 1963*b*. Intragroup genetic compatibility in the *Bufo americanus* species group of toads. Texas J. Sci. 15:15–34.

———. 1964*a*. Evidence bearing on relationships of the *Bufo boreas* group of toads. Texas J. Sci. 16:181–192.

———. 1964*b*. Isolating mechanisms and interspecies interactions in anuran amphibians. Quart. Rev. Biol. 39:334–344.

———. 1966. Genetic compatibility in the *Bufo valliceps* and closely related groups of toads. Texas J. Sci. 18:333–351.

Flindt, R., and H. Hemmer. 1967. Nachweis natürlicher Bastardierung von *Bufo calamita* und *Bufo viridis.* Zool. Anz. 178:419–429.

Gallardo, J. M. 1965. The species *Bufo granulosus* Spix (Salientia: Bufonidae) and its geographic variation. Bull. Mus. Comp. Zool. 134(4):107–138.

Hopkins, D. M. (ed.). 1967. The Bering Land Bridge. Stanford Univ. Press, Stanford, Calif.

Keith, Ronalda. 1968. A new species of *Bufo* from Africa, with comments on the toads of the *Bufo regularis* complex. Amer. Mus. Novitates, no. 2345, pp. 1–22.

Kennedy, J. P. 1962. Spawning season and experimental hybridization of the Houston toad, *Bufo houstonensis.* Herpetologica 17:239–245.

Noble, G. K. 1924. Contributions to the herpetology of the Belgian Congo based on the collection of the American Museum Congo expedition, 1909–1915—with abstracts from the field notes of Herbert Lang and James P. Chapin. Bull. Amer. Mus. Nat. Hist. 49: 147–347.

Szarski, H. 1960. Plazyi gady Ameryki Pólnocnej Przegl. Zool. 4:170–179.

Thornton, W. A. 1955. Interspecific hybridization in *Bufo woodhousei* and *Bufo valliceps.* Evolution 9: 455–468.

Volpe, E. P. 1956. Experimental F_1 hybrids between *Bufo valliceps* and *Bufo fowleri.* Tulane Studies Zool. 4(2):61–75.

———. 1959. Hybridization of *Bufo valliceps* with *Bufo americanus* and *Bufo terrestris.* Texas J. Sci. 11:335–342.

12. Biogenic Amines

JOSE M. CEI

Universidad Nacional de Cuyo
Mendoza, Argentina,

VITTORIO ERSPAMER

Università degli Studi
Rome, Italy,

and

M. ROSEGHINI

Università degli Studi
Rome, Italy

Introduction

Very strong biochemical and metabolic activity occurs in the batrachian skin, where many different substances are produced. There are many important physiological and ecological functions related to the skin, such as passage of water and electrolytes, essential respiratory activity, thermoregulation, secretion of defensive venoms, mimicry, and others. The complicated structure of the secretive layer of the integument emphasizes its intensive cellular activity, especially in such topographically specialized areas of the body as the parotoid, inguinal, lateral, or tibial glands.

During the past ten years, about four hundred amphibian species have been collected throughout the world, and methanol or acetone extracts of the dried or fresh skins of these creatures have been subjected to chemical and biological screening. From this extensive investigation emerged the generalization that the amphibian skin is a formidable storehouse of aromatic biogenic amines (with their precursors and metabolites) and of polypeptides active on smooth muscle and on external secretions. Several new amines and polypeptides have been identified and their structures elucidated. Amines and polypeptides are, of course, only a small part of the biogenic molecules occurring in the amphibian skin. Our screening was limited to amines and polypeptides because of their considerable pharmacological interest, on the one hand, and because of the research aims of our working group, on the other.

While analytic data were accumulating, it soon appeared that amine and polypeptide spectra varied conspicuously among the different genera, species, and perhaps subspecies as well. To be specific, it ap-

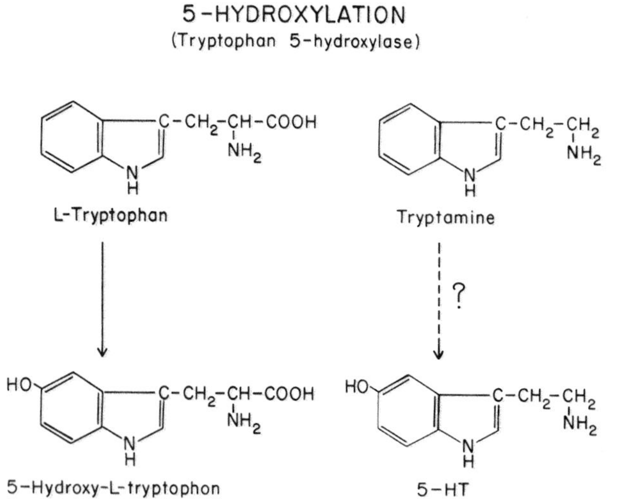

5-HYDROXYLATION
(Tryptophan 5-hydroxylase)

L-Tryptophan → 5-Hydroxy-L-tryptophon

Tryptamine → ? → 5-HT

Fig. 12–1. Enzymatic events that occur in the toad skin: 5-HYDROXYLATION (Tryptophan 5-hydroxylase).

DECARBOXYLATION
(L-Aromatic Acid Decarboxylase (s))

5-Hydroxy-L-tryptophan → 5-Hydroxytryptamine

L-Tryptophan → Tryptamine

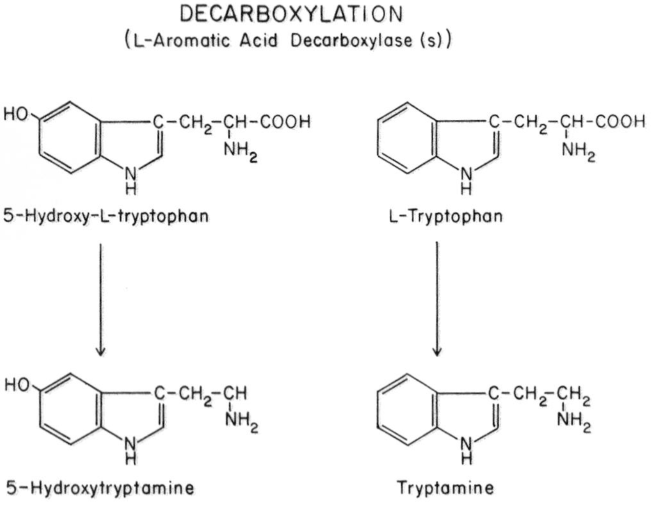

Fig. 12–2. Enzymatic events that occur in the toad skin: DECARBOXYLATION (L-aromatic acid decarboxylase[s]).

peared that the spectrum of biogenic amines and polypeptides could be profitably used, at least as a subsidiary tool, in biochemical taxonomy and for assessing evolutionary relationships. Among the amphibian species and subspecies considered in this extensive study, sixty-five belong to the genus *Bufo*. Three species come from Europe, nine from Africa, five from Asia, and the others from the Americas.

The skin of these *Bufo* species provides a spectacular representation of indolealkylamines. There is no animal or vegetable tissue known that can compete with the toad skin in variety and quantity of indo-

N-METHYLATION
(N-Methyl Transferase)

5-HT → N-Methyl-5-HT → Bufotenine (N, N-Dimethyl-5-HT) → Bufotenidine

5-Methoxytryptamine (5-MT) → N-Methyl-5-MT → O-Methylbufotenine (N, N-Dimethyl-MT)

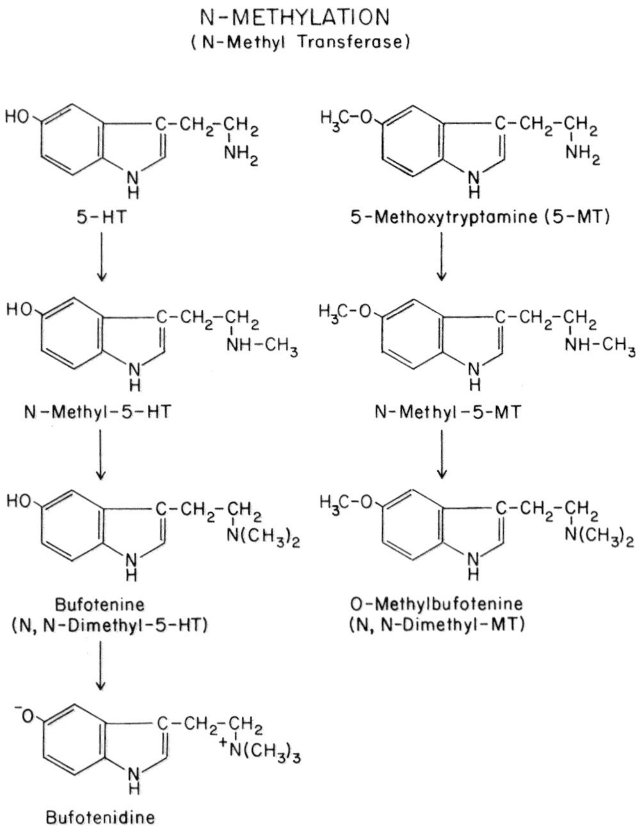

Fig. 12–3. Enzymatic events that occur in the toad skin: N-METHYLATION (N-methyl transferase).

lealkylamines and of their precursors and metabolites. Approximately twenty indole derivatives have been identified so far in extracts of toad skin, and this number will grow.

It is obvious that the occurrence of a given indole derivative implies the occurrence of the enzyme systems catalyzing its biosynthesis. In the case of toad indolealkylamines at least the following enzyme systems are required for their biosynthesis and metabolism: *tryptophan-5-hydroxylase*, the enzyme responsible for the 5-hydroxylation of tryptophan, that is, responsible for the biosynthesis of 5-hydroxytryptophan, the obligatory precursor of all 5-hydroxy- and 5-methoxyindolealkylamines; *aromatic L-amino acid decarboxylase*(s), capable of decarboxylating tryptophan to tryptamine and 5-hydroxytryptophan to 5-HT; *N-methyl transferase*, responsible for the transformation of the primary into the secondary and tertiary amine, and then into the quaternary ammonium base; *5-hydroxyindole-O-methyl transferase*, which causes the methylation of the hydroxy group

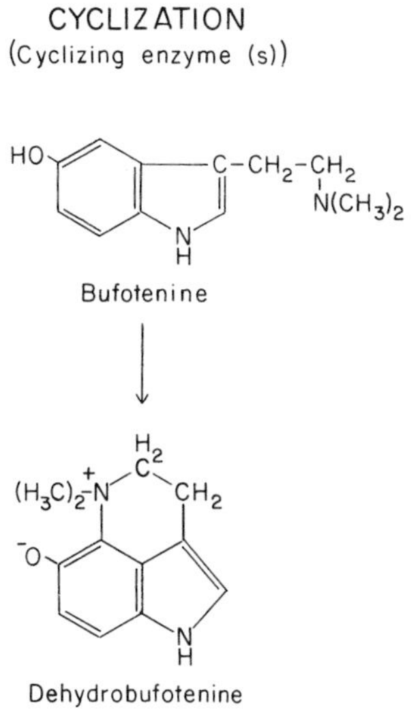

CYCLIZATION
(Cyclizing enzyme (s))

Bufotenine

Dehydrobufotenine

Fig. 12–4. Enzymatic events that occur in the toad skin: CYCLIZATION (cyclizing enzyme[s]).

of the nucleus, forming 5-methoxyindoles; *cyclizing enzymes,* involved in the biosynthesis of dehydrobufotenine, proceeding by dehydrogenation of bufotenine to a *para*-quinonid structure, which then cyclizes by nucleophilic addition of the -$N(CH_3)_2$ group (Robinson et al. 1961); one or two, or even more, *conjugases,* capable of linking sulphuric acid to the phenolic hydroxy group (*O-sulphoconjugase* or *5-sulphoconjugase*) on the side, and of linking again sulphuric acid or other acids to the pyrolic >NH group, in position 1, on the other side (*1-conjugase[s]*); and finally *monoamineoxidase,* responsible for the oxidative deamination of primary and secondary indolealkylamines. Figures 12-1 to 12-8 give a synopsis of the enzymatic events that occur or are likely to occur in the toad skin.

It is probable that several of these enzyme systems are highly specific. But enzymes are proteins, that is, precise chemical structures manufactured in the cell from regular subunits by a complex coding process. Keeping this concept in mind, it is obvious that each amine and each precursor or metabolite may be considered a trait by which we can trace species evolution, and, at the same time, we can get information about the relationships existing among closely related species. Thus, our data might contribute to biochemical taxonomy, which, in its broadest sense, has possibly even greater validity than traditional taxonomy based on morphological characters, such as somatic or osteological structures, size, or coloration.

However, it must be stressed that present data on the spectra of biogenic indolealkylamines should be considered only preliminary to a more extensive investigation and that all inferences drawn from these data should be accepted with prudence. For example, considerable hindrance to the theoretical elaboration of our analytical data is the conspicuous quantitative differences in the amine content found in different individuals of the same species, even when collected at the same time and at the same place. These differences may sometimes even simulate changes in the amine spectrum, due to the occasional absence or presence of a given amine. It is obvious that this physiological handicap may be largely overcome by screening pools of numerous specimens or by numerous individual screenings, which has been done, in our study, whenever possible. Our present data are also unable to demonstrate any kind of reliable sex differences.

In some previous works (Capurro and Silva 1959; Hunsaker, Alston, Blair, and Turner 1961; Wittliff 1962, 1964; Porter 1962, 1964, 1966; Porter and Porter 1967) paper chromatographic analysis of the parotoid-gland secretion of toads has been used in assessing intragroup or intergroup relationships. Research reported in the present chapter, together with that in preceding papers (Cei and Erspamer 1966; Cei, Erspamer, and Roseghini 1967, 1968) and other papers in preparation, represent an effort to check the actual value of this kind of biochemical investigation in as many amphibian species as possible.

The species considered in this chapter, together with data concerning number and state of the skins and date and place of collection, are indicated in Appendix I. Acetone was the solvent routinely used for extracting the skins. In fact, 80 percent acetone is the most suitable solvent for a complete recovery of tissue indolealkylamines. For details on the simple extraction procedure and for a description of paper chromatographic analysis and biological methods, reference has been made to other papers (Erspamer, Roseghini, and Cei 1964; Erspamer, Vitali, Roseghini, and Cei 1964, 1967). The problem of whether the amine spectrum found in dried skins may be considered qualitatively and quantitatively comparable with that occurring in fresh skins has been discussed and will be discussed elsewhere (Cei, Erspamer, and Roseghini 1967).

A number of indole compounds found in the toad skin have been identified in the present report with O-conjugates and 1-conjugates of amines with sul-

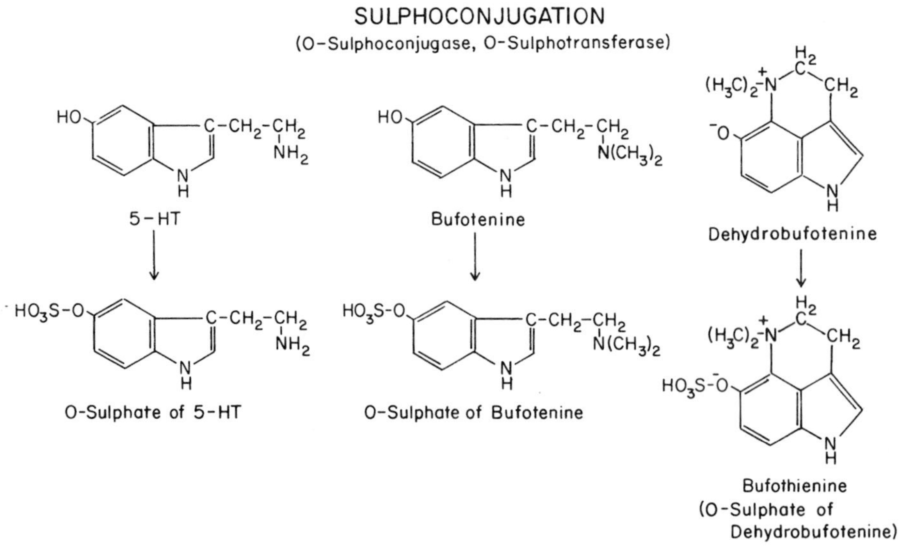

SULPHOCONJUGATION
(O-Sulphoconjugase, O-Sulphotransferase)

HO C-CH$_2$-CH$_2$ NH$_2$ N H
5-HT
HO C-CH$_2$-CH$_2$ N(CH$_3$)$_2$ N H
Bufotenine
H$_2$ C (H$_3$C)$_2$-N$^+$ CH$_2$ $^-$O N H
Dehydrobufotenine

HO$_3$S-O C-CH$_2$-CH$_2$ NH$_2$ N H
O-Sulphate of 5-HT
HO$_3$S-O C-CH$_2$-CH$_2$ N(CH$_3$)$_2$ N H
O-Sulphate of Bufotenine
H$_2$ C (H$_3$C)$_2$-N$^+$ CH$_2$ HO$_3$S-O N H
Bufothienine
(O-Sulphate of Dehydrobufotenine)

Fig. 12–5. Enzymatic events that occur in the toad skin: O-SULPHOCONJUGATION (O-sulphoconjugase, O-sulphotransferase).

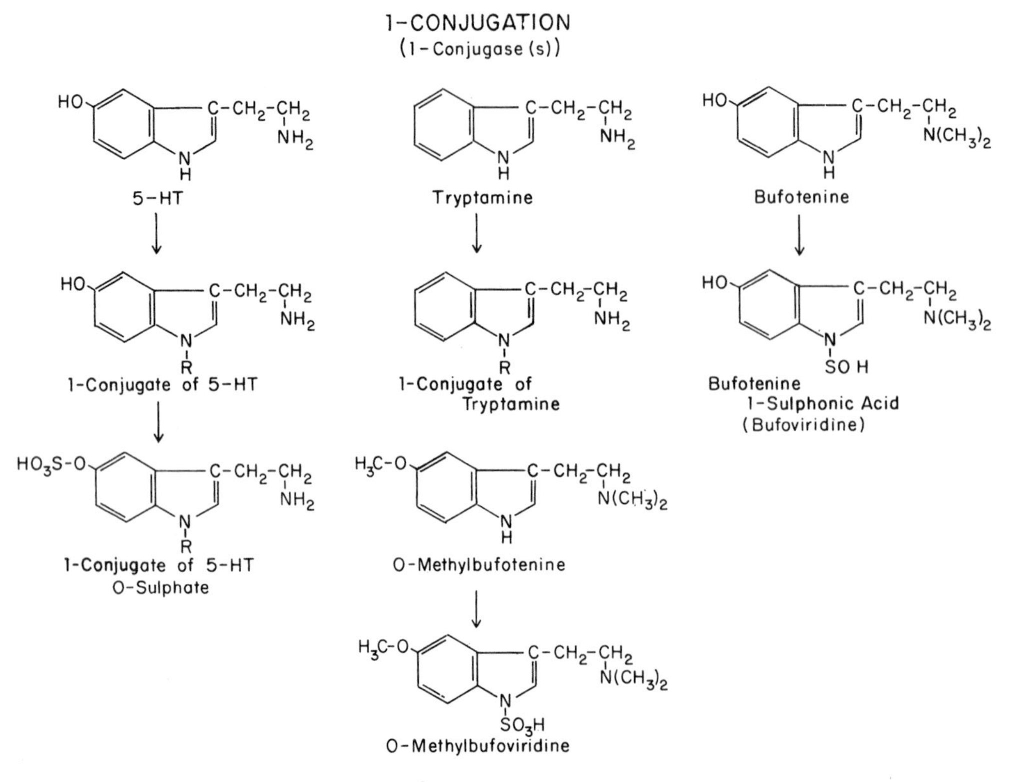

Fig. 12–6. Enzymatic events that occur in the toad skin: 1-CONJUGATION (1-conjugase[s]).

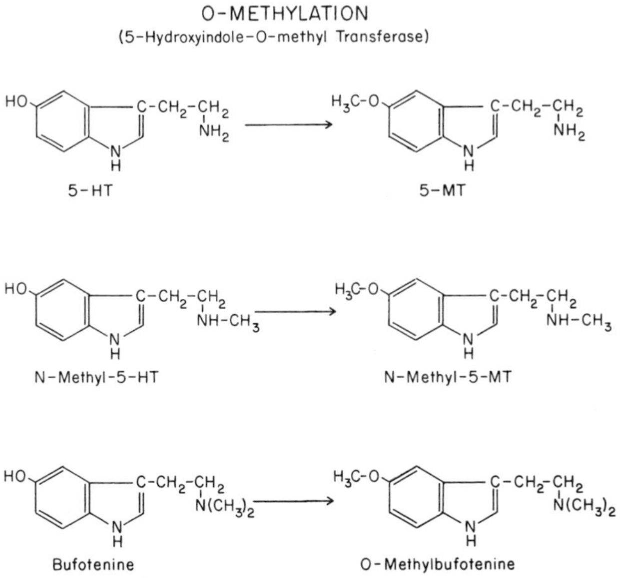

O-METHYLATION
(5-Hydroxyindole-O-methyl Transferase)

5-HT → 5-MT

N-Methyl-5-HT → N-Methyl-5-MT

Bufotenine → O-Methylbufotenine

Fig. 12–7. Enzymatic events that occur in the toad skin: O-METHYLATION (5-hydroxyindole-O-methyl transferase).

phuric acid or other acids. It should be stated that this identification is only tentative. In fact, in spite of considerable positive evidence, only the synthesis of the conjugated compounds shown in the diagrams will prove or disprove their identity with the natural compounds found in the toad skin. This reservation does not apply to bufothionine, whose identification with the O-sulphate of dehydrobufotenine is certain. Prudence is more necessary with bufoviridine, which until now was considered bufotenine O-sulphate; it is possibly bufotenine 1-sulphonic acid.

Taxonomic Application of the Biochemical Data

South American Toads

We will begin by discussing the neotropical stocks of the South American continent, whose ancestral relationships and tentative evolutionary history have been analyzed in another chapter (Chap. 6). These involve early Tertiary genocenters, both in Guiana-Brazilian and Patagonian shields. Two different, perhaps opposing, or diverging stocks are evident on the basis of geographic distribution and morphological arguments, besides the more recent tests of genetic compatibility. One is a western, or Andean stock, which can be provisionally designated as a "southwestern" section of the "thin-skulled line," corre-

OXIDATIVE DEAMINATION
(Monoamineoxidase)

R = -H_2, -HCH_3, [-$(CH_3)_2$]
Hydroxyindolealkylamines (Primary, Secondary, [Tertiary] Amine)

5-Hydroytryptophol

5-Hydroxyindoleacetic Acid

R = -H_2, -HCH_3, [-$(CH_3)_2$]
Methoxyindolealkylamines (Primary, Secondary, [Tertiary] Amine)

5-Methoxytryptophol

5-Methoxyindoleacetic Acid

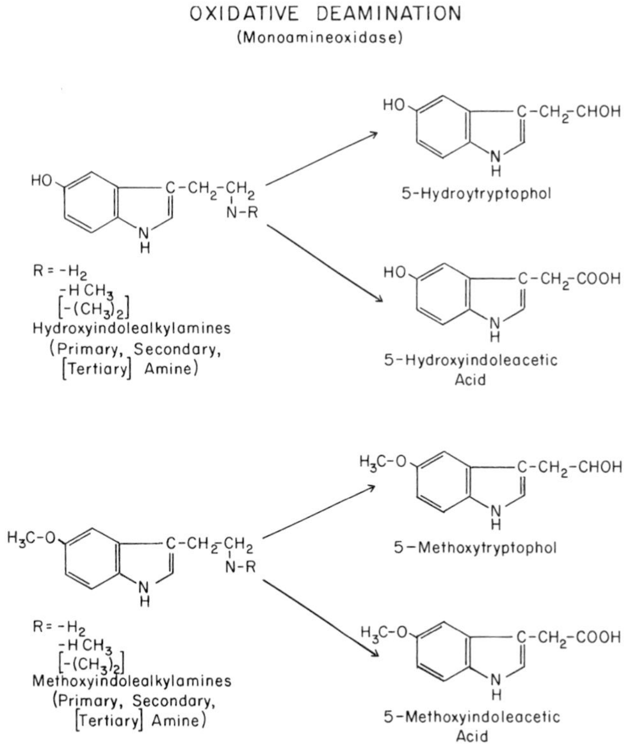

Fig. 12–8. Enzymatic events that occur in the toad skin: OXIDATIVE DEAMINATION (Monoamineoxidase).

sponding to the *spinulosus* complex. Second, there are stocks in South America, probably related to the "southern line" of *valliceps*-like features, such as the *marinus* and *granulosus* groups, which are widely distributed on the entire continental mass, eastward from the Cordilleran range, which is the realm of the *spinulosus* stock.

We have examined most of the taxonomic units of the *spinulosus* complex. In spite of the general similarities within this group, advanced speciation can be recognized in many of its forms, and the ecological and paleogeographic features of its radiation have been discussed by one of us in another chapter (see Chap. 6). The screening of indolealkylamines, which imply specific enzymatic activities, presents a common physiognomy of all the taxonomic units. All species of the complex present a lively biosynthesis of indolealkylamines. However, 5-HT, together with N-methyl-5-HT, is generally absent in the skin, owing to the intensity of N-methylating processes, followed by cyclization and conjugation processes.

From this common pattern some distinctive features emerge for some of the species of the *spinulosus* complex. *B. flavolineatus,* for example, is characterized by an unusually intense and complete N-methylating activity leading to the formation of large amounts of not only bufotenine but also bufotenidine, and by a relatively reduced cyclizing activity, with the consequence of reduced amounts of dehydrobufotenine and bufothionine. Something similar may be observed for *B. trifolium.*

B. chilensis, in turn, is characterized not only by the regular occurrence in its skin of the usual O-sulphate of dehydrobufotenine (bufothionine), but also by the occurrence of the O-sulphate of bufotenine, a derivative found so far only in the *calamita* group from Europe and in some groups of the western section of the American "northern line," but completely lacking in all the representatives of the "southern line." The O-sulphate of bufotenine has also been traced in the "*B. spinulosus* population" of Cajamarca and exceptionally (in one skin out of ten) in the *B. spinulosus* population of San Isidro, Mendoza. Additional research on more specimens is necessary before making definitive assumptions about different characteristics in the amine spectrum of *B. atacamensis* and *B. limensis.*

However, an independent position is demonstrated by our screening for the small *B. variegatus* from the cold forest of Valdivia. It is surely a very ancient Patagonian form, probably from some early Tertiary austral stock of primitive toads, but its amine spectrum differentiates it from any other form of the southwestern section. Its high content of bufotenine is peculiar, and the occurrence of dehydrobufotenine and bufothionine is negligible.

An ancient neotropical branch of the "southern line" has been suggested on the basis of morphological, osteological, and embryological arguments (genetic compatibility). Undoubtedly it is impossible to characterize, with our biochemical evidence, the primeval steps of its phyletic radiation, because of the intervention of the peculiar pattern of speciation, the environmental challenges, and the probable effects of convergent evolution. However, at the level of the present speciation and in accordance with the relatively limited boundaries of the "groups," there is a remarkable agreement between the phenotypical metabolic expression, which the biochemical evidence reflects, and taxonomic status.

For example, *B. marinus* shows a clear uniformity in the amine spectra for all the populations, from Mexico to Surinam, in the Amazonian environment. But the smaller, high Amazonian *B. poeppigi* shows as a "good specific character" a lowered activity of the enzyme system that leads to the formation of dehydrobufotenine. Likewise, a similarity is found in the spectra of *B. marinus* and *B. ictericus,* thus strengthening the evidence that these two, now geographically disjunctive, forms, from the Guiana-Amazonion range and the eastern forest coastal range, are related. They were surely closely related prior to the geological events leading to the dryness of the present central area of Chaco and caatingas. Considerable similarity of the amine spectra indicates a probable late common branching of *B. paracnemis* and *B. arenarum,* perhaps in agreement with some recent paleontological findings (Casamiquela, in press).

The so-called *granulosus* group is a much discussed taxonomic complex. *B. granulosus* Spix has been fractionated into a number of subspecific units, whose geographic distributions, still poorly known, cover much of the continental area, from the Panamian lowlands to the pampean southern range. For morphological, physiological, and probably genetic reasons, many of these so-called morphological subspecies may actually be good biological species, and in a number of cases their evident overlapping can support such a logical assumption. At any rate it is possible that — as in other neotropical forms — the speciation and differentiation of some of these populational groups could have occurred in a relatively short space of geological time. For example, the geographic distribution of *B. major* resembles the geographic xeric central range of the giant toad *B. paracnemis,* which extends from the San Francisco valley to the Argentine Chacoan woods. Screening of the amine spectra of the presently available species demonstrates that the enzyme activity is fundamentally the same in *B. granulosus goeldi,* in *B. major,* in *B. fernandezae,* and in *B. pygmaeus.* Only some quantitative differences can be traced. For example, dehydrobufotenine is always less abundant in the Chacoan toad *B. major* than in other examined forms of the complex. The spectra of indolealkylamines are strikingly similar for the *granulosus* stock and *B. crucifer* stock, which may be a very primitive toad (Chap. 18).

On morphological grounds, a strong relationship has also been suggested between the *granulosus* group and the Caribbean forms of the *peltacephalus* group. The cranial morphology and other arguments have been discussed. However, if one compares the amine spectrum of the skin of *B. peltacephalus* from

Cuba, a striking difference can be noted, indolealkylamines being represented in the *B. peltacephalus* skin only by small amounts of 5-HT and trace amounts of dehydrobufotenine. If some relationship between the insular radiation and an ancestral common neotropical stock is assumed, geographic isolation may have played a very important role in the biochemical evolution. The reduction or shortage of some enzymic activities could thus depend on either some genetic selective factor or genetic isolation (genetic drift?). Similar events — under different environmental factors — have perhaps occurred in the case of *B. atacamensis*, undoubtedly a derivative form from a primitive *chilensis* stock, whose reduced amine spectrum might be strictly conditioned by its complete isolation along the few rivers of the azoic Atacama Desert.

Central and North American Toads

The evolutionary radiation of American toads is enormous, but in the last years some tentative arrangements of their fundamental main lines have been suggested. Tihen (1962) has used osteological arguments, and Blair (1963, 1964, 1966), has pointed out genetic compatibility in interspecific crosses. Two primeval main lines have been tentatively suggested by Blair for the North American toads. They have been named "northern line" (or "thin-skulled line") and "southern line" (or "broad-skulled line").

The most representative elements of the northern line, more precisely of its western section, are the toads of the *boreas* group and related groups (Chap. 7). The amine spectrum in the skin of the *boreas* toads is rather homogeneous, which indicates a close taxonomic affinity among the species of this group, but it shows no peculiar characteristics. The spectrum is very similar to that presented by the European *B. bufo bufo*. *B. canorus*, however, an isolated form from the high summits of Sierra Nevada, California, differs from *B. boreas* in the very high activity of N-methyl transferase, cyclizing enzymes, and o-sulphoconjugase, the last-named leading to an unusual accumulation of bufothionine.

Although they belong to the same western section of the northern line, the other groups of this section present one of the most characteristic spectra of indolealkylamines that can be encountered in toads, which sometimes permits an immediate diagnosis of species. All toads belonging to the *punctatus*, *debilis*, *marmoreus*, and *alvarius* groups show high activity in the indolealkylamine biosynthesis and a pronounced activity of N-methylating enzymes. Cyclizing enzymes are highly active in the *punctatus* and *debilis* groups, while absent or extremely inactive in the *marmoreus* and *alvarius* groups. But the most striking biochemical feature of these groups of toads is the occurrence in their skin of very vigorous conjugation processes. Conjugation may be observed not only for dehydrobufotenine, where this is present, but, what is far more characteristic, also for bufotenine and eventually for 5-HT. In the *punctatus* and *debilis* groups only the 1-conjugate of bufotenine (bufoviridine or bufotenine 1-sulphonic acid) is apparently present. (Trace amounts of bufotenine O-sulphate may have escaped our attention.) In the *marmoreus* and *alvarius* groups, both bufoviridine and, in considerably larger amounts, the O- or 5-sulphate of bufotenine are present. It may be seen that in its indolealkylamine spectrum the *marmoreus* group is indistinguishable from the European *calamita* group, whose early relationship with some primeval representatives of the northern line has been discussed (Blair 1963).

The apparently peculiar behavior of *B. kelloggi*, with its enormous accumulation of dehydrobufotenine and lack of conjugated amines, cannot be considered seriously, because only one specimen was available for the screening. On the contrary, the exceptional position of *Bufo alvarius* may be emphasized on the basis of a careful study. The skin of *Bufo alvarius*, the only one among the skins of the sixty-five toads examined, uniquely possesses a formidable O-methyl transferase activity, which leads to the accumulation of enormous amounts of N, N-dimethyl-5-methoxytryptamine (O-methylbufotenine), and small amounts of N-methyl-5-methoxytryptamine. O-methylbufotenine constitutes sometimes as much as 15 percent of the dry weight of the parotoid and tibial glands of *B. alvarius*. The methoxytryptamines may give origin, following attack by monoamine-oxidase, to the pertinent deaminated metabolites 5-methoxytryptophol and 5-methoxyindole acetic acid, which can regularly be found in the extracts of the *B. alvarius* skin.

The *americanus* group is a well-known set of eastern and in some cases (*B. hemiophrys*, *B. microscaphus*) western species, whose evolutionary trends and probable phyletic history have been thoroughly discussed by Blair (1963). It is a very distinctive group in spite of its morphological and biological radiation, which is probably connected with the late Pliocene and Pleistocene events. The indolealkylamine spectra accord well with the general relationship of the group and its intrageneric differentiation. A reduced cyclizing activity is the rule, resulting in lack or extreme scarcity of dehydrobufotenine. The only exception is the isolated western populations of *B. hemiophrys*. Conjugases of any kind are completely lacking. In addition, specific quantitative enzymatic trends are evident for all the examined forms: *B. americanus, B. terrestris, B. woodhousei, B. microscaphus*, or *B. hemiophrys*, not to mention the intraspecific and geographical variation.

The small eastern toad *B. quercicus* from the forested southeastern United States is perhaps a derivative from the eastern stock. Only moderate amounts of 5-HT and of dehydrobufotenine can be found in its skin, indicating reduced activity of the pertinent enzymes. Blair states in his discussion that *B. quercicus* appears at first glance "a miniature of *B. woodhousei*." Due to the strong agreement existing between the amine spectra and the general arrangement of the *americanus* group as a "low crested" representative of the "northern line," other evidence could be added about the postulated origin of *B. quercicus* from the *americanus* group and about the allocation of this diminutive species of the "northern line."

The "southern line," as treated in the tentative arrangement by Blair, and considering the previous osteological considerations of Tihen (1962), is primarily represented by the *valliceps* stock and its allies. *Bufo valliceps*, of the southern states and Mexico, is closely related to a number of Mexican and Guatemalan forms previously discussed under their specific status by Porter (1964). Our data pertain only to *Bufo valliceps*, in whose skin the only representative of idolealkylamines is 5-HT, indicating the absence of methylating, cyclizing, and conjugating enzymes. (Compare App. I, Table 3.) A great evolutionary radiation seems to have occurred in the heavy-skulled toads of the "southern line," both in Central and in South America. As discussed in other chapters of this book, the primeval relationships of the most remote branches of the American bufonid lines are still unclear, and paleogeographic problems are also related to the more ancient radiating steps of all these groups. Different kinds of evidence may support some remote relationships between *valliceps* and the neotropical stocks, such as the *marinus, granulosus*, and *typhonius* groups. But, on more objective grounds, we must point out that if the biochemical methods and criteria — in our case the amine spectrum — can be very useful for a better screening of the interspecific and intraspecific relationships, this support could only be of relative utility in some special or general cases of phylogenetic reconstruction.

In spite of the above-mentioned reservations in the use of biochemical affinities as a mean of phyletic comparison, we find that the other evidence of taxonomic affinity between *B. cognatus* and *B. speciosus* is also strengthened by the amine spectra of the two species. This is a very elegant example of total agreement between biochemical and morphological characters, and the genetical value of metabolic pathways of indolealkylamines as phenotypical expression of the genetic code is here strikingly focussed. (The *cognatus*-group toads are specialized for a life in desert environments and range over the very extensive Sonoran area, from Texas to Mexico.)

Sulphoconjugating enzymes are lacking in all presently studied Central American toads (*B. luetkeni, B. coccifer*, and *B. coniferus*) except in *B. canaliferus*, from Mexico to Guatemala, in which this enzymic set is present and very active. True evolutionary trends within genetically defined groups can thus be ascertained, in terms of specific genetic information. The information is to some extent independent of the environmental effects, but is subject, however, to general geographic variation.

The *guttatus* and *typhonius* groups, probably of ancient neotropical origin and most likely only recent immigrants to Central America, are still poorly known. It is also difficult to compare their peculiar amine spectra to the previously discussed spectra of the related forms of the *valliceps* group. Considerable activity of cyclizing enzymes and of o-sulfoconjugase may be reported for the *typhonius* toads from Panama. Similarly, a formidable activity of N-methylating and cyclizing enzymes, leading to the accumulation of enormous amounts of dehydrobufotenine (up to 18,000 μg/g dry skin), has been ascertained in Costa Rican specimens of *B. haematiticus*. A very independent ancient radiation of many of these forms can thus be postulated from comparative biochemical evidence.

Eurasian Toads

Eurasian stocks must be divided into three distinct groups.

Toads belonging to the Indonesian or Malayan group (*B. asper* from Malaya and *B. juxtasper* from Borneo) are virtually lacking indolealkylamines; the traces that are present indicate extremely torpid enzyme activity. The same picture may be encountered in the sympatric, related bufonid genus *Pseudobufo subasper* from Malaya.

On the other hand, toads belonging to the continental Euro-Indian group, although including species from very different localities (*B. melanostictus, B. bufo, B. stomaticus*), present a rather uniform amine spectrum, which indicates a lively activity of all enzyme systems involved in the biosynthesis of indolealkylamines, with the exception of 1-conjugase.

Finally, toads belonging to the third group (*B calamita* and *B. viridis*), with European and Mediterranean ranges, are sharply characterized by absence or extreme scarcity of cyclizing enzymes (hence absence of dehydrobufotenine and bufothionine) accompanied by very lively N-methylating and conjugating (both 1- and 5-conjugating) activities, leading to the accumulation of large amounts of bufoviridine and, to a lesser extent, of the O-sulphate of bufotenine. It should be stressed here that the same conjugated derivatives may be found in some American toads of the western section of the so-called northern line.

African Toads

There is remarkable uniformity among all the hitherto examined African toads (*B. mauritanicus, B. regularis, B. kisoloensis, B. berghei, B. funereus, B. garmani, B. rangeri, B. maculatus, B. carens*), despite the important diversity of their chromosomal constitution, which, as reported by Bogart (1968), is $2n = 20$ for *B. regularis, B. gutturalis, B. garmani, B. rangeri, B. brauni, B. latifrons,* and $2n = 22$ for *B. mauritanicus, B. gariepensis, B. rosei,* and *B. carens*. We are dealing with a general paleotropic stock, whose species are perfectly capable (with the exception of *B. carens*) of synthesizing 5-HT but are completely lacking N-methyl transferase activity and hence cyclizing activity. Conjugases, on the contrary, both 5 (O)-conjugase and 1-conjugase, are highly active in all species. African toads are further characterized, again with the exception of *B. carens*, by the content of other indole derivatives (emerald green spot, see Appendix I, Table 1) and by the content of a peculiar hydroxyphenylalkylamine, *kisoloensin*, the study of which is in progress. Among the examined African toads, *B. carens* occupies a unique position, since it contains large amounts of a 1-conjugated derivative of tryptamine, which is a quite unusual indolealkylamine in the amphibian skin.

Needless to say, screening of other species is necessary before drawing acceptable conclusions on the evolutionary relationships existing between the different species of the African stock. For the present we may hold that our biochemical data do not disagree with Bogart's assumption (Chap. 10) that all the African toads possessing 20 chromosomes may likely be derived from a common ancestor having 22 chromosomes.

Conclusion

The amine and polypeptide spectra of the amphibian skin vary among the representatives of different families, but also among different genera, species, and subspecies. The specific spectrum of biogenic amines and polypeptides can be profitably used in biochemical taxonomy and for assessing evolutionary relationships, as in our present screening of sixty-five forms belonging to the genus *Bufo*. The skin of the examined *Bufo* offers a spectacular array of indolealkylamines; approximately twenty indole derivatives have been identified so far. The occurrence of a given indole derivative necessarily implies the occurrence of the enzyme systems catalyzing its biosynthesis; it is probable that the enzyme systems are highly specific, since they are proteins, that is, precise chemical structures manufactured in the cell from regular subunits by a complex coding process. Thus, each amine and each precursor or metabolite may be considered to be a trait by which we can trace species evolution and get information about the relationships existing among more or less closely related species or species group. In this way, and by employing the quantitative differences in the amine content of different individuals of the same species, a number of paleotropical, palearctic, Oriental, nearctic, and neotropical toads have been studied and their specific amine spectra screened and discussed.

It has been mentioned that all the hitherto examined African toads are remarkably uniform, despite their chromosomal diversity. This paleotropical stock is only capable of synthesizing 5-HT; however it is completely lacking N-methyl transferase activity and cyclizing activity, but not conjugase activity (5(O)-conjugase and 1-conjugase). Among the African toads, *B. carens* occupies a unique position, since it contains large amounts of a 1-conjugated derivative of tryptamine, a quite unusual indolealkylamine in the amphibian skin.

The Eurasian stocks can be divided into three distinct groups: a Malayan group virtually lacking indolealkylamines; a Euro-Indian group (*B. melanostictus, B. bufo, B. stomaticus*), presenting a rather uniform amine spectrum, which indicates a lively activity of all enzyme systems involved in the biosynthesis of indolealkylamines, with the exception of 1-conjugase; a third group, the *calamita* group (including *B. viridis*), sharply characterized by lack or extreme scarcity of cyclizing enzymes, accompanied by very lively N-methylating and conjugating activities, leading to the accumulation of large amounts of bufoviridine and of O-sulphate of bufotenine. The same conjugated derivatives may be found in some American toads of the "northern line" of Blair.

Two primeval main lines of the American toads have been tentatively proposed by Blair; the "northern line" and the "southern line." The amine spectra indicate a close taxonomic affinity among the species of the first section, with the *boreas* group being very similar to the European *B. bufo*. Other representatives of this section, belonging to the *punctatus, debilis, marmoreus,* and *alvarius* groups, show a general liveliness of the indolealkylamine biosynthesis and a pronounced activity of N-methylating enzymes. Cyclizing enzymes are highly active in the *punctatus* and *debilis* groups, while lacking or very torpid in the *marmoreus* and *alvarius* groups. Another striking biochemical feature of these toads is the occurrence of vigorous conjugation processes, producing some specific spectra that, as in the case of *marmoreus* group, are indistinguishable from the European *calamita* group, whose early relationships with some primeval representatives of the "northern line" have been discussed (Blair 1963). The exceptional position of *B. alvarius* from the Arizona desert may also be emphasized, the only one among the sixty-five toads examined having the unique property of a formidable O-methyl transferase activity, which leads to the accumulation of enormous amounts of N,N-dimethyl-5-methoxytryptamine.

The indolealkylamine spectra accord well with the general relationship of the *americanus* group and its intragroup differentiation, which is probably related to late Pliocene and Pleistocene events. The agreement existing between the amine spectra and the general taxonomic arrangement may add evidence to the postulated origin of *B. quercicus* from the *americanus* group.

The "southern line," as tentatively presented by Blair, has been discussed following our present biochemical methods and criteria. The absence of methylating, cyclizing, and conjugating enzymes is the rule in *B. valliceps,* in whose skin only 5-HT is reported. The great taxonomic affinity between *B. cognatus* and *B. speciosus* is supported by their amine spectra. Sulphoconjugating enzymes are lacking in the *cognatus* and *coccifer* groups and in *B. coniferus* and *B. luetkeni* of the *valliceps* group, but they are present in the *canaliferus* group. A very independent ancient radiation of the *guttatus* and *typhonius* groups may be postulated from the comparative biochemical evidence.

There is a western or Andean stock that is provisionally designated as a "southwestern" section of the thin-skulled line, corresponding to the *spinulosus* complex. Screening of the indolealkylamines reveals a common physiognomy for all populations studied, but from this common pattern some distinctive biochemical features emerge. This is the case for *B. flavolineatus* and *B. trifolium* from Peru, *B. chilensis* and *B. atacamensis* from Chile, and *B. variegatus* from the cold Valdivian forest.

Finally, an ancient neotropical branch of the "southern line" may be identifiable from the phenotypical metabolic expression of the enzymatic activity of the skin. For example, the clear uniformity of the amine spectra for all the populations of *B. marinus,* from Mexico to Surinam, has been stressed, as well as the similarity between *B. marinus* and *B. icterícus,* now geographically disjunct forms. The considerable similarity of the amine spectra also indicates a late common branching of *B. paracnemis* and *B. arenarum.* Likewise, the many populations of the poorly understood *granulosus* complex show a common enzyme activity, but quantitative differences of significance may be traced, which are related to the probable speciation in the group. There is a noticeable difference, on the other hand, between the *granulosus* group and the morphologically related forms of the *peltacephalus* group.

REFERENCES

Blair, W. F. 1963. Evolutionary relationships of North American toads of the genus *Bufo*: A progress report. Evolution 17:1–16.

———. 1964. Evidence bearing on relationships of the *Bufo boreas* group of toads. Texas J. Sci. 16:181–192.

———. 1966. Genetic compatibility in the *Bufo valliceps* and closely related groups of toads. Texas J. Sci. 18:333–351.

Bogart, J. P. 1968. Chromosome number difference in the amphibian genus *Bufo*: The *Bufo regularis* species group. Evolution 22:42–45.

Capurro, L. F., and F. C. Silva. 1959. Valor taxonómico del estudio cromatográfico del veneno de las parótidas de *Bufo spinulosus* y *Bufo variegatus*. Inv. Zool. Chil. 5:189–197.

Casamiquela, R. M. 1968. Sobre un nuevo *Bufo* fosil de la Provincia de Buenos Aires (Argentina). Rev. Asoc. Paleontol. Argentina (*in press*).

Cei, J. M. 1967. Remarks on the geographical distribution and phyletic trends of South American toads. Texas Mem. Mus. Publ. Pearce-Sellards Series 13:1–21.

———, and V. Erspamer. 1966. Biochemical taxonomy of South American amphibians by means of skin amines and polypeptides. Copeia 1966:74–78.

———, V. Erspamer, and M. Roseghini. 1967. Taxonomic and evolutionary significance of biogenic amines and polypeptides occurring in amphibian skin. I. Neotropical Leptodactylid frogs. Syst. Zool. 16: 328–342.

———. 1968. Taxonomic and evolutionary significance of biogenic amines and polypeptides occurring in amphibian skin. II. Toads of genera *Bufo* and *Melanophryniscus*. Syst. Zool. 17:3.

Erspamer, V., M. Roseghini, and J. M. Cei, 1964. Indole-, imidazole-, and phenyl-alkylamines in the skin of thirteen *Leptodactylus* species. Biochem. Pharmacology 13:1083–1093.

Erspamer, V., T. Vitali, M. Roseghini, and J. M. Cei. 1964. The identification of new histamine derivatives in the skin of *Leptodactylus*. Arch. Bioch. Biophys. 105:620–629.

———. 1967. 5-Methoxy- and 5-hydroxyindoles in the skin of *Bufo alvarius*. Biochem. Pharmacology 16: 1149–1164.

Hunsaker, D., R. E. Alston, W. F. Blair, and B. L. Turner. 1961. A comparison of the ninhydrin positive and phenolic substances of parotoid gland secretions of certain *Bufo* species and their hybrids. Evolution 15: 352–359.

Porter, K. R. 1962. Evolutionary relationships of the *Bufo valliceps* group in Mexico. Ph.D. Thesis, Univ. Texas.

———. 1964. Chromatographic comparisons of the parotoid gland secretions of six species of the *Bufo valliceps* group, p. 451–456. *In* C. A. Leone (ed.), Taxonomic Biochemistry and Serology. Ronald Press, New York.

———. 1964. Distribution and taxonomic status of seven species of Mexican *Bufo*. Herpetologica 19: 229–247.

———, and W. F. Porter. 1967. Venom comparisons and relationships of twenty species of New World toads (genus *Bufo*). Copeia 2:298–307.

Robinson, B., G. F. Smith, A. H. Jackson, D. Shaw, B. Fryman, and V. Deulofeu. 1961. Dehydrobufotenin. Proc. Chem. Soc. 310–311.

Tihen, J. A. 1962. Osteological observations of New World *Bufo*. Amer. Midl. Nat. 67:157–183.

Wittliff, J. L. 1962. Parotoid gland secretions in two species groups of toads (genus *Bufo*). Evolution 16:143–153.

———. 1964. Venom constituents of *Bufo fowleri*, *Bufo valliceps* and their natural hybrids analysed by electrophoresis and chromatography, p. 457–464. *In* C. A. Leone (ed.), Taxonomic Biochemistry and Serology. Ronald Press, New York.

13. Evidence from Parotoid-Gland Secretions

BOBBI S. LOW

University of British Columbia
Vancouver, British Columbia, Canada

Introduction

In any evolutionary study, it is desirable that all available lines of evidence be considered. Nonmorphological data have proved particularly valuable in cases where morphological criteria are inadequate or unavailable. Physiological (Moore 1951; Volpe 1952), ethological (A. P. Blair 1942), mating call (W. F. Blair 1956*a*, *b*), and genetic compatibility (W.F. Blair 1963*a*, *b*, etc., 1966) criteria have proved invaluable in elucidating amphibian relationships. In the genus *Bufo,* the existence of an easily obtained dermal secretion makes it possible to apply comparative biochemical techniques. The secretion is produced by modified dermal glands that are present over most of the skin area but concentrated in the neck region (the parotoid glands).

Because the venom contains many pharmacologically active principles, including indoles, nonindole alkaloids, and bufadienolides, interest in its composition was at first not systematic, but pharmacological. Jensen and Chen (1930; 1936), Deulofeu (1948), Deulofeu and Duprat (1944), Chen and Chen (1933*a*–*h*), and Chen, Jensen, and Chen (1933*a*–*c*) surveyed the venom of numerous species of toads, particularly for comparative pharmacology.

Recently, Hunsaker et al. (1960), Porter and Porter (1967), Porter (1962, 1964*a*), Witliff (1962, 1964), and Low (1967, 1968) have examined with some success the chemical composition of bufonid skin secretions for evolutionary purposes. The secretion is well suited for use in evolutionary comparisons. The venom constitution is genetically determined (Wittliff 1962; Low 1967). Probably excretory in original function (although undoubtedly selected to avoid predators), the venom contains products of secondary metabolic pathways. Thus, it is evolutionarily less conservative than the protein groups common to fundamental pathways and more suitable for study at lower taxonomic levels. Cei, Erspamer, and Roseghini (1968) have demonstrated that analysis of the indolealkylamines can yield valuable information on the enzyme systems present. Because a single tissue in a restricted taxonomic framework is involved, the probability of confusion by apparent but not real similarities, or convergence (Anfinson, 1959; Mayr 1964), is reduced.

The choice of a biochemical approach and the selection of the most appropriate methods of comparison are very important for the success and usefulness

Reference		Arenobufagin	Argentinogenin	Bufalin	Bufarenogenin	Bufotalin	Bufotalidin	Bufotalinin	Cinobufagin	Cinobufotalin	Gamabufotalin	Hellebrigenol	Jamacobufagin	Marinobufagin	Resibufagin	Telocinobufagin	"Bufagins A, B, C"	Gamma sitosterol	Desacetyl-Bufotalin	Desacetyl-Cinobufagin	Desacetyl-Cinobufotalin
1/	*B. alvarius*						1														
2/	*B. arenarum*	3					2	2			1			2		2					
3/	*B. arenarum*	0.67	0.28	0.37	0.21		0.46	0.24				.024		1.57	0.42	0.61					
1/	*B. blombergi*																				
2/	*B. b. bufo*					3	2	1						2		1					
3/	*B. b. bufo*					4.50	0.58	0.53						1.5		0.31					
4/	*B. b. bufo*					+	+							+		+					
2/	*B. b. formosus*	1		(+)							1			2	(+)	1					
5/	*B. b. formosus*	+	+	+		+	+	+	+	+	+			+	+	+			+	+	+
2/	*B. marinus*						1				1	1		3		2					
6/	*B. marinus*		0.01	0.7			0.06				0.02	0.02	0.07	10.0							
2/	*B. mauritanicus*						1	1						1	3						
7/	*B. mauritanicus*	+		2.0			0.1	0.1						10.0	+	+					
8/	*B. mauritanicus*			+			+	+						+	+						
9/	*B. melanostictus*																+				
10/	*B. melanostictus*			(+)		+	+							+	(+)						
11/12/	*B. paracnemis*		+	+			+	+			+	+			+	+		+			
13/	*B. regularis*	+									+										
2/	*B. regularis* (Erythrea)	(+)					2	3				1		1	2						
2/	*B. regularis* (SW Afr.)	2					2	3				1		1	2						
2/	*B. regularis* skin	2									1				1						
1/	*B. spinulosus*	1									1				1						
2/	*B. spinulosus*	2						(+)			2			2	2	1					
1/	*B. valliceps*	2					1					1		3		2					
1/	*B. val.* skin	1												1		1					
14/	*Ch'an Su*	+		1.25-		0.6-	0.17-		1.0-	1.0-	.85-				1.7-	0.5-			0.10-	0.3-	0.12-
				1.5		0.8	0.2		1.2	1.2	1.0				2.0	0.6			0.15	0.4	0.15

Designations are as follows:

1 = present in small amounts
2 = present in moderate amounts
3 = present in large amounts
\+ = present; amount unspecified
(+) = probably present, but identification tentative

If decimal figures are given, they represent percent of venom identified as given compound.

References:

1/ Barbier et al. 1961
2/ Schröter, Tamm, Reichstein, and Deulofeu 1958
3/ Schröter, Tamm, and Reichstein 1958
4/ Urscheler et al. 1955
5/ Iseli et al. 1965
6/ Barbier et al. 1959
7/ Linde and Meyer 1958
8/ Bolliger and Meyer 1957
9/ van Gils 1938
10/ Iseli et al. 1964
11/ Zelnik and Ziti 1962
12/ Zelnik et al. 1964
13/ Bharucha et al. 1961
14/ Ruckstall and Meyer 1957

Table 13–2. Paired Affinities Based on the Bufogenin Venom Compositions Summarized in Table 13–4.

1/	*B. alvarius*																									
2/	*B. arenarum*	17																								
3/	*B. arenarum*	10	46																							
1/	*B. blombergi*	0	0	0																						
2/	*B. bufo*	20	57	37	0																					
4/	*B. bufo*	20	57	37	0	100																				
5/	*B. bufo*	25	43	27	0	80	80																			
2/	*B. formosus*	0	50	45	0	22	22	25																		
6/	*B. formosus*	7	40	47	0	33	33	27	40																	
2/	*B. marinus*	20	30	37	0	43	43	50	38	25																
7/	*B. marinus*	14	43	42	0	20	20	22	30	29	50															
2/	*B. mauritanicus*	25	43	40	0	50	50	33	25	27	29	22														
8/	*B. mauritanicus*	17	63	70	0	50	50	38	63	47	33	27	57													
9/	*B. mauritanicus*	20	38	50	0	43	43	29	38	33	25	33	80	71												
10/	*B. melanostictus*	0	0	0	0	0	0	0	0	0	0	0	0	0	0											
11/	*B. melanostictus*	20	22	37	0	43	43	50	38	33	25	33	60	50	67	0										
12/13/	*B. paracnemis*	11	37	58	0	27	27	18	37	41	40	45	30	45	40	0	27									
14/	*B. regularis*	10	33	9	0	0	0	0	33	13	17	13	0	13	0	0	0	10								
2/	*B. reg.* (Erythrea)	20	50	60	0	38	38	25	33	31	38	30	67	50	57	0	38	36	0							
2/	*B. regularis*	20	50	60	0	38	38	25	33	31	38	30	67	50	57	0	38	36	0	100						
2/	*B. regularis* skin	0	29	18	0	0	0	0	50	20	14	11	17	25	14	0	14	20	67	29	29					
1/	*B. spinulosus*	0	29	18	0	0	0	0	50	20	14	11	17	25	14	0	14	20	67	29	29	100				
2/	*B. spinulosus*	0	71	45	0	38	38	25	71	40	25	18	43	68	38	0	22	36	33	50	50	50	50			
1/	*B. valliceps*	20	57	50	0	43	43	50	38	24	67	33	29	50	25	0	25	27	17	57	57	14	14	38		
1/	*B. val.* skin	0	50	30	0	33	33	40	50	20	33	11	17	43	14	0	14	9	25	29	29	20	20	50	60	
15/	*Ch'an Su*	8	29	29	0	21	21	23	39	80	21	19	14	36	21	0	31	31	17	20	20	25	25	31	21	15

Note: References as in Table 13-1.

of the method. In the genus *Bufo,* general analyses of the venom composition are valid indicators of evolutionary divergence. More restricted studies of the bufadienolide composition have shown essentially no evolutionary correlations. Tables 13-1 and 13-2 summarize the bufadienolide composition and paired affinities (Ellison, Alston, and Turner 1962) based on these compositions. The venom of *B. regularis,* for example, shows no compounds in common when two studies are compared. It is important, therefore, that the same system of analysis be used for all species, that taxonomic designations be as accurate as possible, and that appropriate systems be chosen for study.

Methods and Rationale

Fourteen hundred individual samples of the parotoid-gland secretion were taken from toads of thirty-nine species from eighty-one localities (Table 13-3). The samples were analyzed by two-dimensional paper chromatography, with multiple staining techniques, as previously described (Low 1967, 1968). Samples were chromatographed on Whatman #3 mm. paper in butanol:acetic acid:water (3:1:1). Next, the chromatograms were rotated 90°, and a solvent of 100:1 8 percent aqueous NaCl:gacial acetic acid (Jepson 1963) was used. Chromatograms were examined under ultraviolet light (360 mu) and stained sequentially (Smith 1963) with ninhydrin, Ehrlich reagent, sulfanilic acid, and bromcresol green. Alkali reagent, rhodamine B, and ninhydrin-acetic acid were used as corroborative tests on selected chromatograms. Table 13-4 and figure 13-1 show the compounds characterized. Tentative identifications of 5-hydroxytryptamine (No. 7), bufotenine (No. 50), and urea (No. 13) were made by concurrent chromatography with pure standards in the solvent systems used. Because individual samples were used, not enough material was available to per-

Table 13–3. Summary of the Rf's and Staining Characteristics of the Compounds in the Venoms Examined*

mpound No.	Rf_1	Rf_2	Visible	Ultraviolet	Ninhydrin	Ehrlich	Bromcresol G.	Other
1	22.0	25.1		lt b				
2	22.0	45.15		lt b				
3	28.0	59.8		lt b				
4	27.3	87.4		lt b				
5	49.97	1.5	faint	light				
6	49.97	31.0	faint	light		faint		
7	49.97	44.2	y-brn	dk & g†		pu – gr-brn	pos.	
8	49.97	51.6		b				
9	49.97	62.5		b-g				
10	49.97	71.3		b-g		p-or		
11	47.5	81.5		b/dk				
12	73.7	0.75		lt			pos.	Rhodamine B
13	64.9	93.0			y			
14	22.0	48.5		p				
15	28.0	59.8		p	p-or			
16	49.97	51.6		p				
17	55.0	95.2			pu			
18	49.9	29.0		p				
20	54.0	50.0				p-or		
21	67.0	65.5		lt				
22	65.0	71.5		lt b				
24	69.7	43.2		dk		y		
26	45.0	57.8				pu		
27	63.0	90.6		lt				
29	82.5	83.2		lt b				
30	79.2	83.2		lt		p		
31	11.2	96.5			pu			
32	49.9	77.0		brn				
35	61.0	2.7		lt b				
36	28.0	32.5				pu		
37	64.1	36.9		lt				
39	65.3	29.7		lt				
41	75.9	95.3			p			
43	73.5	41.4		b		pu		
44	85.0	5.0		dk				
45	90.5	70.2			pu			
46	39.0	97.3		b	p-pu			
47	45.6	97.5			p-pu			
49	30.0	93.0			p-pu			
50	71.7	58.3	brn	dk & p†		pu		

Color abbreviations are as follows:

lt = light	g† = green halo
dk = dark	gr = gray
b = blue	or = orange
b-g = blue-green	p = pink
bl = black	p† = pink halo
brn = brown	pu = purple
g = green	y = yellow

) indicates the occasional presence of a second color reaction. Thus, b/pu indicates that the color is usually blue, occasionally purple.
-) denotes a color change. Thus, pu – gr-brn indicates an initial purple color changing to gray brown.

Table 13–4. Summary of Compounds Found in Species Examined*

	1	2	3	4	5	6	7	8	9	10	11	12	13	14	15	16	17	18	20	21	22	24	26	27	29	30	31	32	35	36	37	39	41	43	44	45	46	47	49	50
B. crucifer					4	4	4		4	4	4		4													2	4										4			2
B. arenarum					4	4	4	2	4	4			3										2				4										4			1
B. marinus			2		4	4	2		4	4	4	2	2													1	4										4			4
B. spinulosus												4	1						1		4				4		4							2			4		4	4
B. blombergi			4		4	4	4		4	4	4	4				4																					3			
B. haematiticus																																								
B. bocourti		2	3		3	3	2	2	2	3	2	2	4																								4	3		3
B. valliceps (Tex.)		3	2	1	4	4	4	3	4	4	3	1	3		1												2	3									4			
B. luetkeni					4	4	4	4	4	4	4		4														4										4			
B. ibarrai			4	2	4	4	4		4	4	4		4		1																									
B. cavifrons					4	4	4		4	4	4		4													4											4			
B. coniferus					4	4	4	1	4	4	4	1	2															2									4			
B. canaliferus		2	2		4	4	4		4	4	4		2										2						4								4			
B. occidentalis			4		2	4	4	4	4	4			4																								4			4
B. alvarius												4	3						3						4	4			4		4								4	4
B. cognatus					4	4	4	4	4	4			4						4												2									
B. speciosus					4	4	4	4	4	4		4	4						2												2									
B. boreas												4	4						4		4				4	4			4		4								4	4
B. exsul												4	4						4		4				4	4			4		4	4							4	4
B. americanus					4	4	4		4	4			2						4							4													4	4
B. hemiophrys					4	4	4	2	4	4		2							4							2														4
B. microscaphus					4	4	4		4	4															4	2				1									4	4
B. terrestris					4	4	4	4	4	4			2						4																					4
B. quercicus					4	2	4		4	2			2		2																									4
B. coccifer					4	4	4		4	4	4		3						4								1										4			
B. viridis			4		3		1				3	3	3				1			1		3			4	4								2						4
B. stomaticus							1			2			4						4			3		2	3	3			3		4			2						4
B. bufo			4		1		1			3	3	3					1		3	1	1	3			4	4								2						4
B. melanostictus							3				4		4				4			3					4	4								4	4	4			4	4
B. mauritanicus			2		4	4	4		4	4	4	3				4																					3			
B. superciliaris		3			4	4	4	4	4	4	4	2	4																											
B. garmani		4		4	4	4	4		4	4	4	4	4																											
B. rangeri		3	4		4	4	4	4	4	4	4	4	4				2																							
B. brauni		3			4	4	4	4	4	4	4	4	4																											
B. regularis																																								
El Mahalla			3		4	4	4	4	4	4	4	4	4		1		1																							
Ogbomosho			2		4	4	4	4	4	4	4	4	4																											

(Table continued opposite)

*Occurrence is given in percent class intervals:

1 = occurred in 1–25% of samples examined
2 = occurred in 26–50% of samples examined
3 = occurred in 51–75% of samples examined
4 = occurred in 76–100% of samples examined

Table 13–4. (Continued)

	1	2	3	4	5	6	7	8	9	10	11	12	13	14	15	16	17	18	20	21	22	24	26	27	29	30	31	32	35	36	37	39	41	43	44	45	46	47	49	50
					4	4	4	4	4	4	4	4	4																											
. Johns	3	3	3		4	4	4	4	4	4	4	4	4		1																									
naritzburg		3	3		4	4	4	4	4	4	4	4	4																											
ıço Marques		2	4		4	4	4	4	4	4	4	4	4																											
Trichardt	3	3	4		4	4	4	4	4	4	4	4	4																											
		3	3		4	4	4	4	4	4	4	4	4		1																									
		4	3	4	4	4	4	2	4	4	4	4	4	1																										
ai (1)	3	3			4	4	4	2	4	4	4	4	4																											
ai (2)			2		4	4	4		4	4	4	4	4	4	1	4	2	2																						
Mts.			3		4	4	4		4	4	4	3	3	4		3	2																							
e-Kampala		1	2		4	4	4		4	4	4	4	4	4		4	2																							
as			2		4	4	4		4	4	4	4	4	4		4	2	1																						
atus		3	3		4	4	4	4	4	4			1		1																									
s		3	3	4	4	4	4	4	4	4									3																					
i		4	4		4	4	4		4	4			2																											
rgi			4		4	4	4		4	4																														
epensis					4	4	4		4	4	4		4																											
ngae					4	4	4		4	4	4		2																											

structural analyses. However, much useful sys-
tic information may be obtained from chroma-
phic investigations alone, without structural
ses. As Turner and Alston (1959) and Alston
Turner (1962, 1963) have shown, the com-
ds, even without being identified (as long as
ccurrence in a species is validated), represented
nportant pool of variation for systematic com-
ons.

me form of similarity coefficient (Sokal and
th 1963) must be derived for quantitative com-
on of species. When pooled population samples
ompared, the paired-affinity concept of Ellison,
n, and Turner (1962) is suitable. The paired
ty represents the ratio of compounds shared by
wo species to the total number of compounds
nt in both species, and expresses the chromato-
ic, and presumedly the biochemical, affinity
een the species. When individual samples are
able and when interpopulation comparisons are
able, an affinity weighted for the percent of each
lation possessing the compounds provides more
mation than an affinity based on simple presence

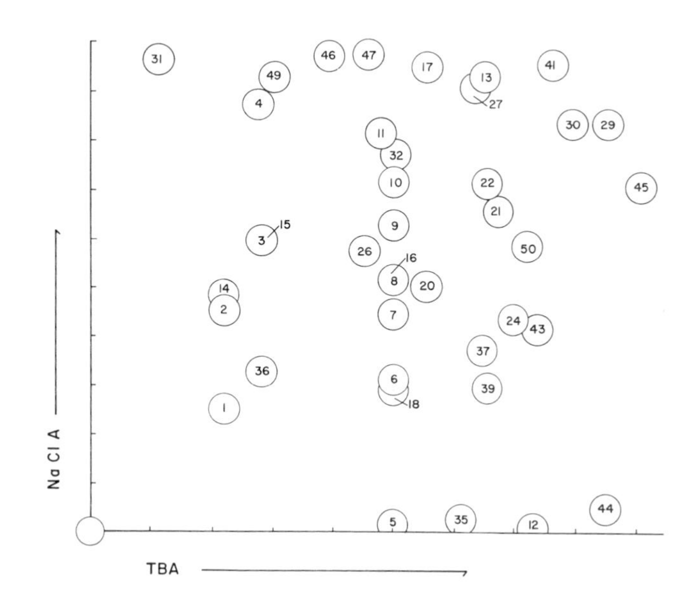

Fig. 13–1. Schematic representation of the compounds found in the *Bufo* venoms examined, and their Rf's in the solvent systems used. See table 13–2 for staining reactions of the compounds.

Table 13–5. Weighted Affinities between *Bufo* Species in Mexico, Central America, and South America

B. bocourti																
B. spinulosus	29															
B. marinus	71	41														
B. arenarum	62	30	75													
B. crucifer	64	31	87	84												
B. blombergi	60	20	73	62	68											
B. haematiticus	62	24	73	72	74	80										
B. canaliferus	64	14	74	74	75	74	74									
B. cavifrons	62	15	75	75	90	72	78	79								
B. coniferus	64	18	78	77	81	77	86	81	86							
B. ibarrai	60	3	69	65	75	76	73	75	79	76						
B. luetkeni	64	25	79	87	90	68	82	75	84	84	75					
B. valliceps	69	—	74	76	75	68	76	77	73	83	74	81				
B. occidentalis	74	26	71	76	72	65	68	67	70	70	71	77	72			
B. coccifer	60	21	75	78	84	72	78	79	86	86	76	84	76			
B. cognatus	53	6	54	72	65	55	72	60	69	68	70	76	65	77		
B. speciosus	57	18	57	70	63	64	81	58	67	69	68	74	66	70	70	92

or absence of compounds. A weighted affinity may be calculated:

$$WA = \frac{\text{total shared occurrence of compounds in species } A \,\&\, B}{\text{total percent occurrence in } A + \text{total percent occurence in } B}$$

This relationship is then expressed as percent affinity. The percent-occurrence figures for all compounds occurring in a species are a quantitative representation of the degree of venom polymorphism in that species. The percent-occurrence figure may represent the actual percent occurrence of any compound in the population or class intervals of percent occurrence. In the present study, four class intervals are considered. Although the use of four classes is arbitrary, Low (1968) has shown that the distortion is minimal: between 1 and 3 percent in most populations. In fact, when fewer than ten individuals from any population are sampled, the use of exact percent-occurrence figures introduces a "false accuracy" bias. A summary of the occurrence of all compounds in the populations considered is given in table 13-4.

The weighted affinities were computer generated (BUFϕ 1; program library, University of British Columbia). The results are also suitable for treatment in the TAXϕN cluster analysis. However, Porter and Porter (1967) have shown that the variation in clustering results from biochemical data in *Bufo* depends heavily on the weighting or linkage criteria. Clusters derived from any similarity coefficient index of venoms would not be realistic in an evolutionary sense, since all data are of the same type and re sent a small portion of the genetic makeup.

When it is possible to relate data from all syst in a form suitable for cluster analysis, the results be meaningful; until that time, there is no ratio for constructing new species groups. The comm accepted groups of Blair (1966) will be used as a frame of reference for the biochemical data.

New World *Bufo*

New World *Bufo* have followed two major lin evolution (Blair 1963*a*), and, as Blair has noted where in this volume, it is probable that the g originated in the American tropics. The greates versity of venom patterns is found there. The so ern radiation is principally South and Central An can in distribution and includes the *valliceps*- and *rinus*-like species. The northern radiation, inclu *americanus*- and *boreas*-like toads, is largely lim to North America, although the montane *spinul* complex of western South America appears to member of this line. In the southern radiation, major pattern trends are evident: the possessio bufotenine and 5-OH tryptamine (*marinus* gr and *B. crucifer*) and the lack of bufotenine (the *liceps* group and allies) (Table 13-4). In the nort radiation, there is a trend toward loss of 5-OH t tamine in the venom, although some species con both 5-OH tryptamine and bufotenine. Chrom graphically, there is a small number of species, cluding *B. coccifer, B. speciosus,* and *B. cogn* which are intermediate in venom composition.

Table 13–6. Weighted Affinities between New World *Bufo* Species

	B. alvarius	*B. boreas*	*B. exsul*	*B. americanus*	*B. hemiophrys*	*B. microscaphus*	*B. terrestris*	*B. quercicus*
B. bocourti	22	23	22	48	56	44	56	52
B. spinulosus	55	61	58	29	21	36	18	18
B. marinus	24	22	21	63	67	61	64	62
B. arenarum	12	11	10	62	66	59	72	64
B. crucifer	19	20	19	67	65	64	65	63
B. blombergi	11	10	10	52	60	54	55	51
B. haematiticus	17	16	15	60	69	57	72	60
B. canaliferus	16	15	14	57	54	54	60	57
B. cavifrons	20	21	20	70	63	62	63	60
B. coniferus	9	8	8	61	65	58	68	62
B. ibarrai	9	11	10	60	58	57	64	65
B. luetkeni	8	10	10	57	60	54	70	57
B. valliceps	10	9	9	52	58	50	63	55
B. occidentalis	20	21	20	63	67	60	78	65
B. coccifer	17	19	18	70	69	57	75	60
B. cognatus	24	27	26	72	77	58	88	62
B. speciosus	32	32	30	65	75	57	80	60

unidentified indole, no. 20 (Tables 13-3, 13-4), is present only in the northern line and in intermediate species.

South American Radiation

The South American species sampled are few in number but evolutionarily important. Only Iquitos (Peru) samples (*B. poeppigi*) were used for comparing venom patterns on a quantitative basis, because the sample size was insufficient for the other populations. *B. marinus*' chromatographic affinities are highest with the other southern-line, bufotenine-containing species, *B. arenarum* (75%) and *B. crucifer* (87%) (Table 13-6). Bertini and Rathe (1962) and Bertini and Cei (1961, 1962) have noted common seroproteins and hemoglobins in the blood of *B. marinus* and *B. arenarum.*

Cei (1956, 1959) noted morphological and physiological polymorphism in *B. arenarum,* and Bertini and Cei (1962) described variations in the electrophoretic patterns of populations in Argentina. Samples used in the present study, from a locality where Cei found extensive polymorphism, were all from the unspotted morph and showed very little chromatographic variability (Table 13-5). The venom contained both bufotenine and 5-OH tryptamine like the venoms of *B. crucifer* (84% affinity), *B. marinus* (75%), *B. bocourti* (62%), and *B. paracnemis.* Although Chen and Chen (1933*a*) and Chen, Jensen, and Chen (1933*a*) described two bufotenines from the venom of *B. arenarum,* no recent workers have been able to verify this (Erspamer 1954; Michl and Kaiser 1963; Low 1967).

The chromatographic affinities of *B. crucifer* are strongest with *B. marinus* (87%) and *B. arenarum* (84%), of the bufotenine-containing species, and with *B. cavifrons* (90%) and *B. luetkeni* (90%), of the 5-OH tryptamine–containing species. Blair (Chap. 11) has demonstrated the low compatibility of *B. crucifer* with other New World species. Morphologically, it is strongly reminiscent of the *valliceps* group of toads; chromatographically, it shows affinities with species of both New World lines. The chromatographic pattern, however, definitely contains bufotenine. It appears likely from the present evidence that *B. crucifer* is an old species, having diverged early in the evolution of the southern species.

The *guttatus* group is represented in my material by the large Colombian *B. blombergi* and the very small Costa Rican *B. haematiticus.* Previous biochemical work has been restricted to the general description of indolethylamine derivatives and cholesterol in the venom of *B. blombergi* (Henderson, Welles, and Chen 1960). Both species possess the cryptic "dead-leaf" pattern, much like the African species *B. superciliaris.* Hybridization data are largely lacking, since the testes of both species contain some unidentified compound that inhibits fertilization (Chap 17). Tihen (1962*a*) split the group: he placed *B. haematiticus* in a monotypic group and included *B. blombergi* in the South American (*marinus*) section of his large *valliceps* group. The venom patterns, both of the 5-OH tryptamine types, show an 80 percent affinity with each other. Other affinities suggest relationship with the *valliceps* radiation of Blair (Table 13-6). The affinities of *B. haematiticus* with *valliceps*-like species range as high as 86 percent with the Costa Rican *B. coniferus.* It is perhaps significant that both *B. blombergi* and *B. haematiticus* show strong affinities with the African *B. superciliaris.*

The polymorphic (Cei 1961) Andean *B. spinulosus* possesses a peculiar venom pattern (Table 13-4); the venom contains bufotenine and lacks 5-OH tryptamine. A few individuals examined possessed the indole no. 20. Presence of bufotenine and loss of 5-OH tryptamine are characteristic of the North American *boreas* group and *B. alvarius,* and the occurrence of no. 20 is largely restricted to North American species. The chromatographic affinities of *B. spinulosus* are generally quite low, the highest being with the *boreas* group and *B. alvarius* (Table 13-6). These range from 61 percent with *B boreas* to 55 percent with *B. alvarius.* Cei, Erspamer, and Roseghini (1968) have found strong biochemical affinity with the *boreas* group. In addition to the loss of 5-OH tryptamine in both groups, they noted the occurrence of bufotenidine, which is restricted to the *boreas* and *spinulosus* groups. With the *americanus* group, chromatographic affinities range from 18 percent with *B. terrestris* to 36 percent with *B. microscaphus.* Affinities between *B. spinulosus* and other South and Central American species are low, ranging from 14 percent with *B. canaliferus* to 41 percent with *B. marinus* (Table 13-5). In addition to the strong chromatographic affinity with the *boreas* group, Tihen (1962*a*) found the skull of *B. spinulosus* "strongly reminiscent" of that of *B. boreas* but attributed the similarity to parallelism. In fact, current information suggests that while the *spinulosus* group is a distinctive complex, not closely related to any species examined, it shows real affinity with the *boreas* group.

Table 13–7. Weighted Affinities between Some *Bufo* Species of the United States and Canada

B. alvarius							
B. boreas	92						
B. exsul	87	95					
B. americanus	47	46	44				
B. hemiophrys	33	33	31	84			
B. microscaphus	41	38	36	82	76		
B. terrestris	27	27	26	84	88	70	
B. quercicus	21	19	18	71	69	68	76

Central and North American Radiation

B. bocourti, from Guatemala, is another distinctive species that (Chap. 11) may belong to the northern, thin-skulled line of toads, along with *B. spinulosus* and the *boreas* group. The venom of *B. bocourti,* like that of the *marinus* group, contains both bufotenine and 5-OH tryptamine. A unique, unidentified compound, no. 47 (Table 13-3) is also present. The highest affinities of *B. bocourti* venom (Table 13-5) are with the venoms of *B. marinus* (71%) and *B. occidentalis* (74%). Blair (1963*a*) noted external morphological similarities between *B. bocourti* and *B. occidentalis.*

B. occidentalis from Mexico shows genetic (Chap. 11) and osteological (Tihen 1962*a*) affinities with the *valliceps* group. Chromatographically, *B. occidentalis* shows the 5-OH tryptamine pattern of the *valliceps* group. Its highest affinities are with *B. luetkeni* (77%; Table 13-5), *B. valliceps* (72%), *B. arenarum* (76%), *B. bocourti* (74%), *B. crucifer,* (72%), and *B. cavifrons* (70%).

B. valliceps is the most widespread of a group of well-differentiated toads ranging principally in the New World tropics. Its cranial morphology has been reviewed by Baldauf (1958) and by Tihen (1962*a*). Blair (1963*a*, 1966) and Porter (1964*a, b, c*) considered *B. valliceps* to be most closely related to *B. mazatlanensis, B. gemmifer, B. cristatus,* and *B. cavifrons.* The chromatographic affinities are highest with *B. coniferus* (83%; Table 13-5) and *B. luetkeni* (81%) of the species examined. The chromatographic pattern is representative of the Central American radiation of toads, characterized by the presence of 5-OH tryptamine and the lack of bufotenine. Although Chen and Chen (1933*e*) reported "vallicepobufotenine" in the venom, no other workers have found any evidence of bufotenine (Table 13-4; Cei, Erspamer, and Roseghini 1968).

The chromatographic evidence supports inclusion

of *B. luetkeni* in the *valliceps* group (Table 13-5). In addition to its affinity with *B. valliceps* (81%), *B. luetkeni* shows strong affinity with *B. crucifer* (90%) and *B. arenarum* (87%), a situation suggestive of a common origin for both the major southern-line radiations.

Blair (1963*a*, 1966) considered *B. canaliferus* distinct enough from the *valliceps* group to be placed in a separate group. Porter (1962, 1964*b*) found the mating call distinct from other members of the *valliceps* group. Although Porter (1962, 1964*a*), found no evidence of indoles in the four *B. canaliferus* he examined, the venom does contain several indoles, the prinicipal one being 5-OH tryptamine (Tables 13-3, 13-4). The chromatographic affinities of *B. canaliferus* are highest with the *valliceps* group (Table 13-5), and in venom characters *B. canaliferus* is no more divergent from the *valliceps-luetkeni* pattern than other species.

The high-altitude *B. cavifrons* ranges through Chiapas, Oaxaca, and Veracruz (Porter 1964*c*) and is a representative member of the *valliceps* group (Tihen 1962*a*; Blair 1963*a*, 1966; Porter 1964*a-c*) in osteology, morphology, and call. On the basis of distribution, Porter (1964*c*) suggested that *B. cavifrons* might represent a recent altitudinal adaptation of *valliceps* stock. One-dimensional chromatographic data were inconclusive (Porter 1964*a*). The two-dimensional chromatographic affinities with *B. luetkeni, B. crucifer* (90%), and *B. arenarum,* as well as recent work by Porter (1967), suggest the possibility that *B. cavifrons* may represent an early divergence within the *valliceps* group. The strongest *valliceps*-group affinity is with *B. coniferus* (86%), which supports Porter's recent (1967) suggestion that *B. cavifrons* is more closely related to *B. coniferus.*

The Guatemalan *B. ibarrai* shows superficial morphological similarities to *B. coccifer* (Stuart 1954*a*, *b*), but neither the genetic compatibility data (Blair 1966) nor the chromatographic information suggests any real relationship (Table 13-5). These lines of evidence show affinity of *B. ibarrai* with the *valliceps* group. The chromatographic pattern is of the *valliceps* type, and a high affinity is shown with the *valliceps*-like venom of *B. cavifrons* (Table 13-5). Like some other New World tropical species, *B. ibarrai* has evolved a venom type strikingly similar to certain African species (Table 13-4).

Chromatographically, *B. coniferus* is no more divergent than the other members, showing an 83 percent affinity with *B. valliceps* (Table 13-5) and 84 percent with *B. luetkeni.* The chromatographic data also do not reflect the east-west dichotomy apparent in genetic compatibility in the *valliceps* group (Blair 1966).

B. coccifer appears to represent an early divergence and constitutes a monotypic species group (Blair 1963*a*). The osteology is similar to that of the *valliceps* group. On the basis of call characters and external morphology, Porter (1964*b*, 1965) suggested that *B. coccifer* is only distantly related to the species of the *valliceps* group. The venom is intermediate in character between the venoms of southern and northern radiation toads, and Low (1967) grouped *B. coccifer* with other species groups as a chromatographic "intermediate line." One-dimensional chromatographic work (Porter 1962, 1964*a*) distinguished only one indole, but it is now apparent that at least three indoles are present (Tables 13-3, 13-4), including 5-OH tryptamine. *B. coccifer* venom lacks bufotenine and contains the indole no. 20 (Table 13-4; Fig. 13-1). Like the *cognatus* group (also considered chromatographically intermediate by Low), chromatographic affinities are moderate to high with members of both the northern and southern radiations. With the *americanus* group, affinities range from 75 percent with *B. terrestris* to 57 percent with *B. microscaphus* (Table 13-4). Affinities with the *boreas* group are less than 20 percent. Affinities with the *valliceps* group agree generally with Blair's (1966) systematic treatment, with certain important exceptions. Blair (1966) reported viable hybrids in a cross of a male *B. coccifer* and a female *B. canaliferus,* contrasting this with a lack of metamorphosis in crosses involving male *B. coccifer* and female *B. valliceps* and *B. luetkeni.* The chromatographic pattern is more similar to that of *B. luetkeni* (84%; Table 13-5) than to those of *B. valliceps* (76%) or *B. canaliferus* (79%). Affinities are also high with *B. coniferus* and *B. cavifrons* (86%).

The *cognatus* group is represented by *B. cognatus* and *B. speciosus.* Osteological characters are similar to those of the *americanus* group. While early genetic compatibility data suggested relationship with species of the southern radiation (Blair 1963*a*), more recent compatibility information (Blair 1966) argued for a position intermediate between the two main New World radiations. As already noted, the venoms of both species are intermediate in character. The two species are closely related and very similar in all characters examined to date. One-dimensional chromatographic studies of the venoms were inconclusive

Table 13–8. Weighted Affinities between Old World Species Sampled

	B. bufo	*B. viridis*	*B. stomaticus*	*B. melanostictus*	*B. g. gariepensis*	*B. g. inyangae*	*B. lönnbergi*	*B. mauritanicus*	*B. superciliaris*	*B. garmani*	*B. rangeri*	*B. brauni*	*B. maculatus*	*B. pusillus*	*B. peretti*
B. bufo															
B. viridis	84														
B. stomaticus	62	58													
B. melanostictus	48	56	46												
B. g. gariepensis	23	30	21	28											
B. g. inyangae	25	29	16	25	90										
B. lönnbergi	29	27	10	9	72	81									
B. mauritanicus	35	34	8	17	71	78	74								
B. superciliaris	27	33	19	27	81	83	66	71							
B. garmani	28	34	18	26	78	79	63	71	86						
B. rangeri	39	45	17	29	78	73	70	72	90	82					
B. brauni	29	35	19	26	79	80	64	72	98	89	93				
B. maculatus	23	24	12	10	66	73	83	65	81	67	81	79			
B. pusillus	30	19	16	7	58	64	76	60	73	70	73	71	87		
B. perreti	27	31	15	13	71	79	89	67	75	75	78	73	87	78	
B. regularis															
El Mahalla	38	44	18	28	77	78	71	75	87	79	93	90	80	69	71
Ogbomosho	34	41	19	26	80	82	71	79	91	82	92	94	77	70	71
Ilesha	30	36	19	27	83	84	67	75	93	84	89	96	74	66	67
Umtali	35	41	18	25	75	76	69	74	93	84	96	95	86	75	77
Entebbe-Kampala	34	40	17	29	73	73	64	82	76	78	82	79	62	56	67
Port St. Johns	33	39	17	24	72	72	66	71	89	81	92	92	82	72	74
Pietermaritzburg	35	41	18	25	76	77	70	75	94	85	97	96	84	76	78
Lourenço Marques	38	44	18	25	76	77	73	75	91	83	97	94	81	74	78
Louis Trichardt	35	41	17	24	71	71	68	70	88	81	94	91	79	72	75
Nyeri	33	39	17	24	72	72	66	71	84	93	88	87	75	77	76
Ol Tukai (1)	28	34	18	26	78	79	63	71	91	88	87	94	72	65	72
Ol Tukai (2)	33	39	17	28	71	71	62	80	72	74	78	75	61	52	62
Vumba Mts.	38	44	15	27	73	77	70	82	74	73	81	75	65	58	70
Mazeras	34	40	17	29	73	73	64	82	73	75	80	76	60	54	64

(Hunsaker et al. 1960; Wittliff 1962; Porter 1962, 1964*a*). Even with two-dimensional resolution, only a one-compound difference is detectable (Table 13-5); the weighted affinity is 92 percent. With species in the northern radiation, affinities are higher with the *americanus* and *quercicus* groups than with the *boreas* group (Table 13-6). With species in the southern radiation, affinities are comparable to those with the *americanus* group (Table 13-5).

The *boreas* group is distributed in montane regions of western North America and is represented here by two of the four species (*B. boreas* and *B. exsul*). *B. alvarius* shows many striking similarities to the *boreas* group. These species share a variety of characteristics: lack of vocal sacs (Inger 1958), loss of call (Blair and Pettus 1954), smooth skin, enlarged venom glands on the hind legs, and tubercles at the jaw articulation (Blair 1963*a*). Five-hydroxytryptamine is lacking in the venom (Table 13-4; Cei, Erspamer, and Roseghini 1968). Chromatographic affinities within the groups are high (Table 13-7): *boreas-exsul,* 95 percent; *boreas-alvarius,* 92 percent; and *alvarius-exsul,* 87 percent. These species are distinct biochemically, and affinities with other New World groups are low (Tables 13-6; 13-7).

B. exsul, first described as a species by Myers (1942), has been variously treated as a species (Schuierer 1962, 1963) and as a subspecies of *B. boreas* (Tihen 1962*a*). Blair (1965) has suggested that, since the population size of this endemic form is so small, differentiation may have occurred as late as the end of the Wisconsin. The very high venom ities between *B. exsul* and other *boreas*-group ıbers, particularly *B. boreas*, lend support to this ;estion.

he venom pattern of *B. alvarius* is strikingly like e of *B. boreas* and *B. exsul* (particularly in the of 5-OH tryptamine), and the affinities with e two species are significantly higher ($p < .001$) ı affinities with any other group.

he *punctatus* and *debilis* groups are not repreed in the present sample. Blair (Chap. 11) has ;ested affinity of these groups with the *boreas* ıp, and biochemical analysis of the venom lends ɔort to this hypothesis. Cei, Erspamer, and Roseıi (1968) have shown that the *boreas, alvarius, ulosus, punctatus,* and *debilis* groups lack 5-OH tamine and frequently contain bufoviridine or ɔviridine 1-sulfonic acid, characteristics peculiar hese species.

'he *americanus* group is a well-defined North erican group, with a dichotomy between eastern ;-calling species and western short-calling species air 1963*a*, 1965). Intragroup relationships have n studied by numerous workers using a variety of racters (Underhill 1962; Stebbins 1951; A. P. Blair 2, 1955; W. F. Blair 1957*a*, *b*, 1959, 1962, 1963*a*; Tihen 1962*a*); it is probably the best-studied cies group of the bufonids. Four species were resented here: *B. americanus, B. hemiophrys, errestris,* and *B. microscaphus.* The venom of all r species contained both bufotenine and 5-OH ɔtamine (Table 13-4). Similar results were reted by Michl and Kaiser (1963) and Erspamer 54) for *B. americanus,* and by Cei et al. (1968). 3. *americanus* ranges widely throughout the east- United States and Canada. It shows affinities of ɔercent with the other eastern species of the *amernus* group tested, *B. terrestris* and *B. hemiophrys* able 13-7). Affinity with the western *B. microphus* is 82 percent. Affinities are considerably low- (<50%) with the *boreas* group, and there is no omatographic evidence to support the suggestion Sanders and Cross (1963), based on chromosomal dies, that *B. americanus* evolved from *B. boreas.* e biochemical evidence indicates that, although h are northern radiation species, there is no close ationship.

3lair (1957*a*, 1963*a*, 1965) and Underhill (1962) 'e suggested that *B. hemiophrys* is closely allied B. *americanus,* perhaps representing a Pleistocene ct. Certainly the chromatographic evidence (Ta- 13-7) indicates a strong relationship.

B. terrestris, from the southeastern United States, shows affinities of 88 percent with *B. hemiophrys,* 84 percent with *B. americanus,* and 70 percent with the western *B. microscaphus* (Table 13-7). Recently, Sanders (1961) proposed a hybrid origin for *B. terrestris* — "Mexican" (i.e., *valliceps* group) and "American" (*americanus* group) ancestors — and Tihen (1962*b*) suggested that not only *B. terrestris* but all Florida species may have been influenced by gene exchange in the past. *B. terrestris* does show moderate affinities with certain *valliceps*-group species (Table 13-6). The venom pattern, however, is typically *americanus*-group type, containing bufotenine, 5-OH tryptamine, and indole no. 20. Anderson (1949) has shown the importance of introgression in evolution, and, despite the usual low fertility of interspecific crosses in amphibians, the possibility of introgression cannot be entirely eliminated. However, the osteological and biochemical affinities of *B. terrestris,* and the osteological peculiarities of other Florida species, do not warrant the assumption of introgressive hybridization without further evidence, nor are they inexplicable by other means. The simplest explanation, of course, is one of diversity of ancestral forms, some of which were intermediate between the two present lines.

The western *B. microscaphus* received conflicting systematic treatment prior to Blair's (1962) application of nonmorphological criteria. A. P. Blair (1955) suggested that the morphological similarity of *B. microscaphus* to the *boreas* group might reflect introgression, while W. F. Blair (1963*a*) suggested that *B. microscaphus* simply showed less divergence from the ancestral type. Certainly the venom pattern shows no close affinity with any of the *boreas* group; affinities range from 36 to 41 percent (Table 13-7). *B. microscaphus* is the only member of the *americanus* group whose venom lacks the indole no. 20 (Table 13-4), and it possesses a compound unique among the species examined (No. 36). Its affinities with other members of the *americanus* group reflect this divergence (Table 13-7).

Blair (1963*a*) demonstrated that *B. woodhousei,* which has an extensive range and occurs sympatrically with several other species (both *americanus* and *valliceps* groups), shows greater intragroup incompatibility than any other *americanus*-group species. He suggested a longer period of divergence for *B. woodhousei.* Of the *americanus*-group species examined by Cei, Erspamer, and Roseghini (1968), *B. woodhousei* was the only species lacking bufotenine (in eleven samples from Austin, Texas). Three individuals from Mesa, Arizona, possessed bufotenine. *B. woodhousei* may form sterile hybrids with members of the *valliceps* group, particularly *B. valliceps* (Thornton 1955; Chap. 11), but it is highly unlikely, however, that the material available to Cei et al. represented hybrid individuals. The venom of *B. woodhousei* appears to be the most divergent of the *americanus* group.

The relationships of tiny *B. quercicus* are difficult to assess. Morphologically, it appears to be a member of the northern radiation (Blair 1963*a*), although osteologically, it seems to belong to the southern radiation (Baldauf 1959; Tihen 1962*a*). The call is quite distinctive (Blair (1963*a*), and hybridization data are few and inconclusive (Chap. 11). The chromatographic pattern is northern in character, containing bufotenine, 5-OH tryptamine, and indole no. 20 (Table 13-4). Affinities with the *americanus* group range from 68 to 76 percent (Table 13-7), while affinities with the *valliceps* group are generally lower (Table 13-6). The balance of the evidence at this time suggests that *B. quercicus* is indeed a northern-line species, perhaps with a long history of divergence.

Old World *Bufo*

Eurasian Radiation

There is a paucity of samples of European and Asian species for comparison, but the available species are significant in the evolutionary picture of the genus. Representatives of four species from widely separated localities were available (Table 13-8): *B. bufo* (Italy), *B. viridis* (Israel), *B. stomaticus* (India), and *B. melanostictus* (Thailand). The first three species are thin-skulled toads showing various degrees of genetic affinity with the thin-skulled species of the New World (Chap. 11), and the last is a thick-skulled species with probable affinity to the New World southern radiation. More species are necessary to clarify the picture, but it is probable that the proliferation of biochemical types characteristic of the New World tropics does not occur in the Old World.

Both bufotenine and 5-OH tryptamine were present in the venom of all four species, and the indole no. 20 was present in all but *B. melanostictus* (Table 13-4). Erspamer (1954) and Michl and Kaiser (1963) reported both bufotenine and 5-OH tryptamine in the venoms of *B. bufo* and *B. viridis*, as did Cei et al.

Table 13–9. Weighted Affinities between New World and Eurasian Species

	bufo	*viridis*	*stomaticus*
alvarius	50	52	71
boreas	49	48	70
exsul	46	45	67
americanus	42	38	46
hemiophrys	45	35	37
microscaphus	41	40	36
terrestris	34	29	37
quercicus	26	34	30
coccifer	30	28	28
spinulosus	43	41	35
bocourti	40	46	27
marinus	43	47	26
arenarum	16	22	19
crucifer	31	37	29
canaliferus	26	29	21
cavifrons	33	39	30
coniferus	25	29	14
ibarrai	33	40	20
luetkeni	21	26	18
valliceps	26	32	15
occidentalis	34	38	30
haematiticus	33	36	17
cognatus	22	20	37
speciosus	27	28	18

(1968). The affinities of these species with e other are summarized in table 13-8.

B. bufo is a widely distributed species, and name as presently used probably refers to seve biological entities. The present samples were fr Italy. Chen and Chen (1933*a*, *b*, *c*), Jensen and C (1930), and Goto (1963*a*, *b*) have examined the s secretion, primarily for pharmacological purpo Michl and Kaiser (1963) have reviewed the bu dienolide constituents. In my material, the venon *B. bufo* showed an 84 percent affinity with the v om of *B. viridis.* The blood protein composition *B. bufo, B. viridis,* and *B. calamita* are more sim to one another than to any other Eurasian spe (Chap. 14).

B. viridis, from Israel, shows high chromatogra affinity with *B. bufo.* Inger (Chap. 8) has show high morphological index between these species. next highest affinity, with *B. stomaticus,* is consi ably lower (58%).

B. stomaticus, a thin-skulled species from Asia, shows relationship to the *viridis* and *bufo* groups of Europe, and to the *boreas* and *spinulosus* groups of the New World. Affinities with *B. bufo* and *B. viridis* and 62 and 58 percent respectively, and affinities with the *boreas*-group species range from 67 to 71 percent (Table 13-9). The available chromatographic evidence indicates that *B. stomaticus* is a member of the same evolutionary line as the *bufo, viridis,* and *boreas* groups and occupies a position intermediate between the New World and European groups.

B. melanostictus, ranging through India, Burma, Ceylon, and the Malayan peninsula (Stejneger 1907), bears morphological resemblance to the *valliceps*-group species of the New World. It is broad skulled, unlike the other Eurasian species available. The chromatographic pattern is qualitatively similar to the bufotenine-containing species of the New World southern radiation (Table 13-4); the venom contains bufotenine and 5-OH tryptamine and lacks the indole no. 20. All chromatographic affinities are low (Tables 13-8; 13-9).

African Radiation

The speciation of *Bufo* in Africa is extensive, possibly exceeded only by that of the New World tropics. Tihen (1960) described two main osteological lines. One, the *angusticeps* complex, lacked dermal ornamentation, possessed an open occipital groove, and showed reduction in ossification of dermal elements. The other, the *regularis complex* (in which he included *B. mauritanicus*), possessed an occipital groove covered dorsally by bone, tended toward dermal ornamentation, and showed less reduction of ossification. He considered the latter condition primitive (Tihen 1960, 1962*a*).

On morphological grounds, Poynton (1964) has recognized five species groups in southern Africa alone.

Chromosomal work by Bogart (1968, Chap. 10) has demonstrated further the complexity of bufonid speciation in Africa. The established diploid number for all New World, European, and Asian species is 22. In the African species he examined, Bogart found a diploid number of 22 for *B. mauritanicus, B. gariepensis, B. rosei, B. lönnbergi,* and *B. carens.* Other African species, including the *regularis* group and related species, have a diploid number of 20, and Bogart suggested that this was probably a derived condition.

Table 13–10. Weighted Affinities between *B. regularis* from Different Localities

El Mahalla, Egypt													
Ogbomosho	96												
Ilesha	94	97											
Port St. Johns	92	91	88										
Pietermaritzburg	94	95	92	96									
Lourenço Marques	94	95	92	93	98								
Louis Trichardt	89	90	87	97	94	94							
Umtali	95	94	91	97	99	97	93						
Nyeri	85	86	83	87	91	89	86	90					
Ol Tukai (1)	84	87	90	93	90	88	92	89	86				
Ol Tukai (2)	82	80	77	75	76	76	72	78	75	74			
Vumba Mts.	82	80	77	75	79	79	74	78	77	73	92		
Entebbe-Kampala	81	82	79	77	81	81	76	80	79	78	96	94	
Mazeras	81	82	79	75	78	78	74	77	77	75	98	94	98

Diversity in the venom patterns is not nearly so extensive in Africa as in the New World tropics. The chromatographic patterns of all species examined were of the 5-OH tryptamine type and were similar to the patterns of members of the *valliceps* and *guttatus* groups of the New World. Neither bufotenine nor indole no. 20 was detected (Table 13-4). Although Chen and Chen (1933*c*) described "regularobufotenine" from the venom of *B. regularis,* no other workers have detected bufotenine in samples of *B. regularis* (Cei et al. 1968; Low 1967).

The problem of African speciation is complicated by polymorphism in *B. regularis* (Table 13-10). This taxonomic designation almost certainly refers to several biological entities, but formal treatment is undesirable at this time because the variation in the different systems studied is not yet fully delineated. Populations from different localities vary in call (Tandy, Chap. 9), morphology (Poynton 1964), blood-protein composition (Guttman 1967; Chap. 14), and venom constitution (Low 1968). While the actual genetic difference involved in the venom polymorphism may be quite small, a gradation in pattern types exists, and affinities between different localities show consistent differences in their affinities with other species. Two main pattern types are evident. With the exception of the Ol Tukai (Kenya) population, the pattern is consistent within localities. However, two very similar species (*B. sp.* and *B. gutturalis*) occur sympatrically at Ol Tukai, and the apparent polymorphism there may result from confusion of the two. One pattern occurs in samples from Egypt, Nigeria, South Africa, and Mozambique, the other in Uganda. Both types appear in Kenya and

Rhodesia. The patterns, for the sake of simplicity, will be referred to in the discussion as "*regularis* 1" and "*regularis* 2," although affinities will be given in the tables by locality.

B. mauritanicus, from northern Africa, is osteologically similar to *B. regularis* and to the *guttatus* group of the New World, and Tihen (1960, 1962*a*) has suggested that this condition is the primitive one. The venom composition is also similar (Table 13-4). The difference in chromosome number, however, would preclude placing *B. mauritanicus* in the *regularis* group. Affinities are nonetheless high, ranging between 70 and 80 percent with most of the *regularis*-group species and *B. gariepensis* (Table 13-8). Affinities with *B. maculatus* (65%) and *B. pusillis* (60%) are somewhat lower. *B. mauritanicus* appears to be most closely related to the *regularis* group, and it is possible that *B. regularis* and its allies evolved from a *mauritanicus*-like ancestor.

B. rangeri, a southern African member of the *regularis* group, was originally described as a subspecies of *B. regularis* (Hewett 1935) and only recently recognized as a good species (Poynton 1964). On morphological grounds, Poynton suggested that *B. rangeri* was more closely related to *B. garmani* and *B. pardalis* than to *B. regularis* and *B. pusillus.* The chromatographic affinities of *B. rangeri,* however, are higher with the various "*regularis* 1" localities (87%–97%; Table 13-8) than with *B. garmani* (82%). Affinity with *B. pusillus* is significantly lower (73%).

B. brauni, a small Tanzanian member of the *regularis* group (Loveridge 1957), and the giant West African *B. superciliaris* show striking morphological (Nieden 1910), ecological (Noble 1924; Sanderson 1936), and biochemical similarities. The venom patterns are practically indistinguishable in the present solvent system (Table 13-4) and have an affinity of 97 percent. Both show high affinities with the "*regularis*" complex (Table 13-8) and with *B. rangeri.*

B. garmani, a savannah form found over most of Africa, is morphologically similar to *B. regularis* (Meek 1897) but may be distinguished from it by call and behavioral differences (Poynton 1964; Tandy, Chap. 9). Chromatographically, it may be distinguished from the "*regularis*" complex. Affinities with other members of the *regularis* group range from 74 to 93 percent (Table 13-8).

B. maculatus (Boulenger 1900) appears very similar to *B. regularis,* and its treatment is a history of taxonomic confusion (Low 1867). It is sympatric over its range with *B. regularis,* and the two species are often in chorus together (Schiøtz 1964*a, b*). The chromatographic pattern is distinct from that of *B. regularis* (Table 13-4; 13-8). Blair (Chap. 11) has suggested that *B. maculatus* and *B. pusillus* may be conspecific and comprise a *maculatus* group separate from the *regularis* group. Neither the venom data nor the blood-protein patterns (Chap. 14) suggest conspecific status for these two species. The affinity of *B. maculatus* with *B. pusillus* is 87 percent (Table 13-8), and with the "*regularis* 1" localities, between 72 and 86 percent. *B. pusillus,* like *B. maculatus,* is a savannah form but principally restricted to southern Africa (Poynton 1964). Its highest chromatographic affinity is with *B. maculatus.* It appears that *B. pusillus* may have derived from a *maculatus*-like ancestor, if the two species are not actually conspecific, as suggested by Blair.

There are few data available bearing on the relationships of *B. perreti* (Schiøtz 1963, 1964*a, b*). The chromatographic affinities are tenuous, because only small amounts of venom are available from individuals, and only eight compounds were detected in the present system.

Only one sample of *B. lönnbergi* was available, and chromatographic affinities are not quantitative in this case. The venom appeared to be somewhat more similar to that of *B. perreti* and *B. maculatus* than to that of the *regularis* group.

B. gariepensis, a southern African 22-chromosome form, is the only representative of the *gariepensis* (= *angusticeps* group of Poynton 1964) in the present sample. The samples are from two localities: Grahamstown, South Africa (*B. g. gariepensis*), and the Inyanga Mountains, Rhodesia (*B. g. inyangae*). Blair (Chap. 11) has suggested that the two forms may be differentiated enough to warrant specific status. The chromatographic affinity between them is 90 percent, which represents only a one-compound difference in venom composition. Compound no. 31 appears to be unique to *B. g. gariepensis* (Table 13-4). The blood hemoglobin and transferrin complements of the two forms are probably identical (Chap. 14). The chromatographic affinities of both forms with *B. mauritanicus* and the *regularis* group (Table 13-8) are lower than the intragroup affinity but range up to 83 percent with *B. superciliaris.* Divergence in the venom constitutions of the *regularis* and *gariepensis* groups is as pronounced as are osteological and morphological divergence.

Table 13–11. Weighted Affinities between African and Southern New World Species

	B. marinus	*B. arenarum*	*B. crucifer*	*B. canaliferus*	*B. cavifrons*	*B. coniferus*	*B. ibarrai*	*B. luetkeni*	*B. occidentalis*	*B. valliceps*	*B. blombergi*	*B. haematiticus*	*B. coccifer*	*B. cognatus*	*B. speciosus*	*B. bocourti*	*B. spinulosus*
B. g. gariepensis	71	74	84	72	83	79	84	84	63	72	68	77	83	73	71	55	10
B. g. inyangae	72	71	79	79	84	87	86	79	63	70	74	84	84	74	71	53	4
B. lönnbergi	62	67	63	69	67	69	82	63	71	63	76	67	67	69	67	52	10
B. mauritanicus	75	64	71	76	75	80	73	71	62	71	96	81	75	57	64	71	18
B. superciliaris	67	69	73	73	77	79	78	83	70	80	69	85	74	79	82	67	9
B. brauni	65	65	71	71	75	77	76	81	68	78	72	88	72	77	86	65	14
B. garmani	64	61	70	70	74	73	80	70	57	72	71	79	71	65	74	59	14
B. rangeri	65	62	66	71	69	71	80	75	72	77	76	82	67	61	69	68	13
B. regularis																	
El Mahalla	68	65	69	69	73	75	84	79	74	76	78	86	70	75	83	66	14
Ogbomosho	61	68	72	72	76	78	82	82	74	76	78	89	73	78	87	66	15
Ilesha	68	70	74	69	78	80	79	84	70	73	75	92	75	80	89	62	15
Port St. Johns	64	61	65	70	68	70	79	75	69	78	73	81	66	70	78	67	13
Pietermaritzburg	68	64	68	73	72	74	81	78	73	80	77	85	69	74	82	70	14
Lourenço Marques	68	64	68	73	72	74	83	78	75	77	79	85	69	74	82	70	14
Louis Trichardt	64	60	64	69	68	69	78	74	71	75	75	80	65	69	77	66	13
Umtali	67	63	68	72	71	73	82	77	72	81	76	84	68	73	81	69	14
Nyeri	64	61	65	70	68	70	82	70	64	76	73	78	66	65	73	67	13
Ol Tukai (1)	64	66	70	70	74	76	75	75	62	75	71	84	71	70	79	64	14
Ol Tukai(2)	64	56	64	64	68	67	76	64	57	65	70	72	65	59	68	54	13
Vumba Mts.	68	59	66	68	69	71	78	66	60	66	82	75	69	61	67	58	11
Entebbe-Kampala	65	57	66	68	69	68	75	66	58	66	81	74	67	61	69	58	13
Mazeras	65	57	66	66	69	68	75	66	58	64	81	74	67	61	69	56	13
B. perreti	62	67	63	75	67	69	80	63	71	71	70	67	67	69	67	59	4
B. maculatus	58	68	59	70	62	67	75	70	75	77	65	71	62	76	74	60	3
B. pusillus	51	60	52	62	55	59	70	62	67	70	61	63	63	76	71	54	3

Intercontinental and Intergroup Affinities

ls Jordan (1905) first suggested, species with a h phylogenetic affinity ordinarily show a high graphic affinity. A study of geographically distant cies showing phylogenetic affinity is useful and y yield clues to past patterns of dispersal. The rcontinental affinities of the species examined are marized in tables 13-11, and 13-12.

ffinities of the European species *B. bufo* and *B. dis* with New World species are highest with nbers of the *boreas* group and *B. alvarius*. Affini- range from 45 percent (*viridis-exsul*) to 52 per- t (*viridis-alvarius*). When *B. stomaticus* and *B. ulosus* are included, the picture is clarified. The World representatives of this line have venoms aining bufotenine, 5-OH tryptamine, and indole 20. The *americanus* and *quercicus* groups have a ern like the Old World species, while the *alvarius* and *boreas* groups do not possess indole no. 20. Affinities between *B. stomaticus* and *B. bufo* and *B. viridis* are 62 and 58 percent respectively. *B. stomaticus* shows affinities ranging from 67 to 71 percent with the *boreas* group and *B. alvarius*. The *boreas* group and *B. alvarius* show affinities with *B. spinulosus* ranging from 55 to 61 percent. The chromatographic evidence suggests that the *spinulosus, boreas, americanus, stomaticus, bufo,* and *viridis* groups are members of a well-defined evolutionary line.

The New World affinities of *B. melanostictus* are all less than 50 percent, the highest affinities being with the *boreas* group. The pattern is characterized by the same marker compounds as are the *marinus* and *crucifer* groups. The chromatographic data are difficult to evaluate, without samples from closely related species.

The chromatographic relationships of the European and Asian species to the African species are sur-

Table 13–12. Weighted Affinities between African and Northern New World Species

	alvarius	*boreas*	*exsul*	*americanus*	*hemiophrys*	*microscaphus*	*terrestris*	*quercicus*
B. g. gariepensis	9	11	11	63	61	60	67	65
B. g. inyangae	7	6	6	69	67	66	74	73
B. lönnbergi	0	0	0	65	69	68	69	67
B. mauritanicus	9	8	8	54	63	57	57	54
B. superciliaris	14	16	15	59	68	56	73	59
B. brauni	19	20	19	57	66	54	71	57
B. garmani	19	20	19	57	60	54	60	57
B. rangeri	18	19	18	53	61	50	66	52
B. regularis El Mahalla	19	20	19	56	64	53	70	59
Ogbomosho	20	21	20	58	67	55	72	58
Ilesha	20	21	20	60	69	57	75	60
Port St. Johns	18	19	18	52	60	50	65	55
Pietermaritzburg	19	20	19	55	63	52	69	55
Lourenço Marques	19	20	19	55	63	52	69	55
Louis Trichardt	17	18	18	52	59	49	64	51
Umtali	18	19	18	54	62	51	68	57
Nyeri	18	19	18	52	60	50	60	52
Ol Tukai (1)	19	20	19	57	65	54	65	57
Ol Tukai (2)	17	18	18	52	54	49	54	54
Vumba Mts.	16	15	14	55	58	52	58	55
Entebbe	18	19	18	53	56	50	56	52
Mazeras	18	19	18	53	56	50	56	52
B. peretti	7	6	6	65	63	62	69	67
B. maculatus	3	3	3	60	67	60	76	65
B. pusillus	9	8	8	62	71	56	76	53

prising, particularly in view of past dispersal schemes proposed on the basis of morphology (Darlington 1957 p. 139). If the bufonids arose in the Old World tropics and dispersal was radial from that point, then affinities should be high between species of contiguous areas of the Old World. In fact, a sharp break in venom type and affinity exists between the African and Near Eastern (e.g. *B. viridis*) species. None of the African species examined contained bufotenine or indole no. 20; all had the 5-OH tryptamine pattern common to the *valliceps* and *guttatus* groups of the New World. The European and Asian species examined were all of the bufotenine types. The highest affinities between any of the African species and any of the European and Asian species were those between *B. viridis* and *B. rangeri* (45%) and *B. viridis* and *B. regularis* from Egypt (44%) (Table 13-8).

The most striking intercontinental affinities are those between species of the southern New World radiation and African species. Of 400 relationshi (Table 13-11), 26 are between 50 and 60 percen 148 between 60 and 70 percent, 171 between 70 an 80 percent, 53 between 80 and 90 percent, and above 90 percent. No relationships are below 50 pe cent, if *B. spinulosus,* which has been shown to b related to the northern radiation, is excluded. Speci of the *valliceps* and *guttatus* groups show strong a finities with the *regularis* and *mauritanicus* grou The affinities of the *guttatus*-group species are pa ticularly interesting because of morphological (Cha 11) and osteological (Tihen 1962*a*) similarities wi the African radiation. The highest affinity of *B. lön bergi* with any other species, New or Old World, with *B. mauritanicus* (96%). *B. haematiticus* sho affinities with *B. brauni* (88%), *B. superciliaris* (85% and the "*regularis* 1" localities (78%–91%).

Summary

The evidence from parotoid-gland secretions suggestive and, in conjunction with other data, mak possible several generalizations about evolutiona trends within the genus. The most persuasive e dence points to the New World tropics as the ge graphic origin for *Bufo*. The primitive venom ty contains 5-OH tryptamine but lacks other indol such as bufotenine, which requires further pathwa The primitive venom pattern is found in the *vallice* and *guttatus* groups of the New World, and in t *mauritanicus, regularis,* and *gariepensis* groups the African radiation. A South American radiati closely allied to these species groups, possesses 5-C tryptamine and bufotenine; this radiation inclu *B. marinus, B. paracnemis, B. arenarum,* and *B. c cifer. B. melanostictus* of the Old World posses this venom type and may be related either to t line or to the *valliceps* line of the southern radiat or to both.

The *spinulosus, boreas, alvarius, punctatus,* a *americanus* groups of the New World, and the *maticus, bufo,* and *viridis* groups of the Old Wo comprise a well-defined biochemical line. This appears to be relatively advanced. The venom c tains bufotenine, bufoviridine, and other indoles la ing in the primitive lines. Certain New World gro (*boreas, alvarius,* and *punctatus*) have seconda lost 5-OH tryptamine.

The African species show high affinities with New World southern-radiation species and have same venom type as the *valliceps* and *gutt* ɔs. Affinities with the European and Asian spe- ɛxamined so far are quite low, although Inger p. 8) has found morphological similarities between the African species and the *pentoni* group; biochemical investigation in this field might prove rewarding.

REFERENCES

ı, R. E., and B. L. Turner. 1962. New techniques analysis of complex natural hybridization. Proc. t. Acad. Sci. 48:130–137.

—. 1963. Biochemical systematics. Prentice-Hall, w York.

ɛrson, Edgar. 1949. Introgressive hybridization. ın Wiley and Sons, New York.

sen, C. B. 1959. The molecular basis of evolution. ın Wiley and Sons, New York.

ıuf, Richard J. 1958. Contributions to the cranial ɛrphology of *Bufo valliceps* Wiegmann. Texas J. Sci. :172–186.

—. 1959. Morphological criteria and their use in ɔwing bufonid phylogeny. J. Morph. 104:527–560.

ɛr, M., M. Bharucha, K. K. Chen, V. Deulofeu, E. li, H. Jaeger, M. Kotake, R. Rees, T. Reichstein, O. ıindler, and E. Weiss. 1961. Papierchromatographische Prüfung weiterer Krötensekrete. Helv. Chim. ta 44:362–367.

ɛr, M., H. Schröter, Kuno Meyer, O. Schindler, and Reichstein. 1959. Die Bufogenine des Paratoidsetes von *Bufo marinus* (L.) Schindler. Helv. Chim. ta 42:2486–2505.

ıi, F., and J. M. Cei. 1961. Seroprotein patterns in *Bufo marinus* complex. Herpetologica 17:231–3.

—. 1962. Electrophoretic observations in seroproıs of various populations of *Bufo arenarum* Hensel Argentina. Acta Physiol. Latinoamer. 12(2):222.

ıi, F., and G. Rathe. 1962. Electrophoretic analysis the hemoglobin of various species of anurans. peia 1962:181–185.

ıcha, M., K. K. Chen, E. Weiss, and T. Reichstein. 61. Regularobufagin. Helv. Chim. Acta 44:844–6.

A. P. 1942. Isolating mechanisms in a complex of r species of toads. Biol. Symposia 6:235–249.

—. 1955. Distribution, variation, and hybridization a relict toad (*Bufo microscaphus*) in southwestern ah. Amer. Mus. Novitates no. 1722, pp. 1–38.

, W. F. 1956*a*. Call difference as an isolating meınism in southwestern toads (genus *Bufo*). Texas Sci. 8:87–106.

—. 1956*b*. The mating call of hybrid toads. Texas Sci. 8:350–355.

———. 1957*a*. Mating call and relationships of *Bufo hemiophrys* Cope. Texas J. Sci. 9:99–108.

———. 1957*b*. Structure of the call and relationships of *Bufo microscaphus* Cope. Copeia 1957:208–212.

———. 1959. Genetic compatibility and species groups in U. S. toads (*Bufo*). Texas J. Sci. 11:427–453.

———. 1962. Non-morphological data in amphibian classification. Syst. Zool. 11:72–84.

———. 1963*a*. Evolutionary relationships of North American toads of the genus *Bufo*: A progress report. Evolution 17:1–16.

———. 1963*b*. Intragroup genetic compatibility in the *Bufo americanus* species groups of toads. Texas J. Sci. 15:15–34.

———. 1965. Amphibian speciation, p. 543–556. *In* H. E. Wright, Jr., and David Frey (eds.), The Quaternary of the United States. Princeton University Press, Princeton, New Jersey.

———. 1966. Genetic compatibility in the *Bufo valliceps* and closely related groups of toads. Texas J. Sci. 18:333–351.

———, and D. Pettus. 1954. The mating call and its significance in the Colorado toad (*Bufo alvarius* Girard). Texas J. Sci. 6:72–77.

Bogart, James P. 1968. Chromosome number difference in the amphibian genus *Bufo*: The *Bufo regularis* species group. Evolution 22:42–45.

Bolliger, Rosemary, and Kuno Meyer. 1957. Isolierung und Identifizierung de Steroide des Giftsekretes der Berberkröte (Pantherkröte) *Bufo mauritanicus* L. Helv. Chim. Acta 40:1659–1670.

Boulenger, G. A. 1900. A list of the batrachians and reptiles of the Gaboon (French Congo), with description of new genera and species. Proc. Zool. Soc. (London) 1900:433–456.

Cei, J. M. 1956. Observaciones genéticas preliminares en poblaciones de anfibios argentinos. Biológicas 22: 45–49.

———. 1959. Ecological and physiological observations on polymorphic populations of the toad *Bufo arenarum* Hensel, from Argentina. Evolution 13:532–536.

———. 1961. *Bufo arunco* (Molina) y las formas chilenas de *Bufo spinulosus* Wiegemann. Inv. Zool. Chil. 7:59–81.

———, V. Erspamer, and M. Roseghini. 1968. Taxonomic and evolutionary significance of biogenic amines occurring in amphibian skin. II. Toads of genera *Bufo* and *Melanophryniscus*. Syst. Zool. 17:3.

Chen, K. K., and A. L. Chen. 1933*a*. Notes on the poisonous secretions of twelve species of toads. J. Pharm. Exp. Ther. 47:281–293.

———. 1933*b*. The physiological activity of the principles isolated from the secretion of the common European toad (*Bufo bufo bufo*). J. Pharm. Exp. Ther. 47:307–320.

———. 1933*c*. The physiological action of the principles isolated from the secretion of the South American toad (*Bufo regularis*). J. Pharm. Exp. Ther. 49:503–513.

———. 1933*d*. The physiological action of the principles isolated from the secretion of the Jamaican toad (*Bufo marinus*). J. Pharm. Exp. Ther. 49:515–525.

———. 1933*e*. A study of the poisonous secretions of five North American species of toads. J. Pharm. Exp. Ther. 49:526–542.

———. 1933*f*. The parotid [sic] secretion of *B. bufo gargarizans* as the source of Ch'an Su. J. Pharm. Exp. Ther. 49:543–547.

———. 1933*g*. The active groupings in the molecules of cino- and marino-bufagins and cino- and vulgaro-bufotoxins. J. Pharm. Exp. Ther. 49:548–560.

———. 1933*h*. Similarity and dissimilarity of bufagins, bufotoxins, and digitaloid glucosides. J. Pharm. Exp. Ther. 49:561–579.

Chen, K. K., H. Jensen, and A. L. Chen. 1933*a*. The physiological activity of the principles isolated from the secretion of *Bufo arenarum*. J. Pharm. Exp. Ther. 49:1–14.

———. 1933*b*. The physiological activity of the principles isolated from the secretion of the European toad (*Bufo viridis viridis*). J. Pharm. Exp. Ther. 49:15–25.

———. 1933*c*. The physiological activity of the principles isolated from the secretion of the Japanese toad (*Bufo formosus*). J. Pharm. Exp. Ther. 49:26–35.

Darlington, Philip J. 1957. Zoogeography: The geographical distribution of animals. John Wiley and Sons, Inc., New York.

Deulofeu, V. 1948. The chemistry of the constituents of toad venoms. Fortschr. Chem. Organ. Naturstoffe 5:241–266.

———, and E. Duprat. 1944. The basic constituents of the venom of some South American toads. J. Biol. Chem. 153:459–463.

Ellison, W. L., R. E. Alston, and B. L. Turner. 1962. Methods of presentation of crude biochemical data for systematic purposes, with particular reference to the genus *Bahia* (Compositae). Amer. J. Bot. 49:599–604.

Erspamer, V. 1954. Pharmacology of indolealkylamines. Pharmacol. Rev. 6:425–487.

Goto, T. 1963*a*. Fluoreszierender Stoff aus *Bufo* *garis*. Konstitutionsaufklärung des Pteridins " chrom." Japan. J. Zool. 14:91–95.

———. 1963*b*. Über einen blau fluoreszierenden "Bufo-chrom," seine Isolierung aus der Haut Kröte *Bufo vulgaris* und seine Verhalter in der wicklungsstadien. Japan. J. Zool. 14:83–90.

Guttman, Sheldon I. 1967. Evolution of blood pr in the cosmopolitan toad genus *Bufo*. Ph. D. T Univ. Texas.

Henderson, Francis G., John S. Welles, and K. K. 1960. Parotid [sic] secretions of *Bufo blomberg* *B. peltacephalus*. Proc. Soc. Exp. Biol. Med. 104: 178.

Hewett, J. 1935. Some new forms of batrachians reptiles from South Africa. Rec. Albany Mus. 4: 357.

Hunsaker, Don II, R. E. Alston, W. F. Blair, and Turner. 1960. A comparison of the ninhydrin tive and phenolic substances of parotoid gland s tions of certain *Bufo* species and their hybrids. E tion 15:352–359.

Inger, R. F. 1958. The vocal sac of the Colorado toad (*Bufo alvarius* Girard). Texas J. Sci. 10: 324.

Iseli, E., M. Kotake, E. Weiss, and T. Reichstein. Die Sterine und Bufadienolide der Haut von *formosus* Boulenger. Helv. Chim. Acta 48:1093–

Iseli, E., E. Weiss, T. Reichstein, and K. K. Chen. Papierchromatographische Prüfung der Sekrete *Bufo melanostictus* Schneider und *Bufo asper* enhorst. Helv. Chim. Acta 47:116–119.

Jensen, H., and K. K. Chen. 1930. Chemical studi toad poisons. II. Ch'an Su, the dried venom o Chinese toad. J. Biol. Chem. 87:741–753.

———. 1936. The chemical identity of certain constituents present in the secretion of various sp of toads. J. Biol. Chem. 116:87–91.

Jepson, J. B. 1963. Indoles and related Ehrlich rea p. 183–211. *In*, Ivor Smith (ed.), Chromatogr and electrophoretic techniques. I. Paper chro graphy. 2nd ed. Interscience Publishers, New Y

Jordan, D. S. 1905. The origin of species through tion. Science 22:545–562.

Linde, H., and K. Meyer. 1958. Vergleichende c ische Untersuchung des Parotoidensekretes und Hautdrüsensekretes der Berberkröte *Bufo maur* *cus* Schlegel. Pharmacol. Acta Helv. 33:327–339.

Loveridge, A. 1957. Checklist of the reptiles and an bians of East Africa (Uganda; Kenya; Tanga Zanzibar). Bull. Mus. Comp. Zool. 117 (2):1–

Low, Bobbi S. 1967. Evolution in the genus *Bufo*: dence from parotoid gland secretions. Ph.D. T Univ. Texas.

———. 1968. Venom polymorphism in *Bufo regularis.* Comp. Biochem. Physiol. 26:247–257.

Mayr, Ernst. 1964. The new systematics, pp. 13–32. *In,* Charles A. Leone (ed.), Taxonomic biochemistry and serology. Ronald Press, New York.

Meek, S. E. 1897. List of fishes and reptiles obtained by the Field Columbian Museum East African Expedition to Somaliland in 1896. Field Mus. Nat. Hist. Zool. 8:165–183.

Michl, H., and E. Kaiser. 1963. Chemie und Biochemie der Amphibiengifte. Toxicon 1963(1):175–228.

Moore, John A. 1951. Hybridization and embryonic temperature studies of *Rana temporaria* and *Rana sylvatica.* Proc. Nat. Acad. Sci. 37:862–866.

Myers, G. S. 1942. The black toad of Deep Springs Valley, Inyo County, California. Occas. Papers Mus. Zool. (Univ. Mich.) 460:1–19.

Nieden, Fritz. 1910. Verzeichnis der bei Amani in Deutschostafrika vorkommenden Reptilien und Amphibien. Sitzber Ges. Naturk. Freunde (Berlin) 441–452.

Noble, G. K. 1924. Contributions to the herpetology of the Belgian Congo, based on a collection in the American Museum Congo Expedition 1909–1915. Bull. Amer. Mus. Nat. Hist. 49:147–347.

Porter, Ken R. 1962. Evolutionary relationships of the *Bufo valliceps* group in Mexico. Ph.D. Thesis, Univ. Texas.

———. 1964*a*. Chromatographic comparisons of the parotoid gland secretions of six species in the *Bufo valliceps* group, p. 451–456. *In,* Charles A. Leone (ed.), Taxonomic biochemistry and serology. Ronald Press, New York.

———. 1964*b*. Morphological and mating call comparisons in the *Bufo valliceps* complex. Amer. Nat. 71: 232–245.

———. 1964*c*. Distribution and taxonomic status of seven species of Mexican *Bufo.* Herpetologica 19: 229–247.

———. 1965. Intraspecific variation in mating call of *Bufo coccifer* Cope. Amer. Midl. Nat. 74:350–356.

———. 1967. *Bufo cavifrons* Firschen collected in Nicaragua. Herpetologica 23:66.

———, and Wendy Porter. 1967. Venom comparisons and relationships of twenty species of New World toads (genus *Bufo*). Copeia 1967:298–307.

Poynton, J. D. 1964. The Amphibia of southern Africa. Ann. Natal Mus. 17:1–334.

Ruckstuhl, Jean-Pierre, and K. Myer. 1957. Isolierung und Aufteilung der chloroformlöslichen Bestandteile und chinesischen Krötengiftdroge Ch'an Su. Helv. Chim. Acta 40:1270–1292.

Sanders, Ottys. 1961. Indications for the hybrid origin of *Bufo terrestris* Bonnaterre. Herpetologica 17:145–156.

———, and J. C. Cross. 1963. Relationships between certain North American toads as shown by cytological study. Herpetologica 19:248–255.

Sanderson, Ivan T. 1936. The amphibians of the Mamfe Division, Cameroons. II. Ecology of the frogs. Proc. Zool. Soc. (London) 1936:165–208.

Schiøtz, Arne. 1963. The amphibians of Nigeria. Vidensk. Medd. dansk. naturh. Foren. (Copenhagen) 125:1–92.

———. 1964*a*. A preliminary list of amphibians collected in Sierra Leone. Vidensk. Medd. dansk. naturh. Foren. (Copenhagen) 127:19–33.

———. 1964*b*. The voices of some West African amphibians. Vidensk. Medd. dansk. naturh. Foren. (Copenhagen) 127:35–83.

Schröter, H., C. Tamm, and T. Reichstein. 1958. Beitrag zur Konstitutionsermittlung des Bufotalinins. Helv. Chim. Acta 41:720–735.

———, and V. Deulofeu. 1958. Über das Vorkommen von Bufotalinin. Helv. Chim. Acta 41:140–151.

Schuierer, F. W. 1962. Remarks upon the natural history of *Bufo exsul* Myers, the endemic toad of Deep Springs Valley, Inyo County, California. Herpetologica 17:260–266.

———. 1963. Notes on two populations of *Bufo exsul* Myers and a commentary on speciation within the *Bufo boreas* group. Herpetologica 18:262–267.

Smith, Ivor (ed.). 1963. Chromatographic and electrophoretic techniques. I. Paper chromatography. 2nd ed. Interscience Publ., New York.

Sokal, Robert R., and Peter H. A. Sneath. 1963. Principles of numerical taxonomy. W. H. Freeman & Co., San Francisco.

Stebbins, R. C. 1951. Amphibians of western North America. Univ. California Press, Berkeley.

Stejneger, L. 1907. Herpetology of Japan and adjacent territory. Bull. U.S. Nat. Mus. 58:1–576.

Stuart, L. C. 1954*a*. Description of some new amphibians and reptiles from Guatemala. Proc. Biol. Soc. (Washington) 67:159–178.

———. 1954*b*. Description of a subhumid corrider across northern Central America, with comments on its herpetofauna indicators. Contr. Lab. Vertebrate Biol. (Univ. Mich.) 65:1–26.

Thornton, W. A. 1955. Interspecific hybridization in *Bufo woodhousei* and *Bufo valliceps.* Evolution 9: 455–468.

Tihen, J. A. 1960. Two new genera of African bufonids, with remarks on the phylogeny of related genera. Copeia 1960:225–233.

———. 1962*a*. Osteological observations on New World *Bufo.* Amer. Midl. Nat. 67:157–183.

———. 1962*b*. A review of New World fossil bufonids. Amer. Midl. Nat. 68:1–49.

Turner, B. L., and R. E. Alston. 1959. Segregation and

recombination of chemical constituents in a hybrid swarm of *Baptisia laevicaules* x *B. viridis* and their taxonomic implications. Amer. J. Bot. 46:678–686.

Underhill, James C. 1962. Intraspecific variation in the Dakota toad, *Bufo hemiophrys*, from northeastern South Dakota. Herpetologica 17:220–231.

Urscheler, H. R., C. Tamm, and T. Reichstein. 1955. Die Giftstoffe der europäischen Erdkröte *Bufo bufo bufo* L. Helv. Chim. Acta 38:883–905.

Van Gils, G. E. 1938. Chemical composition of the parotid [sic] secretion of *Bufo melanostictus*. Acta Brevia Neerland. Physiol. Pharmacol. Microbiol. 8:84. (Chem. Abstr. 32:8010.)

Volpe, E. Peter. 1952. Physiological evidence for natural hybridization of *Bufo americanus* and *B. fowleri*. Evolution 6:393–406.

Wittliff, James L. 1962. Parotoid gland secretions in two species groups of toads. Evolution 16:143–153.

———. 1964. Venom constituents of *Bufo fowleri*, *Bufo valliceps*, and their natural hybrids analysed by electrophoresis and chromatography, p. 457–464. *In*, Charles A. Leone (ed.), Taxonomic biochemistry and serology. Ronald Press, New York.

Zelnik, R., and L. M. Ziti. 1962. Thin layer chromatography: Chromatoplate analysis of the bufadienolides isolated from toad venom. J. Chromatography 9: 371–373.

———, and C. V. Guimaraes. 1964. A chromatographic study of the bufadienolides isolated from the venom of the parotid [sic] glands of *Bufo paracnemis* Lutz 1925. J. Chromatography 15:9–14.

14. Blood Proteins

SHELDON I. GUTTMAN

Miami University
Oxford, Ohio

Introduction

In recent years new criteria have proved useful in determining systematic relationships between organisms. The current trend is to use all available evidence to clarify our understanding of these relationships (Mayr et al. 1953; Blair 1962; Alston and Turner 1963; Mayr 1963). The electrophoretic analysis of proteins is an excellent method of detecting evolutionary affinities (Hubby and Throckmorton 1965; Dessauer 1966). This technique is based on the premise that closely related species share a greater number of macromolecules than distantly related forms. For example, Dessauer and Fox (1964) suggested that homologous proteins of some species may be identical, those of certain genera may cluster within a limited electrophoretic mobility range, patterns of some families may lack components, and entire protein patterns of some orders may show distinctive mobilities.

Few studies of blood proteins have been made within the genus *Bufo* for taxonomic or phylogenetic purposes; the majority of the works have only characterized the patterns of a few species. Rodnan and Ebaugh (1957) and Bertini and Rathe (1962) examined the hemoglobins of certain South American *Bufo* (*marinus* group, *B. arenarum, B. granulosus*). *B. bufo, B. viridis,* and *B. calamita* hemoglobin was sampled electrophoretically by Nakamura (1960) and Marchlewska-Koj (1963). The hemoglobins of *B. fowleri, B. valliceps,* and their natural hybrid were compared by Fox et al. (1961).

Serum proteins of South American *Bufo* were extensively analyzed using paper electrophoresis by Cei and his group (Bertini and Cei 1959, 1960, 1962; Cei and Cohen 1964; Cei et al. 1961). Total serum protein patterns of five species of North American toads were studied (Fox et al. 1961 — *B. fowleri, B. valliceps*; Brown 1964 — *B. americanus, B. woodhousei*; Hebard 1964 — *B. boreas, B. terrestris*). Starch-gel electrophoresis was used to separate the proteins of *B. bufo* (Rossi 1960), while Chen (1967) employed polyacrylamide-disc electrophoresis to examine the serum of *B. bufo* and *B. calamita.*

Results of the above-mentioned studies of blood proteins are not comparable, because each worker examined only a few species and their methods differed. The purpose of the present investigation was

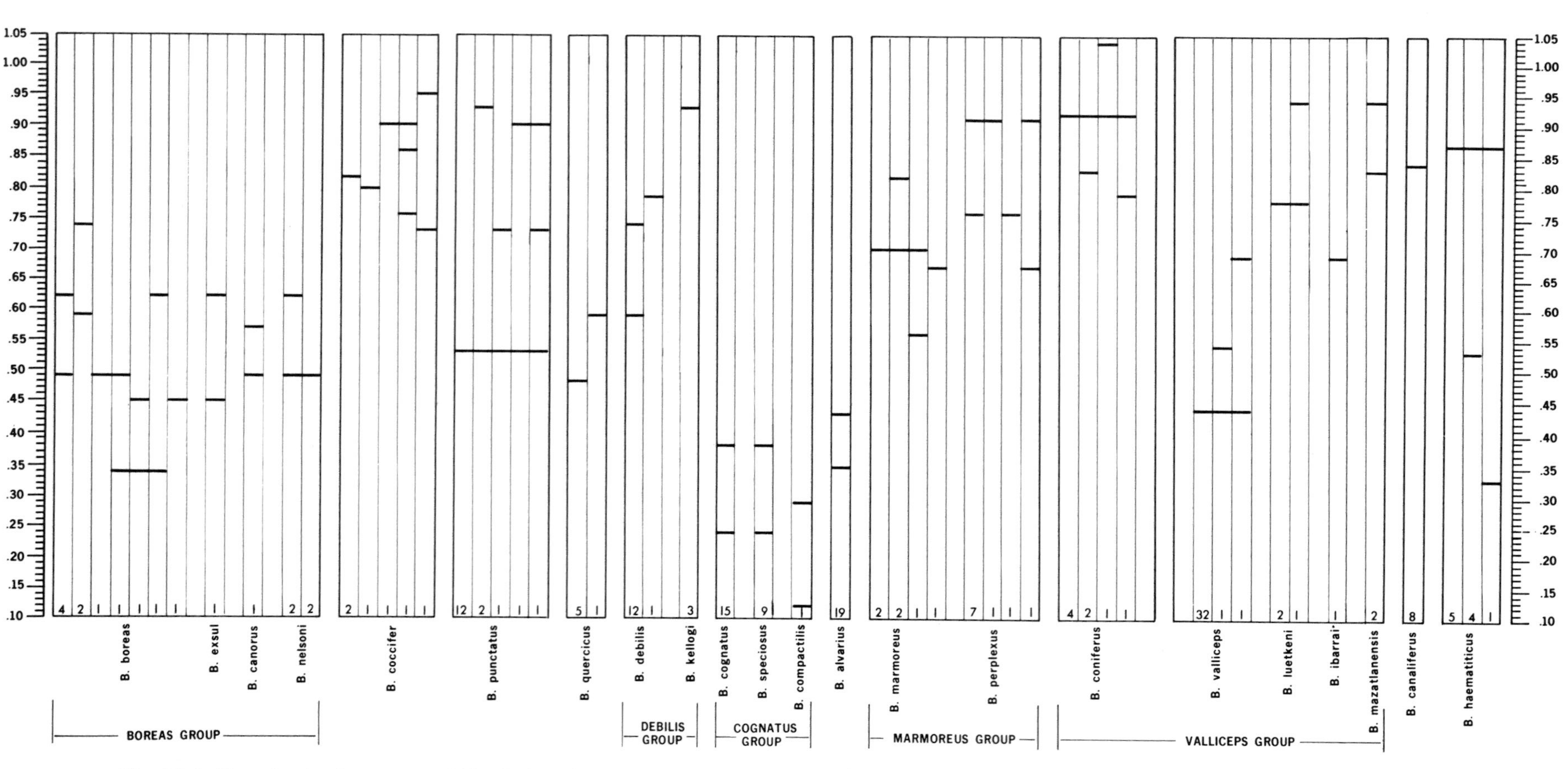

Fig. 14–1. Transferrin phenotypes of North American species groups. The vertical axis indicates relative mobility; the number of individuals of a species with each phenotype is indicated along the horizontal axis. If two or more species of a group were sampled, the group name appears along the horizontal axis.

to examine the blood proteins of as many species as possible, in order to supplement evidence on evolutionary relationships derived from other criteria.

Materials and Methods

The blood of fifty-seven species and some artificial and natural hybrids of *Bufo* was examined by vertical starch-gel electrophoresis. Hemoglobins and transferrins were studied.

Blood was drawn from the left carotid artery into a pipette rinsed with 4 percent EDTA and centrifuged. Plasma was stored at –5° C. Red blood cells were washed three times in 1 percent saline, then lysed by adding a volume of distilled water five times the volume of packed cells, followed by freezing at –5° C. The hemoglobin solution was subjected to electrophoresis immediately after lysis.

The starch-gel electrophoretic technique (Smithies 1959; Kristjansson and Hickman 1965), method of sample preparation, separation of fractions, and staining were described by Guttman (1967*a*, *b*). Briefly, the transferrin procedure involved precipitation of non-iron-binding fractions due to their insolubility in 0.6 percent rivanol (2 Ethoxy – 6,9 diamino acridine lactate) after saturation of the ferric sites on the transferrin molecules by adding 0.15 percent ferric ammonium citrate. The supernatant was subjected to electrophoresis. Transferrins were also located by autoradiography using the procedure of Giblett et al. (1959) to ensure that determinations made by rivanol precipitation were correct.

It was impossible to directly compare migration on different gels, because of temperature fluctuation, different lots of starch, pH variation, and so on. Therefore, relative mobilities were calculated, which were obtained by measuring the distance that each band migrated from the origin to the center of the component and by dividing this mobility by the mobility of a human blood sample present on each gel. The mobilities of toad transferrins and hemoglobins were compared to those of human transferrin C and hemoglobin A respectively. If the relative mobility of one component was within .005 mm. of the other, the transferrins or hemoglobins were considered to have identical mobilities. Samples were reexamined in adjacent slots when any doubt arose about their identity.

Males were sampled almost exclusively, because they were bled at the same time that they were used for artificial-hybridization tests. However, some females were examined and no sexual dimorphism was observed.

Hybrids and the Mechanism of Inheritance of Some Blood Proteins

If parental phenotypes are known or if the parental species are monomorphic or if large samples of the parental types are made so that essentially all morphs are known, then a phenotypic examination of hybrid blood can reveal the mechanism of inheritance of a particular protein.

The majority of the samples were male toads used in artificial crosses. Therefore, the first situation must be eliminated. However, one of the latter two requirements was satisfied in many situations so that some generalizations can be made about the inheritance of transferrins and hemoglobins in *Bufo*.

Transferrins

All *B. valliceps* tested, with the exception of two toads, possessed a single transferrin component (Fig. 14-1). These two individuals showed two bands, one in common with the other *B. valliceps*. This common transferrin was present in all hybrids (Figs. 14-3 and 14-4), both natural and artificial, when *B. valliceps* was either known or assumed to be a parent. In most other species sampled, at least one individual of a particular species possessed a single band. This suggests that alleles determine single transferrin components and that toads with two bands are heterozygotes. This theory was confirmed when hybrids having *B. valliceps* as one of the parents showed one band typical of the other parental species, although more than 75 percent of the individuals of the other species had two bands. For example, 80 percent of the *B. perplexus* sampled had two transferrins; all hybrids of *B. perplexus* had only one of these two components.

In some species (*B. brauni*, *B. carens*, *B. cognatus*, *B. speciosus*, *B. alvarius*) (Figs. 14-1 and 14-2) no fewer than two transferrin components were found. If sampling was adequate, as with the last three species, one is led to believe that each allele determined multiple transferrins. If hybrids with one of the above species as a parent have two bands attributable to this parental type, this hypothesis is confirmed. The hybrid between *B. perplexus* and *B. alvarius* (Fig. 14-4) illustrated such a case. The fastest-migrating band was one of the two found in 80 percent of the *B. perplexus* tested. The other two bands were the

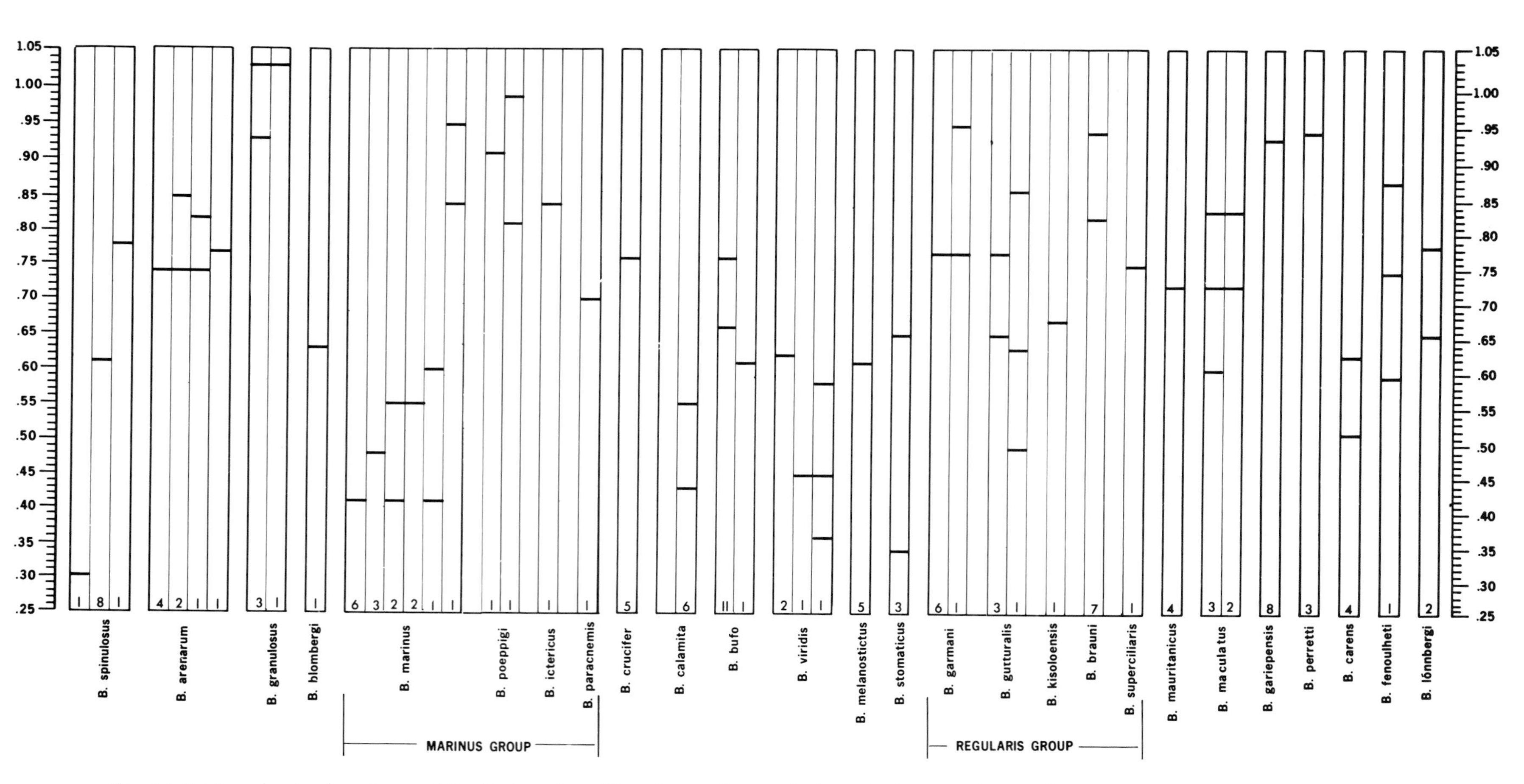

Fig. 14–2. Transferrin phenotypes of South American, Eurasian, and African species groups. The vertical axis indicates relative mobility; the number of individuals of a species with each phenotype is indicated along the horizontal axis. If two or more species of a group were sampled, the group name appears along the horizontal axis.

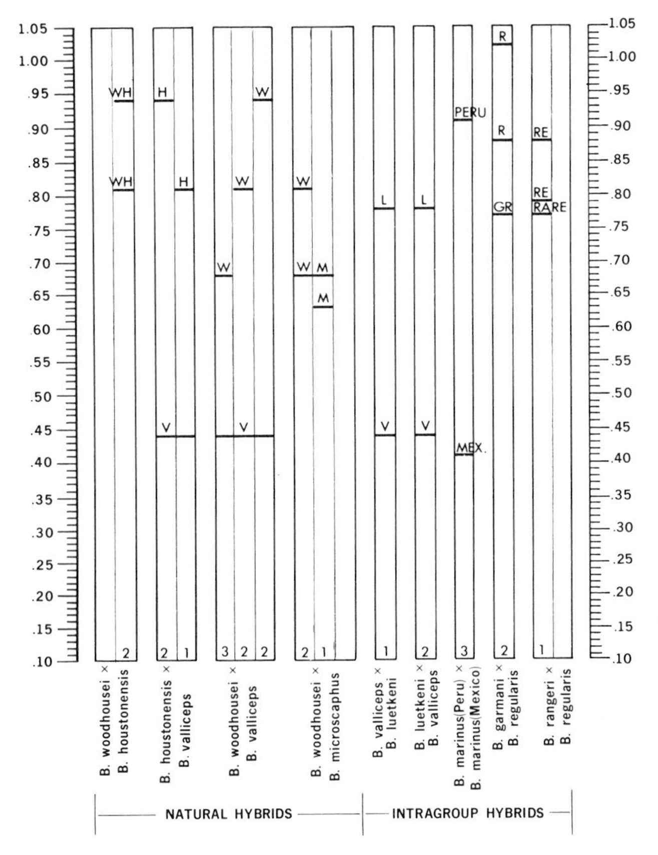

Fig. 14–3. Transferrin patterns of some natural and artificial intragroup hybrids. The vertical axis indicates relative mobility; the number of hybrids with each phenotype is indicated along the horizontal axis. The letters above the bands indicate the first letter(s) of the name of the parental species that possessed that component (Figs. 14–1 and 14–2). If both parents had the same band, the first letters of both species names appear.

Table 14–1. Assumed Natural Hybrids

Cross	No. of Individuals	Location Collected
B. woodhousei x *B. houstonensis*	1	Bastrop, Texas
B. houstonensis x *B. valliceps*	3	Bastrop, Texas
B. woodhousei x *B. valliceps*	7	Austin, Texas
B. woodhousei x *B. microscaphus*	3	St. George, Utah
B. spinulosus x *B. arenarum*	1	Mendoza, Argentina

two transferrins noted in *B. alvarius*. Similar inheritance patterns were noted in hybrids from the following crosses: *B. luetkeni* x *B. compactilis*, *B. valliceps* x *B. cognatus*, and *B. cognatus* x *B. perplexus* (Fig. 14-4). Transferrin inheritance in *B. regularis* and *B. rangeri* (Guttman 1967*b*) is similar to this pattern.

Hemoglobins

Two hemoglobins, one from each parental species, were not typical in *Bufo* hybrids (Fig. 14-7). This situation did prevail in crosses between *B. houstonensis* x *B. valliceps*, *B. woodhousei* x *B. houstonensis*, *B. woodhousei* x *B. valliceps*, and *B. garmani* x *B. regularis*. When both parents possessed hemoglobins with the same mobility, it could not be ascertained whether each parent made an equal contribution to the genetic constitution of the hybrids. This was the situation in the *B. woodhousei* x *B. bufo* and *B. woodhousei* x *B. viridis* crosses. If the hybrids from the latter two crosses did have a component from each parent, it could be assumed that in all the above crosses the toads inherited one codominant allele from each parental type.

The majority of the hybrid toads possessed either one hemoglobin from one parent or a "hybrid" hemoglobin unique to either species. Hybrids with only one hemoglobin band resembling one of the parents resulted from the following crosses: *B. perplexus* x *B. bocourti*, *B. speciosus* x *B. arenarum*, and *B. arenarum* x *B. valliceps*. Wittliff (1964) noted that parotoid-secretion patterns of *B. fowleri* x *B. valliceps* hybrids resembled either parent. Chromatography of indoles yielded patterns like *B. fowleri*. When the venom was subjected to electrophoresis, the patterns of five of the seven hybrids he examined resembled *B. valliceps*. The serum of hybrid trout had electrophoretic patterns characteristic of one of the parental types (Sanders 1964). Manwell et al. (1963) found that both the green sunfish (*Lepomis cyanellus*) and the bluegill (*L. macrochirus*) possess two distinct hemoglobins. A cross between these two species produced ten hybrids, eight of which had the four parental bands and two with only three of the four. It appeared that in the latter two hybrids there was a suppression of the faster-migrating hemoglobin from the *L. cyanellus* parent. Although Wittliff (1964) attributed his findings to genic dominance and recessiveness, this hypothesis does not appear to be applicable here. However, no explanation for this phenomenon is presently available.

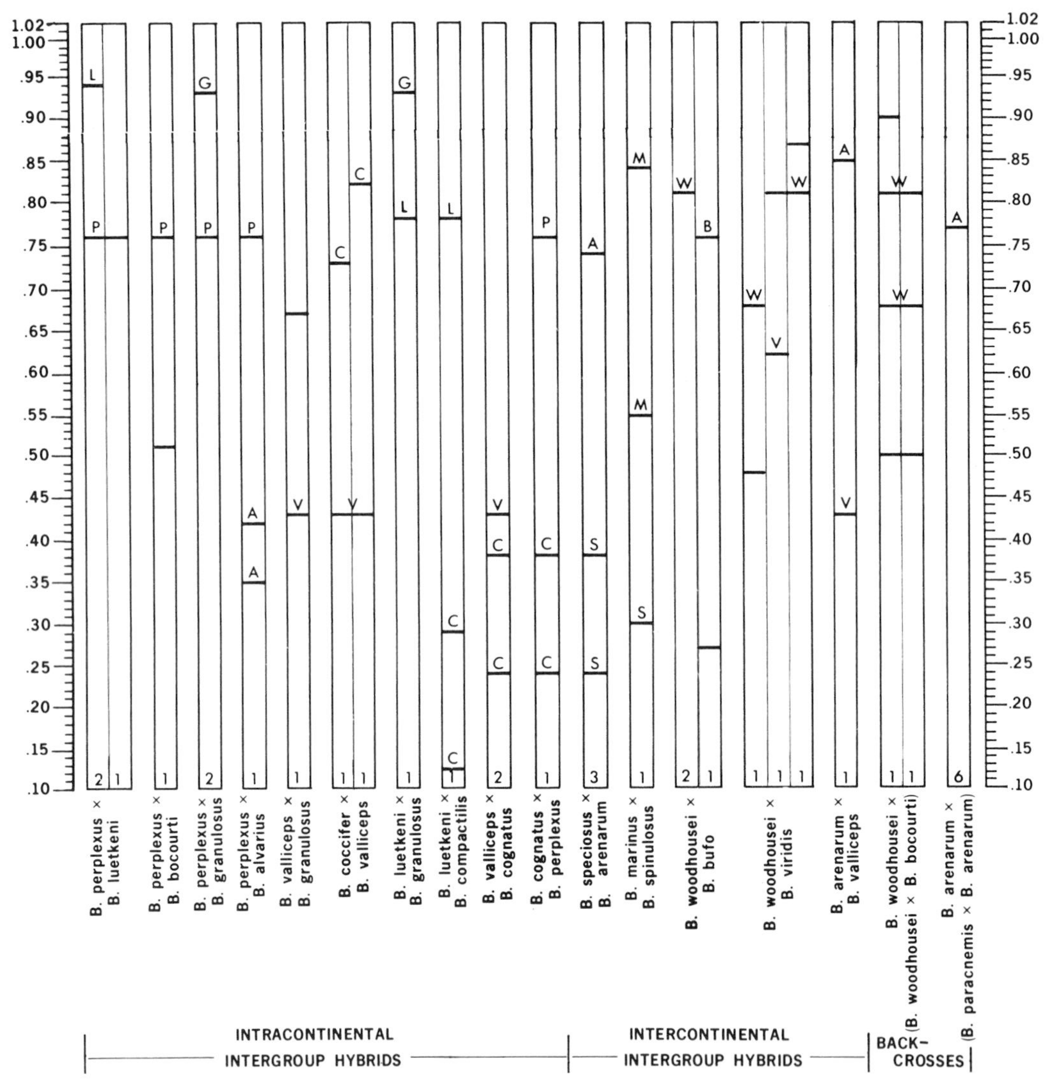

Fig. 14–4. Transferrin phenotypes of some intergroup hybrids and backcrosses. The vertical axis indicates relative mobility; the number of hybrids with each phenotype is indicated along the horizontal axis. The letters above the bands indicate the first letter of the name of the parental species that possessed the component (Figs. 14–1 and 14–2). Transferrins that could not be assigned to either parent do not have letters above them.

"Hybrid" proteins were found in several crosses: *B. perplexus* x *B. luetkeni, B. coccifer* x *B. valliceps, B. marinus* x *B. spinulosus* (Fig. 14-7). In the former two crosses a component from one parent (*B. perplexus* and *B. valliceps* respectively) appeared in addition to the "hybrid" hemoglobin. Manwell et al. (1963) found that the F_1 hybrids from most sunfish crosses possessed hemoglobins of both parents. However, in three crosses the hybrids always had 25 to 40 percent of their hemoglobins with unique electrophoretic properties. Maxwell et al. (1963) and other workers who found "hybrid" proteins (Allen 1962; Cahn et al. 1962; Levinthal et al. 1962) attributed "hybrid" components to polypeptide-chain recombinations.

Polymorphism

Gratzner and Allison (1960) stated that the number of polymorphic hemoglobin systems reported were few and that these were almost entirely confined to mammals. Dessauer et al. (1962) reported that the number and mobilities of toad transferrins were stable within a species. These statements were generally due to any of three weaknesses associated with studies made prior to 1963 (and in some instances true today). First, electrophoretic techniques used were not adequate to detect minor mobility variations, although they may have been present. Moving boundary and paper electrophoresis are markedly inferior to current starch-gel and acrylamide-gel techniques. Second, few investigations were made on blood of lower vertebrates. Most of the research was, and still is, done on the blood of mammals, notably *Homo sapiens*, and birds. Third, the few studies performed with fish, amphibians, and reptiles were generally survey papers, and the number of individuals of any single species sampled was low, in many instances, one.

Of the thirty-seven possible species where different hemoglobin morphs could be found, that is, where two or more individuals were sampled, nineteen species were polymorphic. The transferrins of two or more individuals of each of forty-six species of toads were examined and thirty of these species showed two or more transferrin types. Variation in both proteins was noted in thirteen species.

Variation Resulting from Hybridization

Some polymorphism can be attributed to interspecific hybridization and subsequent introgression. Interspecific hybridization appears to be a major factor in the transferrin and hemoglobin variation in *B. regularis* and *B. rangeri* in the vicinity of Port St. Johns, South Africa (Guttman 1967*b*) and in the *americanus* group in the United States (Guttman 1969).

Interpopulation Variation Resulting from Possible Deficiencies in Taxonomy

Although the hemoglobins of *B. spinulosus* from the two localities sampled were not distinct (Fig. 14-6), the transferrins of individuals from Chile, Argentina, and Peru were different (Fig. 14-2). Blair (Chap. 11) regards *B. spinulosus* as a "super-species." Some South American workers (Cei 1960, 1962; Vellard 1959) believe that there are at least four subspecies in Chile and six in Peru. Cei and Erspamer (1966) noted marked differences in the spectrum of indolealkylamines between populations from different geographic areas. Cei (Chap. 6) now recognizes the existence of several species in the *spinulosus* group. Martin (Chap. 4) found at least three distinct skull types in what is presently called

Table 14–2a. Artificial Hybrids Resulting from Crosses within a Species Group

Cross			No. of Individuals	Cross No.
B. woodhousei Bastrop, Texas	x	*B. houstonensis* Bastrop, Texas	1	B 65-116
B. valliceps Austin, Texas	x	*B. luetkeni* Motagua Valley, Guatemala	1	B 63-96
B. luetkeni Motagua Valley, Guatemala	x	*B. valliceps* Austin, Texas	1	B 64-62
B. luetkeni Motagua Valley, Guatemala	x	*B. valliceps* Austin, Texas	1	B 64-80
B. rangeri Swellendam, South Africa	x	*B. regularis* Ogbomosho, Nigeria	1	B 65-379
B. poeppigi Iquitos, Peru	x	*B. marinus* Veracruz, Mexico	3	B 65-545
B. garmani Kruger National Park, South Africa	x	*B. regularis* Pietermaritzburg, South Africa	2	B 65-429

(Table continued next page)

Table 14–2b. Artificial Hybrids from Intracontinental Intergroup Crosses

Cross			No. of Individuals	Cross No.
B. perplexus Morelos, Mexico	x	*B. luetkeni* Salama, Guatemala	4	B 65-171
B. perplexus Morelos, Mexico	x	*B. bocourti* Motagua Valley, Guatemala	2	B 65-154
B. perplexus Morelos, Mexico	x	*B. granulosus* Chiriqui, Panama	2	B 65-157
B. perplexus Morelos, Mexico	x	*B. alvarius* Mesa, Arizona	1	B 65-170
B. valliceps Austin, Texas	x	*B. granulosus*	1	B 64-149
B. coccifer Esparta, Costa Rica	x	*B. valliceps* Austin, Texas	4	B 65-61
B. luetkeni Motagua Valley, Guatemala	x	*B. granulosus* Chiriqui, Panama	1	B 65-141
B. luetkeni Motagua Valley, Guatemala	x	*B. compactilis* Mexico D. F., Mexico	1	B 65-147
B. valliceps Austin, Texas	x	*B. cognatus* Lubbock, Texas	2	B 64-154
B. cognatus Trans Pecos, Texas	x	*B. perplexus* Morelos, Mexico	1	B 65-233
B. arenarum Mendoza, Argentina	x	*B. spinulosus* Mendoza, Argentina	1	B 65-88

Table 14–2c. Artificial Hybrids from Intercontinental Intergroup Crosses

Cross			No. of Individuals	Cross No.
B. speciosus Hornsby Bend, Texas	x	*B. arenarum* Mendoza, Argentina	3	B 65-128
B. marinus Costa Rica	x	*B. spinulosus* Lima, Peru	1	B 65-227
B. woodhousei Bastrop, Texas	x	*B. bufo* Germany	3	B 65-119
B. woodhousei Austin, Texas	x	*B. viridis* Israel	2	B 65-13
B. arenarum Mendoza, Argentina	x	*B. valliceps* Austin, Texas	1	B 64-301

Table 14–2d. Backcrosses of Artificial Hybrids

Cross			No. of Individuals	Cross No.
B. woodhousei	x	B 64-9 (*B. woodhousei* x *B. bocourti*)	2	B 65-7
B. arenarum Mendoza, Argentina	x	B. 64-309 (*B. paracnemis*, Chaco Central, Argentina x *B. arenarum*, Mendoza, Argentina)	6	B 65-76

B. spinulosus; these correspond with results of this study. He believes that these forms deserve specific status.

Intrapopulation Polymorphism

The greatest polymorphism was found in the *regularis* and *americanus* groups. Although in a few other species some of this variation may be due to interspecific hybridization, most of it cannot be attributed to this factor.

Guttman (1967*b*) found that thirteen *B. regularis* from El Mahalla el Kubra, Egypt, possessed eleven hemoglobin components in eight phenotypes. Four transferrins were present in four phenotypes in the same individuals. Within the *americanus* group, similar variation was noted (Guttman 1969). For example, six transferrin morphs were identified in the sixteen *B. microscaphus* sampled from Virgin, Utah.

Fox et al. (1961) stated that one iron-binding protein is characteristic of *B. valliceps*. The only animals they examined were collected within five miles of New Orleans, Louisiana. Although the transferrin of this species from Austin, Texas, and Yucatan, Mexico, is identical with the component they found in Louisiana, two other Mexican toads of this species possessed additional bands (Fig. 14-1). A faster-migrating transferrin was noted in one individual from Veracruz, while one toad from Tamaulipas had a slower-moving component.

Another toad with a high degree of transferrin polymorphism is *B. coccifer*. The seven animals collected from two localities in Costa Rica possessed six phenotypes. It is interesting to note that the four toads tested for hemoglobin all possessed the same component.

All *B. arenarum* sampled were collected in the vicinity of Mendoza, Argentina. Four molecular species of transferrin were found in the ten toads examined (Fig. 14-2); two hemoglobin types were noted (Fig. 14-6). Bertini and Cei (1960) reported interpopulation differences and stated that the Mendoza population represented a distinct physiological race. Ecological and physiological polymorphism was noted in this population by Cei (1959). He found that 25 percent of the individuals had a pattern of yellow spots and 60 percent possessed a "hypnotic reflex."

Other species showed intrapopulation polymorphism that could not be attributed to interspecific hybridization. *B. garmani, B. viridis, B. nelsoni, B. boreas, B. quercicus, B. punctatus, B. debilis, B. marmoreus, B. perplexus, B. coniferus, B. luetkeni, B. marinus*, and *B. haematiticus* had two or more transferrin phenotypes present in any one population (Figs. 14-1 and 14-2). Intrapopulation variation in hemoglobin components was noted in *B. garmani, B. punctatus, B. kelloggi, B. speciosus, B. cognatus, B. marmoreus, B. perplexus, B. valliceps, B. caniliferus, B. coniferus*, and *B. granulosus* (Figs. 14-5 and 14-6).

Possible Mechanisms Supporting Polymorphism

Most explanations of intrapopulation polymorphism of protein systems, when the variants are present in high frequency, are based on the balanced polymorphic situation of the human hemoglobin A/S heterozygotes postulated by Allison (1954). Selection for heterozygotes also proposes to explain the high frequencies of human hemoglobins C, D, and E. However, these are the only blood proteins where the selective advantages of a particular combination of alleles are known. Although Ashton (1965) postulated the probable existence of a balanced polymorphism in order to explain the transferrin alleles of

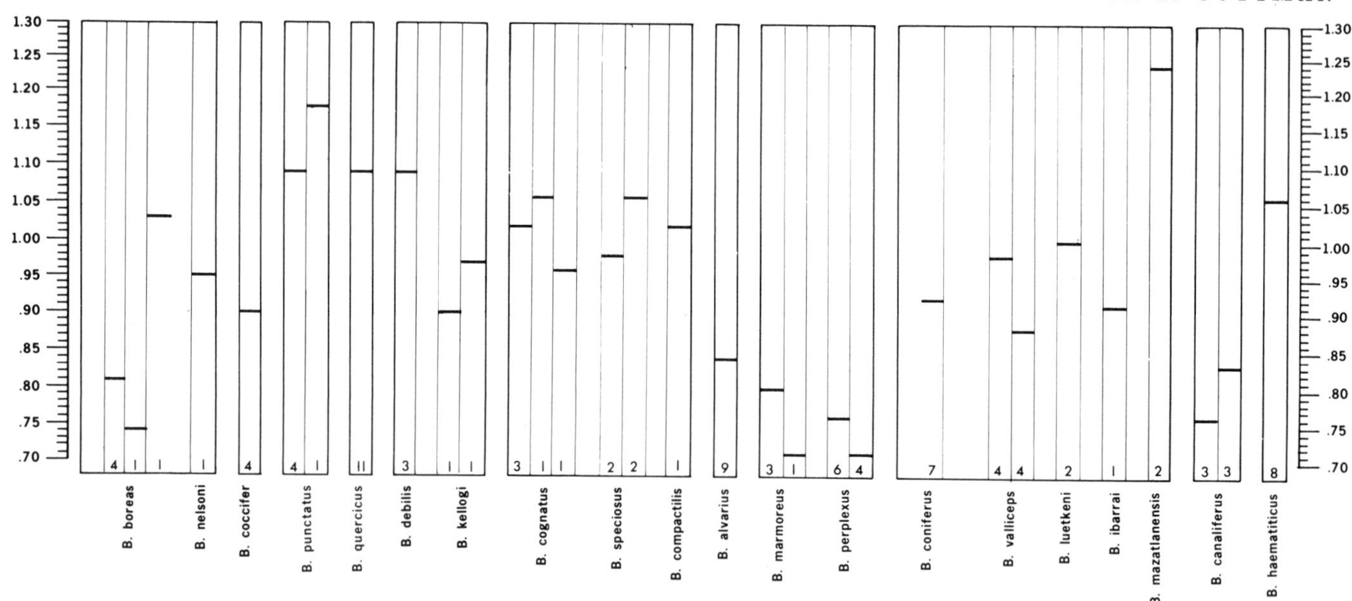

Fig. 14–5. Hemoglobin phenotypes of North American species groups. The vertical axis indicates relative mobility; the number of individuals of a species with each phenotype is indicated along the horizontal axis.

cattle, it has not been shown that different transferrin types differ in their functional ability, that is, to transport iron.

The presence of a diversity of alleles in a given population might be selected under fluctuating environmental conditions. With a variety of alleles at certain essential loci, a population would be endowed with a plasticity not possessed by an essentially homozygous population.

Interpopulation variation could exist if the species is wide ranging, with different populations occupying different habitats. For example, the *B. americanus* found in Ontario, Canada, are not subjected to the same climatic conditions as are conspecific toads from North Carolina or Oklahoma, and different environmental selection pressures probably exist.

Interspecific hybridization (Guttman 1967*b*, 1969; Brown and Guttman 1970) contributes to this variation.

Since the purpose of this study was to examine representative samples of all available species, large numbers of any one species were not tested. Consequently, significant gene frequencies could not be calculated, population interactions could not be studied, and a foundation for selecting the mechanism(s) could not be laid.

Intragroup and Intergroup Relationships

Because of the high degree of polymorphism present in the blood proteins of most species of the genus *Bufo*, great care must be exerted in determining interspecific relationships based on these proteins. However, comparisons of those species with little or no variation yield interesting affinities.

Although their transferrins were different, *B. viridis* and some *B. bufo* possessed the same hemoglobin (Fig. 14-6). Neither the transferrin (Figs. 14-1 and 14-3) nor the hemoglobin (Figs. 14-5 and 14-6) components present in other *B. bufo* or *B. calamita* appeared in the polymorphic *boreas* or *americanus* groups. However, two *B. viridis* possessed iron-binding components that were evident in *boreas*-group toads (Figs. 14-1 and 14-2). Chromatograms of venom from *B. viridis* and *B. bufo* showed a 91 percent affinity with each other (B. Low, Chap. 13). Marchlewska-Koj (1963) also found that the hemoglobin of these two species was the same; she noted also that *B. calamita* had an identical component. Similar patterns were obtained from the serum of *B. bufo* and *B. calamita* (Chen 1967). This evidence suggests that *B. calamita*, *B. viridis*, and *B. bufo* are more closely related to each other than they are to

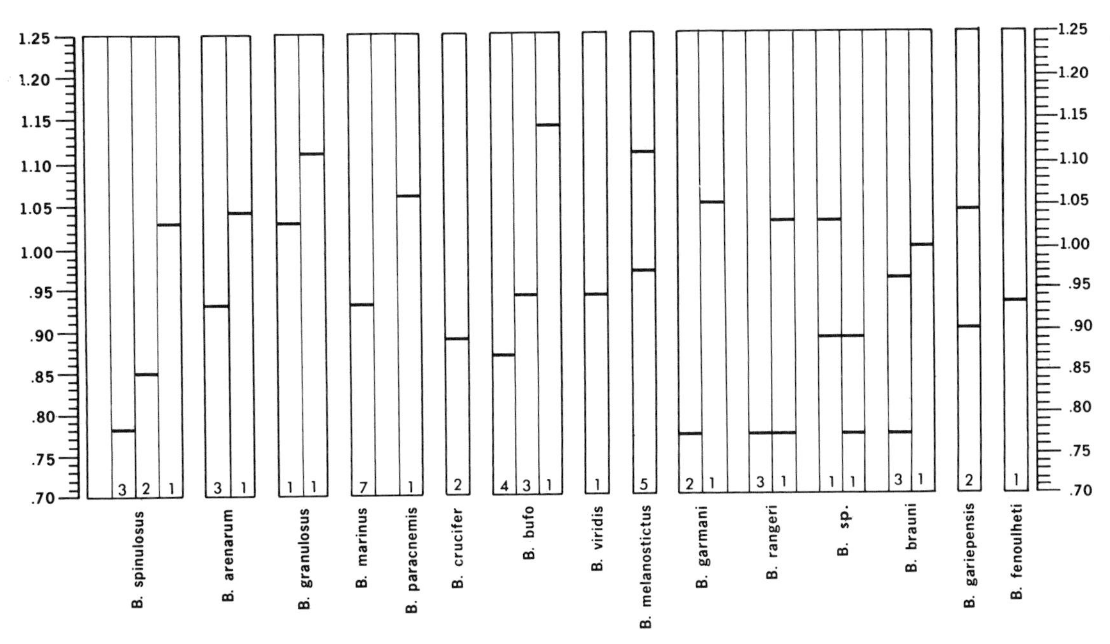

Fig. 14–6. Hemoglobin phenotypes of South American, Eurasian, and African species groups. The vertical axis indicates relative mobility; the number of individuals of a species with each phenotype is indicated along the horizontal axis.

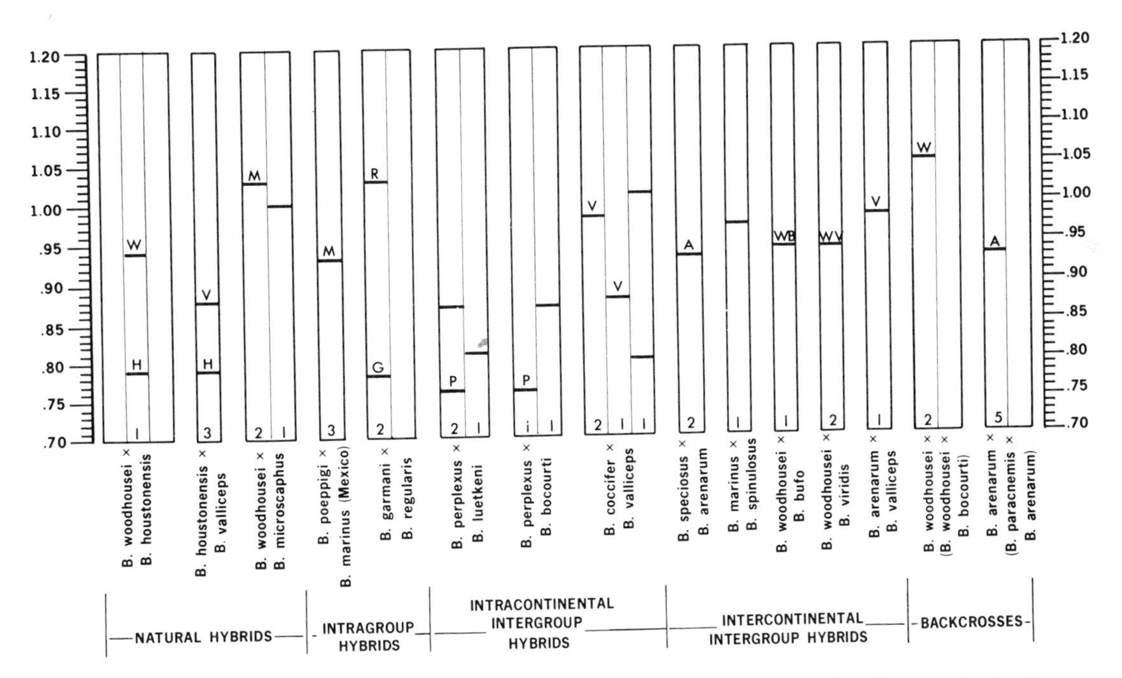

Fig. 14–7. Hemoglobin patterns of some natural and artificial hybrids. The vertical axis indicates relative mobility; the number of hybrids with each phenotype is indicated along the horizontal axis. The letters above the bands indicate the first letter of the name of the parental species that possessed that component (Figs. 14–5 and 14–6). If both parents had the same band, the first letters of both species names appear. Hemoglobins that could not be assigned to either parent do not have letters above them.

the *boreas* or *americanus* groups. Of the former species, *B. viridis* shows a greater affinity to the New World toads (*boreas* group) than the other two forms show.

Within the *boreas* group, *B. boreas* and *B. nelsoni* are more similar to each other than either is to the remaining members of the group. Both transferrins of *B. nelsoni* were present in *B. boreas* (Fig. 14-1); they were found in greater frequency than any of the four other components noted in the latter toad. A transferrin from *B. canorus* and one from *B. exsul* were found in the other species of the group. Karlstrom (1958, 1962) and Mullally and Powell (1958) believe it possible that *B. boreas* and *B. canorus* hybridize. However, natural hybrids have not been found. Schuierer's (1962) comments on speciation within the *boreas* group parallel patterns suggested by transferrin similarities. He postulated that an ancestor in the late Tertiary of western North America gave rise first to a *B. canorus* type toad as a montane isolate, then to *B. exsul*, now endemic to Deep Springs Valley, California; later, as a result of postpluvial dessication, *B. nelsoni* branched off. The line ended with *B. boreas*. Thus, in his scheme, *B. boreas* and *B. nelsoni* are evolutionarily more closely related than the former species is to the remainder of the group.

Evidence of the relationships of *B. debilis* and its allies, *B. kelloggi* and *B. retiformis*, to *B. punctatus* and *B. quercicus*, based on morphologic and genetic compatibility data, is inconclusive. The hemoglobin band in eleven *B. quercicus*, four *B. punctatus*, and three *B. debilis* was identical (Fig. 14-5). One of the two transferrin morphs found in *B. quercicus* was noted in four *B. debilis* (Fig. 14-1). The transferrin band in *B. kelloggi* was seen in three *B. punctatus*. These protein data suggest that the members of this complex of species (*B. punctatus*, *B. quercicus*, and the *debilis* group) are closely related.

Within the *cognatus* group, all *B. cognatus* and *B. speciosus* possessed the same transferrins, which were different from the two found in *B. compactilis* (Fig. 14-1). Thus, the former two species appear more similar to each other than either is to the latter. Although Blair (1963) indicated that the *cognatus* group is a clear-cut group based on morphological criteria, *B. cognatus* is distinct from the other two species in color pattern and size of the cranial crests. The mating call of *B. speciosus* is distinct from the calls of the other two species. High intragroup genetic incompatibility was noted (Blair 1961). Wittliff (1962) was unable to distinguish the chromatographic pattern of parotoid-gland secretions of *B. cognatus* and *B. speciosus*; this pattern differed from that of *B. compactilis*. Martin (1964) found that *B. cognatus* and *B. speciosus* are divergent osteologically.

The *cognatus* group and *B. alvarius* were the only New World toads examined in which single alleles appeared to determine multiple transferrins (Fig. 14-1). The transferrins were not alike, which might indicate convergence. However, evidence that these species are closely related is available (Chap. 11). The integumental pteridine patterns of *B. alvarius* and *B. cognatus*, as reported by Bagnara and Obika (1965), are similar.

Analyses of protein patterns suggest affinities within the African *regularis* group. Evidence on the possibility that *B. regularis* is a mixture of two species was discussed by Guttman (1967*b*). The hemoglobin of *B. rangeri* and *B. garmani* was the same electrophoretically (Fig. 14-6). *B. brauni* usually possessed two bands; the slower-migrating one was the same as the one found in the above-mentioned two species. Poynton (1964) stated that *B. garmani* shows closer affinities to *B. rangeri* than to *B. regularis*. Chromatographic patterns of venom from *B. rangeri*, *B. garmani*, and *B. brauni* are similar (Chap. 13), and consequently, they provide further evidence for the close relationship between these species.

Poynton (1964) placed *B. pusillus* in the *regularis* group. However, tadpoles resulting from reciprocal crosses between *B. maculatus* and members of the *regularis* group did not metamorphose (Chap. 11). *B. maculatus* and *B. pusillus* are considered synonymous by Blair (Chap. 11). An affinity of 87 percent was found between the venom patterns of the latter two species (Chap. 13). The transferrins of these toads were unlike those of any of the *regularis* group. In addition, the transferrins of *B. maculatus* were markedly different from the one component found in *B. pusillus* (Fig. 14-2). A single allele appears to determine multiple components in *B. maculatus;* in *B. pusillus*, an allele specifies a single iron-binding component. Absence of any similarities in blood patterns indicate that these are distinct types.

REFERENCES

Allen, S. L. 1962. Hybrid enzymes and isozymes. Science 138:714.

Allison, A. C. 1954. Protection afforded by sickle-cell trait against subtertian malarial infection. Brit. Med. J. 1:290–294.

Alston, R. E., and B. L. Turner. 1963. Biochemical systematics. Prentice-Hall, Inc., Englewood Cliffs, New Jersey.

Ashton, G. C. 1965. Cattle serum transferrins: A balanced polymorphism? Genetics 52:983–997.

Bagnara, J. T., and M. Obika. 1965. Comparative aspects of integumental pteridine distribution among amphibians. Comp. Biochem. Physiol. 15:33–49.

Bertini, F., and J. M. Cei. 1959. Electroferogrames de proteines seriques en el genero *Bufo*. Actas I° Congr. Zool. Sudamer. (La Plata) 4:161.

———. 1960. Observaciones electroforéticas en seroproteínas de poblaciones argentinas de *Bufo arenarum* Hensel. Rev. Soc. Arg. Biol. 36:355–362.

———. 1962. Seroprotein patterns in the *Bufo marinus* complex. Herpetologica 17:231–239.

Bertini, F., and G. Rathe. 1962. Electrophoretic analysis of the hemoglobins of various species of anurans. Copeia 1962:181–185.

Blair, W. F. 1961. Further evidence bearing on intergroup and intragroup genetic compatibility in toads (genus *Bufo*). Texas J. Sci. 13:163–175.

———. 1962. Non-morphological data in anuran classification. Syst. Zool. 11:72–84.

———. 1963. Evolutionary relationships of North American toads of the genus *Bufo*: A progress report. Evolution 17:1–16.

Brown, L. E. 1964. An electrophoretic study of variation in the blood proteins of the toads, *Bufo americanus* and *Bufo woodhousei*. Syst. Zool. 13:92–95.

———, and S. I. Guttman. 1970. Natural hybridization between the toads *Bufo arenarum* and *Bufo spinulosus* in Argentina. Amer. Midl. Nat. 83:160–166.

Cahn, R. D., O. N. Kaplan, L. Levin, and E. Zwilling. 1962. Nature and development of lactic dehydrogenases. Science 136:962–969.

Cei, J. M. 1959. Ecological and physiological observations on polymorphic populations of the toad *Bufo arenarum* Hensel, from Argentina. Evolution 13:532–536.

———. 1960. Geographic variation of *Bufo spinulosus* in Chile. Herpetologica 16:243–250.

———. 1962. Batracios de Chile. Universidad de Chile, Santiago.

———, F. Bertini, and G. C. Gallopin. 1961. La ratio albumina/globulinas y su probable significado ecológico en los anfibios Sudamericanos. Rev. Soc. Arg. Biol. 37:215–225.

Cei, J. M., and R. Cohen. 1964. Electrophoretic patterns and systematic relations in South American toads. Rutgers Univ. Serol. Mus. Bull. 30:6–8.

Cei, J. M. and V. Erspamer. 1966. Biochemical taxonomy of South American amphibians by means of skin amines and polypeptides. Copeia 1966:74–78.

Chen, P. S. 1967. Separation of serum proteins in different amphibian species by polyacrylamide gel electrophoresis. Experientia 23:1–8.

Dessauer, H. C. 1966. Taxonomic significance of electrophoretic patterns of animal sera. Rutgers Univ., Serol. Mus. Bull. 34:4–8.

———, and W. Fox. 1964. Electrophoresis in taxonomic studies illustrated by analyses of blood proteins, p. 625–647. *In*, C. A. Leone (ed.), Taxonomic biochemistry and serology. Ronald Press, New York.

———, and Q. L. Hartwig. 1962. Comparative study of transferrins of amphibians and reptiles using starch-gel electrophoresis and autoradiography. Comp. Biochem. Physiol. 5:17–29.

Fox, W., H. C. Dessauer, and L. T. Maumus. 1961. Electrophoretic studies of blood proteins of two species of toads and their natural hybrids. Comp. Biochem. Physiol. 3:52–63.

Giblett, E. R., C. G. Hickman, and O. Smithies. 1959. Serum transferrins. Nature 183:1589–1590.

Gratzer, W. B., and A. C. Allison. 1960. Multiple hemoglobins. Biol. Rev. 35:459–506.

Guttman, S. I. 1967*a*. Evolution of blood proteins within the cosmopolitan toad genus *Bufo*. Ph.D. Thesis, Univ. Texas.

———. 1967*b*. Transferrin and hemoglobin polymorphism, hybridization and introgression in two African toads, *Bufo regularis* and *Bufo rangeri*. Comp. Biochem. Physiol. 23:871–877.

———. 1969. Blood protein variation in the *Bufo americanus* species group of toads. Copeia 1969:243–249.

Hebard, W. B. 1964. Serum-protein electrophoretic patterns of the Amphibia, p. 649–657. *In*, C. A. Leone (ed.), Taxonomic biochemistry and serology. Ronald Press, New York.

Hubby, J. L., and L. H. Throckmorton. 1965. Protein differences in *Drosophila*. II. Comparative species genetics and evolutionary problems. Genetics 52:203–215.

Karlstrom, E. L. 1962. The toad genus *Bufo* in the

Sierra Nevada of California. Univ. Calif. Publ. Zool. 62:1–104.

Kristjansson, F. K., and C. G. Hickman. 1965. Subdivision of the allele Tf^D for transferrins in Holstein and Ayrshire cattle. Genetics 52:627–630.

Levinthal, C., E. Signer, and K. Fetherolf. 1962. Reactivation and hybridization of reduced alkaline phosphatase. Proc. Nat. Acad. Sci. 48:1230–1237.

Manwell, C., C. M. A. Baker, and W. Childers. 1963. The genetics of hemoglobin in hybrids. I. A molecular basis for hybrid vigor. Comp. Biochem. Physiol. 10: 103–120.

Marchlewska-Koj, A. 1963. Electrophoretic investigations of hemoglobins in Amphibia. Folia Biol. (Warsaw) 11:167–172.

Martin, R. F. 1964. Osteological morphology and the phylogeny of certain North American toads (genus *Bufo*). M. A. Thesis, Univ. Texas.

Mayr, E. 1963. Animal species and evolution. Belknap Press. Cambridge, Mass.

———, E. G. Linsley, and R. L. Unsinger. 1953. Methods and principles of systematic zoology. McGraw-Hill, New York.

Mullally, D. P., and D. H. Powell. 1958. The Yosemite toad: Northern range extension and possible hybridization with the western toad. Herpetologica 14:31–33.

Nakamura, E. 1960. Das mehrfache Hämoglobin. III. Amphibien und Reptilien. Lymphatologia 4:52–59.

Poynton, J. C. 1964. The Amphibia of southern Africa: A faunal study. Ann. Natal Mus. 17:1–334.

Rodnan, G. P., and F. G. Ebaugh. 1957. Paper electrophoresis of animal hemoglobins. Proc. Soc. Exp. Biol. Med. 95:397–401.

Rossi, A. 1960. Obsservazioni preliminari sul frazionamento delle proteine del siero nei due sessi di *Bufo bufo* (L.) con il metodo dell' elettroforesi sul gel d'amido. Ricera Sci. 30:141–144.

Sanders, B. G. 1964. Electrophoretic studies of serum proteins of three trout species and the resulting hybrids within the family Salmonidae, p. 673–679. *In*, C. A. Leone (ed.), Taxonomic biochemistry and serology. Ronald Press, New York.

Schuierer, F. W. 1962. Notes on two populations of *Bufo exsul* Myers and a commentary on speciation within the *Bufo boreas* group. Herpetologica 18:262–267.

Smithies, O. 1959. Zone electrophoresis in starch gels and its application to studies of serum proteins. Adv. Prot. Chem. 14:65–113.

Vellard, J. 1959. Estudios sobre batracios andinos. V. El género *Bufo*. Mem. Mus. Hist. Nat. (Xavier Prado) 8:1–48.

Wittliff, J. L. 1962. Parotoid gland secretions in two species groups of toads. Evolution 16:143–153.

———. 1964. Venom constituents of *Bufo fowleri, Bufo valliceps* and their natural hybrids analyzed by electrophoresis and chromatography. p. 457–464. *In*, C. A. Leone (ed.), Taxonomic biochemistry and serology. Ronald Press, New York.

Wright, A. H., and A. A. Wright. 1949. Handbook of frogs and toads. Comstock Publ. Co., Ithaca, New York.

15. Evolution of Vocalization in the Genus *Bufo*

WILLIAM F. MARTIN

The University of Texas at Austin
Austin, Texas

Introduction

Adult males of most anuran species produce communicatory sounds. In *Bufo*, these signals are usually regarded either as mating calls, which function as premating isolating mechanisms by selectively attracting conspecific mates (A. P. Blair 1941; W. F. Blair 1956 *et seq.*), or as release calls, involved in the identification of sexual partners (Aronson 1944; A. P. Blair 1947). The function of mating calls has been experimentally demonstrated for numerous anuran species, particularly members of the Hylidae (reviewed by W. Martin, unpubl.), but there is very little evidence on the function(s) of these sounds in *Bufo*. The problem was first studied experimentally by A. P. Blair (1942), who obtained inconclusive results for distance attraction in two species of *Bufo*. However, Bogert (1960) presented evidence that the mating call of *B. terrestris* may be involved in the distant orientation and attraction of toads forming breeding assemblages. Awbrey (1965) synthesized *Bufo* mating calls electronically. He tested the range of different call parameters to which gravid female *B. valliceps* would respond when electronically produced signals were played to them through loud speakers. He also conducted discrimination experiments in which the natural vocalizations of the two species were presented to gravid females, and their choice (indicated by their migration to one or the other of the two broadcasting speakers) was recorded. Thirteen *B. woodhousei woodhousei* females responded without preference to calls of their own subspecies as well as to calls of *B. w. fowleri*. When Awbrey presented calls of *B. valliceps* and a *B. woodhousei* x *B. valliceps* hybrid to the same *B. w. woodhousei* females, they did not respond to either, but "all quickly responded to the *B. w. woodhousei* call given immediately afterward." This single discrimination experiment, in which several females were tested at the same time, constitutes the only experimental evidence that the mating call in *Bufo* is a species-specific sex attractant. Therefore, the mating calls of most species of *Bufo* do not have functional definitions. Only a contextual definition of the mating calls is justified at this time: the vocalization most commonly given by adult male toads without tactile stimulation in breeding aggregations. There is much circumstantial evidence that suggests that these calls function in attracting conspecific mates, and this assumption is made in this report. However, it is debated whether the call serves only to attract males and females to the chorus (Bogert 1960) or if females actually

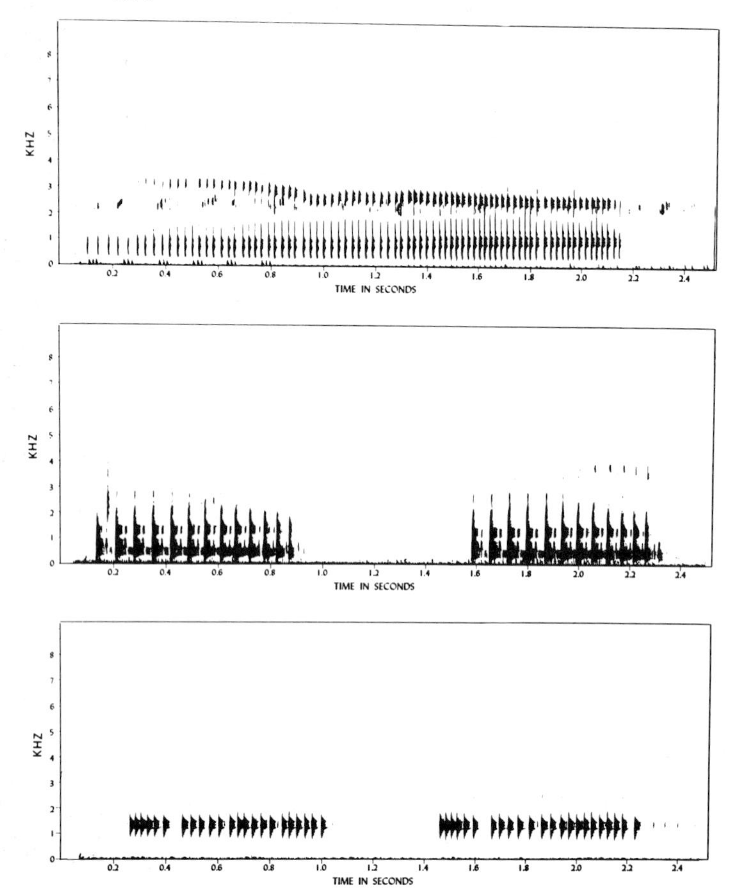

Fig. 15–1. Sound spectrograms of three representative Type I mating calls. (*A*) *B. kerinyagae*, recorded by R. Keith near Nanyuki, Kenya, air and water temperature, 16.1 ° C. (*B*) *B. regularis*, recorded by R. Keith at Amboseli, Kenya, temperature unknown. (*C*) *B. calamita*, recorded by H. Schneider, Boblingen, Germany. Air 4° C; water 6° C. Effective filter band width 300 Hz.

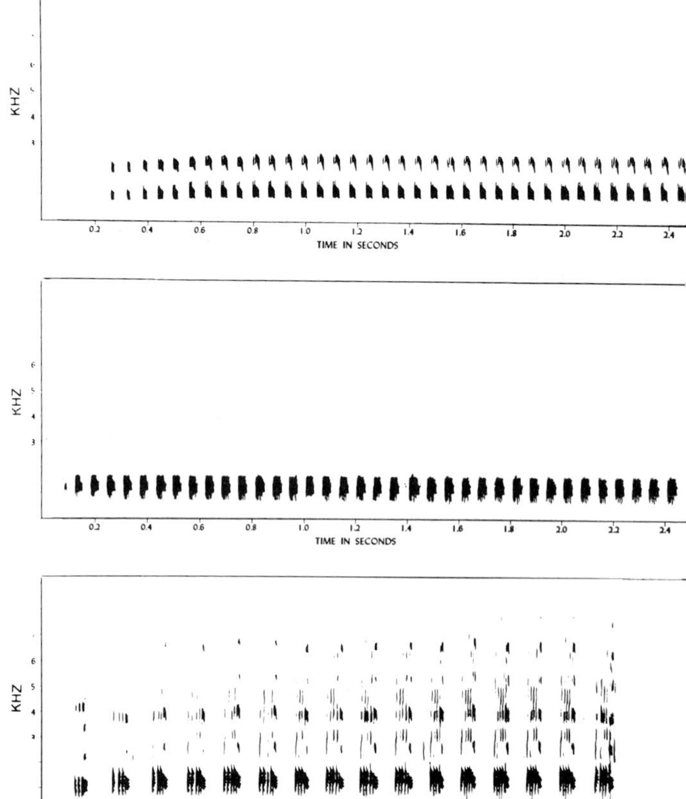

Fig. 15–2. Sound spectrograms of three representative Type II mating calls. (*A*) *B. arenarum*, recorded by W. Birkhead near Canti, Cordoba Prov., Argentina, Air 22° C; water 21° C. (*B*) *B. crucifer*, recorded by B. Lutz, Açude da Solidao, Tijuca Mts., Brazil. (*C*) *B. perplexus*, recorded by K. Porter and C. Nelson near Zampango del Rio, Guerrero, Mexico. Air 23° C. Toad 22° C. Effective filter band width 300 Hz. See also Fig. 15-4f–j.

home on the mating calls of individual conspecific males (W. F. Blair 1956). The distinction is important, for if the former is true, the mating call could not be an effective isolating mechanism when more than one species chorus together.

The release call has been the subject of previous experimentation. Aronson (1944) studied the release calls of *B. americanus, B. w. fowleri,* and *B. terrestris.* The signal consists of two components, a vibration of the flanks and a vocalization. He found that the release call, given by an amplexed male, will cause the amplexing male to release him. The vibration component was effective alone. A. P. Blair (1947) demonstrated that the release call was effective in several other species of *Bufo* and occurred across species lines in some species.

This investigation studies the evolutionary origin and diversification of the vocalizations of members of the genus *Bufo.* Particular attention is paid to the mating calls, for they show a higher degree of species specificity than the release calls do.

The data presented here will be useful taxonomically, but it should be emphasized that it is the evolution of vocalization that is being discussed and not the evolution of the genus *Bufo* per se. Practically, this means that I will draw strongly from the independent taxonomic data of other authors in tracing changes in vocal mechanics, and when those data

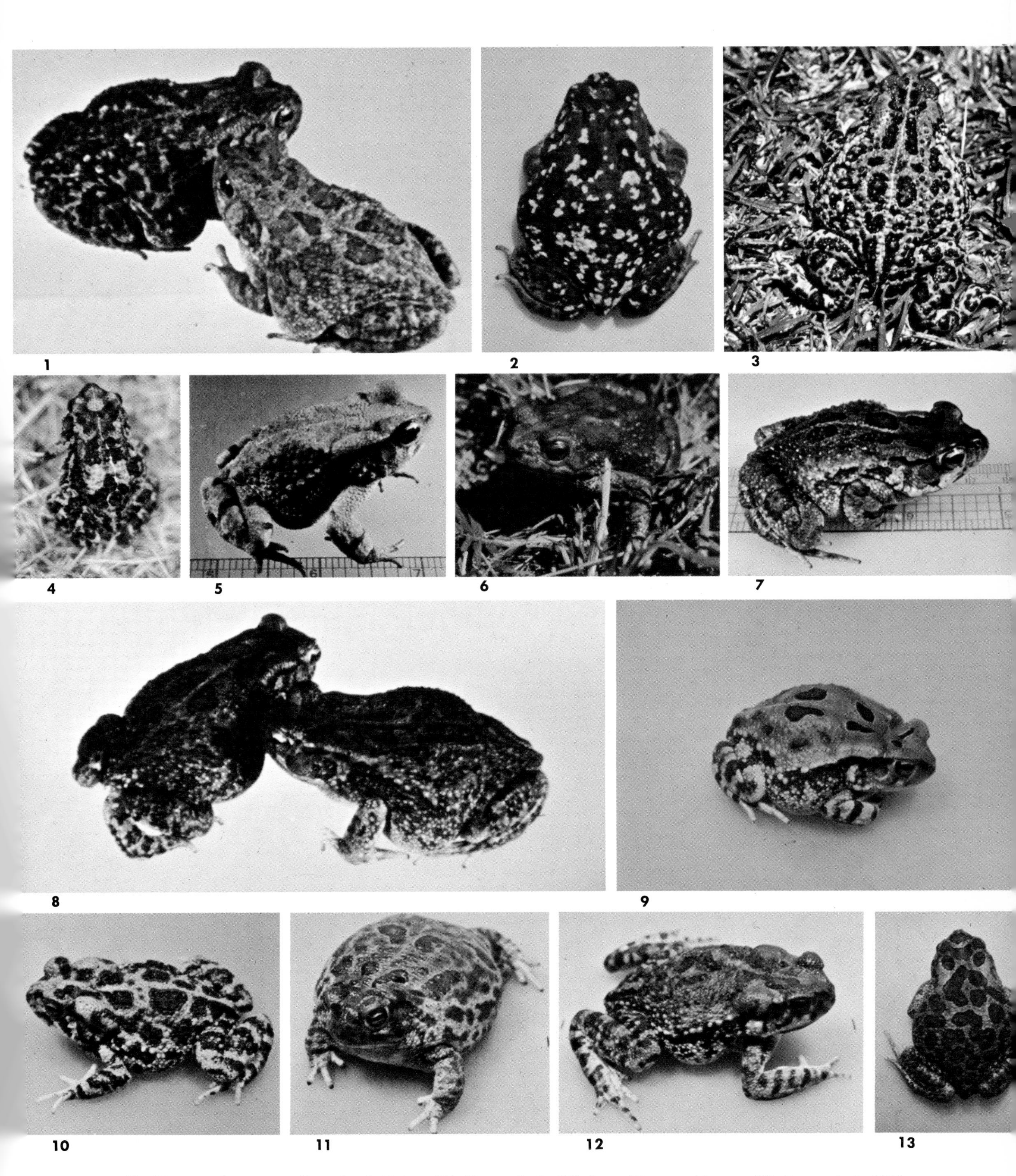

V. "Twenty-chromosome" toads of Africa: (1) *B. regularis* (E) from Mazeras, Kenya, (2) *B. regularis* (W) from Ogbomosho, Nigeria, (3) *B. kerinyagae*, (4) *B. perreti*, (5) *B. brauni*, (6) *B. maculatus*, (7) *B. rangeri*, and (8) *B. garmani*; (9) F_1 hybrid of ♀ *B. brauni* × ♂ *B. garmani*; (10) F_1 of ♀ *B. garmani* × ♂ *B. occidentalis*; (11) F_1 of ♀ *B. rangeri* × ♂ *B. marinus*; (12) F_1 of ♀ *B. garmani* × ♂ *B. marinus*; and (13) the 22-chromosome toad *B. mauritanicus*.

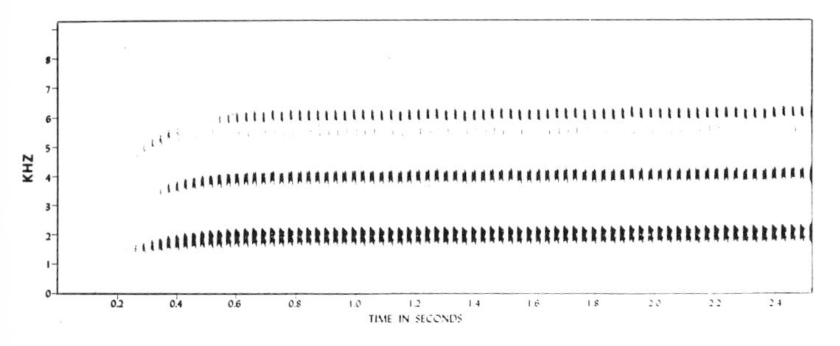

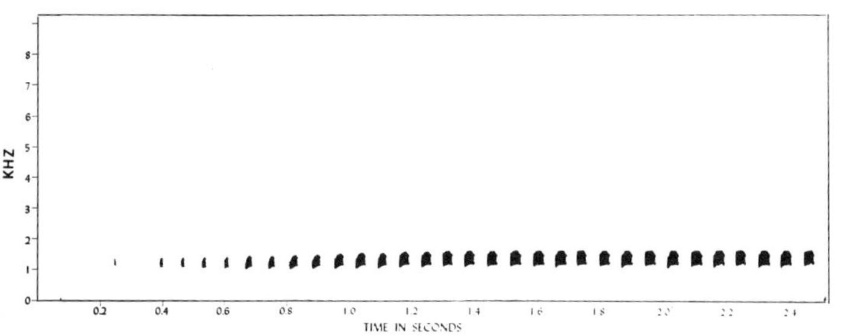

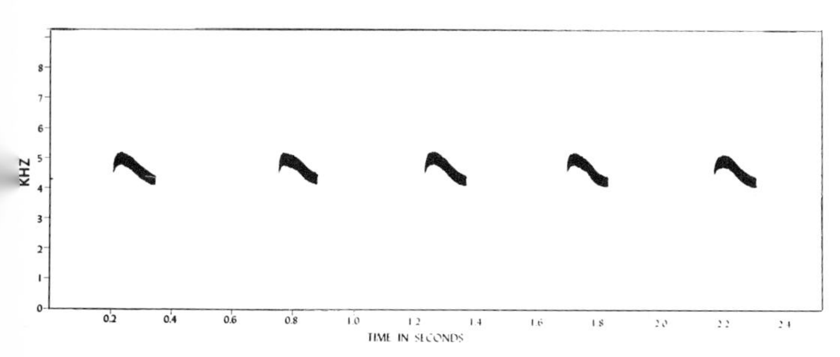

Fig. 15–3. Sound spectrograms of three representative Type III mating calls. (*A*) *B. americanus*, recorded by W. F. Blair near Beavers Bend State Park, Oklahoma, U.S.A. Air 19.4° C; water 20.8° C. (*B*) *B. viridis* recorded by H. Schneider near Tübing, Germany, Air 17° C; water 19° C. (*C*) *B. quercicus*, recorded by C. Johnson 10 mi. N.W. Waycross, Georgia, U.S.A. Air temperature 25° C; water 26° C. Effective filter band width 300 Hz.

suggest a convergence in vocal mechanics, it is considered as such. My data actually testify to an affinity between supposedly convergent forms, and, unless evidence to the contrary from vocal tract anatomy or mechanics is presented, they should be considered as convergent.

Structure of *Bufo* Mating Calls

The Dominant Frequency

When the frequency composition of a *Bufo* mating call is examined, a "dominant" frequency band can be identified (W. F. Blair 1956, *et. seq.*) or one that is more energetic than other frequencies present (Figs. 15-1, 15-2, 15-3). Harmonically related overtones of this dominant frequency are often seen, which are also extremely energetic sometimes. The dominant frequency varies from less than 0.5 KHz in *B. regularis* (Fig. 15-1B) to the 5.3 KHz in *B. quercicus* (Fig. 15-3C).

Frequency Modulation

Changes in the frequency of the dominant within the mating call are extremely common and are pronounced in leptodactylids, particularly the genus *Leptodactylus*. Pronounced frequency modulation in *Bufo* is seen only in the mating call of *B. quercicus* (Fig. 15-3C). This is apparently an example of convergence. However, there are two patterns of less pronounced change in dominant frequency that are widespread in the genus. One pattern is associated with each pulse of the call (Figs. 15-2A, 15-3A,B,C), and the other from one pulse to the next (Figs. 15-2A, 15-3A). The latter is usually associated with the start of the call.

Amplitude Modulation

Within the mating calls of all species studied, except possibly *B. quercicus*, there is a repetitive variation in the amplitude of the dominant frequency. This amplitude modulation, or pulsation, is of three distinct types, as seen in figure 15-4. In Type I (Fig. 15-4K–L) the rise time is fast and is usually followed by a damped oscillation. The rate of this pulsation often changes smoothly within the call (Fig. 15-4K). The Type II modulation is illustrated in figure 15-4F–J. The mating call consists of periodic trains of these complex pulses, and each of the complex pulses contains a repetitive amplitude modulation comparable to the Type I modulation and produced by the same mechanism. The Type III pulsation is presented in figure 15-4A–E. The mating call consists of long, periodic trains of relatively slow, rise-time pulses that are produced by the same mechanism as the slower of the two modulation rates seen in Type II calls.

Vocalizations of *Bufo* contain both active and passive components of variation. Call duration, intercall interval, pulse rate and duration in Type II and III calls, and frequency modulation, which involves contraction of the dilator laryngis muscles, are active components, involving muscle contraction with the periodicity of the mating call variable. The periodicity of passive components is determined by the resonance of some mechanical vibrator in the moving

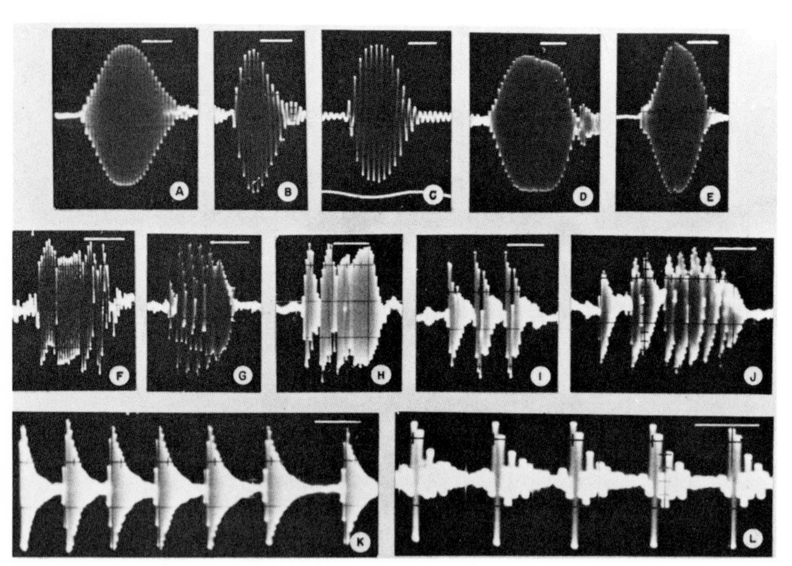

Fig. 15–4. Oscillograms from mating calls of 12 species of *Bufo*. Line on each trace indicates time interval noted below. First row; single Type III pulses, all 5 milliseconds. (*A*) *B. americanus*, (*B*) *B. ibarrai*, (*C*) *B. microscaphus*, (*D*) *B. houstonensis*, (*E*) *B. punctatus*. Second Row; single Type II pulses. (*F*) *B. paracnemis* (20 msec), (*G*) *B. luetkeni* (20 msec), (*H*) *B. cognatus* (10 msec), (*I*) *B. stomaticus* (10 msec), (*J*) *B. marmoreus* (20 msec). Third Row; trains of several Type I pulses. (*K*) *B. calamita* (20 msec), (*L*) *B. brauni* (10 msec). See also Figs. 15-1–15-3.

air stream. Pulse rate in Type I calls, intrapulse amplitude modulation in Type II calls, and dominant frequency and its harmonics in the calls of all species are passive components. All these passive components can be modulated by active muscle contraction.

Methodology

One cannot draw conclusions about evolutionary change in the mating calls of toads simply by studying these signals, because it is difficult to recognize homologous call parameters. Similar sounds can be made by different methods, which might involve anatomically different resonators and contractions of different muscles. Conversely, rather dissimilar sounds can be produced by basically the same mechanism. For these reasons, it is necessary to understand the vocal mechanics involved. The following techniques were used in examining call structure and in studying vocal mechanics.

Detailed analyses of the mating calls of forty-seven species of *Bufo* and several nonbufonids were performed. The release call was also studied in many of these species as well as in several others for which mating calls were not available. All these calls were recorded on magnetic tape and most of the mating-call recordings are in the bioacoustics library of The University of Texas. The recordings were analyzed on a Tektronix 502a oscilloscope, and permanent records were made by photography. The recordings were also analyzed on the Kay Electric Model 6061a sound spectrograph using standard techniques. Oscilloscopic analysis is necessary in order to examine amplitude modulation and waveform, while the sound spectrograph is indispensible for determining the frequency composition of the sound and for studying frequency modulation.

Anatomical studies constituted an important part of this investigation. Larynges of adult males were examined under a binocular dissecting microscope after being cut in the mid-saggital plane between the arytenoid cartilages. This allows efficient study of some aspects of the laryngeal kinetics. The effect of opening the larynx on the tension and the configuration of the vocal cords is easily revealed in this way. The effects of laryngeal muscle contraction on the laryngeal membranes were studied by electrically stimulating fresh preparations. Histological sections of the larynges of six species of *Bufo*, *Odontophrynus americanus*, *Ceratophrys ornata*, and *Pleurodema bibroni* were also prepared.

In order to identify the various membranes responsible for particular characteristics of the call, flared glass tubes (inside diameter, 1.5mm.) were placed in an incision in the posterior top of each lung of freshly pithed adult males of forty-four species. The tubes were then forced forward until they contacted the posterior edge of the cricoid cartilage, where they were securely fastened by clamping the gossamer lung tissue to the glass tube with vinyl-coated copper wire. The vinyl prevented cutting the lung tissue with the wire. Fine rubber tubing connected the glass tubes to an air source. A bleeder valve and aneroid pressure gauge connected in the tubing circuitry permitted controlled variation of air pressure across the laryngeal membranes. However, the pressure on the aneroid gauge is not the actual pressure drop across the laryngeal vibrators, because the gas is moving through the system. Bernoulli's principle dictates that where the cross-sectional area of the system is small, air velocity will be high and air pressure low (Liepmann and Roshko 1957). The tubing circuitry and laryngeal orifice are obviously not of uniform cross-sectional area and, for this reason, all pressures presented here are *relative pressures.* They are a function of the actual pressure differential across the laryngeal vibrators. The artificial air source effectively replaces the pulmonary air supply. The laryngeal vibrators were activated, producing a sound of considerable amplitude, which was tape recorded and analyzed with the oscilloscope and sound spectrograph. Since the lower jaw and associated vocal sac were removed in most experiments, the preparation limited the measurements to variables affecting the sound produced by laryngeal structures and air pressure. Direct observation of the vibrating membranes within the larynx, mechanical damping of these structures with a small glass probe, and extirpation of these vibrators made it possible to determine which membranes were associated with the different characteristics of the sounds produced. These artificially activated signals were then quantitatively compared with the mating call of the respective species.

Finally, laryngeal ablations on live male *Bufo* were used to confirm the observations made associating structure and function. A control mating or release call was first recorded, then the structure of interest in the larynx was damaged or removed. All operations were performed with the animals in hypothermia. Postablation calls were recorded after the animal had revived. A summary of the species used in the experiments is presented in Appendix J.

Origins of Bufonid Vocal Mechanics

The larynges of the bufonid genera *Bufo, Nectophrynoides,* and *Pedostibes,* as well as the atelopodid genus *Atelopus,* possess numerous anatomical characteristics that testify to the phylogenetic affinity of these taxa and that distinguish them from all other Anura that have been examined. As a result of these anatomical differences, many aspects of the vocal mechanics are different from those of other anurans. It is my purpose here to examine these differences and to present a theory about how the transition in vocal mechanics took place with the origin of the Bufonidae. Comparisons are focused on leptodactylids because of growing evidence that they represent the remnants of the bufonid ancestry.

There are a few basic similarities in the larynges of most adult male anurans. The cartilagenous skeleton of the larynx consists of a cricoid ring (sometimes incomplete) to which the anterior extensions of the lungs attach. Articulating with this cricoid cartilage are paired arytenoid cartilages (Fig. 15-5). The opening between the anterior edges of the arytenoid cartilages is the glottis. Muscles originating on the cricoid cartilage and hyoid apparatus open and close the glottis by pivoting the arytenoid cartilages around their articulations (see Figs. 15-5 through 15-7 and 15-9 for the configuration of these muscles). Various accessory cartilages attach to the arytenoids in some taxa, as is described in detail by Blume (1930). Paired vocal cords are attached to the inside surfaces of the arytenoid cartilages. In at least hylids, leptodactylids, bufonids, and *Atelopus*, the posterior edges of the lateral portions of the vocal cords are attached to the arytenoid and to the cricoid cartilages. The attachments to the cricoid cartilage are less elastic than the connections to the arytenoid cartilages. Therefore, when the larynx is opened the vocal cords are stretched around the posterior edges of the arytenoids, which move medially as the anterior edges move laterally (Fig. 15-4). All but three species that I have examined — *Eupemphix*

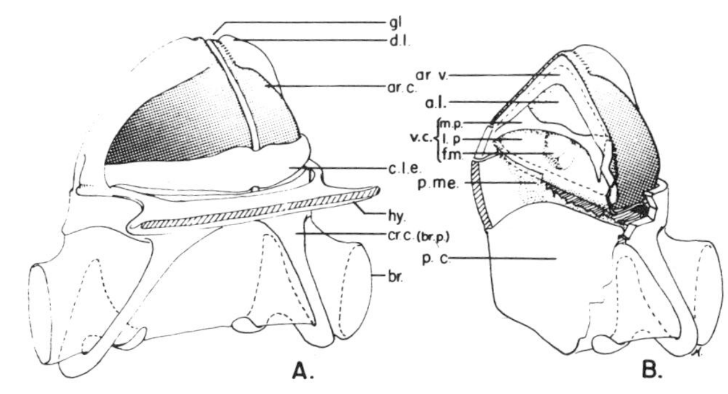

Fig. 15–5. Schematic diagrams of the *Bufo* larynx and hyoid. (*A*) Ventral view with portions of the hyoid removed. The top of the larynx as shown in this drawing projects anteriorly into the mouth cavity. (*B*) Drawn from the same aspect, but the larynx has been sectioned in a parasagittal plane. Several muscles that move the larynx within the mouth cavity and relative to the hyoid have been omitted in these drawings. The constrictor laryngis anterior muscles are also not shown. *v.c.*, vocal cord, *f.m.*, fibrous mass, *l.p.*, lateral portion of vocal cord, *m.p.*, medial portion of vocal cord, *ar.c.*, arytenoid cartilage, *br.* bronchus (left), *c.l.e.*, constrictor laryngis externus, *cr.c.*, cricoid cartilage, *br.p.*, bronchial process of cricoid cartilage, *d.l.*, dilator laryngis muscle, *hy.*, hyoid apparatus, *p.c.*, posterior chamber, *p.me.*, posterior membrane, *gl.*, glottis.

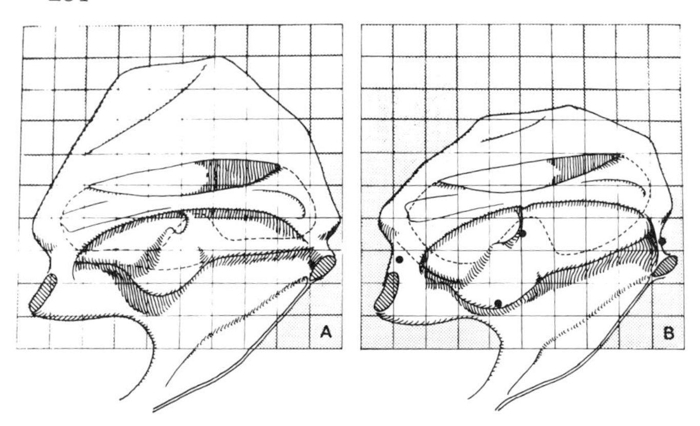

Fig. 15–6. The left half of the larynx of *O. occidentalis* at rest (*A*) and with the arytenoid cartilage pivoted, opening the glottis (*B*). Length of vocal cord medial free edges, 6.4 mm. average (n=7), which was 12.6% of average body size.

pustulosus, Physalaemus cuvieri, and *Edalorhina perezi* (Fig. 15-7E) — have medial free edges on these vocal cords (Fig. 15-7) with free edges projecting posteriorly into the air stream coming from the lungs during vocalization. In all anuran larynges examined, the larynx remains closed when not acted upon by air pressure or muscle contraction. Elasticity of tissue within the larynx forces the anterior tips of the arytenoid cartilages together, and the vocal cords contribute to this elasticity. This elasticity is important, because the arytenoid cartilages vibrate passively during sound production in most bufonids and some leptodactylids. The arytenoids are forced apart by pulmonary air pressure and are returned to the closed position by elasticity during each vibratory cycle. Since the vocal cords are deformed by this opening and closing, their vibration is *mechanically linked* with vibration of the arytenoid cartilages. This linkage can be seen in figure 15-6, where drawings of the larynx of *Odontophrynus occidentalis* are presented. These drawings were made through a dissecting microscope equipped with an occular micrometer, the grid of which is shown. The cricoid cartilage was held in the same position for both drawings, so that movement and tissue deformation is made obvious by comparing the two drawings with reference to the grid. The black spots indicate the points of greatest deformation of tissue. During sound production, the vocal cords balloon anteriorly within the arytenoid lumen, which further stretches the frenular ligaments, increasing the linkage.

The larynges of bufonids and of *Atelopus* consistently differ from those of other anurans in the following characters, which are important to their mechanics of vibration.

A. In larynges of all members of the bufonid-*Atelopus* radiation examined, except those of *Melanophryniscus,* posterior membranes (Fig. 15-5*b*, 15-7L) are present, whereas these structures are unknown in other anurans.

B. In all but ten *Bufo* and in the three species of *Atelopus* examined, fleshy arytenoid valves are present on the anterior edges of the arytenoid cartilages (Figs. 15-5B, 15-7L). *Leptodactylus bolivianus* is the only other anuran examined that has flaps in this position (Fig. 15-7H), which is probably a case of convergence, since these structures are not present in nine other species of this genus that were examined.

C. In bufonids and *Atelopus,* the lateral portions of the vocal cords are expanded, forming planar membranes that are suspended over the arytenoid lumena, and are under considerable tension (Fig. 15-5B, 15-7L). This situation does not exist in *Eleutherodactylus, Crossodactylus, Lithodytes, Pleurodema, Leptodactylus, Calyptocephalella, Telmatobius, Physalaemus, Edalorhina,* or *Eupemphix,* or in anurans of other families, where the lateral portions of the vocal cords, if present, are usually reduced in free surface area. The lateral portions of the vocal cords of *Odontophrynus* (Fig. 15-6) and *Melanophryniscus* (Fig. 15-7J,K) are somewhat expanded, but resonance of these portions of the cords is prevented by cartilaginous buttresses extending from the inside surfaces of the arytenoids in both genera and by chondrified frenular ligaments in *M. stelzneri* (Fig. 15-7K).

D. Many anurans, including *Calyptocephalella gayi,* ten species of *Leptodactylus,* three species of *Pleurodema, Lithodytes lineatus, Odontophrynus americanus, O. occidentalis, Melanophryniscus stelzneri, M.* sp. (Argentina), *Edalorhina perezi, Eupemphix pustulosus, Physalaemus cuvieri, Eleutherodactylus diastema, Ceratophrys ornata, Lepidobatrachus llanensis,* all species of *Bufo,* three species of *Atelopus, Nectophrynoides,* and *Pedostibes,* have condensations of fibrous tissue or cartilage attached to the lateral portions of the vocal cords (Fig. 15-7). The fibrous masses are in firm contact with the posterior edges of the arytenoid cartilages in all these species except *Edalorhina perezi, Eupemphix pustulosus, Physalaemus cuvieri,* and all members of the bu-

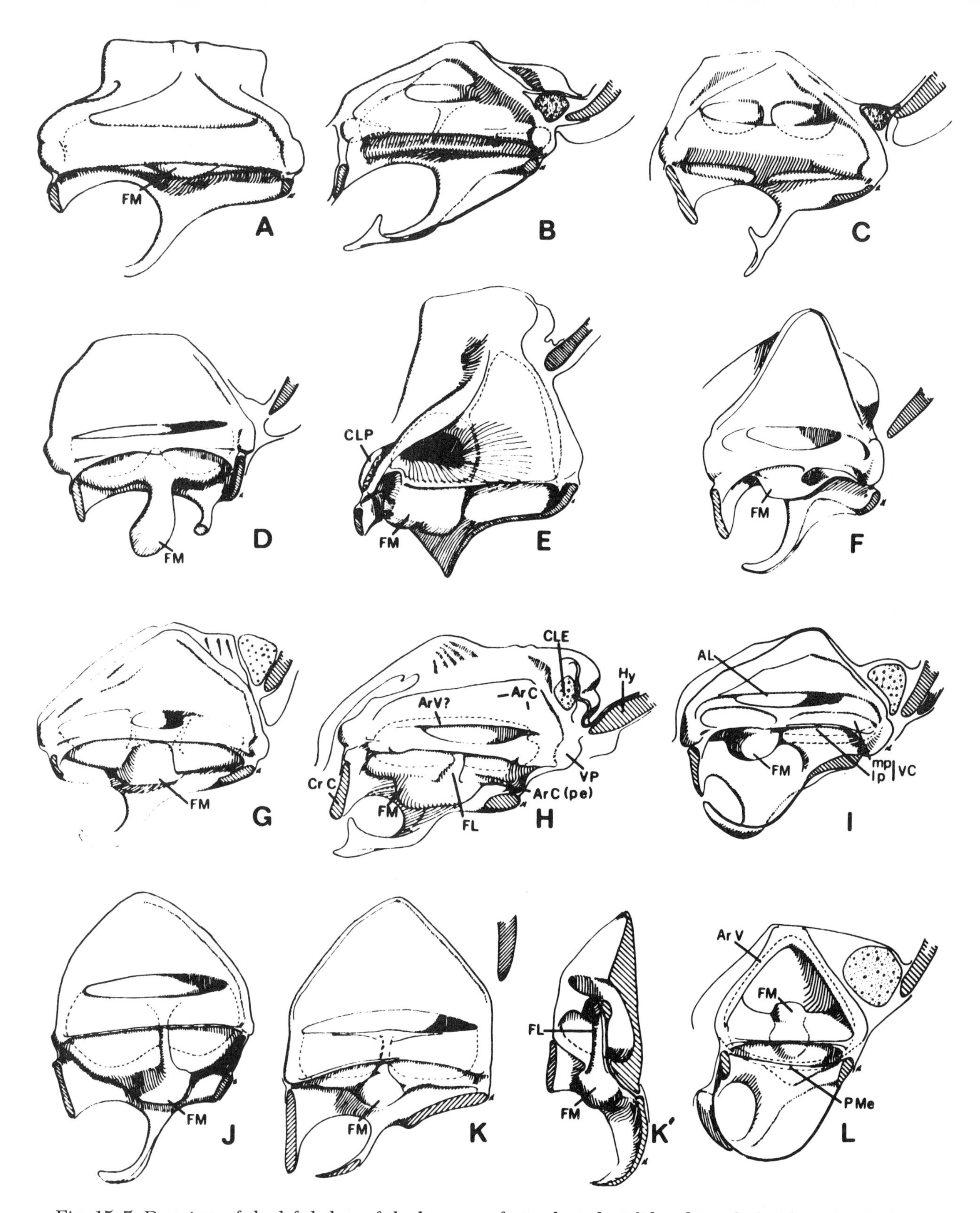

Fig. 15–7. Drawings of the left halves of the larynges of nine leptodactylid and two bufonid species. Left is dorsal, top projects into mouth cavity. *AL*, arytenoid lumen; *ArC*, arytenoid cartilage (*pe*, posterior edge); *ArV*, arytenoid valve; *CLP*, constrictor laryngis posterior muscle; *CrC*, cricoid cartilage; *FL*, frenular ligament; *FM*, fibrous mass; *Hy*, hyoid; *Pme*, posterior membrane; *VC*, vocal cord (*mp*, medial portion, *lp*, lateral portion), *VP*, vocal pulvinarae.

The numerator of fractions following species names below equals length (mm.) of the free edge of the vocal cord; denominator=snout-vent length (mm.). (*A*) *Calyptocephalella gayi* (12.5/124=10%), (*B*) *Eleutherodactylus bufoniformis* (4.5/60=7.5%), (*C*) *E. augusti* (6.7/62=10.8%), (*D*) *E. diastema* (1.7/20=8.5%), (*E*) *Edalorhina perezi* (2.05/28.5=7.2%), (*F*) *Leptodactylus sibilator* (3.3/48=6.9%), (*G*) *L. melanotus* (4/32.5=12.3%), (*H*) *L. bolivianus* (7/90=7.8%), (*I*) *Pleurodema bufonema* (2.5/35.8=7.0%), (*J*) *Melanophryniscus sp.* (Argentina), (*K*) *Melanophryniscus stelzneri* (dorsal portion of left half of larynx), (*L*) *Atelopus ignescens*.

fonid-*Atelopus* radiation. In the bufonid-*Atelopus* complex, they are freely suspended in the lateral portions of the vocal cords (Fig. 15-5, 15-7L) and are free to vibrate during sound production. Fibrous masses are not present in four species of *Eleutherodactylus* (Fig. 15-7B,C), *Crossodactylus guadichaudi, Eupsophus nodosus, Telmatobius montanus,* or *Thoropa miliaris.*

E. In most leptodactylids, as well as in representatives of most other anuran families, frenular ligaments attach the medial portions of the vocal cords to the fibrous masses or to the bases of the lateral portions of the vocal cords when fibrous masses are absent (Fig. 15-7). Only remnants of the frenular ligaments are seen in members of the bufonid-*Atelopus* complex and then only in species with pronounced fibrous condensations on the vocal cords, such as *B. bocourti, B. asper, B. crucifer, B. ictericus, B. paracnemis, B. perreti,* and at least some members of the *regularis* group. In leptodactylids, dilation of the larynx stretches the frenular ligaments (Fig. 15-6), which provides another elastic element for closing the larynx and mechanically coupling vibration of the arytenoid cartilages and medial portions of the vocal cords. *Melanophryniscus stelzneri* is divergent in this respect, because the frenular ligaments are chondrified and stiffly attached to the fibrous condensations. Levers are thus formed that pivot around the fulcra at the bases of the arytenoid cartilages. When the larynx is opened, these levers pivot around their fulcra, causing the anterior portions of the vocal cords to move together in the laryngeal lumen. *Melanophryniscus* sp. (Argentina) does not have chondrified frenular ligaments. The elastic element is absent in the bufonid-*Atelopus* larynx, since the frenular ligaments are missing in most species and reduced to remnants in others. In both leptodactylids and bufonids the ends of the medial portions of the vocal cords also insert on the cricoids, providing another element for elastic return (Fig. 15-6).

F. In all anurans examined, except members of the bufonid-*Atelopus* radiation, *Melanophryniscus, Oreophrynella,* and *Pseudacris feriarum,* the medial portions of the vocal cords insert on cushions of flexible parenchymatous cartilage or of fibrous tissue called vocal pulvinarae (Trewavas 1933). The vocal pulvinarae provide flexibility at the tips of the arytenoid cartilages, so that contraction of the constrictor posterior muscles, which insert on the pulvinarae, can change the tension on the vocal cords (Trewavas 1933; p. 507). Contraction of these muscles can also change the position and configuration of the anterior portions of the vocal cords, as demonstrated by electrical stimulation of these muscles. In *Odontophrynus,* the constrictor posterior muscles are slightly atrophied, being present dorsally, but attaching to the ventral pulvinarae by thin ligaments. Even greater atrophy is seen in the genera *Edalorhina, Eupemphix,* and *Physalaemus,* where ventral pulvinarae are completely absent, and the constrictor posterior muscles are present only as weak slips, with attachment to the dorsal tips of the arytenoids (Fig. 15-7E).

In summary, posterior membranes and arytenoid valves are present in all members of the bufonid-*Atelopus* complex except *Melanophryniscus* and in a few species of *Bufo* where arytenoid valves have been secondarily lost. These structures are absent in all leptodactylids examined except *Leptodactylus bolivianus,* where a convergence on the bufonid-*Atelopus* arytenoid valve is seen. The lateral portions of the leptodactylid vocal cords tend to be reduced in free surface area, whereas they are expanded in bufonids and *Atelopus. Odontophrynus* and *Melanophryniscus* are intermediate in this character, but their expanded lateral portions are prevented from vibration. Fibrous masses are widely distributed in the Leptodactylidae and in the bufonid-*Atelopus* radiation, but these structures have changed position. In the Leptodactylidae and *Melanophryniscus* they are in firm contact with the posterior edges of the arytenoid cartilages, but in bufonids and *Atelopus* they have come to lie in the newly expanded lateral portions of the cords. The leptodactylid genera *Edalorhina, Eupemphix,* and *Physalaemus* have freely suspended fibrous masses. Frenular ligaments occur widely in leptodactylids (as well as in other anuran families) and are present in *Melanophryniscus* but are present only as remnants in a few species of *Bufo* examined. Finally, the mechanism for changing tension, position, and configuration of the vocal cords by muscle contraction is absent in all members of the bufonid-*Atelopus* radiation, including *Melanophryniscus,* but present in all leptodactylids and most other anurans that have been examined. *Odontophrynus* is intermediate in this, and *Edalorhina, Eupemphix,* and *Physalaemus* are even closer to the bufonid-*Atelopus* condition in this character.

All species of the genera *Leptodactylus, Pleurodema, Edalorhina, Eupemphix, Physalaemus,* and *Melanophryniscus* whose karyotypes have been examined have a diploid chromosome number of 22, as do all but several derived species of African *Bufo* (Bogart, personal communication) and all species of *Atelopus* examined. *Odontophrynus occidentalis* also has a diploid complement of 22 (Saez and Brum-Zorrilla 1966). Beçak et al. (1966) demonstrated that *Odontophrynus americanus* is a tetraploid derivative (2n=44). Bogart (1967) compared the chromosomes of *Bufo marinus* with *Odontophrynus* and two species of *Leptodactylus* and found that the centromeric positions are similar in most of the eleven homologous chromosomes. He believes that the dichotomy between 22-chromosome leptodactylids and the line that eventually lead to the Bufonidae probably took place prior to the origin of the genus *Odontophrynus* from the line leading to *Bufo.* Reig (Chap. 3) demonstrates that the extremely rare genus *Macrogenioglottus* (karyotype of 2n=22) has strong morphological affinities with the bufonid-*Atelopus* radiation. He concludes that similarities shared by *Odontophrynus* in four of seven sets of characters represent convergences, although these similarities are often quite distinct. This conclusion is unwarranted on the basis of the evidence he presents, and the similarities provide additional evidence for an affinity of *Odontophrynus* with the bufonid-*Atelopus* radiation, as suggested by Bogart (1967). Also, these 22-chromosome species show evidence of intermediacy between the leptodactylid and bufonid laryngeal conditions, and the laryngeal anatomy of *Melanophryniscus* is intermediate. All this evidence suggests that the laryngeal conditions found in the 22-chromosome leptodactylid species are most likely to reflect those of the ancient ancestry of the bufonid-*Atelopus* radiation.

The larynges of *Eupemphix, Edalorhina,* and *Physalaemus* are similar to one another but they are extremely different from other leptodactylids and bufonids in the configuration of vocal membranes (Fig. 15-7E). Some of these differences are apparently adaptations for producing sound during inhalation and exhalation. The vocal mechanics in these three genera are quite specialized, and have seemingly little to do with the ancestry of vocal mechanics of the bufonid-*Atelopus* complex. However, the chromosomal evidence suggests an affinity with the bufonid-*Atelopus* radiation, and, as mentioned above, the partial atrophy of the vocal cord tensor mechanism and the free suspension of the fibrous masses support the affinity. This evidence indicates that these species represent a single offshoot, divergent in vocal mechanics, that took place with the transition from leptodactylid ancestors to the bufonids and *Atelopus.*

The leptodactylid genera *Leptodactylus, Pleurodema,* and *Odontophrynus* have bulbous fibrous masses at the bases of the arytenoid cartilages that are connected to the medial portions of the vocal cords by frenular ligaments (Figs. 15-6, 15-7F–I). The fibrous masses and frenular ligaments are attached to the lateral portions of the vocal cords by stretchable fibrous connective tissue that holds them against the lateral portions of the vocal cords but allows them to move relative to one another. This is a very generalized and apparently a very old vocal configuration, as evidenced by a configuration very much like it in the primitive *Calyptocephalella gayi* (Fig. 15-7A). The description also fits the larynx of *Lithodytes lineatus* (2n=18; Bogart 1968), and it is quite easy to evolve the bizarre configuration of *Eleutherodactylus diastema* (2n=18, Bogart 1958) from it (Fig. 15-7D). Reig (Chap. 3) suggests an affinity of the Ceratophrynidae with the early radiation of the Bufonidae, and vocal cord configuration is certainly consistent with this suggestion. The larynges of the ceratophrynid species *Ceratophrys ornata* and *Lepidobatrachus llanensis* have this configuration, except that the fibrous masses are much larger, extending anteriorly toward the medial portions of the vocal cords until the delineation between fibrous mass and frenular ligament becomes quite arbitrary in *L. llanensis.* Finally, the association of this vocal cord configuration with the ancestry of the bufonid-*Atelopus* radiation is confirmed by the larynx of the bufonid *Melanophryniscus.* The configuration of vocal cords (Fig. 15-7K) is distinctly like that of *Leptodactylus, Pleurodema,* and *Odontophrynus,* yet this species has lost the vocal pulvinarae and constrictor posterior muscles. The latter condition is almost diagnostic of the bufonid-*Atelopus* radiation. *Melanophryniscus stelzneri* has a Bidder's organ (Griffiths 1954) that is diagnostic of the bufonid-*Atelopus* radiation. Other anatomical data (Griffiths 1963), biochemical data (Cei and Erspamer 1966), and hybridization data (Chap. 11) further confirm this association.

I have not been able to artificially activate the larynges of *Melanophryniscus* or *Pleurodema,* because of their small size and fragile lung tissue. However, this approach has been possible with *Odontophrynus* and *Leptodactylus.* In discussing these results we must review a few basic facts about anuran sound production.

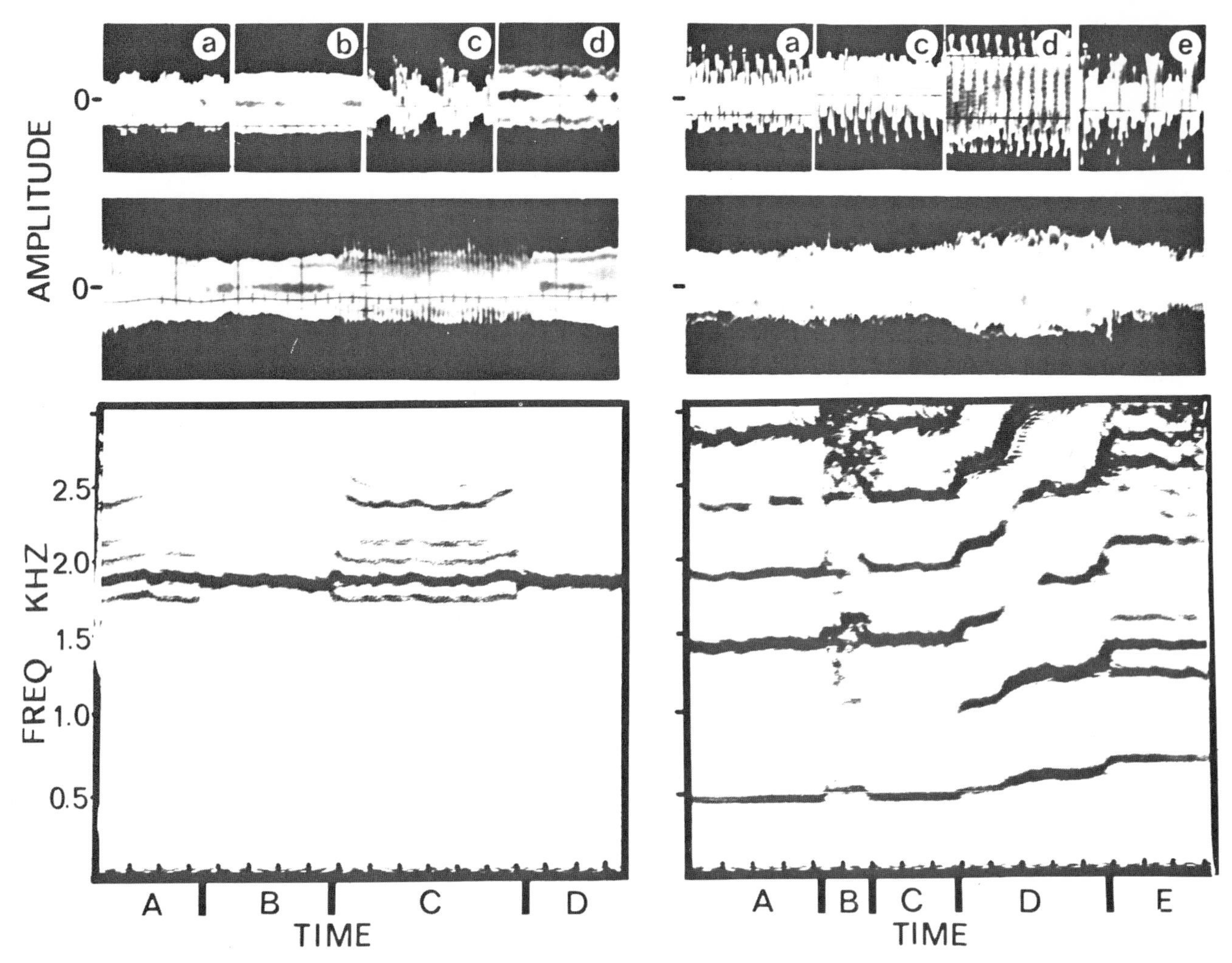

Fig. 15–8. Oscillograms and sonagrams of sounds of artificially activated larynges of *Bufo cognatus* (left) and *Odontophrynus occidentalis* (right). Top oscillograms are 20 msec. samples from the time intervals indicated by corresponding letters below the sonagrams. The sonagrams and the oscillograms immediately above them are on the same time scale (marks on bottom of sonagrams at 50 msec. intervals). Arytenoids artifically damped during intervals B and D for *B. cognatus.*

Sound is produced during exhalation by activating self-exciting vibrators[1] in the larynx with the pulmonary air that is compressed by contracting the internal and external oblique muscles. The vibration of these laryngeal structures causes a periodic variation in the size of the opening from the lungs into the mouth cavity, which in effect breaks the unidirectional flow of air from the lungs into periodic puffs of air that occur at an audible frequency. These puffs of air enter the mouth cavity and vocal sac where they are filtered and their acoustic component is radiated across the surface of the vocal sac into the air. The larynx of *Odontophrynus* extracts this acoustic energy from the compressed pulmonary air by the action of two strongly coupled resonant systems — the vocal cords and the arytenoid cartilages — as in all primitive *Bufo.* Vibration of the arytenoid cartilages opens and closes the glottis with each vibratory cycle, which periodically impedes airflow from the lungs into the mouth cavity. Since it is this air flow that activates the vocal cords, their vibration is interrupted every time the glottis constricts. The temporal

[1] Such vibration differs from forced vibration, for if there is no vibration (of the vocal cords) there is no excitation. Forced vibration results when a vibrator is acted upon by a periodic energy source. In these systems removal of the vibrator does not eliminate the periodic exciting force. The vocal sac exhibits forced vibration, being excited by the laryngeal output. (*See* Bishop 1965 for a discussion of these phenomena.)

pattern of opening and closing the glottis is therefore superimposed on the vibration of the vocal cords by causing the amplitude of vibration of the latter to vary. In other words, vibration of the arytenoid cartilages *amplitude modulates* the vibration of the vocal cords. This is a "Burst-Pulsed" sound system as described by Watkins (1967: p. 33). A carrier oscillator (the vocal cords) radiates a carrier frequency that is amplitude modulated by a second (modulating) oscillator (the arytenoid cartilage system). The relationship is of course reflected in the acoustic output of the larynx. Narrow-band spectrographic analysis of the sound produced during artificial activation of the *Odontophrynus* larynx, while vibration of the arytenoid cartilages was observed with a dissecting microscope, produces a pattern such as that seen in figure 15-8 (interval E). Not surprisingly, the vibration of the vocal cords and of the arytenoid cartilages is not sinusoidal. As a result, each harmonic component of the modulating waveform forms sum and difference products with higher harmonics of the carrier frequency (Watkins 1967).

The mechanical linkage between the arytenoid-cartilage system and the vocal cords was previously described. Acoustic linkage undoubtedly exists as well. Theoretically, the two vibrators will reciprocally influence each other's rates of vibration, since they have elastic elements in common. However, the arytenoid-cartilage system is far more massive than the vocal cords and moves therefore with greater inertia. The velocity of vibration of the arytenoid-cartilage system is therefore less likely to change than is the velocity of vocal cord vibration when the two begin to resonate simultaneously. I have examined 70 narrow-band spectrograms of transitions from vibration of the vocal cords alone to vibration of the arytenoid cartilages plus vocal cords. Changes in the resonant frequency of the vocal cords were observed in sixty-two of these just as the arytenoids started to vibrate. (See transition from interval D to E, Fig. 15-8.) I have not as yet been able to detect reciprocal influences in any of my records.

The mechanical properties of the vocal cords in both *Odontophrynus* and *Bufo* largely determine the dominant frequency of the mating call and its harmonics. Barrio (1964*b*) published mating and release calls of *Odontophrynus occidentalis* and reported that the dominant frequency of the mating call fell between 0.4 and 0.8 KHz. In his spectrogram of the mating call of this species, harmonically related overtones of the dominant frequency are also present,

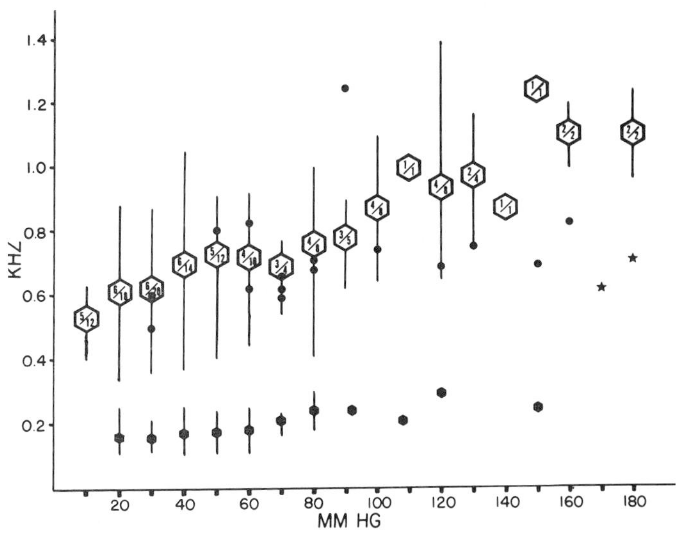

Fig. 15–9. Rates of vibration of the vocal cords and arytenoid cartilages of *Odontophrynus occidentalis* as a function of artificial activation pressure (mmHg). *Open hexagons*, mean vocal cord frequency. *Numerator of the fraction within the hexagon*, number of individuals. *Denominator*, number of frequency determinations. *Small solid hexagons*, means of arytenoid-cartilage vibration rates (37 determinations; 6 animals). The *lines* indicate ranges. *Solid dots*, single vocal cord vibration rates. *Black stars*, vibration rates of damaged vocal cords (see text). Rates determined by measuring harmonic intervals (Watkins 1967).

the fourth falling at 2.8 KHz, which would place the carrier fundamental around 0.7 KHz, which is also close to the dominant frequency of the release call. The fundamental frequencies of vocal cord resonance of seven adult male *O. occidentalis* are presented as a function of activation pressure in figure 15-9. The mean for all these frequencies (regardless of pressure) was 724 Hz (standard deviation 206; standard error 18.0; n=130), which agrees closely with the dominant frequencies for the mating and release call presented by Barrio (1964*b*). These data reflect the tendency of the cords to stop resonating at high pressures. Similar data are presented for *B. marinus*, *B. valliceps*, and *B. cognatus* (Fig. 15-10). Vocal cord resonant frequency data were obtained after removing the arytenoid valves. The histograms showing the distribution of mating call dominant frequencies are compiled by measuring single calls of many individuals. The tendency for the vocal cords to stop resonating at high pressures is not seen. Instead, they respond asymptotically to pressure, the asymptote being near the mean frequency of the dominant for

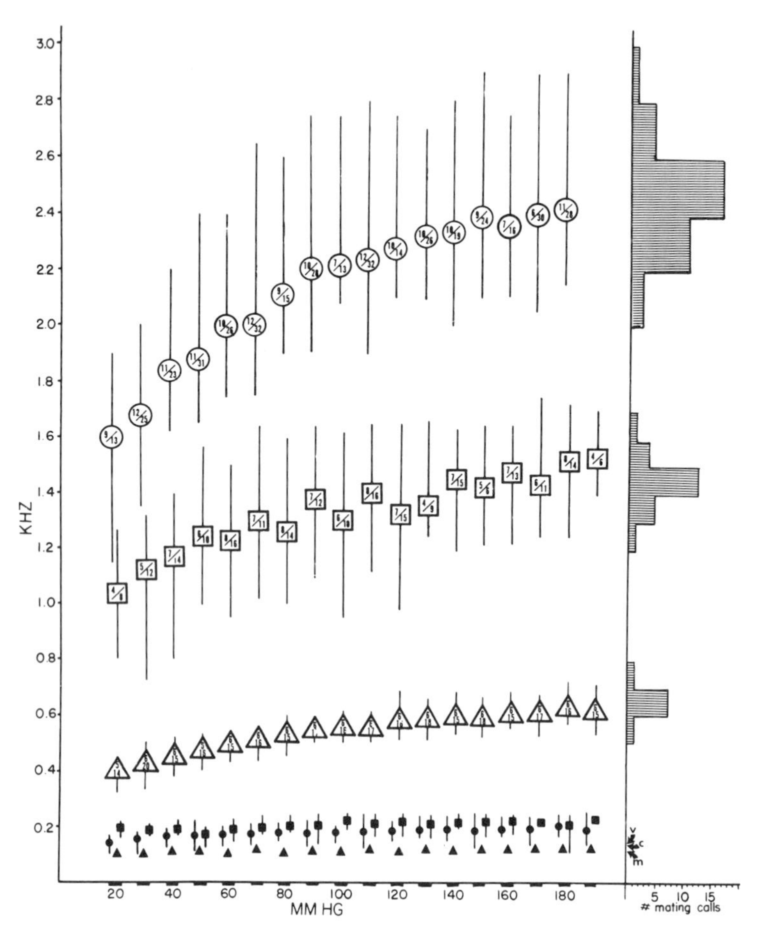

Fig. 15–10. The fundamental frequencies of vibration of the vocal cords (*large symbols*) and arytenoid cartilages (*small, solid symbols*) of 6 *B. marinus* (*triangles*), 8 *B. valliceps* (*squares*), and 12 *B. cognatus* (*circles*) as a function of activation pressure (mmHg) and compared with the dominant frequencies (*histograms*) and mean passive amplitude-modulation rates (*small arrows*) of field-recorded mating calls.

the mating call. Notice that the higher the frequency range of the vocal cord vibration of a particular species, the more sensitive the vocal cords are to changes in pressure. This is a very strong trend in all my data for this genus. However, the mean vocal cord resonant frequencies of all three species span approximately the same percent of an octave between pressures of 20 and 180 mmHg: 78.5 percent for *B. marinus*, 74 percent for *B. valliceps*, and 75 percent for *B. cognatus*. The vocal cords of *O. occidentalis* are more responsive to pressure (Fig. 15-9), spanning 91 percent of an octave between 20 and 180 mmHg. Also, the variance in vocal cord resonant frequencies increases with increasing frequency range in *Bufo*. The averages of the standard deviations of the vocal cord frequencies at all pressures were 41.3 for *B. marinus*, 186.3 for *B. valliceps*, and 202.2 for *B. cognatus*. This trend is also confirmed by other data for this genus. Further data for *Bufo* are presented in table 15-1, where vocal cord fundamental frequencies recorded at 180 mmHg are compared with the dominant frequencies of field-recorded mating calls. Finally, the importance of the vocal cords in establishing the dominant frequency of the mating call was demonstrated in *B. cognatus* and *B. woodhousei* by recording mating calls after damaging the vocal cords. The pronounced dominant frequency was removed from the call by this operation.

Vocal cord resonance in *Odontophrynus* is more variable than in *Bufo*. In *Odontophrynus*, the vocal cords can shift from vibration in one register to another, much as the human cords shift when yodeling. I frequently observed sudden shifts in vibration frequency of the vocal cords, which were often accompanied by slight changes in the size of the glottal aperture. These shifts can result in changes in the fundamental frequency of resonance of the vocal cords as much as 0.6 KHz, but most of the shifts were of less than 0.25 KHz as are those depicted in figure 15-8, interval D. The variability resulting from such shifts is included in figure 15-9. The lowest mean frequency of vibration of the vocal cords (at 20 mmHg) and the dominant frequency of mating calls of *O. occidentalis* (Fig. 15-9) are most comparable to those of *B. marinus* (Fig. 15-10). However, the *O. occidentalis* vocal cords show far more variance independent of pressure than those of *B. marinus*, due to this mechanical instability. The average of the standard deviations of vocal cord resonant frequencies at all pressures between 20 and 160 mmHg for *O. occidentalis* was 156, as compared with 41.3 between 20 and 180 mmHg for *B. marinus*. Also, the resonant frequency of the vocal cords of *O. occidentalis* shows a greater increase in response to pressure increase than does *B. marinus*. For these two reasons, vocal cord resonance in *Odontophrynus* is more variable than in species of *Bufo* whose vocal cords resonate in comparable frequency ranges.

It was mentioned previously that the lateral portions of the vocal cords are not free to resonate in *O. occidentalis* as they are in *Bufo*, which would suggest that it is primarily the medial portions of the cords that are involved in determining the frequency of the dominant for the mating call. By spreading the arytenoids with forceps during artificial activa-

Table 15–1. A Comparison of the Resonant Frequencies of the Vocal Cords in the Larynges of Eight Species of *Bufo* with the Dominant Frequencies of the Field-Recorded Mating Calls*

Species	Artificial				Natural			
	mmHg	Locality	N	Frequency (KHz) Avg. and Range	Frequency (KHz) Avg. and Range	N	Locality	Reference
alvarius	150–180	Phoenix; Tucson	5	0.97 (0.94–1.03)	1.00 (0.95–1.15)	3	Scottsdale; Tucson	—
*cognatus***	180	N.M.; Ariz.	6	2.26 (1.79–2.28)	2.57	13	N.M.; Ariz.	—
crucifer	52–180	Brazil	6	1.22 (1.00–1.37)	1.40 (1.3–1.5)	2	Brazil	Bokermann (Pers. comm.)
luetkeni	180	Guatemala; C.R.	2	1.46 (1.30–1.62)	1.70 (1.65–1.95)	4	Nicaragua	Porter 1966
marinus	180	Guatemala; C.R.	6	0.50 (0.40–0.60)	0.60	2	Guerrero, Mexico	—
regularis	180	Kenya, Ug.; Kamp.; Gha.; Egypt	3	0.44 (0.33–0.60)	0.45 (0.40–0.55)	10	Kenya; Ug.; Kamp.; Nigeria	Tandy (Chap. 9)
marmoreus	180	Mexico	3	1.31 (1.2–1.4)	1.82 (1.65–2.0)	11	Mexico	Porter 1966
*valliceps***	180	Austin, Texas	4	1.32 (1.10–1.50)	1.46 (1.25–1.94)	8	Austin, Texas	Blair 1956 Porter 1964

*These data were selected from similar data on many other species. The selections were made on the basis of availability of mating calls from areas geographically close to the locality where the experimental animals were collected and on sample size.
**Different animals from those used in Figure 15–10.

tion, I was able to excite the medial portion of a single vocal cord of each of the seven *O. occidentalis* to radiate extremely weak tones. These frequencies are also plotted in figure 15-9. They fall within the range of responses of both vocal cords, although their frequency of resonance changed less with changes in activation pressure. At the termination of experiments with *O. occidentalis* Number 6, holes were cut in the lateral portions of the vocal cords, and resonance of the vocal cord medial portion excited again. The resonant frequency of the cords was decreased only slightly by this operation (Fig. 15-9), suggesting that tension in the medial portion of the vocal cords is largely independent of tension in the lateral portions. The tension of the medial portion of the vocal cords is potentially subject to change by contraction of the constrictor posterior muscles, which insert on the pulvinarae (Fig. 15-6). The resonant frequency of the vocal cords in the absence of such muscle contraction, however, matches the dominant frequency of the mating call. The vocal pulvinarae and associated constrictor posterior muscles apparently make only fine adjustments in vocal cord position and tension. The absence of their influence explains the sudden shifts in register recorded during artificial activation that are not evidenced in the mating call of this species.

The mechanics of vibration of the vocal cords in members of the bufonid-*Atelopus* radiation differ in another important way from those in *Odontophrynus*. In *Bufo*, at least, the weights of the fibrous masses are involved in determining the frequency of resonance of the vocal cords. The fundamental frequencies of resonance of the vocal cords of seventy-nine *Bufo* representing twenty-nine species and four hybrid combinations were determined by artificial activation at 180 mmHg relative pressure. A single vocal cord and associated fibrous mass were removed from each of these larynges and weighed. The square root of the inverse of each mass $(1/m)^{1/2}$ (the relationship predicted between mass and fundamental frequency of resonance of a circular membrane; see Kinsler and Frey 1962: p. 89) is plotted against the fundamental frequency of resonance of the vocal

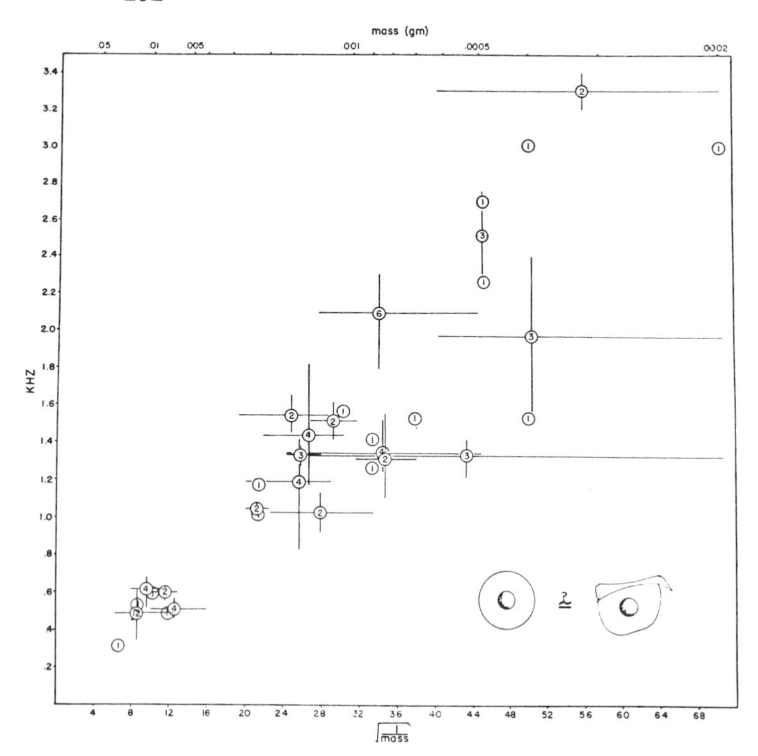

Fig. 15–11. Fundamental frequencies of vibration of *Bufo* vocal cords as a function of vocal cord mass. Circles indicate means for each species, numbers, the sample size, and lines, the ranges. Frequencies measured from photographed oscilloscope traces. Species listed in order of position on frequency scale from bottom to top. Numbers indicate sample size O–l KHz: *B. paracnemis* (1), *B. regularis* (12), *B. regularis* x *B. brauni* (1), *B. marinus* (4), *B. mauritanicus* (1), *B. brauni* x *B. garmani* (1), *B. asper* (2), *B. regularis* x *B. garmani* (4). 1 KHz: *B. arenarum* (1), *B. bufo* (2), *B. alvarius* (2), *B. bocourti* (1), *B. boreas* (4). 1.2 KHz: *B. coniferus* (1), *B. stomaticus* (2), *B. marmoreus* (3), *B. crucifer* (3), *B. valliceps* (4), *B. perplexus* (1). 1.4 KHz: *B. melanostictus* (4), *B. canorus* (2), *B. mazatlanensis* (1), *B. spinulosus* (1), *B. luetkeni* (2), *B. arenarum* x *B. spinulosus* (1). 1.8 KHz: *B. viridis* (3). 2.0 KHz: *B. cognatus* (6), *B. compactilis* (1). 2.4 KHz: *B. hemiophrys* (3), *B. americanus* (1). 2.8 KHz: *B. coccifer* (1), *B. atacamensis* (1). 3.2 KHz: *B. punctatus* (2).

cords in figure 15-11. The linear correlation coefficient for this plot is $\pm$ 0.835, calculated with n=79 to include intraspecific variability. Why does mass correlate so well with the fundamental frequency of resonance when other variables are not directly considered? Powell and Roberts (1923) demonstrated that the addition of a mass at the center of a circular diaphragm lowers the resonant frequency of the diaphragm by an amount five times greater than the lowering produced by adding the same mass uniformly distributed. The fibrous masses contribute greatly to the weight of the vocal cords and they are freely suspended in the center of the vocal cords' lateral portions.

When the arytenoid valves are left intact during artificial activation of the *Bufo* larynx, it rarely happens that the arytenoid cartilages fail to vibrate. The arytenoid cartilages did not fail to vibrate during a single attempt to measure their vibration rate in twelve *B. cognatus* (202 determinations), only for short periods in two of eight *B. valliceps* (98 determinations), and only twice for *B. marinus* (291 determinations) (see Fig. 15-10). In contrast, the vocal cords in all seven specimens of *O. occidentalis* resonated without the intact arytenoid cartilages resonating (Fig. 15-9) for 76 percent of the total recorded activity. On the other hand, removal of the arytenoid valves largely extinguishes arytenoid-cartilage vibration in *Bufo*, whereas the larynx of *O. occidentalis* lacks these structures but still erratically exhibits the vibration. In six of the seven *O. occidentalis* from which data were collected (Fig. 15-9), the arytenoid cartilages suddenly started to vibrate during at least one pressure spectrum (Fig. 15-8; interval E), often in response to slight changes in activation pressure, but equally as often with no known cause. Just as suddenly, the arytenoid cartilages often ceased to vibrate although the vocal cords continued their resonance. The arytenoid valves in the bufonid-*Atelopus* complex have apparently stabilized the vibration of the arytenoid cartilages relative to the condition in *Odontophrynus*. These flaps seal the larynx when it is closed. The arytenoids burst apart, opening the glottis when the laryngeal pressure reaches a certain threshold. The elastic vocal cords are deformed by this motion, which provides energy for the return of the arytenoids to the closed position. Another cycle is initiated when laryngeal pressure again reaches a threshold. The absence of arytenoid valves in *Odontophrynus* allows air to escape from the larynx without vibration of the arytenoid cartilages, thus permitting the vocal cords to resonate without being modulated by the arytenoid system.

The mating call of *O. occidentalis* is Type II, with an active pulse rate of about 6.6 pulses per second (see Barrio 1964). The passive pulsation within each active pulse occurs at a rate of around 80 pulses per second, which is lower than the arytenoid-cartilage vibration rates recorded during artificial activation (Fig. 15-9). This discrepancy presumably results from laryngeal muscle tonus in live, naturally calling animals, which probably decreases the frequency of the arytenoid resonant by increasing the pressure threshold at which they break apart. The same thing is often seen in *Bufo*. In figure 15-10, arytenoid-cartilage vibration rates are consistently higher than passive amplitude-modulation rates in mating calls of *B. valliceps* and *B. cognatus*. However, the rates match quite nicely in the larger *B. marinus*. The arytenoid-cartilage pressure curve for *O. occidentalis* (Fig. 15-9) is based on 37 determinations on 6 animals. In figure 15-10, these curves represent 74 determinations on 6 *B. marinus*, 98 on 8 *B. valliceps*, and 101 on 11 *B. cognatus*. The mean passive amplitude-modulation rates (arrows) are based on measurements of calls of 9 *B. marinus*, 10 *B. cognatus*, and 10 *B. valliceps*.

Mating calls in the genus *Leptodactylus* are extremely different from those of *Bufo* and *Odontophrynus*. Frequency modulation is common in mating calls of *Leptodactylus* and is often quite pronounced (see Fouquette 1960; Barrio 1965, 1966). *L. mystaceus, L. bufonius, L. prognathus, L. anceps*, and *L. sibilator* (Barrio 1965) have whistlelike, frequency-modulated calls that are probably produced by vibration of the vocal cords without vibration of the arytenoid cartilages. This hypothesis remains largely untested, although the limited data collected agree with this theory. I attempted to artificially activate the larynges of two *L. bolivianus*, one *L. ocellatus*, one *L. bufonius*, and two *L. sibilator* and obtained repeatable results only with the last species. In the others, it was extremely difficult to excite sustained resonance of the larynx during artificial activation. When periodic frequencies were observed in the acoustic output, they were very weak and showed no correlation with the structure of field-recorded mating calls. This suggests that contraction of laryngeal muscles is required in order to bring the vocal cords into the appropriate position and under the correct tension for normal sound production. The experiments with *L. sibilator* suggest that active changes in resonant frequencies of the vocal cords are required during the call. The duration of this call is about 0.2 seconds and it is extremely frequency modulated. For instance, the spectrograms of two calls of this species published by Barrio (1965) start with a dominant frequency of about 1 KHz, rise to 3 KHz in one call and 3.2 KHz in the other. Within 0.2 seconds, the vibrator that generates the dominant frequency must have changed in resonant frequency by a factor of at least three in both calls. The larynx of *L. sibilator* was artificially activated at pressures between 20 mmHg and 150 mmHg in 10 mmHg steps. At 20 mmHg the cords resonated with a fundamental frequency of 0.75 KHz. By 40 mmHg, the resonance reached 1 KHz and then increased only slightly as pressure was elevated in 10 mmHg steps to 150 mmHg, where the cords resonated at 1.4 KHz.

Pressure changes alone cannot account for the pronounced frequency modulation seen in the mating call of this species. When at rest, the vocal cords are apparently stretched to a tension such that they resonate at the lowest frequency of the frequency-modulated signal. During the short mating call, this resonant frequency is apparently elevated by contracting laryngeal muscles, probably the constrictor posteriors that operate the vocal pulvinarae at the ends of the medial portions of the vocal cords (Fig. 15-7F–H).

No evidence of vibration of the arytenoid cartilages is seen in the mating call of *L. sibilator* (Barrio 1965; Fig. 15-3). Sporadic vibration of the arytenoid cartilages was observed in one of the two *L. sibilator* tested by artificial activation. It appears that the structure of the mating call in *Leptodactylus* is not determined by laryngeal anatomy to the extent that it is in *Bufo* or even in *Odontophrynus*.

Dominant frequencies of the mating calls in *Leptodactylus, Odontophrynus*, and *Bufo* are apparently homologous in that they are produced by homologous structures, the vocal cords. Furthermore, intrapulse amplitude modulation in mating calls of *Odontophrynus* and *Bufo* are homologous, since both are produced by vibration of the arytenoid cartilages. The mating call of *O. occidentalis* is even broken into periodic pulses that are presumably produced by thoracic or laryngeal muscle contraction as in all primitive (Type II; see next section) *Bufo*. Both *Odontophrynus occidentalis* and *O. americanus* have release calls (Barrio 1964), and their structure suggests that they are produced by continuous contraction of the thoracic musculature. The same applies to the mating call of *O. americanus*. Exactly the same variability exists in the genus *Bufo* (discussed later). There are species of *Bufo* that have pulsed (Type II) mating calls and unpulsed (Type I) release calls, as does *O. occidentalis*, and these include the primitive (Chap. 18) *B. crucifer*. *Odontophrynus* species have very *Bufo*-like calls, and they are produced with larynges that are leptodactylidlike. The mechanics of sound production have been outlined for both genera, but *Odontophrynus* has less anatomical stability of laryngeal output than does *Bufo*. The transition in the mechanics of vibration from an *Odontophrynus*-like ancestor to the configuration of bufonids and *Atelopus* is the accumulation of modifications that lead to this stability. In other words, the vocal mechanics in the bufonid-*Atelopus* radiation have evolved toward greater peripheral tuning of the vocal tract.

The tension, length, and mass of the medial portions of the vocal cords in *Odontophrynus* are anatomically adjusted to produce the dominant frequency of the mating call, even though the laryngeal muscles and vocal pulvinarae are still present (although slightly atrophied). In the genus *Leptodactylus*, the laryngeal muscles modulate the tension, position, and configuration of these portions of the cords. In *Melanophryniscus* the larynx can be viewed as a further stage of peripherally fixing the resonant frequency of the medial portion of the vocal cords. Here, the vocal pulvinarae and the associated constrictor posterior muscles have been lost, foreshadowing the condition in other members of the bufonid-*Atelopus* radiation. In the transition from *Melanophryniscus*-like ancestors to *Bufo*, a further anatomical tuning of the larynx took place. The arytenoid cartilages and lateral portions of the vocal cords changed in configuration, so that the fibrous masses lost contact with the posterior edges of the arytenoids. Thus, the masses were freely suspended and became involved in determining the resonant frequency of the vocal cords. The factors determining the resonant frequency of the vocal cords, in effect, became shifted from the medial to the newly expanded lateral portions of the vocal cords. The vibration of these portions of the cords was already mechanically linked with the vibration of the arytenoid cartilages. This is true because the lateral portions of the vocal cords have their functional attachments to the cricoid cartilages, with the result that movement of the arytenoids stretches the lateral portions of the vocal cords.

The mechanical linkage between medial portions of the vocal cords and arytenoid cartilages provided by the frenular ligaments was lost, for when the fibrous condensations were suspended in the vocal cords the ligaments could not be stretched by opening the larynx. The frenular ligaments atrophied, and only remnants are to be found in a few species of primitive *Bufo*, such as *B. crucifer*. The shift to greater importance of mass in determining vocal cord resonant frequencies also made these structures less sensitive to changes in vibration rate due to changes in activation pressure.

The arytenoid valves further stabilized the laryngeal output, for vibration of the arytenoid cartilages is assured by their presence if the larynx is not actively opened during sound production, as apparently happens in a few species (Type IIIa) that will be discussed later. The evolutionary change involved here is more than a simple refinement of the leptodactylid condition, because vibration of the arytenoid cartilages in *Bufo* is largely dependent on the presence of the arytenoid valves, while in *Odontophrynus*, the arytenoids can vibrate through a large amplitude in the absence of these structures. Some change in the configuration of the bufonid-*Atelopus* larynx must have taken place that prohibits this. Perhaps the fibrous masses in *Odontophrynus* balance the mass of the arytenoid cartilages around their axes of rotation, and the displacement anteriorly of these masses in bufonids and *Atelopus* prevents vibration of the arytenoids when arytenoid valves are not present. Conversely, since the fibrous masses are somewhat elastic, they might contribute to holding the larynx tightly shut, thus allowing resonance to ensue and then to sustain itself.

Damage to the posterior membranes does not affect the frequency composition of field-recorded mating calls in *B. cognatus*, although calls of animals with damaged posterior membranes often have a sporadic start, with the first few thoracic muscle contractions failing to activate the larynx. Eventually, stable resonance is achieved, and the call continues normally. Also, the posterior membranes do not vibrate during artificial activation. These observations suggest that these membranes deflect the flow of pulmonary air between the vocal cords. The fibrous mass in *Odontophrynus*, as well as in many other leptodactylids, could serve this function, since the membranes are often displaced in the larynx so that they come to lie in the openings from the lungs into the larynx (Figs. 15-6, 15-7D–F, H, I).

Radiation of Vocalization within the Genus *Bufo*

The mating calls of all species of *Bufo* that vocalize are characterized by well-defined dominant (or carrier) frequencies and patterns of active and passive amplitude modulation. These variables show a restricted variation within species populations. I believe that species discrimination by conspecific receivers rests largely on these variables. If this is true, concommitant changes in the peripheral and central components of the receiver apparatus must have taken place during the evolutionary divergence of these calls, since species discrimination presumably existed during this divergence. When the physics and physiology of sound reception in *Bufo* are better understood and when comparative data are available on the mechanisms of signal discrimination, we can perhaps better understand the evolution of sound communication in this large genus by coupling these data with those presented in this section.

Dominant Frequency

The mechanism by which the frequency component of the mating call with the maximum acoustical energy (the dominant frequency) is established is fairly uniform within the genus. In all species examined, the vocal cords are intimately involved, but numerous factors can influence their rate of vibration. It is my purpose here to discuss several anatomical and behavioral variables, any one of which could be changed through natural selection and thus result in changes in the dominant frequency of the mating call by influencing the acoustical output of the vocal cords. For some of these factors, there are cogent theoretical reasons to suspect that they will be involved, but experimental evidence is still lacking (e.g., vocal cord tension).

The Larynx. Theoretically, the size, tension, and mass of the vocal cords will be involved in determining their resonant frequency at any given activation pressure. Mass is of disproportionately great importance in determining the resonant frequency of the vocal cords (Fig. 15-11), presumably because the vocal cords are center-weighted by the fibrous masses (see Powell and Roberts 1923). In figure 15-10 data are presented on the resonant frequency of the vocal cords for *B. marinus, B. valliceps,* and *B. cognatus*. The difference in resonance of the vocal cords between *B. marinus* and the others is largely the result of differences in mass, but mass does not account for the pronounced difference between *B. cognatus* and *B. valliceps*. Mechanical properties of the vocal cords other than mass are obviously involved in this difference. The vocal cords of these two species are almost exactly the same size (species Number 12 and 40; Fig. 15-13), which focuses attention on vocal cord tension, although comparisons have yet to be made.

Activation Air Pressure. It was shown earlier that the resonant frequency of the vocal cords of *B. valliceps* and *B. cognatus* (Fig. 15-10) varies with the air pressure used in artificial activation. The response of the vocal cords to air pressure is usually asymptotic, the asymptote being close to the dominant frequency of field-recorded mating calls. This is not always the situation. When the tension, mass, and size of the vocal cords are such that the cords tend to resonate in a high frequency range, high activation pressures commonly cause them to vibrate at frequencies far above the dominant frequency of the mating call. However, at lower pressures, the vocal cord resonance matches the dominant frequency of the mating call, which may mean that these species call at lower pressures than others. Direct recordings of activation pressures developed during vocalization have yet to be made. It should be emphasized that it is the pressure difference between the lungs and mouth cavity that is important.

The Vocal Sac and Mouth Cavity. The vocal cords were mentioned earlier as "self-exciting vibrators" in the sense of Bishop (1965). They extract periodic acoustical energy from a nonperiodic energy source (the pressurized, moving air stream from the lungs). The acoustic energy that is transduced by the vocal cords then acts on the mouth cavity and vocal sac, forcing them to resonate and radiate acoustical energy across the surface of the vocal sac into the air. Resonance of these membranous cavities is excited much as resonance of a stretched membrane or glass bottle can be excited by holding a tuning fork close to them, but there are important differences. Resonance of the mouth cavity and vocal sac causes a periodic fluctuation in pressure within the closed mouth cavity, which can theoretically act back on the vocal cords, influencing their resonant frequency. Wood (1940: p. 77) calls this a *coupled system*, and suggests that the term "forced vibration" should be reserved for cases where the responding system does not modify by its reaction the periodic driving force, as in the tuning fork example. A compromise resonant frequency of the whole system can be established by coupling that is slightly different from the resonant frequencies of the vocal cords, mouth cavity, or vocal sac alone. The absence of the mouth cavity and vocal sac during artificial activation may account for the tendency of vocal cord resonant frequencies to be slightly lower than the dominant frequencies of natural mating calls (Table 15-1). In species where the vocal cords are less massive (due to extreme reduction in the weight of the fibrous masses), one might expect that the vocal sac or mouth cavity will play a larger role in determining

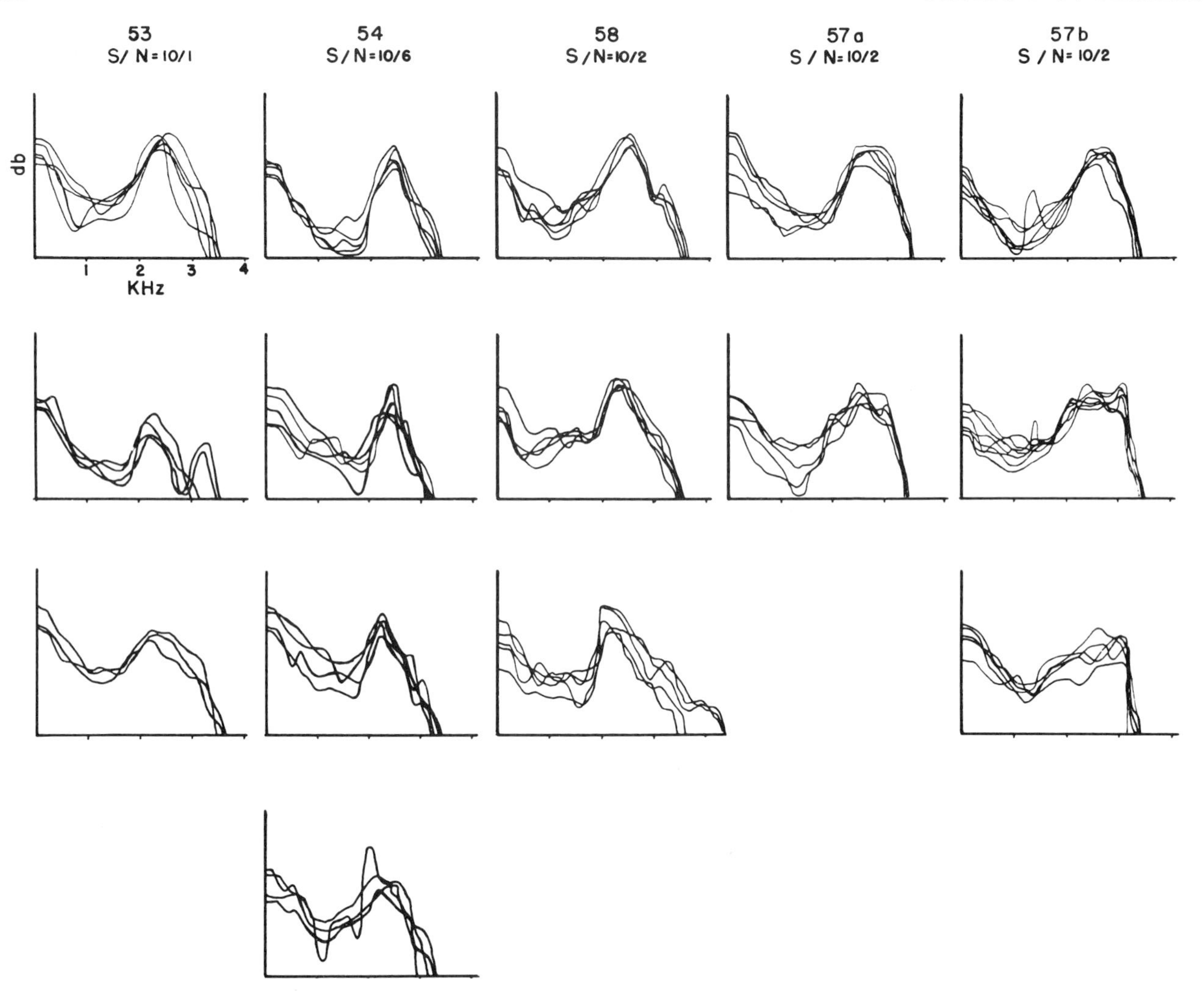

Fig. 15–12. The effect of vocal sac deflation on the spectral composition of 5 mating calls of 4 *B. cognatus*. $S/N=$ signal-to-noise ratio (mating call vs. background-noise amplitude).

the dominant frequency of the mating call. The less massive cords will vibrate with less inertia and therefore be more subject to a shift in resonant frequency resulting from the presence of the other resonators.

In *B. cognatus* at least, the vocal sac is not involved in determining the frequency of the mating call dominant (Fig. 15-12). This species has an extremely long mating call, often lasting a minute. Composite frequency versus amplitude spectra ("sections") are presented from five mating calls of four animals. The top row of the figure consists of composite spectra from portions of the mating calls produced with the vocal sac normally inflated. After these were recorded, part of the air was squeezed out of the vocal sac of each animal. Each row of composite spectra corresponds to an increase in deflation. Air did not move back into the vocal sac, reinflating it, because thoracic muscle contraction is pulsatile in this species, and air apparently returns to the lungs between pulses. Comparisons of absolute sound intensities between composite sections are not meaningful, because of the method of analysis, but it can be seen that (1) the peak of dominant energy does not shift as the vocal sac is deflated and (2) the call is somewhat "detuned" by the deflation, with acoustical energy becoming more and more evenly spread over the frequency spectrum. The primary source of this spreading energy is the sum-and-difference side bands produced by arytenoid-cartilage vibration (see Fig. 15-8, intervals A and C). This detuning is important

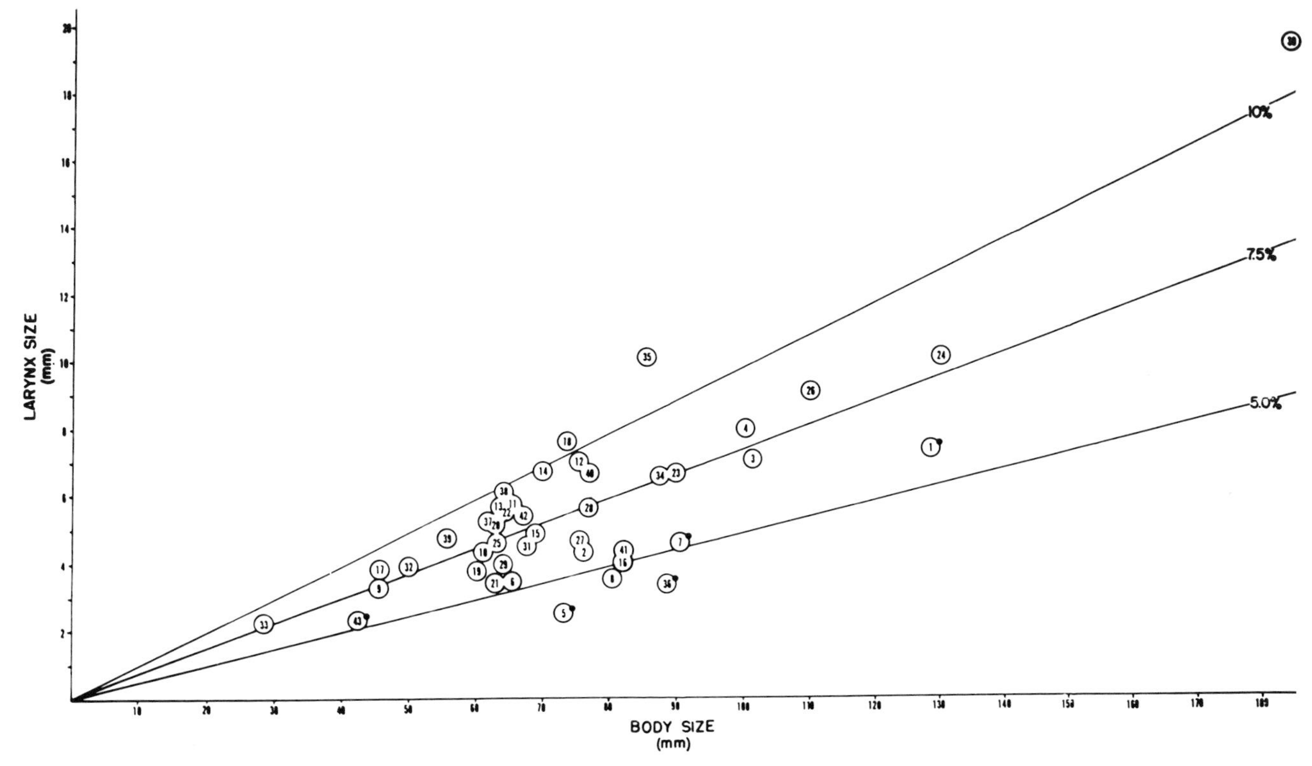

Fig. 15–13. Relationship between body and larynx size within the genus *Bufo*. All measurements on freshly killed animals. *Black spots*, mating call lost or weak. *Number before species name*, species number; *number after*, sample size. 1-*alvarius* (4), 2-*americanus* (1), 3-*arenarum* (1), 4-*asper* (2), 5-*atacamensis* (2), 6-*bocourti* (1), 7-*boreas* (6), 8-*bufo* (5), 9-*canaliferus* (1), 10-*canorus* (2), 11-*coccifer* (1), 12-*cognatus* (8), 13-*compactilis* (1), 14-*coniferus* (4), 15-*crucifer* (5), 16-*gariepensis* (1), 17-*granulosus* (1), 18-*regularis* (1), 19-*haematiticus* (1), 20-*hemiophrys* (3), 21-*ibarrai* (1), 22-*maculatus* (1), 23-*luetkeni* (3), 24-*marinus* (5), 25-*marmoreus* (3), 26-*mauritanicus* (1), 27-*mazatlanensis* (1), 28-*melanostictus* (5), 29-*microscaphus* (2), 30-*paracnemis* (1), 31-*perplexus* (2), 32-*punctatus* (1), 33-*quercicus* (1), 34-*rangeri* (1), 35-*regularis* (13), 36-*spinulosus* (2), 37-*stomaticus* (2), 38-*terrestris* (3), 39-*typhonius* (1), 40-*valliceps* (9), 41-*viridis* (6), 42-*woodhousei* (1), 43-*variegatus* (1).

because it suggests that the vocal sac is highly tuned when inflated, so that it selectively radiates acoustic energy in a narrow band around the resonant frequency of the vocal cords. With few exceptions, the mating calls of *Bufo* have one pronounced dominant frequency. Higher harmonics of the vocal cord fundamental are often present, but when care is taken against overloading the recording and analysis equipment, these higher harmonics are usually found to be considerably less energetic than the fundamental. This is not the situation in several species of African *Bufo* (Chap. 9), for here the vocal sac apparently resonates at frequencies higher than the fundamental frequency of vocal cord vibration. Therefore, one or more dominant peaks at higher frequencies are produced (Fig. 15-1A,B), which are often more energetic than the fundamental frequency of the vocal cords.

Body Size. Blair (1964) demonstrated that small species of *Bufo* tend to have higher dominant frequencies in their mating calls than larger species. Larynx size averages 7.45 percent of body size for forty-three species of *Bufo* (Fig. 15-13). However, some allometric evolution of larynx size relative to body size has taken place within the genus. The larynx is less than 5 percent of body size in the narrow-skulled species *B. bufo*, *B. atacamensis*, and *B. spinulosus*. The last two species apparently do not have functional mating calls, and the call of the first is relatively weak and aberrant (next section). On the other hand, *B. gutturalis*, *B. regularis*, and *B. paracnemis* show ratios in excess of 10 percent. These three species have well-developed mating calls with extremely low dominant frequencies. Other than these exceptions it can be seen in figure 15-13 that larynx size is dependent on body size, which is one reason why body size is inversely related to the dominant frequency. The smaller the animal, the smaller the larynx, and therefore, the smaller the vocal cords within. Also, the smaller the larynx, the smaller the fibrous masses that can be suspended on the vocal cords. Small body size, therefore, influences the resonant frequency of the vocal cords by factors of both size and mass. When coupling with the vocal sac and mouth cavity are involved in determining the dominant frequency of the mating call, body size may be involved, since the volume of both vocal sac and mouth cavity will be limited to some extent by body size.

The evolutionary divergence of the dominant frequency of the mating call in *Bufo* is, therefore, relatively complicated, involving change in response to selection for body size as well as in response to direct selection on the call. Direct selection can operate through change in the weight of the fibrous masses, through allometric evolution of larynx size relative to body size, and theoretically through changes in vocal cord tension. Evolutionary changes in pulmonary air pressure during vocalization can also result in shifts in the dominant frequency of the mating call, as can changes in resonance of the vocal sac and mouth cavity. Minor shifts can also result when amplitude modulation by the arytenoid cartilages is extinguished, as happens in several species. The great heterogeneity in dominant frequencies of mating calls in the genus undoubtedly results mostly from indirect selection on body size and from direct selection on the dominant frequency operating through changes in the vocal cord mass.

Zweifel (1968) found variation in the dominant frequency of the mating call within a single breeding population of *B. woodhousei fowleri* that correlated inversely with body size. Blair (1956) documented a difference in dominant frequency of the mating call between *B. woodhousei fowleri* and another subspecies, *B. w. woodhousei*. The former, smaller subspecies had the higher dominant frequency. Porter (1964) demonstrated clinal variation in body size and dominant frequency in *B. valliceps*. Large-bodied, low dominant-frequency populations are found in Texas. Generally, body size decreases and dominant frequency increases in population as one goes south. Porter (1964) demonstrated similar variation in body size and dominant frequency in *B. mazatlanensis*. On the other hand, Porter (1965) found a great difference in dominant frequency between two disjunct populations of *B. coccifer* that could not result from selection on body size, since no difference in body size could be detected between adult males of the two populations. Porter (1968) found an even more striking case where selection on body size cannot explain divergence in dominant frequency. Toads from a disjunct population of smaller *B. hemiophrys* from Wyoming have lower dominant frequencies than do larger individuals from the main population. If mating calls prove to be predominant premating isolating mechanisms in *Bufo* and if the dominant frequency provides an important discriminatory cue, detailed analysis of the components of variation in vocal mechanics that lead to these divergences would be very profitable, for this is the kind of variation in a premating isolating signal of which species are formed, as explained by Blair (1964).

Patterns of Amplitude Modulation

I will discuss here the taxonomic distribution of the patterns of active and passive amplitude modulation in connection with phyletic conclusions from other evidence.

The primary power supply for sound production of *Bufo* is the internal and external oblique muscles. Without their contraction, the pulmonary air is not pressurized enough to activate laryngeal vibrators. However, it is possible for these muscles to contract and for the larynx to remain silent. Contraction of constrictor laryngis muscles can impede airflow through the larynx, and contraction of the dilator muscles can deactivate the laryngeal vibrators by moving them out of the air stream, as occurs during nonvocal exhalation. The contraction of these muscles can thus turn the activity of the larynx on and off. Their action is reflected in the acoustic output as *active amplitude modulation.* Electromyographic recordings from the internal and external oblique muscles of *B. valliceps*, *B. woodhousei*, *B. americanus*, and *B. arenarum* during release calling confirm that their contraction is pulsatile and periodic, and their pulse rate matches in rate and time the pulsatile amplitude modulation of the acoustically radiated release call (Martin and Gans, in preparation). The pulse rate of the release call approximates that of the mating call in most species of the *americanus* group (Brown and Littlejohn, Chap. 16). Pulsatile thoracic muscle contraction during release calling was also confirmed in *B. coniferus*. All passive vibrators as well as all laryngeal muscles were surgically removed from the vocal tract of a male *B. coniferus*, yet the flanks of the animal still exhibited pulsatile release vibration, the pulse rate being close to that of mating calls recorded at comparable temperatures. At 24.5° C the vibration rate was 40 pps. At 24° C, the mating call of this species has a pulse rate of 39 to 40 pulses per second (Porter 1966). This evidence, coupled with the observation that mating calls of *Bufo* are extremely long (if thoracic muscle contraction is continuous and not pulsatile, call duration is limited by lung volume), suggests that pulsatile thoracic muscle contraction during mating calling is widespread in the genus. The myographic recordings also show that laryngeal muscles can contract with each pulse. A species in which thoracic muscle contraction is continuous while laryngeal muscle contraction is pulsatile has not been found. If such a species is found, its call should be relatively short in duration.

Passive amplitude modulation is produced by vibration of the arytenoid cartilages. The modulation is usually fast in rise time, reflecting the sudden and energetic bursting open of the glottis. Also characteristic of this amplitude modulation is that it changes in rate during the call or occurs aperiodically (Figs. 15-1, 15-4K, L), which presumably results in part from changing or fluctuating pulmonary pressure. In *Bufo*, all species that exhibit this pulsation in their vocalizations have arytenoid valves. Artificial-activation experiments on representatives of forty-eight species of *Bufo* were used to demonstrate the effect of arytenoid-cartilage vibration on the vocal cord output. Also, the arytenoid valves were removed from eighteen animals of seven species after recording control release calls. None of the 2281 postablation calls analyzed exhibited fast rise-time amplitude modulation within pulses, whereas such modulation was present in most of the control calls. A mating call of *B. valliceps* was recorded after the arytenoid valves had been removed, and, as expected, the fast rise-time amplitude modulation was not present in the call.

The Primitive Vocal Pattern: Type II

Oscillograms of single pulses from Type II mating calls are shown in figure 15-4F–J, and sonagrams are shown in figure 15-2. The pulses in figure 15-4F–J are produced in a periodic train that constitutes the mating call. Each pulse in figure 15-4F–J is produced by a single set of coordinated muscle contractions, which probably involves the thoracic musculature in most species. These active pulses are further modulated by vibration of the arytenoid cartilages. The vibration of the vocal cords is therefore modulated by both a passive and an active mechanism. The spectrograms of figure 15-2 were printed with the analyzing filter band width at 300 Hz. The same calls, analyzed with a filter band width = 50 Hz, would reveal the arytenoid-cartilage modulation as sum-and-difference side bands around the dominant frequency instead of as vertical bars. This happens because the arytenoid-cartilage vibration rates are usually higher than 50 pulses per second (see Watkins 1967).

When other kinds of data are consulted, it is found that virtually every species that possesses other characters thought to be "primitive" also possesses this pattern of amplitude modulation (Chap. 18). Every species in South America (the probable place of origin of *Bufo*) for which a mating call has been recorded has this pattern except *B. haematiticus*. The Type II pattern is by far the most widely distributed in the genus, which suggests that this pattern has existed for a relatively long time. Twenty-seven species representing sixteen species groups exhibit this call type (Appendix J). Of these sixteen species groups, only the *valliceps* and *guttatus* groups have members with another call type. This vocal pattern is also seen in the mating call of *Odontophrynus occidentalis* (but not *O. americanus*, which is Type I).

There are six minor variations on this basic Type II pattern, but none are widespread in the genus.

The vocal pattern of *B. canaliferus* is basically of Type II. However, the homologue of the Type II pulse is not the pulsation at 12.44 pps reported by Porter (1964). To press a point, each of Porter's "pulses" actually represents a call, because each of them contains several Type II pulses (Fig. 15-14A). These Type II pulses occur at rates between 66.6 and 75 pps at a water temperature of 29° C (4 animals). *B. canaliferus* has a release vibration produced by thoracic muscle contraction, and the rate of this vibration is very close to the higher mating-call pulse rate (75–87 pps at 29.6° C; one animal) instead of the lower (12.44 pps) rate. The higher mating call pulse rate reflects the active pulsation produced by thoracic muscle contraction. This species is classified as Type II, because arytenoid-valve modulation is present within the active pulses (Fig. 15-14A). However, it is common for arytenoid modulation to disappear from these pulses within a "call" (Porter's "pulse"). This species therefore demonstrates Type IIIb tendencies, which will be discussed later.

The vocal pattern of *B. melanostictus* is only slightly divergent from the Type II pattern. The pulse rate is extremely slow at some localities (5 to 7 pps in a call recorded by R. G. Zweifel in Bogor, Indonesia) and usually variable. Each pulse, however, contains only arytenoid-valve modulation, which is quite variable in rate. The rate and duration of thoracic muscle contraction are not as rigorously controlled as in most Type II species.

The vocal pattern of *B. marmoreus* is also only slightly divergent. The pulse rate is unusually slow (8–9.5 pps; 9 animals), and pulse duration is unusually long (60–80 ms; the same 9 animals). Analysis of thirty-three pulses from the calls of these nine animals showed that all but one pulse consisted of three bursts of arytenoid-cartilage pulses, the first two being relatively short and the last being longer (Fig. 15-4J). The call is divergent only in the consistency of this pattern, which is not present in pulses of the closely related *B. perplexus.* The mechanism by which this consistent pattern is produced in *B. marmoreus* is unknown.

The vocal patterns of members of the *debilis* group (*B. debilis, B. kelloggi,* and *B. retiformis*) are slightly more divergent than in *B. marmoreus.* In some calls of all these species, alternate pulses are identical in amplitude, frequency, and duration, whereas adjacent pulses differ slightly in these characters. The segment from an unusual mating call of *B. debilis* (Fig. 15-4B) suggests that the elements producing adjacent pulses are independent of one another. This call is atypical compared to normal calls of this species, for the pulses of longer duration drop out and they do this without influencing the rate, amplitude, or duration of the shorter pulses. This curious alternation probably results from alternate contraction of vocal muscles somewhere in the system. For instance, the internal and external oblique muscles might alternate. Differences in frequency, duration, and amplitude of adjacent pulses would result if the contractions of these muscles differed slightly in force. The differences in force of contraction would cause differences in the pulmonary pressure developed, which influences all three of the variables that are different. *B. retiformis* has the highest active pulse rate in the genus (224 pps; Bogert 1962). Alternate muscle contraction could explain how pulsation at this rate occurs actively. The pulse rate is so high in these species that the arytenoid valves evidence themselves only at the first of some pulses by imparting an extremely fast rise time to them (Fig. 15-14B). However, the arytenoid valves are sometimes deactivated, and these species, therefore, like *B. canaliferus,* show Type IIIa tendencies.

L. E. Brown (1969) recorded the vocalization of two laboratory male *B. blombergi.* In the absence of a contextual definition of these sounds, they are still of interest from a structural point of view. Most of the calls recorded were of Type II. However, many calls were modulated only by the arytenoid valves and lacked active pulsation. Within three calls, the active pulsation stopped, and the call continued with only passive arytenoid-valve modulation. Because most of the calls were of Type II, this species is classified as such, but the calls without active pulsation testify to tendencies toward the Type I pattern discussed below. Call structure and observations of calling animals in the laboratory indicate that this species might be a good candidate for Type II pulsation involving pulsatile contraction of laryngeal but not thoracic muscles.

B. perreti exhibits tendencies toward the Type I pattern of modulation. In this species, the mating call sometimes changes from passive modulation alone (Type I) to passive and active modulation (Type II) (Fig. 9-4B). The call is therefore intermediate between these two call types.

The Type I Modulation Pattern

B. regularis, B. gutturalis, B. garmani, B. rangeri, B. brauni, B. kerinyagae of the *B. regularis* group, *B. gariepensis, B. pentoni, B. vittatus, B. calamita,* and *B. bufo* lack active pulsation but have passive arytenoid-cartilage modulation in their mating call. A continuous, positive pressure must be applied to the pulmonary air supply during these calls, and air must pass unidirectionally through the larynx into the mouth cavity and vocal sac. This unidirectional flow of air through the larynx causes the duration of the call to be ultimately limited by lung volume. In species that call with the nares closed, as do most *Bufo,* mouth cavity and vocal sac volume may be limiting. Type II and III species are not necessarily limited in this way if the thoracic muscle contraction is pulsatile, for air can be returned to the lungs between pulses. The longest duration for a call of Type I that has been recorded is 2.1 seconds for *B. kerinyagae* (Keith 1968) (see my Fig. 15-1A). Calls that lack active pulsation are classified as Type I.

The call of *B. bufo* is slightly divergent from the

Type I pattern. Arytenoid modulation is present in the mating call, but this modulation often disappears. The call continues to the end with only the vocal cords vibrating. Schneider (1967) demonstrated that the percentage of the call that is "noise-like" (amplitude modulated by arytenoid-cartilage vibration) is highly variable. As discussed later, this species may be losing its mating call.

There is much evidence that the origin of call Type I in the narrow-skulled *B. bufo* and *B. calamita* is independent of the origin of this mechanism in the African species (Chap. 18). Very little is known of the affinities of the narrow-skulled *B. gariepensis,* but Blair (Chap. 18) argues against an affinity with other narrow-skulled species. Low (Chap. 13) presents chromatographic evidence for an affinity with *B. mauritanicus.* Even less is known of the affinities of *B. pentoni* and *B. vittatus.*

The 22-chromosome *B. mauritanicus* is apparently a remnant of the stock that gave rise to the 20-chromosome radiation in Africa (Chap. 18). This species has the primitive Type II modulation pattern. The mating call of *B. perreti* (2n=20) often starts with only arytenoid-cartilage modulation, but active pulsation soon commences. The presence of Type II modulation in this 20-chromosome toad suggests that active pulsation was lost after the reduction in chromosome number. However, *B. gariepensis* (2n=22) has a Type I mating call, and the larynx of this species, as well as the larynx of *B. mauritanicus,* is identical to the *regularis*-group larynx (large fibrous mass relative to larynx size; enormous arytenoid valves; massive arytenoid cartilages). We may then be dealing with one radiation of the Type I mechanism, which took place before the reduction in chromosome number. If so, the call of *B. perreti* is convergent on the Type II mechanism.

It is possible that the Type I mechanism evolved by extreme deceleration of Type II call, so that a single Type I call is homologous to a Type II pulse. An intergradation between call Types I and II can be shown, with Type II pulse rate becoming slower and pulse duration becoming longer until they appear as Type I call rate and call duration respectively. See Chap. 9).

This simple hypothesis is incorrect for the origin of Type I in the 20-chromosome toads and *B. gariepensis.* The mating call of *Odontophrynus americanus* is Type I, as are the long release calls of this species and *O. occidentalis,* but the mating call of *O. occidentalis* is Type II. Furthermore, *B. bocourti* and *B. variegatus* (primitive narrow-skulled species), *B. crucifer, B. ictericus* (primitive broad-skulled species), and *B. mauritanicus* can produce long release calls by the Type I mechanism. Virtually all male toads produce short-duration release calls, each call resulting from a single, forceful thoracic muscle contraction. The "long" calls referred to are those in which contraction of the thoracic musculature is continuous and drawn out, producing a release call that is pulsed by arytenoid-cartilage vibration and having a structure like a Type I mating call. These long calls are not to be confused with the release vibrations of *americanus*-group species (Chap. 16), *B. spinulosus, B. canaliferus, B. melanostictus, B. arenarum, B. periglenes, B. coccifer, B. punctatus, B. granulosus,* and *B. coniferus,* which are produced by pulsatile thoracic muscle contraction. Of the five species with long release calls, *B. bocourti* and *B. variegatus* have apparently lost their mating calls (next section), but the others have Type II mating calls. I have seen a sound spectrogram of an aberrant mating call of the very primitive *B. crucifer* that switched at the end to the Type I mechanism. *B. blombergi* also occasionally exhibited such switching (*see* Brown 1969), although the calls were recorded in the laboratory. *B. perreti* also switches, but from Type I to Type II. The pattern of continuous muscle contraction required for producing calls of Type I is apparently a very old trait in *Bufo,* probably extending back to leptodactylid ancestors, and this contraction pattern exists side-by-side with the Type II pattern in many species. There is no need to derive the pattern anew by decelerating the Type II call in the 20-chromosome toads, particularly since *B. mauritanicus* exhibits this pattern in its release call and since both patterns can be present side-by-side in the mating call of the 20-chromosome *B. perreti.* The origin of the Type I pattern in the 20-chromosome *Bufo* is therefore best considered to have occurred by a shift to the previously existing pattern of continuous contraction.

The Type III Modulation Pattern

Four species of the *valliceps* group (*B. valliceps, B. mazatlanensis, B. luetkeni,* and *B. gemmifer*) have the primitive Type II modulation pattern, but *B. coniferus, B. ibarrai,* and *B. cavifrons* have lost passive amplitude modulation from the call. Similarly, passive amplitude modulation does not occur in the mating calls of *B. haematiticus, B. coccifer,* and *B. viridis* (Fig. 15-3B). Arytenoid valves are present in

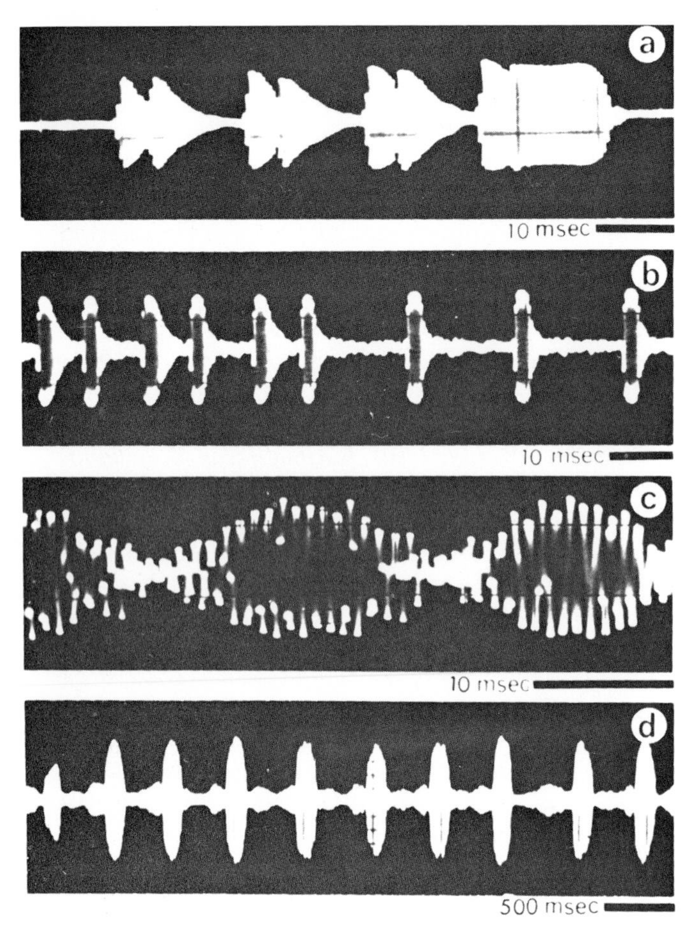

Fig. 15–14. Aberrant mating calls of four species of *Bufo*. (*A*) Single "pulse" from the call of *B. canaliferus* recorded by K. R. Porter east of Arriaga, Chiapas, Mexico. Air 35° C; water 29° C. (*B*) Segment from the first of an aberrant mating call of *B. debilis* recorded by the author near Tatum, New Mexico. Air 20° C.; substrate 20.8° C. (*C*) Segment of mating call of *B. alvarius* recorded by the author near Tucson, Arizona. (*D*) Segment of the mating call of a *B. quercicus* recorded by C. Johnson 10 miles N.W. Waycross, Georgia (air 25° C; water 26° C). Notice the time scale.

the larynges of *B. coniferus, B. ibarrai,* and *B. coccifer* but are reduced in the last two species, particularly in *B. coccifer. B. haematiticus* also exhibits reduced arytenoid valves, although the remnant tissue in this species has a configuration that is different from that in *B. ibarrai* and *B. coccifer.* These partial reductions may have deactivated the vibration of the arytenoid cartilages. The arytenoid cartilages did not vibrate when the larynges of one *B. haematiticus,* one *B. ibarrai,* and two *B. coccifer* were artificially activated but they did vibrate when the larynges of one *B. coniferus* and four *B. viridis* were activated. In at least *B. coniferus* and *B. viridis,* deactivation of the arytenoid cartilages must result from muscle contraction. Electrical stimulation of the dilator laryngis muscles during artificial activation can deactivate the arytenoid cartilages without deactivating the vocal cords (Martin, in preparation).

B. alvarius may be losing its mating call (Blair and Pettus 1954), which, in any case, is undoubtedly divergent. This species calls with its nares open, which means that air passes in one direction through the larynx. The only way that air can be returned to the lungs between pulses is for the air in the mouth cavity to be pressurized above lung pressure while the larynx is open. With the nares open, the mouth cavity air cannot be so pressurized. The unidirectional airflow in *B. alvarius* is confirmed by call structure, for the vocal cords are activated continuously throughout the call (Fig. 15-14C). The calls are pulsed at rates between 40 and 50 pps, and each pulse is of extremely slow rise time. This pulsation is not the result of vibration of the arytenoid cartilages, since the rise time of each pulse is too slow. Also, the pulse rate is about one-third the vibration rate of the arytenoid cartilages during artificial activation, and vibration of the arytenoid cartilage usually results in complete amplitude modulation. The pulsation is probably produced actively, and, since this species has arytenoid valves, it is tentatively classified as Type IIIa although it is obviously divergent.

The six members of the *americanus* group as well as *B. punctatus* exhibit patterns of mating call modulation that are the same as normal Type IIIa species. However, arytenoid valves are not present in the larynges of these species, and artificial activation fails to elicit vibration of the arytenoid cartilages. These species are classified as Type IIIb, with the subscript ("b") indicating the absence of arytenoid valves in their larynges.

The call of *B. quercicus* diverges from this modulation pattern. This species lacks arytenoid valves and arytenoid-cartilage modulation in the mating call, but it is divergent from other Type IIIb species, because each "call" is a single, high-frequency chirp that is not repeatedly amplitude modulated by body-wall contraction. However, these "calls" are repeated periodically (Fig. 15-3C, 15-14D), which suggests that they may actually be homologous to Type IIIb pulses. *B. quercicus* would then be a species with an extremely decelerated mating call.

The Type IIIa vocal pattern apparently arose from the primitive Type II pattern by deactivation of the vibration of the arytenoid cartilages through laryngeal muscle contraction. This probably involved the dilator laryngis muscle. The atrophy of the arytenoid valves seen in the larynges of *B. ibarrai, B. coccifer,* and *B. haematiticus* represents further stages in the loss of this call character.

When considering the evidence for a transition from Type II to Type IIIa within the ancestry of the *valliceps* group, it is significant that this group is rather diversified compared to most other species groups. The impression is of an old group of moderately divergent remnants of an earlier North and Central American radiation that gave rise to several other groups. The three Type IIIa members of the *valliceps* group imply the presence of this modulation pattern in the early radiation. *B. canaliferus* is a divergent Type II remnant of this early radiation (Chap. 18) that shows Type IIIb tendencies (see previous section). On the other hand, *B. occidentalis* is a purely Type II remnant of this early radiation (see Chap. 18).

Since strong affinity between *B. viridis* and the *valliceps* group is not supported by any evidence, it seems that they independently acquired the Type IIIa modulation. *B. viridis* shows reciprocal metamorphosis only with Type II species groups (*cognatus* and *boreas*) and with *B. alvarius* (Type IIIa divergent). In characters other than passive amplitude modulation, the calls of *B. canorus* and *B. viridis* are virtually identical, which suggests affinities between the *boreas* group and *B. viridis.*

Cei, Erspamer, and Roseghini (Chap. 12) present biochemical evidence that testifies to the independent and ancient neotropical origin of the *guttatus* and *typhonius* groups. *B. haematiticus* is a member of the *guttatus* group, which suggests that the origin of Type IIIa vocalization in this species also originated independently. The closely related *B. blombergi* (*guttatus* group) and *B. sternosignatus* (*typhonius* group) have Type II mating calls, which also supports the independent origin of the Type III pattern in *B. haematiticus.*

Other evidence attests to the phylogenetic affinities of all Type IIIb species, with the possible exception of *B. punctatus.* Karyotype data (Chap. 10) support the inclusion of *B. punctatus* within this group, and certain aspects of laryngeal structure other than loss of the arytenoid valves indicate this. Furthermore, *B. punctatus* has a release vibration produced by thoracic muscle contraction, the rate of which matches the pulse rate of the mating call. Pulse rates of the mating call at between 60 and 70 pps were recorded at 27° C. A single release vibration at this same temperature had a contraction rate of 65 pps. This is also true of most *americanus*-group species (Chap. 16). Blair (Chap. 18) points out that the *americanus* group shows strong affinities with both the *valliceps* and *boreas* groups but concludes on the basis of hybridization and chromosomal evidence that the ancestry of the group was from the *valliceps*-group line. This conclusion fits with the evidence from modulation patterns, for it is in the *valliceps* group that several Type IIIa species are found. On the other hand, the only *boreas*-group species that has a mating call (*B. canorus*) has the primitive Type II pattern. Hybridization data also point to a close affinity of *B. coccifer* with the *americanus* group. This species shares one chromosomal constriction with the *americanus* group and one with the *boreas, debilis,* and possibly the *valliceps* groups. Both osteology and parotoid-gland biochemistry point to stronger affinities with the *valliceps* group. *B. coccifer* therefore seems to be somewhat intermediate between the *americanus* and *valliceps* groups, but it has similarities with the *boreas* and possibly with the *debilis* groups. The partial atrophy of the arytenoid valves in *B. coccifer* is consistent with this placement and suggests that the Type IIIb system of the *americanus* group, *B. quercicus,* and possibly *B. punctatus* arose by loss of the arytenoid valves in a divergent line from the primitive *valliceps* radiation of which *B. coccifer* is a remnant.

Atrophy of Components of Airborne Sound Communication Systems

Members of two species groups of narrow-skulled toads (*spinulosus* and *boreas* groups) produce weak chirping sounds in breeding aggregations without tactile stimulation. These sounds have been variously described as being comparable to the voicings of a brood of young domestic goslings (Storer 1925; *B. boreas*) or somewhat like the cheep of young chicks (Cei 1962; *B. atacamensis*). Cei (personal communication) has observed the reproductive behavior of several *spinulosus*-group species, and Blair (personal communication) has observed members of the *boreas* group mating. Both workers felt that these weak, nonspecific calls do not function in attraction to the extent that the more developed mating calls of other species presumably do. Blair and Pettus (1954) also

Table 15–2. Atrophy of Sound Communication Components

Species	Sound Producer				Vocalizations		Sound Receiver				
	V.S.	Ar.V.	P.Me.	V.C.	R.C.	M.C.	Tymp.	E.T.	Ple.	Col.	Op.+M.
I											
Alvarius Gp.											
alvarius	A-R	P	P	P	P	P(R?)	P	P	P	P	P
Boreas Gp.											
boreas	A-P	P	P	P	P	R	P	P	P	P	P
exsul		P	P	P	P	R	P	Presence of tympanum usually			
nelsoni		P	P	P	P	R	P	implies presence of these			
Spinulosus Gp.								structures as well.			
flavolineatus		P	P	P	P		P				
spinulosus	A	P	P	P	P	R	P	P	P	P	P
atacamensis		P	P	P	P	R	P	P	P	P	P
chilensis	A	P	P	P	P	R	P	P	P	P	P
II											
Bocourti Gp.											
bocourti		P	P	P	P	A	A	A	A	A	P
Variegatus Gp.											
variegatus	A	P	P	P	P	R	A	A	A	A	P
Unknown Gps.											
periglenes	A	P	P	P	P	A	A	A-R	A	A	P
lönnbergi	A	P	P	P	P	P?	A	A	A	A	P
anotis		P	P	P	P		A	A	A	A	P
Atelopus											
ignescens	P	P	P	P			A	A	A	A	P
varius	P	P	P	P	P		A	A	A	A	P
sp. (Colombia)	P	P	P	P	P		A	A	A	A	P
Nectophrynoides											
tornieri	A	P	P	P			A	A	A	A	P
Nectophryne Gp.											
afra	A						A	A	A	A	P
Melanophryniscus											
stelzneri	P	A*	A*	P		P	A	A	A	A	P
sp.		A*	A*	P			A	A	A	A	P
III											
Gariepensis Gp.											
rosei	A	P	R-A	R-A	A	A	A	A	A	A	P
Unknown Gp.											
holdridgei	A	P	A	A	A	A	A	A	A	A	P

P–Present; R–Reduced; A–Absent.

*Not the result of atrophy but represents primitive condition (see first section and Fig. 15–7J,K).

(V.S.) = vocal sac; (Ar.V.) = arytenoid valve; (P.Me.) = posterior membrane; (V.C.) = vocal cord; (R.C.) = release call; (M.C.) = mating call; (Tymp.) = tympanum; (E.T.) = eustachean tube; (Ple.) = plectrum; (Col.) = columella; (Op. + M) = opercular muscle and cartilage.

advocate a reduced function for the weak and aberrant mating call of *B. alvarius* (divergent Type IIIa). *B. bufo* (divergent Type I) and *B. haematiticus* (divergent Type IIIa, recorded in the laboratory) have relatively weak and abnormal calls, which might also be best compared with these other calls of apparently reduced function. Other than reduction or loss of the vocal sacs and small larynx relative to body size in *B. bufo* (Fig. 15-13), there are no signs of atrophy of the sound-producing or -receiving structures in these species.

B. bocourti, B. variegatus, B. periglenes, B. lönnbergi, B. anotis, Atelopus ignescens, A. varius, A.sp. (Colombia), *Nectophrynoides tornieri, Melanophry-*

niscus stelzneri and *M. rubriventris* have all lost the middle-ear apparatus except for the operculae and opercular muscles. However, there is no evidence in any of these species that the sound-producing structures are atrophying, although there is reduction or loss of the vocal sac in some species (Table 15-2). Furthermore, well-developed mating calls have been recorded for *M. stelzneri* (Barrio 1964*a*) and possibly *B. lönnbergi* (Chap. 9).

B. rosei and *B. holdridgei* also show atrophy of the receiver apparatus, but, in addition, the laryngeal vocal structures have atrophied (Table 15-2). The all-important vocal cords either lack medial portions or have been reduced to small ridges in all five adult male *B. rosei* that I examined. In a single adult male *B. holdridgei* the vocal cords and the posterior membranes remained only as small ridges on the inside surfaces of the arytenoid cartilages. Arytenoid valves were still present in both these species (Table 15-2).

There are several other bufonid species that I have not examined, which reportedly exhibit atrophy of structures involved in sound communication. Parker (1936) reported that the African species *B. preussi*, and the Asian *B. surdus*, *B. fissipes*, *B. taitanus*, and *B. jordani* lack tympanae, tympanic cavities, and columellae. Savage and Kluge (1961) stated that the Costa Rican *Crepidius epioticus* lacks columellae and ostia pharyngea. The Costa Rican *B. simus*, *B. fastidiosus*, and *B. coerulescens* (Savage 1966), the African *B. micranotis*, *B. osgoodi*, *B. ushoranus* (Loveridge 1925, 1932), and *B. melanopleura* (Schmidt and Inger 1959), and the South American *B. quechua*, *B. ockendeni*, and *B. fissipes* lack tympanae, although the condition of the other middle-ear structures is unknown. Liu (1935) examined fifty-three currently recognized bufonid species and found that vocal sacs are absent in *B. andrewsi*, *B. bankorensis*, all five subspecies of *B. bufo* studied, *B. boreas halophilus*, *B. minshanicus*, *B. superciliaris*, *B. spinulosus*, and *B. variegatus* (see also Inger, Chap. 8).

The members of the first group of species mentioned above (Table 15-2) have apparently undergone a divergence in reproductive behavior, with a deemphasis on vocal communication. However, this behavior is based on observational evidence, and very little anatomical atrophy is seen. The condition in the second group of species poses several interesting physiological problems. Here are eleven species that are capable of producing sound, yet the middle ears have been lost. Because of the poor understanding of acoustical impedence matching in anurans, it cannot be concluded that this loss of the middle ear renders these species deaf to their own vocalizations. Due to the great differences in the density of air and animal tissue, airborne sound impinging on the latter is largely reflected instead of being transmitted into the animal. The middle-ear chain is traditionally considered to perform the function of impedence matching, allowing airborne sound waves to cross the interface with efficiency. If both the emitter and receiver are in water, the middle ear is not necessary, for the density of water is approximately the same as that of animal tissue. If the anuran lungs function as the swim bladder does in many fishes, the animals could even be sensitive to far-field waterborne sound. Furthermore, the middle-ear apparatus would not be necessary when the animals are in direct mechanical contact, as during release calling. The inner ear could still serve as the signal receiver, because the operculae and opercular muscles have been retained (Table 15-2). The opercular muscles originate from the suprascapular process of the pectoral girdle. Therefore, a direct mechanical link for low-frequency vibrations exists from the vibrating flanks of the clasped male through the thumbs, arms, pectoral girdle, opercular muscles, and operculae of the clasping male to his perilymph. However, the loss of both the middle ear and the vocal sac, the latter of which should theoretically decrease the amplitude of the radiated airborne sound, since the laryngeal vibrators are more poorly coupled with the acoustic medium, testifies to a lack of vocalization function in long-distance communication at least. *B. variegatus*, *B. periglenes*, and *B. lönnbergi* exhibit this condition. *B. periglenes* and *B. bocourti* have both been observed in mating aggregations where no mating calls were heard and amplexed pairs were found. Savage concluded that *B. periglenes* does not have a mating call, and Nelson (1966) reached the same conclusion for *B. bocourti*. It may be significant that *B. periglenes*, *B. lönnbergi*, and *Crepidius epioticus* are brightly colored and sexually dichromatic. It is obvious that vocal communication has been lost in the third group, which includes *B. rosei* and *B. holdridgei*. These two losses are undoubtedly independent. It is possibly significant that all the species I have measured that tend toward atrophy of the auditory and communicatory system components have small larynges relative to body size (Fig. 15-13).

There are three trends among the species that show atrophy of sound communication components that should be mentioned. With the exception of *B. alvarius*, there are no known broad-skulled toads that show atrophy of the auditory apparatus, and other evidence shows *B. alvarius* to be close to the narrow-skulled *boreas* group (Chap. 18). Second, all the species in which the middle ear has atrophied are small in body size. Finally, with very few exceptions, the species that show evidence of reduced function of vocalization are species with relatively restricted distributions, the majority in montane habitats. This is true of all the African species (Chap. 9), the five Costa Rican forms, members of the *spinulosus* group and *B. variegatus* (Chap. 6), many species of *Atelopus*, and, to a lesser extent, of members of the *boreas* group (Chap. 7). *B. alvarius* is restricted to the Sonoran Desert.

If we assume that (1) the mating calls of most bufonids are important premating isolating mechanisms and (2) that the evidence presented in this section testifies to a reduction of this function in these forms, then the tendency for species with restricted distributions to show this atrophy might result from decreased contact with other anuran species, a result of their remote distributions. Blair (1958) discussed this idea for *B. alvarius* and several species of the *Rana boylei* group. He suggested that relaxed selection pressure on species-specific, long-distance communication might allow the atrophy of their auditory communication systems. It is impossible to estimate how many times these losses have occurred in *Bufo*, due to the limited knowledge of the taxonomic positions of many rather rare forms. However, it is obvious that the tendency is greatest among the cold-adapted, narrow-skulled species, as well as among some intermediate forms.

Summary

The configuration of vocal structures in the bufonid-*Atelopus* radiation is fairly uniform and very different from other anurans, including leptodactylids. In this report, the problems of the origin and radiation of the unique bufonid-*Atelopus* vocal mechanics were investigated.

Independent taxonomic evidence focuses attention on a complex of 22-chromosome leptodactylids as the probable remnants of the stock giving rise to the bufonid-*Atelopus* radiation. A survey of leptodactylid laryngeal morphology shows that all but three species of the 22-chromosome complex examined have similar larynges, which are also similar to the larynx of the primitive leptodactylid, *Calyptocephalella gayi*, and to members of the Ceratophrynidae. The association of this configuration with the ancestry of the bufonid-*Atelopus* radiation is further confirmed by the laryngeal anatomy of the primitive bufonid genus *Melanophryniscus*, which is intermediate in laryngeal anatomy between leptodactylids and the other members of the bufonid-*Atelopus* radiation. Activation of the larynges with an artificial compressed-air source shows that laryngeal structure is less involved in determining mating call characteristics in members of the 22-chromosome genus *Leptodactylus* than in *Bufo*, and *Odontophrynus occidentalis* ($2n=22$) is intermediate in this respect. Furthermore, the *O. occidentalis* mating call is essentially a Type II (primitive) *Bufo* call, homologous in all passive components to *Bufo* calls, and probably homologous in active modulation. The transition in vocal mechanics from the leptodactylid condition (exemplified by *Leptodactylus* and *Odontophrynus*) results from the accumulation of changes that increase the degree to which laryngeal structure determines call parameters. This is, in effect, a form of peripheral filtering at the emitter end of the communication link, comparable to the peripheral tuning at the inner ear in several anuran species. The transition to the bufonid-*Atelopus* condition involved (1) loss of the constrictor posterior muscles and associated vocal pulvinarae, which together provide a mechanism for changing the tension, position, and configuration of the vocal cords by muscle contraction; (2) stabilization of the arytenoid-cartilage vibration by developing unique valvelike structures on the arytenoid cartilages; and (3) mass-tuning of the bufonid-*Atelopus* vocal cords by changing the position of the fibrous masses, so that they come to lie freely suspended in the newly expanded lateral portions of the vocal cords and, therefore, become involved in determining the vocal cord resonant frequency. One thing that has been lost in this transition is a mechanism for producing extreme frequency modulation in the mating call. Therefore, the transition may also be viewed as a switch from the transmission of information in patterns of frequency modulation to stabilized patterns of amplitude modulation. This transition had already taken place within the Leptodactylidae, since members of the genus *Odontophrynus* are A.M. anurans, while members of the genus *Leptodactylus* tend to be F.M. animals.

The fibrous mass, which is a legacy from the leptodactylid ancestry of the bufonid-*Atelopus* radiation, plays an important role in *Bufo*, for the vocal cords produce the dominant (or carrier) frequency in this genus. Species-specific variation in vocal cord mass is largely responsible for determining their fundamental frequency of resonance. Hence, to a large extent, this one parameter of the vocal tract is responsible for determining the frequency of the dominant of the mating call. However, the air pressure developed during vocalization, the resonant properties of the vocal sac and mouth cavity, and the size and tension of the vocal cords are all potentially involved. Indirect evolutionary pressures on the dominant frequency of the mating call are also exerted through selection on body size, because body size limits to some extent the size of the vocal cords, the size of the fibrous masses that can be suspended on the vocal cords, and the dimensions of the vocal sac and mouth cavity.

The mating calls of *Bufo* can be classified on the basis of the patterns of amplitude modulation, which reflect the amplitude modulation of the vibration of the vocal cords. The vibration of the arytenoid cartilages causes a *passive* amplitude modulation of vocal cord vibration, while pulsatile muscle contraction imparts the *active*, temperature-sensitive pulsation that is characteristic of New World (and some Old World) species. Most of the species examined exhibit both the active and the passive amplitude modulation. All the species that are considered by the other authors as primitive have this modulation pattern, as does *Odontophrynus occidentalis* (but not *O. americanus*). From this apparently primitive pattern, which is called Type II, a pattern lacking active pulsation within the call (Type I), as well as another pattern where passive modulation is lacking (Type III), have also evolved. There are also divergences from these basic types that occur sometimes. The Type III pattern seems to have arisen from the Type II pattern, first by deactivation of arytenoid-cartilage vibration by actively dilating the larynx during vocalization (Type IIIa), and later by loss of the arytenoid valves (Type IIIb), which extinguished arytenoid-cartilage vibration. The Type I pattern may have arisen in some instances by extremely decelerating a Type II call, so that single Type II pulses masquerade as Type I calls. However, it is concluded that the Type I pattern in the African radiation arose by incorporation of a continuous (nonpulsatile) pattern of muscle contraction that is present in the release calls of these species, in the release calls of several closely related species with Type II mating calls and, sporadically, in the mating calls of some Type II species. Indeed, this continuous contraction pattern may go all the way back to the leptodactylid ancestry, since the release call of *O. occidentalis* is Type I, and the mating and release calls of *O. americanus* are Type I.

Many species of the bufonid-*Atelopus* radiation exhibit reduction in behavioral or anatomical components of their sound communication systems. The first condition is one in which very little anatomical atrophy is seen except for reduction of larynx size relative to body size and occasional loss of the vocal sac. However, extensive field observations on breeding animals suggest that vocalization does not play a significant part in attracting a mate. The second condition presents several physiological problems: the vocal tract remains capable of sound production, yet the middle ears are atrophied. Finally, a condition is seen in two species where both the middle ears and the vocal tract are atrophied. Species showing these losses tend to be members of the narrow-skulled radiation and to have remote distributions, particularly in mountainous country. Species exhibiting atrophy of the middle ears are usually small in body size.

REFERENCES

Aronson, L. R. 1944. The mating pattern of *Bufo americanus*, *Bufo fowleri*, and *Bufo terrestris*. Amer. Mus. Novitates, no. 1250, pp. 1–15.

Awbrey, F. T. 1965. An experimental investigation of the effectiveness of anuran mating calls as isolating mechanisms. Ph.D. Thesis, Univ. Texas.

Barrio, A. 1964*a*. Peculiaridades del canto nupcial de *Melanophryniscus stelzneri* (Weyenbergh) (Anura, Brachycephalidae). Physis 24:435–437.

———. 1964*b*. Characteres eto-ecologicos diferenciales entre *Odontophrynus americanus* (Dumeril et Bibron) y *O. occidentalis* (Berg) (Anura, Leptodactylidae). Physis 24:385–390.

———. 1965. Afinidades del canto nupcial de las especies cavicolas del genero *Leptodactylus* (Anura-Leptodactylidae). Physis 25:401–410.

———. 1966. Divergencia acustica entre el canto nupcial de *Leptodactylus ocellatus* (Linne) y *L. chaquensis* (Anura, Leptodactylidae), Physis 26:275–277.

Beçak, M. L., W. Beçak, and M. N. Rabello. 1966. Cytological evidence of constant tetraploidy in the bisexual South American frog *Odontophrynus americanus*. Chromosoma 19:188–193.

Bishop, R. E. D. 1965. Vibration. Cambridge Univ. Press, Cambridge.

Blair, A. P. 1941. Variation, isolation mechanisms, and hybridization in certain toads. Genetics 26:398–417.

———. 1942. Isolating mechanisms in a complex of four species of toads. Biol. Symposia 6:235–249.

———. 1947. The male warning vibration in *Bufo*. Amer. Mus. Novitates, no. 1344, pp. 1–7.

Blair, W. F. 1956. Call difference as an isolation mechanism in southwestern toads (genus *Bufo*). Texas J. Sci. 8:87–106.

———. 1958. Mating call in the speciation of anuran amphibians. Amer. Nat. 92:27–51.

———. 1964. Evolution at populational and interpopulational levels; isolating mechanisms and interspecies interactions in anuran amphibians. Quart. Rev. Biol. 39:333–344.

———, and D. Pettus. 1954. The mating call and its significance in the Colorado River toad (*Bufo alvarius* Girard) Texas J. Sci. 9:72–77.

Blume, W. 1930. Studien am Anurenlarynx. Gegenbaurs Morph. Jahrb. 65:307–464.

Bogart, J. P. 1967. Chromosomes of the South American amphibian family Ceratophridae with a reconsideration of the taxonomic status of *Odontophrynus americanus*. Canad. J. Genet. Cytol. 9:531–542.

———. (In press.) Los cromosomas de anfibios anuros del género Eleutherodactylus. Comunicación en el IV Congreso Latinoamericano de Zoologia, Caracas, Venezuela. November 1968.

Bogert, C. M. 1960. The influence of sound on the behavior of amphibians and reptiles, p. 137–320. *In*, W. W. Lanyon and W. N. Tavolga (eds.), Animal Sounds and Communication. A.I.B.S. Publ. 7.

———. 1962. Isolation mechanisms in toads of the *Bufo debilis* group in Arizona and western Mexico. Amer. Mus. Novitates, no. 2100, pp. 1–35.

Brown, L. E. 1969. Variations in vocalizations produced by the giant South American toad, *Bufo blombergi*. Amer. Midl. Nat. 81:189–197.

Cei, J. M. 1962. Batracios de Chile. Universidad de Chile, Santiago.

———, and V. Erspamer. 1966. Biochemical taxonomy of South American amphibians by means of skin amines and polypeptides. Copeia 1966:74–78.

Fouquette, M. J. 1960. Call structure in frogs of the family Leptodactylidae. Texas J. Sci. 12:201–215.

Griffiths, I. 1954. On the "otic element" in Amphibia Salientia. Proc. Zool. Soc. (London) 124:35–50.

———. 1963. The phylogeny of the Salientia. Biol. Rev. 38:241–292.

Keith, R. 1968. A new species of *Bufo* from Africa, with comments on the toads of the *Bufo regularis* complex. Amer. Mus. Novitates, no. 2345, pp. 1–22.

Liepmann, H. W., and A. Roshko. 1957. Elements of gasdynamics. Wiley, New York.

Liu, C. C. 1935. Types of vocal sac in the Salientia. Proc. Boston Soc. Nat. Hist. 41:19–40.

Loveridge, A. 1925. Notes on east African batrachians collected 1920–1923, with the description of four new species. Proc. Zool. Soc. (London) 1925:763–791.

———. 1932. Eight new toads of the genus *Bufo* from east and central Africa. Occas. Papers Boston Soc. Nat. Hist. 8:43–54.

Nelson, C. E. 1966. Notes on some Mexican and Central American amphibians and reptiles. Southwestern Naturalist 11:123–124.

Parker, H. W. 1936. Dr. Karl Jordan's expedition to south-west Africa and Angola: Herpetological collections. Nov. Zool. 40:115–146.

Porter, K. R. 1964. Morphological and mating call comparisons in the *Bufo valliceps* complex. Amer. Mid. Nat. 71:232–245.

———. 1965. Intraspecific variation in mating call of *Bufo coccifer* Cope. Amer. Mid. Nat. 74:250–256.

———. 1966. Mating calls of six Mexican and Central American toads (genus *Bufo*). Herpetologica 22:60–67.

———. 1968. Evolutionary status of a relict population of *Bufo hemiophrys* Cope. Evolution 22:583–594.

Powell, J. H., and J. H. T. Roberts. 1923. On the frequency of vibration of circular diaphragms. Proc. Phys. Soc. 35:170–182.

Saez, F. A., and N. Brum-Zorilla. 1966. Karyotype variation in some species of the genus *Odontophrynus* (Amphibia-Anura). Caryologica 14:55–63.

Savage, J. M. 1966. An extraordinary new toad (Bufo) from Costa Rica. Rev. Biol. Trop. 14:153–167.

———, and A. G. Kluge. 1961. Rediscovery of the strange Costa Rica toad, *Crepidius epioticus* Cope. Rev. Biol. Trop. 9:39–51.

Schmidt, Karl P., and R. F. Inger. 1959. Exploration du Parc National de l'Upemba. Inst. Parcs Nationaux Congo Belge. Fascicle 56. Brussels

Schneider, H. 1967. Die Paarungsrufe einheimischer Froschlurche (Discoglossidae, Pelobatidae, Bufonidae, Hylidae). Z. Morph. Okol. Tiere 57:119–136.

Storer, T. I. 1925. A synopsis of the Amphibia of California. Univ. Calif. Publ. Zool., Vol. 21.

Trewavas, E. 1933. The hyoid and larynx of the Anura. Phil. Trans. Royal Soc. (London) 222:401–527.

Watkins, W. A. 1967. The harmonic interval: fact or artifact in spectral analysis of pulse trains, p. 15–43. *In*, W. N. Tavolga (ed.), Marine Bioacoustics, Oxford and New York, Pergamon Press.

Wood, A. B. 1940. Acoustics. Dover, New York.

Zweifel, R. G. 1968. Effects on temperature, body size, and hybridization on mating calls of toads, *Bufo a. americanus* and *Bufo woodhousei fowleri*. Copeia 1968:269–284.

16. Male Release Call in the *Bufo americanus* Group

LAUREN E. BROWN
Illinois State University
Normal, Illinois,
and
M. J. LITTLEJOHN
University of Melbourne
Parkville, Victoria, Australia

Introduction

W. F. Blair (1963), on the basis of laboratory artificial hybridization experiments, included six nominal species within the *Bufo americanus* group: *B. americanus, B. hemiophrys, B. houstonensis, B. microscaphus, B. terrestris,* and *B. woodhousei.* This assemblage has long been of special interest to students of evolution and is perhaps unique among amphibians in that more instances of both intra- and intergroup natural hybridization are known than in any other comparable taxonomic group (Anderson, Liner, and Etheridge 1952; Ballinger 1966; A. P. Blair 1941, 1955; W. F. Blair 1956*a*, 1957; Brown 1967; Cory and Manion 1955; Henrich 1968; McCoy, Smith, and Tihen 1967; Neil 1949; Thornton 1955; Volpe 1952, 1959, 1960; Zweifel 1968; and others).

Although more may be known about the *americanus* group than any other species group within the genus *Bufo,* the stage of speciation reached by most of the allopatric forms remains an enigma. Only one species, *B. woodhousei,* is to any extent sympatric with other members of the group and thus exposed to the absolute test of species validity. Yet this species occasionally forms natural hybrids (which are fertile; Chap. 11) with every member of the *B. americanus* group with which it is sympatric. Morphological, biochemical, cytological, ecological, and behavioral differences occur among the six forms, but the significance of these differences for species status is not clear.

Males of the genus *Bufo* and of other anuran genera produce a distinct acousticomechanical signal that is known as the release call, or warning vibration. This signal is a sex identification mechanism that ensures rapid release when one male is clasped by another (Aronson 1944). Rapid releases are presumably advantageous since energy is conserved, gamete wastage is prevented, and the amount of time that the sexually preoccupied males are exposed to predation is reduced (W. F. Blair 1968).

Certain assumptions can be made about the release call that suggest it is considerably more valuable as a phylogenetic tool than the mating call is. First, the release call is not known to function as an isolating mechanism or sex attractant and therefore should not be exposed to the same divergent adaptive selection pressures that influence mating call structure. Second, there should be selection for relative uniformity in the release call structure among sympatric species, since male-male sexual pairings are unproductive regardless of whether they are intra- or inter-specific. Consequently, this character should be conservative in its evolution.

Our aims have been:

1. To survey the release calls of the *B. americanus* species group;
2. To determine the effects of temperature and body size on the release call structure;
3. To ascertain similarities and differences that may exist between the release calls and mating calls of the members of this group;
4. To assess the value of the structure of the release call as an aid in identifying natural hybrids;
5. To attempt a reconstruction of the evolutionary history of the group on the basis of a presumably conservative behavioral character — the male release call.

Materials and Methods

The release calls were recorded on tape at 19 cm. per second, using Stancil-Hoffman M9, Uher 4000 Report S, and Sony 101 tape recorders with Altec 633A, Beyer M69, or Sony microphones respectively. Each individual was held near the microphone, or in direct contact with it, and release calls were elicited by applying pressure to the back and sides of the animal with the fingers. It was necessary to make two series of recordings for each individual, since the release calls can consist of two components — release chirp and release vibration — which have quite different sound intensities. Both are usually produced during a single call. The first series was made at the normal recording level and the animal's head was held about 2–3 cm. away from the microphone in order to record the louder release chirp. In the second series, the recording level was increased to overload (about +2 to +3 db), and the animal's ventral surface was placed in contact with the microphone in order to pick up the release vibration, which has a much lower intensity.

Tapes were subsequently analyzed on a Sona-Graph Sound Spectrograph Model 6061-A (Kay Electric Co.) and a double-beam cathode-ray oscilloscope (Cossor Model 1049). The data obtained from these analyses included dominant frequency, call duration, pulse repetition rate, and note structure.

Cloacal temperatures were recorded to the nearest 0.1° C with a Schultheis quick-reading thermometer immediately before and after the taping of the release calls of each animal. These initial and final temperatures showed little difference and were averaged. The size of each animal was measured to the nearest mm. by flattening the dorsal surface of the specimen onto a ruler and noting the distance between the tip of the snout and center of the cloaca.

Release calls of ninety-one sexually mature males were recorded: fourteen *B. americanus* (9 from Dawson Lake, Illinois; 3 from Lake Rosseau, Ontario, Canada; 2 from Minnesota, exact locality not known); five *B. hemiophrys* (2 from Ada, Minnesota; 3 from Minnesota, specific locality not known); fifteen *B. houstonensis* (12 from Bastrop, Texas; 3 from Houston, Texas); ten *B. microscaphus* (all from Virgin, Utah); eleven *B. terrestris* (10 from Bay Co. and Gulf Co., Florida; 1 from Gainesville, Florida); and thirty-four *B. woodhousei* (20 from Austin, Texas; 4 from Bastrop, Texas; 3 from Pflugerville, Texas; 2 from Gonzales, Texas; 1 from Hornsby's Bend, Texas; 1 from Montgomery Co., Texas; and 3 from St. George, Utah). The release calls of two B. *houstonensis* x *B. woodhousei* natural hybrids from Bastrop, Texas, were also recorded.

To determine the effect of temperature on the structure of the release call, five *B. woodhousei* males from Austin, Texas, were subjected to a series of eight water baths in temperatures ranging from 15 to 33° C. Release calls were tape-recorded after the animals equilibrated to each water temperature.

The other individuals (except *B. houstonensis*) were equilibrated to 24–25°C in a temperature-controlled room or water bath before release calls were recorded in order to ensure that temperature differences would not invalidate interspecies comparisons. Release calls of *B. houstonensis* were recorded before the technique of temperature standardization had been developed and there is thus a greater range of cloacal temperatures for this species. However, this variability provided another opportunity to analyze the effect of temperature on the release call.

For effective comparison of mating calls and release calls it was necessary that both signals be recorded at the same temperature. Fifteen *B. wood-*

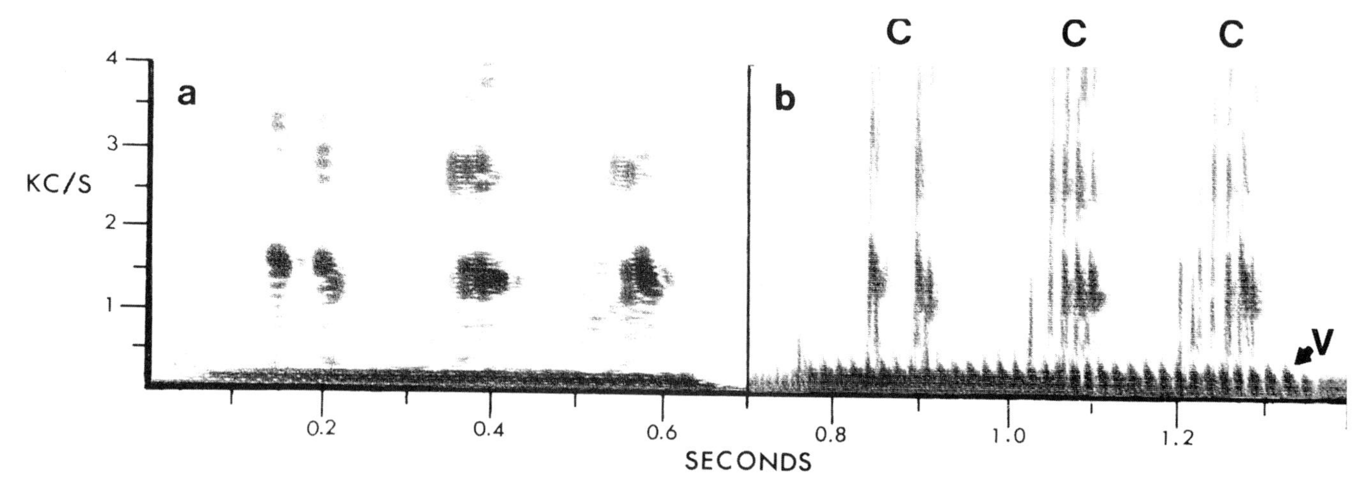

Fig. 16–1. Audiospectrograms of release calls of a male *Bufo terrestris*, one of a series collected in Bay and Gulf counties, Florida, and recorded at a cloacal temperature of 23.8° C. (*a*) Narrow band display (45 cps filter setting); (*b*) Wide band display (300 cps filter setting). *C* indicates release chirps; *V* indicates release vibration.

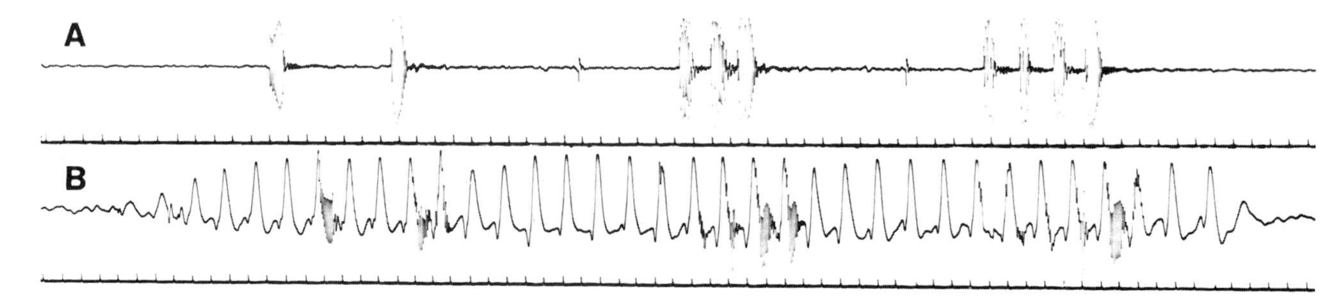

Fig. 16–2. Oscillograms of release calls of a male *Bufo terrestris*, one of a series collected in Bay and Gulf counties, Florida, and recorded at a cloacal temperature of 23.8° C. (*A*) Release chirps; (*B*) Release vibration with superimposed (low recording level) release chirps. The time marker below each tracing indicates 10 msec. intervals.

housei males from Austin, Texas, were sampled in the field in the following manner. First, three to five mating calls of each individual were recorded; the animal was then captured and its cloacal temperature quickly taken. This was followed by immediately recording five or six release calls. A final cloacal temperature was then read. The difference between the two cloacal temperatures was not more than 1°C and usually less than 0.5°C. Mating calls were analyzed in the same manner as the release calls, except that durations were determined to 0.1 second with a stopwatch while the recordings were played back at half speed.

Analyses of variance (randomized block design) were carried out to determine the effect of cloacal temperature on the release call of the five temperature-treated *B. woodhousei*. Correlation coefficients were computed in order to analyze the effect of cloacal temperature on the release call of *B. houstonensis*. Likewise, the relationship of snout-vent length to structure of the release call was determined by computing correlation coefficients. For all other comparisons, T tests were utilized.

Release Vocalizations

Although there is considerable variation in the pulse rates, dominant frequencies, and durations of the release vibration and release chirp, the basic structure of each major component is similar among all members of the *B. americanus* group. The release calls of this group are quite different from those of other species of *Bufo* that have been examined (Brown and Littlejohn, unpublished data). Audiospectrograms of a typical release vibration and release chirp of a *B. terrestris* from Gainesville, Florida, are shown in figure 16-1. Oscillograms of a release

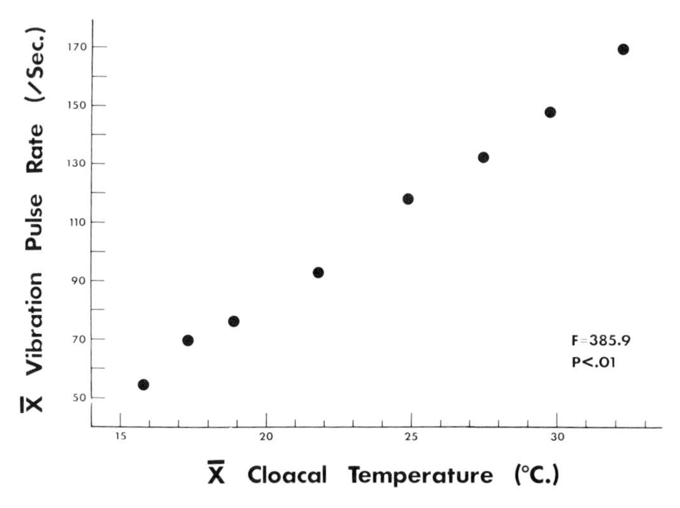

Fig. 16–3. The effect of temperature on pulse rates of the release vibration of five temperature-treated *B. woodhousei.*

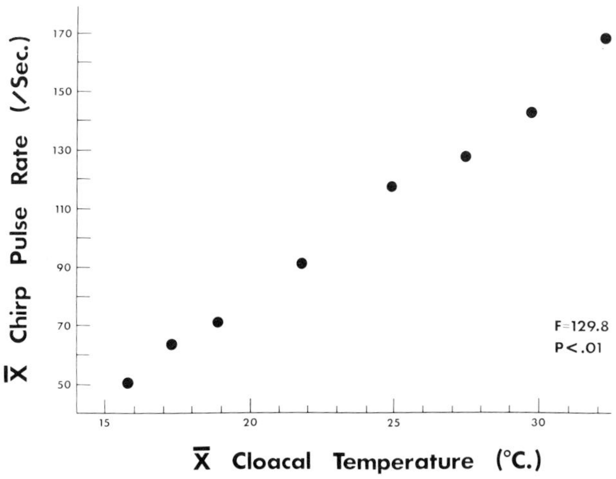

Fig. 16–4. The effect of temperature on pulse rates of the release chirp in four temperature-treated *B. woodhousei.*

vibration and release chirp from the same animal are shown in figure 16-2.

The release vibration is a barely audible, pulsed sound that is usually under two seconds in duration. Its dominant frequency (= fundamental) is very low and cannot be determined through spectrographic analysis. Oscilloscopic analysis suggests that most of the energy is contained in this low fundamental frequency. As the release vibration progresses, its pulse rate decreases.[1] This component is accompanied by body vibrations that presumably determine its fundamental frequency.

The vocalized release chirp contains more energy and is shorter than the release vibration. During production of the release chirp there is often a partial inflation of the vocal sac. The dominant frequency of the chirp can be determined by audiospectrographic analysis; it is considerably higher than that of the release vibration. The release chirp is the more variable of the two components. Its dominant frequency often increases and/or decreases considerably as the chirp progresses. Typically, the release chirp is pulsed, although sometimes no pulsation is visible on the audiospectrograms. The unpulsed release chirps occur when there is no coincidental release vibration. Thus, it is quite possible that pulsed release chirps represent an amplitude modulation by the release-vibration fundamental frequency or by the same mechanism that generates the release vibration.

The release chirp is usually produced in conjunction with the release vibration, although each component can be produced independently. Sometimes as many as four or more release chirps are given during the production of a single release vibration.

The Effect of Temperature on the Release Call

Pulse Rate of Release Vibration

Variations in temperature have a striking effect on the pulse rates of the release vibrations of *B. woodhousei*, *B. houstonensis*, and presumably of the

Table 16–1. Correlation of Cloacal Temperature and Variables of the Release Calls of *B. houstonensis*

Relationship	r	P
Cloacal temperature and release-vibration pulse rate	+.791	<.001
Cloacal temperature and release-chirp pulse rate	+.781	<.001
Cloacal temperature and release-chirp dominant frequency	+.654	<.01
Cloacal temperature and release-vibration duration	–.679	<.01
Cloacal temperature and release-chirp duration	–.083	<.80,>.70

[1] This decrease made it difficult to determine pulse rates. The release chirps often concealed some portions of the release vibration, making the calculation of a single pulse rate for an entire release vibration impossible. Therefore, pulse rates were calculated for each visible segment of the release vibration (over equivalent time spans) and subsequently averaged.

other members of the *B. americanus* group. There is a positive correlation between cloacal temperature and pulse rate of the release vibration for the five temperature-treated *B. woodhousei* from Austin, Texas (Fig. 16-3). The value of *F* is highly significant. The pulse rates at the highest cloacal temperatures ($\overline{X}$ = 32.3° C) are approximately three times greater than those at the lowest temperatures ($\overline{X}$ = 15.8° C). For *B. houstonensis* there is a highly significant positive correlation between cloacal temperature and the pulse rate of the release vibration (Table 16-1).

Pulse Rate of Release Chirp

There is a positive correlation between cloacal temperature and pulse rate of the release chirp for the temperature-treated *B. woodhousei* (Fig. 16-4). The value of *F* is highly significant. Pulse rates of the release chirp increase with temperature in a similar manner to the release-vibration pulse rates. One of the *B. woodhousei* gave only a few pulsed release chirps (although many nonpulsed chirps were given). The data from this specimen were therefore not included in this analysis of variance. Cloacal temperature has a highly significant positive relationship with the pulse rate for the release chirp of *B. houstonensis* (Table 16-1).

Duration of Release Vibration

Cloacal temperature is negatively correlated with release-vibration duration for the five temperature-treated *B. woodhousei* (Fig. 16-5). The *F* value is highly significant. For *B. houstonensis*, the negative correlation coefficient of these two variables is highly significant (Table 16-1).

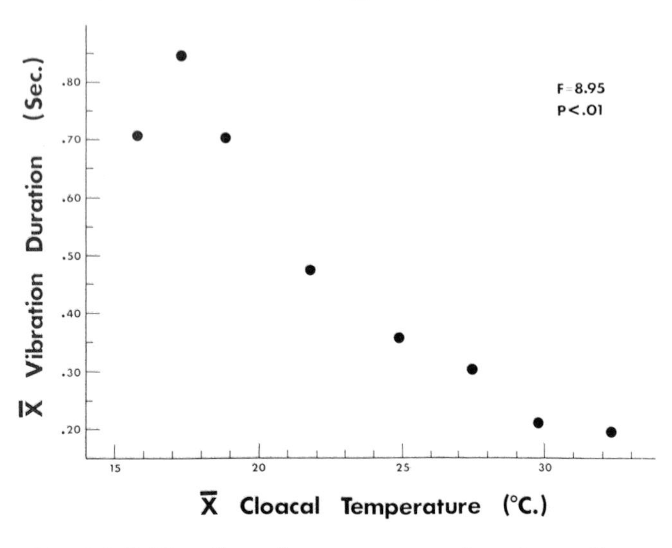

Fig. 16–5. The effect of temperature on the release-vibration duration of five temperature-treated *B. woodhousei.*

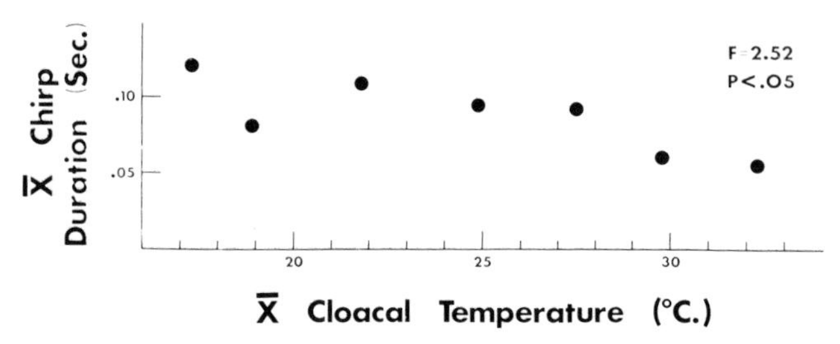

Fig. 16–6. The effect of temperature on duration of the release chirp in five temperature-treated *B. woodhousei.*

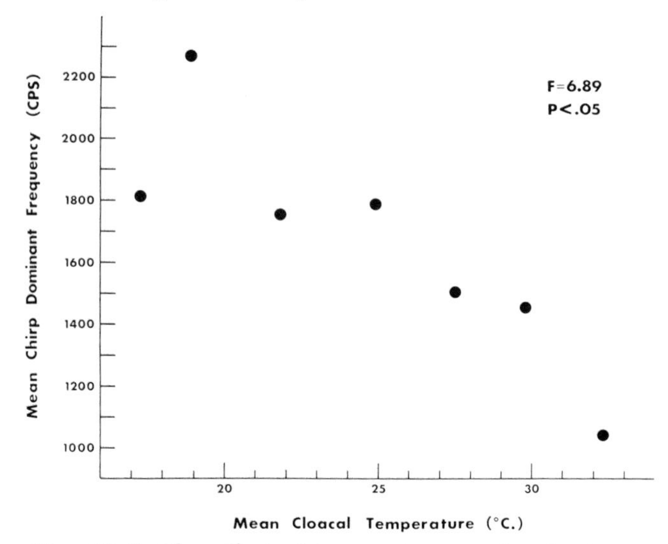

Fig. 16–7. The effect of temperature on the dominant frequency of the release chirp in five temperature-treated *B. woodhousei.*

Duration of Release Chirp

For the five temperature-treated *B. woodhousei*, the release chirp duration is negatively correlated with cloacal temperature (Fig. 16-6). *F* is barely significant. The negative correlation coefficient for the *B. houstonensis* sample is not significant (Table 16-1). These inconsistencies may reflect the considerable variability in the duration of the release chirp.

Dominant Frequency of Release Chirp

Cloacal temperature is negatively correlated with the dominant frequency of the release chirp of the five temperature-treated *B. woodhousei* (Fig. 16-7). The *F* value is significant. The dominant frequency of the *B. houstonensis* release chirp, however, shows

Table 16–2. Correlation of Snout-Vent Length and Variables of the Release Calls of Members of the *B. americanus* Species Group

Species	Relationship	r	P
B. americanus	Snout-vent length and release-vibration pulse rate	–.192	<.60, >.50
B. microscaphus	Snout-vent length and release-vibration pulse rate	+.713	<.05
B. terrestris	Snout-vent length and release-vibration pulse rate	–.609	<.10, >.05
B. woodhousei	Snout-vent length and release-vibration pulse rate	+.232	<.40, >.30
All specimens	Snout-vent length and release-vibration pulse rate	+.258	<.10, >.05
B. americanus	Snout-vent length and release-chirp pulse rate	–.297	<.40, >.30
B. microscaphus	Snout-vent length and release-chirp pulse rate	+.913	<.001
B. terrestris	Snout-vent length and release-chirp pulse rate	–.447	<.20, >.10
B. woodhousei	Snout-vent length and release-chirp pulse rate	+.246	<.40, >.30
All specimens	Snout-vent length and release-chirp pulse rate	+.252	<.10, >.05
B. americanus	Snout-vent length and release-chirp dominant frequency	–.473	<.20, >.10
B. microscaphus	Snout-vent length and release-chirp dominant frequency	+.038	>.90
B. terrestris	Snout-vent length and release-chirp dominant frequency	–.135	<.80, >.70
B. woodhousei	Snout-vent length and release-chirp dominant frequency	+.016	>.90
All specimens	Snout-vent length and release-chirp dominant frequency	–.377	<.01
B. americanus	Snout-vent length and release-vibration duration	+.123	<.70, >.60
B. microscaphus	Snout-vent length and release-vibration duration	+.144	<.70, >.60
B. terrestris	Snout-vent length and release-vibration duration	+.538	<.20, >.10
B. woodhousei	Snout-vent length and release-vibration duration	–.371	<.20, >.10
All specimens	Snout-vent length and release-vibration duration	+.195	<.20, >.10
B. americanus	Snout-vent length and release-chirp duration	+.485	<.10, >.05
B. microscaphus	Snout-vent length and release-chirp duration	–.476	<.20, >.10
B. terrestris	Snout-vent length and release-chirp duration	+.537	<.20, >.10
B. woodhousei	Snout-vent length and release-chirp duration	+.236	<.40, >.30
All specimens	Snout-vent length and release-chirp duration	+.262	<.05

a highly significant positive correlation with cloacal temperature (Table 16-1).

The Effect of Body Size on the Release Call

The effect of body size on release call structure is less obvious than the effect of temperature, which may be due to small sample sizes and the limited ranges of variation in snout-vent length (except in *B. americanus* and *B. woodhousei*). No regression analyses of snout-vent length were carried out for *B. hemiophrys*, due to the small sample size (N = 5). Since the release calls of *B. houstonensis* were recorded over a greater range of cloacal temperatures than were those of the other species, no regression analyses of snout-vent length were carried out for this species.

The effects of snout-vent length on the various components of the release calls of *B. americanus*, *B. microscaphus*, *B. terrestris*, and *B. woodhousei* at approximately 25° C are shown in table 16-2. In addition, correlation coefficients were calculated for the combined data of these four species ("all specimens").

Pulse Rate of Release Vibration

There is a significant positive correlation between snout-vent length and pulse rate of the release vibration for *B. microscaphus*. The correlation coefficients for the other species and "all specimens" are not significant.

Pulse Rate of Release Chirp

B. microscaphus also shows a highly significant positive correlation between snout-vent length and the pulse rate for the release chirp. The other species and "all specimens" show no significant relationship.

Dominant Frequency of Release Chirp

A highly significant negative correlation is found between snout-vent length and dominant frequency of the release chirp when the data of all specimens are combined. There is no significant correlation between these variables when each of the species is analyzed separately.

Duration of Release Vibration

None of the correlation coefficients for snout-vent

Table 16–3. Release Call and Mating Call Characteristics of Fifteen Male *B. woodhousei* from Austin, Texas

Individual No.	Cloacal Temperature		Pulse Rate			Dominant Frequency		Duration		
	Initial	Final	Vibration	Chirp	Mating Call	Chirp	Mating Call	Vibration	Chirp	Mating Call
1	18.7	19.5	73.8	66.7	108.1	1250	1313	1.208	0.114	2.18
2	19.0	19.3	78.3	64.1	118.3	—	953	0.623	0.055	2.57
3	18.5	19.5	72.1	72.9	109.5	1750	1250	0.543	0.061	1.63
4	19.1	19.8	76.7	74.1	118.3	2219	1282	0.780	0.065	1.85
5	18.8	19.3	82.4	90.6	108.4	2500	1350	0.757	0.106	2.72
6	18.5	18.8	85.8	85.8	111.4	3500	1375	0.918	0.079	2.65
7	18.5	18.4	74.2	71.4	107.8	2500	1362	0.678	0.077	2.23
8	18.0	18.0	68.9	67.9	107.8	1875	1325	0.677	0.101	2.06
9	18.6	18.4	72.9	70.5	117.2	2438	1282	0.988	0.059	2.15
10	18.3	18.6	76.0	75.0	112.2	1625	1213	0.832	0.061	1.93
11	20.2	20.4	72.8	74.1	124.0	1375	1266	1.170	0.151	2.03
12	19.6	19.7	75.8	74.0	118.0	1531	1375	0.658	0.159	2.23
13	20.4	20.6	88.6	85.8	132.0	2500	1263	0.718	0.111	1.80
14	20.8	21.5	86.9	90.1	134.0	1438	1387	0.705	0.095	2.60
15	20.0	20.5	88.7	90.0	128.0	1125	1313	1.062	0.067	2.04

length and duration of the release vibration is significant.

Duration of Release Chirp

Only when all specimens are combined is a significant positive correlation found between snout-vent length and duration of the release chirp; the correlation coefficients for the various species analyzed separately are not significant.

It is worth emphasizing the results of the regression analyses of snout-vent length for *B. americanus* and *B. woodhousei*. These two species have the largest sample sizes and considerable ranges of variation in snout-vent length (Table 16-5). Yet in neither species does snout-vent length show a significant correlation with any of the variables of the release call. A. P. Blair (1947) reported that three *B. w. fowleri* had a mean pulse rate for the release vibration of 93.3 pulses per second, while the comparable figure for five *B. w. woodhousei* was 93.2 pulses per second. Although he did not give the snout-vent lengths of these animals, it is probable that the *B. w. fowleri* were smaller than the *B. w. woodhousei*. The high values for the pulse rates of the release vibration that he obtained for both subspecies are consistent with the results of our correlation analyses. We examined only one *B. w. fowleri*; the pulse rates for this individual do not differ from those of the *B. w. woodhousei*. Thus, on the basis of present evidence, body size does not affect the pulse rates of the release vibration of *B. woodhousei*.

Comparison of Mating Calls and Release Calls

Bufo woodhousei

The field-recorded mating and release calls of the fifteen *B. woodhousei* from Austin, Texas, are quite different in all respects that were considered (Table 16-3). The pulse rates of the release vibration and release chirp are both significantly lower than the mating call pulse rate ($P < .001$ in each case). There is also a highly significant difference ($P < .01$) between the dominant frequencies of the release chirp and mating call, the latter being the lower. Both the release vibration and the release chirp are shorter in duration than the mating call ($P < .001$ in each case).

Other Species

Extrapolated values of some characteristics of the mating call for members of the *B. americanus* group are listed in table 16-4. Since pulse rates of the mating call, and sometimes durations, were positively correlated with temperature, it was necessary to adjust these variables to the same temperatures at which the release calls were recorded before comparison. However, none of the dominant-frequency values listed for mating call was adjusted, since temperature apparently has only a slight effect on dominant frequency — at least for *B. americanus, B. woodhousei*, and *B. hemiophrys* (Zweifel 1968; Porter 1968).

Adjusted estimates of pulse rates and durations

Table 16–4. Mating Call Characteristics of Members of the *B. americanus* Species Group*

Species	Temperature (° C)	Pulse Rate (Pulses/Sec.)	Dominant Frequency (CPS)	Duration (Sec.)
B. americanus	25.1	55.0	1400–1900	4.2
B. hemiophrys	25.0	71.7, 73.7	1590, 1760	7.0, 5.8
B. houstonensis	21.9	32.0	2072	10.5
B. terrestris	24.0, 27.0	73.9	2161	4.5
B. woodhousei	24.9	168.6	1800–2100	1.13

*The values listed for *B. americanus* and *B. woodhousei* are estimates calculated using the predictive equations and data of Zweifel (1968). The *B. hemiophrys* data are estimated from the predictive equation and Fig. 2 of Porter (1968). The first *B. hemiophrys* value in each column is an estimated prediction for an isolated Wyoming population; the second value is for animals from Manitoba and the Dakotas. The *B. houstonensis* data are means of 14 individuals recorded by Brown (1967); likewise, the *B. terrestris* data represents means of eleven individuals recorded by W. F. Blair (1956*b*). See the text for further explanation.

of the mating call for *B. americanus* and *B. woodhousei* were calculated using the predictive equations of Zweifel (1968).

Porter (1968) found that pulse rate is the only variable of the *B. hemiophrys* mating call that is correlated with temperature. His estimated regression coefficient (b = +3 pulses/° C) was used to adjust the mean *B. hemiophrys* pulse rates to 25° C. The means of dominant frequency and duration were estimated from his figure 2. Porter compared the mating calls of *B. hemiophrys* from the main distribution of the species (Dakotas and Manitoba) with mating calls of individuals from a relictual population in Wyoming. He suspected that there might be reinforcement of the mating call due to the proximity of *B. americanus* in the Dakotas and Manitoba but not in Wyoming where *B. americanus* does not occur.

The values of the variables in *B. houstonensis* mating call listed in table 16-4 are means calculated from raw data for fourteen individuals recorded by Brown (1967) at temperatures around 22°C. It was not possible to record accurate cloacal temperatures of these individuals, because the temperatures increased very rapidly as soon as they were handled. The temperature listed is the mean of water and air temperatures.

There are not enough data available on the mating call characteristics of *B. terrestris* to permit regression analyses. The values for this species (Table 16–4) are the means of raw data reported by W. F. Blair (1956*b*) for eleven individuals. Calls of these males were recorded at an air temperature of 24° C and a water temperature of 27° C.

W. F. Blair (1957) has described the mating calls of nine *B. microscaphus*. It is not possible to apply temperature adjustments to his raw data, because the temperature range was so narrow (air temperatures: 14.0 and 17.0°C; water temperatures: 16.0 and 17.0° C).

PULSE RATE

With the exception of *B. woodhousei*, pulse rates of the release vibration and mating call of each species are similar (Tables 16-4, 16-5). The same generalization can be made when pulse rates of the release chirp and mating call are compared. The predicted pulse rates of the release vibration and release chirp at 21.9°C for *B. houstonensis* are 36.7 and 39.5 pulses per second respectively. These are reasonably close to the pulse rate of the mating call at the same temperature (32.0 pulses/sec). The release vibration and release chirp pulse rates differ somewhat from the mating call pulse rate in *B. terrestris*. However, the cloacal temperatures are not known for the eleven *B. terrestris* males recorded by W. F. Blair (1956*b*), and a precise comparison with the release call is not possible. In general, pulse rates of the release call and the mating call in the four species differ only slightly.

DOMINANT FREQUENCY

There is a general similarity between the dominant frequencies of the release chirp and the mating call for each species except *B. woodhousei*. The snout-vent lengths of individuals analyzed for the release call undoubtedly differ from those of males used in the analysis of mating call. These differences in length may provide a partial explanation for the slight differences in frequency between the release chirp and mating call, since size is sometimes correlated with dominant frequency (see Zweifel 1968). The fundamentals for the release vibration are considerably lower than the mating call dominant frequencies for all species.

DURATION

The mating call is considerably longer than the release vibration and the release chirp for every species.

Comparison of the Release Vibration and Release Chirp

Pulse Rate

Although the pulse rate of the release chirp is significantly greater than that of the release vibration for three species (*B. americanus*, $P < .05$; *B. houstonensis*, $P < .001$; *B. terrestris*, $P < .01$), the differences are only slight. This may be due to differences in the methods of calculation of pulse rates (see fn. 1). The two pulse rates do not differ significantly for the other three species (*B. hemiophrys*, $.70 > P > .60$; *B. microscaphus*, $P > .90$; *B. woodhousei*, $.70 > P > .60$).

Dominant Frequency

The dominant frequencies of the release chirp for all species are much greater than those of the release vibration. The means of the dominant frequencies of the release chirp range from 1350 cps for *B. woodhousei* to 2048 cps for *B. terrestris*. In both, the dominant frequencies of the release vibration are too low to determine from audiospectrographic sections.

Duration

For all species, the release vibration is significantly longer than the release chirp (*B. americanus*, $P < .001$; *B. hemiophrys*, $P < .02$; *B. houstonensis*, $P < .001$; *B. microscaphus*, $P < .001$; *B. terrestris*, $P < .001$; *B. woodhousei*, $P < .001$). The duration of *B. houstonensis* release vibration is significantly correlated with temperature, whereas the duration of the *B. houstonensis* release chirp is not. However, when the effect of temperature on the duration of the release vibration and of the release chirp of *B. woodhousei* was analyzed, it was found that the F values were significant in each case.

Interspecific Comparison

The interspecific overlap of ranges in variation of the dominant frequencies of the release chirp, duration of the release vibration, and duration of release chirp (Table 16-5) precludes their use in determining phylogenetic affinities. The more limited intraspecific variation in the pulse rates of the release vibration makes this attribute considerably more valuable in determining evolutionary relationships. The pulse rate of the release chirp is not utilized in the following interspecies comparisons, because it is most likely generated by amplitude modulation of the fundamental of the release vibration. Consequently, interspecies variation in pulse rate is essentially the same for both of these variables.

B. woodhousei has the highest mean pulse rate for the release vibration (Table 16-5) and is the most clearly differentiated member of the *B. americanus* group. The difference between the means for the pulse rate of the release vibration of *B. woodhousei* and *B. microscaphus* (the species with the second highest pulse rate mean) is large, 28.2 pulses per second, and there is no overlap in ranges of variation. One *B. hemiophrys* has a pulse rate for the release vibration within the range of the pulse rate of *B. woodhousei*. This individual, however, is quite atypical of *B. hemiophrys*.

The mean for the pulse rate of the release vibration of *B. hemiophrys* is close to that of *B. microscaphus* (the difference being only 4.1 pulses/sec), but there is little overlap in ranges of variation. The pulse rate for the release vibration of one *B. hemiophrys* falls within the range for *B. microscaphus*. Likewise, the pulse rate for the release vibration of a single *B. microscaphus* is found within the range of *B. hemiophrys*.

There are notable differences between the pulse rates for the release vibration of *B. terrestris* (the species with the next highest pulse rate) and *B. hemiophrys*. The means for the pulse rates of the release vibrations of these two species differ by 17.3 pulses per second. The pulse rate of the release vibration for a single *B. terrestris* falls within the range of *B. hemiophrys*. Similarly, the pulse rate of only one *B. hemiophrys* falls within the range of *B. terrestris*.

The pulse rates of the release vibration of *B. americanus* and *B. terrestris* are moderately different (means differ by 9.0 pulses/sec). There is some overlap in the ranges for the two species but it results largely from the record of one *B. americanus*. This individual was recorded at the highest cloacal temperature of any of the *B. americanus*. Only one other *B. americanus* had a pulse rate for the release vibra-

Table 16–5. Release-Call Characteristics of Members of the *B. americanus* Species Group.

Species or Hybrid	N	Cloacal Temperature (°C)		Snout-Vent Length (mm.)		Vibration Pulse Rate (Pulse/Sec)		Chirp Dominant Frequency (CPS)		Vibration Duration (Sec.)		Chirp Duration (Sec.)	
		$\overline{X}$	Range	$\overline{X}$	Range	$\overline{X}$	Range	$\overline{X}$	Range	$\overline{X}$	Range	$\overline{X}$	Range
B. americanus	14	25.1	23.8–25.4	78.9	58–98	51.8	42.4–61.3	1406	938–2750	0.723	0.388–1.283	0.115	0.082–0.174
B. hemiophrys	5	25.0	24.3–26.2	59.6	54–63	78.1	67.5–99.0	1788	937–2313	0.438	0.185–0.638	0.100	0.071–0.118
B. houstonensis	15	22.9	21.3–25.8	64.8	57–73	40.2	28.9–53.8	1596	1094–2375	0.658	0.240–1.150	0.139	0.077–0.218
B. microscaphus	10	24.5	24.0–25.2	61.9	60–67	82.2	74.8–87.8	1597	907–3150	0.355	0.272–0.470	0.079	0.058–0.103
B. terrestris	11	24.2	23.8–25.5	60.4	54–67	60.8	57.3–69.4	2048	1188–3146	0.514	0.218–0.757	0.090	0.065–0.136
B. woodhousei	19	24.9	23.6–25.8	81.6	65–95	110.4	89.2–131.2	1350	688–2313	0.450	0.155–0.750	0.087	0.029–0.150
B. houstonensis X *B. woodhousei*	2	22.5	21.7–23.3	69.0	66–72	65.6	65.0–66.1	1144	1125–1157	0.604	0.548–0.660	0.077	0.070–0.083

tion within the range of *B. terrestris* and then only slightly.

The predicted pulse rate for release vibration for *B. houstonensis* at 25° C (47.3 pulses/sec) and the mean value for *B. americanus* (Table 16-5) differ by only 4.5 pulses per second. The difference between the mean pulse rate for the release vibration of the three *B. houstonensis* that were recorded at cloacal temperatures nearest to 25° C (50.5 pulses/sec) and the mean pulse rate of all the *B. americanus* is only 1.3 pulses per second. Considerable overlap occurs in the ranges for the pulse rate of the release vibration of *B. americanus* and *B. houstonensis*. This overlap would probably be greater if all the *B. houstonensis* were recorded at cloacal temperatures of 25° C.

Pulse Rate of the Release Vibration as an Aid in the Identification of Natural Hybrids

Brown and Guttman (1970) found that a natural hybrid *Bufo arenarum* x *Bufo spinulosus* had a pulse rate for the release call intermediate between those of the parental species. Additional morphological and biochemical evidence further confirmed the hybrid identification. Natural hybrids between *B. woodhousei* and *B. houstonensis* have been found near Bastrop, Texas (Brown 1967). Since the parental species are very similar in morphology, there are no apparent structural characteristics that can be used to positively identify these hybrids. However, the mating call of the hybrid is intermediate between those of the parental species in pulse rate, dominant frequency, and duration (Brown 1967). The release calls of two *B. woodhousei* x *B. houstonensis* natural hybrids were recorded to determine whether the pulse rate of the release vibration could also be of use in identifying these hybrids.

Predictive equations from the regression analyses of temperature and pulse rate were utilized to obtain values for the pulse rates of the release vibration of the parental species at the temperatures of the hybrids. For *B. woodhousei*, means were calculated for the predicted values of the five temperature-treated individuals from Austin, Texas. Since the *B. houstonensis* were recorded at a number of different temperatures, only one predictive equation was used.

One *B. woodhousei* x *B. houstonensis* hybrid recorded at a cloacal temperature of 21.7° C had a pulse rate for the release vibration of 65.0 pulses per second. The mean of the predicted pulse rates of the *B. woodhousei* at 21.7° C is 95.1 pulses per second. For *B. houstonensis* the predicted value of the pulse rate at 21.7° C is 36.0 pulses per second. The second natural hybrid had a pulse rate of 66.1 pulses per second at a cloacal temperature of 23.3° C. The predicted pulse rates for *B. woodhousei* and *B. houstonensis* are 105.8 and 41.4 pulses per second respectively. Thus, the intermediate pulse rates provide an additional means for identifying *B. woodhousei* x *B. houstonensis* natural hybrids.

Discussion

If a phylogeny of the *B. americanus* group were based solely on interspecific variations in the pulse rate of the release vibration, it would imply the following relationships. At one extreme, with the highest pulse rate, is the distinctly differentiated *B. woodhousei*. At the other extreme, with the lowest pulse rates, are *B. americanus* and *B. houstonensis*. Intermediate between these two extremes are *B. terrestris*, *B. microscaphus*, and *B. hemiophrys*. *B. hemiophrys* and *B. microscaphus* appear closely related and show the closest relationship to *B. woodhousei*. *B. terrestris* stands between the two apparently closely related pairs *B. americanus*–*B. houstonensis* and *B. hemiophrys*–*B. microscaphus*. It appears, however, more closely allied to *B. americanus* and *B. houstonensis*. Thus, the *B. americanus* group could, on the basis of the pulse rates of the release vibration, be separated into three subgroups, the first containing *B. americanus*, *B. houstonensis*, and *B. terrestris*, the second including *B. hemiophrys* and *B. microscaphus*, and the third having *B. woodhousei* as its sole representative. A dendrogram based on the release-vibration pulse rates is shown in figure 16-8.

A. P. Blair (1947), using a kymograph and a mechanical coupling attached to toads, was able to determine the release-vibration pulse rates of males of seven North American species of *Bufo*. From the results of this work he recognized three classes based on release-vibration pulse rates:

I. Species with a high release-vibration pulse rate – *B. w. woodhousei* (N = 5; $\bar{X}$ = 92.3; range = 84–104) *B. w. fowleri* (N = 3; $\bar{X}$ = 93.3; range = 80–104)

II. Species with an intermediate release-vibration pulse rate – *B. terrestris* (N = 9; $\bar{X}$ = 51.3; range = 42–64)

III. Species with a low release-vibration pulse rate – *B. cognatus* (N = 2; $\bar{X}$ = 6.5; range = 5–9.5) *B. valliceps* (N = 12; $\bar{X}$ = 13.0; range = 9.5–16)

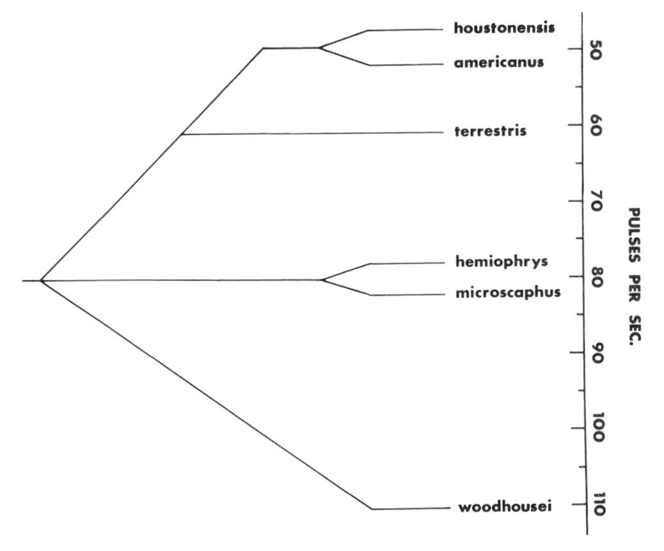

Fig. 16–8. A dendrogram of the *B. americanus* species group based on the pulse rates of the release vibration. A pulse-rate scale is shown horizontally. The line of each species ends at a position along this scale which corresponds to the species release-vibration pulse rate.

B. marinus (N = 3; $\overline{X}$ = 6.9; range = 5–9)
B. alvarius (N = 1; $\overline{X}$ = 5.0)
B. speciosus (only gross observations made)

A. P. Blair possibly obtained his data at a lower temperature (he gave no temperature data) than we did, for the pulse-rate means listed for both *B. woodhousei* and *B. terrestris* are somewhat lower than ours. However, the difference in means for the pulse rate of these two species is similar to that which we found.

Using evidence from genetic compatibility experiments and other sources, W. F. Blair (1963: p. 30) proposed the following evolutionary history for the *B. americanus* species group:

> In brief, *B. woodhousei* seems likely to represent the earliest speciation by reason of its sympatry with all of the other species except *B. hemiophrys*, by morphological reasons of distribution of throat pigments in breeding males, and possibly by reason of the more rapidly pulsed and less finely tuned call. A subsequent subdivision may have separated the complex of eastern, long-calling species (*B. americanus, B. terrestris, B. houstonensis*) from the two western, short-calling species (*B. hemiophrys, B. microscaphus*). Finally, *B. houstonensis* appears to be the youngest species of the group, if indeed it has achieved reproductive isolation from *B. americanus* from whose range it is disjunct.

The evidence of the pulse rates for the release call closely agrees with W. F. Blair's phylogenetic conclusions. If one were to base a phylogeny on the pulse rates of the mating call, it would probably be similar to that suggested by W. F. Blair and by the pulse rates for the release vibration.[2] The main exception is the degree of differentiation attained by *B. woodhousei.* The pulse rates for the release vibration suggest that this species is the most divergent member of the *B. americanus* group. Pulse rates of the mating call suggest, however, that *B. woodhousei* is much more distantly related to the other members of the group than the pulse rates of the release call suggest. At 25° C the difference between the pulse rates of the release vibration of *B. houstonensis* (predicted) and *B. woodhousei* (mean) is 63.1 pulses per second. These are the species with the lowest and highest pulse rates for the release vibration. In contrast, the predicted pulse rates (Table 16-4) of the mating call for *B. americanus* (with a pulse rate close to that of the related *B. houstonensis*) and *B. woodhousei* show a much greater difference of 113.6 pulses per second. The very high pulse rate of the *B. woodhousei* mating call accounts for the mating call range being greater than that of the release vibration.

The *B. woodhousei* mating call, because of its importance as an isolating mechanism, is highly divergent. This is to be expected, since *B. woodhousei* is the only species that is sympatric with other members of the *B. americanus* group. The release call is probably not exposed to strong selection for interspecific divergence. No advantage is conferred on either species in male-male interspecific pairings by a lack of release-call communication. The failure of such pairs to rapidly cease amplexus would deter the male of each species from maximizing his contribution to the gene pool of a subsequent generation.

Summary

The bufonid male release call serves as a sex identification mechanism between male toads. In the species of the *B. americanus* group it consists of two major components: the release vibration and the release chirp. We have attempted to determine the de-

[2] At present it is not possible to make accurate correlations of the mating call pulse rates of *B. terrestris* and *B. microscaphus* to 25° C; thus, more confident phylogenetic inferences cannot be made from their mating call structure.

gree to which interspecific variation in the male release call reflects the evolution of the *B. americanus* species group. Secondary aims were to determine the effects of temperature and snout-vent length on the release call; to compare release calls with mating calls; and to examine the release calls of natural hybrids.

Temperature has a marked effect on the release call. For *B. woodhousei* and *B. houstonensis* the pulse rates for both the release vibration and the release chirp show positive correlations with cloacal temperature (highly significant *F* and *r* values). For *B. houstonensis* the dominant frequency of the release chirp has a highly significant positive correlation with temperature. In addition, the dominant frequency of the release chirp of *B. woodhousei* is negatively correlated with temperature (significant *F* value). Temperature is negatively correlated with the duration of the release vibration in *B. woodhousei* and *B. houstonensis* (highly significant *F* and *r* values). The duration of the release chirp is also negatively correlated with temperature; the *F* value for *B. woodhousei* is significant, but the correlation coefficient for *B. houstonensis* is not significant.

Possible effects of body size on the release call are less evident because of narrow ranges of size variation. Nonetheless, snout-vent length does not appear to have as striking an influence as temperature. Pulse rates of both the release vibration and release chirp of *B. microscaphus* show significant positive correlations with snout-vent length; the correlation coefficients for the other species are not significant. There is a significant negative correlation between the dominant frequency of the release chirp and snout-vent length only when the data for all species are combined. Snout-vent length does not appear to affect the duration of the release vibration. It does, however, have a significant effect on the duration of the release chirp, but only when the data from all specimens are combined.

The mating calls and release calls of *B. woodhousei* differ significantly in duration, pulse rate, and dominant frequency. Mating calls and release calls of the other species do not show such clear-cut differences, although detailed comparisons were not made; however, in all species, durations of the release vibration and release chirp are shorter than those of mating calls.

Natural hybrids between *B. houstonensis* and *B. woodhousei* were easily identified by pulse rates of the release vibration, which were intermediate between those of the parental species. Morphological characteristics of the parental species overlap considerably in variation, and consequently hybrids cannot be identified strictly on this basis.

The pulse rate for the release vibration was the only useful attribute in determining evolutionary relationships. A phylogeny based on these pulse rates agrees closely with the evolutionary history proposed by W. F. Blair (1963). The pulse rates for the release vibration suggest that *B. americanus* and *B. houstonensis* are closely related, as apparently are *B. hemiophrys* and *B. microscaphus*. *B. terrestris* occupies an intermediate position between these two pairs but is closer to *B. americanus* and *B. houstonensis*. The most clearly differentiated member of the group is *B. woodhousei*. *B. microscaphus* and *B. hemiophrys* appear most closely related to *B. woodhousei*. A phylogeny of the *B. americanus* group based on the mating call pulse rates is similar. The major exception is *B. woodhousei*, which has an extremely high pulse rate for the mating call. This is perhaps because *B. woodhousei* is the only species of the group that is sympatric with other members of the group. Under sympatric conditions there should be a premium on a highly differentiated mating call because of the importance of this signal as an isolating mechanism. In contrast, the pulse rate of the release vibration in *B. woodhousei* is not nearly as divergent as that of the mating call. Conservative variation of the release call is presumably advantageous because of the possible utilization of this signal in interspecies communication.

REFERENCES

Anderson, P. K., E. A. Liner, and R. E. Etheridge. 1952. Notes on amphibian and reptile populations in a Louisiana pineland area. Ecology 33:274–278.

Aronson, L. R. 1944. The sexual behavior of Anura VI — The mating pattern of *Bufo americanus, Bufo fowleri*, and *Bufo terrestris*. Amer. Mus. Novitates, no. 1250, pp. 1–15.

Ballinger, R. E. 1966. Natural hybridization of the toads *Bufo woodhousei* and *Bufo speciosus*. Copeia 1966: 366–368.

Blair, A. P. 1941. Variation, isolation mechanisms, and hybridization in certain toads. Genetics 26:398–417.

———. 1947. The male warning vibration in *Bufo*. Amer. Mus. Novitates, no. 1344, pp. 1–7.

———. 1955. Distribution, variation, and hybridization in a relict toad (*Bufo microscaphus*) in southwestern Utah. Amer. Mus. Novitates, no. 1722, pp. 1–38.

Blair, W. F. 1956*a*. The mating calls of hybrid toads. Texas J. Sci. 8:350–355.

———. 1956*b*. Call difference as an isolation mechanism in southwestern toads (genus *Bufo*). Texas J. Sci. 8: 87–106.

———. 1957. Structure of the call and relationships of *Bufo microscaphus* Cope. Copeia 1957:208–212.

———. 1963. Intragroup genetic compatibility in the *Bufo americanus* species group of toads. Texas J. Sci. 15:15–34.

———. 1968. Amphibians and reptiles, p. 289–310. *In*, T. A. Sebeok (ed.), Animal communication. Indiana Univ. Press, Bloomington.

Brown, L. E. 1967. The significance of natural hybridization in certain aspects of the speciation of some North American toads (genus *Bufo*). Ph.D. Thesis, Univ. Texas.

———, and S. I. Guttman. 1970. Natural hybridization between the toads *Bufo arenarum* and *Bufo spinulosus* in Argentina. Amer. Midl. Nat.: 160–166.

Cory, L., and J. J. Manion. 1955. Ecology and hybridization in the genus *Bufo* in the Michigan-Indiana region. Evolution 9:42–51.

Henrich, T. W. 1968. Morphological evidence of secondary intergradation between *Bufo hemiophrys* Cope and *Bufo americanus* Holbrook in eastern South Dakota. Herpetologica 24:1–13.

McCoy, C. J., H. M. Smith, and J. A. Tihen. 1967. Natural hybrid toads, *Bufo punctatus* x *Bufo woodhousei*, from Colorado. Southwestern Naturalist 12:45–54.

Neil, W. T. 1949. Hybrid toads in Georgia. Herpetologica 5:30–32.

Porter, K. R. 1968. Evolutionary status of a relict population of *Bufo hemiophrys* Cope. Evolution 22:583–594.

Thornton, W. A. 1955. Interspecific hybridization in *Bufo woodhousei* and *Bufo valliceps*. Evolution 9: 455–468.

Volpe, E. P. 1952. Physiological evidence for natural hybridization of *Bufo americanus* and *Bufo fowleri*. Evolution 6:393–406.

———. 1959. Experimental and natural hybridization between *Bufo terrestris* and *Bufo fowleri*. Amer. Midl. Nat. 61:295–312.

———. 1960. Evolutionary consequences of hybrid sterility and vigor in toads. Evolution 14:181–193.

Zweifel, R. G. 1968. Effects of temperature, body size, and hybridization on mating calls of toads, *Bufo a. americanus* and *Bufo woodhousei fowleri*. Copeia 1968:269–285.

17. Characteristics of the Testes

W. FRANK BLAIR

The University of Texas at Austin
Austin, Texas

Introduction

The sacrifice of male toads for artificial crosses has permitted the testes of all the species used in the hybridization experiments to be examined. However, it was not until many animals had been used that systematic observations on the character of the testes were recorded. At first, sketches were made showing the approximate shape and proportion, and the color was recorded. Later, actual measurements as well as records of color were made.

The two main measurable attributes of the testes, color and proportions, do have phylogenetic significance and do reinforce other evolutionary evidence. All the sketches, measurements, and records of color were made in freshly killed animals.

Proportions

In general, the proportions of the testes correspond to the two main evolutionary lines. The narrow-skulled line of toads tends to have short, thick testes, the width of which usually equals 30 percent or more of the length. The broad-skulled line tends to have elongated testes that have a diameter usually much less than 30 percent of the length (Appendix K-1, 2, and 3). Seasonal variations related to the reproductive season probably account for the considerable variation reflected in appendix K-1, 2, but the general pattern holds with only a small number of apparent exceptions. There is some interesting variation within groups, but it is not presently possible to say whether this is genetic or the result of seasonal change.

In the *spinulosus* group of the narrow-skulled line, three species, *B. spinulosus*, *B. flavolineatus*, and *B. chilensis*, showed maximum ratios of more than 30 percent, and two species, *B. trifolium* and *B. atacamensis*, showed ratios of less than 30 percent (Appendix K-1). In the *boreas* group, *B. boreas* and *B. canorus* showed ratios in excess of 30 percent, but one *B. exsul* showed only 28 percent. *B. bocourti* fitted the same pattern, with a range of 28 to 54 percent and a mean of 38 percent in eight individuals. In the *americanus* group, which is somewhat peripheral in the narrow-skulled line, three species showed maximum ratios of more than 30 percent. These were *B. americanus*, *B. hemiophrys*, and *B. microscaphus*. *B. woodhousei* and *B. terrestris* are alike in that they have ratios only slightly below these, but the relict *B. houstonensis*, in the one individual examined, showed a marked departure from this pattern, with a ratio of only 7 percent. Some individuals of *B. debilis* of the *debilis* group exceed 30 percent in this ratio, but *B. retiformis* and *B. kelloggi*, represented by one individual each, fall below it. *B. viridis, B. calamita,* and *B. bufo* of Eurasia and *B. stomaticus* of southeastern Asia fit this pattern. So do *B. periglenes* of Costa Rica, *Melanophryniscus* (probably = *Bufo*) *stelzneri* of South America, and *B. gariepensis, B. inyangae,* and *B. lönnbergi* of Africa. *B. quercicus* of North America fits the same pattern with a range of 33–50 percent among three individuals.

Other narrow-skulled toads, *cognatus* group, show variation of only 8 percent in two *B. compactilis,* 13–31 percent in *B. cognatus,* and 18–45 percent in *B. speciosus.*

Among groups that are intermediate in skull type, *B. punctatus,* among six individuals, showed a range of 8–15 percent. The two species of the *marmoreus* group showed a range from 6 to 28 percent, and *B. alvarius* showed a range of 22–26 percent and a mean of 23 percent. The greatest discrepancy in the narrow-skulled toads was found in the African *B. carens,* in which the range was 6–9 percent in four individuals.

In the broad-skulled line, the South American species *B. crucifer* showed a ratio of 13–18 percent in five individuals. In the remaining South American broad-skulls (*marinus* group) there is a dichotomy, with *B. arenarum* and *B. ictericus* showing ranges up to 38 and 31 percent respectively, and with *B. marinus, B. paracnemis* and *B. poeppigi* having ranges up to 19 percent (8 individuals), 13 percent (4 individuals), and 20 percent (6 individuals), respectively. *B. granulosus,* an intermediate species in several respects, had a testis width-to-length ratio of 8–28 percent in two individuals. *B. typhonius* had a ratio of 14–28 percent in two individuals.

B. haematiticus and *B. holdridgei,* of the broad-skulled line, had ratios of 7 and 20 percent, respectively, in one individual of each species.

The *valliceps* group is somewhat intermediate, with three species showing maximum ratios of more than 30 percent and three showing lesser ratios. *B. valliceps* (8–37%), *B. luetkeni* (7–30%), and *B. coniferus* (8–50%) show high ratios. *B. ibarrai,* in two individuals, showed ratios of 9 and 10 percent. *B. mazatlanensis,* in two individuals, had ratios of 9 and 10 percent. *B. gemmifer* (one individual) showed a ratio of 11 percent. *B. canaliferus,* a close relative of the *valliceps* group, showed a range of 15 to 25 percent, among eight individuals. *B. occidentalis,* with two individuals, showed a range of 16 to 33 percent. *B. melanostictus* of southeastern Asia, among seven individuals, showed a range of 5 to 13 percent. *B. asper* showed a range of 6 to 16 percent among six individuals examined.

Table 17–1. Group of Species Groups and Species of *Bufo* Based on Testis Shape and Color

$\frac{\text{Width}}{\text{Length}} = >30\%$	$\frac{\text{Width}}{\text{Length}} = <30\%$
Black and Yellow Present	Yellow Present
cognatus	*marmoreus*
spinulosus	*punctatus*
Yellow Present	*alvarius*
bocourti	*asper*
boreas	*marinus*
bufo	*crucifer*
americanus	*canaliferus*
coccifer	*melanostictus*
arenarum	*regularis*
valliceps	*maculatus*
mauritanicus	White–Flesh
White–Flesh–Pale Gray Present	*granulosus*
stomaticus	*typhonius*
occidentalis	*carens*
gariepensis	*haematiticus*
lönnbergi	*holdridgei*
periglenes	
Testes Black	
viridis	
calamita	
Testes Reticulate	
debilis	
quercicus	
Melanophryniscus (= *Bufo*?) *stelzneri*	

B. mauritanicus of North Africa, although a broad-skulled toad, has a high ratio, with a range of 18 to 46 percent and a mean of 29 percent in three individuals. All of the ten 20-chromosome African toads examined (*regularis* and *maculatus* groups) had maximum ratios of less than 30 percent.

If all the species groups for which testis data are available are arranged in two groups — those in which at least some maximum ratios of 30 percent or more were found and those in which no species with a maximum of 30 percent or more were found — these two groups, with few exceptions, correspond to the narrow- and the broad-skulled lines (Table 17-1). Among those with ratios of 30 percent and higher, the only exceptions are *B. arenarum, B. ictericus, B. occidentalis,* the *valliceps* group (only 3 of the 6 species examined), and *B. mauritanicus.* In the alternative group, the only exceptions are *B. granulosus, B. punctatus, B. alvarius,* and the *marmoreus* group (all of which are more or less intermediate in skull type and evolutionary position), and *B. carens.* The testes of the large, broad-skulled *B. blombergi* of South America and the comparable *B. superciliaris* of West Africa were not measured, but, from observations of the former and from Noble's (1924) description of the latter, both fall in the group with elongated testes.

Color

Testis color is quite variable (Appendix K-1, 2), even within a species, but it does provide some valuable information. The color descriptions used are subjective ones, and choice of a category is sometimes arbitrary under such a system.

One clearly obvious phenomenon is the occurrence of a black coating of the testes in various members of the narrow-skulled line of toads but its absence in all broad-skulled toads that have been examined. Four of six *B. chilensis* had black testes; the other two had dark gray testes. Fifteen individuals of four other species of the *spinulosus* group had testes that were yellow or of similarly pale color. One of six *B. cognatus* had a black testis, although the five others and individuals of the two other species of the group had yellow, cream, or flesh-colored testes. Seven of eight *B. viridis* had black testes, and the eighth had dark gray ones. Five of six *B. calamita* had black testes, and the sixth had dark gray ones. One of nine *B. bufo* had one black and one yellow testis, and two had dark gray testes. The testes of the other six ranged from yellow to cream to white.

Interesting results were obtained from the hybridization of members of the narrow-skulled line. Male hybrids between female *B. woodhousei* and male *B. viridis*, *B. bufo*, and *B. calamita*, not unexpectedly, had black testes. However, in several crosses of a narrow-skulled and a broad-skulled toad, the F_1 male had black testes. One of two hybrids from the combination of ♀ *B. hemiophrys* x ♂ *B. valliceps* had black testes, and the other had flesh-colored testes. All three hybrids from the combination ♀ *B. perplexus* x ♂ *B. luetkeni*, one from the combination of ♀ *B. speciosus* x ♂ *B. granulosus*, one from the combination of ♀ *B. speciosus* x ♂ *B. arenarum*, and two from the combination of ♀ *B. arenarum* x ♂ *B. marmoreus* had black testes. Two of three hybrids from the combination of ♀ *B. coccifer* x ♂ *B. valliceps* had black testes, and the third had dark gray ones. It thus appears that in some of the narrow-skulled species the determiner for a melanin-coated testis persists but is suppressed; it is expressed only under the heterozygosity of an interspecies hybrid.

Yellow testes occur commonly in both evolutionary lines (Appendix K-3). In species with yellow testes, some individuals often had cream-colored or white testes. Whether this variation is purely seasonal and physiological is uncertain, but I suspect that it probably is. In both lines, there are groups of species in which yellow testes have not been recorded, and, instead, the color ranges from white to flesh-colored to pale gray. In the group with short, compact testes, *B. stomaticus*, *B. gariepensis*, *B. lönnbergi*, and *B. periglenes* of the narrow-skulled line fit this category, as well as the broad-skulled *B. occidentalis*. In the group with elongate testes, *B. granulosus*, *B. typhonius*, *B. haematiticus*, and *B. holdridgei* of the broad-skulled line and *B. carens* of the narrow-skulled line have testes of these colors. One of two *B. typhonius* had cream-colored testes and the other had white, so it is possible that this species fits with the yellow-testis group.

Pale-colored testes with fine gray reticulations have been found in four species, *B. debilis* and *B. kelloggi* (but not in *B. retiformis*, of the *debilis* group), *B. quercicus*, and *Melanophryniscus* (=*Bufo*?) *stelzneri*.

Intragroup Variation

Some variation from species to species in the same species group is based on such small samples that its significance cannot be properly assessed. The *americanus* group shows wide variation in the ratio of width to length of the testes. *B. americanus* and *B. microscaphus* have the most compact testes. *B. terrestris*, *B. hemiophrys*, and *B. woodhousei* tend to have more elongate ones, while the one specimen of *B. houstonensis*, an almost certain Pleistocene relict from *B. americanus* stock, had very elongate testes.

In the *cognatus* group (Appendix K-1), *B. cognatus* and *B. speciosus* have compact testes, but the testes of two individuals of *B. compactilis* examined were much more elongate than the most elongate of those of the ten individuals of the two other species.

As mentioned earlier, two species of the *debilis* group have reticulated testes; the third, *B. retiformis*, had a flesh-colored testis in the one individual examined, which also had relatively elongate testes in comparison with the two other species. This might suggest closer relationship of the latter than of either to *B. retiformis*.

In the *marinus* group, *B. arenarum* and *B. ictericus*, which show close affinity within the group on evidence from hybridization (Chap. 11), have distinctly shorter, broader testes than other members of the group, but *B. paracnemis,* which also shows affinity with these two, has elongate testes closely comparable to those of *B. marinus* and *B. poeppigi.*

There is wide intraspecific variation in the width/length ratio of the testes in the *valliceps* group, but there is an apparent dichotomy, with *B. valliceps, B. luetkeni,* and *B. coniferus* tending to greater compactness than *B. ibarrai, B. mazatlanensis,* and *B. gemmifer.*

The previously mentioned dichotomy in testis color in the *spinulosus* group is unquestionably significant. Of dubious significance is the relatively great elongation of the testes in the single specimen of *B. atacamensis* examined (Appendix K-1).

The *regularis* group shows a great deal of intragroup variation, and, since the sample size is fairly adequate, it is possible to assess this variation with some degree of confidence. The testes tend to be elongate and rounded in cross-section, as in most other broad-skulled toads, and several of the species are indistinguishable in the characters of the testes. *B. rangeri* tends to show the greatest compaction of the testes, which are usually yellow by contrast with the usually white or near-white testes of other members of the group. The most interesting variation is found in the two cryptic species that we have referred to as *B. regularis* (E) and *B. regularis* (W). Observations of the testes are available for the eastern populations from nine localities over a huge range from Kenya to South Africa. The testes are mostly the normal ones of broad-skulled toads, with white or near-white color and elongate form. The greatest departure from this norm in testis shape is found in two individuals from Lourenço Marques, Mozambique, and one of two from the Vumba Mountains, Rhodesia (Appendix K-1). The only marked departure in color was found in the yellow testes of one toad from Pietermaritzburg, South Africa, and one identified as *B. regularis* (E) from Kabete, Kenya. One of two males from Nyeri, Kenya, identified by geographical location as *B. regularis* (E), had the type of testis found in Nigerian *B. regularis* (W) to be described below. Since the eastern and western species seem to be indistinguishable on external characters, it is not possible to say whether this one individual is a *B. regularis* (W) that is sympatric with *B. regularis* (E) in Kenya or whether it indicates overlap of the two populations in testis characters. The distinction is so clear between the two in testis character that one is tempted to accept the former explanation.

B. regularis (W) shows remarkable geographic variation in what is presumably one species. The testis in this species differs from all other species of *Bufo* examined in that it is somewhat ribbonlike rather than cylindrical. Three males from El Mahalla el Kubra, Egypt, had yellow testes, and one from Entebbe, Uganda, had pale yellow testes. In the others, the testes were white to flesh colored. One of two males from Kampala, Uganda, had an irregular mass that defied measurement but that contained sperm that fertilized eggs. The most peculiar type of testis was found in toads from Ilesha and Ogbomosho, Nigeria. In these the flesh-colored to white testes were more or less vesicular, and in some the vesicles were essentially separate from one another.

Hypertrophy of Testes

Enormous hypertrophy of the testes has occurred independently in both the broad-skulled, elongate-testis line and in the narrow-skulled, compact-testis lines. In the latter group, the only examples have been found in *B. gariepensis, B. inyangae,* and *B. rosei* of South Africa (Appendix K-4). In the former group, hypertrophy of the testes has occurred at least twice or there has been intercontinental dispersal to account for similarity of the African *B. superciliaris* and the South American *B. blombergi.* In addition to these two species, *B. haematiticus* and *B. holdridgei* of Central America also have hypertrophied testes.

In both groups the testes are enormously enlarged by comparison with normal testes of their respective type (compact or elongate). The body cavity is literally crammed with the hypertrophied testes. In the tiny (35 mm. snout-vent length) *B. inyangae,* a testis that was measured had dimensions of 6 x 13 mm., and the other testis was equally developed. The testes of the much larger *B. gariepensis* are proportionately large. In the hypertrophy of the elongate testis, there is much coiling, and again the body cavity is crammed with testicular material. A *B. haematiticus* with snout-vent length of 51 mm. had testes 28 mm. in length (left testis measured). A *B. holdridgei* only 45 mm. in snout-vent length had testes 11 mm. in length (left testis measured). The very large *B. blombergi* has testes that are proportionately enlarged. Our only information about the testes of *B. superciliaris* comes from the description of the hypertrophied testes in this species by Noble (1924), who speculated that this hypertrophy is related to reproductive potential. This theory lacks credence, since a very small testes can produce an excess of sperm.

In attempted hybridization of the males of these six species in which the testes are hypertrophied, there were quite different results than in other interspecific crosses in *Bufo*. Normally, there is development to gastrula even when very distantly related eggs and sperm are mixed, and even the sperm of other genera can be found to stimulate development to gastrula. However, in thirty-eight attempted crosses between these males (Appendix K-4) and females of a wide range of species, only eight survived to gastrulation. In two of the eight a few very abnormal neurulae were formed. There was no cleavage in twelve of the attempted crosses, although fertilization was usually high in the controls. In nine other attempted crosses, fewer than 10 percent of the eggs cleaved.

In some instances the jelly of the eggs turned white and the melanin of the eggs turned gray within the first hour. Abnormal cleavage was common, and, except for the cases noted above, development ceased prior to gastrulation. These clues indicate that something contained in the hypertrophied testes is having an effect on the eggs. However, the biological significance of this phenomenon must be cautiously interpreted until additional experiments can be made. There has been no chance to test the effect on the eggs in the in vivo discharge of sperm over eggs; thus there is no certainty that the results seen in the in vitro mixing of eggs and sperm would pertain under natural conditions. Second, it has not been possible to test the effects on homospecific eggs. Furthermore, the chemical agent remains to be identified.

Present evidence does demonstrate that the sperm suspension made from the crushed hypertrophied testes has a unique effect on eggs of numerous heterospecific females. If this effect does pertain in nature, it could function to reduce interspecific competition from larvae of other species when several species use the same breeding pool.

General Conclusions

1. The same dichotomy of the genus *Bufo* into broad-skulled and narrow-skulled toads is also found in general in the shape of the testes, with the narrow-skulled toads tending to have short, broad testes in which the width/length ratio is 30 percent or more. The broad-skulled toads, with few exceptions, have elongate testes in which the width/range ratio is usually much less than 30 percent.

2. Black testes occur in several species groups of the narrow-skulled line but are unknown in the broad-skulled line. Furthermore, F_1 hybrids between species groups of the narrow-skulled line and between them and representatives of the broad-skulled line have, in several instances, black testes. This suggests that a determiner for a melanin-coated testes exists as well as a suppressor, which fails to be effective under the heterozygosity of interspecies hybridization.

3. There is considerable intraspecific variation in color and proportions of the testes, but information is inadequate for determining whether this is a reflection of seasonal and physiological condition or of genetic difference.

4. In some species groups there is marked variation from species to species in color and shape of the testes, and, in the African species *B. regularis* (W), there is striking geographic variation, if this is indeed one biological species.

5. Hypertrophy of the testes has appeared in both the narrow-skulled, compact-testes line and in the broad-skulled, elongate-testes line. In the former line, this hypertrophy is known only in the species *B. gariespensis*, *B. inyangae*, and *B. rosei* of the *gariepensis group*. In the broad-skulled, elongate-testis line, it is known to have occurred in *B. blombergi*, *B. haematiticus*, and *B. holdridgei* of South and Central America and in *B. superciliaris* of Africa. Sperm suspensions from all males that have been tested inhibit normal development in heterospecific mixing of sperm and eggs, but additional work is needed to determine the biological significance of this phenomenon.

REFERENCES

Noble, G. K. 1924. Contributions to the herpetology of the Belgian Congo, based on a collection in the American Museum Congo Expedition 1909–1915. Bull. Amer. Mus. Nat. Hist. 49:147–347.

VI. Miscellaneous Bufo and significantly related forms: (1) *B. gariepensis*, (2) *B. rosei*, (3) *B. lönnbergi*, and (4) *B. vertebralis* of Africa; (5) ♀ *B. periglenes*, (6) ♂ *B. periglenes*, and (7) ♂ *B. holdridgei* of Costa Rica; (8) *Odontophrynus americanus* and (9) *Melanophryniscus* (= *Bufo*?) *stelzneri* from Argentina; (10) *Atelopus oxyrhynchus* from Venezuela; (11) *Dendrobates pumilio* from Costa Rica; (12) *Phyllobates* sp. from Peru; (13) *Pedostibes* from the East Indies; and (14–15) *B. typhonius* from Iparia, Peru.

18. Summary

W. FRANK BLAIR

The University of Texas at Austin
Austin, Texas

Introduction

In this final chapter, we will attempt to collate the multidisciplinary evidence presented in the preceding chapters as it provides answers to the main questions posed in chapter 1. The evidence from the diverse systems provides convincing answers to some of the questions. In other instances the evidence provides only tentative answers. Much is still to be learned of details of the relations of some species of *Bufo*, and there remains the important question of the relationships between the genus *Bufo* and other forms that are customarily treated as bufonid genera. Nevertheless, the broad pattern of evolution of the two hundred or more species of *Bufo* and the broad aspects of the spread of this genus to a nearly cosmopolitan distribution can now be reconstructed.

Where Did the Genus *Bufo* Originate?

The best estimate of place of origin of the genus *Bufo* is the New World and probably South America. The inadequate fossil record (Chap. 2) makes it necessary to turn largely to evidence from living forms. Nevertheless, important clues are provided by such fossil record as there is.

The oldest known frogs are from the Triassic of South America, and leptodactylid frogs were present there at least as early as the early Tertiary (Hecht 1963; Reig, Chap. 3).

The oldest known fossil that is possibly a *Bufo* is from the Lower Oligocene of Patagonia (Chap. 2). If indeed a *Bufo*, this fossil represents the narrow-skulled line, represented today by the *spinulosus* and other groups. A fossil form representing the broad-skulled type of *Bufo* is known from the Upper Miocene of the Magdalena Valley of Colombia. The oldest unquestionable *Bufo* is a member of the narrow-skulled lineage, is known from the Lower Miocene of Florida (Chap. 2), and is apparently ancestral to forms that occur there today. Thus, there is at least an implication that toads of the two divergent skull types existed in South America and the New World by the Late Miocene and perhaps much earlier.

Another important point of evidence is that the family Leptodactylidae, from which *Bufo* evolved, occurs in great diversity in the New World, especially in South America, and, with the exceptions of a questionable African genus (*Heleophryne*), occurs nowhere else except in Australia where there are no native *Bufo*. Poynton (1964) regards this African genus as unquestionably leptodactylid, but it has not been studied by modern methods of analysis.

The geographical conclusions of Metcalf (1923) from the study of anurans and their opalinid parasites, criticized by Darlington (1957) and others, are

Table 18–1. Number of Characters Shared among Ten Attributes

Species Groups													
valliceps													
spinulosus	7												
bocourti	5	3											
marmoreus	5	3	6										
alvarius	4	4	5	6									
cognatus	6	7	3	5	3								
boreas	6	5	2	7	5	8							
punctatus	2	2	3	6	3	5	5						
debilis	6	2	3	6	6	3	3	5					
viridis	5	3	4	4	5	6	4	7	6				
stomaticus	7	9	7	3	5	6	5	1	3	2			
bufo	6	7	3	3	4	8	6	4	4	5	6		
americanus	5	5	5	6	5	7	9	6	3	4	5	6	
calamita	3	3	5	6	4	6	6	7	3	6	2	6	6

Compared: Vocal type, presence or absence of bufoviridin and indole no. 20, presence or absence of chromosomal constrictions A, B, D, E, F, J, K. Narrow-skulled, North and Central American intermediates and *B. valliceps*. Comparisons with *B. debilis*, *B. punctatus*, *B. marmoreus*, and *B. calamita* are biased because data on indole no. 20 are absent in these species.

considerably parallel to the ones that have emerged in this book. This evidence argues against the past occurrence of leptodactylids in Eurasia and against a movement of bufonids from North America to South America in the Tertiary. Such movement would be expected if the bufonids had originated in the Old World and subsequently spread to the New World.

The various South American genera that are of prime importance in tracing the transition from Leptodactylidae to *Bufo* are just now receiving the attention they need. The rare Brazilian *Macrogenioglottus* (Chap. 3) is a highly significant intermediate form. *Melanophryniscus* is intermediate in vocal structures between Leptodactylidae and *Bufo* (Chap. 15), yet diploid hybrids were produced between *Bufo* and *Melanophryniscus* and survived to the larval stage (Chap. 11). This evidence alone is enough to strongly suggest South America as the area of origin of the genus *Bufo*.

Chromosomal evidence also strengthens the case for a South American origin of *Bufo* (Chap. 10). The chromosomal number 22, found in all *Bufo* examined except for one African complex, is common in South American leptodactylids. Furthermore, one chromosomal constriction seems to be shared among some *Bufo*, *Melanophryniscus stelzneri*, and *Odontophrynus americanus*. Other karyotypic evidence also points to a South American origin. Six of twelve secondary constrictions (Chap. 10) are found in at least some members of the broad-skulled and narrow-skulled lines of toads and in at least some of the intermediate forms. These are constrictions A, B, D, J, K, L. *Bufo crucifer* of Brazil, with constrictions A, B, D, J, K, has all but one of these constrictions, and no other species examined by Bogart even approaches this species in this respect (Table 18-1).

Supporting biochemical evidence comes from the transferrins. Among fifty-seven transferrins found in the *Bufo* examined (Chap. 14), only one is found in broad- and narrow-skulled and intermediate toads. This transferrin is the only one found in *B. crucifer*, which is the only broad-skulled toad in which it has been found.

Another argument for a New World origin of *Bufo* comes from the present existence in South America, Central America, and southern North America of intermediate forms that connect the two major evolutionary lines (narrow-skulled and broad-skulled) occurring on all major landmasses inhabited by the genus. *B. arenarum* is such a form when genetic compatibility data are considered (Chap. 11). Additionally, both osteological extremes of skull type are present in the Andean *spinulosus* group (Chap. 4). Other intermediate forms are found in Central America and southern North America.

No such intermediates between narrow- and broad-skulled lines have been found among the Eurasian and African species that have been studied. In fact, there is a sharp dichotomy in genetic compatibility between the broad-skulled and narrow-skulled Eurasian species studied and between African representatives of these two lines (Chap. 11). The world-wide pattern is one that would be expected if *Bufo* arose in the New World and subsequently spread to Eurasia and Africa. The persistence of intermediate forms between lineages that, from the fossil evidence, appear to have been separate since the mid-Tertiary seems incredible. However, the genetic compatibility of African and South American toads is indicative of the remarkable evolutionary conservatism of *Bufo*.

One additional bit of evidence is important. Hybridization data (Chap. 11) have shown that the radiation of *Bufo* in South and North America has been very closely linked, which implies relatively free interchange of *Bufo* between the two continents throughout the Tertiary.

The weight of the evidence summarized above strongly favors the New World as the place of origin of the genus *Bufo*.

Previous discussions about the area of *Bufo* origin have been largely speculative and were based on morphology alone or they have represented efforts to fit this taxon to general zoogeographical theories (e.g., Matthew 1939; Darlington 1957). Baldauf (1959) followed the concepts of Matthew and Schaeffer (1949) in assuming a palearctic origin, although the sparse evidence he cited is subject to almost any interpretation. Darlington (1957), like Matthew, has tried to fit zoogeography to theory. In a remarkably circuitous argument, he theorized that "the genus [*Bufo*] . . . has radiated from the Old World tropics." Although leptodactylids are unknown there, he argued that, since bufonids have been derived from leptodactylids, *then the former presence of leptodactylids is indicated* [italics mine].

We should ask how much evidence has accumulated during the researches on which this book is based that argues in favor of an Old World origin for *Bufo* rather than for a neotropical origin. There is slight evidence, none of which is unequivocal. The splitting of the African *Bufo* into numerous monotypic "species groups" might suggest a longer history for the African fauna than for the New World component, in which intermediate groups interconnect the two main evolutionary lines of *Bufo*. However, this splitting might reflect nothing more than differential rates of evolution and differential opportunities for speciation within the groups during Pliocene and Pleistocene climatic shifts.

Heleophryne in South Africa provides an explanation for the presence of African leptodactylids, if indeed this is a leptodactylid genus. However, the absence of leptodactylids from tropical Africa suggests that *Heleophryne* may be a sweepstakes invader of Africa rather than a relic of a long-existing African leptodactylid stock. Furthermore, it is hard to believe that the leptodactylids were ever in the African tropics, since this is largely a tropical and subtropical family. One would expect tropical or subtropical remnants if the family was ever there. Furthermore, the findings of Metcalf (1923), deprecated by Darlington (1957) and Baldauf (1949), argue against the presence of leptodactylids in Eurasia in the past and against a north-to-south movement of *Bufo* in the New World.

No great quantity of new evidence has become available on the relationship of the other Old World bufonid genera to *Bufo*. The biochemical evidence (Chap. 12) suggests that *Pseudobufo* is probably derived from *Bufo* of the same region. *Pedostibes* of Indonesia has the L chromosomal constriction that is in the Asian *B. melanostictus,* Eurasian *B. calamita,* African *B. mauritanicus,* and South American *spinulosus* group (Chap. 10), which suggests the origin of *Pedostibes* from *Bufo,* if it even deserves to be considered a separate genus. Laryngeal morphology supports this affinity (Chap. 15).

In sum, the accumulated evidence points to the neo-tropics as the area of origin of the genus *Bufo* and does not convincingly reinforce alternative theories of an Old World tropics area or a palearctic area of origin.

What Is the History of the World-Wide Spread of *Bufo*?

If we accept the strong evidence for an origin of *Bufo* from New World leptodactylid ancestors, the second major question of interest is: how was the present, nearly cosmopolitan, distribution achieved?

Major earth features that might be expected to influence the spread of *Bufo* are the following:

1. Isolation of the continental landmasses by water gaps at various times in the Tertiary and Quaternary.
2. Existence of climatic zones that might serve as barriers or alternatively as dispersal routes for groups with restrictive adaptation to climatic type, for instance, warm-tropical adaptation, as in the broad-skulled toads.
3. Shift of the climatic zones under the influence of Pliocene and Pleistocene continental glaciations and interglacial conditions.

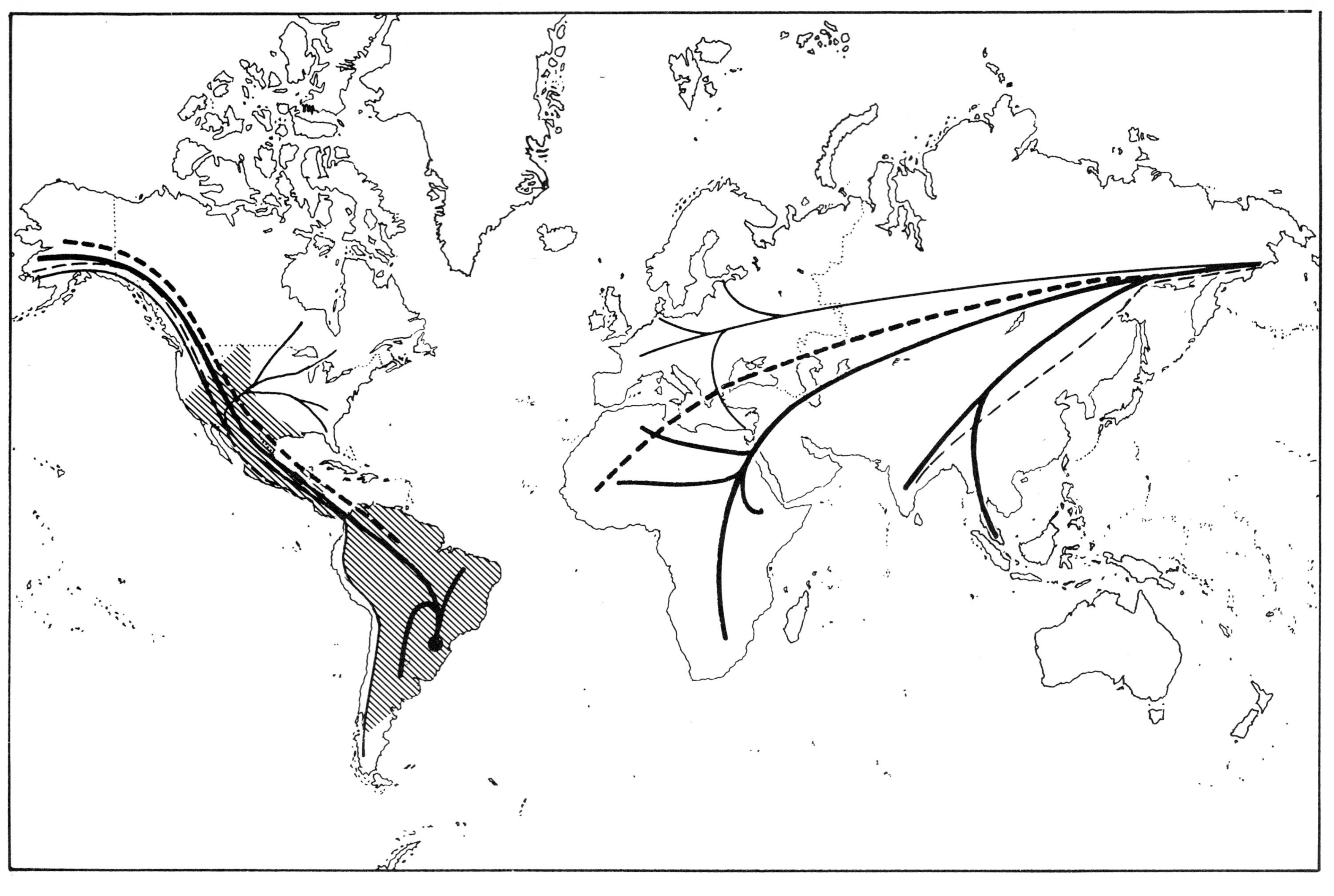

Figure 18–1. Possible lines of intercontinental exchange of *Bufo* as suggested by multidisciplinary evidence. Branchings of the various lines are symbolic of radiation and do not represent specific dichotomies. *Solid, heavy line*, broad-skulled toads (*marinus, valliceps, melanostictus, regularis* groups); *broken, heavy line*, broad-skulled toads (*guttatus* group of South America, *B. superciliaris* of Africa); *solid, light line*, narrow-skulled toads (*spinulosus, boreas, viridis* groups, etc.); *broken, light line, marmoreus* group–*stomaticus* group; *hatched area*, intermediate species known to be present; and *dot, Bufo crucifer*.

4. Existence of south-to-north-oriented mountain ranges, such as the Andean-Sierra Nevadan system, which might act as dispersal routes for cold-adapted types.

5. Existence of east-west-trending mountain chains, as in Eurasia, which would act as lethal traps for warmth-adapted types that might be forced southward by climatic cooling.

6. Continental mass and environmental diversity, which would provide a template for the adaptive radiation of the genus on the respective continents.

The evidence shows that there have been at least three, and possibly more, invasions of the Old World by *Bufo* from the New World.

One of these lines is represented by the cold-adapted and narrow-skulled toads that range from the cold Valdivian forests of the southern Andes northward to Ecuador (*spinulosus* group), recur in the highlands of Guatemala (*B. bocourti*), range through the Sierra Nevada and the Rocky Mountains of western North America to coastal Alaska (*boreas* group), range across Eurasia (*viridis* group, *B. calamita*, and *B. bufo*) to North Africa. Evidence from hybridization (Chap. 11, Fig. 11-2) points to the reality of this evolutionary line. If *Paleophrynus gessneri* (Chap. 2) is a *Bufo*, as seems likely, then the narrow-skulled toads had reached Europe by the middle Miocene.

There is consistent evidence from many sources for the reality of this evolutionary line. One feature of the line is the tendency toward reduction or loss of the mating call, especially in the New World forms (Chap. 15). The relictual species *B. canorus* of the Sierra Nevada of California, which retains the mating call, has a vocalization that is very similar to that of the Eurasian *B. viridis*. Both species have fairly similar color patterns and both show more sexual dimorphism than is usual in *Bufo*. Cei (Chap. 6) has noted the similarity in appearance of *B. rubropunctatus* of the Andean *spinulosus* group and the relictual *B. canorus* of the North American *boreas* group.

The biochemical evidence from the biogenic amines of the skin is highly consistent with the hybridization evidence. Bufotenine is present in all members of this line as well as in most broad-skulled and intermediate groups in the New World. This compound is lacking in all African species examined by either Low (Chap. 13), or Cei, Erspamer, and Roseghini (Chap. 12). The unidentified compound designated indole no. 20 by Low is found only in

Table 18–2. Shared State (presence or absence) of forty Biochemical Compounds of the Skin (Chap. 13)

valliceps												
spinulosus	17											
bocourti	30	24										
marmoreus	—	—	—									
alvarius	21	33	24	—								
cognatus	29	26	32	—	29							
boreas	17	33	21	—	38	27						
punctatus	—	—	—	—	—	—	—					
debilis	—	—	—	—	—	—	—	—				
viridis	24	3	6	—	0	4	27	—	—			
stomaticus	14	29	21	—	33	26	32	—	—	29		
bufo	20	24	25	—	26	25	24	—	—	36	29	
americanus	26	28	30	—	29	32	29	—	—	26	28	24

Narrow-skulled Asian, North, and Central American intermediates and *valliceps* group. Comparisons with *B. debilis*, *B. punctatus*, *B. marmoreus* and *B. calamita* are biased because of the absence of data on indole no. 20 in these species.

members of the narrow-skulled line (*spinulosus* group, *boreas* group, *B. alvarius*, *B. bufo*) and in a few closely related forms (*B. stomaticus*, and the *cognatus*, *quercicus*, *americanus*, and *coccifer* groups). Bufoviridin (Chap. 12) was found only in this line (*B. bocourti*, *B. viridis*, *B. calamita*) and in other closely related forms (*B. punctatus*, *B. alvarius*, and *debilis* and *marmoreus* groups).

Chromosomal evidence (Chap. 10) does not sharply distinguish this line when one considers the occurrence of secondary constrictions among broad-skulled, narrow-skulled, and intermediate toads. Six of the twelve secondary constrictions, as previously mentioned, are found in both broad- and narrow-skulled toads and in some intermediates. Two are found only in intermediates, two only in narrow-skulled and intermediate toads, and one only in broad-skulled and intermediate types. However, constrictions D, E, F, H, I, and J are lacking from the broad-skulled group, with the exception of *B. crucifer*.

Among the fifty-seven transferrins found (Chap. 14), only one was found in broad-skulled, narrow-skulled, and intermediate toads, and four additional ones were shared by the broad-skulled and narrow-skulled lines but were not found in intermediates. Seven were found only in narrow-skulled toads, and five others were found also in intermediate forms. Eight were found only in broad-skulled toads, and an additional seventeen were also found in intermediate toads. Sixteen transferrins were found only in the intermediate toads. Thus, transferrins sharply differentiate the two lines.

The testes are typically compact in the narrow-skulled line, and (Table 17-1) all except *B. alvarius* (which is classed as broad-skulled, Chap. 3) have a maximum width-to-length ratio of more than 30 percent, and most of the intermediate forms have the same type of testes. The broad-skulled toads, with the exception of *B. arenarum* and *B. mauritanicus,* have elongate testes. The narrow-skulled toads encompass a wide range of change in vocal mechanism, ranging from the presumably primitive Type II of the New World broad-skulled toads to Type IIIa of some North American and Eurasian species groups and Type IIIb of the *americanus* group and *B. punctatus* and *B. quercicus.* Furthermore, two Eurasian groups (*bufo, calamita*) have the specialized Type I mechanism that characterizes most African *Bufo.*

The second intercontinental transfer involved broad-skulled toads. These include the South American *marinus* group, which today ranges north to southern Texas, *B. melanostictus* of southern Asia, and, in Africa, *B. mauritanicus* and the large radiation of toads (*regularis* group, *maculatus* group, *B. steindachneri, B. perreti*) in which a pair of chromosomes has been lost. This distribution of broad-skulled toads could result from a crossing of the Bering land bridge to Asia in the Tertiary and subsequent spread from Eurasia to Africa. That these broad-skulled toads had reached Africa by Miocene time is shown by Upper Miocene fossils similar to *B. regularis* from Morocco (Chap. 2). The invasion of Eurasia must have come at a time of extensive tropical climate in the middle latitudes of the northern continents, for the broad-skulled toads seem to have remained consistently a warmth-adapted group. Elimination of this line from the northern middle latitudes could be attributed to the climatic shifts accompanying continental glaciations of the Pliocene or Pleistocene, which would account for the present broadly disjunct distribution of the broad-skulled toads.

Evidence from hybridization (Chap. 11) provides convincing support that the broad-skulled toads of South America, southern Asia, and Africa are broad-skulled because of affinity rather than because of convergence. Both the African complex and the Asian *B. melanostictus* have retained high genetic compatibility between their females and the males of the American broadskulls (Table 11-12), but there is incompatibility in the reciprocals of these crosses. Absence of any evidence of compatibility between the African and Eurasian broad-skulled toads could mean that both are derivatives from an invading broad-skulled ancestral stock or that there were two broad-skulled invasions rather than one.

The biochemical evidence from skin secretions is fairly complex. The skin of most New World broad-skulled, narrow-skulled, and intermediate toads contains bufotenine, as does the skin of the Asian broad-skulled *B. melanostictus.* However, all African species examined (Chap. 12, 13), including the broad-skulled toads, lack this compound. However, bufoviridin (Chap. 12) and indole no. 20 of Low (Chap. 13) are absent from all broad-skulled toads examined.

Other kinds of evidence for the distinctness of the broad-skulled line (chromosomal, blood proteins, vocal type, testes type) have been mentioned above in the discussion of the narrow-skulled line.

A third possible crossing from North America to Asia may be represented by *B. stomaticus* of southern Asia. R. Martin (Chap. 4) has classed this as a narrow-skulled toad. The high genetic compatibility between this and the two members of the Mexican *marmoreus* group (Chap. 11), in contrast with the low compatibility with other Eurasian species, suggests closer affinity between *stomaticus* and the *marmoreus* group than between *B. stomaticus* and its Eurasian congeners (Table 11-11).

Chromosomal evidence also supports the theory of a third crossing, because *B. stomaticus* shares with the *marmoreus* group a secondary constriction (B) that has been found in no other species outside the New World. Also, the call type (Type II) is similar in *B. stomaticus* and in the *marmoreus* group, while it is Type I in two of the Eurasian narrow-skulled toads (*B. bufo* and *B. calamita*) and Type IIIa in the third (*B. viridis*).

A fourth possible dispersal is suggested by the similarity between the large forest toad *B. superciliaris* of western Africa and *B. blombergi* of northwestern South America. In addition to their large size and "dead-leaf" color pattern these toads share hypertrophy of the elongate-type testes. The highest affinity in biogenic amines shown by *B. superciliaris* is with *B. haematiticus,* which, like *B. blombergi* is a member of the *guttatus* group (Chap. 13). However, *B. blombergi* displays highest affinity with *B. mauritanicus,* and *B. haematiticus* also shows high affinity with *B. brauni* of the *regularis* group. This suggests but does not prove that the similarities between *B.*

blombergi and *B. superciliaris* may be the result of convergent evolution in parts of a common ancestral stock.

What Is the Evolutionary History of *Bufo*?

It is possible to trace many aspects of the world deployment of *Bufo* from the available consensus data, but some aspects remain obscure.

Bufo crucifer of Brazil may be the most similar of all living *Bufo* to the ancestral form of the genus. The chromosomal number is the basic one of 2n=22 that is found in all known *Bufo* except an African radiation that has lost one pair (Chap. 10). Five of the six secondary chromosomal constrictions that occur widely in the genus (in broad-skulled, narrow-skulled, and intermediate types) are present in *B. crucifer*. Furthermore, the only transferrin known from broad-skulled, narrow-skulled, and intermediate toads is the only transferrin detected in *B. crucifer*, the only broad-skulled toad in which it has been found. Evidence from hybridization shows that this species is incompatible with narrow-skulled toads and virtually incompatible with broad-skulled toads, which implies marked genetic differentiation of both lines from the *B. crucifer* genotype. Evidence that the ancestral stock from which *Bufo* evolved was broad skulled (Chap. 3) adds strength to the case for *B. crucifer* as a primitive type.

The sparse fossil evidence for a very early dichotomy into broad-skulled and narrow-skulled lines has already been presented. When, in the Tertiary, this might have occurred cannot be established. It had to be earlier than the late Miocene; it was presumably much earlier than that. Reig (Chap. 3) suggests that anuran history has been generally underestimated because of the inadequate fossil record, and I suspect that he is correct.

In South America, the narrow-skulled line of toads became adapted to relatively cold or montane conditions, and this adaptation greatly influenced their subsequent spread to other continents. In addition, this adaptation was probably significant in limiting the radiation of this line in South America relative to that of the broad-skulled line. The distribution is principally Andean, with Pleistocene relicts in the Sierra Córdoba of Argentina. All are members of a single species group or, in the opinion of some coauthors, of two groups. The Valdivian species, *B. variegatus*, considered by some as distinct from the *spinulosus* group, has only the K secondary chromosomal constriction, but the same is true of *B. atacamensis* of the *spinulosus* group. Osteological characters (Chap. 4) suggest that *B. chilensis* and *B. limensis* are closer to the dichotomy of broad-skulled and narrow skulled toads than are other members of the group, including *B. variegatus*. The black testes of *B. chilensis* (Chap. 17) suggest closer affinity of this species than of the others with the narrow-skulled invaders of Eurasia, and there is higher genetic compatibility between it and *B. viridis* than there is between any other member of this group and *B. viridis*.

The South American broad-skulled toads, with primary adaptation to warm, tropical conditions, have occupied much of the landmass of that continent. In so doing, they have adapted to both wet and dry tropical and subtropical environments. The complex that includes *B. marinus* and related species ranges from Patagonia to the Rio Grande of North America. *B. arenarum* of this group is of particular interest because of its intermediacy in genetic compatibility between the narrow- and broad-skulled lines and because of its compatibility with the toads that are intermediate in skull type. Among the species of the *marinus* complex, *B. ictericus* and *B. paracnemis* are most compatible with *B. arenarum*, while *B. marinus* and *B. poeppigi* represent a separate branch within this group (Chap. 11). These last two, although long confused by taxonomists, differ in transferrins (Chap. 14), in parotoid-gland secretions (Chap. 13), and in morphology.

The evolutionary position of *B. typhonius* is uncertain. The limited hybridization data suggest distant relationships with the *marinus* and *granulosus* groups (Chap. 11). In external appearance and osteological characters *B. typhonius* is rather similar to another small species, *B. haematiticus*, which may be its nearest relative. *B. blombergi*, like *B. haematiticus* a member of the *guttatus* group, is the only South American toad examined other than *B. typhonius* in which bufotenine is lacking, which suggests the possibility of closest relationships of *B. typhonius* to the *guttatus* groups.

The evolutionary position of the broad-skulled *guttatus* group is unknown. Some substance in the hypertrophied testes of *B. blombergi*, *B. haematiticus*, and *B. holdridgei* has inhibited development and prevents the assessment of the genetic compatibility of these toads with other species (Chap. 17). High affinity in skin secretions (Chap. 13) between these toads and such African broad-skulled toads as *B. mauritanicus*, *B. superciliaris*, and members of the *regularis* group implies that the *guttatus* group branched from the line of broad-skulled toads that penetrated Africa, but it provides no information about the time of the branching in connection with other dichotomies. *B. haematiticus* shows great osteological similarity to *B. asper* of southeastern Asia (Chap. 4).

One striking feature of *Bufo* evolution in South America is the persistence there, as well as in Central America and southern North America, of species intermediate in genetic compatibility between the broad- and narrow-skulled toads. *B. arenarum,* which has retained compatibility with the *marinus* group, is one of these intermediate species, and it provides a connecting link between the South American *Bufo* and various intermediate types in Central and North America. The second intermediate group in South America is the widely distributed *granulosus* group. Hybridization data indicate that this group must have branched from the line leading from *B. arenarum* to *B. valliceps* and the Central and North American intermediates rather than from a line leading from *B. marinus* and *B. poeppigi.* The B secondary constriction, common in intermediate and narrow-skulled toads, but known only in *B. crucifer* of the broad-skulled toads, occurs in *B. granulosus,* and its presence adds weight to this conclusion.

The broad-skulled toads must have extended northward in at least the western part of North America in the Tertiary and crossed the Bering land bridge to Asia in a period of extensive tropical or subtropical climates in the northern mid-latitudes. Today, *B. marinus* of this line extends northward only to the Rio Grande in Texas. The present northern limit on this tropic-adapted species is presumably a climatically imposed one, and it probably represents the maximum geographic recovery from a southward displacement in the Wisconsin stage of the Pleistocene.

The narrow-skulled line is represented in the highlands of Guatemala by *B. bocourti.* Hybridization and biochemical data, as well as osteology, point to the affinity of this isolated species with the narrow-skulled line, even though there is an apparent gap from Ecuador to Guatemala in the present geographic distribution of this line. The vocal mechanics during release calling (*B. bocourti* has apparently lost its mating call: Chap. 15) are Type I, as in *B. bufo* and *B. calamita* during mating calling, rather than Type III, as in *B. viridis* (Chap. 15). This same line is represented in western North America by the *boreas* group, which is distributed largely in the Rocky Mountains and in the Sierra Nevada and other western ranges. In the *boreas* group, *B. canorus* seems the most primitive, since it has a mating call (very similar to that of the Eurasian *B. viridis*) while *B. boreas, B. exsul,* and *B. nelsoni* have lost the mating call.

Several species intermediate between the broad-skulled and narrow-skulled toads occur in Central America and southern North America. The evolution of these intermediate forms has been very close to that of the narrow-skulled toads, and a remarkable amount of genetic compatibility has been retained between the narrow-skulled and the intermediate species groups.

The *valliceps* group of Central America, Mexico, and the western Gulf of Mexico coast of the United States is like *B. arenarum* in being in many ways intermediate between broad- and narrow-skulled toads although both are osteologically broad skulled. The two complexes are surely members of the same evolutionary branch. There is high compatibility between them (Table 11-10), and both show at least some metamorphosis in both reciprocals of crosses with the *granulosus* and *spinulosus* groups and with *B. alvarius,* which, itself, seems to be an old, relictual species. Closer affinity of *B. arenarum* than of the *valliceps* group with *B. marinus* is suggested by reciprocal metamorphosis between *B. arenarum* and *B. marinus* but not between the members of the *valliceps* group and *B. marinus.* Both chromosomal constrictions present in *B. arenarum* are also found in the *valliceps* group, which shows the largest number found in any group. The *valliceps* group is like *B. arenarum* in lacking bufoviridin and indole no. 20 but it differs from that species in being one of the few New World groups lacking bufotenine.

The vocal mechanics (Chap. 15) in the *valliceps* group are variable, with some species (*B. valliceps, B. mazatlanensis, B. gemmifer,* and *B. luetkeni*) having the primitive Type II vocal mechanism, while others (*B. coniferus, B. cavifrons,* and *B. ibarrai*) are intermediate (Type IIIa) between this type and the Type IIIb mechanism (arytenoid valve lost) of the *americanus* group.

The *valliceps* group is also highly variable in karyotype. *B. coniferus* has only the primitive South American K constriction. This feature, as well as the southernmost geographic position of the species within the group, might suggest that it is the oldest member of the group. *B. valliceps* has a large number

Table 18–3. Characters of Narrow-Skulled and Intermediate Toads that Show Various Degrees of Genetic Compatibility

Narrow and Intermediate-Skulled Toads Hybrid Combinations ♀ x ♂	Skull Narrow (N) Skull Intermediate (I) Skull Broad (B)		Fused (F) Unfused (U)		Highest % F_1 Met.	Highest % Recip. F_1 Met.	Shared Characters of 10 from Table 18-1	Shared Characters of 40 from Table 18-2	Weighted Affinities from Table 13-10
valliceps x *marmoreus*	B	I	F	F	68.3	93.1	5	—	—
valliceps x *cognatus*	B	I-N	F	U	92.0	72.6	6	29	66
valliceps x *alvarius*	B	B	F	F	21.7	14.3	4	21	10
valliceps x *spinulosus*	B	B-N	F	F	72.8	3.7	6	17	—
valliceps x *americanus*	B	N	F	U	5.2	72.2	5	27	52
marmoreus x *cognatus*	I	I-N	F	U	26.2	91.1	5	—	—
marmoreus x *alvarius*	I	B	F	F	26.3	81.7	6	—	—
marmoreus x *boreas*	I	N	F	U	—	66.7	7	—	—
marmoreus x *spinulosus*	I	B-N	F	F	62.3	30.9	3	—	—
marmoreus x *stomaticus*	I	N	F	F	74.0	66.7	4	—	—
cognatus x *boreas*	I-N	N	U	U	30.7	65.0	8	27	32
cognatus x *alvarius*	I-N	B	U	F	26.7	81.1	3	29	32
cognatus x *viridis*	I-N	N	U	F	55.9	22.1	6	24	20
cognatus x *debilis*	I-N	I	U	F	33.8	G	3	—	—
boreas x *alvarius*	N	B	U	F	83.3	9.0	5	38	92
boreas x *punctatus*	N	I	U	F	66.7	47.2	5	—	—
boreas x *viridis*	N	N	U	F	14.8	1.9	4	27	48
boreas x *americanus*	N	N	U	U	6.9	79.9	9	29	46
alvarius x *viridis*	B	N	F	F	6.3	30.2	4	31	52
punctatus x *stomaticus*	I	N	F	F	13.0	9.2	2	—	—
spinulosus x *stomaticus*	B-N	N	F	F	10.0	5.5	8	29	35

Comparisons with *B. debilis*, *B. punctatus*, *B. marmoreus*, and *B. calamita* are biased because data on indole no. 20 in these species are absent.

of secondary constrictions. *B. luetkeni* has only two constrictions, and these apparently correspond to the two constrictions that are found in *B. woodhousei* of the *americanus* group.

Two Mexican species, *B. occidentalis* and *B. canaliferus*, are close to the *valliceps* group and represent separate branches from a *valliceps*-group-like ancestor. Both are compatible in reciprocal crosses with the *valliceps* group. The former is classed as intermediate in skull width and the latter is classed as broad skulled (Chap. 4); in both the proötic is fused as in the *valliceps* group. *B. occidentalis* has the Type II vocalization as in a part of the *valliceps* group. *B. canaliferus* has also Type II (although divergent) but has a tendency to produce parts of the call by the Type IIIa mechanism as in some members of the *valliceps* group.

In their karyotypes, both share the primitive A constriction, which is also present in some members of the *valliceps* group. *B. occidentalis* has the F constriction, which is questionably present in *B. valliceps* but is otherwise restricted to the narrow-skulled line and near relatives of that line. *B. canaliferus* has the C and E constrictions, the first of which is present in *B. valliceps* and the second of which is questionably present there. These two constrictions are otherwise restricted to the narrow-skulled line and its near relatives. In the forty compounds reported from the parotoid glands of *Bufo* (Chap. 13), both species share thirty-one of forty possible conditions of presence or absence of compounds with the *valliceps* group. On the basis of weighted affinities (Chap. 13), those between *B. canaliferus* and the *valliceps* group (75–81) and between *B. occidentalis* and the *valliceps* group (70–77) are of the same order of magnitude as those between members of the *valliceps* group (73–86). Like the *valliceps* group, these two species show high weighted affinities with the

Table 18–4. Concordance in Characters among Narrow- and Intermediate-Skulled Toads

	Vocal Type	Proötic	Skull Width	Bufoviridin	Indole No. 20
	II	Unfused	Narrow	Absent	Present
boreas	yes	yes	yes	yes	yes
cognatus	yes	yes	part	yes	yes
stomaticus	yes	no	yes	yes	yes
americanus	no	yes	yes	yes	yes
quercicus	no	yes	yes	yes	
coccifer	no	no	no	?	yes
bufo	no	part	yes	yes	yes
	III	Fused	Intermediate	Present	Absent
alvarius	yes	yes	yes (B)	yes	no
viridis	yes	yes	yes	yes	yes
valliceps	yes	yes	yes (B)	no	yes
punctatus	yes	yes	yes	yes	?
debilis	yes	yes	yes	yes	?
marmoreus	no	yes	yes	yes	?
calamita	no	yes	no	yes	?
	loss	Fused	Intermediate	Present	Absent
spinulosus	yes	yes	part	no	no
bocourti	yes	yes	yes	yes	yes

B - broad-skulled.

americanus and *cognatus* groups but very low affinities with the *boreas* group and with *B. alvarius.*

All the evidence indicates that *B. occidentalis* and *B. canaliferus* are independent branches from *valliceps*-like toads, which implies that they are not involved in the transition from the broad-skulled *valliceps* group to the narrow-skulled type of toad.

If one starts with the *valliceps* group and examines the transition to narrow-skulled toads, as indicated by genetic compatibility, two pathways of change are suggested (Fig. 11-2). One leads from the *valliceps* group, then through the *cognatus* group to the *boreas* group to *B. punctatus*, or through the *cognatus* group to the Eurasian *viridis* group. The other leads from the *valliceps* group through the *marmoreus* group to *B. alvarius* on through the *marmoreus* group to the Eurasian *stomaticus* group. However, both *B. alvarius* and toads of the *marmoreus* group show high compatibility with the *cognatus* group, and the *marmoreus* group also shows some compatibility in reciprocal crosses with the South American *spinulosus* group, comprised mostly of narrow-skulled toads.

If we look at five key characters (Table 18-4) we can see that the subdivision into two lines is partly corroborated by correspondence of species groups in these characters. The *cognatus* and *boreas* groups share the primitive Type II vocal mechanism, unfused proötic, narrow skull, absence of bufoviridin, and presence of indole 20. The *americanus* group shares all these attributes except that it has evolved to a Type IIIb vocal mechanism. The Eurasian *B. stomaticus* fits with this group in all characters except its fused proötic.

Only *B. viridis* shows all the alternative characters of the second group. However, *B. alvarius* and *B. punctatus* are consistent in all five characters except that the first has indole no. 20, and the presence or absence of this compound in *B. punctatus* is unknown. *B. bocourti* fits here in four characters, but the vocal mechanism cannot be determined, because the mating call has been lost.

The *marmoreus* group fits with the second set of species groups in fused proötic, intermediate skull width, and presence of bufoviridin. It shares vocal Type II with the first group and shows fairly high genetic compatibility with *B. cognatus* of that group, and its compatibility is even higher with *B. stomaticus* of that group. The mixture of characters in the *marmoreus* group is consistent with its position of crossability with both group I and group II toads (Fig. 11-2).

The *debilis* group fits with the second group in its Type III vocalization, fused proötic, intermediate skull width, and presence of bufoviridin. The condition for indole no. 20 is unknown. In the ten states compared in table 18-1, it shares six with the *viridis* and *alvarius* members of the group II, and five with *B. punctatus,* and three with *B. bocourti.* Five states are shared with the *bufo* and *americanus* groups, three with the *boreas* and *cognatus* groups, and only two with the *stomaticus* group of group I.

The *americanus* group is a North American radiation that is clearly definable. Its origins must have occurred near the time that the line branched that led to the *cognatus* and *boreas* groups, because there are almost equal indications of affinity with the *valliceps* group and the *boreas* group. There is metamorphosis of reciprocal hybrids between members of the *americanus* group and members of both the *valliceps* and *boreas* groups. In both instances the survival is low and approximately equal when the male parent is of the *americanus* group. However, a hybrid male between *B. luetkeni (valliceps* group) and *B. microscaphus (americanus* group) was backcrossed and found to be fertile. This result and the presence in both *B. woodhousei* and *B. luetkeni* of the A and possibly the B secondary chromosomal constriction point to the *valliceps*-group-like level of origin of the *americanus group.* However, the *americanus* group shares nine of ten possible states with the *boreas* group, versus only five with the *valliceps* group (Table 18-1). There is little difference in biochemical compounds in the skin. The *americanus* group shares twenty-eight states with the *valliceps* group, and thirty with the *boreas* group. Some members of the *valliceps* group have the Type IIIa vocal mechanism that is a precursor of the Type IIIb that characterizes the *americanus* group. *B. praevius* from the lower Miocene of Florida (Chap. 2) may well be an ancestor of the *americanus* group.

B. quercicus seems to be a derivative of the *americanus* group, or probably of the line, as represented by the Miocene *B. praevius,* which presumably led to that group. It is chromosomally like the *americanus* group in having the A or D and questionably the B constrictions. The vocal mechanism is Type IIIb, as in the *americanus* group. The external appearance is that of a diminutive member of this group. F_1 hybrids have been produced between females of three species of the *americanus* group and males of *B. quercicus.* F_1 hybrids between female *B. quercicus* and male *B. americanus* survived to the larval stage.

B. coccifer is difficult to place. Genetic compatibility data suggest that it is most closely related to the *americanus* group, which is the only group with which it has been found to form hybrids that metamorphose in both reciprocals of the cross. The vocal mechanism is Type IIIa, but the arytenoid valve shows sign of atrophy, which would be a state leading to the IIIb mechanism of the *americanus* group. The A secondary chromosomal constriction is present, as in the *americanus* group, and the H constriction is present, as in the *boreas, debilis,* and possibly the *valliceps* groups. Alternatively, both the parotoid-gland secretions and osteology show greater affinity with the *valliceps* group than with the *americanus* group (Chap. 13).

Evolutionary relations of the Eurasian toads are less clear than those of the New World toads. *B. melanostictus* is clearly an Old World derivative of a broad-skulled line that crossed the Bering land bridge from east to west and moved on into Africa. *B. melanostictus* has the primitive A chromosomal constriction, which traces all the way back to the South American *B. crucifer* and which occurs in both broad- and narrow-skulled lines. It also has the L constriction, which is also in the *spinulosus* group of South America and in the narrow-skulled *B. calamita* of Eurasia and broad-skulled *B. mauritanicus* of North Africa. Absence of this constriction from the North American radiation and from the South American broad-skulled toads suggests a very early crossing of ancestral forms of all these toads to Eurasia. *B. melanostictus* is like the New World broad-skulled toads, since it has a Type II vocal mechanism and bufotenine but no bufoviridin or indole no. 20 in the skin. The external appearance is rather similar to *B. valliceps,* including the yellow throat color of adult males. There is very strong osteological similarity to the *valliceps* group (Chap. 4).

The four narrow-skulled Asian or Eurasian species about which we have information seem distinct from one another. As mentioned above, *B. viridis* and *B. stomaticus* represent parts of two separate evolutionary lines that can be traced from North America to Eurasia. It is significant that *B. stomaticus* and the two members of the *marmoreus* group show higher genetic compatibility (Table 18-3) in both reciprocals of crosses than do members of the *marmoreus group* and other New World species or *B. stomaticus* and the other Old World species. In fact, F_1 hybrids were inviable between female *B. stomaticus* and males of *B. viridis, B. bufo,* and *B. calamita* (Table

18-5). In addition, *B. stomaticus* shows high to low compatibility in crosses of its females with males of two North American groups (*marmoreus*, 66% metamorphosis, and *punctatus*, 9% metamorphosis) with which *B. viridis* forms only inviable hybrids. Conversely, *B. stomaticus* females form only inviable hybrids in crosses with North American narrow-skulled toads (*boreas, cognatus, alvarius,* and *debilis* groups) in which at least some metamorphosing hybrids in the comparable cross involving females of *B. viridis* are produced. *B. viridis* and *B. stomaticus* are similar only in producing inviable hybrids in crosses of their females with males of the *americanus* group (Table 18-5).

All the genetic compatibility data point to rather distant affinities among these four narrow-skulled species. Inviable hybrids result from crosses of female *B. calamita* with males of *B. viridis* and *B. bufo* (Table 18-5). These results compared with the results from crosses of male *B. viridis, B. calamita,* and *B. stomaticus* with the North American species implies that *B. bufo* is farther removed from the New World radiation than the other three are. Morphological characters (Chap. 8) support this conclusion.

In terms of the five characters treated in table 18-4, *B. bufo* fits in group I, along with *B. stomaticus,* differing only in the specialized Type I vocalization by which it resembles the African radiation. Of the ten attributes in which *Bufo* are compared (Table 18-1), *B. bufo* shows the highest affinity with the *cognatus* group (group 1), with which it shares eight states. Its second highest affinity (7 shared states) is with the intermediate South American *spinulosus* group. *B. bufo* also shows equally high affinities (6 shared states) with the New World toads of the *valliceps* and *boreas* groups, as it does with *B. viridis* (5), *B. calamita* (6), and *B. stomaticus* (6).

In the biochemical characters treated in table 18-2, *B. bufo* shows the highest affinity with *B. viridis,* sharing thirty-six of forty possible states.

B. calamita does not fit clearly in either group I or group II as based on the characters compared in table 18-4. The vocal mechanism is the specialized Type I, as in *B. bufo.* The skull is narrow, but the proötics are fused. Bufoviridin is present. In the ten characters compared in table 18-1, *B. calamita* shows the highest affinity with *B. punctatus* (7 shared states), and it shares six with *B. bufo* (group I) and with the New World *cognatus, boreas,* and *americanus* groups (group I) and with two group II species groups, *viridis* and *marmoreus.*

The southeastern Asian *B. asper* is difficult to relate to the other toads. It is quite removed from most other *Bufo* with which it has been tested for genetic compatibility. No hybrid between female *B. asper* and a long series of species (Chap. 11) developed beyond the larval stage, and most failed at gastrulation. In crosses of male *B. asper,* the only diploid hybrids that metamorphosed (25%) were in a cross with a female of the *americanus* group. This suggests that *B. asper* is closer to the *americanus* group than are the extreme, broad-skulled South American toads, such as *B. marinus,* the broad-skulled *B. melanostictus* of Asia, or the African broad-skulled toads. However, *B. asper* is very similar to *B. haematiticus* in skull characters and in its nearly unique vertebral characters (Chap. 4).

The skull is of the broad type, with unfused proötic. Bufoviridin is absent. In skin biochemistry *B. asper* differs strikingly from the other *Bufo* studied, since it virtually lacks indolealkylamines. The same is true of *B. juxtasper* and *Pseudobufo subasper* Chap. 7), which suggests that *P. subasper* may be derived from the *B. asper* group.

The relation of the evolution of the African toads to that of the toads of other continents is difficult to assess. For one thing, the African *Bufo* have been more isolated than the toads of other continents, which is reflected in the lack of compatibility in reciprocal crosses between African toads and toads of other continents (Tables 11-9, 11-10, 11-11) and in the very high incompatibility between African and Eurasian toads. No viable, normal diploid hybrids have been produced between toads of these two landmasses (Fig. 11-3). The only survivors of attempted crosses have been polyploids or gynogenetic diploids.

The one African group—a large one—that can be related with confidence to *Bufo* evolution on other continents is the large complex of 20-chromosome toads of the *regularis* group and *B. maculatus, B. steindachneri,* and *B. perreti.* As discussed earlier, the high compatibility between females representative of this complex of broad-skulled toads and South American broad-skulled toads, as well as biochemical evidence, places these toads as derivatives of an early dispersal of broad-skulled toads from the New World. In addition to having a different chromosome number (Chap. 10), these toads have

Table 18–5. Stage Reached or Percentage Metamorphosis

♀	♂										
	viridis	*stomaticus*	*calamita*	*bufo*	*boreas*	*cognatus*	*alvarius*	*americanus*	*marmoreus*	*debilis*	*punctatus*
viridis		69	5	6	2	22	24	N	L	8	L
stomaticus	L		L	L	L	N	L	G	66	L	9
calamita	L*			G				G			
bufo											
boreas	15	4	6	5							
cognatus	56	2	33	L							
alvarius	6		L	G							
americanus	48	53	52	15							
marmoreus	L	74	N	G							
debilis											
punctatus	84	13		L							

Crosses among the four narrow-skulled Eurasian species and crosses with North American narrow-skulled toads.

*Flindt & Hemmer (1967)

evolved a peculiar vocal mechanism (Type I) that they share with the narrow-skulled *B. bufo* and *B. calamita* of Eurasia and *B. gariepensis* of Africa. *B. perreti* is Type I and II.

Whether the broad-skulled *B. mauritanicus*, with 22 chromosomes, and *B. superciliaris*, with unknown karyotype, represent branches of the same invasion that led to the 20-chromosome toads is uncertain. The remarkable similarity between *B. superciliaris* and members of the South American *guttatus* group has already been mentioned. There has been little testing of the genetic compatibility of B. *mauritanicus* with other toads, because of inability to obtain females. There are indications, however, that *B. mauritanicus* is indeed close to the 20-chromosome toads and thus represents a less-differentiated descendent of the invasion that was responsible for the 20-chromosome toads. There was small percentage metamorphosis of the hybrids when males of *B. mauritanicus* were crossed with females of *B. regularis* (W) and *B. maculatus* (Chap. 11). Hybrids between females of the *americanus* group and *B. mauritanicus* males failed at gastrulation as would happen if the *americanus*-group female were crossed with a *regularis*-group, *marinus*-group, or *B. melanostictus* male. Pentaploid hybrids resulted from a cross of a female *B. viridis* with a male *B. mauritanicus* (Chap. 10).

B. mauritanicus has the Type II vocal mechanism typical of South American broad-skulled toads and some North American and Asiatic groups. The *regularis*, *maculatus*, and *gariepensis* groups have a specialized Type I vocal mechanism, but the 20-chromosome *B. perreti* changes from Type I to Type II during the mating call.

The chromosomal evidence could indicate the 20-chromosome toads and *B. mauritanicus* branched from a common ancestor. The latter has the primitive L secondary constriction that also occurs in the *spinulosus* group and in *B. melanostictus* and *B. calamita* of Eurasia. Fusion of the K and L constrictions could have resulted in the R constriction characteristic of the *regularis* group (Chap. 10), but, if so, the K was subsequently lost in *B. mauritanicus*. The K constriction is typical of New World broad-skulled toads and also of the South American *spinulosus* group and the Asiatic narrow-skulled *B. stomaticus*.

Strong osteological and biochemical similarities point to close relationship between *B. mauritanicus* and the 20-chromosome toads. All are similar, as is *B. superciliaris*, in lacking bufotenine in the skin. On the basis of weighted affinities (Chap. 13) in parotoid-gland secretions, the relationship is fairly close. Weighted affinities between *B. mauritanicus* and the 20-chromosome toads range from 60 percent with *B. maculatus* (= *B. pusillus* of Chap. 8) to 82 percent with some populations of *B. regularis*. This compares with a range of 72 to 98 percent within the *regularis* group and of 73 to 98 percent between *B. superciliaris* and the *regularis* group.

The 20-chromosome toads have undergone considerable differentiation. *B. perreti* is virtually incompatible with the toads of the *regularis* group, and *B. maculatus* is incompatible. *B. maculatus* lacks the R chromosomal constriction typical of the *regularis* group. It does have one constriction that is peculiar to itself and to *B. brauni* and *B. garmani*. In parotoid-gland secretions, *B. maculatus* and *B. perreti* show high affinities with the toads of the *regularis* group (Table 13-9), but they are of the same order of magnitude as those between the 22-chromosome, narrow-skulled African toads and the *regularis* group. The vocal mechanism is Type I in all of the 20-chromosome toads, but *B. perreti* changes in the course of a call from Type I to the Type II that is found in *B. mauritanicus* (Chap. 15).

The narrow-skulled, 22-chromosome toads of Africa are difficult to relate to other *Bufo*. The one that is distinctly different from other *Bufo* seems to be *B. carens*. No hybrid ever developed to a feeding larva in crosses of female *B. carens* with three members of the *regularis* group, *B. maculatus*, and *B. perreti* among the 20-chromosome toads or with the African narrow-skulled *B. gariepensis*. The same was true of crosses of female *B. carens* with four Eurasian narrow-skulled toads and the broad-skulled *B. melanostictus*, and of the cross of a male *B. carens* with a female *B. viridis*. Likewise, crosses between female *B. carens* and males of fourteen New World narrow-skulled toads and six species of broad-skulled toads never developed to feeding larva. The same was true in the cross of a male *B. carens* with the broad-skulled *B. valliceps*.

B. carens is highly differentiated in its skin compounds. The 5-HT, characteristic of the 20-chromosome toads and *B. mauritanicus*, is lacking in *B. carens*, which also differs in the lack of N-methyl transferase activity and in the high activity of conjugases. *B. carens* is unique in having a 1-conjugated derivative of tryptamine (Chap. 12). It is also unique in the morphology of the parasphenoid bone (Chap. 4), which further reinforces the conclusion that this species has evolved far from other species of *Bufo*. The vocal mechanism is Type I, as in the 20-chromosome toads, but this is of little significance, since this type appears also in the Eurasian *B. bufo* and *B. calamita*.

The *gariepensis* group also remains puzzling in relationships. No chromosomal constrictions were found. Genetic compatibility has had no adequate testing. Females have been unavailable, and the inhibitory substance in the hypertrophied testes of the males makes it impossible to test their genetic compatibility with other species groups. Superficially, *B. gariepensis* shows similarity to *B. viridis*. The skull is of the narrow type, with fused proötic. The data from parotoid venom are interesting but of little help. Weighted affinities of *B. gariepensis* and *B. inyangae* are relatively high and of the same order of magnitude with other African *Bufo*, both broad and narrow skulled, and with most New World groups of both skull types. However, affinities are strikingly low with the South America *spinulosus* group, the North American *boreas* group and *B. alvarius*, and the Eurasian narrow-skulled *B. bufo*, *B. viridis*, and *B. calamita* and the broad-skulled *B. melanostictus*. Like the broad-skulled African toads, the *gariepensis* group lacks bufotenine and indole no. 20 (Chap. 13), and 5-HT pattern is shared with them. The biochemical evidence thus argues strongly against the deriving of this group from the line of narrow-skulled toads represented by *B. viridis*.

B. lönnbergi shows no evidence of close relationship with any other species. This narrow-skulled toad has the Type II, primitive vocal mechanism. In the karyotype, the only secondary constriction identifiable is the widely distributed A constriction. In crosses between males of *B. lönnbergi* and females of three species of the regularis group and of *B. marinus* there was failure of the hybrids prior to hatching. *B. lönnbergi* closely parallels the *gariepensis* group in its low affinity with the Eurasian toads and with the *spinulosus* and *boreas* groups and *B. alvarius* when parotoid-gland secretions are compared.

There is very little information on the relationships of the small African *B. vertebralis* and *B. lughensis*. The only evidence comes from hybridization. In crosses of female *B. vertebralis* with males of African

B. regularis (W) and *B. gariepensis,* the hybrids failed at gastrulation, and the same results were obtained in crosses with the Eurasian *B. bufo,* North American *americanus* and *valliceps* groups, and the South American broad-skulled *B. marinus* and *B. arenarum.* It is probably significant that development proceeded to larva in crosses with the *viridis, boreas, cognatus, marmoreus,* and *debilis* groups and with *B. punctatus.*

Major Unanswered Questions

Although the multidisciplinary approach has provided a broad view of the world-wide radiation of the genus *Bufo,* there remain a number of troublesome and interesting questions.

1. What is the meaning of the similarity of both broad-skulled and narrow-skulled African toads in their skin venoms? The broad-skulled African toads are identifiable as a radiation from immigrant broad-skulled toads from the New World. No such clear relationship of African and other narrow-skulled toads has been detected. Is it possible that the African narrow-skulled toads are independently derived from the broad-skulled ancestor that provided the basis for the extensive African radiation of broad-skulled toads? The low biochemical similarity of the African toads of both skull types with Eurasian toads and with several New World narrow-skulled toads suggests this possibility. More probably, the biochemical similarity between the African toads and most New World narrow-skulled and broad-skulled toads suggests that the African narrow-skulled toads could be the result of a dispersal of narrow-skulled toads wholly independent of the Asian and Eurasian toads represented in this study. The biochemical similarities of the African toads are, however, still unexplained. Answers to these questions are dependent on adequate examination of African narrow-skulled species by the methods represented here.

2. The number of entries of broad-skulled toads into Africa remains to be clarified. There are striking similarities between *B. superciliaris,* of presently unknown karyotype, and the 20-chromosome toads, but there are even stronger similarities between this species and toads of the South American *guttatus* group.

3. The number of interchanges of narrow-skulled toads between North America and Asia needs clarification. There seems to have been at least two. Were there more? Ultimate availability of gravid female *B. bufo* and *B. calamita* should contribute to the answer, and studies of other Asian species will do likewise.

4. What, if any, is the significance of the occurrence of montane isolate species of small narrow-skulled toads on all continents? Are these remnants of another dispersal of narrow-skulled toads, or do they represent relictual species of the same dispersal that brought *B. viridis* to Eurasia?

5. The relationships of the various bufonid genera need to be examined by the modern techniques of analysis that we have used for the genus *Bufo.* The same may be said in general for the higher categories of anurans. Most of the progress in anuran classification and evolutionary relations of recent years has been made at the level of species and species complexes. However, we continue to use an archaic system of higher categories based on little more than morphology. A modern approach to the relations of the higher anuran categories will certainly result in some major changes in the classificatory system and may hopefully clarify some presently existing puzzles in anuran phylogeny and biogeographical relations.

REFERENCES

Baldauf, R. G. 1959. Morphological criteria and their use in showing bufonid phylogeny. J. Morph. 104: 527–560.

Darlington, P. J., Jr. 1957. Zoogeography: The geographical distribution of animals. John Wiley and Sons, New York.

Hecht, M. K. 1963. A reevaluation of the early history of the frogs. Part II. Syst. Zool. 12:20–35.

Matthew, W. D. 1939. Climate and Evolution. Second ed. New York Acad. Sci., New York.

Metcalf, M. M. 1923. The origin and distribution of the Anura. Amer. Nat. 57:385–411.

Poynton, J. D. 1964. The amphibia of Southern Africa. Ann. Natal Mus. 17:1–334.

Appendix A. Species Examined in Preparation of Chapters 4, 10, 11, 13, 14, and 15

Species	Number of Specimens					
	Chap. 4	Chap. 10	Chap. 11	Chap. 13	Chap. 14	Chap. 15*
Crucifer group						
B. crucifer						
Rio de Janeiro, Brazil	4	6	24	4	1	
Campo Grande, São Paulo Brazil			3		4	
Brazil					1	6
Marinus group						
B. marinus						
Puerto Rico			5			
Colombia	1		6			
Lake Catemaco, Veracruz		2	14			
Mexico	3	3	3			
Costa Rica		2	12	4	3	
Turrialba, Costa Rica		1	4			
Central America			1			
Siguirres, Costa Rica			6			1
Nautla, Veracruz			1			
Iparia, Peru		6	1			
Guatemala		1	4		2	5
Tamaulipas, Mexico				1		
Chiapas, Mexico					1	
Veracruz, Mexico		1			8	7
El Salvador					1	
Rio Grande City, Texas						1
Guerrero, Mexico						2
Canal Zone						3
B. poeppigi						
Iquitos, Peru		3	12	18	2	
Iparia, Peru		10	8			
B. paracnemis						
Fernandez, Santiago del Estero, Argentina	1	2	2		1	
El Colorada Formosa, Argentina			4			
San Miguel de Tucuman, Argentina			2			1
B. ictericus						
Misiones, Argentina	1	1	3		1	
Campo Grande, Brazil			1			
Teresopolis, Brazil			1			1
São Jose de Rio Preto, Brazil			1			
No locality			2			
B. arenarum						
Mendoza, Argentina	4	6	34	25*	10	
Fernandez, Santiago del Estero, Argentina	3	1	4	8		
San Miguel de Tucuman, Argentina	1	1	7	4		2
Guayamallen, Argentina			4			
Poterillos, Argentina			1			
No locality			1			
Buenos Aires, Argentina	1					
Argentina	4					
Mendoza (lab raised)				10		
Cordoba Prov., Argentina						3

* Asterisked items refer to individuals in which only mating calls were examined, along with those individuals examined anatomically. These do not include African species. Most of the mating calls were examined by both Tandy and Martin.

Appendix A (Continued)

Species	Number of Specimens					
	Chap. 4	Chap. 10	Chap. 11	Chap. 13	Chap. 14	Chap. 15
Guttatus group						
B. haematiticus						
Canal Zone	1					
Río Reventazon, Costa Rica	4	3	15	14	10	
Puerto Viejo, Costa Rica		3				
No locality						1
B. blombergi						
Colombia	5	1	4	6*	1	2
B. holdridgei						
Costa Rica	3	2	1			1
Typhonius group						
B. typhonius						
Canal Zone	2	1	1			
Iparia, Peru		3	4			2
Granulosus group						
B. humboldti						
Chiriqui, Panama	3		13		4	3
Canal Zone		4	8			
San Jose de Guiavare, Colombia			1			
Girardot, Colombia		1	3			
Villavicencio, Colombia		2	2			
B. major						
Resistencia, Argentina	1					
B. fernandezae						
Resistencia, Argentina	2		15			
Santa Cruz, Uruguay			2			
B. d'orbignyi						
No locality	1					
B. pygmaeus						
No locality			4			
Spinulosus group						
B. spinulosus						
Mendoza, Argentina	1	4	11	7	8	2
Malargüe, Mendoza, Argentina			5			
Poterillos, Mendoza, Argentina			5			
Guayamallen, Argentina			1			
San Isidro, Mendoza, Argentina		2	9			1
No locality	4					
B. atacamensis						
Atacama, Chile	2	1	5			2
B. chilensis						
Til Til, Chile	3		7			
Santiago, Chile	2					
Zapallar, Chile	3	2	4			
Chile			3		1	1
B. limensis						
Lima, Peru	1		5		1	
B. trifolium						
Palca, Peru	3		2			
B. flavolineatus						
Junin Plateau	4		4			
B. variegatus						
Bariloche, Argentina	3	4	4			1

Appendix A (Continued)

Species	Number of Specimens					
	Chap. 4	Chap. 10	Chap. 11	Chap. 13	Chap. 14	Chap. 15
Bocourti group						
B. bocourti						
San Cristobal, Chiapas, Mexico			2			
Sal Caja, Guatemala			13			
Motagua Valley, Guatemala			3			
Tecpan, Guatemala			2			
San Marcos, Guatemala			1			
Jalapa, Guatemala			1			
Quezeltenango, Guatemala			25			1
Guatemala	13	2	8	2		5
Valliceps group						
B. valliceps						
Austin, Texas	9	10	153		27	48
Columbus, Texas			3			
Bee Creek, Texas			1			
Fentress, Texas			1			
Westphalia, Texas			1			
Houston, Texas			2			
Red Rock, Texas			1			
San Marcos, Texas			1			
Bastrop, Texas		3	1			
La Ventana, Veracruz			3			
Cuatalapan, Veracruz			1			
Fortin, Veracruz		1	1	2	1	
Juan Diaz Covarrubias, Veracruz			1			
Zanatepec			1			
Mexico			2			
Huautla, Puebla			3			
Padilla, Tamaulipas			1			
Merida, Yucatan		1	4		5	
Quezeltenango, Guatemala			2			
Retalhuleu, Guatemala			4			
Guatemala		1	1			
Sinaloa, Mexico	1					
Tampico, Mexico				1		
El Encinal, Mexico				1		
Sinton, Texas				48		
Tamaulipas, Mexico				2		
B. cavifrons						
Volcan San Martin	1			1		
Bastonal, Veracruz			2			1
No locality	1					
B. mazatlanensis						
Mazatlan, Mexico		2	2			
Tepic, Nyarit, Mexico			6			
Ixtlan, Nyarit, Mexico			5			
San Blas, Mexico			2			
Alamos, Sonora, Mexico			2			1
Sonora, Mexico			1			6
Mexico	2		4		2	
B. gemmifer						
Acapulco, Mexico			3			
Tecpan de Guerrero, Mexico			4			
Las Cruces, Guerrero, Mexico			1			
Guerrero, Mexico	1		1			4

Appendix A (Continued)

Species	Number of Specimens					
	Chap. 4	Chap. 10	Chap. 11	Chap. 13	Chap. 14	Chap. 15
B. ibarrai						
Jalapa, Guatemala	3		15	7	1	
Guatemala			1			4
B. luetkeni						
Motagua Valley, Guatemala			26			
Toculutan, Guatemala			1			1
Baja Verapaz, Guatemala			2			
Choluteca, Honduras			8			1
Liberia, Costa Rica			4	3*		1
Guatemala	5	4			3	
Managua, Nicaragua						4
B. coniferus						
Moravia de Turrialba, Costa Rica	3	6	49	16	8	3
Liberia, Costa Rica			1			
Cartaga, Costa Rica			2			
Costa Rica	3		3			8
Canaliferus group						
B. canaliferus						
Arriga, Chiapas			11			
Tonala, Chiapas			2			
Oaxaca			1			
Tapanatepec, Oaxaca			2			
Suchitepequez, Guatemala			1			
Retalhuleu, Guatemala			26	9	8	1
Tehuantepec, Oaxaca			1			
No locality			7			
Chiapas, Mexico	12	5				2
Guatemala						2
Occidentalis group						
B. occidentalis						
Cuautla, Morelos			2			
Chihuahua			1			
Oaxaca	2	2	2		6	
Tapascolula, Mexico			1			
Juchatengo, Mexico			1			
Mexico	2	3	4			
Puebla, Mexico						1
Marmoreus group						
B. marmoreus						
Zanatepec, Oaxaca			1			
Tehuantepec, Oaxaca			10			1
Oaxaca			2		1	1
Tapanatepec, Oaxaca			7			
Las Cruces, Chiapas			1			
Cintalapa, Chiapas						9
Tierra Colorada, Guerrero	1		10		4	
Santiago, Colima			7			1
Matias Romero, Oaxaca			2			
Mexico	8	5	15			
Puebla, Mexico					2	1
B. perplexus						
Zampingo, Guerrero			1			1
Juchatengo, Oaxaca			1			
Cuautla, Morelos			3			3

Appendix A (Continued)

Species	Number of Specimens					
	Chap. 4	Chap. 10	Chap. 11	Chap. 13	Chap. 14	Chap. 15
Amayuca, Morelos			2			1
Izucar de Matamoras, Puebla	1	6	19		8	2
Tlaltzapan, Morelos			14			
Tonatico, Mexico			1			
Ocotito, Guerrero			1		1	
Guerrero			1			
Yanatepec, Morelos	1		1		4	1
Mexico			3			
Alvarius group						
B. alvarius						
Avra Valley, Arizona			2			
Eloy, Arizona			2			
Mesa, Arizona		1	20			
Phoenix, Arizona			3		3	2
Tempe, Arizona			1			
Tucson, Arizona		3	8	11*	14	1
Arizona	5		4			1
Sonoita, Sonora			1			
No locality			6			1
Cognatus group						
B. cognatus						
Tulia, Texas			1			
Lubbock, Texas		4	10			
Spearman, Texas		2	9			
Trans-Pecos, Texas			1			
Tucson, Arizona			3			45
Green Valley, Arizona						1
Safford, Arizona			1			6
Avra Valley, Arizona			1			
Douglas, Arizona			2			
Portal, Arizona			4			1
Mesa, Arizona			12		6	4
Wheaton, Minnesota			15			
Minnesota			1		1	
Durango, Mexico			2			
Ciudad Obregon, Sonora			4	14	4	
Sonora, Mexico		1	1			
No locality			4			
Deming, New Mexico						5
Arizona			12			
Lamb County, Texas						12
Hobbs, New Mexico						4
Rodeo, New Mexico						22
B. compactilis						
Jiquilpan, Jalisco		3	6		1	
Guadalajara, Mexico			2			1
Teotihuacan, D.F.			2			
Federal District, Mexico			2			
B. speciosus						
Austin, Texas		4	12			
Utley, Texas			24			
Luling, Texas			1			
Manor, Texas			1			
Lockhart, Texas			1			

Appendix A (Continued)

Species	Number of Specimens					
	Chap. 4	Chap. 10	Chap. 11	Chap. 13	Chap. 14	Chap. 15
Hornsby Bend, Texas			5			
Kirkland, Texas			1			
Childress, Texas			1			
Cotulla, Texas			1			
Big Spring, Texas			1			
Lubbock, Texas			2			15
Sweetwater, Texas						5
Skidmore, Texas			3			
Pflugerville, Texas		4	15	8	8	
Port Aransas, Texas			3			
Benavides, Texas			1			
Falfurrias, Texas		1	2		1	
Artesia, New Mexico			1			
Sabinas, Coahuila			1			
No locality			1			
Boreas group						
B. boreas						
Bakersfield, California			5			
Mary Lake, California		3	1			1
Mesa Lake, Colorado			1			
Boulder, Colorado			1		1	
Fort Collins, Colorado			15			
Fruita, Colorado			1			
Snowy Range, Wyoming			4		6	1
Medicine Bow Mts., Wyoming			8	1		
Moscow, Idaho		3	15		2	2
Clear Lake, Oregon			1			
East Lake, Oregon			1			
Johnson Lake, British Columbia, Canada		1	5	8	3	
Colorado			11			
California			4			
No locality						1
B. nelsoni						
Beatty, Nevada		4	3		4	
B. exsul						
Deep Springs Valley, California		2	7	2	1	
B. canorus						
Mary Lake, California		8	20			2
California			4		1	1
Punctatus group						
B. punctatus						
Bee Creek, Texas			6			
Medina, Texas			3			
Austin, Texas			1			
San Marcos, Texas			2			
Blanco, Texas			7			
Cedar Park, Texas			1			
Kerrville, Texas			1			
Pontotoc, Texas			3			
Wimberly, Texas		4	10			
Enchanted Rock, Texas		4	1			
Langtry, Texas			1			
Limpia Canyon, Texas			2			
Dripping Springs, Texas			4			

Appendix A (Continued)

Species	Number of Specimens					
	Chap. 4	Chap. 10	Chap. 11	Chap. 13	Chap. 14	Chap. 15
Valentine, Texas		1	3		2	
Bee Caves, Texas			1			
Marble Falls, Texas			2			
Texas	4		3			
Portal, Arizona			1			
Santa Rita Mts., Arizona			2			1
Mesa, Arizona			5			
Las Vegas, Nevada			6			
Jalapa, Aguascalientes, Mexico			3			
Mexico			3			2
Jollyville, Texas			1			
Arizona	3		6			2
No locality			3			
Zacatecas, Mexico	1					
Travis County, Texas		2			14	
Tucson, Arizona					1	
Debilis group						
B. debilis						
Big Spring, Texas			4			
Brownwood, Texas			2			
Luling, Texas			1			
Brackettville, Texas			1			
Fentress, Texas			3			
Liberty Hill, Texas		4	18		11	
Valentine, Texas			5		1	
Marble Falls, Texas			1			
Lubbock, Texas			5		1	
Red Rock, Texas			3			
Kingsland, Texas			2			
Lampasas, Texas			7			
Pedernales River, Texas			1			
Tatum, New Mexico						3
Mexico			2			
No locality			4			
Texas	4	4				
Coahuila, Mexico	1					
B. kelloggi						
Hermosillo, Mexico			1			
Tepic, Nyarit, Mexico			1			
Navajoa, Sonora			2			
Sonora, Mexico	1	3	6		3	2
No locality			2			1
B. retiformis						
Sells, Arizona			6			
No locality		4	10			
Hermosillo, Mexico	2		3			
Sonora, Mexico						1
Americanus group						
B. americanus						
Boxley, Arkansas			3			1
Fayetteville, Arkansas			8			
Exeter, Missouri			1			
Woodruff, Kansas			1			
Broken Bow, Oklahoma			4			

Appendix A (Continued)

Species	Number of Specimens					
	Chap. 4	Chap. 10	Chap. 11	Chap. 13	Chap. 14	Chap. 15
Wilburton, Oklahoma			3			
Ouachita Mts., Oklahoma			2			
Tulsa, Oklahoma			1			
Bethel, Oklahoma			8			
Muskogee, Oklahoma			1			
La Grange, Indiana		1	1			
Urbana, Illinois			2			
Charles County, Maryland			1		2	
Jackson County, Georgia			1			
Ohio			3			
Itasca, Minnesota			2			
Oklahoma			2			
Illinois			2			
Minnesota		3	15		2	
James Bay, Ontario			9	9		
Parry Sound District, Ontario		4	1			
Montreal, Canada			1			
Ontario, Canada				1	7	
No locality						1
B. terrestris						
Jackson County, Georgia			1			
Newton, Georgia			3			
Savannah, Georgia		2	2			
Gainesville, Florida			39			
Hogtown Sink, Florida			1			
De Leon Springs, Florida			2			
Florida			7			
No locality			2			2
Bay County, Florida			3	11	3	2
Georgia		2	4			
Westonia, Mississippi			1			
Jasper County, South Carolina					5	
McIntosh County, Georgia					2	
B. houstonensis						
Houston, Texas		2	4		2	
Bastrop, Texas		5	2		7	2
B. hemiophrys						
Wheaton, Minnesota			16			1
Mahnomey, Minnesota			2			
Minnesota		5	26	16	7	3
Delta, Manitoba			10			
Grank Forks, North Dakota			2			
Laramie, Wyoming			2			
No locality			2			
B. microscaphus						
Birch Creek, Utah			1			
North Creek, Utah			1			
Zion, Utah		3	7	12	16	
Sand Cove Reservoir, Utah			3			
St. George, Utah			16			3
Utah			7			
Chihuahua			1			
No locality			2			
Baker Reservoir, Utah						1

Appendix A (Continued)

Species	Number of Specimens					
	Chap. 4	Chap. 10	Chap. 11	Chap. 13	Chap. 14	Chap. 15
B. woodhousei						
Douglas, Arizona			1			
Mesa, Arizona			9		15	
Moapa, Nevada			1			
Brookshire			1			
Beaumont, Texas			1			
Montgomery, Texas			1			
Fentress, Texas			1			
Austin, Texas		4	31		2	2
Sealey, Texas			1			
Wichita Falls, Texas			1			
Weatherford, Texas						1
Houston, Texas			1			
Red Rock, Texas			1			
Waller County, Texas			5			
Karnack, Texas			6			
Bastrop, Texas		3	4		6	
Utley, Texas			1			
Hornsby Bend, Texas			3			
Alabama			1			
Eastern Texas		2	11			
Virgin River, Utah			1			
O'Neill, Nebraska			1			
Shreveport, Louisiana			1			
Broken Bow, Oklahoma			4			
Fort Collins, Colorado			1			
No locality			4			
Fredericksburg, Texas					1	
St. George, Utah					7	6
Coccifer group						
B. coccifer						
Tapanatepec, Oaxaca			3			
Zanatepec, Oaxaca			2			
Juchitan, Oaxaca			2			
Cuscutlan, El Salvador			3			
San Vicente, El Salvador			12			
Costa Rica		1	2	19		
Guanacaste, Costa Rica			2			
Esparta, Costa Rica			5		2	1
Liberia, Costa Rica			3		4	1
Retalhuleu, Guatemala			3	5		
Choluteca, Honduras			1			
Central America	6		16			
San Miguel, El Salvador			2			
El Salvador	2	3		3		
Quercicus group						
B. quercicus						
Miami, Florida			3	9		
Florida		1	5			
Mississippi			5			
No locality			2			1
Bay County, Florida		5			3	
Gulf County, Florida					7	
Long County, Florida					1	

Appendix A (Continued)

Species	Number of Specimens					
	Chap. 4	Chap. 10	Chap. 11	Chap. 13	Chap. 14	Chap. 15
Waycross, Georgia						5
Ludowici, Georgia					1	
Periglenes group						
B. periglenes						
Monteverde, Costa Rica	1		2			2
Viridis group						
B. viridis						
Israel	9	8	42	10	3	3
No locality	4		3		1	1
Germany						1
Calamita group						
B. calamita						
The Hague, Holland			1		1	
Holland			1			
France			2			
Germany						1
Europe	6	6	29		5	1
Bufo group						
B. bufo						
Firenze, Italy	2		6	8	4	
Germany		1	4			
Holland			3			
Boille St. Paul, France		4	18			3
No locality	3		25		9	3
B. japonicus						
Tokyo	2	1	1		1	
B. asiaticus						
Masan, Korea	1					
Stomaticus group						
B. stomaticus						
Calcutta, India	3	6	21	8	3	2
Orissa, India						1
Mysengh, East Pakistan			7			
No locality			1			
Melanostictus group						
B. melanostictus						
Kuala Lumpur, Malaysia	1		6			1
Thailand	1		5	9	5	
Mysengh, East Pakistan			5			
Orissa, India						1
India		4	10			5
Ceylon						2
Bogor, Indonesia						1
No locality		2	1			2
Asper group						
B. asper						
Malaysia	2	1	22			3
Mauritanicus group						
B. mauritanicus						
Morocco	4	4	9	10	4	2
Regularis group						
B. brauni						
Amani, Tanzania	2	6	8	20*	7	1

Appendix A (Continued)

Species	Number of Specimens					
	Chap. 4	Chap. 10	Chap. 11	Chap. 13	Chap. 14	Chap. 15
B. garmani						
Letaba, Kruger National Park, South Africa	2	2	6	9	4	
Umtali, Rhodesia			1	2	2	
Louis Trichardt, South Africa				3		
Ol Tukai, Kenya		1	1	3	1	
No locality			1			
East Africa						1
B. rangeri						
Swellendam, South Africa	4	4	7	28	7	
Hicport, South Africa						1
Mussel Bay, South Africa	1		3	8	3	
Port St. John, South Africa	1	2	5	8	5	
Bloemfontein, South Africa			1			
Grahamstown, South Africa			4			1
No locality	2		1			2
B. pardalis						
South Africa	1		2			
B. regularis (E)						
Pietermaritzburg, South Africa			3	10	3	
Port St. Johns, South Africa		2	11	58	12	
Lourenço Marques, Mozambique		2	6	33	5	
Umtali, Rhodesia	2	1	4	34	1	
Louis Trichardt, South Africa	1		2	17	1	
Nyeri, Kenya	1	2	8	39	5	
Mazeras, Kenya	1	2	4		4	
Vumba Mts., Rhodesia		1	5	23	5	
Ol Tukai, Kenya	1	1	6	13	4	6
Nairobi, Kenya			2			
Kabete, Kenya			1			
No locality			3			
Transkei, South Africa	1					
B. regularis (W)						
Achi, Nigeria			1			
Calabar, Nigeria			1			
Ogbomosho, Nigeria	5	2	20	15	7	
Ilesha, Nigeria			2	2		
Kampala, Uganda			5			2
Entebbe, Uganda		2	2	20	6	
El Mahalla el Kubra, Egypt	1	6	18	41	14	6
Cape Coast, Ghana		2	3			
Ghana						2
Ibadan, Nigeria		1	3			
No locality		2				
Nigeria	3		7			
B. sp.						
Mtito Andei, Kenya		2	1		1	1
Ol Tukai, Kenya		4	1		3	
B. kisoloensis						
Nyeri, Kenya		1	1			1
B. kerinyagae						
Kabete, Kenya		1	3			1
Nanyuki, Kenya						3
Maculatus group						
B. maculatus						
Kampala, Uganda		1	1	3		

Appendix A (Continued)

Species	Number of Specimens					
	Chap. 4	Chap. 10	Chap. 11	Chap. 13	Chap. 14	Chap. 15
Nigeria			6			
Achi, Nigeria			1			
Idanre, Nigeria			4			
Ilesha, Nigeria			2	3	2	
Ogbomosho, Nigeria			1			
Mukeza, Tanzania	1		1	3	1	
Entebbe, Uganda	2		1	7	1	
Mazeras, Kenya			5	9	2	
Ibadan, Nigeria		6	2			1
No locality	1		1			
Calabar, Nigeria			1			
B. pusillus						
Letaba, Kruger National Park, South Africa		1		3	1	
Umtali, Rhodesia				3		
Perreti group						
B. perreti						
Idanre, Nigeria	7	1	11	20	3	
Gariepensis group						
B. gariepensis						
Grahamstown, South Africa	3		5	5	3	
B. inyangae						
Inyanga Mts., Rhodesia	2	2	4	2	5	
B. rosei						
Table Mts., South Africa	12	1	4			5
Lönnbergi group						
B. lönnbergi						
Kabete, Kenya			1			1
Nanyuki, Kenya	2	2	4	1	2	
Vertebralis group						
B. vertebralis						
Victoria West, South Africa			2			
Inyanga, Rhodesia					1	
Lughensis group						
B. lughensis						
Baraqui, Kenya			2			
Carens group						
B. carens						
M'Kuze, Zululand	1		6		3	
South Africa	4	2	4		1	1

Appendix B. Eurasian Species of Bufo: Definition of Characters Examined

Skull

1. Skull length measured mid-dorsally from anterior end of internasal suture to opening of foramen magnum.
2. Skull width relative to skull length. Width was measured at the ventral end of the suspensorium.
3. Skull height relative to skull length. Skull height was measured from the ventral articular surface to the supraoccipital surface. Projections caused by dorsal bony crests were not included in this dimension.
4. Shape of palatal fenestra (width divided by length). Maximum width was measured between lateral edge of parasphenoid process and lateral border of fenestra. Maximum length was measured between palatine and transverse ramus of pterygoid.
5. Ratio of depth to width of braincase measured at anterior end of occipital arterial canal. Measurements were made on the exterior surfaces of the bones. The width dimension was taken between the anterior ends of the occipital canals.
6. Ratio of width to length of parasphenoid process. The length was measured between the anterior tip of the process and the midline of the transverse arm of the parasphenoid. Maximum width of the anterior process was used.
7. Rugosity of dorsal surface of the skull. Nasals, frontoparietals, squamosals, and proötics were examined. These surfaces varied from smooth to pitted to rugose. The last two states were distinguished on the basis of whether the discontinuities in the surface were depressed or raised. Some skulls were both pitted and rugose.
8. Separation of frontoparietals. The bones are joined by a hairline suture; separated by a very narrow, parallel-sided space; or widely separated in the anterior third or two-thirds to form a fontanelle that is not parallel sided. The second condition may be an artifact of drying. Only the third condition is regarded as "frontoparietals separated."
9. Dorsal exposure of sphenethmoid. If the frontoparietals do not abut the nasals completely, a varying portion of the sphenethmoid is exposed between the two sets of bones. The amount of exposure of the sphenethmoid is a measure, therefore, of the degree of fusion of frontoparietals and nasals. The character was recorded in the following states: not exposed dorsally; exposed in a small area of the dorsal midline; exposed for about half the width of the snout; exposed the full width of the snout.
10. Occipital canal. The canal for the occipital artery runs from the supraoccipital region to the posteromedial corner of the orbit and is roofed over by bone to varying degrees The character was recorded in the following states: canal completely exposed, hence appearing as a groove in the dorsal surface of the skull; ¼ to ½ of canal covered by bone; canal covered by bone for its entire length. In a few instances a very narrow sliver of bone arched over the canal at one of its ends. These skulls are arbitrarily classified as having exposed canals.
11. Dorsal otic plate of squamosal. The dorsal portion of the squamosal may form a broad plate that overlies the proötic to varying extents. In some skulls, however, the squamosal merely overlies the lateral edge of the proötic and does not form an "otic plate."
12. Frontoparietal-proötic fusion. These two bones may be completely separated or completely fused. In the first instance the sutures are visible in the wall of the orbit and dorsally. Presumably an intermediate could occur, one in which only the orbital suture or only the dorsal suture was visible. This state was not encountered in the Eurasian toads examined. Even in skulls having the frontoparietal overlying the proötic and meeting the squamosal, a dorsal suture between frontoparietal and proötic was distinct if a suture was visible in the wall of the orbit.
13. Overlap of quadratojugal and maxilla. The amount of overlap was determined while viewing the skull dorsolaterally. In effect, only the dorsolateral overlap of the quadratojugal on the maxilla was considered. The amount of overlap was gauged in terms of the pterygoidal fenestra: no overlap; overlap slight, tip of quadratojugal not reaching anterior half of pterygoidal fenestra; overlap extensive, tip of quadratojugal in anterior half of fenestra; overlap complete, tip of quadratojugal reaching pterygoid. The first state (no overlap) is generally associated with a very thin quadratojugal.
14. Hyomandibular foramen. The ramus hyomandibularis of the seventh cranial nerve runs laterally across the anterior face of the proötic and passes through the suprapterygoid fenestra (Tihen 1962). The nerve may cross the fenestra through the cartilaginous wall, or it may pass through a notch in the lateral edge of the proötic, or it may run through a complete circle of bone near the end of the proötic. These were the three states recorded.
15. Suprapterygoid fenestra. As defined by Tihen (1962), this is the space bounded by the proötic, the squamosal, and the pterygoid. The character was recorded as present if this space was occupied

by cartilaginous material and as absent if occluded by bone.

16. Pterygoid-parasphenoid angle. When a *Bufo* skull is viewed from the rear, an angle (or angles) is formed between the pterygoid and the parasphenoid, or between the quadratic and parasphenoidal arms of the pterygoid or at both points. The character was recorded simply as a comparison of angles: intrapterygoid angle greater than, equal to, or smaller than the pterygoid-parasphenoid angle.
17. Transverse parasphenoid ridge. Bufonids may have a ridge running across the transverse posterior axis of the parasphenoid. This character appears in the following states: absent, in which case the parasphenoid is flat; shallow, obtuse ridge; deep, broad, obtuse ridge; narrow, deep, sharp ridge.
18. Palatine denticulation. The ventral edge of the palatine may be smooth, slightly undulate, or notched (denticulate). Care must be taken not to confuse natural serrations with notches made accidentally in cleaning the skull.

Vertebral Column

19. Relative length of third transverse process. The length measured from tip to tip and divided by skull length. This character gives an indication of the relative width of the axial skeleton.
20. Relative length of the seventh transverse process, the presacral process. The length was measured from tip to tip and divided by the length of the third transverse process. This character gives a measure of the shape of the axial skeleton. The smaller the ratio, the more the axial skeleton tapers posteriorly.
21. Relative width of the sacral diapophysis. The maximum width of the bony outer end of the sacral diapophysis was divided by skull length.
22. Size of the foramen of the third spinal nerve. The opening was treated as an ellipse and its area divided by the length of the third centrum. All measurements for this character were made with an ocular micrometer.
23. Vertebral crests. The majority of vertebral columns examined have a single low dorsal crest formed by the neural spines. Some, however, have two low parallel dorsal ridges. If one imagines splitting each neural spine and pressing each half laterally, one can visualize the appearance of the double-crested columns. Since the first two vertebrae never have distinct neural crests, only vertebrae three to seven were used to categorize columns in this character.

Muscles

24. Humerodorsalis (Haines 1939). This muscle originates on the humerus and extends out the third and fourth fingers to varying distances. In the species of *Bufo* examined, a slip of this muscle to the second finger appears only in odd individuals and will not be discussed further. The first of the two main types or states has branches that extend as tendons to the terminal phalanges of the third and fourth fingers and a short accessory branch that inserts on the outer side of the fourth metacarpal. The second type has the usual long slip to the third finger but only the short accessory slip to the fourth. This muscle is called extensor digitorum communis longus in Gaupp (1896) and Kändler (1924).
25. Supinator manus (Haines 1939). All species of *Bufo* examined have this muscle, called abductor indicis longus by Gaupp (1896) and Kändler (1924). Not all species, however, have the branch from the humerus. The presence or absence of this branch, which becomes enlarged in adult males (Kändler 1924), were the two states recorded.
26. Extensor digitorum communis longus of the tarsus (Dunlap 1960). This muscle has a varying number of slips to the toes. The species of *Bufo* are more variable in this regard than is indicated by Dunlap. The character states were defined by the digits on which the muscle inserted: toes 2 and 3, 2 to 4, 3 and 4, and 2 to 5.
27. Adductor longus. In the great majority of toads examined, the adductor longus is distinct; that is, one can recognize two diagonally oriented thigh muscles: one, the pectineus, inserting on the middle of the femur, and the second, the adductor longus, continuing distally and fusing with the adductor magnus. In some bufonids only one muscle can be detected in this area and it inserts on the middle of the femur. Dunlap (1966) has shown that in larval *Rana pipiens* the adductor longus and pectineus develop from a common premuscular mass and that they separate by larval stage IX. We may assume that the developmental pattern is much the same in bufonids. The bufonids having only a single muscle inserting on the femur have probably undergone reduction in the size of this muscle mass as well as failure of the differentiation process. I have arbitrarily defined this state as "adductor longus absent." Dunlap (1960) refers to this state in *Atelopus cruciger* in the same terms.

External features

28. Tympanum. Two states were recognized: tympanum present and visible; tympanum absent.
29. Cranial crests. Crests are here defined as cornified, raised edges, curved or straight. Usually, but not invariably, the skin of the crests is fused to the underlying bones, which usually are also raised. Six

states were recognized: no crests; canthal crest only; supratympanic crest only; orbital crests only, including at least one in the preorbital, supraorbital, or postorbital position; orbital crests plus supratympanic; orbital crests plus parietal; orbitals plus supratympanic plus parietal.

30. Tibia gland. To distinguish an ordinary glandular wart, which all these species have, from a more specialized structure, a tibia gland was defined as a mass of glands larger than the diameter of the eye. In some cases the tibia gland was conspicuously raised. In others, its presence had to be confirmed by dissection. Two states were recognized: present and absent.
31. Tarsal ridge. This structure is a distinctly raised, continuous, sharp ridge extending proximally from the inner metatarsal tubercle. A series of separated tubercles on the inner edge of the tarsus was not considered a tarsal ridge. Present and absent were the states recorded.
32. Subarticular tubercles. Only the tubercles under the digital joints were considered. Usually these subarticular tubercles are large and single in species of *Bufo*. In some toads the distal tubercles may be cardioid or divided, sometimes under all digits but more commonly under the third finger and fourth toe. This character was recorded in four states: all subarticular tubercles single; only third finger with divided subarticular tubercles; only fourth toe with divided tubercles; third finger and fourth toe with divided tubercles. Cardioid tubercles were treated as single.

Vocal apparatus

33. Vocal sac present or absent. To ensure that males were mature, only those with nuptial pads were examined.
34. Gular pigmentation. Melanophores are scattered through the muscle and connective tissue investing the vocal sac in some species. To see them, the gular skin must be cut. This character was recorded in two states, present or absent.

Characters 33 and 34 were combined in the analysis. These secondary sex characters are clearly interrelated. No males lacking vocal sacs have pigment in the gular musculature. The "new" character has three states: no vocal sac; vocal sac present, gular pigment lacking; vocal sac and gular pigment present.

Appendix C. Eurasion Species of Bufo: Material Examined

Taxa examined are listed in alphabetic order. Trinomials are ignored and each subspecies is listed as a separate taxon. The number following the name is the total number. In parentheses, in order, are the number of whole preserved specimens, the number of skeletons, and the number viewed in X-ray. Where more than 10 were studied, the number is given simply as 10+. In the second set of parentheses are listed the abbreviations of the institutions providing the material.

abatus Ahl 1 (1,0,1) (ZMB)
asper Gravenhorst 10+ (10+,8,2) (FMNH, USNM)
bankorensis Barbour 10+ (10+,7,1) (FMNH)
beddomi Günther 1 (1,0,1) (BM)
biporcatus Gravenhorst 4 (4,1,0) (FMNH)
bufo Linnaeus 10+ (10+,3,0) (FMNH)
calamita Laurenti 6 (5,1,0) (FMNH)
celebensis Schlegel 1 (1,0,0) (FMNH)
claviger Peters 2 (2,0,1) (FMNH, USNM)
dhufarensis Parker 4 (4,0,2) (BM, FMNH)
divergens Peters 10+ (10+,3,0) (FMNH)
dodsoni Boulenger 9 (9,2,0) (FMNH)
fergusoni Boulenger 10+ (10+,2,0) (FMNH)
galeatus Günther 2 (2,0,0) (FMNH)
gargarizans Cantor 10+ (10+,6,0) (FMNH)
himalayanus Günther 5 (5,0,2) (BM, SNF, USNM)
juxtasper Inger 10+ (10+,2,0) (FMNH)
latastei Boulenger 1 (1,0,1) (BM)
macrotis Boulenger 10+ (10+,0,3) (AMNH, FMNH, USNM)
mauritanicus Schlegel 10+ (10+,2,0) (FMNH)
melanostictus Schneider 10+ (10+,9,1) (FMNH)
minshanicus Stejneger 10+ (10+,1,0) (USNM)
olivaceus Blanford 10 (10,0,0) (CAS, MNS)
orientalis Werner 10+ (10+,5,0) (FMNH)
parietalis Boulenger 2 (2,0,0) (BM)
parvus Boulenger 10+ (10+,3,2) (FMNH)
pentoni Anderson 3 (3,0,0) (SNF)
philippinicus Boulenger 10+ (10+,7,1) (FMNH)
quadriporcatus Boulenger 10+ (10+,1,0) (FMNH)
raddei Strauch 10+ (10+,5,0) (FMNH)
stejnegeri Schmidt 1 (1,0,0) (FMNH)
stomaticus Lütken 10 (10,2,1) (AMNH, CAS, FMNH)
stuarti Smith 2 (2,0,2) (BM)
sulphureus Grandison and Daniel 2 (2,0,2) (BM)
sumatranus Peters 1 (1,0,1) (ZMB)
surdus Boulenger 2 (2,0,1) (BM, MNS)
tibetanus Zarevsky 10+ (10+,2,0) (FMNH)
viridis Laurenti 10+ (10+,3,1) (FMNH)

The following taxa were not available for study: *Bufo brevirostris* Rao, *burmanus* Andersson, *chlorogaster* Daudin, *hololius* Günther, *luristanicus* Schmidt, *microtympanum* Boulenger, *oblongus* Nikolsky, *pageoti* Bourret, *persicus* Nikolsky, *tienhoensis* Bourret, *valhallae* Meade-Waldo.

Several forms of *Bufo bufo* were not studied, partly for lack of time, but mainly because they would not add much to an analysis of species groups.

Bufo kelaarti Günther is a peculiar species questionably assigned to this genus. Its relationship to *Bufo* will be the subject of a separate paper.

Appendix D. Mathematical Analysis of Species Karyotypes

Introduction

r: the ratio of the short arm divided into the long arm.

type: the type of constriction. Metacentric constrictions (*m*) have an *r* of 1.0 to 1.6; submetacentric constrictions (*sm*) have an *r* of 1.7 to 2.9; subtelocentric constrictions (*st*) have an *r* of 3.0 to 6.9; an *r* of 7.0 or above is telocentric (*t*).

%: the percentage length of the chromosome compared to the longest chromosome having a percentage length of 100.

C: refers to secondary constrictions. The ratio presented is determined as if the secondary constriction was the primary constriction; *s* or *l* follows this ratio if the secondary constriction is on the short arm or the long arm.

Species		Chromosomes										
		1	2	3	4	5	6	7	8	9	10	11
B. americanus	r	1.4	1.2	1.5	1.9	1.1	1.2	1.3	1.2	1.2	2.4	1.9
	type	m	m	m	sm	m	m	m	m	m	sm	sm
	%	99.4	95.2	82.9	74.5	62.0	54.1	29.6	25.2	20.9	18.4	12.3
	C	3.4 l	1.1 s				1.2 l					
B. houstonensis	r	1.4	1.0	1.4	1.8	1.2	1.2	2.0	1.3	1.8	1.1	1.2
	type	m	m	m	sm	m	m	sm	m	sm	m	m
	%	100	96.3	86.0	70.8	57.5	44.0	24.6	22.2	21.6	21.2	15.6
	C	3.6 l	1.3									2.2 s
B. microscaphus	r	1.1	1.4	1.4	2.4	1.0	1.1	1.6	2.2	1.3	1.1	1.7
	type	m	m	m	sm	m	m	m	sm	sm	m	sm
	%	100	88.5	77.4	63.0	52.0	44.2	22.6	19.5	19.5	15.0	13.3
	C	1.0 l	1.0 s									
B. terrestris	r	1.1	1.2	1.2	2.2	1.2	1.7	1.0	1.3	1.2	1.3	1.2
	type	m	m	m	sm	m	sm	m	m	m	m	m
	%	100	87.5	76.6	60.0	47.5	40.0	24.4	21.9	20.3	18.1	12.5
	C	1.5 s	1.9 s									
B. hemiophrys	r	1.3	1.1	1.6	2.2	1.2	1.2	2.6	1.2	1.6	1.1	1.5
	type	m	m	m	sm	m	m	sm	m	m	m	m
	%	100	99.4	74.4	64.7	58.3	47.4	23.1	20.8	16.7	13.5	9.6
	C		1.6 s									
B. woodhousei	r	1.1	1.4	1.5	2.4	1.2	1.3	1.0	2.8	1.6	1.1	1.2
	type	m	m	m	sm	m	m	m	sm	m	m	m
	%	100	93.5	77.4	67.7	54.2	42.6	21.0	21.0	14.5	13.5	10.3
	C	1.2 l										
B. luetkeni	r	1.0	1.2	1.6	1.8	1.0	1.1	1.5	1.2	2.4	1.1	1.7
	type	m	m	m	sm	m	m	m	m	sm	m	sm
	%	100	89.8	79.7	71.2	61.0	57.6	29.8	26.4	25.4	19.7	13.6
	C	1.6 s										
B. valliceps	r	1.1	1.2	1.6	1.4	1.0	1.2	1.5	1.6	3.0	1.2	1.2
	type	m	m	m	m	m	m	m	m	st	m	m
	%	100	96.6	78.4	71.8	64.0	50.4	27.0	23.4	22.6	17.0	14.6
	C	1.4 s 5.6 s		3.1 l					5.0 s			
B. coniferus	r	1.2	1.3	1.7	1.2	1.2	1.8	1.4	1.8	3.6	1.2	1.6
	type	m	m	sm	m	m	sm	m	sm	st	m	m
	%	100	86.0	66.0	65.0	56.7	43.3	30.0	25.0	20.7	14.0	10.7
	C							4.3 s				
B. cognatus	r	1.2	1.2	1.2	1.6	1.2	1.4	1.3	2.0	1.4	1.2	1.4
	type	m	m	m	m	m	m	m	sm	m	m	m
	%	100	88.2	76.0	71.6	57.2	44.7	23.5	22.0	20.6	12.7	10.8
	C	1.0 l										

Appendix D (Continued)

Species		Chromosomes 1	2	3	4	5	6	7	8	9	10	11
B. compactilis	r	1.2	1.4	1.4	1.8	1.0	1.2	2.2	1.3	1.6	1.6	1.6
	type	m	m	m	sm	m	m	sm	m	m	m	m
	%	100	92.7	73.3	67.3	56.2	41.4	24.9	23.9	20.3	17.5	15.5
	C	1.0 l										1.0 l
B. boreas	r	1.3	1.1	1.7	1.8	1.0	1.4	1.3	2.6	1.5	1.3	1.8
	type	m	m	sm	sm	m	m	m	sm	m	m	sm
	%	98.7	92.6	83.4	82.8	67.8	45.4	24.2	23.7	23.0	13.9	13.0
	C	1.6 l										
B. nelsoni	r	1.2	1.2	1.2	1.8	1.1	1.8	1.4	2.4	1.2	1.2	1.1
	type	m	m	m	sm	m	sm	m	sm	m	m	m
	%	100	97.5	84.6	82.1	59.2	37.8	26.1	25.4	21.1	16.9	14.4
B. canorus	r	1.2	1.3	1.4	1.9	1.2	1.7	1.6	1.2	1.4	1.2	1.2
	type	m	m	m	sm	m	sm	m	m	m	m	m
	%	100	95.2	88.8	81.0	64.6	46.4	27.2	26.5	22.1	19.0	14.0
	C		1.6 l		2.5 s	1.0 l	1.5 l					
B. marmoreus	r	1.2	1.2	1.2	1.8	1.2	1.5	1.3	2.2	1.2	2.0	2.0
	type	m	m	m	sm	m	m	m	sm	m	sm	sm
	%	100	95.4	77.2	77.2	69.5	54.4	30.9	28.1	25.3	15.8	14.7
	C	1.4 l										
		1.7 l										
B. perplexus	r	1.0	1.2	1.4	1.6	1.2	1.0	1.2	2.2	1.2	1.4	1.3
	type	m	m	m	m	m	m	m	sm	m	m	m
	%	100	95.0	78.0	72.4	60.6	54.2	33.0	27.6	27.6	22.2	18.8
	C		1.5 s									
B. punctatus	r	1.1	1.3	1.2	2.5	1.2	1.4	2.2	1.2	1.4	1.1	1.5
	type	m	m	m	sm	m	m	sm	m	m	m	m
	%	100	92.7	83.0	69.6	59.8	40.0	22.4	20.0	18.5	15.5	13.0
	C	1.3 s		2.8 l								
B. debilis	r	1.4	1.2	1.5	2.0	1.0	1.2	1.3	2.5	1.2	1.4	1.6
	type	m	m	m	sm	m	m	m	sm	m	m	m
	%	100	86.8	81.7	65.4	57.6	43.8	22.4	22.2	19.4	16.1	14.0
	C	1.6 s								1.2 l		
B. kelloggi	r	1.2	1.4	1.4	2.2	1.2	1.3	1.6	1.6	1.4	1.1	2.1
	type	m	m	m	sm	m	m	m	m	m	m	sm
	%	100	86.0	68.4	60.9	56.8	47.8	20.6	17.6	17.5	14.2	12.0
	C	2.0 s								1.8 l		
B. retiformis	r	1.1	1.2	1.4	2.2	1.1	1.2	2.3	1.2	1.6	1.2	1.2
	type	m	m	m	sm	m	m	sm	m	m	m	m
	%	100	86.3	77.8	63.2	56.6	43.8	24.0	21.2	19.0	16.4	13.8
	C	1.4 l	1.4 l	1.0 l		1.5 l						
		2.2 l	1.1 l									
		3.0 s	3.4 s									
		6.0 s										
B. quercicus	r	1.1	1.1	1.4	1.9	1.0	1.3	2.1	1.2	1.0	1.2	1.1
	type	m	m	m	sm	m	m	sm	m	m	m	m
	%	100	96.0	74.7	68.7	58.2	45.4	22.2	19.8	15.8	13.7	10.5
	C		1.6 l									
B. alvarius	r	1.1	1.2	1.3	1.3	1.2	1.6	1.3	2.8	1.5	1.4	1.1
	type	m	m	m	m	m	m	m	sm	m	m	m
	%	100	83.5	78.6	69.1	55.3	53.2	30.9	28.0	25.1	18.9	14.5
	C	2.2 s	1.9 l				3.7 s					

Appendix D (Continued)

Species		Chromosomes										
		1	2	3	4	5	6	7	8	9	10	11
B. canaliferus	r	1.4	1.1	1.2	1.8	1.2	1.4	1.4	1.2	1.5	1.1	1.4
	type	m	m	m	sm	m	m	m	m	m	m	m
	%	100	93.7	84.3	71.8	62.1	55.6	33.8	32.6	27.3	20.5	17.1
	C	3.4 l	1.8 s									
B. bocourti	r	1.4	1.0	1.4	1.8	1.0	1.1	1.4	1.4	1.2	1.2	1.6
	type	m	m	m	sm	m	m	m	m	m	m	m
	%	100	98.0	87.1	75.2	65.8	58.9	38.6	34.2	32.7	25.7	22.3
	C	1.8 l						2.1 s				
B. occidentalis	r	1.2	1.3	1.4	1.2	1.2	1.6	1.8	1.6	1.6	1.0	1.8
	type	m	m	m	m	m	m	sm	m	m	m	sm
	%	97.8	94.2	76.4	65.8	56.2	50.4	27.2	24.8	22.4	18.0	16.6
	C	1.5 s		1.2 l								
B. holdridgei	r	1.2	1.4	1.4	1.6	1.2	1.2	1.2	2.2	1.2	1.2	1.8
	type	m	m	m	m	m	m	m	sm	m	m	sm
	%	100	90.2	84.1	76.8	63.4	56.1	34.6	31.7	31.2	21.5	16.6
	C	2.2 s										
B. haematiticus	r	1.3	1.4	1.5	1.7	1.0	1.2	1.4	1.1	1.7	1.3	1.4
	type	m	m	m	sm	m	m	m	m	sm	m	m
	%	100	91.8	73.4	68.8	61.0	47.4	33.8	29.0	25.8	23.2	19.6
	C	1.2 s			3.4 s		2.0 s					
B. variegatus	r	1.2	1.4	1.3	2.2	1.0	1.5	1.5	1.4	1.1	1.7	1.2
	type	m	m	m	sm	m	m	m	m	m	sm	m
	%	100	97.5	82.8	75.0	63.0	47.9	32.9	30.1	26.6	21.3	21.0
	C							3.0 s				
B. granulosus	r	1.2	1.3	1.4	1.6	1.4	1.3	1.4	2.7	1.6	1.2	1.8
	type	m	m	m	m	m	m	m	sm	m	m	sm
	%	100	94.8	86.5	74.1	67.2	54.6	31.2	26.8	26.5	15.5	15.0
	C	1.8 s	6.6 l		14.0 l					2.3 l		
B. spinulosus	r	1.0	1.6	1.3	2.0	1.2	2.1	2.2	1.2	1.7	1.1	2.0
	type	m	m	m	sm	m	sm	sm	m	sm	m	sm
	%	100	87.8	74.8	73.4	52.4	42.0	26.5	24.6	21.2	15.3	11.8
	C			4.4 l								
				1.6 l								
B. marinus	r	1.1	1.1	1.6	2.0	1.4	1.5	2.5	1.7	1.3	1.0	2.0
	type	m	m	m	sm	m	m	sm	sm	m	m	sm
	%	100	93.5	79.7	73.2	61.0	40.6	28.4	26.0	26.0	16.3	12.2
	C							6.0 s				
B. atacamensis	r	1.1	1.2	1.4	1.4	1.6	1.2	1.9	1.2	1.2	1.1	1.0
	type	m	m	m	m	m	m	sm	m	m	m	m
	%	100	88.3	78.8	70.8	64.2	45.0	31.6	30.0	26.7	22.9	14.2
	C					1.1 l		5.2 s				
B. paracnemis	r	1.1	1.2	1.8	2.0	1.4	1.4	1.4	1.2	1.6	1.2	1.4
	type	m	m	sm	sm	m	m	m	m	m	m	m
	%	100	96.2	78.3	68.2	58.6	48.6	29.7	26.2	22.4	15.4	14.9
	C							2.8 s				
B. poeppigi	r	1.4	1.0	1.4	2.7	1.5	1.5	1.6	1.3	1.5	1.3	1.3
	type	m	m	m	sm	m	m	m	m	m	m	m
	%	98.8	95.8	85.1	78.9	70.5	45.4	30.2	25.6	23.5	15.2	11.8
	C	1.5 l						3.6 s				

Appendix D (Continued)

Species		Chromosomes										
		1	2	3	4	5	6	7	8	9	10	11
B. arenarum	r	1.4	1.2	1.6	2.2	1.2	1.4	1.9	1.2	1.2	1.4	1.0
	type	m	m	m	sm	m	m	sm	m	m	m	m
	%	100	87.7	74.8	68.0	64.6	45.4	29.1	28.2	24.0	16.0	14.0
	C	2.3 s						4.4 s				
B. crucifer	r	1.2	1.3	1.9	1.3	2.0	1.4 1.2	2.1	1.4	1.7	1.0	1.4
	type	m	m	sm	m	sm	m m	sm	m	sm	m	m
	%	100	83.2	69.1	63.8	52.3	49.9 32.7	24.4	22.5	21.0	14.4	13.6
	C	1.0 l 2.0 l				5.2 s		4.5 s				
B. bufo	r	1.2	1.2	1.5	1.9	1.0	1.8	2.4	1.4	1.0	1.6	1.6
	type	m	m	m	sm	m	sm	sm	m	m	m	m
	%	100	95.6	80.9	75.9	72.0	52.9	25.0	23.5	23.5	19.1	18.2
	C	1.3 l 3.8 s		1.0 l								
B. viridis	r	1.2	1.2	1.2	2.0	1.0	1.7	2.6	1.0	1.4	1.2	1.5
	type	m	m	m	sm	m	sm	sm	m	m	m	m
	%	100	93.2	80.1	71.8	58.9	55.8	26.6	26.0	22.8	18.6	15.5
	C	1.5 l	1.5 l	1.4 l								
B. stomaticus	r	1.2	1.2	1.4	1.6	1.0	1.7	1.4	1.4	1.2	1.2	1.1
	type	m	m	m	m	m	sm	m	m	m	m	m
	%	100	91.1	77.3	69.6	59.5	56.4	32.4	28.2	26.0	19.6	15.0
	C	1.7 l	1.9 l 2.9 l	1.4 l 3.0 s			6.6 s	3.3 s				
B. calamita	r	1.2	1.3	1.4	2.0	1.1	1.3	1.1	1.7	1.2	1.4	2.2
	type	m	m	m	sm	m	m	m	sm	m	m	sm
	%	100	96.0	79.8	66.8	63.4	48.5	27.8	26.4	22.4	20.7	17.2
	C		1.5 l			3.4 l						8.5 l
B. melanostictus	r	1.5	1.0	1.6	1.8	1.1	1.8	1.6	1.1	1.0	1.2	2.1
	type	m	m	m	sm	m	sm	m	m	m	m	sm
	%	100	91.2	83.8	70.6	65.4	54.4	34.9	31.2	27.6	23.9	22.8
	C	1.0 l										11.5 l
B. regularis	r	1.3	1.3	1.6	1.6	1.2	1.6	1.2	1.1	2.2	1.4	
El Mahalla el	type	m	m	m	m	m	m	m	m	sm	m	
Kubra, Egypt	%	100	91.5	83.6	68.2	53.0	50.8	28.5	24.3	23.4	22.2	
	C		1.4 s				1.2 l					
B. regularis	r	1.2	1.2	1.6	2.0	1.3	1.4	1.4	1.3	1.8	1.2	
Nyeri, Kenya	type	m	m	m	sm	m	m	m	m	sm	m	
	%	100	89.4	81.0	65.8	61.0	52.4	32.4	29.0	26.6	26.3	
	C						1.3 l	4.4 l				
B. regularis	r	1.3	1.2	1.8	1.9	1.2	1.3	1.0	1.1	2.0	1.6	
Mazuras,	type	m	m	sm	sm	m	m	m	m	sm	m	
Kenya	%	100	86.1	66.2	59.6	45.7	41.4	27.2	24.8	22.5	19.2	
	C				2.3 s		1.5 l					
B. regularis	r	1.3	1.2	1.3	1.9	1.4	1.0	1.2	1.4	1.1	1.8	
Ol Tukai,	type	m	m	m	sm	m	m	m	m	m	sm	
Kenya	%	100	96.0	84.7	70.6	61.3	51.6	32.2	26.2	23.4	21.0	
	C						2.1 l					
B. regularis	r	1.2	1.2	1.5	1.8	1.2	1.4	1.2	1.0	1.4	2.4	
Umtali,	type	m	m	m	sm	m	m	m	m	m	sm	
Rhodesia	%	100	91.0	84.5	69.9	57.2	51.2	35.4	28.8	27.0	24.0	
	C		1.3 l				1.0 l					

Appendix D (Continued)

Species		Chromosomes										
		1	2	3	4	5	6	7	8	9	10	11
B. regularis	r	1.3	1.9	1.4	1.6	1.4	1.7	1.4	1.4	1.2	1.4	
Vumba	type	m	sm	m	m	m	sm	m	m	m	m	
Mountains,	%	100	94.0	87.4	81.4	64.1	55.3	32.7	27.0	25.4	23.9	
Rhodesia	C		2.2 s		1.3 l		1.0 l					
B. regularis	r	1.2	1.5	1.6	1.4	1.1	1.3	1.3	1.4	1.0	1.8	
Lourenço Mar-	type	m	m	m	m	m	m	m	m	m	sm	
ques, Mozambique	%	100	98.7	82.7	74.4	64.7	56.1	36.8	32.7	29.5	28.2	
B. regularis	r	1.0	1.3	1.5	1.7	1.4	1.2	1.1	1.4	2.4	1.1	
Ghana	type	m	m	m	sm	m	m	m	m	sm	m	
	%	100	93.0	87.2	72.1	61.6	56.3	36.0	24.4	23.2	23.2	
B. regularis	r	1.2	1.2	1.3	1.6	1.0	1.1	1.2	1.0	1.2	2.1	
Port St. Johns,	type	m	m	m	m	m	m	m	m	m	sm	
South Africa	%	100	95.6	82.3	73.4	58.6	52.1	31.0	27.0	26.1	22.1	
B. gutturalis	r	1.3	1.3	1.6	1.6	1.2	1.1	1.3	1.2	2.4	2.0	
	type	m	m	m	m	m	m	m	m	sm	sm	
	%	100	95.0	76.3	72.2	59.6	54.0	28.5	23.2	20.8	19.2	
	C		2.1 s			1.8 s						
B. garmani	r	1.2	1.2	1.4	1.4	1.2	1.4	1.2	1.2	1.8	1.2	
	type	m	m	m	m	m	m	m	m	sm	m	
	%	100	96.1	87.2	75.1	61.4	52.0	36.3	30.0	23.5	23.5	
	C	5.5 s		3.0 s	3.0 s		2.0 s					
		1.5 l										
B. rangeri	r	1.7	1.2	1.7	1.2	1.1	1.3	1.5	1.7	1.2	2.4	
	type	sm	m	sm	m	m	m	m	sm	m	sm	
	%	100	98.3	82.6	75.6	67.4	50.4	30.9	24.3	22.8	19.6	
	C				1.5 s							
B. maculatus	r	1.5	1.1	1.3	1.5	1.2	1.6	1.2	2.5	1.1	1.2	
	type	m	m	m	m	m	m	m	sm	m	m	
	%	100	82.8	79.0	68.6	54.8	52.8	33.3	25.0	22.6	21.4	
	C		1.6 l	6.8 s			1.1 l					
B. brauni	r	1.2	1.2	1.3	1.8	1.2	1.5	1.2	1.4	2.5	1.2	
	type	m	m	m	sm	m	m	m	m	sm	m	
	%	100	93.6	78.1	75.4	59.1	45.0	31.8	27.7	24.9	24.9	
	C				1.8 s		1.3 l					
B. mauritanicus	r	1.4	1.4	1.6	1.3	1.0	1.0	1.4	1.2	1.0	1.3	1.4
	type	m	m	m	m	m	m	m	m	m	m	m
	%	100	92.2	74.6	74.6	61.7	46.8	29.8	25.4	20.3	19.7	19.7
	C											2.0 s
B. gariepensis	r	1.2	1.2	2.3	1.3	1.7	1.4	2.1	1.3	1.3	1.2	1.2
	type	m	m	sm	m	sm	m	sm	m	m	m	m
	%	100	86.2	72.8	65.8	57.7	40.6	29.2	27.7	22.8	18.4	13.2
B. lönnbergi	r	1.3	1.4	1.4	2.1	1.0	1.4	1.9	1.6	2.0	2.4	1.2
	type	m	m	m	sm	m	m	sm	m	sm	sm	m
	%	100	90.2	84.2	67.7	67.4	50.5	32.1	26.6	26.6	22.7	17.8
	C		3.5 l									

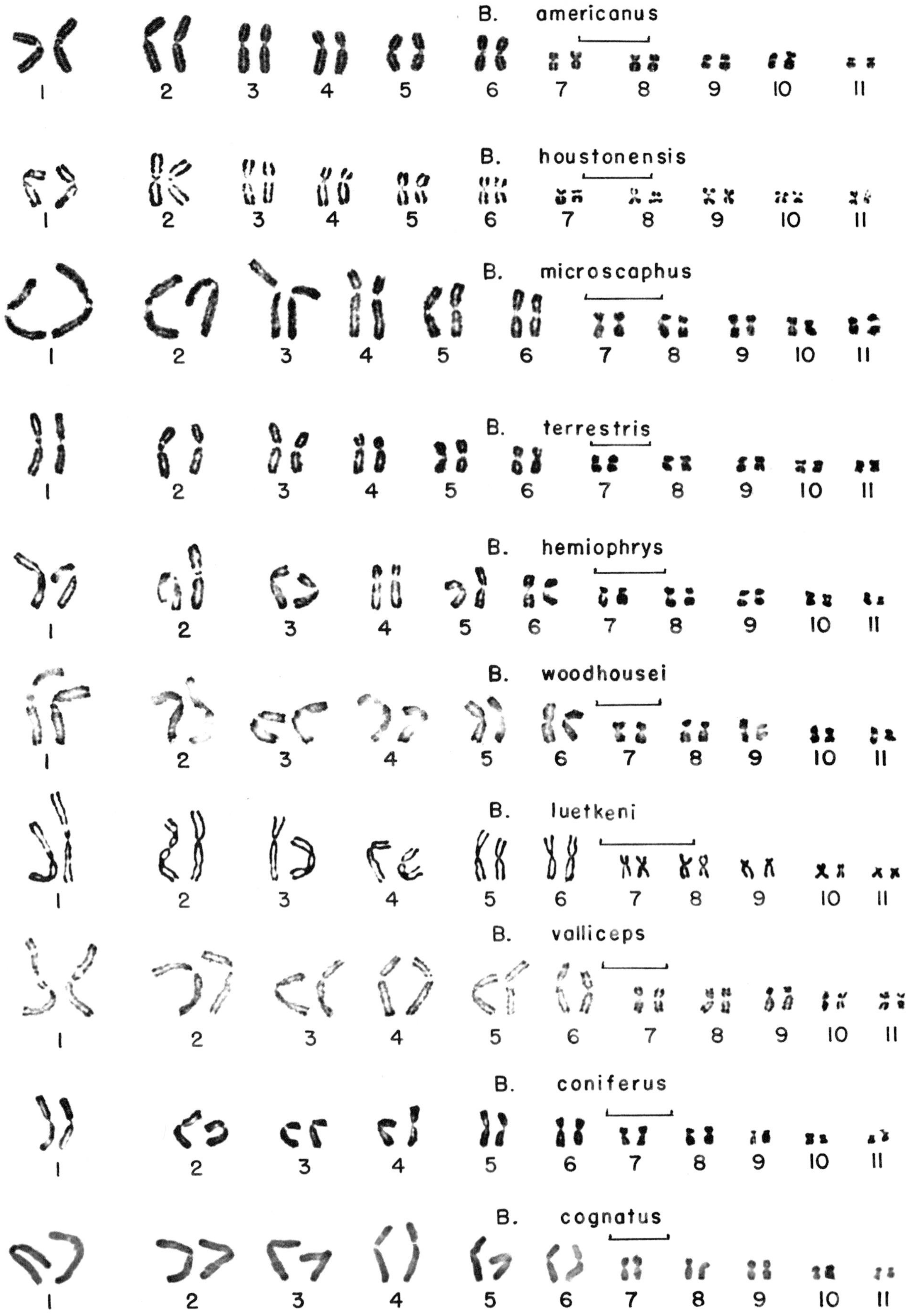

Appendix E–1. Karyotypes of Species of *Bufo*

B. americanus, Minnesota; *B. houstonensis*, Houston, Texas; *B. microscaphus*, Zion National Park, Utah; *B. terrestris*, McIntosh County, Georgia; *B. hemiophrys*, Minnesota; *B. woodhousei*, Austin, Texas; *B. luetkeni*, Guatemala; *B. valliceps*, Austin, Texas; *B. coniferus*, Moravia de Turrialba, Costa Rica; *B. cognatus*, Spearman, Texas.

The scale represents 10 micra in each case.

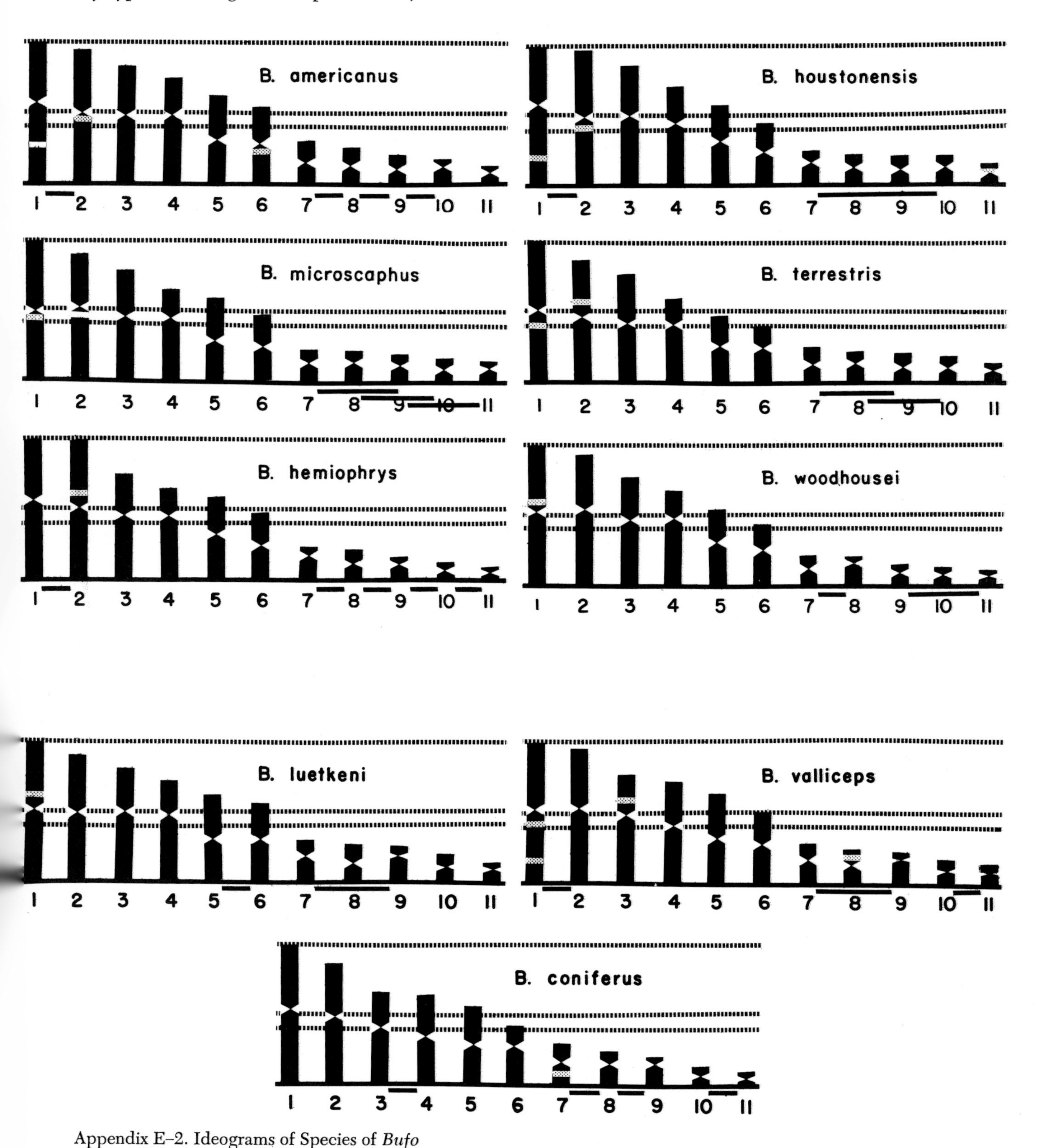

Appendix E–2. Ideograms of Species of *Bufo*

The dotted lines represent 40, 50, and 100 percent of the longest chromosome. Secondary constrictions are represented by voids. Lines connecting karyotype numbers below the base line indicate that these chromosomes are indistinguishable in length. Ideograms were constructed from the results presented in Appendix D.

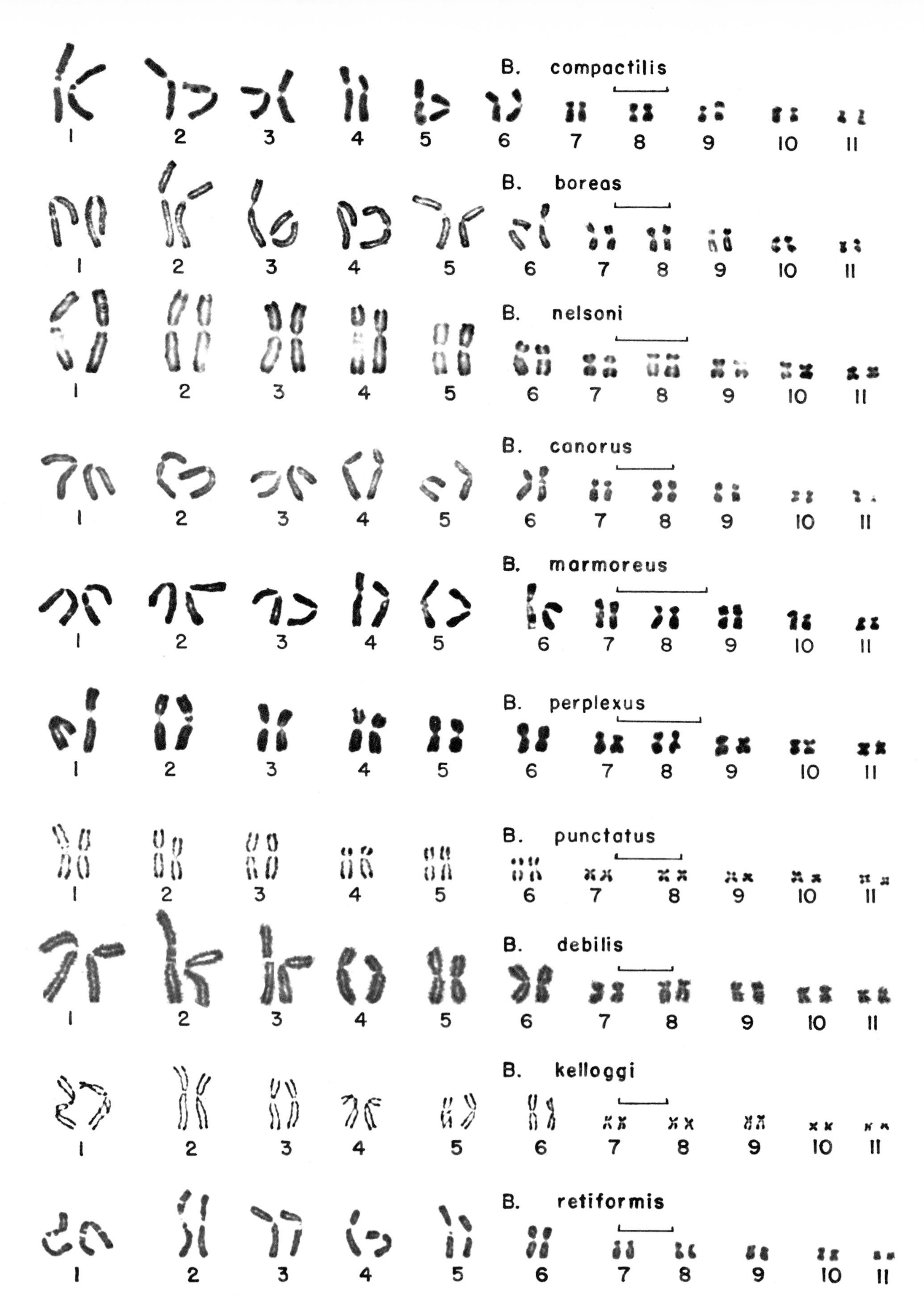

Appendix E–3. Karyotypes of Species of *Bufo*

B. compactilis, Jalisco, Mexico; *B. boreas*, Moscow, Idaho; *B. nelsoni*, Beatty, Nevada; *B. canorus*, Mary Lake, California; *B. marmoreus*, Guerrero, Mexico; *B. perplexus*, Morelos, Mexico; *B. punctatus*, Travis County, Texas; *B. debilis*, Liberty Hill, Texas; *B. kelloggi*, Sonora, Mexico; *B. retiformis*, Sonora, Mexico.

The scale represents 10 micra in each case.

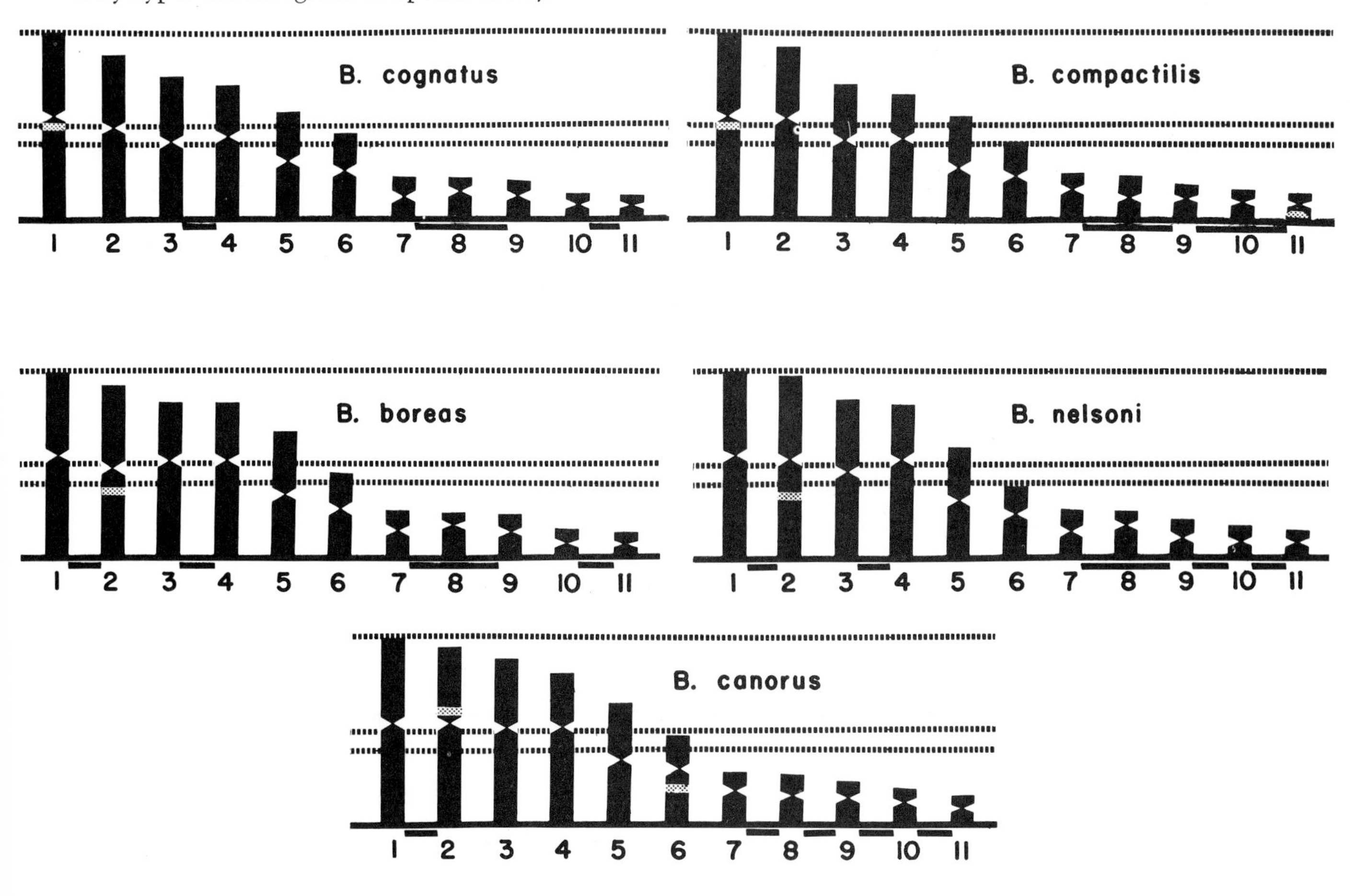

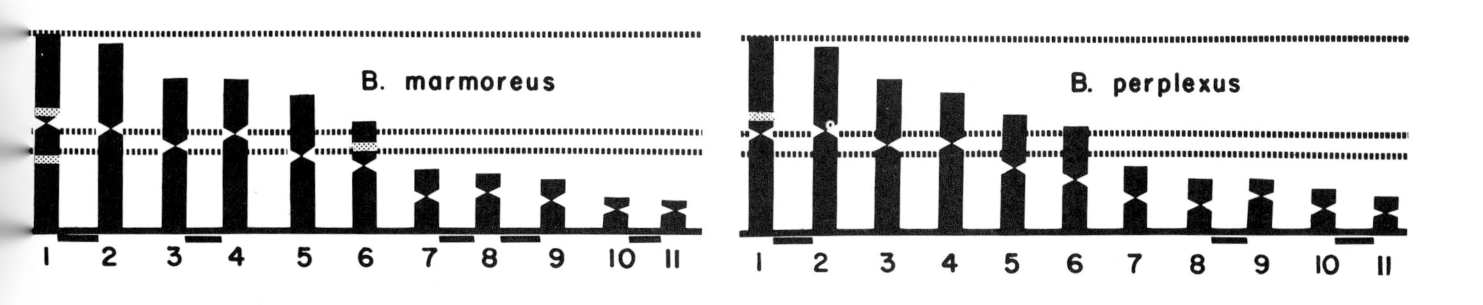

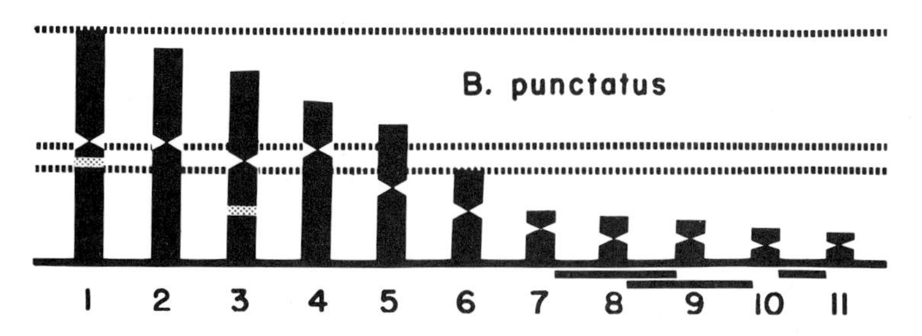

Appendix E–4. Ideograms of Species of *Bufo*

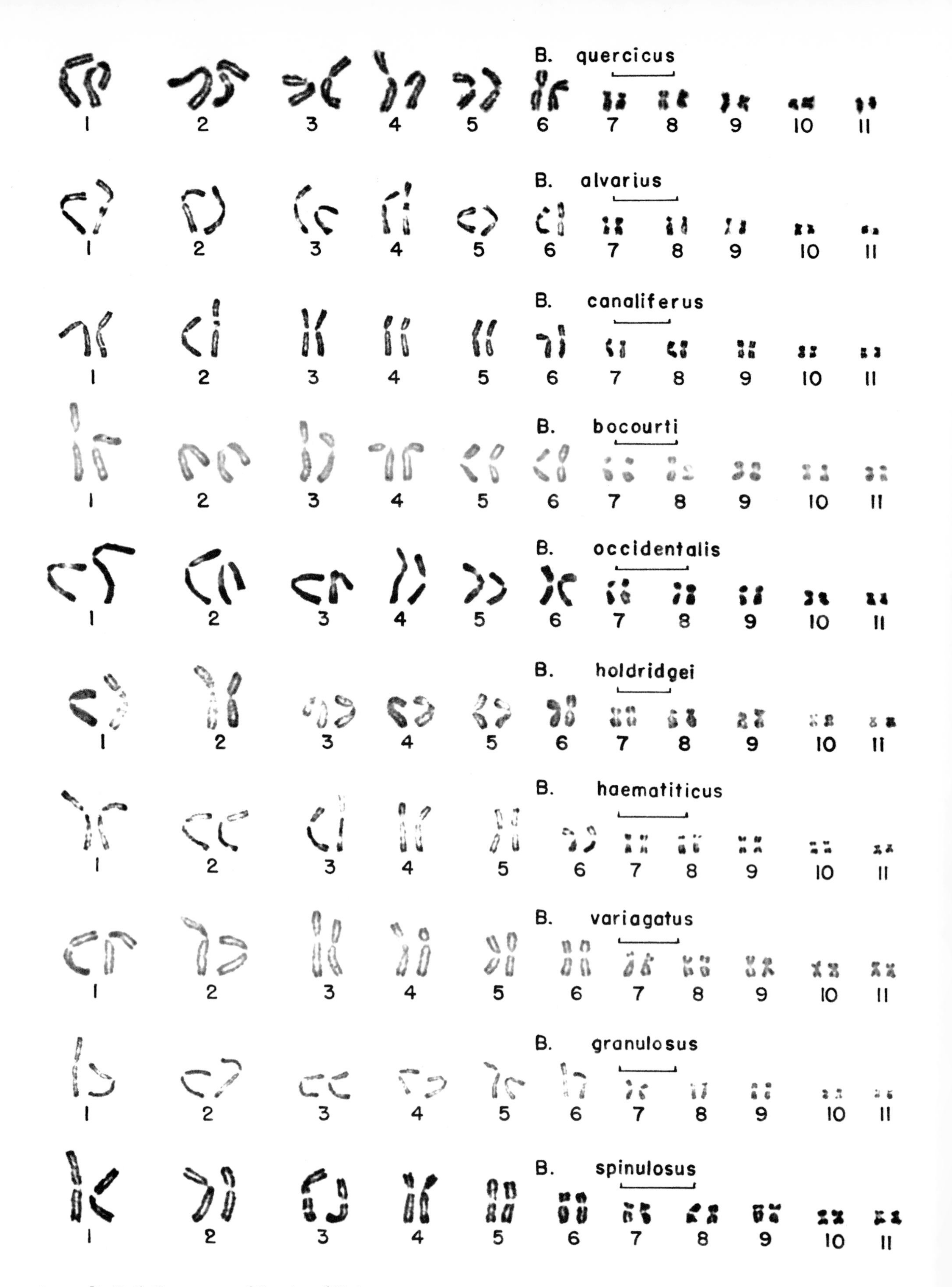

Appendix E–5. Karyotypes of Species of *Bufo*

B. quercicus, Bay County, Florida; *B. alvarius*, Tucson, Arizona; *B. canaliferus*, Rotalhuleu, Guatemala; *B. bocourti*; *B. occidentalis*, Oaxaca, Mexico; *B. holdridgei*, Costa Rica; *B. haematiticus*, Rio Reventazon, Costa Rica; *B. variegatus*, Argentina; *B. granulosus*, Canal Zone, Panama; *B. spinulosus*, Mendoza, Argentina.

The scale represents 10 micra in each case.

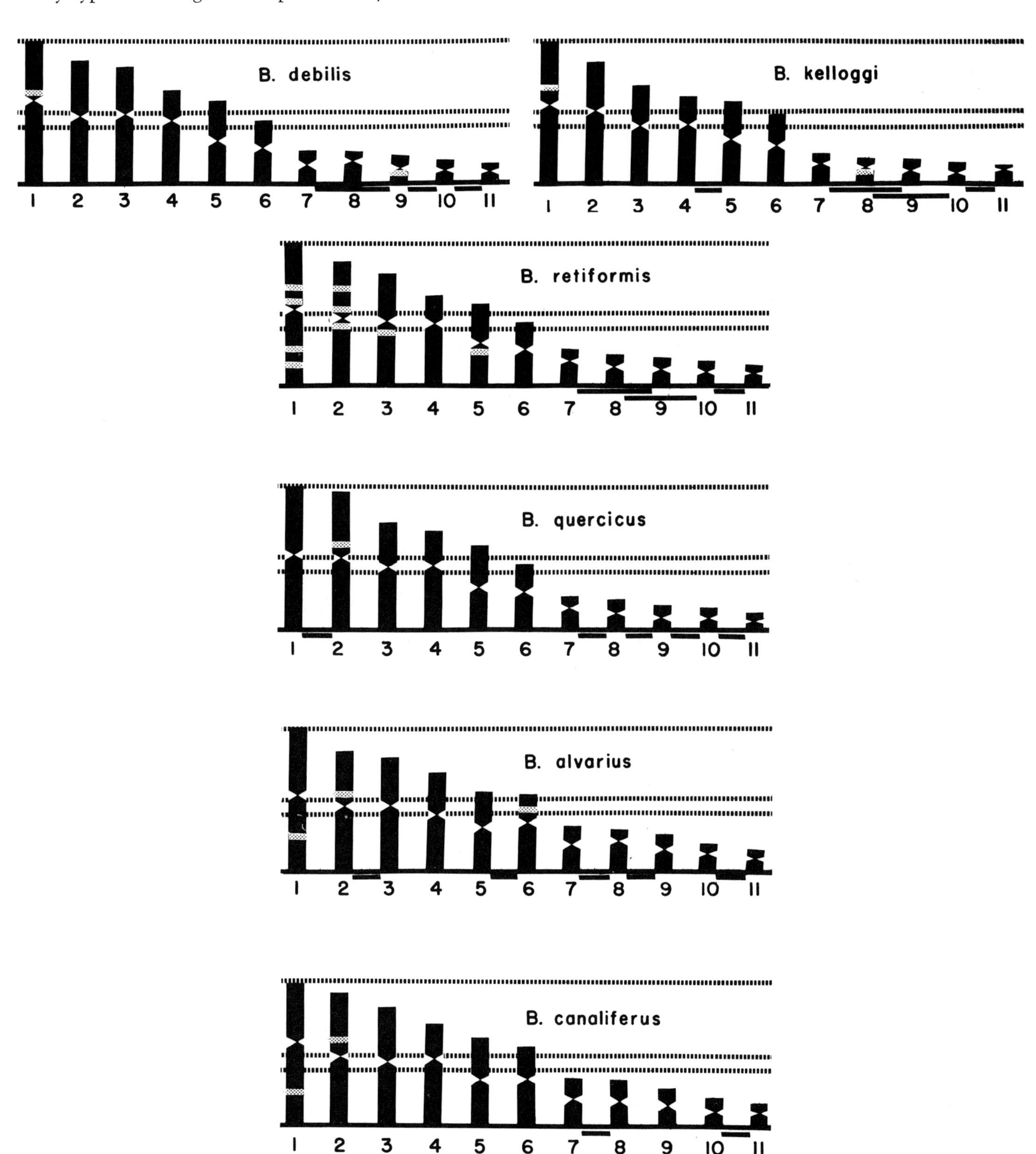

Appendix E-6. Ideograms of Species of *Bufo*

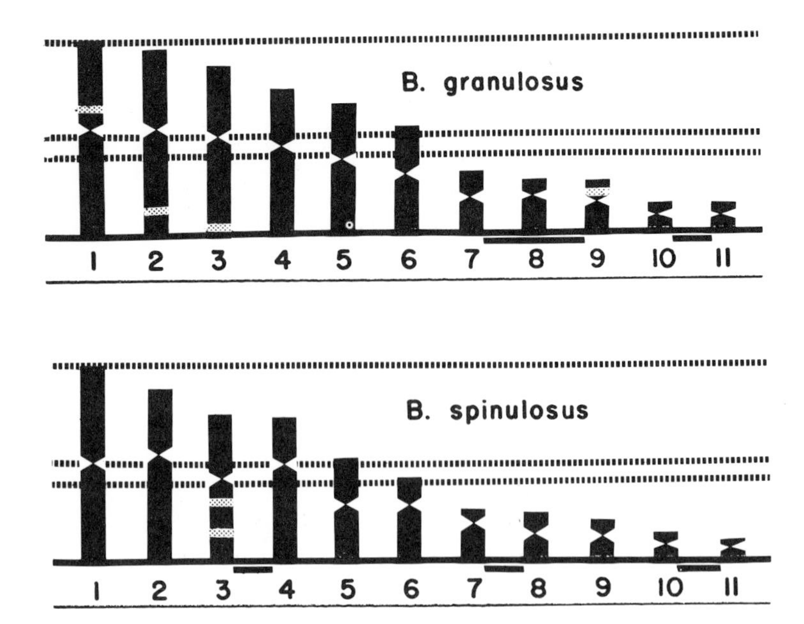

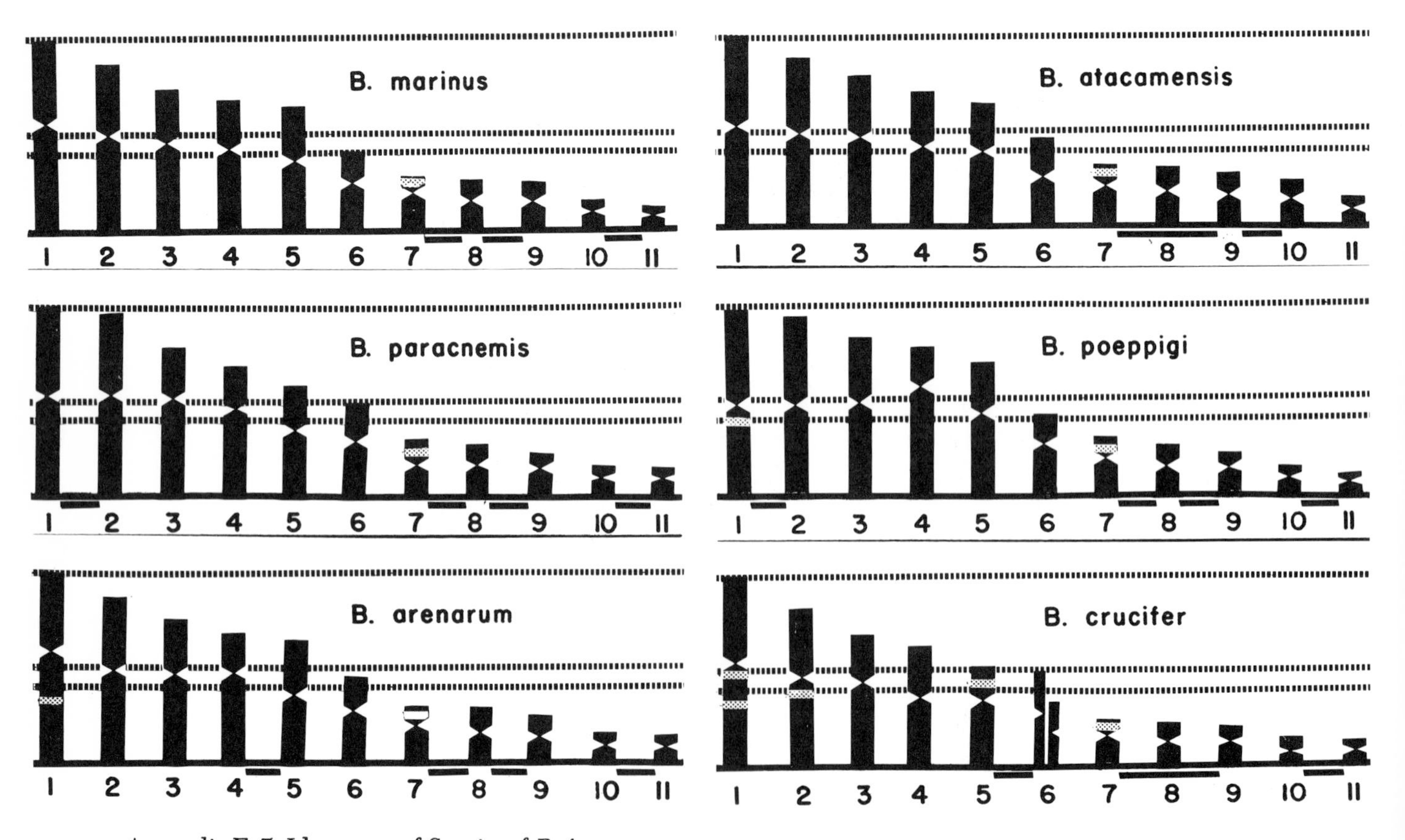

Appendix E–7. Ideograms of Species of *Bufo*

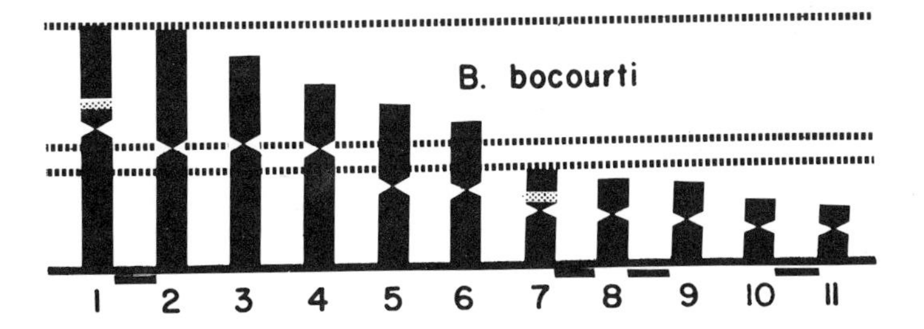

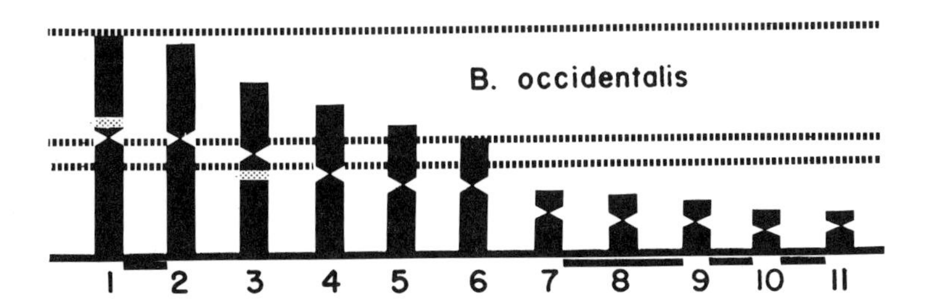

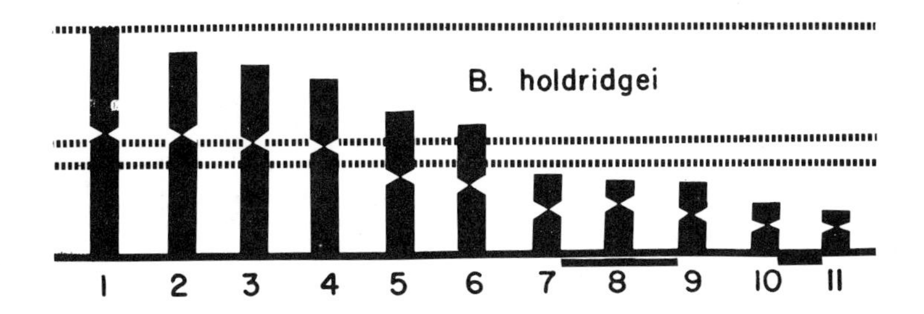

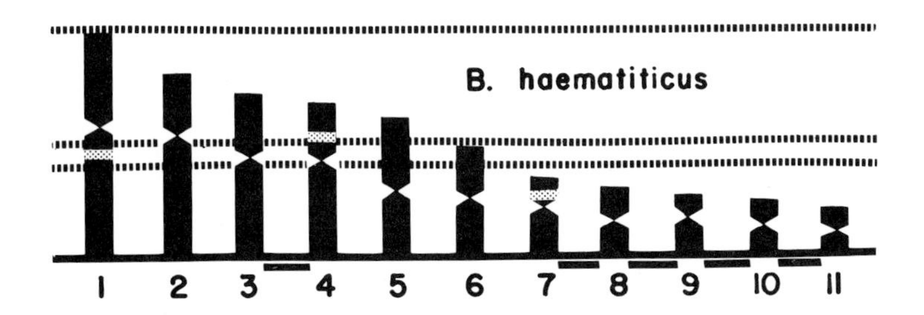

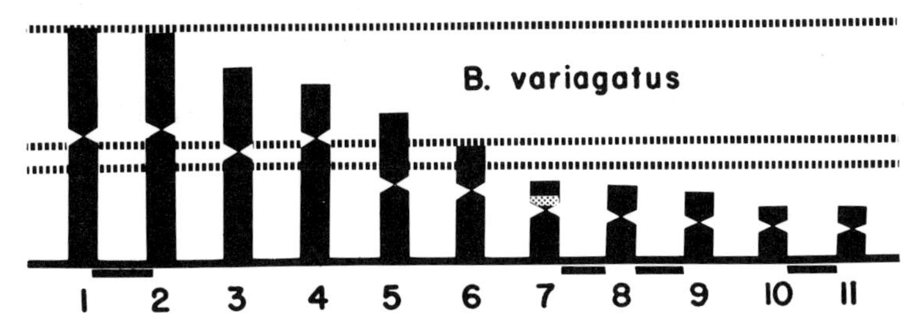

Appendix E-8. Ideograms of Species of *Bufo*

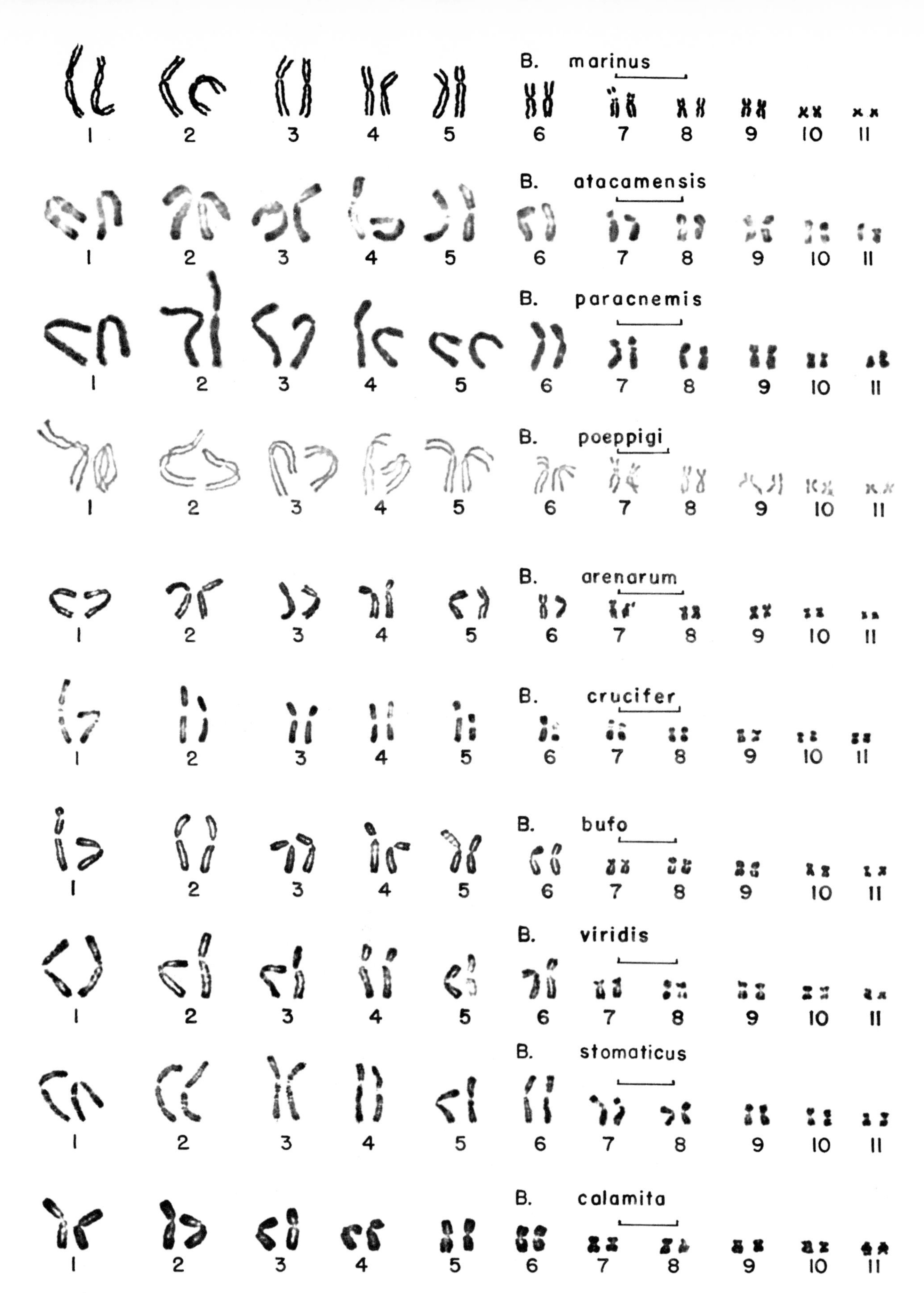

Appendix E–9. Karyotypes of Species of *Bufo*

B. marinus, Mexico; *B. atacamensis; B. paracnemis*, Argentina; *B. poeppigi*, Iparia, Peru; *B. arenarum*, Mendoza, Argentina; *B. crucifer*, Rio de Janeiro, Brazil; *B. bufo*, Ferenzini, Italy; *B. viridis*, Israel; *B. stomaticus*, India; *B. calamita*, The Hague, Netherlands.

The scale represents 10 micra in each case.

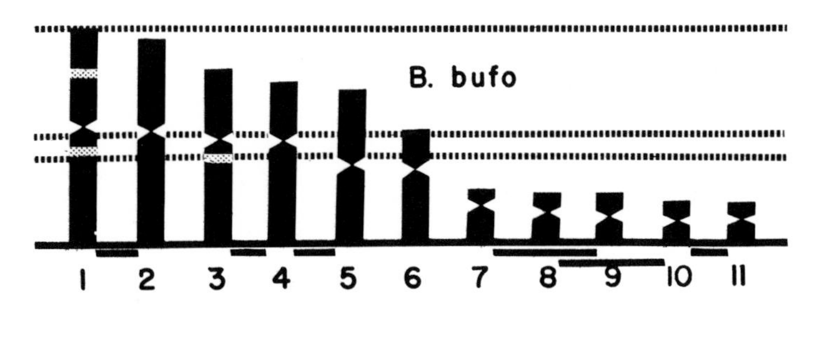

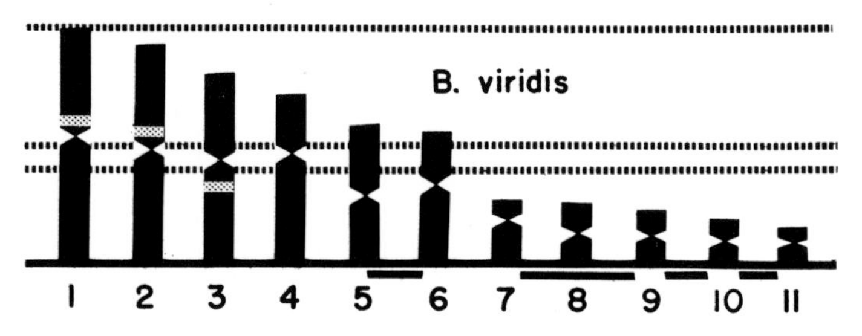

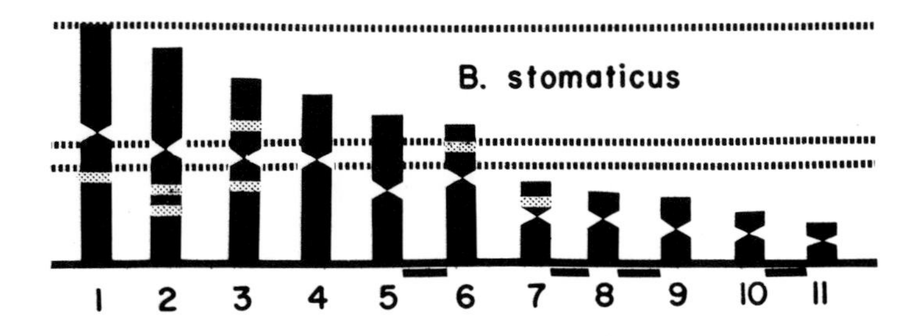

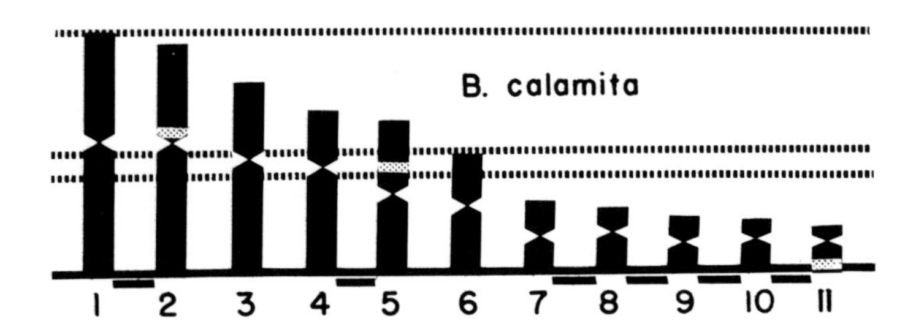

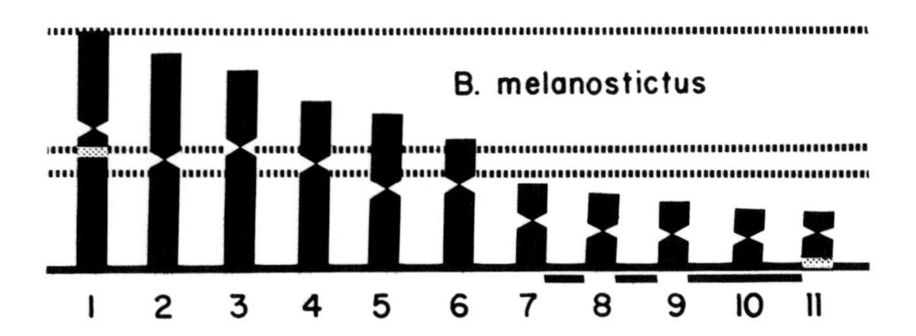

Appendix E–10. Ideograms of Species of *Bufo*

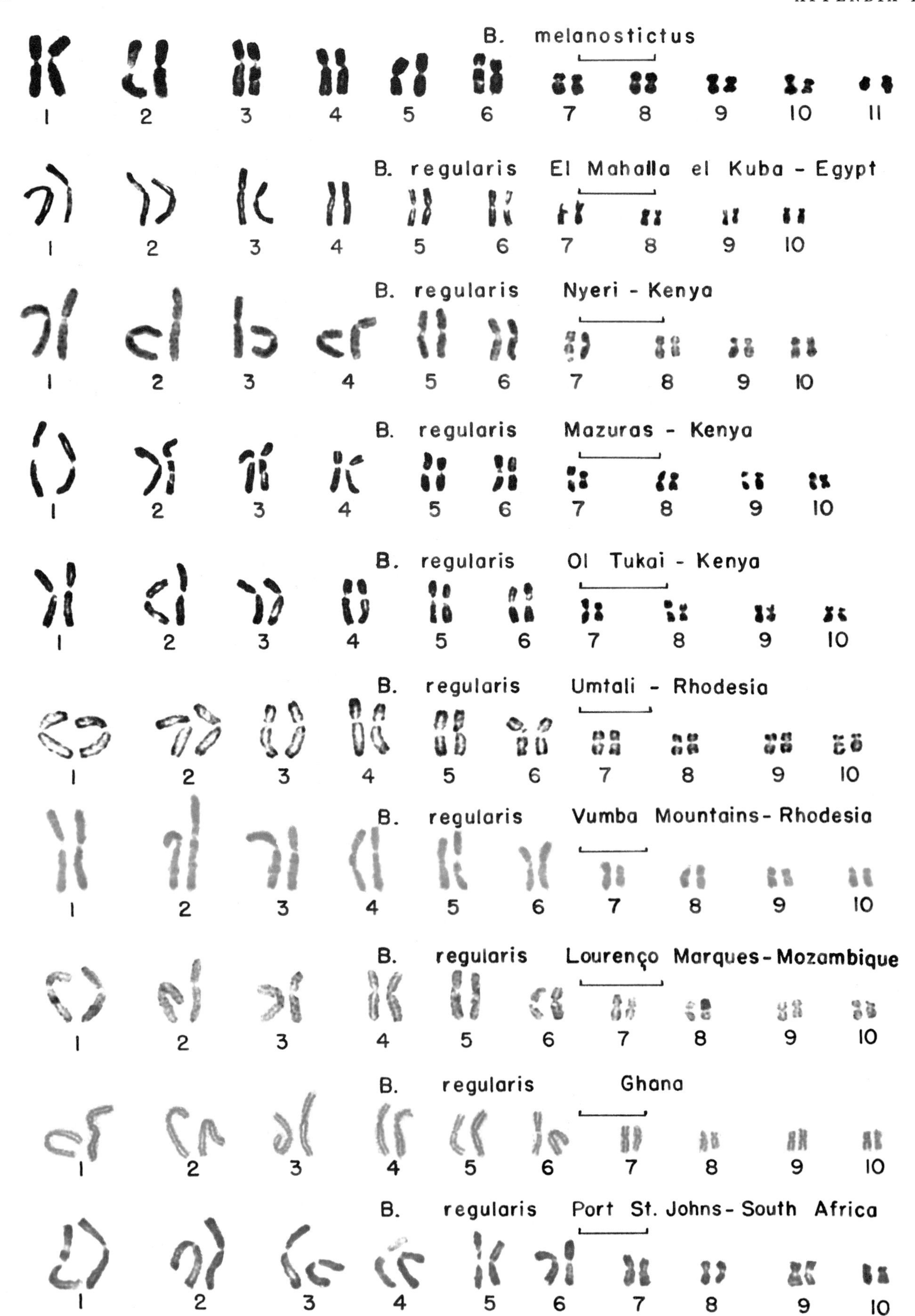

Appendix E–11. Karyotypes of Species of *Bufo*
B. melanostictus, Thailand; populations of *B. regularis* in Africa.

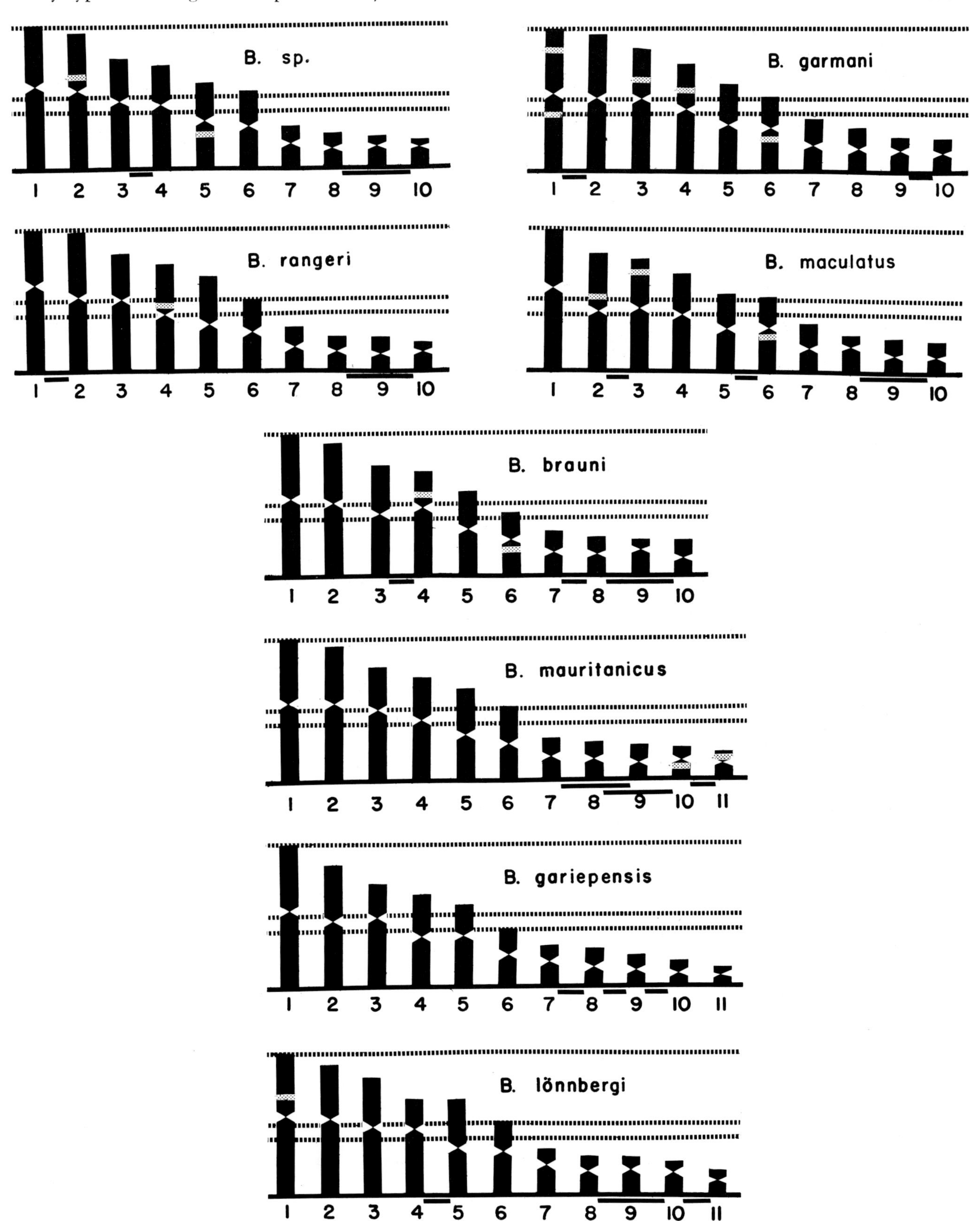

Appendix E–12. Ideograms of Populations of *B. regularis* and *B. kerinyagae*

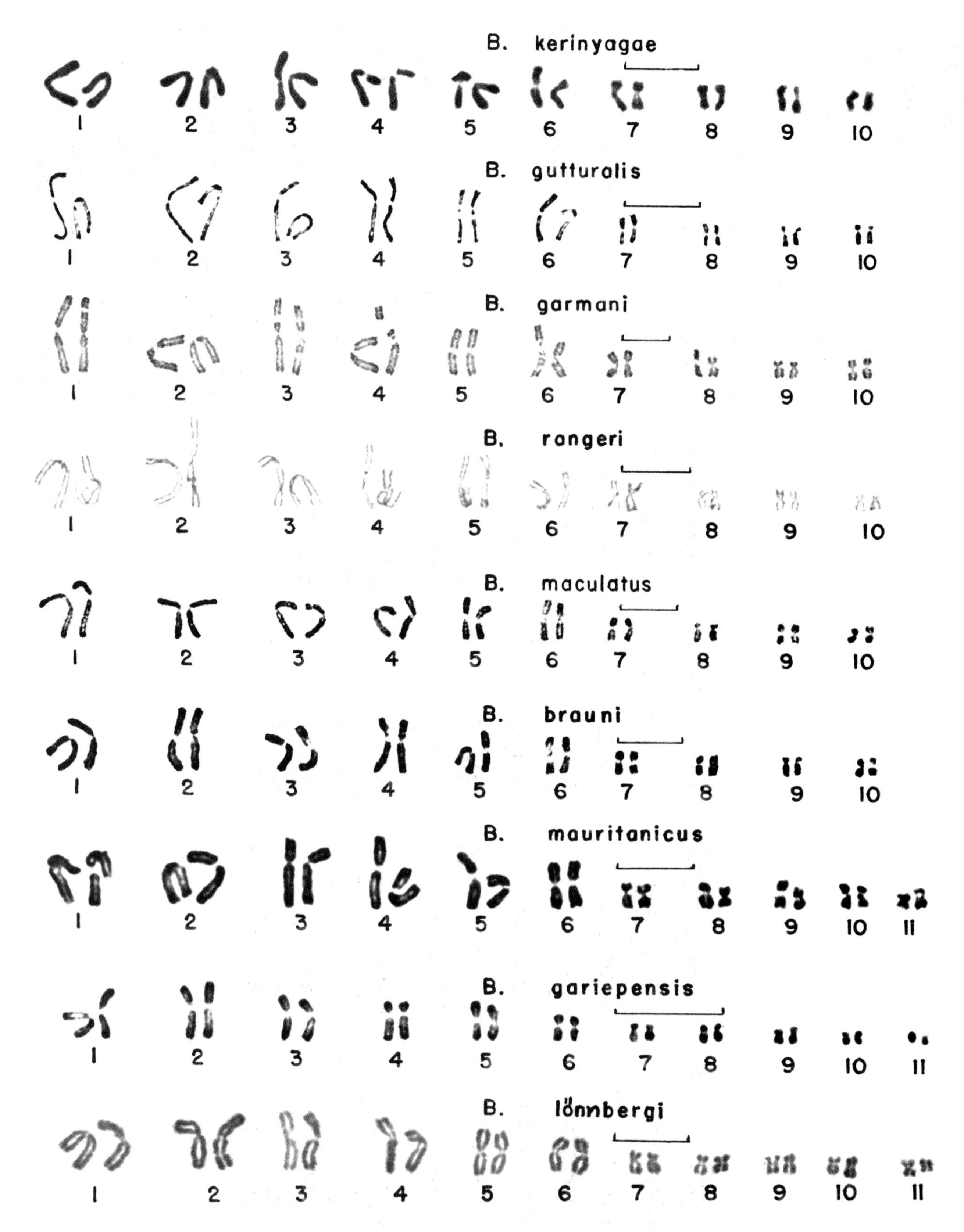

Appendix E–13. Karyotypes of Species of *Bufo*
B. kerinyagae, Kenya; *B. maculatus,* Mtito Andei, Kenya; *B. garmani,* Kruger National Park, South Africa; *B. rangeri,* Swellendam, South Africa; *B. maculatus,* Mazeras, Kenya; *B. brauni,* Amani, Tanzania; *B. mauritanicus,* Morocco; *B. gariepensis,* Grahamstown, South Africa; *B. lönnbergi,* Nanyuki, Kenya.
The scale represents 10 micra in each case.

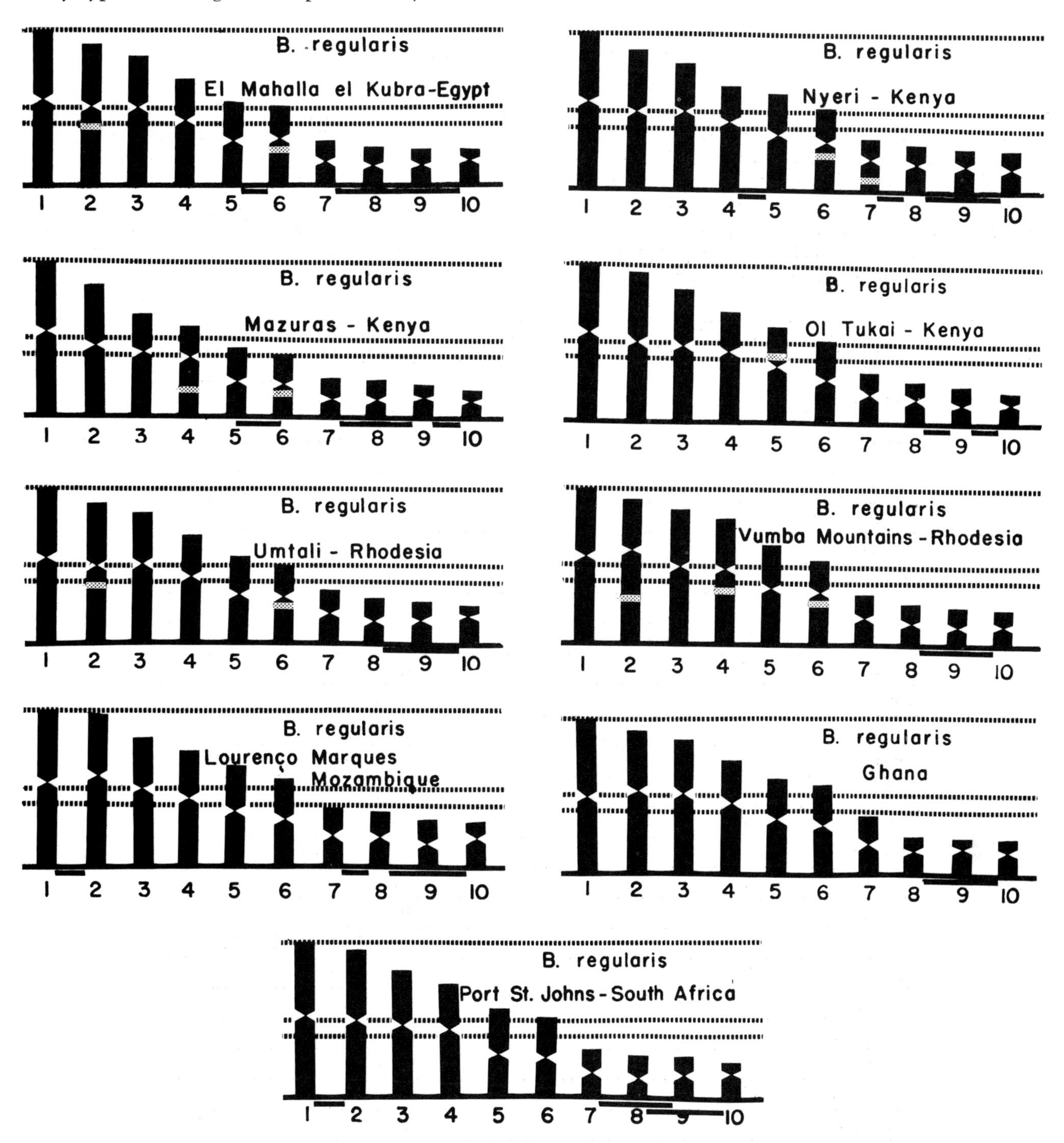

Appendix E–14. Ideograms of Species of *Bufo*

Appendix F. Distribution and Secondary Constrictions in Species of *Bufo*

An *X* indicates that the species is found in the various continents: North America (*NA*), South America (*SA*), Central America (*CA*), Europe (*EU*), Asia (*AS*), or Africa (*AF*). The secondary constrictions were classified *A* through *L* (see text). The classified secondary constrictions found in each species are presented in the column, Secondary Constrictions. If the particular secondary constriction can not be definitely assigned to the species in question, (?) follows the classified constriction. Unclassified secondary constrictions found in the various species are defined as being metacentric (*m*), submetacentric (*sm*), subtelocentric (*st*), or telocentric (*t*) on a certain chromosome number on either the long (*l*) or short (*s*) arm. Thus, *sm(11)s* would indicate that the species has a submetacentric secondary constriction on the short arm of chromosome 11.

Species	NA	SA	CA	EU	AS	AF	Secondary Constrictions
americanus	X						D, C, I
houstonensis	X						D, C, I sm(11)s
microscaphus	X						A, D
terrestris	X						A, D
hemiophrys	X						A, D
woodhousei	X						A, B?
luetkeni	X		X				A, B?
valliceps	X						A, C, G, K, H?, F?, E?
coniferus			X				K
cognatus	X						A
speciosus	X						A
compactilis	X						A, m(11)l
boreas	X						A or D
canorus	X						A or D, I
marmoreus	X						B?, J, sm(1)s
perplexus	X						A or D
punctatus	X						A, D?, sm(3)l
debilis	X						A, m(9)l
kelloggi	X						B, H?, sm(8)l
retiformis	X						A, B, D, E, F, H, J, st(1)s
quercicus	X						A or D, B?
alvarius	X						B, E, F?, J
canaliferus			X				A, C, E
bocourti			X				B, K
occidentalis	X						A, F
holdridgei			X				B
haematiticus		X	X				A, G, K
variegatus		X					K
granulosus		X	X				B, K, st(2)l, t(3)l
spinulosus		X					E, F, K, L[*], st(3)l

[*] *L* in *B. (spinulosus) chilensis.*

Appendix F (Continued)

Species	NA	SA	CA	EU	AS	AF	Secondary Constrictions
marinus	X	X	X				K
atacamensis		X					K
paracnemis		X					K
poeppigi		X					A, K
arenarum		X					B, K
crucifer		X					A, B, D, J, K
bufo				X	X		A or D, F, st(1)s
viridis				X	X	X	A, D, F
stomaticus					X		B, E, F, G, K, sm (2)l
calamita				X			A or D, L
melanostictus					X		A, L
regularis						X	R (see text)
gutturalis						X	R?, sm(5)l, sm(2)s
garmani						X	R, st(4)s, st(3)s, sm(2)s, st(2)l?, m(2)l
rangeri						X	R, sm(2)s, m(4)s
maculatus						X	R, sm(2)l, st(3)5
brauni						X	R, st(3)s, sm(4 or 3)s
mauritanicus						X	L, sm(10)l
gariepensis						X	
lönnbergi						X	A

Appendix G. Combinations of Species of *Bufo* Which Produce Diploid Progeny and Those Species Which Have Been Found to Produce Only Polyploid Offspring

Species	2n, 2n&3n, or 2n&5n	Only 3n or 5n
americanus ♀	*quercicus* B68-103 *luetkeni* B68-110	
asper ♂	*terrestris* B67-225	
houstonensis ♀	*hemiophrys* B66-14, B66-15 *microscaphus* B67-39	
houstonensis ♂	*woodhousei* B65-116 *hemiophrys* B66-39	
microscaphus ♂	*houstonensis* B67-39	
terrestris ♀	*quercicus* B67-242 *asper* B67-225	
terrestris ♂	*woodhousei* B68-43 *hemiophrys* B68-55	
hemiophrys ♀	*punctatus* B66-43 *houstonensis* B66-39 *cognatus* B66-40 *terrestris* B68-55 *speciosus* B68-61 *valliceps* B68-57 *luetkeni* B68-11 *stomaticus* B68-62	*spinulosus* B68-54 (3n)
hemiophrys ♂	*houstonensis* B66-14, B66-15	
woodhousei ♀	*houstonensis* B65-116 *occidentalis* B66-46 *nelsoni* B66-49 *perplexus* B66-48 *quercicus* B66-193 *terrestris* B68-43 *bocourti* B65-7 *stomaticus* B68-41 *Melanophryniscus stelzneri* B68-46	*viridis* B65-13 (3n)
luetkeni ♀	*perplexus* B66-161 *speciosus* B66-171 *spinulosus* B66-166	
luetkeni ♂	*hemiophrys* B68-111 *americanus* B68-110 *marmoreus* B66-285	
valliceps ♀	*alvarius* B65-311 *cognatus* B67-27 *coniferus* B66-263	*viridis* B66-267 (3n)
valliceps ♂	*hemiophrys* B68-57 *coccifer* B65-61 *arenarum* B66-142 *melanostictus* B67-117 *melanostictus* B67-91 *regularis* B65-501	*stomaticus* B67-47 (3n) *rangeri* B65-382 (5n)
coniferus ♂	*valliceps* B66-263	
cognatus ♀	*marmoreus* B65-229 *perplexus* B65-233 *debilis* B65-241	

Appendix G (Continued)

Species	2n, 2n&3n, or 2n&5n	Only 3n or 5n
cognatus ♂	*alvarius* B65-265 *speciosus* B66-37, B67-85 *hemiophrys* B66-40	*viridis* B67-27 (3n)
speciosus ♀	*marmoreus* B66-33, B65-355 *occidentalis* B65-356 *cognatus* B66-37, B67-85 *mazatlanensis* B67-86 *canorus* B67-312 *bocourti* B67-315 *perplexus* B65-359 *granulosus* B66-275 *arenarum* B65-128 *stomaticus* B67-84	*quercicus* B66-277 (3n) *canaliferus* B66-276 (3n)
speciosus ♂	*luetkeni* B66-171 *hemiophrys* B68-61	*canaliferus* B66-175 (3n)
boreas ♂	*canorus* B67-167 *melanostictus* B67-102	
nelsoni ♂	*punctatus* B66-64 *woodhousei* B66-49	
canorus ♀	*boreas* B67-167 *bocourti* B67-205 *spinulosus* B67-207	
canorus ♂	*speciosus* B67-312 *stomaticus* B67-232	
marmoreus ♀	*luetkeni* B66-285	
marmoreus ♂	*speciosus* B65-355, B66-33 *cognatus* B65-229 *stomaticus* B67-135	
coccifer ♂	*valliceps* B65-61	
perplexus ♀	*alvarius* B65-170, B66-248 *spinulosus* B66-240	*granulosus* B65-157 (3n)
perplexus ♂	*speciosus* B65-359 *punctatus* B65-289 *cognatus* B65-233 *woodhousei* B66-48 *luetkeni* B66-173 *arenarum* B66-360 *stomaticus* B67-135	*regularis* B66-153 (5n) *melanostictus* B67-122 (3n)
punctatus ♀	*perplexus* B65-289 *nelsoni* B66-64 *debilis* B66-69 *kelloggi* B66-312 *spinulosus* B66-43	
punctatus ♂	*hemiophrys* B66-43	
debilis ♂	*cognatus* B65-241 *punctatus* B66-69	*viridis* B67-21 (3n) *garmani* B66-121 (3n)
kelloggi ♂	*punctatus* B66-312 *arenarum* B66-356 *viridis* B67-17	

Appendix G (Continued)

Species	2n, 2n&3n, or 2n&5n	Only 3n or 5n
quercicus ♂	*woodhousei* B66-193 *terrestris* B66-242 *americanus* B66-103	*speciosus* B66-277 (3n)
alvarius ♀	*cognatus* B65-265	
alvarius ♂	*valliceps* B65-311 *perplexus* B65-170, B66-248 *viridis* B67-36, B67-24 *melanostictus* B67-114	*marinus* B65-561 (3n)
canaliferus ♀	*perplexus* B66-173 *arenarum* B66-179, B66-183 *spinulosus* B66-185	*speciosus* B66-175 (3n)
canaliferus ♂	*arenarum* B66-349	*speciosus* B66-276 (3n)
bocourti ♂	*speciosus* B67-315 *canorus* B67-205 *woodhousei* B65-7	
occidentalis ♂	*woodhousei* B66-46 *speciosus* B65-356 *garmani* B66-130	
granulosus ♀	*stomaticus* B68-39	*perplexus* B65-157 (3n)
granulosus ♂	*speciosus* B66-275 *arenarum* B66-359 *regularis* B67-62	
spinulosus ♂	*punctatus* B66-28 *luetkeni* B66-166 *perplexus* B66-240 *canorus* B67-207 *canaliferus* B66-185 *arenarum* B67-153 *flavolineatus* B67-173 *flavolineatus* B67-174 *viridis* B67-13	*hemiophrys* B68-54 (3n) *marinus* B65-555 (5n) *garmani* B66-117 (3n) *stomaticus* B67-46 (3n)
flavolineatus ♀	*spinulosus* B67-173 *spinulosus* B67-174	
marinus ♀		*alvarius* B65-561 (3n) *spinulosus* B65-555 (5n) *viridis* B65-492 (5n)
marinus ♂	*arenarum* B66-137, B66-147 *poeppigi* B68-21 *crucifer* B67-181, B67-354 *rangeri* B65-373 *garmani* B65-436 *regularis* B65-528, B67-75 *melanostictus* B67-92	*ictericus* B65-352 (3n) *viridis* B67-1 (5n)
atacamensis ♀	*stomaticus* B67-280	
paracnemis ♂	*arenarum* B64-309	
poeppigi ♀	*marinus* B68-21	
arenarum ♀	*valliceps* B66-142 *kelloggi* B66-356 *perplexus* B66-360	

Appendix G (Continued)

Species	2n, 2n&3n, or 2n&5n	Only 3n or 5n
arenarum ♂	*marinus* B66-137, B66-147 *granulosus* B66-359 *canaliferus* B66-349 *spinulosus* B67-153	
arenarum ♂	*speciosus* B65-128 *canaliferus* B66-183, B66-179 *ictericus* B65-346, B65-356 *paracnemis* B64-309 *melanostictus* B67-103 *rangeri* B65-377 *brauni* B65-479 *regularis* B65-477 *garmani* B66-114	
crucifer ♀		*asper* B67-355 (3n)
crucifer ♂	*marinus* B67-181, B67-354	*melanostictus* B67-265 (3n)
ictericus ♀	*arenarum* B65-346, B65-356	*marinus* B65-352 (3n)
melanostictus ♀	*alvarius* B67-114 *boreas* B67-102 *valliceps* B67-91, B67-117 *arenarum* B67-103 *marinus* B67-92	*perplexus* B67-122 (3n) *crucifer* B67-265 (3n) *asper* B67-261 (3n)
bufo ♂	*woodhousei* B65-119 *viridis* B67-2	
viridis ♀	*kelloggi* B67-17 *alvarius* B67-24, B67-36 *spinulosus* B67-13 *calamita* B65-536 *bufo* B67-2 *stomaticus* B67-12	*debilis* B67-21 (3n) *boreas* B67-14 (3n) *marinus* B67-1 (5n) *mauritanicus* B67-6 (5n)
viridis ♂		*woodhousei* B65-13 (3n) *valliceps* B66-267 (3n) *cognatus* B67-27 (3n) *regularis* B65-492 (5n)
stomaticus ♀	*perplexus* B67-135 *canorus* B67-232 *marmoreus* B67-133	*valliceps* B67-47 (3n) *spinulosus* B67-46 (3n) *asper* B67-224 (3n)
stomaticus ♂	*hemiophrys* B68-62 *woodhousei* B68-41 *granulosus* B68-39 *atacamensis* B67-280 *viridis* B67-12	*speciosus* B67-84 (3n)

Appendix G (Continued)

Species	2n, 2n&3n, or 2n&5n	Only 3n or 5n
calamita ♂	*viridis* B65-536	
asper ♂	*terrestris* B67-225	*crucifer* B67-355 (3n) *stomaticus* B67-224 (3n) *melanostictus* B67-261 (3n)
regularis ♀	*valliceps* B65-501 *granulosus* B67-62 *arenarum* B65-477 *marinus* B65-528, B67-75 *rangeri* B66-100 *perreti* B65-435 *garmani* B65-474, B66-97 *brauni* B65-475, B66-99 *maculatus* B65-520 *garmani?* B65-529 *mauritanicus* B66-252	*perplexus* B66-153 (5n) *viridis* B65-492 (5n) *kerinyagae* B65-524 (3n)
regularis ♂	*rangeri* B65-372, B65-379 *garmani* B66-123, B66-125 *brauni* B65-456, B65-517 *gutturalis* B66-304	
gutturalis ♀	*regularis* B66-304	
garmani ♀	*occidentalis* B66-130 *arenarum* B65-114 *marinus* B65-436 *rangeri* B65-426 *regularis* B66-123, B66-125 *perreti* B65-434 *brauni* B66-107	*debilis* B66-121 (3n) *spinulosus* B66-117 (3n)
garmani ♂	*regularis* B65-474, B66-97 *maculatus* B66-97 *brauni* B65-475	
rangeri ♀	*marinus* B65-373 *arenarum* B65-377 *regularis* B65-372, B65-379	*valliceps* B65-382 (5n)
rangeri ♂	*garmani* B65-426 *regularis* B66-100 *brauni* B65-472	
perreti ♂	*regularis* B65-435 *garmani* B65-434	
maculatus ♀	*garmani* B65-538	
maculatus ♂	*brauni* B65-512, B65-513 *regularis* B65-520	
brauni ♀	*arenarum* B65-479 *maculatus* B65-512, B65-513 *regularis* B65-517, B65-456 *rangeri* B65-472 *garmani* B65-475	
brauni ♂	*regularis* B66-99 *garmani* B66-107	
Mazeras, Kenya *kerinyagae* ♂	*regularis* Ogbomosho, Nigeria B65-455	Entebbe, Uganda B65-524 *regularis* (3n)

Appendix G (Continued)

Population	2n, 2n&3n, or 2n&5n	Only 3n or 5n
mauritanicus ♂	*regularis* B66-252	*viridis* B67-6 (5n)

Populations of Bufo Regularis : Combinations Producing Diploids and Polyploids

Population		
El Mahalla el Kubra, Egypt ♀	Port St. Johns B66-91, B66-98 Kampala, Uganda B66-257 Vumba Mountains B66-93 3n ♂ of B66-93 backcrossed to ♀ from Ghana. The progeny B67-65 had abnormal chromosome numbers. Ogbomosho, Nigeria B66-79 Lourenço Marquis, Mozambique B66-95 ♂ of B66-95 backcrossed to Ghana female. There were abnormal numbers of chromosomes in the progeny B67-66.	
Nyeri, Kenya ♂	Entebbe, Uganda B65-529 Port St. Johns B66-6	
Ol Tukai, Kenya		
(1) ♂		Ghana B67-70 (3n)
(2) ♂	Ghana B67-71	
Umtali, Rhodesia ♀	Louis Trichardt B65-428	
Vumba Mountains ♂	El Mahalla el Kubra B66-93 3n ♂ of B66-93 backcrossed to ♀ from Ghana. The progeny B67-65 had abnormal chromosome numbers.	
Lourenço Marques, Mozambique ♂	El Mahalla el Kubra B66-95 ♂ of B66-95 backcrossed to Ghana ♀. There were abnormal numbers of chromosomes in the progeny B67-66.	
Ghana ♀	B66-95 (El Mahalla x Lourenço Marques) B67-66 — abnormal numbers 3n B66-93 (El Mahalla x Vumba Mountains) B67-65 — abnormal numbers Ol Tukai (2) B67-71	Ol Tukai (1) B67-70 (3n)
Port St. Johns, South Africa ♀	Entebbe, Uganda B66-1 Pietermaritzburg B66-5 Nyeri, Kenya B66-6	
Port St. Johns, South Africa ♂	El Mahalla el Kubra B66-91 El Mahalla el Kubra B66-98	
Louis Trichardt, South Africa ♂	Umtali, Rhodesia B65-428	
Ogbomosho, Nigeria ♂	Entebbe, Uganda B65-522 El Mahalla el Kubra B66-79	
Entebbe, Uganda ♀	Ogbomosho, Nigeria B65-522 Nyeri, Kenya B65-529	
Entebbe, Uganda ♂	Port St. Johns B66-1	
Pietermaritzburg, South Africa ♂	Port St. Johns B66-5	
Kampala, Uganda ♂	El Mahalla el Kubra B66-257	

Appendix H. Results of Interspecific Hybridization of *Bufo*

Combination		Cross Number	% Fertilized	% Fertilized Hatched	Number Larvae	% Metamorphose	Calculated % Original Metamorphose	Stage Reached
♀	♂							
CRUCIFER	*SPINULOSUS*							
crucifer x *flavolineatus*		67-194	5.1	—	—	—	—	G–N
	MARINUS							
crucifer x *arenarum*		67-182	93.6	20.3	12	—	—	G–L
crucifer x *marinus*		67-181	90.7	23.1	49	2.0	0.5	M
crucifer x *marinus*		67-354	67.7	13.6	27	25.9	3.5	M
	GUTTATUS							
crucifer x *haematiticus*								
	BOCOURTI							
crucifer x *bocourti*		67-177	81.8	—	—	—	—	G
crucifer x *bocourti*		67-179	93.3	—	—	—	—	G
	BOREAS							
crucifer x *boreas*		67-183	79.7	—	—	—	—	G
crucifer x *canorus*		67-357	64.4	—	—	—	—	G
	MARMOREUS							
crucifer x *marmoreus*		67-351	90.9	—	—	—	—	G
crucifer x *perplexus*		67-184	100.0	—	—	—	—	G–N
	COGNATUS							
crucifer x *cognatus*		67-192	42.8	—	—	—	—	G
crucifer x *speciosus*		67-358	41.6	—	—	—	—	G
	DEBILIS							
crucifer x *retiformis*		67-353	61.5	—	—	—	—	G
	VALLICEPS							
crucifer x *coniferus*		67-187	92.3	—	—	—	—	G–N
crucifer x *luetkeni*		67-186	91.5	—	—	—	—	G–N
crucifer x *valliceps*		67-176	82.7	—	—	—	—	G
crucifer x *valliceps*		67-178	100.0	—	6	—	—	G–L
crucifer x *valliceps*		67-193	60.7	5.9	15	—	—	N–L
	CANALIFERUS							
crucifer x *canaliferus*		67-188	91.2	4.5	6	—	—	G–L
	COCCIFER							
crucifer x *coccifer*		67-189	98.7	—	—	—	—	G–N
	AMERICANUS							
crucifer x *americanus*		67-185	97.1	—	—	—	—	G
	VIRIDIS							
crucifer x *viridis*		67-190	96.8	—	—	—	—	G–N
	BUFO							
crucifer x *bufo*		67-352	70.0	—	—	—	—	G
	MELANOSTICTUS							
crucifer x *melanostictus*		67-356	15.0	—	—	—	—	G
	ASPER							
crucifer x *asper*		67-355	78.4	2.5	2	50.0	1.2	M
MARINUS	*SPINULOSUS*							
arenarum x *chilensis*		64-415	45.6	52.4	71	—	—	L
arenarum x *chilensis*		64-424	92.0	67.4	91	1.2	0.8	M
arenarum x *chilensis*		67-153	10.9	100.0	119	7.6	7.6	M
arenarum x *flavolineatus*		67-157	5.3	66.7	2	100.0	66.7	M
arenarum x *spinulosus*		65-88	82.8	68.3	236	41.1	28.1	M
marinus x *limensis*		65-227	30.2	100.0	112	40.2	40.2	M
poeppigi x *spinulosus*		65-555	70.5	7.3	60	1.7	.02	M
poeppigi x *spinulosus*		68-23	92.5	44.1	65	—	—	L

M = metamorphosed L = stopped as larva N = stopped as neurula G = stopped as gastrula x = chemical inhibition
S♂ = male proved sterile in F_2 or backcross F♂ = male proved fertile in F_2 or backcross

Combination ♀	♂	Cross Number	% Fertilized	% Fertilized Hatched	Number Larvae	% Metamorphose	Calculated % Original Metamorphose	Stage Reached
MARINUS	*GRANULOSUS*							
arenarum x *fernandezae*		64-432	100.0	66.2	214	33.2	22.0	M
arenarum x *humboldti*		66-359	93.5	100.0	155	90.5	84.6	M
arenarum x *humboldti*		64-417	27.4	85.7	32	59.4	50.9	M
marinus x *pygmaeus*		64-278	69.0	17.2	64	—	—	L
poeppigi x *humboldti*		66-213	31.8	78.6	113	—	—	L
	MARINUS							
arenarum x *marinus*		64-416	98.3	—	—	—	—	G
arenarum x *marinus*		66-147	52.4	93.7	90	41.1	38.5	M
arenarum x *marinus*		66-137	30.0	90.1	35	25.7	23.1	M
arenarum x *paracnemis*		65-85	95.7	1.5	40	40.0	0.6	M
arenarum x *paracnemis*		66-341	23.8	—	—	—	—	G–N
arenarum x *poeppigi*		65-84	84.1	—	8	25.0	—	M
ictericus x *arenarum*		65-346	99.0	100.0	104	83.6	83.6	M
ictericus x *paracnemis*		65-353	51.0	19.2	35	45.7	8.8	M
ictericus x *poeppigi*		65-352	49.2	16.7	23	13.0	2.2	M
marinus x *arenarum*		64-126	51.1	86.9	149	2.7	2.3	M
marinus x *arenarum*		64-287	70.1	4.2	174	—	—	L
marinus x *ictericus*		64-123	77.9	95.0	257	29.2	27.7	M
marinus x *marinus*		64-274	76.0	47.5	118	—	—	L
marinus x *marinus*		64-294	80.5	77.8	92	—	—	L
marinus x *paracnemis*		64-127	47.8	96.9	181	15.5	15.0	M
poeppigi x *arenarum*		68-22	96.5	84.6	250	14.8	11.9	M
poeppigi x *marinus*		65-545	82.8	29.2	114	44.7	13.0	M
poeppigi x *marinus*		66-214	50.0	48.0	92	—	—	L
poeppigi x *marinus*		67-213	60.0	4.8	3	66.7	3.2	M
poeppigi x *marinus*		68-21	100.0	48.8	224	32.8	32.8	M
poeppigi x *marinus*		68-25	80.1	70.3	257	40.8	52.7	M
poeppigi x *marinus*		68-26	92.5	81.6	248	87.8	81.2	M
poeppigi x *poeppigi*		65-562	23.6	92.3	50	60.0	55.4	M
paracnemis x *arenarum*		64-309	84.6	100.0	80	65.0	65.0	M F♂
paracnemis x *ictericus*		64-310	33.3	85.7	6	75.0	64.3	M
	CRUCIFER							
arenarum x *crucifer*		66-342	25.0	—	—	—	—	G
poeppigi x *crucifer*		65-559	31.1	—	3	—	—	L
poeppigi x *crucifer*		67-201	40.0	—	3	—	66.7	M
poeppigi x *crucifer*		67-212	30.8	—	—	—	—	G–N
	GUTTATUS							
arenarum x *haematiticus*		66-355	5.4	—	—	—	—	G
ictericus x *blombergi*		65-350	45.0	—	—	—	—	G
	BOCOURTI							
arenarum x *bocourti*		64-318	10.3	—	—	—	—	G–N
arenarum x *bocourti*		64-429	100.0	—	—	—	—	G–N
arenarum x *bocourti*		67-156	27.0	—	—	—	—	G–N
marinus x *bocourti*		64-279	73.5	—	—	—	—	G
	BOREAS							
arenarum x *boreas*		66-144	41.3	46.5	70	—	—	L
arenarum x *boreas*		67-158	19.4	—	—	—	—	G
arenarum x *boreas*		64-419	66.7	—	—	—	—	G
marinus x *boreas*		64-129	57.4	—	—	—	—	G
poeppigi x *boreas*		66-212	88.6	—	—	—	—	G
poeppigi x *canorus*		67-215	24.5	—	—	—	—	G

M = metamorphosed L = stopped as larva N = stopped as neurula G = stopped as gastrula x = chemical inhibition
S♂ = male proved sterile in F_2 or backcross F♂ = male proved fertile in F_2 or backcross

Appendix H (Continued)

Combination	Cross Number	% Fertilized	% Fertilized Hatched	Number Larvae	% Metamorphose	Calculated % Original Metamorphose	Stage Reached
♀ *MARINUS* ♂ *ALVARIUS*							
arenarum x *alvarius*	64-423	96.4	68.5	118	31.3	21.4	M
marinus x *alvarius*	64-283	44.8	84.6	72	—	—	L
marinus x *alvarius*	64-269	62.5	50.0	65	—	—	L
poeppigi x *alvarius*	66-215	28.3	68.4	70	—	—	L
poeppigi x *alvarius*	65-561	32.7	29.4	35	14.3	4.2	M
MARMOREUS							
arenarum x *marmoreus*	64-315	42.8	91.7	136	2.9	2.7	S♂
arenarum x *perplexus*	66-360	90.6	100.0	148	—	—	L
arenarum x *perplexus*	66-133	44.7	100.0	63	1.6	1.6	M
marinus x *marmoreus*	64-131	91.4	15.1	56	—	—	L
poeppigi x *marmoreus*	66-210	78.4	22.5	26	—	—	G–L
poeppigi x *perplexus*	66-205	91.3	26.2	—	—	—	G–N
poeppigi x *perplexus*	67-214	76.0	60.5	29	—	—	L
COGNATUS							
arenarum x *cognatus*	64-430	55.5	50.0	55	—	—	L
arenarum x *cognatus*	64-304	79.4	—	—	—	—	G
arenarum x *speciosus*	64-316	21.7	100.0	8	—	—	L
arenarum x *speciosus*	64-433	68.9	25.8	108	—	—	L
marinus x *cognatus*	64-128	68.0	—	—	—	—	G–N
marinus x *cognatus*	64-286	84.6	—	3	—	—	L
marinus x *speciosus*	64-275	75.2	35.9	14	—	—	L
poeppigi x *speciosus*	66-209	60.5	—	—	—	—	G–N
DEBILIS							
arenarum x *debilis*	64-422	97.4	—	—	—	—	N
arenarum x *debilis*	66-145	22.0	—	2	—	—	L
arenarum x *kelloggi*	66-356	98.0	15.7	48	16.0	15.7	M
arenarum x *retiformis*	67-437	50.9	21.8	26	—	—	L
marinus x *debilis*	64-130	8.1	—	—	—	—	G
poeppigi x *debilis*	66-207	83.9	—	—	—	—	G
VALLICEPS							
arenarum x *coniferus*	66-353	98.3	3.6	4	—	—	G–L
arenarum x *coniferus*	67-434	43.4	80.4	137	13.8	11.1	M
arenarum x *coniferus*	67-436	55.7	64.1	170	12.3	7.9	M
arenarum x *coniferus*	67-438	67.8	75.0	130	2.3	1.7	M
arenarum x *ibarrai*	64-425	98.0	—	—	—	—	N
arenarum x *ibarrai*	65-81	80.2	72.3	247	0.4	0.3	N
arenarum x *luetkeni*	64-303	25.7	11.1	4	—	—	L
arenarum x *luetkeni*	65-83	87.5	80.3	185	—	—	L
arenarum x *luetkeni*	64-314	24.0	100.0	196	—	—	L
arenarum x *luetkeni*	64-317	93.9	100.0	196	—	—	L
arenarum x *mazatlanensis*	66-358	95.1	100.0	258	—	—	L
arenarum x *valliceps*	66-142	61.3	100.0	38	28.9	28.9	M
arenarum x *valliceps*	64-301	84.0	66.7	111	78.9	52.6	M
marinus x *luetkeni*	64-124	42.7	5.3	22	—	—	L
poeppigi x *coniferus*	66-220	45.9	75.3	54	—	—	L
poeppigi x *coniferus*	68-30	—	—	—	—	—	G
poeppigi x *coniferus*	67-216	11.8	—	—	—	—	G
poeppigi x *luetkeni*	68-27	96.2	3.6	27	—	—	L
poeppigi x *valliceps*	68-29	86.7	24.2	216	—	—	L
CANALIFERUS							
arenarum x *canaliferus*	66-349	98.1	86.5	145	—	—	L
arenarum x *canaliferus*	67-435	64.8	?	100	—	—	L

M = metamorphosed L = stopped as larva N = stopped as neurula G = stopped as gastrula x = chemical inhibition
S♂ = male proved sterile in F_2 or backcross F♂ = male proved fertile in F_2 or backcross

Combination ♀	♂	Cross Number	% Fertilized	% Fertilized Hatched	Number Larvae	% Metamorphose	Calculated % Original Metamorphose	Stage Reached
MARINUS	*CANALIFERUS*							
arenarum x *canaliferus*		67-154	13.0	100.0	77	—	—	L
arenarum x *canaliferus*		67-155	12.5	100.0	18	—	—	L
arenarum x *canaliferus*		66-345	58.8	3.3	8	—	—	L
marinus x *canaliferus*		64-273	40.0	77.8	59	—	—	L
marinus x *canaliferus*		64-284	33.3	21.0	51	—	—	L
	OCCIDENTALIS							
arenarum x *occidentalis*		66-131	57.1	100.0	156	—	—	L
	COCCIFER							
arenarum x *coccifer*		64-319	2.8	—	—	—	—	G–N
arenarum x *coccifer*		64-427	92.6	—	—	—	—	G
poeppigi x *coccifer*		66-204	91.3	—	—	—	—	G–N
	AMERICANUS							
arenarum x *americanus*		64-431	93.2	—	—	—	—	G
arenarum x *hemiophrys*		67-431	55.2	—	—	—	—	G
arenarum x *microscaphus*		64-428	12.7	—	—	—	—	G
arenarum x *terrestris*		65-92	78.3	—	—	—	—	G–N
arenarum x *woodhousei*		64-420	88.7	—	—	—	—	G
arenarum x *woodhousei*		64-302	44.7	—	—	—	—	G
marinus x *americanus*		64-281	98.2	—	—	—	—	G
marinus x *microscaphus*		64-280	30.4	—	—	—	—	G
marinus x *woodhousei*		64-125	18.3	—	—	—	—	G
marinus x *woodhousei*		64-293	71.0	—	—	—	—	G
poeppigi x *hemiophrys*		65-557	59.6	—	—	—	—	G
poeppigi x *microscaphus*		66-206	86.5	—	—	—	—	G
poeppigi x *terrestris*		66-208	78.3	—	—	—	—	G
	QUERCICUS							
arenarum x *quercicus*		66-350	90.9	50.0	9	—	—	G–L
poeppigi x *quercicus*		65-546	65.1	—	—	—	—	G–N
poeppigi x *quercicus*		66-217	45.6	—	—	—	—	G
	VIRIDIS							
arenarum x *viridis*		64-421	95.5	—	—	—	—	G
poeppigi x *viridis*		65-558	64.6	—	—	—	—	G–N
	CALAMITA							
poeppigi x *calamita*		65-554	74.1	—	—	—	—	G
	BUFO							
arenarum x *bufo*		65-82	30.4	—	—	—	—	G
arenarum x *bufo*		66-138	17.2	—	—	—	—	G
arenarum x *bufo*		66-146	47.3	—	—	—	—	G
arenarum x *bufo*		66-348	98.3	—	—	—	—	G
poeppigi x *bufo*		65-550	58.6	—	—	—	—	G–N
	STOMATICUS							
arenarum x *stomaticus*		66-343	74.2	—	8	—	—	L
arenarum x *stomaticus*		66-352	97.0	93.8	261	—	—	L
	MELANOSTICTUS							
arenarum x *melanostictus*		66-141	56.1	—	—	—	—	G
arenarum x *melanostictus*		66-347	?	—	—	—	—	G
poeppigi x *melanostictus*		66-203	—	—	—	—	—	G
	ASPER							
arenarum x *asper*		67-430	31.2	—	4	—	—	L
poeppigi x *asper*		67-211	38.7	—	—	—	—	G–N

M = metamorphosed L = stopped as larva N = stopped as neurula G = stopped as gastrula x = chemical inhibition
S♂ = male proved sterile in F_2 or backcross F♂ = male proved fertile in F_2 or backcross

Combination ♀	♂	Cross Number	% Fertilized	% Fertilized Hatched	Number Larvae	% Metamorphose	Calculated % Original Metamorphose	Stage Reached
MARINUS	*MAURITANICUS*							
arenarum x *mauritanicus*		66-351	98.3	—	—	—	—	G
arenarum x *mauritanicus*		66-357	98.2	—	—	—	—	G
marinus x *mauritanicus*		64-267	57.6	89.5	67	—	—	L
	GARIEPENSIS							
poeppigi x *inyangae*		65-552	—[1]	—	—	—	—	—
	LÖNNBERGI							
poeppigi x *lönnbergi*		65-553	68.7	—	—	—	—	G–N
	REGULARIS							
arenarum x *brauni*		66-143	65.9	—	—	—	—	G
arenarum x *rangeri*		67-432	14.9	—	—	—	—	G
arenarum x *regularis* (W)		65-91	100.0	—	—	—	—	G
arenarum x *regularis* (E)		66-126	32.7	—	—	—	—	G
arenarum x *regularis* (E)		66-136	53.9	—	—	—	—	G
ictericus x *regularis* (W)		65-349	43.1	—	—	—	—	G
marinus x *regularis* (W)		64-271	88.9	—	—	—	—	G
poeppigi x *brauni*		65-548	81.7	—	—	—	—	G
poeppigi x *garmani*		65-547	65.1	—	—	—	—	G
poeppigi x *rangeri*		65-543	92.2	—	—	—	—	G
poeppigi x *regularis* (W)		66-219	61.1	—	—	—	—	G
poeppigi x *regularis* (E)		65-544	85.7	—	—	—	—	G
poeppigi x *regularis* (E)		65-556	65.6	—	—	—	—	G
poeppigi x *regularis* (E)		65-560	36.6	—	—	—	—	G
poeppigi x *regularis* (W)		68-28	88.2	—	—	—	—	G
	MACULATUS							
ictericus x *maculatus*		65-351	27.4	—	—	—	—	G
poeppigi x *maculatus*		65-551	72.5	—	—	—	—	G–N
TYPHONIUS	*GRANULOSUS*							
typhonius x *fernandezae*		68-32	4.5	—	—	—	—	G
typhonius x *fernandezae*		68-37	4.5	—	—	—	—	G
	MARINUS							
typhonius x *marinus*		67-410	48.6	11.1	2	—	—	G–L
	BOCOURTI							
typhonius x *bocourti*		67-411	46.9	—	—	—	—	G–N
	MARMOREUS							
typhonius x *marmoreus*		67-414	35.0	—	—	—	—	G
	VALLICEPS							
typhonius x *coniferus*		67-408	51.3	10.0	2	—	—	G–L
typhonius x *valliceps*		67-409	21.6	—	—	—	—	G
typhonius x *valliceps*		68-49	27.2	—	—	—	—	G
	AMERICANUS							
typhonius x *americanus*		67-413	66.7	—	—	—	—	G
typhonius x *woodhousei*		68-44	88.5	—	—	—	—	G
	REGULARIS							
typhonius x *regularis*[2]		67-412	30.5	—	—	—	—	G
BOCOURTI	*SPINULOSUS*							
bocourti x *spinulosus*		65-95	27.9	—	—	—	—	G–N
bocourti x *spinulosus*		65-112	37.5	—	—	—	—	N

M = metamorphosed L = stopped as larva N = stopped as neurula G = stopped as gastrula x = chemical inhibition
S♂ = male proved sterile in F_2 or backcross F♂ = male proved fertile in F_2 or backcross

[1] Inadvertently discarded. [2] Location unknown.

Combination ♀	♂	Cross Number	% Fertilized	% Fertilized Hatched	Number Larvae	% Metamorphose	Calculated % Original Metamorphose	Stage Reached
BOCOURTI	GRANULOSUS							
bocourti x *humboldti*		65-101	12.5	—	—	—	—	G
bocourti x *humboldti*		65-113	6.3	—	—	—	—	G
	MARINUS							
bocourti x *arenarum*		65-105	90.0	—	—	—	—	G–N
bocourti x *poeppigi*		65-103	17.0	—	—	—	—	G
	BOREAS							
bocourti x *boreas*		65-96	13.0	—	—	—	—	G
bocourti x *boreas*		68-16	—	—	—	—	—	G
	PUNCTATUS							
bocourti x *punctatus*		65-97	14.5	—	—	—	—	G
	MARMOREUS							
bocourti x *marmoreus*		68-11	11.6	20.0	1	—	—	L
	COGNATUS							
bocourti x *cognatus*		65-98	27.8	10.0	2	—	—	L
	VALLICEPS							
bocourti x *valliceps*		65-94	25.8	—	—	—	—	G
bocourti x *valliceps*		65-100	34.0	—	—	—	—	G
bocourti x *valliceps*		65-109	39.1	—	—	—	—	G
	COCCIFER							
bocourti x *coccifer*		65-102	5.1	—	—	—	—	G
bocourti x *coccifer*		65-111	12.9	—	—	—	—	G
bocourti x *coccifer*		65-114	1.8	—	—	—	—	G
	AMERICANUS							
bocourti x *terrestris*		65-107	57.4	—	—	—	—	G
bocourti x *woodhousei*		65-93	6.7	—	—	—	—	G
bocourti x *woodhousei*		65-99	41.4	—	—	—	—	G
bocourti x *woodhousei*		65-108	43.9	—	—	—	—	G
	VIRIDIS							
bocourti x *viridis*		68-8	12.5	—	—	—	—	G
	CALAMITA							
bocourti x *calamita*		68-9	7.0	—	—	—	—	G
	BUFO							
bocourti x *bufo*		65-104	9.8	—	—	—	—	G
	STOMATICUS							
bocourti x *stomaticus*		68-10	9.3	—	—	—	—	G
	MELANOSTICTUS							
bocourti x *melanostictus*		68-12	4.3	—	—	—	—	G
	ASPER							
bocourti x *asper*		68-15	18.6	—	—	—	—	G
	REGULARIS							
bocourti x *rangeri*		68-7	4.3	—	—	—	—	G
bocourti x *regularis* (W)		65-106	83.9	—	—	—	—	G
SPINULOSUS	SPINULOSUS							
chilensis x *spinulosus*		67-415	94.2	88.0	174	0.1	0.1	M
flavolineatus x *chilensis*		67-173	66.7	85.0	110	10.2	8.5	M
flavolineatus x *spinulosus*		67-174	100.0	30.4	89	11.2	3.4	M
spinulosus x *chilensis*		67-440	74.4	48.6	73	4.1	2.0	M
	GRANULOSUS							
chilensis x *humboldti*		64-399	41.2	—	—	—	—	N

M = metamorphosed L = stopped as larva N = stopped as neurula G = stopped as gastrula x = chemical inhibition
S♂ = male proved sterile in F_2 or backcross F♂ = male proved fertile in F_2 or backcross

Appendix H (Continued)

Combination ♀	♂	Cross Number	% Fertilized	% Fertilized Hatched	Number Larvae	% Metamorphose	Calculated % Original Metamorphose	Stage Reached
SPINULOSUS	*MARINUS*							
atacamensis x *marinus*		67-296	17.1	—	—	—	—	G
chilensis x *arenarum*		64-386	100.0	41.7	164	14.6	6.1	M
chilensis x *arenarum*		67-420	100.0	51.5	136	8.8	4.5	M
chilensis x *marinus*		67-422	92.8	—	—	—	—	G
chilensis x *paracnemis*		64-407	64.2	—	—	—	—	N
spinulosus x *poeppigi*		65-294	45.2	—	—	—	—	G
	CRUCIFER							
atacamensis x *crucifer*		67-278	28.4	—	—	—	—	G
	PERIGLENES							
spinulosus x *periglenes*		67-441	87.0	—	6	—	—	L
	BOCOURTI							
chilensis x *bocourti*		64-395	14.5	—	—	—	—	G
chilensis x *bocourti*		67-419	97.9	16.7	31	—	—	L
spinulosus x *bocourti*		67-443	81.4	31.4	20	5.0	1.6	M
	BOREAS							
chilensis x *boreas*		64-387	51.7	—	—	—	—	N
chilensis x *boreas*		64-389	82.8	—	—	—	—	N
chilensis x *boreas*		64-401	28.6	—	—	—	—	G
atacamensis x *boreas*		67-290	5.3	—	—	—	—	G
atacamensis x *canorus*		67-294	76.7	24.2	36	—	—	G–L
chilensis x *boreas*		67-418	79.1	39.6	61	—	—	L
chilensis x *canorus*		67-204	29.2	28.6	6	33.3	9.5	M
spinulosus x *boreas*		65-299	15.8	—	—	—	—	G–N
spinulosus x *boreas*		65-292	73.4	78.7	141	—	—	L
flavolineatus x *boreas*		67-175	76.9	34.0	30	—	—	L
flavolineatus x *canorus*		67-171	60.5	78.2	118	—	—	L
	ALVARIUS							
chilensis x *alvarius*		64-404	19.5	—	1	100.0	—	M
atacamensis x *alvarius*		67-277	17.0	3.2	2	—	—	G–L
atacamensis x *alvarius*		67-295	19.1	—	1	—	—	G–L
	PUNCTATUS							
atacamensis x *punctatus*		67-284	60.4	62.5	21	—	—	G–L
atacamensis x *punctatus*		67-289	6.9	—	—	—	—	G–N
spinulosus x *punctatus*		65-295	3.5	—	—	—	—	G–N
spinulosus x *punctatus*		65-301	5.1	—	—	—	—	G–N
	MARMOREUS							
atacamensis x *marmoreus*		67-298	4.4	—	—	—	—	G–L
spinulosus x *marmoreus*		65-293	91.2	89.1	193	34.7	30.9	M
spinulosus x *perplexus*		65-169	90.6	—	—	—	—	G
	COGNATUS							
chilensis x *cognatus*		64-397	71.0	9.1	2	—	—	L
chilensis x *speciosus*		64-410	23.1	—	—	—	—	G
atacamensis x *cognatus*		67-286	16.0	37.5	3	—	—	L
atacamensis x *speciosus*		67-293	51.9	3.7	4	—	—	G–L
chilensis x *cognatus*		67-421	64.3	88.9	8	12.5	11.1	M
spinulosus x *cognatus*		65-300	15.4	87.5	7	14.3	12.5	M
	DEBILIS							
chilensis x *debilis*		64-403	55.5	—	—	—	—	G–N
atacamensis x *retiformis*		67-283	69.2	—	7	—	—	L
spinulosus x *retiformis*		67-446	22.5	5.0	6	—	—	L

M = metamorphosed L = stopped as larva N = stopped as neurula G = stopped as gastrula x = chemical inhibition
S♂ = male proved sterile in F_2 or backcross F♂ = male proved fertile in F_2 or backcross

Appendix H (Continued)

Combination ♀	♂	Cross Number	% Fertilized	% Fertilized Hatched	Number Larvae	% Metamorphose	Calculated % Original Metamorphose	Stage Reached
SPINULOSUS	*VALLICEPS*							
chilensis x *ibarrai*		64-408	25.0	7.7	7	14.3	1.1	M
chilensis x *luetkeni*		64-406	62.7	—	17	—	—	L
chilensis x *valliceps*		64-398	66.7	—	—	—	—	G–N
chilensis x *valliceps*		64-388	72.7	78.1	105	4.8	3.7	M
atacamensis x *coniferus*		67-299	39.5	—	—	—	—	G
atacamensis x *coniferus*		67-300	30.0	—	—	—	—	G
atacamensis x *valliceps*		67-287	100.0	33.3	6	—	—	N–L
spinulosus x *coniferus*		67-445	66.7	—	3	—	—	L
	COCCIFER							
spinulosus x *coccifer*		65-296	63.0	—	—	—	—	G
	AMERICANUS							
chilensis x *americanus*		64-409	57.1	—	—	—	—	N
chilensis x *microscaphus*		64-405	18.9	—	—	—	—	G
chilensis x *woodhousei*		64-390	71.1	—	—	—	—	G–N
chilensis x *woodhousei*		64-391	61.3	—	—	—	—	G–N
chilensis x *woodhousei*		64-392	78.8	—	—	—	—	G–N
spinulosus x *hemiophrys*		67-447	65.3	—	—	—	—	G–N
	VIRIDIS							
chilensis x *viridis*		64-402	32.6	—	—	—	—	N
chilensis x *viridis*		67-417	91.1	84.3	143	—	—	L
flavolineatus x *viridis*		67-172	38.7	50.0	102	—	—	L
	CALAMITA							
spinulosus x *calamita*		65-298	56.6	—	—	—	—	G–N
	BUFO							
atacamensis x *bufo*		67-292	48.5	—	2	—	—	G-L
spinulosus x *bufo*		65-297	41.9	—	—	—	—	G
	STOMATICUS							
spinulosus x *stomaticus*		67-442	35.5	52.4	43	4.6	2.4	M
	ASPER							
atacamensis x *asper*		67-276	21.6	—	—	—	—	G
	REGULARIS							
chilensis x *regularis*[3]		64-411	27.8	—	—	—	—	G–N
chilensis x *regularis*[4]		64-413	8.5	—	—	—	—	G
spinulosus x *rangeri*		67-448	33.8	—	—	—	—	G–N
	MACULATUS							
chilensis x *maculatus*		64-393	53.8	—	—	—	—	G–N
chilensis x *maculatus*		64-394	55.8	—	—	—	—	G
GRANULOSUS	*SPINULOSUS*							
pygmaeus x *spinulosus*		65-59	60.0	100.0	223	27.8	27.8	M
	MARINUS							
fernandezae x *arenarum*		64-344	75.9	66.7	60	65.0	50.0	M
fernandezae x *ictericus*		68-36	85.7	—	—	—	—	G–N
fernandezae x *marinus*		64-133	low	—	—	—	—	G
fernandezae x *poeppigi*		68-40	11.4	—	—	—	—	G
humboldti x *poeppigi*		65-57	98.3	—	—	—	—	G
pygmaeus x *arenarum*		64-214	73.3	13.6	28	32.1	4.4	M
	CRUCIFER							
fernandezae x *crucifer*		68-33	66.7	—	—	—	—	G

M = metamorphosed L = stopped as larva N = stopped as neurula G = stopped as gastrula x = chemical inhibition
S♂ = male proved sterile in F_2 or backcross F♂ = male proved fertile in F_2 or backcross

[3] Location unknown. [4] Location unknown.

Combination ♀	♂	Cross Number	% Fertilized	% Fertilized Hatched	Number Larvae	% Metamorphose	Calculated % Original Metamorphose	Stage Reached
GRANULOSUS	*TYPHONIUS*							
fernandezae x *typhonius*		68-32	4.3	—	—	—	—	G
fernandezae x *typhonius*		68-37	1.0	—	—	—	—	G
	BOCOURTI							
pygmaeus x *bocourti*		64-213	8.9	—	—	—	—	G–N
humboldti x *bocourti*		65-56	70.5	—	—	—	—	G–N
	BOREAS							
fernandezae x *boreas*		68-31	47.6	30.0	14	—	—	L
	ALVARIUS							
fernandezae x *alvarius*		64-353	low	—	—	—	—	G
	PUNCTATUS							
fernandezae x *punctatus*		64-347	13.9	—	—	—	—	G–N
	COGNATUS							
fernandezae x *cognatus*		64-134	low	—	—	—	—	N
fernandezae x *cognatus*		64-352	low	—	—	—	—	G
fernandezae x *speciosus*		64-350	low	—	—	—	—	N
	DEBILIS							
pygmaeus x *debilis*		64-210	12.3	—	—	—	—	G
	VALLICEPS							
fernandezae x *coniferus*		68-35	15.8	—	—	—	—	G–N
fernandezae x *luetkeni*		64-132	low	—	—	—	—	G–N
fernandezae x *valliceps*		64-22	6.4	6.2	3	—	—	L
fernandezae x *valliceps*		64-348	3.9	—	2	50.0	—	M
pygmaeus x *valliceps*		64-211	62.8	2.3	1	—	—	L
	COCCIFER							
fernandezae x *coccifer*		64-343	24.4	—	—	—	—	G–N
fernandezae x *coccifer*		64-346	25.9	—	—	—	—	G–N
	AMERICANUS							
fernandezae x *woodhousei*		64-52	50.0	1.2	1	—	—	L
fernandezae x *woodhousei*		64-351	low	—	—	—	—	G
fernandezae x *woodhousei*		64-212	low	—	—	—	—	G
	QUERCICUS							
fernendezae x *quercicus*		64-345	3.1	—	—	—	—	G
	VIRIDIS							
humboldti x *viridis*		65-58	4.4	20.0	2	—	—	L
	CALAMITA							
fernandezae x *calamita*		68-38	46.5	52.5	63	—	—	L
	STOMATICUS							
fernandezae x *stomaticus*		68-34	29.2	—	—	—	—	G–N
fernandezae x *stomaticus*		68-39	54.2	100.0	32	—	—	L
	REGULARIS							
humboldti x *regularis* (W)		65-60	90.0	—	2	—	—	L
BOREAS	*SPINULOSUS*							
boreas x *spinulosus*		63-43	100.0	94.8	249	63.0	59.7	M
canorus x *chilensis*		67-204	29.2	28.6	6	33.3	9.5	M
canorus x *flavolineatus*		67-163	7.1	—	—	—	—	G
canorus x *trifolium*		67-161	3.1	50.0	1	—	—	L
canorus x *trifolium*		67-207	33.6	75.6	31	25.8	19.5	M
	GRANULOSUS							
boreas x *humboldti*		65-185	55.5	48.0	70	37.1	17.8	M

M = metamorphosed L = stopped as larva N = stopped as neurula G = stopped as gastrula x = chemical inhibition
S♂ = male proved sterile in F_2 or backcross F♂ = male proved fertile in F_2 or backcross

Combination ♀	♂	Cross Number	% Fertilized	% Fertilized Hatched	Number Larvae	% Metamorphose	Calculated % Original Metamorphose	Stage Reached
BOREAS	*MARINUS*							
boreas x *arenarum*		65-189	45.0	3.7	4	25.0	0.9	M
boreas x *poeppigi*		65-212	44.8	—	—	—	—	G
	CRUCIFER							
canorus x *crucifer*		67-183	79.8	—	—	—	—	G
canorus x *crucifer*		67-357	69.0	—	—	—	—	G
	BOCOURTI							
boreas x *bocourti*		65-184	41.4	37.5	107	8.4	3.1	M
canorus x *bocourti*		67-162	33.3	—	—	—	—	G
canorus x *bocourti*		67-205	71.7	53.5	58	24.1	12.9	M
	BOREAS							
boreas x *canorus*		63-55	47.8	93.8	230	83.5	78.2	M
boreas x *canorus*		63-57	93.5	97.3	269	58.5	53.1	M
boreas x *exsul*		65-192	30.0	94.4	17	76.5	72.2	M
boreas x *exsul*		65-217	12.0	100.0	20	10.0	—	M
boreas x F_1 (*boreas* x *canorus*)		64-442	36.2	64.7	18	50.0	32.3	M
	ALVARIUS							
boreas x *alvarius*		63-51	98.2	90.9	255	80.0	72.7	S♂
boreas x *alvarius*		63-52	98.5	70.3	244	68.8	52.7	M
boreas x *alvarius*		65-222	20.4	100.0	42	83.3	83.3	M
	PUNCTATUS							
boreas x *punctatus*		63-41	low	100.0	21	66.7	66.7	M
boreas x *punctatus*		65-221	25.0	75.0	22	27.3	20.5	M
	MARMOREUS							
boreas x *marmoreus*		63-45	100.0	86.8	246	73.1	63.4	M
boreas x *perplexus*		63-50	98.4	82.2	251	71.3	58.6	M
canorus x *marmoreus*		67-398	82.7	62.8	43	18.6	11.7	M
	COGNATUS							
boreas x *cognatus*		65-190	52.4	?	75	1.3	<1.3	M
boreas x *cognatus*		64-444	34.6	—	—	—	—	L
boreas x *compactilis*		63-58	97.2	100.0	269	65.0	65.0	M
boreas x *speciosus*		65-215	17.6	33.3	8	—	—	L
canorus x *cognatus*		67-165	50.0	—	—	—	—	G
canorus x *cognatus*		67-166	2.1	100.0	1	—	—	L
canorus x *cognatus*		67-396	80.4	7.3	16	—	—	L
canorus x *speciosus*		67-399	37.7	57.7	15	—	—	G–L
	DEBILIS							
boreas x *debilis*		65-211	67.6	58.0	95	11.6	6.7	M
boreas x *debilis*		65-223	18.9	100.0	29	6.9	6.9	M
boreas x *debilis*		65-225	5.9	—	20	—	—	L
canorus x *retiformis*		67-397	58.8	22.5	17	—	—	L
	VALLICEPS							
boreas x *ibarrai*		65-187	44.3	37.1	100	35.0	13.0	M
boreas x *mazatlanensis*		63-53	97.9	97.9	246	60.2	58.9	M
boreas x *valliceps*		63-40	100.0	4.4	22	54.5	2.4	M
boreas x *valliceps*		65-220	6.2	100.0	23	21.7	21.7	M
	OCCIDENTALIS							
boreas x *occidentalis*		63-44	98.0	—	—	—	—	G
	COCCIFER							
boreas x *coccifer*		65-188	79.5	—	—	—	—	G

M = metamorphosed L = stopped as larva N = stopped as neurula G = stopped as gastrula x = chemical inhibition
S♂ = male proved sterile in F_2 or backcross F♂ = male proved fertile in F_2 or backcross

Combination ♀	♂	Cross Number	% Fertilized	% Fertilized Hatched	Number Larvae	% Metamorphose	Calculated % Original Metamorphose	Stage Reached
BOREAS	*AMERICANUS*							
boreas x *americanus*		63-39	98.1	61.5	211	—	—	L
boreas x *americanus*		63-56	88.5	81.1	262	0.4	0.3	M
boreas x *hemiophrys*		63-46	98.2	81.8	147	—	—	L
boreas x *microscaphus*		65-193	57.1	50.0	2	—	—	L
boreas x *microscaphus*		65-213	20.6	42.8	8	—	—	L
boreas x *microscaphus*		65-226	30.0	—	9	—	—	L
boreas x *terrestris*		65-191	36.4	83.3	25	8.0	6.9	M
boreas x *woodhousei*		63-42	98.5	69.1	145	—	—	L
canorus x *hemiophrys*		67-170	52.2	—	—	—	—	G–N
	VIRIDIS							
boreas x *viridis*		65-186	60.3	25.7	73	57.5	14.8	M
canorus x *viridis*		67-169	33.3	—	—	—	—	G
canorus x *viridis*		67-206	63.3	61.3	70	2.8	1.7	M
	CALAMITA							
boreas x *calamita*		63-37	100.0	55.9	144	10.4	5.8	M
boreas x *calamita*		63-59	100.0	76.6	251	—	—	L
boreas x *calamita*		65-224	27.6	100.0	73	—	—	L
	BUFO							
boreas x *bufo*		63-49	97.3	52.8	39	10.2	5.4	M
boreas x *bufo*		63-54	61.5	100.0	102	—	—	L
canorus x *bufo*		67-164	26.3	—	—	—	—	G
canorus x *bufo*		67-208	46.2	—	—	—	—	G
	STOMATICUS							
canorus x *stomaticus*		67-394	47.2	17.6	13	23.0	4.0	M
	MELANOSTICTUS							
canorus x *melanostictus*		67-395	51.8	—	—	—	—	G
	ASPER							
canorus x *asper*		67-393	26.0	—	—	—	—	G
	REGULARIS							
boreas x *regularis* (W)		65-216	18.4	28.5	8	—	—	L
	MACULATUS							
boreas x *maculatus*		65-219	25.5	—	—	—	—	G
	PERRETI							
boreas x *perreti*		65-218	24.4	60.0	22	—	—	L
ALVARIUS	*SPINULOSUS*							
alvarius x *spinulosus*		65-258	57.8	—	—	—	—	G
	MARINUS							
alvarius x *arenarum*		65-263	78.3	87.0	147	0.7	0.6	M
alvarius x *arenarum*		64-358	38.3	4.3	1	—	—	L
alvarius x *poeppigi*		65-264	65.3	—	—	—	—	G
alvarius x *marinus*		65-286	91.2	—	—	—	—	G
	GUTTATUS							
alvarius x *blombergi*		65-290	92.4	—	—	—	—	G
	BOCOURTI							
alvarius x *bocourti*		62-269	85.9	—	—	—	—	G
	BOREAS							
alvarius x *boreas*		65-256	7.8	10.0	13	—	—	L
alvarius x *boreas*		65-260	53.6	77.7	42	—	—	L
alvarius x *boreas*		65-282	63.6	88.1	237	—	—	L
alvarius x *boreas*		64-356	1.0	—	—	—	—	G

M = metamorphosed L = stopped as larva N = stopped as neurula G = stopped as gastrula x = chemical inhibition
S ♂ = male proved sterile in F_2 or backcross F ♂ = male proved fertile in F_2 or backcross

Combination ♀	♂	Cross Number	% Fertilized	% Fertilized Hatched	Number Larvae	% Metamorphose	Calculated % Original Metamorphose	Stage Reached
ALVARIUS	*BOREAS*							
alvarius x *boreas*		64-321	11.8	25.0	24	—	—	L
alvarius x *exsul*		65-275	5.1	100.0	4	—	—	L
alvarius x *exsul*		65-288	46.0	100.0	223	9.0	9.0	M
	PUNCTATUS							
alvarius x *punctatus*		64-357	4.9	—	—	—	—	G
alvarius x *punctatus*		65-257	80.9	92.1	107	—	—	L
	MARMOREUS							
alvarius x *marmoreus*		65-261	69.7	78.3	18	66.7	52.2	M
alvarius x *marmoreus*		65-283	92.8	87.7	139	64.7	56.7	M
alvarius x *perplexus*		65-270	83.7	87.8	73	93.1	81.7	M
alvarius x *perplexus*		65-289	65.2	90.0	227	29.5	23.5	M
	COGNATUS							
alvarius x *cognatus*		65-255	75.0	100.0	6	50.0	50.0	M
alvarius x *cognatus*		65-265	92.9	93.5	253	73.5	68.7	M
alvarius x *cognatus*		64-354	6.3	—	—	—	—	G
alvarius x *speciosus*		65-272	95.6	98.5	255	82.3	81.1	M
	DEBILIS							
alvarius x *debilis*		65-266	98.1	86.8	166	—	—	L
	VALLICEPS							
alvarius x *ibarrai*		64-359	28.1	—	—	—	—	G
alvarius x *luetkeni*		64-360	24.6	21.4	3	66.7	14.3	M
alvarius x *valliceps*		65-262	53.5	23.3	7	—	—	L
alvarius x *valliceps*		65-281	90.7	25.6	95	—	—	L
	COCCIFER							
alvarius x *coccifer*		65-269	85.9	—	—	—	—	G
	AMERICANUS							
alvarius x *americanus*		65-280	92.1	—	—	—	—	G–N
alvarius x *hemiophrys*		65-279	93.4	—	—	—	—	G–N
alvarius x *microscaphus*		65-273	74.1	—	—	—	—	G–N
alvarius x *terrestris*		65-259	78.6	—	—	—	—	G–N
alvarius x *woodhousei*		64-355	6.5	—	—	—	—	G
alvarius x *woodhousei*		65-274	68.7	—	—	—	—	G–N
	QUERCICUS							
alvarius x *quercicus*		65-302	3.4	—	—	—	—	G–N
	VIRIDIS							
alvarius x *viridis*		65-267	65.0	64.1	225	9.8	6.3	M
	CALAMITA							
alvarius x *calamita*		65-277	4.8	100.0	64	—	—	L
alvarius x *calamita*		65-287	72.7	75.0	164	—	—	L
	BUFO							
alvarius x *bufo*		65-276	43.3	—	—	—	—	G
	REGULARIS							
alvarius x *regularis* (W)		65-278	97.8	—	—	—	—	G
	MACULATUS							
alvarius x *maculatus*		65-284	86.5	—	—	—	—	G
	PERRETI							
alvarius x *perreti*		65-285	91.0	—	—	—	—	G

M = metamorphosed L = stopped as larva N = stopped as neurula G = stopped as gastrula x = chemical inhibition
S♂ = male proved sterile in F_2 or backcross F♂ = male proved fertile in F_2 or backcross

Combination ♀	♂	Cross Number	% Fertilized	% Fertilized Hatched	Number Larvae	% Metamorphose	Calculated % Original Metamorphose	Stage Reached
PUNCTATUS	*SPINULOSUS*							
punctatus x *spinulosus*		66-68	15.9	7.1	1	—	—	L
punctatus x *spinulosus*		66-67	19.0	—	—	—	—	G
punctatus x *spinulosus*		66-28	76.1	12.5	20	25.0	3.1	M
	GRANULOSUS							
punctatus x *fernandezae*		64-50	42.5	—	—	—	—	G–N
punctatus x *fernandezae*		64-182	92.8	—	—	—	—	G–N
punctatus x *humboldti*		66-317	2.2	—	—	—	—	G
	MARINUS							
punctatus x *arenarum*		64-179	100.0	—	—	—	—	G–N
punctatus x *marinus*		62-75	36.8	—	—	—	—	G
	CRUCIFER							
punctatus x *crucifer*		66-311	36.2	—	—	—	—	G
	BOCOURTI							
punctatus x *bocourti*		64-166	72.9	62.8	156	91.0	57.1	M
punctatus x *bocourti*		64-167	90.4	74.5	151	87.4	79.0	M
	BOREAS							
punctatus x *boreas*		64-323	34.4	61.0	80	75.0	45.7	M
punctatus x *exsul*		65-251	11.8	100.0	20	35.0	35.0	M
punctatus x *nelsoni*		66-63	95.6	96.6	145	48.9	47.2	M
punctatus x *nelsoni*		66-64	100.0	84.8	116	9.5	8.1	M
	ALVARIUS							
punctatus x *alvarius*		64-215	50.0	—	7	—	—	L
punctatus x *alvarius*		64-322	16.7	16.7	12	25.0	4.2	M
	MARMOREUS							
punctatus x *marmoreus*		61-36	95.5	95.3	300	72.0	68.6	M
punctatus x *marmoreus*		64-48	41.8	63.6	76	52.6	33.4	M
punctatus x *marmoreus*		64-180	97.3	100.0	273	94.1	94.1	M
punctatus x *perplexus*		64-176	89.1	71.9	201	90.5	65.1	M
punctatus x *perplexus*		64-177	88.5	93.7	88	97.7	91.5	M
punctatus x *perplexus*		64-178	100.0	100.0	288	95.5	95.5	M
punctatus x *perplexus*		65-254	53.7	100.0	91	85.7	85.7	M
	COGNATUS							
punctatus x *cognatus*		61-33	97.9	76.6	36	22.2	17.0	M
punctatus x *cognatus*		64-184	100.0	97.6	82	90.2	88.0	M
punctatus x *cognatus*		64-185	100.0	86.7	62	98.4	85.3	M
punctatus x *cognatus*		64-327	69.7	88.7	107	69.2	61.4	M
punctatus x *compactilis*		66-313	26.8	—	—	—	—	G
punctatus x *speciosus*		61-39	52.9	74.0	20	40.0	29.6	M
punctatus x *speciosus*		64-181	100.0	64.5	160	71.9	46.4	M
	DEBILIS							
punctatus x *debilis*		60-2	27.5	—	11	—	—	G–L
punctatus x *kelloggi*		66-312	12.8	60.0	22	47.3	28.4	M
punctatus x *retiformis*		67-382	22.5	64.7	22	40.9	26.5	M
	VALLICEPS							
punctatus x *coniferus*		66-310	25.9	—	—	—	—	G
punctatus x *coniferus*		67-383	3.8	—	—	—	—	G
punctatus x *coniferus*		64-329	63.0	—	—	—	—	G
punctatus x *gemmifer*		64-174	93.9	28.2	65	1.5	0.4	M
punctatus x *gemmifer*		64-175	97.9	80.8	48	—	—	L
punctatus x *ibarrai*		64-326	50.0	75.6	91	40.6	30.7	M

M = metamorphosed L = stopped as larva N = stopped as neurula G = stopped as gastrula x = chemical inhibition
S♂ = male proved sterile in F_2 or backcross F♂ = male proved fertile in F_2 or backcross

Combination ♀	♂	Cross Number	% Fertilized	% Fertilized Hatched	Number Larvae	% Metamorphose	Calculated % Original Metamorphose	Stage Reached
PUNCTATUS	*VALLICEPS*							
punctatus x *luetkeni*		66-315	6.1	—	—	—	—	G
punctatus x *luetkeni*		64-168	48.9	14.9	27	—	—	G
punctatus x *luetkeni*		64-169	92.4	77.5	88	—	—	L
punctatus x *mazatlanensis*		66-314	16.9	—	—	—	—	G
punctatus x *valliceps*		61-34	87.0	—	—	—	—	G
punctatus x *valliceps*		61-38	85.6	—	—	—	—	G
punctatus x *valliceps*		62-74	66.4	1.3	1	—	—	G–L
punctatus x *valliceps*		64-183	77.6	—	—	—	—	G–N
	CANALIFERUS							
punctatus x *canaliferus*		66-309	43.9	—	—	—	—	G
punctatus x *canaliferus*		60-104	15.5	—	13	—	—	G–L
punctatus x *canaliferus*		67-384	12.5	—	—	—	—	N
	OCCIDENTALIS							
punctatus x *occidentalis*		66-62	31.7	—	—	—	—	G
	COCCIFER							
punctatus x *coccifer*		64-324	54.7	—	—	—	—	G
	AMERICANUS							
punctatus x *americanus*		60-1	42.3	21.4	74	—	—	G–L
punctatus x *hemiophrys*		60-8	1.6	—	1	—	—	G–L
punctatus x *microscaphus*		66-200	12.3	—	—	—	—	G
punctatus x *terrestris*		66-199	31.8	—	—	—	—	G
punctatus x *terrestris*		60-9	3.6	—	—	—	—	G–N
punctatus x *woodhousei*		61-37	97.7	—	—	—	—	G–N
	QUERCICUS							
punctatus x *quercicus*		67-385	6.9	—	—	—	—	G
punctatus x *quercicus*		69-9	10.9	75.0	16	16.7	12.5	M
	VIRIDIS							
punctatus x *viridis*		65-252	37.3	100.0	50	84.0	84.0	M
	CALAMITA							
punctatus x *calamita*		64-47	8.0	—	—	—	—	G–N
punctatus x *calamita*		64-170	42.4	20.0	9	—	—	L
punctatus x *calamita*		64-171	78.9	11.3	15	—	—	L
	BUFO							
punctatus x *bufo*		64-49	53.7	69.0	42	—	—	L
punctatus x *bufo*		64-164	27.4	—	1	—	—	L
punctatus x *bufo*		64-165	74.1	—	1	—	—	L
punctatus x *bufo*		66-71	42.2	11.4	4	—	—	L
punctatus x *bufo*		69-10	3.2	—	—	—	—	G
	STOMATICUS							
punctatus x *stomaticus*		67-381	19.5	30.4	7	42.8	13.0	M
	ASPER							
punctatus x *asper*		67-380	30.9	—	—	—	—	G
	REGULARIS							
punctatus x *regularis* (W)		64-216	8.2	—	—	—	—	G–N
punctatus x *regularis* (E)		66-65	91.3	—	—	—	—	G
punctatus x *regularis* (E)		66-66	36.7	—	—	—	—	G
	PERRETI							
punctatus x *perreti*		65-253	81.8	—	—	—	—	G

M = metamorphosed L = stopped as larva N = stopped as neurula G = stopped as gastrula x = chemical inhibition
S♂ = male proved sterile in F_2 or backcross F♂ = male proved fertile in F_2 or backcross

Combination ♀	♂	Cross Number	% Fertilized	% Fertilized Hatched	Number Larvae	% Metamorphose	Calculated % Original Metamorphose	Stage Reached
MARMOREUS	*SPINULOSUS*							
marmoreus x *spinulosus*		64-45	90.2	72.0	260	1.5	1.1	M
marmoreus x *spinulosus*		64-203	50.5	15.7	13	—	—	L
perplexus x *spinulosus*		65-169	90.5	—	—	—	—	G
perplexus x *spinulosus*		66-240	47.2	21.4	62	29.1	6.2	M
	GRANULOSUS							
marmoreus x *humboldti*		66-286	66.4	47.5	68	35.3	16.8	M
perplexus x *humboldti*		65-157	95.8	67.6	89	13.5	9.1	M
	MARINUS							
marmoreus x *arenarum*		64-44	100.0	—	—	—	—	G–N
marmoreus x *marinus*		64-192	96.0	—	—	—	—	G
perplexus x *arenarum*		65-161	92.1	8.5	8	—	—	G
perplexus x *marinus*		65-159	93.5	—	—	—	—	G
	GUTTATUS							
marmoreus x *haematiticus*		66-282	23.5	—	—	—	—	G
perplexus x *haematiticus*		66-245	1.6	—	—	—	—	G
perplexus x *haematiticus*		66-249	21.8	—	—	—	—	G
perplexus x *haematiticus*		66-261a[5]	—	—	—	—	—	x
perplexus x *haematiticus*		66-261b[6]	—	—	—	—	—	x
	BOCOURTI							
marmoreus x *bocourti*		67-404	68.5	54.8	180	60.5	33.2	M
marmoreus x *bocourti*		67-405	18.8	4.0	24	100.0	4.0	M
marmoreus x *bocourti*		64-43	80.5	93.5	258	—	—	L
perplexus x *bocourti*		65-154	100.0	100.0	140	—	—	S♂
	BOREAS							
marmoreus x *boreas*		64-42	20.0	—	—	—	—	G
marmoreus x *boreas*		67-301	1.2	—	—	—	—	G
marmoreus x *canorus*		67-303	17.2	—	6	—	—	L
perplexus x *boreas*		65-244	90.7	—	—	—	—	G
perplexus x *boreas*		66-237	13.4	—	—	—	—	G
	ALVARIUS							
perplexus x *alvarius*		66-248	90.5	84.0	86	31.4	26.3	M
	PUNCTATUS							
marmoreus x *punctatus*		64-41	21.9	7.1	16	—	—	L
marmoreus x *punctatus*		66-289	53.3	35.0	32	—	—	L
perplexus x *punctatus*		65-148	90.0	100.0	245	—	—	L
perplexus x *punctatus*		65-167	90.7	10.2	27	—	—	L
	COGNATUS							
marmoreus x *cognatus*		64-97	75.7	64.1	199	4.5	2.9	M
marmoreus x *speciosus*		62-71	45.2	9.1	24	25.0	2.3	M
perplexus x *cognatus*		65-153	100.0	70.0	139	37.4	26.2	M
perplexus x *speciosus*		65-163	97.9	79.2	58	3.4	2.7	M
perplexus x *speciosus*		66-58	1.6	100.0	1	—	—	L
perplexus x *speciosus*		66-59	48.6	47.0	8	37.5	17.4	M
perplexus x *speciosus*		66-238	40.7	31.8	39	35.8	11.4	M
	DEBILIS							
marmoreus x *debilis*		62-72	44.6	58.6	47	—	—	L
marmoreus x *kelloggi*		66-284	6.3	—	—	—	—	G–N
marmoreus x *retiformis*		67-302	34.2	—	—	—	—	G
marmoreus x *retiformis*		67-402	95.0	80	220	—	—	L

M = metamorphosed L = stopped as larva N = stopped as neurula G = stopped as gastrula x = chemical inhibition
S♂ = male proved sterile in F_2 or backcross F♂ = male proved fertile in F_2 or backcross

[5] No fertilization. [6] No fertilization

Combination ♀	♂	Cross Number	% Fertilized	% Fertilized Hatched	Number Larvae	% Metamorphose	Calculated % Original Metamorphose	Stage Reached
MARMOREUS	*DEBILIS*							
perplexus x *debilis*		65-158	82.2	100.0	58	6.9	6.9	M
perplexus x *debilis*		66-246	97.5	95.5	140	—	—	L
	VALLICEPS							
marmoreus x *coniferus*		67-308	10.5	—	—	—	—	G
marmoreus x *coniferus*		66-280	49.4	—	—	—	—	G
marmoreus x *luetkeni*		66-285	54.1	95.5	144	81.5	77.8	M
marmoreus x *mazatlanensis*		63-71	16.7	100.0	77	77.9	77.9	M
marmoreus x *valliceps*		62-36	45.8	—	—	—	—	G–N
marmoreus x *valliceps*		64-189	80.0	3.7	3	—	—	L
marmoreus x *valliceps*		67-406	63.0	—	—	—	—	G
marmoreus x *valliceps*		67-407	50.6	—	—	—	—	G
perplexus x *coniferus*		66-228	24.6	—	—	—	—	G
perplexus x *coniferus*		66-232	17.2	—	—	—	—	G
perplexus x *coniferus*		66-235	27.1	—	—	—	—	G
perplexus x *coniferus*		66-244	67.3	—	—	—	—	G
perplexus x *coniferus*		66-223	12.6	—	—	—	—	G
perplexus x *coniferus*		66-251	20.0	—	—	—	—	G
perplexus x *ibarrai*		65-155	55.2	100.0	50	58.0	58.0	M
perplexus x *luetkeni*		65-171	85.5	98.3	169	94.7	93.1	S♂
perplexus x *valliceps*		66-56	17.1	—	—	—	—	G
perplexus x *valliceps*		65-151	100.0	1.7	3	—	—	L
perplexus x *valliceps*		66-57	37.5	—	—	—	—	G
perplexus x *valliceps*		66-239	88.5	28.9	48	37.5	10.8	M
	CANALIFERUS							
marmoreus x *canaliferus*		64-195	42.0	—	2	—	—	L
perplexus x *canaliferus*		66-224	44.9	—	—	—	—	G
perplexus x *canaliferus*		66-230	23.1	—	—	—	—	G
perplexus x *canaliferus*		66-231	73.4	—	—	—	—	G
	OCCIDENTALIS							
marmoreus x *occidentalis*		63-73	—	—	—	—	—	G
	COCCIFER							
marmoreus x *coccifer*		66-281	63.0	—	—	—	—	G
perplexus x *coccifer*		65-156	40.3	—	—	—	—	G
	AMERICANUS							
marmoreus x *americanus*		62-37	41.7	—	—	—	—	G–N
marmoreus x *americanus*		63-61	79.3	—	—	—	—	G–N
marmoreus x *americanus*		61-58	89.5	—	—	—	—	G
marmoreus x *hemiophrys*		67-306	25.2	—	—	—	—	G
marmoreus x *microscaphus*		66-287	3.5	—	—	—	—	G
marmoreus x *terrestris*		66-288	39.5	—	—	—	—	G
marmoreus x *woodhousei*		64-191	26.8	—	—	—	—	G
perplexus x *americanus*		65-150	100.0	—	—	—	—	G
perplexus x *hemiophrys*		65-165	97.3	—	—	—	—	G
perplexus x *microscaphus*		65-172	96.1	—	—	—	—	G
perplexus x *terrestris*		65-174	92.7	—	—	—	—	G
perplexus x *woodhousei*		65-173	98.2	—	—	—	—	G
	QUERCICUS							
marmoreus x *quercicus*		66-279	54.1	—	—	—	—	G
perplexus x *quercicus*		66-233	48.0	17.4	17	—	—	L
perplexus x *quercicus*		66-225	5.6	—	—	—	—	G
perplexus x *quercicus*		66-229	2.9	—	—	—	—	G

M = metamorphosed L = stopped as larva N = stopped as neurula G = stopped as gastrula x = chemical inhibition
S♂ = male proved sterile in F_2 or backcross F♂ = male proved fertile in F_2 or backcross

Combination ♀	♂	Cross Number	% Fertilized	% Fertilized Hatched	Number Larvae	% Metamorphose	Calculated % Original Metamorphose	Stage Reached
MARMOREUS	VIRIDIS							
perplexus x *viridis*		65-152	94.3	88.0	134	—	—	L
	CALAMITA							
marmoreus x *calamita*		63-73	—	—	—	—	—	G
perplexus x *calamita*		65-160	90.2	—	—	—	—	G–N
	BUFO							
marmoreus x *bufo*		67-307	24.0	—	—	—	—	G
perplexus x *bufo*		65-162	94.1	—	—	—	—	G
	STOMATICUS							
marmoreus x *stomaticus*		67-400	77.4	100.0	209	35.0	35.0	M
marmoreus x *stomaticus*		67-403	100.0	100.0	23	74.0	74.0	M
	MELANOSTICTUS							
marmoreus x *melanostictus*		67-305	52.0	—	—	—	—	G
	ASPER							
marmoreus x *asper*		67-304	26.7	—	—	—	—	G
	MAURITANICUS							
perplexus x *mauritanicus*		66-253	13.6	—	—	—	—	G
	GARIEPENSIS							
perplexus x *inyangae*		66-242	20.5	—	—	—	—	G
perplexus x *inyangae*		66-243	62.2	—	—	—	—	G
	REGULARIS							
marmoreus x *regularis* (W)		64-205	95.7	—	—	—	—	G
perplexus x *regularis* (W)		65-166	94.9	—	—	—	—	G
	MACULATUS							
perplexus x *maculatus*		65-245	95.9	—	—	—	—	G
	PERRETI							
perplexus x *perreti*		65-246	94.4	—	—	—	—	G
COGNATUS	SPINULOSUS							
cognatus x *spinulosus*		65-243	51.6	—	—	—	—	G–N
speciosus x *variegatus*		67-79	8.7	5.3	1	—	—	L
	GRANULOSUS							
cognatus x *humboldti*		65-231	10.4	50.0	24	8.3	4.2	M
speciosus x *humboldti*		66-275	11.5	50.0	103	21.0	10.5	M
	MARINUS							
cognatus x *arenarum*		64-242	6.9	—	—	—	—	G–N
cognatus x *arenarum*		64-254	35.6	14.3	9	—	—	L
cognatus x *marinus*		64-240	38.2	—	—	—	—	G–N
cognatus x *marinus*		64-264	54.3	—	—	—	—	G
speciosus x *arenarum*		65-128	87.9	82.3	72	25.0	20.6	S♂
speciosus x *marinus*		64-372	89.9	—	—	—	—	G–N
	CRUCIFER							
speciosus x *crucifer*		65-360	76.0	—	—	—	—	G
	GUTTATUS							
speciosus x *blombergi*		65-126	56.5	—	—	—	—	G
	BOCOURTI							
cognatus x *bocourti*		64-241	7.6	55.5	20	10.0	5.5	M
cognatus x *bocourti*		64-261	88.7	100.0	195	33.3	33.3	M
cognatus x *bocourti*		67-452	8.8	92.8	45	17.6	16.3	M
speciosus x *bocourti*		64-106	65.7	54.7	87	2.3	2.4	M
speciosus x *bocourti*		67-315	80.5	41.5	153	5.2	2.1	M

M = metamorphosed L = stopped as larva N = stopped as neurula G = stopped as gastrula x = chemical inhibition
S♂ = male proved sterile in F_2 or backcross F♂ = male proved fertile in F_2 or backcross

Combination ♀	♂	Cross Number	% Fertilized	% Fertilized Hatched	Number Larvae	% Metamorphose	Calculated % Original Metamorphose	Stage Reached
COGNATUS	*BOREAS*							
cognatus x *boreas*		64-237	14.8	—	8	75.0	—	M
cognatus x *boreas*		64-250	0.5	100.0	1	100.0	100.0	M
cognatus x *boreas*		64-257	42.6	100.0	129	—	—	L
speciosus x *boreas*		67-318	13.8	55.5	14	21.4	11.9	M
speciosus x *boreas*		64-105	15.6	61.9	13	7.7	4.8	M
speciosus x *boreas*		64-107	16.7	—	—	—	—	N
speciosus x *boreas*		65-55	17.0	77.8	42	—	—	L
speciosus x *canorus*		67-312	67.2	92.8	139	33.1	30.7	M
speciosus x *exsul*		65-363	10.0	33.3	1	—	—	L
speciosus x *nelsoni*		66-38	2.0	—	—	—	—	G
	ALVARIUS							
cognatus x *alvarius*		64-247	33.0	3.4	5	—	—	L
cognatus x *alvarius*		64-260	43.4	69.4	88	38.5	26.7	M
speciosus x *alvarius*		64-368	40.0	75.8	155	29.0	22.0	M
	PUNCTATUS							
cognatus x *punctatus*		64-239	7.9	—	—	—	—	G–N
cognatus x *punctatus*		64-366	0	0	1	—	—	L
speciosus x *punctatus*		64-374	5.5	—	—	—	—	G
	MARMOREUS							
cognatus x *marmoreus*		65-229	39.6	58.0	187	85.5	49.5	M
cognatus x *marmoreus*		65-239	82.5	100.0	233	91.1	91.1	M
cognatus x *perplexus*		65-228	10.4	57.0	12	41.7	23.8	M
cognatus x *perplexus*		65-233	10.7	25.0	182	85.2	21.3	M
speciosus x *marmoreus*		65-355	25.9	28.6	26	56.2	15.1	M
speciosus x *marmoreus*		66-33	77.0	59.7	232	82.5	50.2	M
speciosus x *perplexus*		65-359	83.0	84.5	233	61.7	52.2	M
speciosus x *perplexus*		66-270	17.4	2.8	8	100.0	2.8	M
	COGNATUS							
cognatus x *compactilis*		64-375	36.8	100.0	235	25.1	25.1	M
cognatus x *speciosus*		64-235	—	100.0	9	44.4	44.4	M
cognatus x *speciosus*		64-236	71.4	5.7	20	95.0	5.4	M
cognatus x *speciosus*		64-251	33.9	100.0	144	68.0	68.0	F♂
speciosus x *cognatus*		56-26	—	—	190	65.6	—	M
speciosus x *cognatus*		57-8	high	high	300	20.3	—	M
speciosus x *cognatus*		60-21	100.0	100.0	180	6.1	6.1	M
speciosus x *cognatus*		61-88	100.0	100.0	300	2.3	2.3	M
speciosus x *cognatus*		66-37	49.4	72.0	50	42.0	30.2	M
speciosus x *cognatus*		67-85	89.0	84.5	67	25.4	21.4	M
speciosus x *compactilis*		61-90	100.0	100.0	180	37.8	37.8	M
speciosus x *speciosus*		60-103	89.7	—	130	4.6	—	M
	DEBILIS							
cognatus x *debilis*		65-241	84.0	94.6	235	35.8	33.8	M
cognatus x *debilis*		64-238	33.4	—	4	75.0	—	M
cognatus x *debilis*		64-253	13.1	75.0	31	12.9	9.7	M
cognatus x *retiformis*		67-451	47.5	—	2	—	—	L
speciosus x *debilis*		57-6	—	—	—	—	—	L
speciosus x *debilis*		64-367	74.5	8.5	35	—	—	L
speciosus x *kelloggi*		67-87	68.5	46.0	6	—	—	L
speciosus x *retiformis*		67-310	39.2	—	—	—	—	G–N
	VALLICEPS							
cognatus x *coniferus*		67-449	84.7	—	—	—	—	G
cognatus x *luetkeni*		64-249	13.1	—	—	—	—	M

M = metamorphosed L = stopped as larva N = stopped as neurula G = stopped as gastrula x = chemical inhibition
S♂ = male proved sterile in F_2 or backcross F♂ = male proved fertile in F_2 or backcross

Combination	Cross Number	% Fertilized	% Fertilized Hatched	Number Larvae	% Metamorphose	Calculated % Original Metamorphose	Stage Reached
♀ COGNATUS ♂ VALLICEPS							
cognatus x *luetkeni*	64-256	76.6	100.0	226	72.6	72.6	M
cognatus x *valliceps*	64-233	62.2	14.3	36	—	—	L
cognatus x *valliceps*	64-259	59.6	32.1	10	—	—	L
cognatus x *valliceps*	64-364	45.3	—	19	15.8	—	M
speciosus x *coniferus*	67-317	40.4	—	—	—	—	G–L
speciosus x *coniferus*	66-274	67.5	—	—	—	—	G
speciosus x *ibarrai*	64-369	96.8	82.0	250	22.4	18.4	M
speciosus x *luetkeni*	64-370	31.1	47.4	93	54.8	26.0	M
speciosus x *mazatlanensis*	67-86	57.1	63.6	128	42.2	26.8	M
speciosus x *valliceps*	57-3	53.5	13.0	3	—	—	M
speciosus x *valliceps*	57-7	4.2	54.5	60	10.0	5.4	M
speciosus x *valliceps*	61-89	97.8	100.0	180	2.8	2.8	M
CANALIFERUS							
cognatus x *canaliferus*	64-248	4.4	—	—	—	—	G–N
cognatus x *canaliferus*	64-258	34.0	100.0	40	10.0	10.0	M
speciosus x *canaliferus*	66-272	1.7	50.0	3	33.3	16.7	M
speciosus x *canaliferus*	67-319	4.6	—	—	—	—	G
speciosus x *canaliferus*	66-276	67.2	—	—	—	—	G
OCCIDENTALIS							
speciosus x *occidentalis*	65-356	6.2	66.7	36	22.2	14.6	M
COCCIFER							
cognatus x *coccifer*	64-365	57.1	—	—	—	—	G
speciosus x *coccifer*	64-371	49.4	—	—	—	—	G
AMERICANUS							
cognatus x *americanus*	64-262	82.2	62.7	51	—	—	L
cognatus x *americanus*	64-245	22.5	—	—	—	—	G–N
cognatus x *hemiophrys*	65-232	23.2	—	—	—	—	G–N
cognatus x *microscaphus*	64-246	33.3	—	—	—	—	G–N
cognatus x *terrestris*	65-230	19.7	35.7	11	—	—	L
cognatus x *woodhousei*	64-232	7.2	—	—	—	—	G
cognatus x *woodhousei*	64-252	50.0	64.5	40	—	—	L
compactilis x *americanus*	62-38	16.7	—	—	—	—	G–N
speciosus x *hemiophrys*	65-362	88.9	—	—	—	—	G–N
speciosus x *hemiophrys*	66-35	48.0	—	—	—	—	G
speciosus x *houstonensis*	66-26	73.0	—	—	—	—	G–N
speciosus x *microscaphus*	65-364	55.5	—	—	—	—	G–N
speciosus x *microscaphus*	66-271	10.3	—	—	—	—	G
speciosus x *terrestris*	57-17	high	—	334	—	—	L
QUERCICUS							
speciosus x *quercicus*	65-354	32.2	42.1	65	—	—	L
speciosus x *quercicus*	66-277	51.0	32.3	20	45.0	14.5	M
VIRIDIS							
cognatus x *viridis*	65-235	37.5	77.8	214	71.9	55.9	M
CALAMITA							
cognatus x *calamita*	65-236	71.4	87.5	235	37.4	32.7	M
BUFO							
cognatus x *bufo*	65-237	8.2	—	1	—	—	L
speciosus x *bufo*	67-311	31.9	—	—	—	—	G–L
STOMATICUS							
cognatus x *stomaticus*	67-450	41.5	—	—	—	—	L
speciosus x *stomaticus*	67-84	70.0	33.3	29	6.9	2.3	M

M = metamorphosed L = stopped as larva N = stopped as neurula G = stopped as gastrula x = chemical inhibition
S♂ = male proved sterile in F_2 or backcross F♂ = male proved fertile in F_2 or backcross

Combination ♀	♂	Cross Number	% Fertilized	% Fertilized Hatched	Number Larvae	% Metamorphose	Calculated % Original Metamorphose	Stage Reached
COGNATUS	*MELANOSTICTUS*							
speciosus x *melanostictus*		67-313	40.2	—	—	—	—	G
	ASPER							
speciosus x *asper*		67-316	51.0	—	—	—	—	G
	MAURITANICUS							
cognatus x *mauritanicus*		64-263	69.6	—	—	—	—	G
cognatus x *mauritanicus*		64-265	62.5	—	—	—	—	G
	GARIEPENSIS							
speciosus x *rosei*		65-358	24.1	—	—	—	—	G
	REGULARIS							
cognatus x *regularis* (W)		64-244	32.0	—	—	—	—	G–N
speciosus x *rangeri*		65-361	58.8	—	—	—	—	G
speciosus x *regularis* (W)		65-365	52.0	—	—	—	—	G
	MACULATUS							
cognatus x *maculatus*		65-240	47.4	—	—	—	—	G
	PERRETI							
cognatus x *perreti*		65-238	78.3	—	—	—	—	G
DEBILIS	*MARINUS*							
debilis x *marinus*		59-36	70.0	—	—	—	—	G
	BOREAS							
debilis x *boreas*		66-30	5.6	—	—	—	—	G
	ALVARIUS							
debilis x *alvarius*		59-41	—	—	—	—	—	G
	PUNCTATUS							
debilis x *punctatus*		59-35	—	—	—	—	—	G
	COGNATUS							
debilis x *speciosus*		59-40	80.0	—	—	—	—	G
	DEBILIS							
kelloggi x *debilis*		66-290	69.7	23.9	11	45.4	10.8	M
kelloggi x *debilis*		64-312	47.2	—	—	—	—	G
retiformis x *debilis*		62-60	low	—	—	—	—	N
	VALLICEPS							
debilis x *valliceps*		59-27	—	—	—	—	—	G
debilis x *valliceps*		59-31	30.0	—	—	—	—	G
debilis x *valliceps*		59-39	83.3	—	—	—	—	G
debilis x *valliceps*		59-42	—	—	—	—	—	G
kelloggi x *valliceps*		64-311	67.0	—	—	—	—	G
	AMERICANUS							
debilis x *americanus*		59-34	50.0	—	—	—	—	G
debilis x *americanus*		59-43	—	—	—	—	—	G–N
debilis x *hemiophrys*		59-38	low	—	—	—	—	G–N
debilis x *woodhousei*		59-28	—	—	—	—	—	G–N
debilis x *woodhousei*		59-30	—	—	—	—	—	G–N
debilis x *woodhousei*		59-33	30.0	—	—	—	—	G–N
retiformis x *woodhousei*		62-59	high	—	—	—	—	G
retiformis x *woodhousei*		62-81	9.5	—	—	—	—	G

M = metamorphosed L = stopped as larva N = stopped as neurula G = stopped as gastrula x = chemical inhibition
S♂ = male proved sterile in F_2 or backcross F♂ = male proved fertile in F_2 or backcross

Combination ♀	♂	Cross Number	% Fertilized	% Fertilized Hatched	Number Larvae	% Metamorphose	Calculated % Original Metamorphose	Stage Reached
VALLICEPS	*SPINULOSUS*							
luetkeni x *spinulosus*		66-166	9.3	100.0	70	72.8	72.8	M
valliceps x *spinulosus*		64-120	19.0	63.6	87	66.7	42.4	M
	GRANULOSUS							
luetkeni x *fernandezae*		64-84	97.2	24.3	76	—	—	L
luetkeni x *humboldti*		65-141	60.0	66.7	128	57.8	38.5	M
valliceps x *fernandezae*		63-28	75.0	88.2	345	50.2	44.3	M
valliceps x *fernandezae*		64-118	67.8	56.1	32	3.1	1.7	M
valliceps x *pygmaeus*		64-117	26.6	9.5	9	—	—	L
valliceps x *pygmaeus*		64-148	89.7	85.0	—	—	—	L
valliceps x *pygmaeus*		64-149	90.4	9.3	50	14.0	1.7	M
	MARINUS							
luetkeni x *arenarum*		63-26	100.0	93.0	300	91.7	85.3	M
luetkeni x *ictericus*		64-85	95.4	—	20	10.0	—	M
luetkeni x *marinus*		64-71	44.8	—	—	—	—	G–L
luetkeni x *marinus*		64-160	95.1	—	—	—	—	G
mazatlanensis x *marinus*		63-89	79.7	—	—	—	—	G
valliceps x *arenarum*		64-64	65.4	86.8	93	60.2	52.2	M
valliceps x *ictericus*		64-92	12.3	50.0	12	75.0	37.5	M
valliceps x *ictericus*		64-93	5.2	—	8	50.0	—	M
valliceps x *marinus*		62-4	73.0	—	—	—	—	G–N
valliceps x *marinus*		63-27	100.0	—	—	—	—	G–N
valliceps x *paracnemis*		63-30	88.0	33.9	52	—	—	L
	CRUCIFER							
luetkeni x *crucifer*		67-219	94.2	—	—	—	—	G
mazatlanensis x *crucifer*		67-246	52.0	—	—	—	—	G
valliceps x *crucifer*		65-366	9.7	—	—	—	—	G
valliceps x *crucifer*		66-291	77.2	—	—	—	—	G–N
	TYPHONIUS							
valliceps x *typhonius*		68-49	27.2	—	—	—	—	G
	GUTTATUS							
valliceps x *blombergi*		65-304	13.2	—	—	—	—	G
valliceps x *blombergi*		65-310	6.3	—	—	—	—	G–N
valliceps x *haematiticus*		66-261 (a-d)	—	—	—	—	—	—[7]
valliceps x *haematiticus*		66-262	17.6	—	—	—	—	2 cell
valliceps x *haematiticus*		66-268	4.1	—	—	—	—	8 cell
	BOCOURTI							
luetkeni x *bocourti*		64-70	59.6	100.0	234	—	—	L
luetkeni x *bocourti*		67-237	30.4	95.0	62	—	—	L
mazatlanensis x *bocourti*		63-97	54.3	96.0	137	—	—	L
valliceps x *bocourti*		63-103	75.9	1.7	13	—	—	L
valliceps x *bocourti*		63-93	30.7	92.8	243	4.9	4.5	M
valliceps x *bocourti*		64-143	31.9	78.4	139	—	—	L
valliceps x *bocourti*		64-144	94.8	—	10	—	—	L
	BOREAS							
luetkeni x *boreas*		64-69	41.6	—	—	—	—	G–L
luetkeni x *canorus*		67-220	74.0	7.9	28	—	—	L
luetkeni x *canorus*		67-228	62.4	3.8	10	—	—	L
mazatlanensis x *boreas*		63-82	70.2	1.5	5	—	—	L
mazatlanensis x *canorus*		67-251	23.8	—	—	—	—	G
valliceps x *boreas*		57-51	—	—	200	—	—	L
valliceps x *boreas*		57-17	—	—	13	—	—	L

M = metamorphosed L = stopped as larva N = stopped as neurula G = stopped as gastrula x = chemical inhibition
S♂ = male proved sterile in F_2 or backcross F♂ = male proved fertile in F_2 or backcross

[7] Chemical inhibition; no fertilization.

Combination	Cross Number	% Fertilized	% Fertilized Hatched	Number Larvae	% Metamorphose	Calculated % Original Metamorphose	Stage Reached
♀ VALLICEPS ♂ BOREAS							
valliceps x *boreas*	65-307	26.7	—	—	—	—	G–N
valliceps x *boreas*	65-312	67.0	84.5	65	—	—	L
valliceps x *exsul*	65-371	6.9	7.7	4	—	—	L
ALVARIUS							
mazatlanensis x *alvarius*	64-219	84.2	81.2	239	—	—	L
valliceps x *alvarius*	65-311	11.6	92.5	51	23.5	21.7	M
valliceps x *alvarius*	65-303	33.0	27.3	16	12.5	3.4	M
PUNCTATUS							
luetkeni x *punctatus*	64-66	23.3	17.6	23	—	—	L
luetkeni x *punctatus*	64-82	45.7	62.5	10	—	—	L
mazatlanensis x *punctatus*	64-222	32.5	—	—	—	—	G
valliceps x *punctatus*	64-91	4.3	60.0	8	—	—	L
MARMOREUS							
luetkeni x *marmoreus*	64-158	71.2	—	59	93.2	—	M
luetkeni x *perplexus*	66-161	10.0	10.0	100	9.7	97.0	M
luetkeni x *perplexus*	66-170	31.0	10.0	113	81.5	81.5	M
mazatlanensis x *marmoreus*	63-83	35.2	87.1	227	78.4	68.3	M
mazatlanensis x *perplexus*	64-220	93.6	88.6	178	—	—	L
valliceps x *marmoreus*	61-74	82.9	100.0	300	1.7	1.7	M
valliceps x *perplexus*	62-79	58.5	9.7	57	8.8	0.8	M
valliceps x *perplexus*	65-369	1.6	—	—	—	—	G–N
COGNATUS							
luetkeni x *cognatus*	64-159	56.0	62.1	124	91.1	56.6	S♂
luetkeni x *compactilis*	65-147	12.7	87.5	77	49.3	48.1	M
luetkeni x *speciosus*	66-171	90.3	53.0	100	84.0	44.5	M
luetkeni x *speciosus*	64-65	51.7	75.5	174	21.8	16.5	M
mazatlanensis x *cognatus*	64-221	22.9	—	—	—	—	G
mazatlanensis x *cognatus*	64-223	6.1	100.0	45	42.2	42.2	M
mazatlanensis x *speciosus*	63-84	62.0	96.8	230	87.4	92.0	S♂
valliceps x *cognatus*	57-25	—	—	—	—	—	G–L
valliceps x *cognatus*	57-72	high	—	—	—	—	M
valliceps x *cognatus*	64-154	91.4	82.8	253	34.4	28.5	S♂
valliceps x *compactilis*	62-7	77.8	62.8	472	29.9	18.8	M
valliceps x *speciosus*	56-8	—	—	600	61.8	—	M
valliceps x *speciosus*	57-11	—	—	300	—	—	L
valliceps x *speciosus*	64-155	40.8	—	80	8.7	—	M
DEBILIS							
luetkeni x *debilis*	64-67	8.0	33.3	7	—	—	L
mazatlanensis x *debilis*	63-85	25.0	64.4	158	—	—	L
valliceps x *debilis*	57-53	low	—	1	—	—	L
valliceps x *debilis*	57-55	low	—	—	—	—	G–N
valliceps x *debilis*	57-67	low	—	3	—	—	L
valliceps x *retiformis*	67-288	5.8	—	—	—	—	N
VALLICEPS							
luetkeni x *coniferus*	67-223	71.7	—	—	—	—	G
luetkeni x *coniferus*	67-217	93.8	—	—	—	—	G
luetkeni x *gemmifer*	64-86	96.5	100.0	255	43.1	43.1	M
luetkeni x *ibarrai*	65-133	97.5	7.5	49	—	—	L
luetkeni x *mazatlanensis*	64-161	90.6	62.5	180	67.2	42.0	F♂
luetkeni x *valliceps*	64-62	35.8	34.2	77	74.0	25.3	M
luetkeni x *valliceps*	64-80	98.0	95.9	247	88.6	85.0	F♂
luetkeni x *valliceps*	64-163	86.3	68.2	30	66.7	45.9	M

M = metamorphosed L = stopped as larva N = stopped as neurula G = stopped as gastrula x = chemical inhibition
S♂ = male proved sterile in F_2 or backcross F♂ = male proved fertile in F_2 or backcross

Combination ♀	♂	Cross Number	% Fertilized	% Fertilized Hatched	Number Larvae	% Metamorphose	Calculated % Original Metamorphose	Stage Reached
VALLICEPS	*VALLICEPS*							
luetkeni x *valliceps*		67-222	64.3	10.0	54	3.7	3.7	M
mazatlanensis x *coniferus*		67-247	50.6	—	—	—	—	G
mazatlanensis x *coniferus*		67-251	23.8	—	—	—	—	G–N
mazatlanensis x *luetkeni*		63-98	100.0	100.0	9	42.8	33.3	F ♂
mazatlanensis x *valliceps*		61-62	98.1	92.4	380	70.0	64.7	M
mazatlanensis x *valliceps*		63-92	66.7	95.4	165	87.6	83.6	F ♂
valliceps x *cavifrons*		62-76	78.0	53.1	148	53.4	28.3	M
valliceps x *cavifrons*		62-80	1.4	100.0	7	71.4	71.4	M
valliceps x *coniferus*		66-250	25.0	75.0	10	50.0	37.5	M
valliceps x *coniferus*		64-339	1.1	—	16	25.0	—	M
valliceps x *coniferus*		67-38	100.0	100.0	1	100.0	100.0	M
valliceps x *coniferus*		66-263	72.1	29.6	52	84.6	25.0	M
valliceps x *coniferus*		67-123	25.5	84.4	71	42.3	35.7	M
valliceps x *gemmifer*		60-22	100.0	ca 100.0	300	28.3	28.3	F ♂
valliceps x *ibarrai*		64-332	10.0	22.2	27	14.7	3.3	F ♂
valliceps x *ibarrai*		64-338	—	—	5	80.0	—	M
valliceps x *ibarrai*		64-340	1.3	100.0	10	60.0	60.0	M
valliceps x *ibarrai*		64-363	80.5	100.0	182	69.2	69.2	M
valliceps x *luetkeni*		64-63	6.0	20.0	19	15.8	3.2	M
valliceps x *luetkeni*		63-96	66.1	79.5	251	72.5	57.6	F ♂
valliceps x *mazatlanensis*		63-86	48.3	100.0	228	96.1	96.1	M
valliceps x *mazatlanensis*		61-73	25.0	100.0	300	41.0	41.0	F ♂
valliceps x *mazatlanensis*		61-75	low	100.0	28	89.3	89.3	M
	CANALIFERUS							
luetkeni x *canaliferus*		64-63	6.0	20.0	19	15.8	3.2	M
mazatlanensis x *canaliferus*		63-86	48.3	100.0	228	96.1	96.1	M
valliceps x *canaliferus*		61-80	79.5	75.2	400	35.2	26.5	S ♂
	OCCIDENTALIS							
valliceps x *occidentalis*		62-78	39.1	24.0	95	72.6	17.4	M
	COCCIFER							
luetkeni x *coccifer*		65-131	63.5	—	—	—	—	G
luetkeni x *coccifer*		67-227	73.5	—	—	—	—	G
mazatlanensis x *coccifer*		62-62	50.0	—	—	—	—	G
mazatlanensis x *coccifer*		63-88	61.3	<1.0	1	—	—	L
mazatlanensis x *coccifer*		67-255	1.4	—	—	—	—	G
valliceps x *coccifer*		63-95	50.0	5.7	16	—	—	L
valliceps x *coccifer*		65-69	11.1	10.0	5	—	—	L
valliceps x *coccifer*		65-71	1.2	—	—	—	—	G
valliceps x *coccifer*		65-72	76.7	—	—	—	—	G–N
	AMERICANUS							
luetkeni x *americanus*		64-308	100.0	—	—	—	—	G
luetkeni x *houstonensis*		65-145	91.1	—	—	—	—	G
luetkeni x *hemiophrys*		65-138	91.1	2.0	2	—	—	L
luetkeni x *hemiophrys*		67-221	9.3	—	1	—	—	L
luetkeni x *microscaphus*		64-162	35.8	5.2	3	66.7	3.5	M
luetkeni x *terrestris*		65-134	8.3	—	—	—	—	G
luetkeni x *woodhousei*		64-68	52.0	—	—	—	—	G
mazatlanensis x *americanus*		63-80	71.5	—	—	—	—	G
mazatlanensis x *americanus*		67-252	51.5	—	—	—	—	G
mazatlanensis x *hemiophrys*		67-249	67.7	—	—	—	—	G
mazatlanensis x *hemiophrys*		67-253	11.0	—	—	—	—	G
mazatlanensis x *microscaphus*		64-224	9.5	—	—	—	—	G
mazatlanensis x *microscaphus*		64-225	47.9	—	—	—	—	G

M = metamorphosed L = stopped as larva N = stopped as neurula G = stopped as gastrula x = chemical inhibition
S ♂ = male proved sterile in F_2 or backcross F ♂ = male proved fertile in F_2 or backcross

Combination		Cross Number	% Fertilized	% Fertilized Hatched	Number Larvae	% Metamorphose	Calculated % Original Metamorphose	Stage Reached
♀ *VALLICEPS*	♂ *AMERICANUS*							
mazatlanesis x *terrestris*		67-248	20.3	—	—	—	—	G
mazatlanensis x *woodhousei*		63-81	45.6	—	—	—	—	G
valliceps x *microscaphus*		62-6	73.8	—	—	—	—	G–N
	QUERCICUS							
luetkeni x *quercicus*		66-167	55.4	4.3	9	—	—	L
valliceps x *quercicus*		65-316	39.4	89.3	220	—	—	L
valliceps x *quercicus*		64-333	4.6	—	4	—	—	L
	VIRIDIS							
luetkeni x *viridis*		65-143	98.1	88.7	247	—	—	L
valliceps x *viridis*		62-11	low	—	2	50.0	—	M
valliceps x *viridis*		65-46	68.0	—	—	—	—	G–N
valliceps x *viridis*		65-47	22.2	—	—	—	—	G–N
valliceps x *viridis*		65-48	16.9	—	—	—	—	G–N
valliceps x *viridis*		66-267	88.4	73.7	30	3.3	2.4	M
	CALAMITA							
mazatlanensis x *calamita*		63-90	77.3	90.2	123	—	—	L
valliceps x *calamita*		62-10	100.0	19.1	9	—	—	L
	BUFO							
luetkeni x *bufo*		65-144	97.8	—	—	—	—	G
mazatlanensis x *bufo*		63-91	17.6	—	1	—	—	L
valliceps x *bufo*		63-104	60.6	—	—	—	—	G
valliceps x *bufo*		65-49	30.0	—	—	—	—	N
valliceps x *bufo*		65-50	72.7	—	—	—	—	N
valliceps x *bufo*		65-51	70.3	—	—	—	—	G
	MELANOSTICTUS							
luetkeni x *melanostictus*		66-165	100.0	—	—	—	—	G
valliceps x *melanostictus*		63-24	35.8	—	—	—	—	G
valliceps x *melanostictus*		63-29	8.4	—	—	—	—	G
valliceps x *melanostictus*		63-31	9.6	—	—	—	—	G
	ASPER							
luetkeni x *asper*		67-218	64.3	—	—	—	—	G
mazatlanensis x *asper*		67-254	4.0	—	—	—	—	G
	CARENS							
valliceps x *carens*		66-266	69.5	—	—	—	—	G
	REGULARIS							
luetkeni x *regularis* (W)		66-169	45.3	—	—	—	—	G
luetkeni x *regularis* (W)		65-146	91.2	—	—	—	—	G
valliceps x *garmani*		66-120	89.8	—	—	—	—	G
valliceps x *pardalis*		65-120	38.4	—	—	—	—	G
valliceps x *regularis* (W)		66-124	7.8	—	—	—	—	G
valliceps x *regularis* (W)		65-313	70.0	—	—	—	—	G
	MACULATUS							
valliceps x *maculatus*		64-121	31.6	—	—	—	—	G
valliceps x *maculatus*		64-141	94.7	—	—	—	—	G
valliceps x *maculatus*		64-142	67.0	—	—	—	—	G
	PERRETI							
valliceps x *perreti*		65-314	26.4	—	—	—	—	G
	PELTACEPHALUS							
valliceps x *peltacephalus*		61-61	80.8	84.2	400	79.7	67.1	M
valliceps x *peltacephalus*		64-156	15.1	100.0	231	93.9	93.9	M

M = metamorphosed L = stopped as larva N = stopped as neurula G = stopped as gastrula x = chemical inhibition
S♂ = male proved sterile in F_2 or backcross F♂ = male proved fertile in F_2 or backcross

Combination ♀	♂	Cross Number	% Fertilized	% Fertilized Hatched	Number Larvae	% Metamorphose	Calculated % Original Metamorphose	Stage Reached
CANALIFERUS	*SPINULOSUS*							
canaliferus x *spinulosus*		66-185	1.3	100.0	2	50.0	50.0	M
canaliferus x *spinulosus*		66-186	?	?	1	—	—	L
canaliferus x *spinulosus*		66-180	5.0	100.0	1	—	—	L
	MARINUS							
canaliferus x *arenarum*		66-345	58.8	3.3	8	—	—	L
canaliferus x *arenarum*		66-183	22.0	93.3	51	47.0	43.8	M
canaliferus x *arenarum*		66-174	9.1	75.0	27	11.1	8.3	M
canaliferus x *arenarum*		66-179	14.3	57.1	25	24.0	13.7	M
canaliferus x *marinus*		66-184	72.1	29.5	43	2.3	0.7	M
	BOCOURTI							
canaliferus x *bocourti*		64-296	50.0	—	4	—	—	L
	BOREAS							
canaliferus x *boreas*		61-127	4.0	—	—	—	—	G-N
canaliferus x *canorus*		67-370	89.7	54.2	52	—	—	L
	ALVARIUS							
canaliferus x *alvarius*		64-297	20.8	—	—	—	—	G
	PUNCTATUS							
canaliferus x *punctatus*		61-114	100.0	—	—	—	—	G–N
	MARMOREUS							
canaliferus x *marmoreus*		61-110	22.7	75.0	17	52.9	39.7	M
canaliferus x *perplexus*		66-173	69.6	100.0	100.0	46.0	46.0	M
canaliferus x *perplexus*		66-182	21.3	58.9	23.0	60.9	35.9	M
	COGNATUS							
canaliferus x *cognatus*		61-109	6.5	100.0	4	25.0	25.0	M
canaliferus x *cognatus*		61-121	5.4	100.0	13	—	—	L
canaliferus x *speciosus*		61-112	22.0	55.5	77	15.6	8.7	M
canaliferus x *speciosus*		66-175	13.8	85.7	13	23.1	19.8	M
	DEBILIS							
canaliferus x *debilis*		60-102	96.4	—	111	0.7	—	M
canaliferus x *kelloggi*		61-128	86.2	—	—	—	—	G–N
canaliferus x *retiformis*		61-129	low	—	—	—	—	G–N
canaliferus x *retiformis*		67-369	3.6	—	—	—	—	G
	VALLICEPS							
canaliferus x *coniferus*		67-136	78.1	—	—	—	—	G
canaliferus x *coniferus*		67-368	27.3	—	—	—	—	G–N
canaliferus x *gemmifer*		61-115	96.7	27.6	81	76.8	21.2	M
canaliferus x *luetkeni*		64-299	28.9	18.2	4	50.0	9.1	M
canaliferus x *mazatlanensis*		61-104	low	—	72	100.0	—	M
canaliferus x *valliceps*		61-105	—	100.0	1	100.0	100.0	M
canaliferus x *valliceps*		61-111	100.0	18.7	107	77.6	14.5	S♂
canaliferus x *valliceps*		61-116	49.1	51.7	52	50.0	25.8	M
canaliferus x *valliceps*		61-117	—	—	7	28.6	—	M
canaliferus x *valliceps*		67-367	9.3	41.6	17	47.0	19.5	M
	COCCIFER							
canaliferus x *coccifer*		61-107	100.0	55.8	300	6.3	3.5	M
canaliferus x *coccifer*		61-113	—	—	—	—	—	G
canaliferus x *coccifer*		66-189	15.4	61.5	8	—	—	L
canaliferus x *coccifer*		66-190	56.4	100.0	31	—	—	L
	AMERICANUS							
canaliferus x *americanus*		61-119	low	—	—	—	—	G
canaliferus x *americanus*		61-120	low	—	—	—	—	G

M = metamorphosed L = stopped as larva N = stopped as neurula G = stopped as gastrula x = chemical inhibition
S♂ = male proved sterile in F_2 or backcross F♂ = male proved fertile in F_2 or backcross

Combination ♀	♂	Cross Number	% Fertilized	% Fertilized Hatched	Number Larvae	% Metamorphose	Calculated % Original Metamorphose	Stage Reached
CANALIFERUS	*AMERICANUS*							
canaliferus x *microscaphus*		64-300	24.0	—	—	—	—	G
canaliferus x *terrestris*		61-108	low	—	—	—	—	G
canaliferus x *woodhousei*		61-118	low	—	—	—	—	G
	QUERCICUS							
canaliferus x *quercicus*		66-181	4.1	20.0	15	—	—	L
canaliferus x *quercicus*		66-187	12.7	100.0	6	—	—	L
	VIRIDIS							
canaliferus x *viridis*		67-140	100.0	16.9	41	—	—	L
	BUFO							
canaliferus x *bufo*		66-188	5.8	—	—	—	—	G
	MELANOSTICTUS							
canaliferus x *melanostictus*		66-172	45.7	—	—	—	—	G
	ASPER							
canaliferus x *asper*		67-366	12.7	—	1	—	—	G–L
	REGULARIS							
canaliferus x *regularis* (W)		66-196	88.2	—	—	—	—	G
OCCIDENTALIS	*SPINULOSUS*							
occidentalis x *spinulosus*		65-227	82.6	60.5	41	4.9	3.0	M
	GRANULOSUS							
occidentalis x *humboldti*		66-323	25.9	0.5	7	—	—	L
	MARINUS							
occidentalis x *arenarum*		65-323	44.4	12.5	1	100.0	12.5	M
occidentalis x *arenarum*		65-334	4.0	50.0	2	100.0	50.0	M
occidentalis x *poeppigi*		65-329	55.5	—	—	—	—	G–N
	CRUCIFER							
occidentalis x *crucifer*		66-320	13.2	—	—	—	—	G
	GUTTATUS							
occidentalis x *haematiticus*		66-330	7.7	—	—	—	—	G
	BOREAS							
occidentalis x *boreas*		66-328	12.4	—	—	—	—	G–N
occidentalis x *boreas*		65-326	40.0	8.3	4	—	—	L
occidentalis x *canorus*		65-341	14.1	—	—	—	—	G
	ALVARIUS							
occidentalis x *alvarius*		65-325	4.4	—	—	—	—	G–L
occidentalis x *alvarius*		66-332	37.2	49.0	69	—	—	L
	PUNCTATUS							
occidentalis x *punctatus*		65-320	35.0	—	—	—	—	G–N
occidentalis x *punctatus*		65-343	9.4	—	—	—	—	G
occidentalis x *punctatus*		66-324	31.9	5.1	11	—	—	G–N
	MARMOREUS							
occidentalis x *marmoreus*		66-333	43.9	19.6	64	—	—	L
occidentalis x *marmoreus*		65-328	10.4	—	3	—	—	L
occidentalis x *perplexus*		65-338	23.2	—	—	—	—	G
occidentalis x *perplexus*		66-337	17.1	—	—	—	—	G–N
	COGNATUS							
occidentalis x *cognatus*		65-318	0.3	—	—	—	—	N
occidentalis x *cognatus*		65-337	39.6	—	—	—	—	G–L
occidentalis x *cognatus*		65-348	43.4	4.3	31	—	—	L
occidentalis x *speciosus*		65-340	19.0	—	—	—	—	G–L
occidentalis x *compactilis*		66-322	14.3	20.0	7	—	—	L

M = metamorphosed L = stopped as larva N = stopped as neurula G = stopped as gastrula x = chemical inhibition
S♂ = male proved sterile in F_2 or backcross F♂ = male proved fertile in F_2 or backcross

Combination ♀	♂	Cross Number	% Fertilized	% Fertilized Hatched	Number Larvae	% Metamorphose	Calculated % Original Metamorphose	Stage Reached
OCCIDENTALIS	*DEBILIS*							
occidentalis x *debilis*		65-321	82.0	—	—	—	—	G–L
occidentalis x *debilis*		65-344	25.0	—	—	—	—	N–L
occidentalis x *kelloggi*		66-321	16.7	—	—	—	—	G–N
	VALLICEPS							
occidentalis x *coniferus*		66-319	0.9	100.0	9	—	—	G–L
occidentalis x *ibarrai*		66-326	3.7	20.0	20	—	—	L
occidentalis x *luetkeni*		66-325	12.1	85.7	72	—	—	L
occidentalis x *mazatlanensis*		66-327	9.8	81.8	10	—	—	L
occidentalis x *valliceps*		65-324	6.3	50.0	18	—	—	L
occidentalis x *valliceps*		65-347	39.7	93.1	97	27.6	25.7	M
occidentalis x *valliceps*		66-336	24.5	46.1	114	—	—	L
	COCCIFER							
occidentalis x *coccifer*		65-330	15.9	—	—	—	—	G
occidentalis x *coccifer*		66-335	22.8	—	—	—	—	G–N
	AMERICANUS							
occidentalis x *americanus*		65-322	73.2	—	—	—	—	G–N
occidentalis x *hemiophrys*		65-339	20.4	—	—	—	—	G
occidentalis x *microscaphus*		65-342	18.4	—	—	—	—	G
occidentalis x *terrestris*		66-329	32.1	—	—	—	—	G–N
occidentalis x *woodhousei*		66-334	26.4	—	—	—	—	G
	QUERCICUS							
occidentalis x *quercicus*		65-331	29.0	—	—	—	—	G
	VIRIDIS							
occidentalis x *viridis*		65-319	37.3	14.2	6	—	—	G–L
	CALAMITA							
occidentalis x *calamita*		65-332	18.4	—	—	—	—	G
	BUFO							
occidentalis x *bufo*		65-333	low	—	—	—	—	G
	REGULARIS							
occidentalis x *regularis* (W)		66-331	25.5	—	—	—	—	G
COCCIFER	*SPINULOSUS*							
coccifer x *spinulosus*		66-75	14.7	—	—	—	—	G
	GRANULOSUS							
coccifer x *granulosus*		63-78	25.3	14.3	11	—	—	L
coccifer x *granulosus*		65-66	24.6	100.0	25	4.0	4.0	M
	MARINUS							
coccifer x *arenarum*		65-182	47.0	4.2	8	12.5	0.5	M
coccifer x *poeppigi*		65-63	86.4	—	—	—	—	G–N
	BOCOURTI							
coccifer x *bocourti*		65-64	87.5	100.0	195	—	—	L
	BOREAS							
coccifer x *boreas*		65-177	100.0	34.1	43	4.6	1.6	M
	PUNCTATUS							
coccifer x *punctatus*		64-334	1.3	33.3	1	100.0	33.3	M
	MARMOREUS							
coccifer x *marmoreus*		65-176	100.0	45.0	71	4.2	1.9	M
coccifer x *perplexus*		65-175	100.0	34.9	46	8.7	3.0	M

M = metamorphosed L = stopped as larva N = stopped as neurula G = stopped as gastrula x = chemical inhibition
S♂ = male proved sterile in F_2 or backcross F♂ = male proved fertile in F_2 or backcross

Combination ♀	♂	Cross Number	% Fertilized	% Fertilized Hatched	Number Larvae	% Metamorphose	Calculated % Original Metamorphose	Stage Reached
COCCIFER	COGNATUS							
coccifer x *cognatus*		64-335	1.2	33.3	2	—	—	L
coccifer x *cognatus*		65-65	97.3	77.8	188	—	—	L
coccifer x *speciosus*		65-183	16.0	12.5	1	—	—	L
	DEBILIS							
coccifer x *debilis*		65-67	67.2	93.0	80	—	—	L
	VALLICEPS							
coccifer x *coniferus*		67-428	71.9	—	—	—	—	G
coccifer x *coniferus*		64-336	21.6	—	—	—	—	N
coccifer x *ibarrai*		65-68	100.0	87.3	248	0.4	0.3	M
coccifer x *luetkeni*		65-179	63.0	37.9	52	26.9	10.2	M
coccifer x *valliceps*		61-102	94.7	50.0	305	18.7	9.3	M
coccifer x *valliceps*		64-337	2.3	—	—	—	—	G
coccifer x *valliceps*		65-61	9.8	27.3	21	76.2	20.8	M
	CANALIFERUS							
coccifer x *canaliferus*		63-79	40.0	66.7	42	—	—	L
	AMERICANUS							
coccifer x *hemiophrys*		67-429	50.0	—	—	—	—	G
coccifer x *microscaphus*		65-181	19.7	—	—	—	—	G
coccifer x *terrestris*		60-106	89.8	—	450	0.2	—	M
coccifer x *terrestris*		60-107	—	—	6	—	—	L
coccifer x *woodhousei*		63-77	40.3	6.9	119	—	—	L
	VIRIDIS							
coccifer x *viridis*		65-178	3.6	33.3	3	—	—	L
	BUFO							
coccifer x *bufo*		66-76	40.0	—	—	—	—	G
	STOMATICUS							
coccifer x *stomaticus*		67-423	98.5	70.3	145	—	—	L
	MELANOSTICTUS							
coccifer x *melanostictus*		67-424	27.2	—	—	—	—	G
coccifer x *melanostictus*		67-427	100.0	—	—	—	—	G
	ASPER							
coccifer x *asper*		67-426	5.4	—	—	—	—	G–N
	REGULARIS							
coccifer x *regularis* (E)		67-425	88.7	—	—	—	—	G
AMERICANUS	SPINULOSUS							
americanus x *spinulosus*		68-80	38.4	—	—	—	—	G
americanus x *spinulosus*		68-81	68.9	2.5	38	2.6	.05	M
americanus x *spinulosus*		68-82	97.5	47.2	36	—	—	L
hemiophrys x *spinulosus*		68-54	74.2	51.0	46	—	—	L
hemiophrys x *spinulosus*		68-56	97.5	59.7	33	—	—	L
hemiophrys x *spinulosus*		68-79	83.6	67.8	66	—	—	L
woodhousei x *spinulosus*		63-33	100.0	91.2	352	0.8	0.7	M
woodhousei x *spinulosus*		64-8	88.9	93.7	69	1.4	1.3	M
americanus x *variegatus*		69-13	74.0	89.3	139	—	—	L
	GRANULOSUS							
woodhousei x *fernandezae*		64-35	73.6	7.5	64	3.1	0.2	M
americanus x *humboldti*		69-11	65.6	33.3	55	80.0	52.5	M

M = metamorphosed L = stopped as larva N = stopped as neurula G = stopped as gastrula x = chemical inhibition
S♂ = male proved sterile in F_2 or backcross F♂ = male proved fertile in F_2 or backcross

Combination ♀	♂	Cross Number	% Fertilized	% Fertilized Hatched	Number Larvae	% Metamorphose	Calculated % Original Metamorphose	Stage Reached
AMERICANUS	*MARINUS*							
americanus x *marinus*		62-21	14.8	—	—	—	—	G
hemiophrys x *marinus*		63-65	92.8	—	—	—	—	G
terrestris x *marinus*		64-76	72.2	—	—	—	—	G
woodhousei x *arenarum*		63-8	100.0	61.1	142	43.7	26.7	M
woodhousei x *ictericus*		64-19	100.0	90.2	195	59.5	53.7	S♂
woodhousei x *ictericus*		64-20	30.8	54.6	93	28.8	15.7	M
woodhousei x *ictericus*		64-21	91.5	46.3	85	35.3	16.3	M
woodhousei x *marinus*		63-19	100.0	—	—	—	—	G
woodhousei x *marinus*		65-15	100.0	—	—	—	—	G
woodhousei x *marinus*		63-16	100.0	—	—	—	—	G
woodhousei x *marinus*		63-20	100.0	—	—	—	—	G
woodhousei x *marinus*		63-21	100.0	—	—	—	—	G
woodhousei x *poeppigi*		68-47	61.0	—	—	—	—	G
woodhousei x *paracnemis*		64-16	100.0	—	—	—	—	G
woodhousei x *paracnemis*		64-17	55.7	—	—	—	—	G
woodhousei x *paracnemis*		64-18	60.8	—	—	—	—	G
	CRUCIFER							
americanus x *crucifer*		68-70	8.5	—	—	—	—	G
americanus x *crucifer*		68-71	32.4	—	—	—	—	G
americanus x *crucifer*		68-73	15.9	—	—	—	—	G
hemiophrys x *crucifer*		68-68	70.1	—	—	—	—	G
hemiophrys x *crucifer*		68-69	80.4	—	—	—	—	G
hemiophrys x *crucifer*		68-72	5.9	—	—	—	—	G
woodhousei x *crucifer*		68-42	92.4	—	—	—	—	G
	GUTTATUS							
microscaphus x *blombergi*		65-306	35.0	—	—	—	—	N
woodhousei x *blombergi*		65-125	16.0	—	—	—	—	G
	BOCOURTI							
americanus x *bocourti*		64-61	36.9	100.0	46	60.9	60.9	M
terrestris x *bocourti*		64-74	74.6	36.2	217	15.2	5.5	M
woodhousei x *bocourti*		64-9	100.0	54.7	252	13.4	7.3	M F♂
woodhousei x *bocourti*		64-34	91.7	90.9	120	2.5	2.3	M
	BOREAS							
americanus x *boreas*		68-89	73.8	—	2	—	—	L
americanus x *boreas*		68-91	45.0	11.1	19	21.1	2.2	M
americanus x *boreas*		68-92	16.7	20.0	9	—	—	L
hemiophrys x *boreas*		63-62	41.7	93.3	168	77.4	72.2	M
hemiophrys x *boreas*		68-90	79.7	56.4	31	3.2	1.8	M
microscaphus x *boreas*		—	—	—	—	—	—	—
microscaphus x *boreas*		64-57	80.3	8.9	33	54.5	4.8	M
terrestris x *boreas*		—	—	—	—	—	—	—
terrestris x *boreas*		64-73	55.1	39.5	137	70.8	28.0	M
woodhousei x *boreas*		58-1	high	low	600	58.0	—	M
woodhousei x *boreas*		63-34	100.0	98.5	244	81.1	79.9	M
woodhousei x *nelsoni*		66-49	84.0	84.5	161	13.7	11.6	M
	ALVARIUS							
americanus x *alvarius*		64-229	6.7	100.0	3	66.7	66.7	M
americanus x *alvarius*		64-230	6.7	100.0	5	80.0	80.0	M
woodhousei x *alvarius*		59-5	—	—	2	50.0	—	M
woodhousei x *alvarius*		61-24	100.0	25.0	228	43.5	10.9	S♂
woodhousei x *alvarius*		61-29	—	—	8	62.5	—	M

M = metamorphosed L = stopped as larva N = stopped as neurula G = stopped as gastrula x = chemical inhibition
S♂ = male proved sterile in F_2 or backcross F♂ = male proved fertile in F_2 or backcross

Combination ♀	♂	Cross Number	% Fertilized	% Fertilized Hatched	Number Larvae	% Metamorphose	Calculated % Original Metamorphose	Stage Reached
AMERICANUS	*PUNCTATUS*							
americanus x *punctatus*		69-15	14.1	54.6	38	44.7	24.4	M
hemiophrys x *punctatus*		66-43	100.0	83.9	126	42.1	35.3	M
microscaphus x *punctatus*		64-55	88.7	27.3	51	19.6	5.3	M
woodhousei x *punctatus*		64-186	38.2	76.9	70	60.0	46.1	M
woodhousei x *punctatus*		64-188	58.4	86.7	239	62.1	53.8	M
	MARMOREUS							
americanus x *marmoreus*		61-53	100.0	97.2	300	29.7	28.9	M
americanus x *marmoreus*		64-228	—	—	5	—	—	L
americanus x *marmoreus*		66-221	11.3	100.0	16	56.2	56.2	M
hemiophrys x *marmoreus*		62-67	—	—	19	5.3	—	M
microscaphus x *marmoreus*		64-59	94.3	45.4	44	22.7	10.3	M
terrestris x *marmoreus*		61-44	100.0	100.0	300	66.3	66.3	S♂
woodhousei x *marmoreus*		62-64	73.5	90.0	218	11.9	10.7	M
hemiophrys x *perplexus*		66-41	97.5	94.8	131	5.0	4.7	M
terrestris x *perplexus*		67-257	46.3	2.3	1	—	—	G–L
woodhousei x *perplexus*		66-48	77.1	35.2	119	33.6	11.8	M
	COGNATUS							
americanus x *cognatus*		59-19	high	—	450	40.7	—	M
americanus x *cognatus*		59-20	low	—	90	28.9	—	S♂
americanus x *cognatus*		64-231	low	—	20	20.0	—	M
americanus x *speciosus*		62-222	52.6	100.0	139	8.6	8.6	M
hemiophrys x *cognatus*		66-40	68.0	73.5	85	3.5	2.6	M
hemiophrys x *speciosus*		66-42	91.6	97.1	113	—	—	L
hemiophrys x *speciosus*		68-60	87.2	82.4	123	7.3	6.0	M
hemiophrys x *speciosus*		68-61	92.1	65.7	112	8.9	5.9	M
microscaphus x *cognatus*		64-56	73.8	12.9	4	25.0	3.2	M
microscaphus x *cognatus*		64-58	—	—	14	78.6	—	M
terrestris x *cognatus*		58-9	100.0	3.4	3	66.7	2.3	M
terrestris x *speciosus*		58-10	—	—	9	11.1	—	M
terrestris x *speciosus*		64-75	76.3	20.0	34	14.7	2.9	M
woodhousei x *cognatus*		58-2	100.0	—	600	19.5	—	M
woodhousei x *cognatus*		58-5	100.0	—	600	28.2	—	S♂
woodhousei x *cognatus*		58-8	100.0	—	570	14.0	—	M
woodhousei x *compactilis*		63-7	100.0	100.0	245	19.2	19.2	S♂
woodhousei x *speciosus*		58-7	ca 8.0	—	180	20.0	—	M
woodhousei x *speciosus*		59-10	50.0	—	14	78.6	—	S♂
	DEBILIS							
americanus x *debilis*		62-20	52.0	65.4	46	—	—	L
hemiophrys x *debilis*		63-64	90.6	100.0	229	8.3	8.3	M
microscaphus x *debilis*		64-60	low	—	4	—	—	L
terrestris x *debilis*		57-65	mod.	—	31	3.2	—	M
woodhousei x *debilis*		61-23	100.0	68.3	240	0.4	0.3	M
	VALLICEPS							
americanus x *coniferus*		68-97	27.6	—	—	—	—	G
americanus x *coniferus*		68-98	66.1	—	—	—	—	G
americanus x *luetkeni*		68-109	65.0	3.9	—	—	—	L
americanus x *luetkeni*		68-110	27.6	100.0	13	4.6	4.6	M
americanus x *luetkeni*		68-112	11.4	—	4	50.0	<50.0	M
americanus x *gemmifer*		62-25	98.2	50.0	297	7.1	3.6	M
hemiophrys x *coniferus*		68-99	24.4	—	—	—	—	G
hemiophrys x *coniferus*		68-100	51.1	—	—	—	—	G
hemiophrys x *luetkeni*		68-111	74.1	100.0	40	4.0	4.0	M
hemiophrys x *mazatlanensis*		63-68	100.0	80.0	92	90.2	72.2	M

M = metamorphosed L = stopped as larva N = stopped as neurula G = stopped as gastrula x = chemical inhibition
S♂ = male proved sterile in F_2 or backcross F♂ = male proved fertile in F_2 or backcross

Combination		Cross Number	% Fertilized	% Fertilized Hatched	Number Larvae	% Metamorphose	Calculated % Original Metamorphose	Stage Reached
♀	♂							
AMERICANUS	*VALLICEPS*							
hemiophrys x *valliceps*		63-66	100.0	84.0	104	3.7	3.1	M
hemiophrys x *valliceps*		68-57	74.2	41.3	44	9.1	3.8	M
hemiophrys x *valliceps*		68-58	100.0	90.9	40	2.5	2.3	M
hemiophrys x *valliceps*		68-59	37.1	—	—	—	—	G
microscaphus x *valliceps*		64-51	90.2	47.8	204	31.9	15.2	M
terrestris x *luetkeni*		64-72	72.6	57.8	126	67.5	39.0	M
terrestris x *mazatlanensis*		64-101	100.0	46.1	117	46.1	21.2	M
terrestris x *valliceps*		60-14	76.6	—	390	56.9	—	M
woodhousei x *coniferus*		68-45	95.0	—	—	—	—	G
woodhousei x *gemmifer*		61-27	low	—	26	46.1	—	M
woodhousei x *ibarrai*		65-17	95.1	89.6	—	—	—	L
woodhousei x *ibarrai*		65-18	88.1	100.0	152	0.6	0.6	M
woodhousei x *luetkeni*		64-10	96.9	87.1	254	76.4	66.5	M
woodhousei x *mazatlanensis*		62-5	31.2	60.0	206	72.3	43.4	M
	CANALIFERUS							
hemiophrys x *canaliferus*		67-321	—	—	—	—	—	G–L
terrestris x *canaliferus*		61-46	26.7	100.0	4	25.0	25.0	M
woodhousei x *canaliferus*		62-66	60.3	52.6	109	19.3	10.1	M
	OCCIDENTALIS							
hemiophrys x *occidentalis*		63-74	2.6	100.0	1	—	—	L
woodhousei x *occidentalis*		66-46	2.6	—	1	—	—	L
	COCCIFER							
americanus x *coccifer*		66-222	39.6	—	—	—	—	G
americanus x *coccifer*		68-106	56.2	—	—	—	—	G
americanus x *coccifer*		68-107	34.2	—	—	—	—	G
hemiophrys x *coccifer*		63-76	50.9	17.8	5	—	—	L
hemiophrys x *coccifer*		68-105	95.0	—	—	—	—	G
hemiophrys x *coccifer*		68-108	7.6	—	—	—	—	G
terrestris x *coccifer*		67-244	29.2	—	—	—	—	G
woodhousei x *coccifer*		62-63	96.6	63.1	86	46.5	29.3	M
	AMERICANUS							
americanus x *hemiophrys*		61-50	—	—	47	85.2	—	M
americanus x *hemiophrys*		61-51	—	—	278	34.9	—	M
americanus x *houstonensis*		59-22	—	—	—	68.7	—	M
americanus x *houstonensis*		69-14	70.7	75.0	97	—	—	—
americanus x *microscaphus*		56-24	—	—	—	—	—	M
americanus x *microscaphus*		62-32	100.0	100.0	344	24.4	24.4	M
americanus x *terrestris*		61-52	—	—	286	52.5	—	M
americanus x *woodhousei*		61-47	—	—	123	42.0	—	M
americanus x *woodhousei*		61-54	high	100.0	288	45.5	45.5	M
americanus x *woodhousei*		62-28	100.0	71.5	97	47.5	33.9	M
hemiophrys x *americanus*		62-29	64.3	37.8	300	42.0	15.8	M
hemiophrys x *americanus*		68-87	93.9	86.9	140	29.4	25.2	M
hemiophrys x *americanus*		68-88	43.5	—	—	—	—	G
hemiophrys x *houstonensis*		66-39	55.7	100.0	104	23.7	23.7	M
hemiophrys x *microscaphus*		67-320	3.9	50.0	10	10.0	5.0	M
hemiophrys x *terrestris*		68-51	12.9	31.2	16	31.2	9.7	M
hemiophrys x *terrestris*		68-53	74.6	65.9	150	37.3	24.6	M
hemiophrys x *terrestris*		68-55	—	—	173	—	—	M
hemiophrys x *woodhousei*		62-30	97.0	83.6	180	26.6	22.3	M
hemiophrys x *woodhousei*		62-31	—	—	7	85.6	—	M
houstonensis x *americanus*		65-115	39.0	100.0	30	80.0	80.0	M

M = metamorphosed L = stopped as larva N = stopped as neurula G = stopped as gastrula x = chemical inhibition
S♂ = male proved sterile in F_2 or backcross F♂ = male proved fertile in F_2 or backcross

Combination ♀	♂	Cross Number	% Fertilized	% Fertilized Hatched	Number Larvae	% Metamorphose	Calculated % Original Metamorphose	Stage Reached
AMERICANUS	*AMERICANUS*							
houstonensis x *microscaphus*		67-39	—	100.0	37	27.0	27.0	M
houstonensis x *terrestris*		65-122	38.2	100.0	231	77.9	69.0	M
houstonensis x *woodhousei*		65-127	53.7	100.0	156	85.9	85.9	M
microscaphus x *americanus*		67-259	5.6	100.0	11	36.3	36.3	M
microscaphus x *hemiophrys*		67-258	2.2	100.0	3	—	—	L
microscaphus x *terrestris*		64-54	92.5	8.0	60	65.0	52.0	M
microscaphus x *terrestris*		61-99	—	—	419	46.5	—	M
microscaphus x *woodhousei*		64-53	—	—	40	37.5	—	M
terrestris x *americanus*		56-22	ca 100.0	ca 100.0	?	?	?	?
terrestris x *hemiophrys*		60-10	high	98.0	300	97.5	95.5	M
terrestris x *houstonensis*		56-16	100.0	—	—	—	—	M
terrestris x *houstonensis*		56-19	?	?	?	?	?	?
terrestris x *microscaphus*		64-81	—	8.4	—	—	—	M
terrestris x *microscaphus*		64-103	—	—	6	—	—	L
terrestris x *microscaphus*		61-126	86.8	—	409	29.1	—	M
terrestris x *woodhousei*		60-15	81.5	27.7	360	61.1	16.9	M
woodhousei x *americanus*		61-55	100.0	—	258	48.4	—	M
woodhousei x *americanus*		62-12	59.3	6.2	—	—	—	M
woodhousei x *americanus*		62-13	—	17.7	—	—	—	M
woodhousei x *hemiophrys*		59-1	100.0	—	600	24.4	—	M
woodhousei x *hemiophrys*		57-36	—	—	279	31.2	—	M
woodhousei x *houstonensis*		65-116	40.8	100.0	230	93.5	81.3	M
woodhousei x *microscaphus*		63.6	100.0	85.2	213	40.4	34.4	F ♂
woodhousei x *microscaphus*		57-10	?	?	?	?	?	L
woodhousei x *microscaphus*		61-106	low	—	26	69.0	—	M
woodhousei x *terrestris*		64-31	95.7	85.3	158	72.1	61.5	M
woodhousei x *terrestris*		64-32	100.0	98.5	163	91.5	90.0	M
woodhousei x *terrestris*		57-1	low	—	—	—	—	M
woodhousei x *terrestris*		61-28	100.0	100.0	299	73.4	73.4	M
woodhousei x *terrestris*		68-43	74.1	96.7	193	—	—	L
	QUERCICUS							
americanus x *quercicus*		68-101	54.2	80.8	54	—	—	L
americanus x *quercicus*		68-103	33.3	40.0	25	—	—	L
hemiophrys x *quercicus*		68-102	86.8	91.3	90	—	—	L
hemiophrys x *quercicus*		68-104	31.4	—	—	—	—	G
microscaphus x *quercicus*		65-305	20.2	76.5	1	7.7	5.9	M
terrestris x *quercicus*		67-242	2.9	100.0	4	25.0	25.0	L
woodhousei x *quercicus*		66-158	23.2	100.0	5	—	—	L
woodhousei x *quercicus*		66-193	81.5	87.1	187	51.1	49.8	M
	VIRIDIS							
americanus x *viridis*		68-84	7.0	—	—	—	—	G
americanus x *viridis*		68-85	61.3	21.0	18	11.1	2.3	M
americanus x *viridis*		68-86	72.1	35.5	32	15.6	5.5	M
hemiophrys x *viridis*		68-83	92.8	58.9	50	32.1	18.9	M
terrestris x *viridis*		61-45	45.7	100.0	27	48.1	48.1	M
woodhousei x *viridis*		65-13	97.7	95.3	174	23.6	22.5	S ♂
woodhousei x *viridis*		65-14	100.0	98.3	159	43.4	42.7	S ♂
	CALAMITA							
americanus x *calamita*		62-34	41.5	36.4	18	27.8	10.2	M
hemiophrys x *calamita*		63-60	100.0	73.7	272	71.3	52.5	M
terrestris x *calamita*		64-78	62.2	37.5	141	30.5	11.4	M
woodhousei x *calamita*		63-32	17.8	87.5	21	42.8	36.4	M
woodhousei x *calamita*		63-35	85.1	90.0	326	19.3	17.4	S ♂

M = metamorphosed L = stopped as larva N = stopped as neurula G = stopped as gastrula x = chemical inhibition
S ♂ = male proved sterile in F_2 or backcross F ♂ = male proved fertile in F_2 or backcross

Combination ♀	♂	Cross Number	% Fertilized	% Fertilized Hatched	Number Larvae	% Metamorphose	Calculated % Original Metamorphose	Stage Reached
AMERICANUS	*BUFO*							
americanus x *bufo*		68-93	76.5	—	—	—	—	G
americanus x *bufo*		68-94	54.5	22.2	25	—	—	L
americanus x *bufo*		68-95	73.6	28.5	42	4.0	1.1	M
americanus x *bufo*		69-12	79.2	55.3	99	2.0	1.1	M
hemiophrys x *bufo*		68-52	52.5	—	—	—	—	G
hemiophrys x *bufo*		68-96	26.9	—	—	—	—	G
hemiophrys x *bufo*		63-63	31.9	100.0	54	14.8	14.8	M
woodhousei x *bufo*		65-22	65.0	23.1	61	3.3	0.8	M
woodhousei x *bufo*		65-119	87.9	89.6	226	11.5	10.3	S ♂
terrestris x *bufo*		64-79	30.9	19.0	62	9.7	1.9	M
	STOMATICUS							
americanus x *stomaticus*		68-65	22.6	16.7	90	34.0	5.7	M
americanus x *stomaticus*		68-66	6.0	—	—	—	—	G–N
americanus x *stomaticus*		68-67	20.0	64.7	11	81.8	52.9	M
hemiophrys x *stomaticus*		68-62	74.5	100.0	211	18.4	18.4	M
hemiophrys x *stomaticus*		68-63	87.0	83.0	160	10.0	8.3	M
hemiophrys x *stomaticus*		68-64	14.9	—	—	—	—	G–N
woodhousei x *stomaticus*		68-41	64.2	76.7	233	—	—	L
	MELANOSTICTUS							
woodhousei x *melanostictus*		63-11	100.0	—	—	—	—	G
woodhousei x *melanostictus*		63-12	95.7	—	6	—	—	L
woodhousei x *melanostictus*		63-22	100.0	—	—	—	—	G
woodhousei x *melanostictus*		63-23	9.6	—	—	—	—	G
	ASPER							
terrestris x *asper*		67-225	18.4	72.0	23	34.8	25.1	M
	MAURITANICUS							
terrestris x *mauritanicus*		64-87	90.7	—	—	—	—	G
	REGULARIS							
terrestris x *regularis* (W)		64-77	53.7	—	—	—	—	G
woodhousei x *brauni*		66-18	90.9	—	—	—	—	G
woodhousei x *pardalis*		65-121	80.0	—	—	—	—	N
woodhousei x *rangeri*		66-19	100.0	—	—	—	—	G
woodhousei x *regularis* (W)		66-159	94.5	—	—	—	—	G
woodhousei x *regularis* (W)		65-118	86.5	—	—	—	—	G
woodhousei x *regularis* (W)		64-7	96.3	—	—	—	—	G
hemiophrys x *regularis* (E)		65-500	27.3	—	—	—	—	G
woodhousei x *regularis* (E)		66-20	100.0	—	—	—	—	G
	MELANOPHRYNISCUS							
woodhousei x *M. stelzneri*		68-46	77.4	83.3	40	—	—	L
QUERCICUS	*GRANULOSUS*							
quercicus x *humboldti*		69-6	7.9	—	—	—	—	G
	PUNCTATUS							
quercicus x *punctatus*		69-7	15.4	—	—	—	—	G
	COGNATUS							
quercicus x *speciosus*		69-8	11.8	—	—	—	—	G
	VALLICEPS							
quercicus x *valliceps*		69-3	33.0	—	—	—	—	G
	AMERICANUS							
quercicus x *americanus*		69-5	27.6	9.4	3	—	—	L
quercicus x *woodhousei*		69-1	2.4	—	—	—	—	G
quercicus x *woodhousei*		69-2	—	—	—	—	—	G

M = metamorphosed L = stopped as larva N = stopped as neurula G = stopped as gastrula x = chemical inhibition
S ♂ = male proved sterile in F_2 or backcross F ♂ = male proved fertile in F_2 or backcross

Combination ♀	♂	Cross Number	% Fertilized	% Fertilized Hatched	Number Larvae	% Metamorphose	Calculated % Original Metamorphose	Stage Reached
VIRIDIS	SPINULOSUS							
viridis x *chilensis*		67-458	7.6	37.6	33	36.4	13.6	M
viridis x *spinulosus*		67-13	44.7	100.0	174	13.2	13.2	M
viridis x *spinulosus*		64-448	94.4	39.2	62	1.6	0.6	M
	GRANULOSUS							
viridis x *granulosus*		64-458	24.6	7.1	2	—	—	L
viridis x *granulosus*		64-470	40.7	29.2	50	30.0	8.8	M
	MARINUS							
viridis x *arenarum*		64-451	98.2	—	—	—	—	N
viridis x *marinus*		64-454	92.4	—	—	—	—	G
viridis x *marinus*		67-460	7.3	—	—	—	—	G
viridis x *marinus*		67-1	93.7	4.9	1	—	—	L
	CRUCIFER							
viridis x *crucifer*		67-32	35.1	—	—	—	—	G
viridis x *crucifer*		67-33	37.5	—	—	—	—	G
	GUTTATUS							
viridis x *haematiticus*		67-15	63.3	—	—	—	—	B
viridis x *haematiticus*		67-16	40.0	—	—	—	—	B
viridis x *holdridgei*		67-29	—	—	—	—	—	—[8]
	PERIGLENES							
viridis x *periglenes*		67-457	13.6	—	—	—	—	G
	BOCOURTI							
viridis x *bocourti*		64-457	100.0	100.0	263	8.0	8.0	M
	BOREAS							
viridis x *boreas*		67-14	68.6	37.0	100	—	—	L
viridis x *boreas*		64-447	96.7	69.0	135	—	—	L
viridis x *boreas*		64-469	69.0	62.5	67	—	—	L
viridis x *canorus*		64-462	84.1	97.3	200	2.0	1.9	M
	ALVARIUS							
viridis x *alvarius*		67-24	84.4	48.4	174	24.1	11.7	M
viridis x *alvarius*		67-36	69.2	69.3	76	43.5	30.2	M
viridis x *alvarius*		67-37	60.0	72.7	124	13.7	9.9	M
viridis x *alvarius*		67-461	97.8	2.2	2	—	—	N
	PUNCTATUS							
viridis x *punctatus*		65-389	26.0	15.0	3	—	—	L
viridis x *punctatus*		65-533	2.1	—	—	—	—	G
viridis x *punctatus*		65-534	7.1	100.0	6	—	—	L
	MARMOREUS							
viridis x *marmoreus*		67-20	52.7	—	—	—	—	G
	COGNATUS							
viridis x *cognatus*		67-27	42.6	52.2	132	9.3	4.9	M
viridis x *cognatus*		67-459	7.3	75.0	17	29.4	22.1	M
viridis x *cognatus*		64-542	71.1	40.6	73	—	—	N
	DEBILIS							
viridis x *debilis*		67-21	74.3	42.3	73	19.4	8.2	M
viridis x *debilis*		64-455	89.7	98.3	189	—	—	L
viridis x *kelloggi*		67-17	9.3	40.0	9	—	—	L
viridis x *retiformis*		67-455	30.2	12.5	24	—	—	L

M = metamorphosed L = stopped as larva N = stopped as neurula G = stopped as gastrula x = chemical inhibition
S♂ = male proved sterile in F_2 or backcross F♂ = male proved fertile in F_2 or backcross

[8] Chemical inhibition.

Combination ♀	♂	Cross Number	% Fertilized	% Fertilized Hatched	Number Larvae	% Metamorphose	Calculated % Original Metamorphose	Stage Reached
VIRIDIS	*VALLLICEPS*							
viridis x *coniferus*		67-19	3.0	—	—	—	—	G
viridis x *coniferus*		67-26	23.4	—	—	—	—	G
viridis x *ibarrai*		64-467	55.7	—	3	—	—	N
viridis x *luetkeni*		64-464	97.2	—	6	—	—	L
viridis x *valliceps*		67-23	66.1	30.7	39	5.1	1.6	M
viridis x *valliceps*		64-449	97.8	—	3	—	—	L
viridis x *valliceps*		64-465	1.6	—	1	—	—	L
	COCCIFER							
viridis x *coccifer*		64-456	85.4	—	—	—	—	G
	AMERICANUS							
viridis x *americanus*		67-18	52.7	—	—	—	—	G–N
viridis x *microscaphus*		64-460	98.0	—	1	—	—	N
viridis x *woodhousei*		64-446	81.8	—	—	—	—	N
	QUERCICUS							
viridis x *quercicus*		67-34	35.9	42.9	52	—	—	L
viridis x *quercicus*		67-35	80.0	100.0	142	—	—	L
	CALAMITA							
viridis x *calamita*		65-425	27.7	27.8	5	—	—	L
viridis x *calamita*		65-536	54.5	100.0	59	5.1	5.1	M
	BUFO							
viridis x *bufo*		67-461	18.6	—	6	—	—	L
viridis x *bufo*		67-2	83.7	44.4	116	13.8	6.1	M
viridis x *bufo*		67-3	100.0	70.0	75	—	—	L
	STOMATICUS							
viridis x *stomaticus*		67-12	85.1	68.9	177	68.4	47.1	M
	MELANOSTICTUS							
viridis x *melanostictus*		67-4	79.5	—	—	—	—	G
viridis x *melanostictus*		67-5	90.7	—	—	—	—	G
	ASPER							
viridis x *asper*		67-456	20.0	—	—	—	—	G
	MAURITANICUS							
viridis x *mauritanicus*		67-6	—	—	6	33.3	—	M
viridis x *mauritanicus*		67-7	—	—	4	25.0	—	M
	GARIEPENSIS							
viridis x *gariepensis*		65-424	34.1	—	—	—	—	G
viridis x *inyangae*		65-537	8.5	—	—	—	—	G?
	VERTEBRALIS							
viridis x *lughensis*		67-28	14.9	—	2	50.0	—	M
	CARENS							
viridis x *carens*		64-459	98.0	—	—	—	—	G
	REGULARIS							
viridis x *garmani*		65-539	83.3	—	—	—	—	G
viridis x *garmani*		64-463	80.6	—	—	—	—	G
viridis x *rangeri*		65-540	54.0	—	—	—	—	G
viridis x *regularis* (W)		64-466	94.7	—	—	—	—	G
viridis x *regularis* (W)		65-535	6.5	—	—	—	—	G
viridis x *regularis* (W)		67-8	62.5	—	—	—	—	G
viridis x *regularis* (W)		67-9	60.0	—	—	—	—	G
	MACULATUS							
viridis x *maculatus*		64-450	93.8	—	—	—	—	G

M = metamorphosed L = stopped as larva N = stopped as neurula G = stopped as gastrula x = chemical inhibition
S♂ = male proved sterile in F_2 or backcross F♂ = male proved fertile in F_2 or backcross

Combination ♀	♂	Cross Number	% Fertilized	% Fertilized Hatched	Number Larvae	% Metamorphose	Calculated % Original Metamorphose	Stage Reached
CALAMITA	*AMERICANUS*							
calamita x *woodhousei*		60-7	ca 5+	—	—	—	—	G
STOMATICUS	*SPINULOSUS*							
stomaticus x *spinulosus*		67-46	29.6	100.0	18	5.5	5.5	M
stomaticus x *variegatus*		69-17	92.0	91.5	111	—	—	L
	GRANULOSUS							
stomaticus x *humboldti*		69-18	40.0	64.3	12	—	—	L
	MARINUS							
stomaticus x *arenarum*		67-129	27.8	—	—	—	—	G
stomaticus x *marinus*		67-130	18.4	—	—	—	—	G
	CRUCIFER							
stomaticus x *crucifer*		67-245	84.5	—	—	—	—	G
stomaticus x *crucifer*		69-16	38.4	20.0	1	—	—	L
	BOCOURTI							
stomaticus x *bocourti*		67-389	0.9	—	—	—	—	N
stomaticus x *bocourti*		67-240	88.0	100.0	51	—	—	L
	BOREAS							
stomaticus x *boreas*		67-42	75.9	15.3	18	—	—	L
stomaticus x *canorus*		67-239	82.9	100.0	29	—	—	L
stomaticus x *canorus*		67-232	53.6	93.0	80	—	—	L
	ALVARIUS							
stomaticus x *alvarius*		67-390	1.5	100.0	2	—	—	L
	PUNCTATUS							
stomaticus x *punctatus*		67-386	4.4	30.8	15	13.7	4.2	M
	MARMOREUS							
stomaticus x *marmoreus*		67-126	17.3	100.0	15	66.7	66.7	M
stomaticus x *marmoreus*		67-133	20.0	84.2	16	12.5	10.5	L
stomaticus x *perplexus*		67-132	8.8	50.0	9	—	—	L
stomaticus x *perplexus*		67-135	39.3	59.1	13	76.9	45.4	M
	COGNATUS							
stomaticus x *cognatus*		67-131	7.1	—	—	—	—	N
stomaticus x *speciosus*		67-127	2.6	—	—	—	—	G
	DEBILIS							
stomaticus x *retiformis*		67-387	1.9	80.0	5	—	—	L
	VALLICEPS							
stomaticus x *coniferus*		67-256	48.8	—	—	—	—	G
stomaticus x *coniferus*		67-233	82.0	—	—	—	—	G
stomaticus x *valliceps*		67-47	21.4	33.4	1	—	—	L
stomaticus x *valliceps*		67-125	29.0	9.1	1	—	—	L
	CANALIFERUS							
stomaticus x *canaliferus*		67-236	68.2	—	—	—	—	G–N
	COCCIFER							
stomaticus x *coccifer*		67-238	77.0	—	—	—	—	G
	AMERICANUS							
stomaticus x *americanus*		67-128	23.8	—	—	—	—	G–L
stomaticus x *hemiophrys*		67-230	49.0	—	—	—	—	G
stomaticus x *terrestris*		67-231	43.5	—	—	—	—	G

M = metamorphosed L = stopped as larva N = stopped as neurula G = stopped as gastrula x = chemical inhibition
S♂ = male proved sterile in F_2 or backcross F♂ = male proved fertile in F_2 or backcross

Combination		Cross Number	% Fertilized	% Fertilized Hatched	Number Larvae	% Metamorphose	Calculated % Original Metamorphose	Stage Reached
♀	♂							
STOMATICUS	QUERCICUS							
stomaticus x *quercicus*		67-243	37.8	100.0	34	—	—	L
	VIRIDIS							
stomaticus x *viridis*		67-41	90.5	52.7	30	—	—	L
	CALAMITA							
stomaticus x *calamita*		67-388	18.6	80.5	31	—	—	L
	BUFO							
stomaticus x *bufo*		67-43	60.7	2.1	1	—	—	L
	MELANOSTICTUS							
stomaticus x *melanostictus*		67-44	62.7	—	—	—	—	G
	ASPER							
stomaticus x *asper*		67-224	32.7	63.1	12	8.3	5.2	M
stomaticus x *asper*		67-391	13.5	—	—	—	—	G
	REGULARIS							
stomaticus x *regularis* (W)		67-45	25.0	—	—	—	—	G
MELANOSTICTUS	SPINULOSUS							
melanostictus x *atacamensis*		67-270	10.4	43.8	27	—	—	G–L
melanostictus x *spinulosus*		67-269	56.7	71.2	142	—	—	L
melanostictus x *spinulosus*		67-105	—	—	—	—	—	G–N
melanostictus x *spinulosus*		67-106	3.6	100.0	12	8.3	8.3	M
	MARINUS							
melanostictus x *arenarum*		67-103	13.0	22.2	37	21.6	4.4	M
melanostictus x *arenarum*		67-104	3.9	100.0	100	35.0	35.0	M
melanostictus x *marinus*		67-92	100.0	95.5	163	81.6	78.1	M
	CRUCIFER							
melanostictus x *crucifer*		67-260	81.8	55.5	35	40.0	22.2	M
melanostictus x *crucifer*		67-265	37.2	21.0	33	36.3	7.6	M
	BOCOURTI							
melanostictus x *bocourti*		67-274	79.2	7.1	8	—	—	L
	BOREAS							
melanostictus x *boreas*		67-101	8.2	20.4	20	—	—	L
melanostictus x *boreas*		67-102	26.8	100.0	41	2.5	2.5	M
melanostictus x *canorus*		67-268	73.8	76.4	85	—	—	G–L
	ALVARIUS							
melanostictus x *alvarius*		67-264	52.9	87.0	80	—	—	L
melanostictus x *alvarius*		67-113	29.7	45.5	24	—	—	L
	PUNCTATUS							
melanostictus x *punctatus*		67-109	22.8	—	—	—	—	G–N
melanostictus x *punctatus*		67-110	35.7	—	—	—	—	G–N
melanostictus x *punctatus*		67-273	71.8	—	—	—	—	G–N
	MARMOREUS							
melanostictus x *marmoreus*		67-107	20.0	50.0	4	—	—	L
melanostictus x *marmoreus*		67-108	82.3	73.4	20	—	—	G–L
melanostictus x *perplexus*		67-121	89.7	57.4	20	—	—	L
melanostictus x *perplexus*		67-122	87.4	77.1	20	—	—	L
	COGNATUS							
melanostictus x *cognatus*		67-275	62.2	46.2	18	—	—	L
melanostictus x *cognatus*		67-99	2.6	100.0	40	—	—	L
melanostictus x *cognatus*		67-100	100.0	—	—	—	—	G–N

M = metamorphosed L = stopped as larva N = stopped as neurula G = stopped as gastrula x = chemical inhibition
S♂ = male proved sterile in F_2 or backcross F♂ = male proved fertile in F_2 or backcross

Combination		Cross Number	% Fertilized	% Fertilized Hatched	Number Larvae	% Metamorphose	Calculated % Original Metamorphose	Stage Reached
♀	♂							
MELANOSTICTUS	*DEBILIS*							
melanostictus x *kelloggi*		67-111	11.4	—	—	—	—	G
melanostictus x *kelloggi*		67-112	74.2	—	—	—	—	G–N
melanostictus x *retiformis*		67-271	55.2	90.6	109	—	—	L
	VALLICEPS							
melanostictus x *coniferus*		67-88	100.0	79.4	177	10.7	8.5	M
melanostictus x *luetkeni*		67-263	47.2	57.7	111	—	—	L
melanostictus x *valliceps*		67-91	100.0	82.4	210	15.5	12.8	M
melanostictus x *valliceps*		67-117	92.3	63.9	186	17.5	11.2	M
melanostictus x *valliceps*		67-118	61.3	85.2	123	34.9	29.7	M
	CANALIFERUS							
melanostictus x *canaliferus*		67-267	83.6	83.6	151	—	—	L
	COCCIFER							
melanostictus x *coccifer*		67-95	98.7	72.3	52	—	—	L
melanostictus x *coccifer*		67-96	21.0	83.3	12	—	—	L
	AMERICANUS							
melanostictus x *americanus*		67-115	95.8	—	—	—	—	G–N
melanostictus x *americanus*		67-116	26.6	—	—	—	—	G–N
melanostictus x *microscaphus*		67-97	51.0	—	—	—	—	G–N
melanostictus x *microscaphus*		67-98	16.4	—	—	—	—	G–N
melanostictus x *terrestris*		67-90	100.0	14.7	20	—	—	L
	QUERCICUS							
melanostictus x *quercicus*		67-119	17.6	21.0	20	—	—	L
melanostictus x *quercicus*		67-120	44.0	54.6	20	—	—	L
	VIRIDIS							
melanostictus x *viridis*		67-89	100.0	—	—	—	—	G–N
	BUFO							
melanostictus x *bufo*		67-262	35.9	—	—	—	—	G
	STOMATICUS							
melanostictus x *stomaticus*		67-270	10.4	—	—	—	—	G–L
melanostictus x *stomaticus*		67-93	98.5	—	—	—	—	G–N
	ASPER							
melanostictus x *asper*		67-261	46.5	35.0	38	2.6	0.9	M
	REGULARIS							
melanostictus x *regularis* (W)		67-94	100.0	—	—	—	—	G
ASPER	*SPINULOSUS*							
asper x *atacamensis*		67-328	28.6	—	—	—	—	G
	MARINUS							
asper x *arenarum*		67-338	49.0	—	—	—	—	G–N
asper x *marinus*		67-339	9.5	—	—	—	—	G
	BOCOURTI							
asper x *bocourti*		67-323	13.2	—	—	—	—	G
	BOREAS							
asper x *boreas*		67-322	16.7	—	—	—	—	G
asper x *canorus*		67-333	54.9	—	—	—	—	G
	ALVARIUS							
asper x *alvarius*		67-332	18.6	—	—	—	—	G
	PUNCTATUS							
asper x *punctatus*		67-325	20.9	—	—	—	—	G

M = metamorphosed L = stopped as larva N = stopped as neurula G = stopped as gastrula x = chemical inhibition
S ♂ = male proved sterile in F_2 or backcross F ♂ = male proved fertile in F_2 or backcross

Combination ♀	♂	Cross Number	% Fertilized	% Fertilized Hatched	Number Larvae	% Metamorphose	Calculated % Original Metamorphose	Stage Reached
ASPER	*MARMOREUS*							
asper x *marmoreus*		67-326	20.5	—	—	—	—	G
asper x *perplexus*		67-335	53.4	—	—	—	—	G
	COGNATUS							
asper x *cognatus*		67-342	0.8	—	—	—	—	G
asper x *speciosus*		67-331	68.9	—	—	—	—	G
	VALLICEPS							
asper x *luetkeni*		67-334	50.9	—	—	—	—	G
asper x *valliceps*		67-329	28.2	—	—	—	—	G
	CANALIFERUS							
asper x *canaliferus*		67-330	62.8	—	—	—	—	G–N
	COCCIFER							
asper x *coccifer*		67-327	44.3	—	—	—	—	G
	AMERICANUS							
asper x *americanus*		67-336	35.9	—	—	—	—	G
asper x *hemiophrys*		67-324	15.9	—	—	—	—	G
asper x *microscaphus*		67-337	45.3	—	—	—	—	G
	BUFO							
asper x *bufo*		—	—	—	—	—	—	G
	MELANOSTICTUS							
asper x *melanostictus*		—	—	—	—	—	—	G
VERTEBRALIS	*MARINUS*							
vertebralis x *arenarum*		67-144	100.0	—	—	—	—	G–N
vertebralis x *marinus*		67-302	4.4	—	—	—	—	G
	BOREAS							
vertebralis x *boreas*		66-301	2.4	—	—	—	—	G
vertebralis x *boreas*		67-145	100.0	13.6	15	—	—	L
	PUNCTATUS							
vertebralis x *punctatus*		67-146	96.4	16.2	64	—	—	L
	MARMOREUS							
vertebralis x *marmoreus*		67-150	87.0	21.0	39	—	—	L
	COGNATUS							
vertebralis x *cognatus*		67-147	99.0	15.1	75	—	—	L
	DEBILIS							
vertebralis x *debilis*		67-142	32.4	33.3	4	—	—	L
	VALLICEPS							
vertebralis x *valliceps*		66-303	4.0	—	—	—	—	G
vertebralis x *valliceps*		67-148	100.0	—	—	—	—	G
	AMERICANUS							
vertebralis x *hemiophrys*		67-141	89.5	—	—	—	—	G
	VIRIDIS							
vertebralis x *viridis*		67-149	83.1	7.0	100	—	—	L
	BUFO							
vertebralis x *bufo*		66-300	3.3	—	—	—	—	G
vertebralis x *bufo*		67-143	97.2	—	—	—	—	G–N
	GARIEPENSIS							
vertebralis x *gariepensis*		66-299	19.8	—	—	—	—	G
	REGULARIS							
vertebralis x *regularis*		66-298	11.3	—	—	—	—	G

M = metamorphosed L = stopped as larva N = stopped as neurula G = stopped as gastrula x = chemical inhibition
S♂ = male proved sterile in F_2 or backcross F♂ = male proved fertile in F_2 or backcross

Combination ♀	♂	Cross Number	% Fertilized	% Fertilized Hatched	Number Larvae	% Metamorphose	Calculated % Original Metamorphose	Stage Reached
CARENS	*SPINULOSUS*							
carens x *chilensis*		67-364	8.9	—	—	—	—	G
	MARINUS							
carens x *arenarum*		65-414	35.7	—	—	—	—	G–N
carens x *arenarum*		65-415	8.2	—	—	—	—	G
carens x *marinus*		65-392	34.4	—	—	—	—	G–N
	CRUCIFER							
carens x *crucifer*		67-198	87.8	—	—	—	—	G
	GUTTATUS							
carens x *haematiticus*		67-197	6.2	—	—	—	—	B or G
	BOCOURTI							
carens x *bocourti*		67-363	3.3	—	—	—	—	G
carens x *bocourti*		67-195	9.1	—	—	—	—	G
	BOREAS							
carens x *boreas*		65-393	76.0	—	—	—	—	G
carens x *canorus*		67-361	0.5	—	—	—	—	G
	PUNCTATUS							
carens x *punctatus*		65-403	13.7	—	—	—	—	G
	MARMOREUS							
carens x *perplexus*		65-410	11.6	—	—	—	—	G
carens x *perplexus*		65-411	6.0	—	—	—	—	G
	COGNATUS							
carens x *cognatus*		65-418	45.2	—	—	—	—	G–N
carens x *cognatus*		65-419	87.5	—	—	—	—	G–N
carens x *speciosus*		65-401	97.0	—	—	—	—	G–N
	DEBILIS							
carens x *debilis*		65-402	24.1	—	—	—	—	G
	VALLICEPS							
carens x *coniferus*		67-200	40.0	—	—	—	—	G
carens x *valliceps*		65-394	93.7	—	—	—	—	G
	CANALIFERUS							
carens x *canaliferus*		67-199	75.8	4.7	2	—	—	G–L
	COCCIFER							
carens x *coccifer*		67-196	13.7	—	—	—	—	G
carens x *coccifer*		67-365	1.0	—	—	—	—	G
	AMERICANUS							
carens x *hemiophrys*		65-420	53.6	—	—	—	—	G
carens x *hemiophrys*		65-421	90.9	—	—	—	—	G
carens x *microscaphus*		65-399	30.6	—	—	—	—	G
carens x *microscaphus*		65-400	4.3	—	—	—	—	G
	QUERCICUS							
carens x *quercicus*		67-203	40.0	—	—	—	—	G
	VIRIDIS							
carens x *viridis*		65-408	7.9	—	—	—	—	G–N
carens x *viridis*		65-409	38.0	—	—	—	—	G–N
	CALAMITA							
carens x *calamita*		65-391	25.7	—	—	—	—	G
carens x *calamita*		65-398	4.8	—	—	—	—	G
	BUFO							
carens x *bufo*		65-412	5.2	—	—	—	—	G–N
carens x *bufo*		65-413	4.3	—	—	—	—	G

M = metamorphosed L = stopped as larva N = stopped as neurula G = stopped as gastrula x = chemical inhibition
S ♂ = male proved sterile in F_2 or backcross F ♂ = male proved fertile in F_2 or backcross

Combination ♀	♂	Cross Number	% Fertilized	% Fertilized Hatched	Number Larvae	% Metamorphose	Calculated % Original Metamorphose	Stage Reached
CARENS	*STOMATICUS*							
carens x *stomaticus*		67-362	11.0	—	—	—	—	G
	MELANOSTICTUS							
carens x *melanostictus*		67-360	1.7	—	—	—	—	G
	ASPER							
carens x *asper*		67-359	25.9	—	—	—	—	G
	GARIEPENSIS							
carens x *gariepensis*		65-416	—	—	—	—	—	—[9]
carens x *gariepensis*		65-417	—	—	—	—	—	—[10]
	REGULARIS							
carens x *rangeri*		65-395	47.4	—	—	—	—	N
carens x *regularis* (E)		65-390	87.2	—	—	—	—	N
carens x *regularis* (W)		65-423	77.8	—	—	—	—	G
	MACULATUS							
carens x *maculatus*		65-404	13.6	—	—	—	—	G–N
carens x *maculatus*		65-405	67.9	—	—	—	—	G–N
	PERRETI							
carens x *perreti*		65-406	52.4	9.1	2	—	—	L
carens x *perreti*		65-407	24.1	—	—	—	—	G–N
REGULARIS	*SPINULOSUS*							
brauni x *spinulosus*		65-487	28.2	18.2	10	—	—	G–L
brauni x *spinulosus*		65-488	23.1	22.2	6	—	—	L
garmani x *spinulosus*		66-117	83.3	93.3	142	3.5	3.3	M
regularis (W) x *atacamensis*		67-344	23.7	4.3	8	—	—	L
regularis (W) x *variegatus*		67-63	90.8	1.0	13	—	—	L
regularis (E) x *spinulosus*		65-486	16.1	—	5	—	—	L
	GRANULOSUS							
regularis (W) x *granulosus*		67-62	63.7	32.7	46	—	—	L
	MARINUS							
brauni x *arenarum*		65-478	96.0	62.5	46	—	—	L
brauni x *arenarum*		65-479	77.1	88.9	74	—	—	L
brauni x *marinus*		65-484	94.7	88.9	96	—	—	L
brauni x *marinus*		65-485	66.7	92.9	26	—	—	L
garmani x *arenarum*		66-114	23.9	90.9	130	90.8	82.5	M
garmani x *marinus*		65-436	74.7	97.4	237	53.2	51.8	M
garmani x *marinus*		65-442	81.5	100.0	174	56.9	56.9	M
kerinyagae x *marinus*		68-136	48.1	84.6	67	43.3	36.6	M
kerinyagae x *poeppigi*		68-140	38.6	40.9	11	36.4	14.9	M
rangeri x *arenarum*		65-377	49.1	100.0	166	13.8	13.8	M
rangeri x *marinus*		65-373	81.1	97.7	112	24.1	23.5	M
regularis (E) x *arenarum*		65-477	42.9	50.0	71	1.4	0.7	M
regularis (E) x *marinus*		65-483	35.4	30.4	67	1.5	0.5	M
regularis (E) x *marinus*		65-441	2.4	100.0	1	—	—	L
regularis (W) x *marinus*		65-528	90.9	100.0	200	0.5	0.5	M
regularis (W) x *marinus*		68-4	86.5	95.4	152	—	—	L
regularis (W) x *marinus*		67-75	80.7	57.3	243	—	—	L
regularis (E) x *marinus*		68-133	90.6	87.5	76	1.3	1.1	M
regularis (W) x *paracnemis*		68-117	97.9	66.7	190	—	—	L
regularis (W) x *poeppigi*		68-118	100.0	51.9	214	—	—	L

M = metamorphosed L = stopped as larva N = stopped as neurula G = stopped as gastrula x = chemical inhibition
S♂ = male proved sterile in F_2 or backcross F♂ = male proved fertile in F_2 or backcross

[9] No fertilization. [10] No fertilization.

Combination ♀	♂	Cross Number	% Fertilized	% Fertilized Hatched	Number Larvae	% Metamorphose	Calculated % Original Metamorphose	Stage Reached
REGULARIS	*CRUCIFER*							
regularis (E) x *crucifer*		68-134	76.7	100.0	34	—	—	L
regularis (W) x *crucifer*		68-121	90.5	75.2	178	—	—	L
kerinyagae (E) x *crucifer*		68-135	68.6	45.8	15	13.3	6.1	M
	GUTTATUS							
regularis (W) x *haematiticus*		66-261c	—	—	—	—	—	x
	BOCOURTI							
regularis (W) x *bocourti*		68-3	96.2	—	—	—	—	G
regularis (W) x *bocourti*		67-345	30.9	—	—	—	—	G–N
regularis[11] x *bocourti*		67-374	96.6	—	—	—	—	G
	BOREAS							
brauni x *boreas*		65-481	50.9	—	—	—	—	G
brauni x *boreas*		65-482	87.2	—	—	—	—	G
garmani x *boreas*		65-449	34.2	—	—	—	—	G
rangeri x *boreas*		65-380	96.1	—	—	—	—	G–N
regularis (E) x *boreas*		65-480	28.6	7.1	15	—	—	L
	ALVARIUS							
brauni x *alvarius*		65-514	96.6	17.9	38	—	—	L
regularis (W) x *alvarius*		67-72	63.7	13.8	22	—	—	L
regularis (E) x *alvarius*		65-515	12.8	—	5	—	—	L
	PUNCTATUS							
brauni x *punctatus*		65-518	71.9	—	—	—	—	G
garmani x *punctatus*		66-122	20.0	—	—	—	—	G–N
rangeri x *punctatus*		65-386	62.7	—	—	—	—	G
regularis (E) x *punctatus*		65-454	4.7	—	—	—	—	G
regularis (E) x *punctatus*		65-519	2.6	—	—	—	—	G
	MARMOREUS							
rangeri x *americanus*		65-381	58.2	—	—	—	—	G
rangeri x *perplexus*		65-385	62.3	—	—	—	—	G–N
regularis (W) x *perplexus*		66-153	32.5	7.2	5	20.0	4.4	M
regularis (W) x *perplexus*		67-77	66.7	—	3	—	—	L
	COGNATUS							
brauni x *cognatus*		65-509	100.0	—	—	—	—	G
brauni x *cognatus*		65-510	27.8	—	—	—	—	G
garmani x *speciosus*		65-439	88.5	—	10	—	—	L
rangeri x *speciosus*		65-384	74.9	22.2	12	—	—	L
regularis (W) x *cognatus*		65-531	77.5	30.9	31	—	—	G–L
regularis (E) x *cognatus*		65-508	4.6	—	—	—	—	G
regularis (W) x *speciosus*		66-103	13.0	22.2	37	21.6	4.8	M
regularis (E) x *speciosus*		65-440	16.4	33.3	4	50.0	16.6	M
	DEBILIS							
garmani x *debilis*		66-121	54.2	75.9	86	2.3	17.5	M
rangeri x *debilis*		65-388	17.8	—	—	—	—	G
regularis (W) x *debilis*		67-76	38.7	—	—	—	—	G
regularis (W) x *debilis*		66-101	100.0	72.4	53	—	—	G–N
regularis[12] x *retiformis*		67-372	71.9	—	—	—	—	G
	VALLICEPS							
brauni x *valliceps*		65-502	94.3	33.3	31	—	—	L
garmani x *valliceps*		65-446	54.0	100.0	113	—	—	L
kerinyagae x *valliceps*		68-144	36.4	33.3	19	42.1	14.0	M
rangeri x *valliceps*		65-382	70.6	68.7	73	2.7	1.8	M
regularis (W) x *coniferus*		68-1	96.3	96.0	174	—	—	L

M = metamorphosed L = stopped as larva N = stopped as neurula G = stopped as gastrula x = chemical inhibition
S♂ = male proved sterile in F_2 or backcross F♂ = male proved fertile in F_2 or backcross

[11] Location unknown. [12] Location unknown.

Combination ♀	♂	Cross Number	% Fertilized	% Fertilized Hatched	Number Larvae	% Metamorphose	Calculated % Original Metamorphose	Stage Reached
REGULARIS	*VALLICEPS*							
regularis (W) x *coniferus*		67-347	13.6	—	—	—	—	G
regularis (E) x *coniferus*		68-17	69.2	—	—	—	—	L
regularis[13] x *coniferus*		67-376	100.0	92.5	161	—	—	G–L
regularis (W) x *luetkeni*		67-346	24.5	—	—	—	—	G
regularis (E) x *luetkeni*		68-18	58.7	29.6	27	—	—	L
regularis[14] x *luetkeni*		67-377	66.6	—	57	—	—	G–L
regularis (W) x *valliceps*		68-2	95.8	86.9	160	—	—	L
regularis (E) x *valliceps*		65-501	28.6	78.6	31	—	—	L
regularis (E) x *valliceps*		65-447	25.5	33.3	4	—	—	L
regularis (W) x *valliceps*		68-119	81.2	25.6	21	—	—	L
	CANALIFERUS							
regularis[15] x *canaliferus*		67-378	97.7	100.0	143	1.4	1.4	M
	OCCIDENTALIS							
garmani x *occidentalis*		66-120	89.8	—	—	—	—	G
	COCCIFER							
regularis (W) x *coccifer*		67-60	60.0	—	—	—	—	G–N
	AMERICANUS							
brauni x *hemiophrys*		65-498	100.0	—	—	—	—	G
brauni x *hemiophrys*		65-499	50.0	—	—	—	—	G
garmani x *microscaphus*		65-448	43.5	—	—	—	—	G
garmani x *woodhousei*		66-132	4.6	—	—	—	—	G
kerinyagae x *americanus*		68-142	14.3	—	—	—	—	G
kerinyagae x *hemiophrys*		68-143	28.6	—	—	—	—	G
rangeri x *americanus*		65-381	58.2	—	—	—	—	G
regularis (W) x *americanus*		65-532	27.4	—	—	—	—	G
regularis (E) x *hemiophrys*		68-20	36.0	—	—	—	—	G
	QUERCICUS							
regularis (W) x *quercicus*		67-375	87.9	1.5	1	—	—	G–L
	VIRIDIS							
brauni x *viridis*		65-493	100.0	—	—	—	—	G–N
brauni x *viridis*		65-494	50.0	—	—	—	—	G
regularis (W) x *viridis*		67-74	39.4	2.2	15	—	—	L
regularis (E) x *viridis*		65-492	39.0	6.2	7	—	—	L
	CALAMITA							
garmani x *calamita*		65-445	85.7	—	—	—	—	G–N
rangeri x *calamita*		65-378	54.4	—	—	—	—	G–N
regularis (E) x *calamita*		65-453	83.5	3.0	90	—	—	L
	BUFO							
brauni x *bufo*		65-469	18.0	—	—	—	—	G
brauni x *bufo*		65-470	41.2	—	—	—	—	G–N
garmani x *bufo*		66-109	48.9	—	—	—	—	G
kerinyagae x *bufo*		68-141	32.3	—	—	—	—	G
regularis (W) x *bufo*		66-102	83.6	—	—	—	—	G
regularis (E) x *bufo*		65-468	32.8	—	—	—	—	G
	STOMATICUS							
regularis (W) x *stomaticus*		67-61	97.5	—	—	—	—	G–N
regularis[16] x *stomaticus*		67-373	98.2	—	17	—	—	G–L
	MELANOSTICTUS							
regularis (W) x *melanostictus*		66-149	63.7	11.9	40	—	—	L
regularius (E) *x melanostictus*		68-19	63.8	—	—	—	—	G

M = metamorphosed L = stopped as larva N = stopped as neurula G = stopped as gastrula x = chemical inhibition
S♂ = male proved sterile in F_2 or backcross F♂ = male proved fertile in F_2 or backcross

[13] Location unknown. [14] Location unknown. [15] Location unknown. [16] Location unknown.

Combination		Cross Number	% Fertilized	% Fertilized Hatched	Number Larvae	% Metamorphose	Calculated % Original Metamorphose	Stage Reached
♀	♂							
REGULARIS	*ASPER*							
regularis (W) x *asper*		67-343	25.0	28.0	47	—	—	L
regularis[17] x *asper*		67-371	84.3	35.7	85	—	—	L
	MAURITANICUS							
regularis (W) x *mauritanicus*		66-252	91.9	91.2	192	15.6	14.2	M
regularis (W) x *mauritanicus*		67-69	17.6	44.4	40	7.5	3.3	M
	GARIEPENSIS							
brauni x *inyangae*		65-467	64.5	—	—	—	—	G
brauni x *rosei*		65-504	—	—	—	—	—	—[18]
rangeri x *rosei*		65-374	—	—	—	—	—	—[19]
regularis (E) x *gariepensis*		65-397	60.7	—	—	—	—	G
regularis (E) x *inyangae*		65-465	—	—	—	—	—	—[20]
regularis (E) x *rosei*		65-503	—	—	—	—	—	—[21]
	LÖNNBERGI							
brauni x *lönnbergi*		65-458	16.7	—	—	—	—	G
brauni x *lönnbergi*		65-506	100.0	—	—	—	—	G
brauni x *lönnbergi*		65-507	31.0	—	—	—	—	G
regularis (E) x *lönnbergi*		65-457	53.5	60.5	73	—	—	L
regularis (E) x *lönnbergi*		65-505	4.1	—	—	—	—	N
regularis (E) x *lönnbergi*		65-450	72.7	—	—	—	—	N
regularis (E) x *lönnbergi*		65-451	13.8	—	—	—	—	N
regularis (W) x *lönnbergi*		68-116	98.4	9.1	23	—	—	L
	CARENS							
brauni x *carens*		65-490	95.9	—	—	—	—	N
brauni x *carens*		65-491	37.5	13.3	2	—	—	L
garmani x *carens*		65-432	96.8	—	—	—	—	G
regularis (E) x *carens*		65-489	72.2	—	—	—	—	N
regularis (E) x *carens*		65-433	13.1	—	—	—	—	G
	REGULARIS							
brauni x *garmani*		65-475	75.0	88.9	80	68.7	61.1	M
brauni x *garmani*		65-476	10.2	88.9	8	87.5	77.8	M
brauni x sp.		65-463	—	—	—	—	—	—[22]
brauni x *rangeri*		65-472	81.7	83.7	70	34.3	28.7	M
brauni x *rangeri*		65-473	11.1	37.5	3	100.0	37.5	M
brauni x *regularis* (W)		65-456	62.8	96.3	26	57.7	55.6	M
brauni x *regularis* (E)		65-496	—	—	—	—	—	—[23]
brauni x *regularis* (E)		65-497	—	—	—	—	—	—[24]
brauni x *regularis* (E)		65-517	59.3	87.5	26	88.5	77.4	M
brauni x *regularis* (E)		66-305	100.0	82.3	122	78.7	64.8	M
brauni x *regularis* (E)		66-307	80.5	13.8	8	62.5	8.6	M
garmani x *brauni*		66-107	89.1	90.2	137	37.9	34.2	M
garmani x *gutturalis*		66-110	—	—	3	—	—	L
garmani x *rangeri*		65-426	81.6	82.3	234	99.1	81.6	M
garmani x *regularis* (E)		66-135	28.6	—	5	100.0	—	M
garmani x *regularis* (E)		66-125	47.5	100.0	119	84.1	84.1	M
garmani x *regularis* (E)		65-429	86.8	79.7	247	92.3	73.6	M
garmani x *regularis* (W)		65-437	81.2	100.0	252	80.2	80.2	M
garmani x *regularis* (E)		66-111	100.0	81.0	100	81.0	60.5	M
garmani x *regularis* (E)		66-112	100.0	72.0	100	72.0	51.9	M
garmani x *regularis* (E)		66-113	16.7	100.0	107	70.0	70.0	M
garmani x *regularis* (W)		66-108	95.7	100.0	145	87.6	87.6	M
garmani x *regularis* (E)		66-115	11.4	100.0	95	81.1	81.1	M
garmani x *regularis* (W)		66-123	9.3	100.0	73	90.5	90.5	M

M = metamorphosed L = stopped as larva N = stopped as neurula G = stopped as gastrula x = chemical inhibition
S♂ = male proved sterile in F_2 or backcross F♂ = male proved fertile in F_2 or backcross

[17] Location unknown. [18] No fertilization. [19] No fertilization. [20] No fertilization.
[21] No fertilization. [22] No fertilization. [23] No fertilization; male possibly sterile. [24] No fertilization; male possibly sterile.

Appendix H (Continued)

Combination	Cross Number	% Fertilized	% Fertilized Hatched	Number Larvae	% Metamorphose	Calculated % Original Metamorphose	Stage Reached
♀ *REGULARIS* ♂ *REGULARIS*							
garmani x *regularis* (E)	66-129	46.3	92.0	123	78.1	71.9	M
gutturalis x *regularis* (W)	66-304	86.0	74.4	112	49.1	36.5	M
gutturalis x *regularis* (W)	66-308	55.2	12.5	2	50.0	6.3	M
kerinyagae x *rangeri*	68-137	47.0	26.7	9	66.7	17.8	M
rangeri x *regularis* (W)	65-372	53.7	96.4	138	87.7	84.5	M
rangeri x *regularis* (W)	65-379	82.1	84.4	96	75.0	63.3	M
regularis (W) x *brauni*	66-99	74.5	65.5	125	20.0	13.1	M
regularis (E) x *brauni*	65-459	96.8	88.3	253	63.6	56.2	M
regularis (W) x *garmani*	66-97	62.8	52.4	127	92.1	58.0	M
regularis (E) x *garmani*	65-474	44.6	86.5	232	85.8	74.2	M
regularis (E) x *garmani*	65-444	19.7	64.3	32	43.7	28.1	M
regularis (W) x *garmani*	65-529	73.5	100.0	136	32.3	32.3	M
regularis (E) x *gutturalis*	65-462	4.6	—	1	—	—	L
regularis (W) x *rangeri*	65-525	97.0	100.0	112	55.4	55.4	M
regularis (E) x *rangeri*	65-471	8.5	14.3	25	92.0	13.2	M
regularis (E) x *rangeri*	65-427	23.5	33.3	43	69.8	23.3	M
regularis (W) x *rangeri*	66-87	28.2	—	—	—	—	G
regularis (W) x *rangeri*	66-90	23.1	9.5	4	—	—	L
regularis (W) x *rangeri*	66-92	92.0	89.1	141	12.0	10.7	M
regularis (W) x *rangeri*	66-100	78.4	55.0	122	31.1	17.1	M
regularis (W) x *rangeri*	68-120	96.2	84.3	143	7.0	5.9	M
regularis (E) x *regularis* (E)	65-428	29.4	45.0	89	48.3	21.7	M
regularis (E) x *regularis* (W)	65-438	25.9	100.0	114	58.7	58.7	M
regularis (W) x *regularis* (W)	65-521	85.4	97.9	106	87.7	83.5	M
regularis (W) x *regularis* (W)	66-522	94.3	100.0	110	89.1	89.1	M
regularis (W) x *regularis* (E)	65-523	98.3	98.3	118	5.9	5.8	M
regularis (W) x *regularis* (W)	65-526	90.5	100.0	98	92.9	92.9	M
regularis (W) x *regularis* (E)	65-530	72.5	86.2	145	—	—	L
regularis (W) x *regularis* (E)	65-527	97.6	97.5	99	—	—	L
regularis (W) x *regularis* (E)	65-524	91.5	81.5	204	—	—	L
regularis (W) x *rangeri*	68-120	96.2	84.3	143	7.0	15.0	M
regularis (E) x *regularis* (W)	66-1	43.8	48.0	64	34.4	16.5	M
regularis (E) x *regularis* (E)	66-4	41.0	50.0	132	23.5	11.7	M
regularis (E) x *regularis* (E)	66-5	35.8	5.3	5	40.0	2.1	M
regularis (E) x *regularis* (E)	66-6	63.8	86.7	66	19.7	16.1	M
regularis (E) x *regularis* (E)	66-7	36.6	90.9	10	30.0	27.3	M
regularis (W) x *regularis* (E)	66-77	12.6	100.0	35	2.8	2.8	M
regularis (W) x *regularis* (E)	66-259	100.0	97.4	115	—	—	L
regularis (W) x *regularis* (E)	66-78	84.6	94.0	171	—	—	L
regularis (W) x *regularis* (E)	66-98	73.5	60.0	130	0.8	0.5	M
regularis[25] x *regularis*[25]	67-379	100.0	100.0	152	—	—	L
regularis (W) x *regularis* (W)	66-79	83.0	92.3	246	39.0	36.0	M
regularis (W) x *regularis* (W)	66-81	28.3	100.0	113	65.5	65.5	M
regularis (W) x *regularis* (W)	66-257	87.8	100.0	102	85.3	85.3	M
regularis (W) x *reg-rang* (hybrid)	66-82	24.5	—	—	—	—	G
regularis (W) x *regularis* (E)	66-255	90.4	100.0	107	—	—	L
regularis (W) x *regularis* (E)	66-83	65.5	58.3	98	—	—	L
regularis (W) x *regularis* (E)	66-254	92.0	100.0	117	—	—	L
regularis (W) x *regularis* (E)	66-84	48.0	16.7	80	—	—	L
regularis (W) x *regularis* (E)	66-88	54.1	47.8	122	2.4	1.1	M
regularis (W) x *regularis* (E)	66-93	87.3	3.3	116	7.7	2.6	M
regularis (W) x *regularis* (E)	66-89	41.6	12.0	24	—	—	L
regularis (W) x *regularis* (E)	66-95	94.2	69.4	134	1.5	1.0	M
regularis (W) x *regularis* (E)	66-91	81.6	88.1	137	0.7	<1.0	M

M = metamorphosed L = stopped as larva N = stopped as neurula G = stopped as gastrula x = chemical inhibition
S♂ = male proved sterile in F_2 or backcross F♂ = male proved fertile in F_2 or backcross

[25] Location unknown.

Combination		Cross Number	% Fertilized	% Fertilized Hatched	Number Larvae	% Metamorphose	Calculated % Original Metamorphose	Stage Reached
♀ REGULARIS	♂ REGULARIS							
regularis (W) x *regularis* (E)		66-96	18.4	—	8	—	—	L
regularis (W) x *regularis* (E)		67-70	47.5	87.5	>142	0.7	0.6	M
regularis (W) x *regularis* (E)		67-69	17.6	44.4	40	7.5	3.3	M
regularis (W) x *regularis* (E)		67-71	76.2	80.0	164	1.2	1.0	M
regularis (E) x *regularis* (W)		65-455	98.0	100.0	208	91.8	91.8	M
regularis (E) x *regularis* (E)		65-495	3.2	—	—	—	—	G
regularis (E) x *regularis* (E)		65-516	28.0	57.1	57	70.2	40.1	M
regularis (W) x *regularis* (E)		68-113	94.6	97.1	134	3.9	5.7	M
regularis (W) x *kerinyagae*		68-114	100.0	43.0	243	1.2	3.6	M
	MACULATUS							
brauni x *maculatus*		65-512	92.3	100.0	110	—	—	L
brauni x *maculatus*		65-513	45.4	100.0	5	—	—	L
garmani x *maculatus*		65-430	98.4	—	8	—	—	G–L
kerinyagae x *maculatus*		68-138	66.7	—	—	—	—	L
rangeri x *maculatus*		65-375	45.0	90.5	44	—	—	L
regularis (W) x *maculatus*		65-520	98.2	100.0	194	—	—	L
regularis (E) x *maculatus*		66-2	55.6	64.1	121	—	—	L
regularis (E) x *maculatus*		66-3	56.0	35.7	104	—	—	L
regularis (E) x *maculatus*		65-511	1.7	—	—	—	—	G
regularis (E) x *maculatus*		65-431	19.6	50.0	55	—	—	L
regularis (W) x *maculatus*		66-78	—	—	—	—	—	L
	PERRETI							
garmani x *perreti*		65-434	86.7	78.8	221	2.3	1.8	M
rangeri x *perreti*		65-375	45.0	90.5	44	—	—	L
regularis (E) x *perreti*		65-435	18.4	75.0	176	0.6	0.4	M
	MELANOPHRYNISCUS							
regularis (W) x *Melanophryniscus*		68-6	12.5	—	—	—	—	G
MACULATUS	SPINULOSUS							
maculatus x *spinulosus*		64-202	33.3	—	—	—	—	G–N
	GRANULOSUS							
maculatus x *fernandezae*		64-208	98.7	69.2	154	—	—	L
maculatus x *humboldti*		64-378	19.1	—	2	—	—	L
	MARINUS							
maculatus x *arenarum*		64-198	97.7	66.1	77	—	—	L
maculatus x *arenarum*		68-130	15.1	—	5	—	—	L
maculatus x *marinus*		64-206	98.0	86.3	230	0.9	0.8	M
maculatus x *marinus*		64-295	58.1	44.4	40	—	—	L
maculatus x *marinus*		65-202	86.9	97.7	148	—	—	L
maculatus x *marinus*		68-129	14.0	—	24	—	—	L
maculatus x *paracnemis*		68-131	10.7	10.0	40	—	—	L
maculatus x *poeppigi*		68-128	9.2	16.7	9	—	—	L
	CRUCIFER							
maculatus x *crucifer*		68-123	100.0	45.4	115	—	—	L
	BOCOURTI							
maculatus x *bocourti*		64-201	100.0	—	—	—	—	G–N
	BOREAS							
maculatus x *boreas*		64-288	34.2	—	4	—	—	L
maculatus x *boreas*		64-382	54.0	—	—	—	—	G–N
maculatus x *boreas*		65-200	86.0	—	—	—	—	G

M = metamorphosed L = stopped as larva N = stopped as neurula G = stopped as gastrula x = chemical inhibition
S♂ = male proved sterile in F_2 or backcross F♂ = male proved fertile in F_2 or backcross

Combination ♀	♂	Cross Number	% Fertilized	% Fertilized Hatched	Number Larvae	% Metamorphose	Calculated % Original Metamorphose	Stage Reached
MACULATUS	*ALVARIUS*							
maculatus x *alvarius*		64-270	21.6	72.7	25	—	—	L
	PUNCTATUS							
maculatus x *punctatus*		65-541	18.9	—	—	—	—	G
maculatus x *punctatus*		65-199	100.0	1.6	1	—	—	G–L
	MARMOREUS							
maculatus x *marmoreus*		64-207	98.0	6.0	3	—	—	L
maculatus x *perplexus*		65-198	98.1	—	—	—	—	G–N
	COGNATUS							
maculatus x *cognatus*		64-277	86.7	—	—	—	—	G–N
maculatus x *cognatus*		65-207	100.0	—	—	—	—	G
maculatus x *speciosus*		64-199	93.3	—	3	—	—	L
	DEBILIS							
maculatus x *debilis*		64-200	70.1	17.9	20	—	—	N
maculatus x *debilis*		65-201	98.3	—	—	—	—	G
	VALLICEPS							
maculatus x *coniferus*		68-124	32.5	46.1	43	—	—	L
maculatus x *ibarrai*		64-380	21.4	—	—	—	—	G–N
maculatus x *luetkeni*		64-379	22.2	—	—	—	—	G–N
maculatus x *valliceps*		64-197	98.4	43.3	141	—	—	L
maculatus x *valliceps*		65-208	94.4	—	1	—	—	L
maculatus x *valliceps*		68-132	38.6	—	20	—	—	L
	CANALIFERUS							
maculatus x *canaliferus*		64-276	97.9	76.1	72	—	—	L
	COCCIFER							
maculatus x *coccifer*		64-289	92.8	—	2	—	—	L
maculatus x *coccifer*		64-383	50.7	—	5	—	—	L
maculatus x *coccifer*		64-385	17.9	—	18	—	—	L
maculatus x *coccifer*		65-209	97.0	—	—	—	—	G
	AMERICANUS							
maculatus x *americanus*		64-290	100.0	—	—	—	—	G
maculatus x *hemiophrys*		65-210	92.8	—	—	—	—	G
maculatus x *microscaphus*		64-292	48.6	—	—	—	—	G
maculatus x *terrestris*		65-205	98.3	—	—	—	—	G
maculatus x *woodhousei*		64-196	81.0	—	—	—	—	G–N
maculatus x *woodhousei*		64-381	51.3	—	—	—	—	G–N
	QUERCICUS							
maculatus x *quercicus*		68-127	16.7	—	—	—	—	G
	VIRIDIS							
maculatus x *viridis*		65-197	86.8	—	—	—	—	G–N
	CALAMITA							
maculatus x *calamita*		65-204	100.0	—	—	—	—	G
maculatus x *calamita*		64-204	37.5	—	—	—	—	G
	BUFO							
maculatus x *bufo*		64-209	15.5	—	—	—	—	G
maculatus x *bufo*		65-203	98.5	—	—	—	—	G
	MAURITANICUS							
maculatus x *mauritanicus*		64-266	97.9	97.9	246	0.4	<0.1	M

M = metamorphosed L = stopped as larva N = stopped as neurula G = stopped as gastrula x = chemical inhibition
S♂ = male proved sterile in F_2 or backcross F♂ = male proved fertile in F_2 or backcross

Combination ♀	♂	Cross Number	% Fertilized	% Fertilized Hatched	Number Larvae	% Metamorphose	Calculated % Original Metamorphose	Stage Reached
MACULATUS	*REGULARIS*							
maculatus x *brauni*		66-154	41.8	89.4	66	—	—	L
maculatus x *garmani*		65-538	95.6	95.4	122	—	—	L
maculatus x *kerinyagae*		68-125	80.3	68.6	95	—	—	L
maculatis x *pardalis*		65-206	84.7	100.0	190	—	—	L
maculatus x *rangeri*		68-122	95.3	48.8	220	—	—	L
maculatus x *regularis* (W)		65-195	83.7	100.0	210	—	—	L
maculatus x *regularis* (E)		65-542	40.0	100.0	2	—	—	L
maculatus x *regularis* (W)		66-155	52.0	96.2	80	—	—	L
maculatus x *regularis* (W)		70-5	17.9	—	—	—	—	G
maculatus x *regularis* (W)		70-6	22.9	—	—	—	—	G
maculatus x *regularis* (W)		70-9	29.7	3.0	8	—	—	L
maculatus x *regularis* (W)		70-10	98.1	100.0	162	—	—	L
maculatus x *regularis* (W)		70-11	99.0	99.0	160	—	—	L
maculatus x *regularis* (E)		66-156	53.1	88.0	56	—	—	L
maculatus x *regularis* (E)		66-157	4.5	71.4	5	—	—	L
maculatus x *regularis* (E)		68-125	81.0	68.6	95	—	—	L
STEINDACHNERI	*MARINUS*							
steindachneri x *paracnemis*		70-16	88.8	—	—	—	—	G–N
	VALLICEPS							
steindachneri x *valliceps*		70-17	98.0	—	—	—	—	G
	AMERICANUS							
steindachneri x *woodhousei*		70-18	100.0	—	—	—	—	G
	REGULARIS							
steindachneri x *regularis* (W)		70-15	—	—	8	—	—	L
	MACULATUS							
steindachneri x *maculatus*		70-14	100.0	99.1	192	—	—	L
	PERRETI							
maculatus x *perreti*		65-194	100.0	100.0	177	—	—	L
PERRETI	*REGULARIS*							
perreti x *regularis* (W)		70-32	98.2	—	—	—	—	G–N
perreti x *regularis* (W)		70-34	94.1	—	—	—	—	G–N
	MACULATUS							
perreti x *maculatus*		70-33	98.1	—	—	—	—	G–N
perreti x *maculatus*		70-35	72.3	—	—	—	—	G–N
LUGHENSIS	*MARINUS*							
lughensis x *marinus*		66-296	52.8	—	—	—	—	G
	BOREAS							
lughensis x *boreas*		66-294	83.3	—	—	—	—	G
	DEBILIS							
lughensis x *kelloggi*		66-297	7.1	—	—	—	—	G–N
	GARIEPENSIS							
lughensis x *gariepensis*		66-295	21.0	—	—	—	—	G
	REGULARIS							
lughensis x *regularis* (W)		66-293	57.6	—	—	—	—	G

M = metamorphosed L = stopped as larva N = stopped as neurula G = stopped as gastrula x = chemical inhibition
S ♂ = male proved sterile in F_2 or backcross F ♂ = male proved fertile in F_2 or backcross

Appendix I. Spectra of Indolealkylamines in the Toad Skin

Table 1. African Stocks
(Indolealkylamine contents are expressed in μg/g dried or fresh skin)

	5-HT	N-Methylated Derivatives	Cyclized Derivatives	Conjugated Derivatives C_a	C_b	Emerald Green Spot
Regularis group						
regularis						
Congo, VII. 1957						
Pool 220 skins D	540	—	—	(15)	(25)	+++
Uganda, II. 1964						
Pool 6 skins D	2500	—	—	(50)	(110)	+++
Kenya, V. 1967						
Pool 6 skins F	1100	—	—	(60)	(80)	++
Kenya, I. 1968						
Pool 50 skins F	375	—	—	(3)	(35)	++
Nigeria, VIII. 1968						
1 skin D	1600	—	—	—	(30)	+++
rangeri						
Graham Town, South Africa, VIII. 1968						
4 skins D	700-2600	—	—	(30-300)	(160-2400)	+++
latifrons						
Ibadan, Nigeria, VIII. 1968						
Pool 2 skins D	1160	—	—	(?)	(25)	++++
kisoloensis						
Congo, V. 1952						
Pool 12 skins D	600	—	—	—	—	++
berghei						
Congo, V. 1952						
Pool 20 skins D	150	—	—	—	—	++
funereus						
Congo, VIII. 1957						
Pool 129 skins D	540	—	—	(?)	(250)	+++
garmani						
South Africa, XII. 1967	1250 ± 265				371 ± 61	
7 skins F	(650-2600)	—	—	0-15	(120-600)	++++
South Africa, II. 1968						
Pool 24 skins F	1000	—	—	(15)	(250)	++++
South Africa, III. 1968						
Pool 10 skins F	850	—	—	(13)	(200)	++++
Mauritanicus group						
mauritanicus						
Morocco, V. 1967						
Pool 5 skins F	480	—	—	(5)	(500)	(+)
Morocco, VI. 1967						
Pool 12 skins F	650	—	—	(?)	(5)	+++
Morocco, I. 1968						
Pool 14 skins F	675	—	—	(?)	(20)	+++
Carens group						
carens						
South Africa, III. 1968						
Pool 200 skins F	—	—	—	Unknown 1-conjugate of tryptamine (>200–300 μg/g)		

D, dried skin; *F*, fresh skin — not detected; (?), doubtful occurrence.

C_a, supposed O-sulphate of 5-HT; C_b, supposed O-sulphate of 5-HT with an unknown acid radical attached in position 1, i.e., to the pyrrolic > NH group.

Emerald green spot, compound giving, on paperchromatograms, an emerald green spot with the *p*-dimethylaminobenzaldehyde reagent.

Values for C_a and C_b in parentheses are only indicative, and expressed as 5-HT following the *p*-dimethylaminobenzaldehyde reaction. For the emerald green spot the number of + indicates increasing amounts. For example, ++, 100–200 μg/g; ++++, > 400 μg/g.

Appendix I (Continued)

Table 2. Euro-Asiatic Stocks
(Indolealkylamine contents are expressed in μg/g dried or fresh skin)

	5-HT	N-Methylated Derivatives			Cyclized Derivatives (Dehydro-bufotenine)	Conjugated Derivatives		
						5 (O)-conjugated		1-conjugated
		M_1	M_2	M_3		C_1	C_2	C_3
South-Oriental or Indonesian or Malayan group								
B. asper								
Malaya, V. 1966 Pool 3 skins D	—	—	—	—	—	—	—	—
Malaya, VIII. 1967 9 skins D	—	—	—	—	6-30 (in 3 specimens only)	—	—	—
B. juxtasper								
Borneo, XI. 1966 Pool 8 skins D	2-3(?)	—	—	—	—	—	—	—
Pseudobufo subasper								
Malaya, X. 1964. Pool 13 skins D	8-10	—	—	—	—	—	—	—
Euro-Indian (Paleartic) group								
B. melanostictus								
Malaya, VII. 1964 Pool 100 skins D	130	—	20	1250	8-9	—	—	—
India, VIII. 1967 5 skins D	—	—	676±66 (460-800)	1014±78 (830-1250)	(10-30 in two specimens)	276±44 (200-420)	—	—
Pakistan, VIII. 1968. 1 skin D	120	110	480	60	—	—	—	—
B. stomaticus								
India, VIII. 1967 5 skins D	203 ± 99 (30-550)	—	411±130 (115-770)	1545±879 (75-3700)	45±29 (5-160)	77±25 (30-140)	—	—
B. bufo formosus								
Japan, VI. 1962 Pool 9 skins D	65	60	150	140	25	375	—	—
B. b. bufo								
France, VIII. 1967 Pool 7 skins D	10	—	175	215	700	1000	—	—
B. b. spinosus								
Italy, IV. 1964 Pool 7 skins F	20	—	450	50	130	320	—	—
Italy, IV. 1966 Pool 15 skins F	2-12	—	169±24 (45-400)	54±10 (5-150)	98±11 (60-175)	320±33 (90-750)	—	—
Italy, IV. 1967 Pool 10 skins F	8	—	250	80	80	180	—	—
Calamita group								
B. calamita								
Central Europe, VI. 1964 Pool 5 skins F	6-8	15	105	35	—	—	++	110
Central Europe, VI. 1967 Pool 3 skins F	100	200	500	20	—	—	+	750
Central Europe, IX. 1967 Pool 45 skins F	15	35	220	100	—	—	+++	950
B. viridis								
Italy, IV. 1959 Pool 150 skins F	5-10	25	650	30	—	—	?	480
Italy, IV. 1964 Pool 20 skins F	10	?	750	50	—	—	?	110

Table 2. Euro-Asiatic Stocks (Continued)
(Indolealkylamine contents are expressed in μg/g dried or fresh skin)

	5-HT	N-Methylated Derivatives			Cyclized Derivatives (Dehydro-bufotenine)	Conjugated Derivatives		
						5 (O)-conjugated		1-con-jugated
		M_1	M_2	M_3		C_1	C_2	C_3
B. viridis								
Italy, IV. 1967 Pool 12 skins F	30	—	1500	100	—	—	—	300
Italy, IX. 1967 Pool 116 skins F	12	—	1000	80	—	—	—	500
Israel, VIII. 1967 6 skins D	64 ± 19 (6-140)	—	2325±303 (1150-3250)	850±154 (300-1300)	—	—	?	2540± 426 (1040-4000)
Israel, XII. 1967 4 skins F	9 ± 2 (7-12)	—	2475±367 (1750-3500)	175±28 (125-250)	—	—	?	1350± 221 (800-1800)

D, dried skin; *F*, fresh skin; —, not detected; ?, doubtful occurrence.

M_1, N-methyl-5-HT; M_2 bufotenine; M_3, bufotenidine.

C_1, bufothionine; C_2, supposed O-sulfate of bufotenine; C_3, bufoviridine (supposed bufotenine 1 sulfonic acid).

+, ++, +++, presence of the compound in varying amounts; no quantitative data available.

Table 3. North American Stocks
Northern Line. Western Section
(Indolealkylamine contents are expressed in μg/g dried or fresh tissue)

	5-HT	N-Methylated Derivatives			Cyclized Derivatives (Dehydro-bufotenine)	Conjugated Derivatives		
						5 (O)-conjugated		1-con-jugated
		M_1	M_2	M_3		C_1	C_2	C_3
Boreas group								
boreas								
Wyoming (Med. Bow Mnt.), U.S.A., VII. 1964 14 skins D	24.3±2.3 (20-40)	—	520±66 (220-880)	31±2 (20-50)	7.4±1.3 (5-20)	67.5±9.3 (20-120)	—	—
Colorado, Cameron Pass, U.S.A., VII. 1964 4 skins D	—	—	1465+139 (260-2500)	72.5±7 (20-130)	—	492.5±13 (360-570)	—	—
Colorado, U.S.A., IX. 1967 4 skins D	0-10	—	305±42 (190-370)	0-80 (in few specimens)	—	351±282 (60-1200)	—	—
halophilus								
California, Riverside, U.S.A., IX. 1963 1 skin D	—	—	275	50	—	275	—	—
California, May Lake, U.S.A., VIII. 1967 6 skins D	—	—	136.6±89 (0-580)	0-25 (in few specimens)	0.4 (in few specimens)	487.5±217 (60-1350)	—	—
canorus								
California, U.S.A., IX. 1967 18 skins D	23±6.6 (0-110)	0-120 (in few speci-mens)	448±116 (50-1900)	0-30 (in few specimens)	84±18 (6-320)	1943±322 (560-5600)	—	—
Alvarius group								
alvarius								
See Table 3 bis (p. 440)								

Appendix I (Continued)

Table 3. North American Stocks (Continued)

	5-HT	N-Methylated Derivatives			Cyclized Derivatives (Dehydro-bufotenine)	Conjugated Derivatives		
		M_1	M_2	M_3		5 (O)-conjugated C_1	C_2	1-conjugated C_3
Punctatus group								
punctatus								
Arizona, U.S.A., VIII. 1964 6 skins D	—	—	0-20	—	543±136 (125-1000)	465±109 (100-750)	—	—
Arizona, U.S.A., X. 1964 Pool 3 skins D	—	—	—	—	500	400	—	60
Arizona, U.S.A., VIII. 1965 1 skin D	—	—	50	—	900	600	—	5
California, U.S.A., VIII. 1965 Pool 2 skins D	—	—	10	—	500	1200	—	575
Mexico, VIII. 1966 Pool 8 skins D	—	—	20	—	1750	4000	—	3000
Texas, U.S.A., VIII. 1967 Pool 3 skins D	—	130	12	—	1800	840	—	2800
Debilis group								
debilis								
Texas, U.S.A., VII. 1965 9 skins D	—	0-75 (in few specimens)	0-64 (in few specimens)	—	1745±368 (550-3600)	802±167 (390-1700)	—	20-250
Texas, U.S.A., VIII. 1967 1 skin D	90	—	—	—	1800	840	—	2800
kelloggi								
Mexico, Sonora, VIII. 1967 1 skin D	16	13	23	—	7000	—	—	—
retiformis								
Mexico, VIII. 1967 7 skins D	104±26 (32-200)	0-85 (in few specimens)	—	—	2041±244 (280-5000)	389±222 (40-1700)	—	192±64 (25-385)
Marmoreus group								
marmoreus								
Mexico, Guerrero, IX. 1964 Pool 4 skins D	150	—	770	—	—	—	?	480
Mexico, VIII. 1967 Pool 5 skins D	200	?	1390	—	—	—	220	1050
perplexus								
Mexico, Puebla, VII. 1967 Pool 9 skins D	140	—	1400	—	5	—	1300	10
Mexico, Puebla, VII. 1966 Pool 6 skins D	150	—	2400	—	—	—	360	6
Mexico, VIII. 1967 Pool 7 skins D	160	—	1750	—	5	—	800[a]	—
Bocourti group								
bocourti								
Guatemala, VIII. 1967 Pool 12 skins D	—	—	3700	2500	—	—	1050	75
Guatemala, VIII. 1967 6 skins D	—	—	1900±296 (800-2800)	841±158 (300-1250)	—	—	486±213 (100-1450)	—

D, dried skin; *F*, fresh skin; —, not detected; ?, doubtful occurrence.

M_1, N-methyl-5-HT; M_2, bufotenine; M_3, bufotenine.

C_1, bufothionine; C_2, supposed O-sulphate of bufotenine; C_3, bufoviridine (supposed bufotenine 1 sulphonic acid).

[a] occurrence also of the supposed O-sulphate of 5-HT (~ 70 μg) g.

Appendix I (Continued)

Table 3 bis. The content of bufotenine and O-methylbufotenine in the skin of *Bufo alvarius* (in mg/g skin) Large Cutaneous Glands (Coxal and Parotoid Glands) and the Remaining Skin were Separately Investigated

(No. of Skins)	Bufotenine		O-Methylbufotenine	
	Glands	Remaining Skin	Glands	Remaining Skin
(1) USA, Arizona, Tucson, 1963, 11	2.2	1.5	80	3.2
(2) USA, Arizona, Mesa, 1964, 5	0.9, 1.1	0.33, 2.1	100, 160	1.2, 3.0
(9) USA, Arizona, Mesa, 1964, 8	2.3±0.50	1.07±0.19	98±11	2.0±0.2
	(0.9-5)	(0.8-2.2)	(60-160)	(1.0-3.5)
(4) USA, Arizona, Mesa, 1965, 9	0.3-0.5	1.25-5.5	107-146	2.0-3.9
(1) USA, California, 1965, 7	0.1	0.7	51.5	0.6
(3) USA, Arizona, Mesa, 1966, 8	0.1	0.5-0.9	77-107	0.4-0.7

In the pool of the 9 nonglandular skins collected in Arizona during August 1964, the following additional indolealkylamines were detected (in μg/g dry skin): 5-HT 4-6, N-methyl-5-HT 30-40, N-methyl-5-methoxytryptamine 10-15, bufoviridine 15-20, O-sulphate of bufotenine 80-90 and O-methylbufotenine 1-sulphonic acid 90-100 (Erspamer, Vitali, Roseghini and Cei, 1967).

(Reported from Cei, Erspamer and Roseghini, 1968).

Table 4. North American Stocks
Northern line. Eastern Session
(Indolealkylamine contents are expressed in μg/g dried or fresh tissue)

	5-HT	N-Methylated Derivatives			Cyclized Derivatives (Dehydrobufotenine)	Conjugated Derivatives		
						5 (O)-conjugated		1-conjugated
		M_1	M_2	M_3		C_1	C_2	C_3
Americanus group								
americanus								
Hudson, Ohio, U.S.A., IX. 1967 1 skin D	270	—	370	750	—	—	—	—
Minnesota, U.S.A., X. 1967 4 skins D	533±24 (35-750)	80±24 (15-130)	205±101 (50-490)	352±96 (75-500)	18.7±3.7 (15-30)	—	—	—
terrestris								
Florida, U.S.A., VI. 1966 Pool 4 skins D	1600	70	20	1300	25	—	—	—
Savannah, Georgia, U.S.A., X. 1967 1 skin D	1700	—	—	1850	75	—	—	—
w. woodhousei								
Texas, U.S.A., VI. 1964 2 skins D	500–800	40	—	950, 1200	150, 200	—	—	—
Texas, U.S.A., VI. 1965 9 skins D	302±58 (100-600)	50±11 (20-100)	—	306±55 (70-600)	27±6 (7-55)	—	—	—
w. australis								
Mesa, Arizona, U.S.A., IX. 1964 8 skins D	1081±158 (550-2000)	154±30 (80-300)	—	1469±323 (500-3000)	—	—	—	—
microscaphus								
Utah, U.S.A., VI. 1966 Pool 2 skins D	600	160	60	5200	—	—	—	—
hemiophrys								
Dakota, U.S.A., VI. 1965 7 skins D	317±82 (100-600)	96±26 (20-170)	68±22 (10-180)	676±144 (270-1200)	1993±806 (250-6000)	—	—	—
hemiophrys								
Minnesota, U.S.A., IX. 1967 10 skins D	108±38 (10-400)	56.8±19 (5-160)	34±6 (5-60)	893±202 (130-1860)	1098±142 (510-1950)	—	—	—
Quercicus group								
quercicus								
Georgia, U.S.A., IX. 1967 4 skins 4 skins D	84.7±57 (5-250)	—	—	—	104.7±56 (25-270)	—	—	—

D, dried skin; *F*, fresh skin; —, not detected.
M_1, N-methyl-5-HT; M_2, bufotenine; M_3, bufotenidine.
C_1, bufothionine; C_2, supposed O-sulphate of bufotenine; C_3, bufoviridine (supposed bufotenine 1-sulphonic acid).

Appendix I (Continued)

Table 5. Central American Stocks
Southern Line
(Indolealkylamine contents are expressed in μg/g dried or fresh tissue)

	5-HT	N-Methylated Derivatives			Cyclized Derivatives (Dehydro-bufotenine)	Conjugated Derivatives		
						5 (O)-conjugated		1-conjugated
		M_1	M_2	M_3		C_1	C_2	C_3
Cognatus group								
cognatus								
Mexico, Durango, XI. 1963 1 skin D	600	20	—	—	10	—	—	—
Arizona, U.S.A., VIII. 5 skins D	750±246 (200-1600)	48±25 (5-140)	—	—	26±11 (7-75)	—	—	—
Texas, U.S.A., VI. 1965 10 skins D	38±9 (10-90)	traces	0-10	—	—	—	—	—
Arizona, U.S.A., VII. 1965 10 skins D	213±48 (10-460)	64±13 (5-180)	—	—	91±27 (6-260)	—	—	—
Arizona, U.S.A., VIII. 1967 Pool 4 skins D	100	35	25	—	100	—	—	—
speciosus								
Texas, U.S.A., XI. 1963 1 skin D	650	90	—	—	50	—	—	—
Texas, U.S.A., V. 1964 12 skins D	2079±351 (300-4500)	274±34 (50-450)	—	—	98±26 (30-290)	—	—	—
Texas, U.S.A., X. 1967 4 skins D	987±252 (400-1500)	154±23 (110-215)	—	—	58±36 (5-160)	—	—	—
Valliceps group								
coniferus								
Costa Rica, IX. 1967 12 skins D	89±25 (1-300)	225±69 (10-750)	—	—	0-30 (in some specimens)	2-40 (in few specimens)	—	—
luetkeni								
Nicaragua, XI. 1964 4 skins D	405±162 (150-900)	20±1 (8-30)	42±14 (12-75)	—	—	—	—	—
Honduras, X. 1967 3 skins D	(200, 1200, 3400)	(40, 65, 530)	(0, 0, 45)	—	—	—	—	—
valliceps								
Texas, U.S.A., VIII. 1963 Pool 7 skins D	1600	—	—	—	—	—	—	—
Texas, U.S.A., VI. 1964 7 skins D	1200±140 (500-1600)	—	—	—	—	—	—	—
Texas, U.S.A., IX. 1967 3 skins D	(40, 200, 1000)	—	—	—	—	—	—	—
Texas, U.S.A., X. 1967 Pool 8 skins D	700	—	—	—	—	—	—	—
Coccifer group								
coccifer								
San Vicente, Salvador IX. 1964 Pool 12 skins D	120	10	—	1450	—	—	—	—
Costa Rica, X. 1967 4 skins D	38±8 (20-60)	—	—	501±132 (260-870)	8±4.5 (0-20)	—	—	—

Appendix I (Continued)

Table 5. Central American Stocks (Continued)

	5-HT	N-Methylated Derivatives			Cyclized Derivatives (Dehydro-bufotenine)	Conjugated Derivatives		
						5 (O)-conjugated		1-con-jugated
		M_1	M_2	M_3		C_1	C_2	C_3
Canaliferus group								
canaliferus								
Chiapas, Mexico, XI. 1963 1 skin D	3000	120	—	80	—	250	—	—
Guatemala, VII. 1966 Pool 4 skins D	5000	150	—	—	—	375	—	—
Tehuantepec, Mexico, IX. 1967 4 skins D	1972±1346 (400-6000)	45±28 (6-130)	15±10 (2-45)	—	14±2 (8-20)	2225±744 (400-3900)	—	—
Guatemala, X. 1967 5 skins D	3100±693 (2200-5000)	192±69 (55-430)	—	—	0-80 (in few specimens)	4140±795 (2800-7000)	—	—
(*incertae sedis* groups)								
Guttatus group								
haematiticus								
Turrialba, Cost Rica, VIII. 1964 Pool 12 skins D	2100	—	—	—	9000	—	—	—
Costa Rica, X. 1967 1 skin D	300	—	85	85	18000	140	—	—
Periglenes group								
periglenes								
Costa Rica, VIII. 1968 1 skin D	40	—	1160	—	—	—	—	—
Typhonius group								
typhonius								
Darien, Panama, X. 1965 1 skin D	10	—	—	—	450	790	—	—

D, dried skin; *F*, fresh skin; —, not detected.
M_1, N-methyl-5-HT; M_2, bufotenine; M_3, bufotenidine.
C_1, bufothionine; C_2, supposed O-sulphate of bufotenine; C_3, bufoviridine (supposed bufotenine 1-sulphonic acid).

Table 6. Neotropical Stocks
Andean-Patagonian Line
(Southwestern Section of a primeval "Northern Line"?)
(Indolealkylamine contents are expressed in μg/g dried or fresh tissue)

	5-HT	N-Methylated Derivatives			Cyclized Derivatives (Dehydro-bufotenine)	Conjugated Derivatives		
						5 (O)-conjugated		1-con-jugated
		M_1	M_2	M_3		C_1	C_2	C_3
Spinulosus group								
limensis								
Trujillo, Peru, I. 1965 7 skins D	—	—	111±36 (25-300)	350±102 (65-800)	463±34 (320-560)	720±111 (300-1200)	—	—
limensis-like population, Cajamarca, Peru, 2800 m. I. 1965 7 skins D	—	—	127±27 (60-250)	211±52 (80-400)	1835±319 (950-3000)	2100±322 (940-3750)	—	—
flavolineatus								
Junin Lake, Peru, 4200 m. V. 1967 11 skins D	0-30 (in few specimens)	—	2275±389 (460-4200)	2218±280 (750-4100)	140±50 (0-480)	394±145 (0-1700)	—	—

Appendix I (Continued)

Table 6. Neotropical Stocks (Continued)

	5-HT	N-Methylated Derivatives			Cyclized Derivatives (Dehydro-bufotenine)	Conjugated Derivatives		
						5 (O)-conjugated		1-conjugated
		M_1	M_2	M_3		C_1	C_2	C_3
trifolium								
Huanuco, Peru, 1800 m. I. 1963 Pool 3 skins D	—	—	550	100	—	—	—	—
Tarma, Peru, 3500 m. V. 1967 3 skins D	(traces)	—	(450, 550, 700)	(300, 1200, 1250)	(8, 25, 480)	(2. 5, 35, 750)	—	—
spinulosus								
Cuzco, Peru, 3500 m. I. 1963 Pool 18 skins D	—	—	230	30	180	940/1100	—	—
San Pedro, Antofagasta, Chile, 2500 m. X. 1965 6 skins D	—	—	785±264 (400-2000)	61±7 (40-80)	536±275 (150-1900)	2200±440 (500-4400)	—	—
Sosneado, Mendoza, Argentina, 1000 m. III. 1965 9 skins D	scarce (20-150 in few specimens)	—	134±31 (20-320)	33±4 (25-60)	929±172 (360-2200)	840±128 (375-1850)	—	—
Malargue, Mendoza, Argentina 1000 m. III. 1965 6 skins D	(traces)	—	71±26 (20-120)	26±9 (10-70)	850±160 (400-1500)	360±109 (90-950)	—	—
San Isidro, Mendoza, Argentina, 1200 m. X. 1967 10 skins D	— (75 in one specimen)	—	265±22 (160-460)	114±14 (37-175)	1288±247 (600-3200)	3233±353 (2300-5900)	(110 in one specimen)	—
atacamensis								
Huasco River, Chile, Atacama Desert, VII. 1967 9 skins D	(traces)	—	(traces in one specimen)	—	173±65 (6-500)	690±206 (50-1700)	—	—
chilensis								
Zapallar, Chile, I. 1966 Pool 8 skins D	—	—	310	35	570	650	270	—
Valparaiso, Chile I. 1966 Pool 15 skins D	—	—	250	45	1000	1000	240	—
Zapallar, Chile, V. 1967 6 skins D	—	—	1600±305 (250-2300)	150±46 (75-350)	734±112 (450-1100)	2358±218 (1400-2800)	400±90 (100-675)	—
Zapallar, Chile, VIII. 1967 4 skins D	—	—	577±343 (115-1600)	31±7 (15-50)	450±163 (40-800)	1407±443 (680-2700)	70±20 (30-110)	—
Valparaiso, Chile, II. 1968 10 skins D	—	—	952±377 (5-3500)	67±17.7 (5-175)	778±134 (240-1400)	2069±350 (560-4500)	97±31 (8-300)	—
Valparaiso, Chile, II. 1968 Pool 63 skins D	—	—	2300	175	1180	1960	160	—
rubropunctatus								
Llanquihue, Chile, VII. 1968 1 skin F	—	—	175	—	5	2.5	—	—
variegatus group								
variegatus								
Lakar Lake, Argentina, I. 1964 Pool 6 skins D	—	—	8600	750	—	—	—	—
Bariloche, Argentina, IX. 1967 1 skin D	—	—	16200	350	60	140	—	—
Valdivia, Chile, X. 1967 1 skin D	—	—	18500	625	(traces)	140	—	—

D, dried skin; *F*, fresh skin; —, not detected.
M_1, N-methyl-5-HT; M_2, bufotenine; M_3, bufotenidine.

C_1, bufothionine; C_2, supposed O-sulphate of bufotenine; C_3, bufoviridine (supposed bufotenine 1-sulphonic acid).

Table 7. Neotropical Stocks
Neotropical Branch of the Southern Line
(Indolealkylamine contents are expressed in μg/g dried or fresh tissue)

	5-HT	N-Methylated Derivatives			Cyclized Derivatives (Dehydro-bufotenine)	Conjugated Derivatives		
						5 (O)-conjugated		1-conjugated
		M_1	M_2	M_3		C_1	C_2	C_3
"*Marinus* Stock"								
Marinus group								
m. marinus								
Nicaragua, IX. 1964 7 skins D	554±117 (50-1600)	353-149 (25-1200)	—	—	3743±652 (1500-6000)	—	—	—
Costa Rica, V. 1962 Pool 4 skins D	100	50	—	—	2200	320	—	—
Paramaribo, Surinam, V. 1960 Pool 12 skins male specimens D	100	130	4	—	3000	465	—	—
Paramaribo, Surinam, V. 1960 Pool 22 skins female specimens D	30	40	3	—	2700	430	—	—
Paramaribo, Surinam, IV. 1962 Pool 6 skins D	140	100	—	—	6000	30	—	—
Hawaii, X. 1967 1 skin D	70	170	—	—	1600	420	—	—
poeppigi								
Coroico, Bolivia, I. 1963 1 skin D	165	150	—	—	1100	300	—	—
Tingo Maria, Peru, II. 1963 Pool 5 skins D	160	100	—	—	270	300	—	—
Huancabamba, Peru, I. 1965 Pool 3 skins D	970	380	—	—	330	32	—	—
Tumbez, Peru, I. 1965 8 skins D	577±120 (220-1300)	342±56 (180-650)	—	—	535±36 (290-750)	—	—	—
Machala, Ecuador, I. 1965 1 skin D	400	250	—	—	140	10	—	—
ictericus								
Rio de Janeiro, Brazil, II. 1962 Pool 10 skins D	10	20	—	—	4500	310	—	—
São Paulo, Brazil. 1962 Pool F	3.5	4	3	—	1800	45	—	—
Misiones, Argentina, X. 1962 Pool 7 skins D	60	50	15	—	3300	465	—	—
paracnemis								
Tucuman, Argentina, I. 1960 Pool 8 skins D	25	—	2800	220	120	140	—	—
Formosa, Argentina, IV. 1960 Pool 8 skins D	70	80	4300	180	180	360	—	—
Parana River, Corrientes, Argentina, II. 1961 Pool 2 skins D	15	—	5000	650	250	370	—	—
Ibera Lake, Corrientes, Argentina, II. 1961 Pool 5 skins D		—	4500	400	270	370	—	—
Misiones, Argentina, II. 1961 Pool 11 skins D	100	200	2500	250	1500	450	—	—

Appendix I (Continued)

Table 7. Neotropical Stocks (Continued)

	5-HT	N-Methylated Derivatives			Cyclized Derivatives (Dehydro-bufotenine)	Conjugated Derivatives		
						5 (O)-conjugated		1-conjugated
		M_1	M_2	M_3		C_1	C_2	C_3
arenarum								
Cordoba and Tucuman, Argentina, 1960 Several pools, 260 skins D	40-250	40-200	800-2600	20-130	250-700	190-560	—	—
Mendoza, Argentina, VI. 1965 13 skins D	145±20 (40-300)	80±10 (35-160)	604±58 (270-900)	55±6 (20-100)	286±50 (75-715)	254±39 (60-600)	—	—
"*Crucifer* Stock"								
Crucifer group								
crucifer								
Rio de Janeiro, Brazil, IX. 1967 9 skins D	83±27 (4-280)	39±16 (0-160)	16±8 (0-72)	—	423±130 (120-1200)	202±68 (55-560)	—	—
"*Granulosus* Stock"								
Granulosus group								
g. goeldi								
Amazonas, Brazil, I. 1967 10 skins D	81.2±11 (30-155)	17.8±3 (6-35)	6.6±1.3 (0-15)	—	2250±187 (1300-3200)	840±75 (400-1250)	—	—
pygmaeus								
La Criolla, Santa Fe, Argentina, XI. 1964 11 skins D	52±16 (5-190)	99±19 (15-200)	92±16 (25-200)	—	2339±407 (430-4300)	350±64 (95-1000)	—	—
major								
Resistencia, Chaco, Argentina, XI. 1962 Pool 6 skins D	250	130	200	—	550	980	—	—
Resistencia, Chaco, Argentina, I. 1964 Pool 10 skins D	270	120	120	—	1000	940	—	—
Formosa, Argentina, I. 1964 Pool 10 skins D	60	20	140	—	traces	375	—	—
fernandezae								
Buenos Aires, Argentina, VIII. 1962 Pool 15 skins D	60	350	150	—	1200	1100	—	—
Resistencia, Chaco, Argentina, XI. 1962 Pool 6 skins D	150	500	150	—	2400	1350	—	—
Resistencia, Chaco, Argentina, I. 1964 Pool 3 skins D	750	650	80	—	1250	1930	—	—
La Criolla, Santa Fe, Argentina, XI. 1962	194±32 (70-400)	609±86 (220-1200)	238±46 (20-500)	—	2199±252 (840-3600)	1680±184 (420-2750)	—	—
Peltacephalus group								
peltacephalus								
Cuba, III. 1968	110	—	—	—	15	—	—	—

D, dried skin; *F*, fresh skin; —, not detected.
M_1, N-methyl-5-HT; M_2, bufotenine; M_3, bufotenidine.

C_1, bufothionine; C_2, supposed O-sulphate of bufotenine; C_3, bufoviridine (supposed bufotenine 1-sulphonic acid).

Appendix J: Summary of Bioacoustical Investigations

Group and Species	1 Anat. Examined	2 Arytenoid Valve	3 Mod. Pattern Examined	4 M.C. Type	5 Artificial Activation	6 Ablation Exp.
Crucifer Gp.						
crucifer	x	x	x	II	x	
Marinus Gp.						
arenarum	x	x	x	II	x	x
marinus	x	x	x	II	x	x
paracnemis	x	x	x	II	x	
ictericus	x	x	x	II	x	
Guttatus Gp.						
blombergi			x?	I+II		
haematiticus	x	x–	x?	"III"	x	
guttatus						
Typhonius Gp.						
sternosignatus			x	II		
typhonius	x	x			x	
Bocourti Gp.						
bocourti	x	x			x	x
Spinulosus Gp.						
spinulosus	x	x			x	
atacamensis	x	x			x	
chilensis	x	x			x	
variegatus	x	x				
Granulosus Gp.						
granulosus	x	x	x	II	x	
Boreas Gp.						
boreas	x	x			x	
exsul						
nelsoni						
canorus	x	x	x?	II	x	
Alvarius Gp.						
alvarius	x	x	x	?	x	
Punctatus Gp.						
punctatus	x	—	x	III	x	
Marmoreus Gp.						
marmoreus	x	x	x	II	x	
perplexus	x	x	x	II	x	
Cognatus Gp.						
cognatus	x	x	x	II	x	x
compactilis	x	x	x	II	x	
speciosus	x	x	x	II	x	x
Debilis Gp.						
debilis	x	x	x	"II"		x
kelloggi	x	x	x	"II"		
retiformis	x	x	x	"II"		
Valliceps Gp.						
valliceps	x	x	x	II	x	x
mazatlanensis	x	x	x	II	x	
gemmifer			x	II		
luetkeni	x	x	x	II	x	
cristatus						
cavifrons			x?	III		
coniferus	x	x	x	III	x	
ibarrai	x	x–	x	III	x	
Canaliferus Gp.						
canaliferus	x	x	x	"II"		
Occidentalis Gp.						
occidentalis			x	II		
Coccifer Gp.						
coccifer	x	x–	x	III	x	
cycladen			x	III		
Americanus Gp.						
americanus	x	—	x	III	x	
houstonensis	x	—	x	III		
terrestris	x	—	x	III	x	
hemiophrys	x	—	x	III	x	
microscaphus	x	—	x	III	x	
woodhousei	x	—	x	III	x	x
Quercicus Gp.						
quercicus	x	—	x	"III"		
Viridis Gp.						
viridis	x	x	x	III	x	
raddei						
Bufo Gp.						
bufo	x	x	x	I	x	
Calamita Gp.						
calamita	x	x	x	I	x	
Stomaticus Gp.						
stomaticus	x	x	x	II	x	
Melanostictus Gp.						
melanostictus	x	x	x	II	x	x
Asper Gp.						
asper	x	x			x	
Carens Gp.						
carens			x	?		
Mauritanicus Gp.						
mauritanicus	x	x	x	II	x	x

Appendix J (Continued)

Group and Species	Anat. Examined 1	Arytenoid Valve 2	Mod. Pattern Examined 3	M.C. Type 4	Artificial Activation 5	Ablation Exp. 6
Regularis Gp.						
kisoloensis			x	I		
regularis	x	x	x	I	x	x
gutturalis	x	x			x	
garmani			x	I		
rangeri	x	x	x	I	x	
brauni			x	I		
superciliaris			x	I		
Lönnbergi Gp.						
lönnbergi	x	x				
Unknown Affinities						
steindachneri			x	I+II		
vittatus			x	I		
Maculatus Gp.						
maculatus	x	x	x	I	x	
Perreti Gp.						
perreti			x	I+II		
Gariepensis Gp.						
gariepensis	x	x	x	I	x	
inyangae						
rosei	x	x				
Hybrids						
arenarum x *spinulosus*	x	x			x	
brauni x *garmani*	x	x			x	
regularis x *brauni*	x	x			x	
regularis x *garmani*	x	x			x	
Totals	56	48x			I—13 II—27 III—15	48

Column:

1. Anat. Examined: laryngeal anatomy examined in at least one adult male
2. Arytenoid Valve:
   (x)–arytenoid valve present
   (x–)–arytenoid valve present but reduced
   (—)–arytenoid valve absent
3. Mod. Pattern Examined:
   (x)–field-recorded mating call examined
   (x?)–laboratory-recorded mating call examined
4. M.C. Type: mating call classified on basis of amplitude modulation pattern
   (I)–Type I
   (II)–Type II
   (III)–Type III
   (I+II)–Gives calls of both types or switches from one to other
   ("I", "II", or "III")–Type I, II, or III divergent
5. Artificial Activation: larynx of at least one adult male artificially activated
6. Ablation Exp.: some structures in the vocal tract were either damaged or removed, and calls recorded afterwards from one adult male.

Appendix K–1. Characteristics of Testes of Individual Toads, from Sketches of Testes and Color Notes

		Length/Width Ratio of Testes		Testis Color										
Species	No.	% Range	% Mean	Black	Dark Gray	Dark Flesh	Flesh	Yellow	Cream	White	Reticulated	Unclassified	Cystic	Remarks
spinulosus	11	11–37	20				2	6	1	2				
flavolineatus	1	33	33					1						
trifolium	2	14–26	20				1		1					
atacamensis	1	15	15					1						
chilensis	6	18–50	32	4	2									
bocourti	8	28–54	38				3	3	2					
cognatus	6	13–31	21	1			1	3	1					
speciosus	4	18–45	29						4					
compactilis	2	8–9	8				1	1						
punctatus	6	8–15	12				2	3	1					
marmoreus	4	10–28	20		1		1	1	1					
perplexus	6	6–26	14				3	3						
alvarius	5	22–26	23					4	1					
canorus	2	27–50	38					2						
boreas	5	22–50	35				2	3						
exsul	1	28	28					1						
americanus	3	19–37	30						1	2				
terrestris	4	17–28	23		1			2	1					
houstonensis	1	7	7									1		
hemiophrys	4	20–31	25				1	2		1				
microscaphus	4	18–40	27				2			2				
woodhousei	4	10–26	17					1	1	2				
coccifer	7	9–55	22				1	1	2	3				
quercicus	3	33–50	39								1			
debilis	3	12–37	25			1		1			1			
retiformis	1	12	12			1								
kelloggi	1	23	23								1			
viridis	8	13–45	31	7	1									
calamita	6	18–50	30	5	1									
bufo	9	18–50	32	1*	2			3	2	1				*1 Black and 1 Yellow
stomaticus	4	19–40	28				2			1		1		
carens	4	6–9	8							3				
periglenes	1	44	44				1							
Melanophryniscus stelzneri	2	35–45	40								2			
gariepensis	1	43	43							1				
inyangae	1	46	46							1				
lönnbergi	2	36–50	43							1		1		
crucifer	5	13–18	16					1	2	1		1		
arenarum	7	16–18	25				2	2	2	1				
icterius	2	15–31	23				2							
paracnemis	4	10–13	12				1	1	2					
marinus	8	12–19	15				1	1	3	1		2		
poeppigi	6	4–20	13				2		3	1				
granulosus	2	8–28	18				2							
typhonius	2	14–28	21						1	1				

Appendix K–1 (Continued)

Species	No.	Length/Width Ratio of Testes % Range	% Mean	Black	Dark Gray	Dark Flesh	Flesh	Yellow	Cream	White	Reticulated	Unclassified	Cystic	Remarks
				Testis Color										
haematiticus	1	7	7							1				hyper-trophied
holdridgei	1	20	20							1				
valliceps	10	8–37	23			8			2					
ibarrai	2	9–10	9					2						
mazatlanensis	2	6–7	6							2				
gemmifer	1	11	11							1				
luetkeni	8	7–30	19			3	4			1				
coniferus	10	9–50	25				1	6	2			1		
canaliferus	8	15–25	20				2	2	2	2				
occidentalis	2	16–33	24		1					1				
melanostictus	7	5–13	8					2	1	4				
asper	6	12–16	14				1	4	1					
mauritanicus	3	18–46	29				1		1					
regularis (E)	23	4–24	14			1	1	2	5	13			1	
regularis (W)	17	6–27	13			1	2	3	1	3			7	
regularis gp. sp.	2	9–18	13							2				
kisoloensis	1	9	9						1					
pardalis	1	14	14							1				
garmani	4	7–13	10						2	2				
rangeri	9	9–27	14											
kerinyagae	3	15–17	16				1		1			1		
brauni	4	7–16	12						2	2				
maculatus	4	8–15	11				2		1	1				

Appendix K-2. Characteristics of Testes of Individual Toads from Measured Material

Species	Testis in mm. Width	Length	Ratio	Testis Color Black	Dark Gray	Dark Flesh	Flesh	Yellow	Cream	White	Unclassified	Reticulated	Remarks
spinulosus	3	8	37				+						
cognatus	2	10	31					+					
speciosus	2	9	22						+				
boreas	4.5	9	45					+					
americanus	3	9	33						+				
terrestris	2	7	28					+					
hemiophrys	3	11	27				+						
woodhousei	4.5	10	45						+				
terrestris	2	12	17					+					
coccifer	2	9	22							+			
quercicus	1	2	50									+	
bufo	5	11	45		+								
bufo	3	11	27		+								
stomaticus	3	9	33				+						

Appendix K–2 (Continued)

Species	Testis in mm.			Testis Color										Remarks
	Width	Length	Ratio	Black	Dark Gray	Dark Flesh	Flesh	Yellow	Cream	White	Unclassified	Reticulated		
carens	—	26	—							+				
inyangae	6	13	46							+				
crucifer	2	12	17							+				
crucifer	2	15	13						+					
crucifer	2	11	18											
arenarum	3	17	18						+					
paracnemis	3	25	11					+						
paracnemis	4	30	13						+					
marinus	5	26	19											
poeppigi	2	14	14						+					
poeppigi	4	22	18						+					
poeppigi	1	22	4						+					
poeppigi	3	15	20				+							
typhonius	2	7	28						+					
haematiticus	—	28	—							+				S–V length 51mm!!
holdridgei	—	11	—							+				
valliceps	3	10.5	28			+								
valliceps	3	9	33			+								
valliceps	2	10	20			+								
valliceps	3	8	37			+								
luetkeni	2	8	25			+								
coniferus	3	6	50						+					
coniferus	3	7	43				+							
coniferus	2	8	25											
regularis (E)	2.5	22	11					+						
regularis	2	13	15											
regularis	2	13	15						+					
regularis	2	12	17				+							
rangeri	2	17	12											

Appendix K–3. Characteristics of Testes of F_1 Hybrid Toads

♀ Parent & Type	♂ Parent & Type	Length/Width Ratio			Testis Color										Remarks
		No.	Range	Mean	Black (B)	Dark Gray (dG)	Dark Flesh (dF)	Flesh (F)	Yellow (Y)	Cream (C)	White (W)	Reticulated (R)	Unclassified	Cystic (Cy)	
woodhousei (17 W C Y)	*viridis* (31 B G)	3	9–67	34	3										1 is ½ black
woodhousei (17 W C Y)	*bufo* (32 Y G C B W)	1	20	20	1										
woodhousei (17 W C Y)	*calamita* (30 G B)	1	—	—	1										
woodhousei (17 W C Y)	*boreas* (35 Y F)	1	50	50	1										
woodhousei (17 W C Y)	*bocourti* (38 Y F C)	2	35–44	39				2							
woodhousei (17 W C Y)	*compactilis* (8 Y F)	1	—	—				1							eggs in B.O.

♀ Parent & Type	♂ Parent & Type	Length/Width Ratio			Testis Color										Remarks
		No.	Range	Mean	Black (B)	Dark Gray (dG)	Dark Flesh (dF)	Flesh (F)	Yellow (Y)	Cream (C)	White (W)	Reticulated (R)	Unclassified	Cystic (Cy)	
woodhousei (17 W C Y)	*arenarum* (25 F Y C W)	1	75	75			1*								*red
woodhousei (17 W C Y)	*icterICUS* (23 F)	1	71	71			1								
woodhousei (17 W C Y)	*luetkeni* (19 F dF W)	1	14	14			1								
hemiophrys (25 F Y W)	*valliceps* (23 dF C)	2	18	18	1			1							
microscaphus (27 F W)	*valliceps* (23 dF C)	1	60	60				1							
luetkeni (19 F dF W)	*valliceps* (23 dF C)	4	30–92	58		3	1								
luetkeni (19 F dF W)	*mazatlanensis* (6 W)	3	33–45	38				2			1				
luetkeni (19 F dF W)	*arenarum* (25 F Y C W)	1	57	57			1								
luetkeni (19 F dF W)	*granulosus* (18 F)	1	11	11			1								
luetkeni (19 F dF W)	*compactilis* (8 F Y)	1	18	18			1								
mazatlanensis (6 W)	*luetkeni* (19 F dF W)	1	14				1								
mazatlanensis (6 W)	*valliceps* (23 dF C)	2	—	—			1						1*		*ovitestis
mazatlanensis (6 W)	*speciosus* (29 C)	1	22	22			1								
mazatlanensis (6 W)	*marmoreus* (20 dG F Y C)	3	20–32	26		1							1		
valliceps (23 dF C)	*mazatlanensis* (6 W)	3	—	—			3								
valliceps (23 dF C)	*luetkeni* (19 F dF W)	1	28	28		1									
valliceps (23 dF C)	*canaliferus* (20 F Y C W)	2	—	—			2								
valliceps (23 dF C)	*granulosus* (18 F)	2	14–71	42			2								
valliceps (23 dF C)	*arenarum* (25 F Y C W)	4	29–73	44				4							
canaliferus (20 F Y C W)	*valliceps* (23 dF C)	2	—	—							2				
luetkeni (19 dF W)	*microscaphus* (27 F W)	2	33–55	44				1							
perplexus (14 F Y)	*bocourti* (38 Y F C)	2	20–22	21								2			
perplexus (14 F Y)	*speciosus* (29 C)	1	20	20							1				
perplexus (14 F Y)	*luetkeni* (19 F dF W)	3	17–18	18	3										
perplexus (14 F Y)	*valliceps* (23 dF C)	1	25	25							1				
perplexus (14 F Y)	*granulosus* (18 F)	1	44	44	1										
perplexus (14 F Y)	*spinulosus* (20 F Y C W)	1	17	17				1							
marmoreus (20 dF F Y C)	*mazatlanensis* (6 W)	1	31–13*	31–13*											*tapered
cognatus (21 Y C F B)	*perplexus* (14 F Y)	1	83	83									1		
speciosus (29 C)	*arenarum* (25 F Y C W)	1	20	20	1										
canaliferus (20 F Y C W)	*arenarum* (25 F Y C W)	1	44	44							2				
pygmaeus (18 F)	*arenarum* (25 F Y C W)	1	27	27			1								
coccifer (22 W C Y F)	*valliceps* (23 dF C)	3	10–20	16	2	1									
marinus (15 C Y W F)	*limensis* (?)	1	5	5							1				
paracnemis (12 C Y F)	*arenarum* (25 F Y C W)	1	18	18				1							
arenarum (25 F Y C W)	*marmoreus* (20 dG F Y C)	2	45–67	56	2										
arenarum (25 F Y C W)	*spinulosus* (20 F Y C W)	1	12	12							1				
brauni (12 C W)	*garmani* (10 C W)	2	19–20	19							2				
brauni (12 C W)	*regularis* (W) (C Y W F)	1	14	14				1							
garmani (10 C W)	*regularis* (W) (13 C Y W F)	2	30–38	34							2				
garmani (10 W)	*regularis* (E) (14 W C Y)	3	8–18	13							3				
regularis (W) (13 C W Y F)	*regularis* (E) (14 W C Y)	1	40	40							1				

Mean, width/length ratio, and abbreviations of colors of parental species are in parentheses after names.

Appendix K–4. Results of Crosses between Narrow-Skulled Toads with Hypertrophied Testes and other Species

	Experimental		Control		
♂ *gariepensis*	% Fert.	Stage Reached	% Fert.	Stage Reached	Remarks
♀ *regularis* (E)	60.7	G	**	—	**no control
♀ *carens*	—*	—	29.2	H(D)	*Jelly turned white, many sperm present
♀ *carens*	—	—	90.9	G*	*Interspecific
♀ *lughensis*	21.0	–G	83.3	G*	*Interspecific
♀ *vertebralis*	19.8	–G*	10.3	G**	*Cleavage abnormal **Interspecific
♀ *viridis*	34.1	G	27.7	L*	*Interspecific
♂ *inyangae*					
♀ *regularis* (W)	—	—	98.0	A	
♀ *brauni*	—	—	60.0	A	
♀ *brauni*	64.5		64.3	A	
♀ *viridis*	8.5	–G	54.5	A*	*Interspecific
♀ *perplexus*	48.8	–G*	47.2	A**	*only 37.0% of fert. to 64-cell **Interspecific
♀ *perplexus*	62.4	–G*	47.2	A**	*only 10.7% of fert. to 64-cell **Interspecific
♀ *marinus*	—	—	72.5	N*	*Interspecific
♂ *rosei*					
♀ *rangeri*	—	—	66.2	A	
♀ *regularis* (W)	—	—	28.6	H(D)*	*Interspecific
♀ *brauni*	—	—	100.0	G*	*Interspecific
♂ *blombergi*					
♀ *woodhousei*	16.0	G	55.5	A	
♀ *microscaphus*	35.0	N*	19.3	H(D)**	*Abnormal **Hatch, discarded
♀ *valliceps*	6.3	N*	15.6	A	*Abnormal
♀ *valliceps*	13.2	–G*	33.0	A	*Cleavage abnormal
♀ *alvarius*	92.4	G*	23.0	A	*Cleavage abnormal
♀ *speciosus*	56.4	G	87.9	A	
♂ *haematiticus*					
♀ *perplexus*	1.6	–G	97.7	L*	*Interspecific
♀ *perplexus*	21.7	–G	93.9	A	
♀ *perplexus*	—	—	93.9	A	
♀ *marmoreus*	42.8	–B	89.4	A	
♀ *punctatus*	1.8	–G*	**	—	*Cleavage abnormal **No control
♀ *valliceps*	—	—	*	—	*No control
♀ *valliceps*	17.6	–G	*	—	*No control
♀ *valliceps*	4.1	–G*	88.4	A**	*Cleavage abnormal **Interspecific
♀ *occidentalis*	7.7	B	32.0	G*	*Interspecific
♀ *crucifer*	6.4	–G	96.5	A	
♀ *arenarum*	5.1	B*	100.0	A	*Cleavage abnormal
♀ *viridis*	63.3	B	68.6	L*	*Interspecific
♀ *viridis*	40.0	B	51.6	A	
♀ *regularis* (W)	—	—	100.0	L*	*Interspecific
♀ *carens*	6.1	G	87.8	G*	*Interspecific
♂ *holdridgei*					
♀ *viridis*	—	—	66.0	A	

INDEX